PLASTICITY, LIMIT ANALYSIS, STABILITY AND STRUCTURAL DESIGN

An Academic Life Journey from Theory to Practice

PLASTICITY, LIMIT ANALYSIS, STABILITY AND STRUCTURAL DESIGN

An Academic Life Journey from Theory to Practice

Wai-Fah Chen 陳惠發

University of Hawaii, USA

Editor

Lian Duan 段煉

California Department of Transportation, USA

World Scientific

NEW JERSEY · LONDON · SINGAPORE · BEIJING · SHANGHAI · HONG KONG · TAIPEI · CHENNAI · TOKYO

Published by

World Scientific Publishing Co. Pte. Ltd.

5 Toh Tuck Link, Singapore 596224

USA office: 27 Warren Street, Suite 401-402, Hackensack, NJ 07601

UK office: 57 Shelton Street, Covent Garden, London WC2H 9HE

Library of Congress Cataloging-in-Publication Data

Names: Chen, Wai-Fah, 1936–   author. | Duan, Lian, 1954–   editor.

Title: Plasticity, limit analysis, stability and structural design : an academic life journey from
  theory to practice / Wai-Fah Chen, University of Hawaii, USA ;
  editor, Lian Duan, California Department of Transportation, USA.

Description: Singapore ; Hackensack, NJ ; London : World Scientific, [2021] |
  Includes bibliographical references and index.

Identifiers: LCCN 2020053355 | ISBN 9789811229732 (hardcover) | ISBN 9789811231407 (paperback) |
  ISBN 9789811229749 (ebook for institutions) | ISBN 9789811229756 (ebook for individuals)

Subjects: LCSH: Structural engineering. | Geotechnical engineering. |
  Chen, Wai-Fah, 1936– | Civil engineers--United States--Biography.

Classification: LCC TA637 .C44 2021 | DDC 624.1--dc23

LC record available at https://lccn.loc.gov/2020053355

British Library Cataloguing-in-Publication Data

A catalogue record for this book is available from the British Library.

For any available supplementary material, please visit
https://www.worldscientific.com/worldscibooks/10.1142/12083#t=suppl

PREFACE
自序

This book is a personal anthology (陳惠發院士文集) of my utmost academic works and accomplishments with my former students and colleagues and is intended as an enduring record for the engineering community many years to come.

I was born in Nanjing, China (中國南京) in 1936. I studied civil engineering at the National Cheng-Kung University (國立成功大學) in Tainan (台南), Taiwan (台灣), and obtained my BS degree in 1959. In 1961, I moved to Bethlehem, Pennsylvania, and graduated from Lehigh University with an MS in Structural Engineering in 1963. Thereafter, I transferred to Brown University to continue my graduate study and wrote my doctoral dissertation on soil plasticity and limit analysis under the supervision of Professor Daniel C Drucker. I was second to the last of Drucker's PhD students at Brown. My career path and learning process are summarized in Chapter 1: Lifelong Learning (學無止境).

In 1966, my new bride, the former Linlin Hsuan (宣玲玲), and I moved to Bethlehem, where I was an assistant professor to a full professor at Lehigh University for ten years (1966–1976). In 1976, we moved to West Lafayette, Indiana, to start my new position as a professor (1976–1992), the George E Goodwin Distinguished Professor of Civil Engineering (1992–1999), and the Head of Structural Engineering (1980–1999) at Purdue University. After 23 years of services at Purdue, I accepted an appointment at the University of Hawaii as the Dean of Engineering, and so I moved to Honolulu, Hawaii, in 1999. I retired in 2006. My 40-year professional career and academic life journey are summarized in the following three chapters: Chapter 2: Lehigh University, 1966–1976, Beginning to Show (初露頭角), Chapter 3: Purdue University, 1976–1999, Highly Productive Years (得心應手), and Chapter 4: University of Hawaii, 1999–2006, The Pursuit of Excellence (追求卓越).

It was at Lehigh University that I made major contributions to the theory of plasticity and its application to the design of bi-axially loaded columns in engineering structures with the use of the then newly developed plastic theory. Lynn Beedle founded the Lehigh group in plastic design for steel structures, and through his leadership, the group became internationally known for its pioneering work in implementing the plasticity theory to structural design. At Lehigh, I published over 90 papers, some in collaboration with colleagues and doctoral students from the Fritz Engineering Laboratory. Most of these papers dealt with the bi-axially loaded columns, known as beam-columns in frame design, particularly, the theoretical development of three-dimensional plastic behavior of beam-columns. However, the approach was also balanced by a number of papers on the experimental evaluation of H-columns (as used in building frames) and large cylindrical fabricated columns (as used in offshore structures), as well as the tests of some full-scale steel beam-to-column building connections.

In 1975, I published a milestone treatise on *Limit Analysis and Soil Plasticity*, in which I introduced the concept of plasticity and limit theorems to practical applications in soil mechanics. I brought them together for the first time in a unified manner, which at the time was known only to a few in the field. In particular, the limit theorems led directly to limit design — a technique that could predict the stability and load-carrying capacity problems in geotechnical engineering such as slope stability, earth pressures on retaining structures, and bearing capacity of footings. These limit theorems had immediate application to geotechnical problems that were formerly beyond the scope of engineering practice. The geotechnical engineering community was brought up-to-date with this new publication, and the book was well-received with wide attention; it generated huge excitement in the geotechnical engineering profession as reflected in the high regard in various book reviews published in different technical journals around the world. The merit of limit analysis lies in the fact that engineers can now make practical and safe decisions on the design of complex load-bearing components based on relatively simple calculations.

In 1976–1977, I published a two-volume work on *Theory of Beam-Columns* with my doctoral student, T Atsuta. This work was the first ever in this field to discuss systematically the complete theory of in-plane and space beam-columns. This landmark work shows how these theories are applied for the solution of practical design in steel frames.

In 1982, I published another pioneering work on *Plasticity in Reinforced Concrete*, in which I presented a unified approach to mathematical modeling of concrete materials as commonly used in reinforced concrete structural analysis. Also, I was among the first few to show how the limit analysis could be used in the analysis and design of tensile-weak material like concrete or rocks, and I later developed constitutive equations for concrete materials that could be applied effectively to the analysis and design of reinforced concrete structures with the powerful computer-based finite element methods that were very popular at the time. With the theorems of limit analysis, calculations were performed on deep beams and structural joints with ease. The success of this work played a strong role in the development of the American Concrete Institute's *Building Code Requirements for Structural Concrete*, which bases strength estimates of complex joints or deep beams with openings on lower bound limit theorem, now known as the *strut-tie model* in the terminology of the concrete community.

In my later years at Purdue, I became active in a field sometimes known as *advanced analysis*, which attempts to bridge the gap in design between using elastic analysis for a structural system while sizing up the structural members with an inelastic analysis through the use of the so-called "effective length factor", or simply known as the K-factor design as popularly used in engineering practice. Unlike the older second-order analysis used in the seismic and static design, which is used as supplemental checking only, the concept of advanced analysis can now be used as primary design and member size framing without the use of the K-factor procedure.

Research works on advanced analysis are now in full swing to develop nonlinear procedures and software for practical use in design office. The theory and approaches for advanced analysis of plane frames composed of members of compact sections, fully braced out-of-plane, have been well developed, verified and coded by the American Institute of Steel Construction (AISC) in the 2005 AISC "Specifications for Structural Steel Buildings" as well as by other institutions around the world.

In the later part of the 1990s during the Internet bubble time, there was greater awareness and growing realization among the Hawaii State's leaders, public leaders, and private sector leaders, that the engineering workforce would play a central role in shaping the future of Hawaii's high tech economy, and that the research mission at the University of Hawaii (UH) and its College of Engineering, in particular, was really pivotal for this development.

In September 1999, I was recruited by UH to take the Deanship of its College of Engineering to accomplish this noble mission. To this end, the College was shaped to follow a strategic plan that was built around my clear vision to build it into one of the top 50 engineering schools in the country. In my vision paper in 1999, the strategic plan included efforts to strengthen the UH faculty, improve its facilities, and increase enrollment. A lot of progress was made in all these three areas — faculty was added, the research program grew, facilities were improved, and enrollment increased.

In order to achieve this lofty goal, my top priority at the beginning was to make UH the first-ever university in the university system to start merit raises based on performance to retain faculty. Not only did I recruit over 20 new faculty at market price, but we also raised UH's enrollment numbers from 500 undergraduates to 760. Our goal was 800. Also, we increased external fundings to boost UH's graduate programs. At Stanford University, every faculty produced four masters and one PhD graduates per year. The UH faculty only produced half a master per year and very few PhDs. When I came in, we had a budget of $4–5 million. Over the years, the figure has since reached $8 million, and the UH is now working toward $10 million. Finally, in terms of facilities, several multi-million-dollar laboratories were built.

In 2006, I stepped down from the deanship with the satisfaction of having achieved the challenges and goals set up for me by the administration in 1999. In addition, I had left behind a solid roadmap or blueprint for the College to continue reaching its goal of being ranked as one of the top 50 engineering schools in the country. Looking back, I had the great fortune of having the opportunity to continue my life's work in the field that I love (and will always love) as a graduate research faculty at UH Manoa's research campus since 2007.

Chapter 5: Selected Academic Bulletins (學術論文通報) presents 23 academic article bulletins in their original formats to highlight my 10 major research areas including Limit Analysis, Beam-Columns, Semi-rigid Construction, Offshore Structures, Concrete Plasticity, Soil Plasticity, Structural Stability, Concrete Building Construction, Advanced Analysis, and Structural Engineering. My PhD advisor DC Drucker, and my former students, Toshio Atsuta (PhD, 1972), DA Ross (PhD, 1978), GP Rentschler (PhD, 1979), E Mizuno (PhD, 1981), EM Lui (呂汶) (PhD, 1985), XL Liu (劉西拉) (PhD, 1985), Lian Duan (段煉) (PhD, 1990), KH Mosallam (PhD, 1991) and JYR Liew (劉德源) (PhD, 1992), each co-authored with me on some of the abovementioned topics.

Chapter 6: Academic Masterpiece Books (學術專著) documents 23 books along with their peer journal reviews. My former students, Atsuta, Rentschler, Mizuno, AF Saleeb (PhD, 1981), DJ Han (韓大建) (PhD, 1984), EM Lui, XL Liu, IS Sohal (PhD, 1986), S El-Metwally (PhD, 1986), H Zhang (張宏) (PhD, 1991), Mosallam, Liew, SE Kim (PhD, 1996) and YS Kim (PhD, 1998), each co-authored with me on some of these books. Chapter 6 also lists 33 edited books and 57 contributing book chapters.

Chapter 7: Academic Publications (期刊論文目錄) lists 359 peer-reviewed journal articles, 263 articles in conference proceedings and symposiums, and 110 invited lectures and keynotes.

I participated actively in several professional societies. I was most active in the Structural Stability Research Council (SSRC) Task Groups research activities and served as a member of the Executive Committee of SSRC for decades. For a long time, I was a co-technical editor of the American Society of Civil Engineers (ASCE) Journal of Engineering Mechanics and the ASCE Journal of Structural Engineering, among others. I was also actively involved in the AISC Specifications Committee for many years as a member and helped implement research and theory into engineering practice.

I have been a member of the National Academy of Engineering and the prestigious Academician of Academia Sinica (中央研究院院士) since 1995 and 1997, respectively. I am also an ASCE Honorary Member since 1997, and I received the Lifetime Achievement Award from AISC in 2003.

At Lehigh University, I advised several masters and nine doctoral students over a period of ten years. At Purdue University, I advised 45 doctoral students from various countries around the world spanning more than two decades. At the University of Hawaii, I advised one doctoral student, 3 postdoctoral students and several visiting scholars from Japan, Hong Kong, Egypt and China, including Taiwan. In total, I had 55 PhD students, and their dissertations are listed in Chap. 8.

Waifah Chen (陳惠發)
Honolulu, Hawaii (夏威夷檀香山)
May 2020

AUTHOR
作者简介

Wai-Fah Chen at a Glance (陳惠發)

Professional Positions
- Professor and Dean of the College of Engineering, University of Hawaii, 1999–2006
- Professor — George E. Goodwin Distinguished Professor of Civil Engineering, and Head of Structures Area, School of Civil Engineering, Purdue University, 1976–1999
- Assistant Professor to Professor, Lehigh University, 1966 to 1976

Education
- PhD in solid mechanics, Brown University, RI, 1966
- MS in structural engineering, Lehigh University, PA, 1963
- BS in civil engineering, National Cheng-Kung University, Taiwan, 1959

Research Interests
- Constitutive modeling of engineering materials
- Soil and concrete plasticity
- Structural connections and structural stability

Awards and Honors
- 1984 US Senior Scientist Award, Alexander von Humboldt Foundation, Germany
- 1985 TR Higgins Lectureship Award, American Institute of Steel Construction
- 1985 Raymond C Reese Research Prize, American Society of Civil Engineers
- 1988 Distinguished Alumnus Award, National Cheng-Kung University
- 1990 Shortridge Hardesty Award, American Society of Civil Engineers
- 1991 Honorary Fellow, Singapore Structural Steel Society
- 1995 Member, National Academy of Engineering
- 1997 Honorary Member, American Society of Civil Engineers
- 1998 Member, Academia Sinica (Taiwan's National Academy of Science)
- 1999 Distinguished Engineering Alumnus Medal, Brown University
- 2003 Lifetime Achievement Award, America Institute of Steel Construction
- 2003 Featured in the biographical monograph *Giants of Engineering Science* as one of ten leading engineering scientists in the world

Distinctions
- Author or co-author of 23 books
- Author or co-author of 359 peer-reviewed journal articles
- Author or co-author of 263 conference proceedings and symposiums articles
- Author or co-author of 57 contributing chapters
- Editor or co-editor of 33 engineering books including 5 handbooks
- Presented 110 invited lectures or keynote topics
- Served on the editorial boards of 37 technical journals
- Listed on 35 Who's Who publications
- Consulting editor of the *Encyclopedia of Science and Technology*

NAE Election

Elected to the National Academy of Engineering, 1995.

Academic Sinica Election

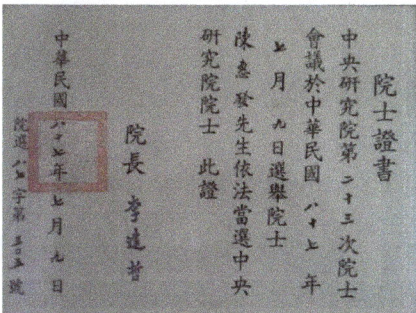

Elected to the Academia Sinica (Taiwan's National Academy of Science), 1998.

ASCE Awards

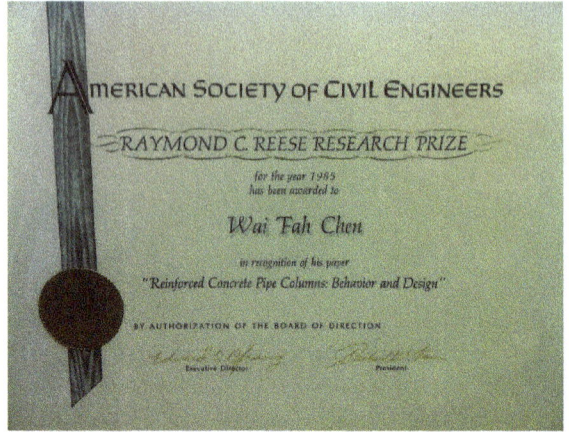

Awarded the Raymond C Reese Research Prize with Xila Liu, American Society of Civil Engineers, 1985.

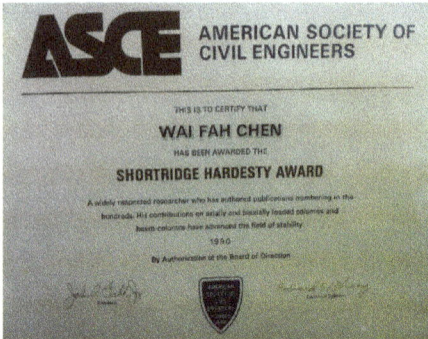

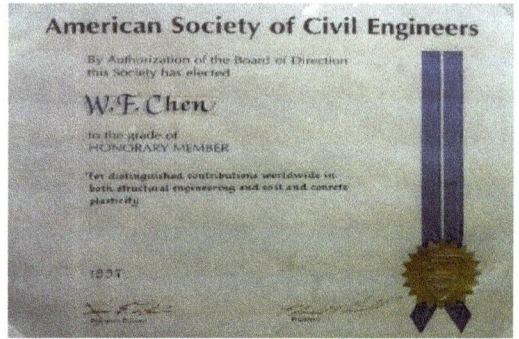

Left: Awarded the Shortridge Hardesty Award, American Society of Civil Engineers, 1990.
Right: Elected as an Honorary Member of the American Society of Civil Engineers, 1997.

AISC Awards

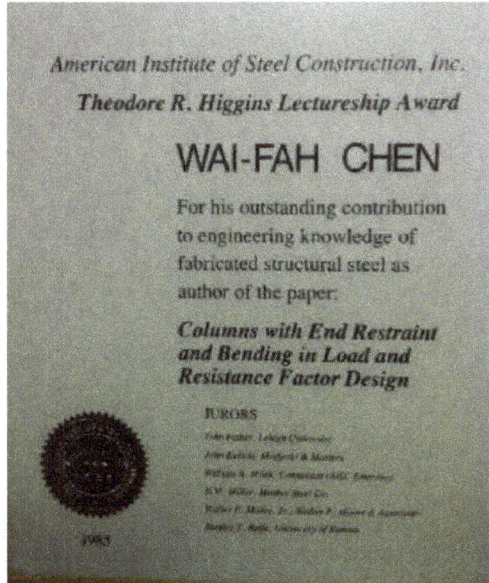

Awarded the TR Higgins Lectureship Award, American Institute of Steel Construction, 1985.

Presented with a Lifetime Achievement Award, America Institute of Steel Construction, 2003.

National Cheng-Kung University Award

From left to right: The National Cheng-Kung University Distinguished Alumnus Award, 1988, with former President Nee (center) and former President Ma (right).

Brown University Alumnus Medal

Distinguished Engineering Alumnus Medal with Dean Rod Clifton, Brown University, 1999.

Top 10 Engineering Scientists

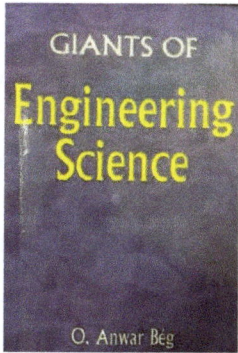

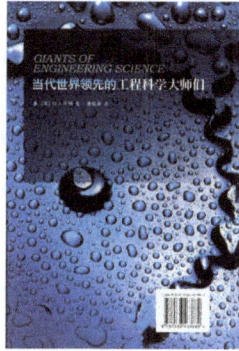

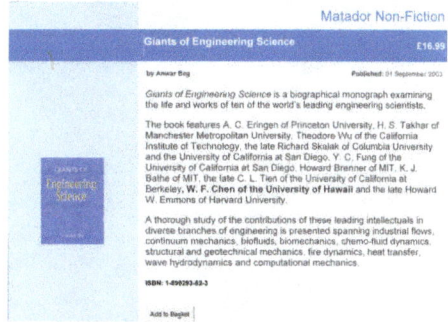

GIANTS OF

Engineering Science

O. Anwar Bég

GIANTS OF
ENGINEERING SCIENCE
当代世界领先的工程科学大师们

Matador Non-Fiction

Giants of Engineering Science £16.99

by Anwar Bég   Published: 01 September 2003

Giants of Engineering Science is a biographical monograph examining the life and works of ten of the world's leading engineering scientists.

The book features A. C. Eringen of Princeton University, H. S. Takhar of Manchester Metropolitan University, Theodore Wu of the California Institute of Technology, the late Richard Skalak of Columbia University and the University of California at San Diego, Y. C. Fung of the University of California at San Diego, Howard Brenner of MIT, K. J. Bathe of MIT, the late C. L. Tien of the University of California at Berkeley, W. F. Chen of the University of Hawaii and the late Howard W. Emmons of Harvard University.

A thorough study of the contributions of these leading intellectuals in diverse branches of engineering is presented spanning industrial flows, continuum mechanics, biofluids, biomechanics, chemo-fluid dynamics, structural and geotechnical mechanics, fire dynamics, heat transfer, wave hydrodynamics and computational mechanics.

ISBN: 1-899293-83-3

Add to Basket

Featured in the biographical monograph *Giants of Engineering Science*, 2003. (In English and Chinese.)

CONTENTS

CHAPTER 1
LIFELONG LEARNING
學無止境

1.1 National Cheng-Kung University — Civil Engineering, 1955–1959
(國立成功大學－土木工程)

In 1955, after I graduated from the Senior High School of the Taiwan Normal University, I passed the national examination and entered the National Cheng-Kung University (NCKU 國立成功大學) in Tainan, Taiwan. I graduated from NCKU with a Bachelor of Science in Civil Engineering (BSCE) degree in 1959. At the time, Taiwan was at a crossroads; land reform and economic development were the focus of the government, a top priority for the regime's survival. With an infusion of American aid, educational reform followed, and higher education became very much Americanized. Purdue University (Purdue) in West Lafayette, Indiana, was selected to be NCKU's counterpart by the United States (US) State Department. Purdue served as our role model for curriculum reform, teacher training, and reeducation. Many of NCKU's existing faculty members were sent to Purdue for advanced degrees. Several Purdue faculty members came over to teach a variety of courses and also guided the top administrators for the university system reform.

Those days, we used the original English textbooks as our official textbooks. Since NCKU had only a few original books from overseas, they were very expensive, and the University set aside a room for us to read these "original English books" in the library. Later, these original textbooks were reprinted in Taiwan at an affordable price. Several of my classes were also taught directly by Purdue professors. There were good interactions between us. It was a struggle, of course, for all of us to learn to read, write and understand English quickly since we were not prepared at all for such a drastic change.

This turned out to be a blessing for many of us who came to the US later for advanced degrees. As a result of this relationship, I had a deep impression of Purdue and knew about its great engineering school in the US since my undergraduate years. Who could predict that 20 years later, I would become a faculty member at Purdue, be promoted steadily to the Head of Structural Engineering in 1980 and become its

Receiving the NCKU (成大) Distinguished Alumnus Award (傑出校友) in 1988. Here, I am with the civil engineering faculty and top administrators (first row, from left: Chairman Tang (譚建國), President Ma (馬哲儒), former President Nee (倪超) and Professor Shi (史惠順).

Four roommates on the NCKU campus, 1958.
Left (from left): Me, Sinclair Wang (王繩康), George Lo (羅金陵), Paul Sun (孫長貴).
Right (from left): Me, Sinclair, Paul, George.

NCKU's President Hwung (黃煌輝) at the 2014 Sinica Meeting.

Exchanging gifts with President Hwung (黃煌輝) at the 2014 Sinica Meeting.

Left: My 1988 Distinguished Alumnus Award at NCKU.
Center: With former President Nee.
Right: With former President Ma.

first George E Goodwin Distinguished Professor of Civil Engineering in its School of Civil Engineering in 1992.

1.2 Lehigh University — Structural Engineering, 1961–1963
(里海大學－結構工程)

1.2.1 *Graduate research assistant*

Lehigh University is located in Bethlehem, Pennsylvania. Bethlehem is a steel town where the Bethlehem Steel Corporation was formed in 1904 and later became the US's second-largest steel company. Lehigh's Fritz Engineering Laboratory (Fritz Lab), named after the original university trustee and Bethlehem Iron Works (later, Bethlehem Steel Corporation) founder, received a major addition in 1955 and was hailed as the world's most impressive structural testing facility at the time. Lynn S Beedle

Lehigh University (left) and the Fritz Engineering Laboratory (right).

was director of the Lab when I enrolled as a graduate student in the department of civil engineering. Beedle was also acting Chairman of the department at the time. We all worked under him.

The stipend for a half-time research assistantship was about $2000 per academic year at the time. Tuition fees for all graduate assistants like us were waived. Most of the research and tests were carried out in the Fritz Lab, which was a five-story steel-framed building housing a 5 million pounds (2,500 US Tons) universal testing machine, the largest one in the world at the time. While the structural research facilities took up the largest space, several other research and instruction laboratories were located throughout the building. Offices, classrooms and various smaller laboratories were located on the second and upper floors.

Two-thirds of the civil engineering faculty members at Lehigh were in structural engineering; steel research activities through the sponsorship of the American Institute of Steel Construction (AISC) for the steel industry were Lehigh's main source of revenue. A revolutionary method for steel design, known as "plastic design", for steel structures was the main theme of research under the leadership of Beedle at the time. Lehigh was also well known for its engineering program, in particular, structural engineering. The nickname for Lehigh's football team was "The Engineers". It seems that my life journey and fate are always associated with engineering-focused schools.

The first one was NCKU, where I received my BSCE degree in 1959. NCKU's original name was Tainan Institute of Engineering (台南工學院). The nickname of Purdue University, where I spent 23 years as a faculty member, was "The Boilermakers". What a coincidence. I stayed in a dormitory during my first year. The food was very good and plentiful. I was impressed with the unlimited supply of milk and ice-cream and the three regular meals. The only problem with dormitory living was that I had to move out to another dormitory during the holiday period so that the university could lock up most of its facilities.

My monthly paycheck was about $200. It was enough for my simple living, and I began to send $50 cash via registered mail back home to support my parents. This practice lasted for the next 45 years.

In the second year, I moved out and rented a house near the campus with three friends. Since everyone's schedule was different, we did not share cooking and grocery shopping. I spent most of my time in my office. My officemate was Hai-Sang Lew, an American-Korea student. My immediate research supervisor was John Hanson, a doctoral student working on his PhD thesis, and I helped him test concrete specimens. His thesis was on the shear strength of reinforced concrete beams. The second year, a new student from Switzerland, John Badoux, joined our research group, and I became his immediate supervisor. I received my Master of Science in Civil Engineering (MSCE) degree in structural engineering from Lehigh in 1963 and decided to transfer to Brown University in Providence, Rhode Island, to focus on theoretical work of solid mechanics. Each of us then went our own ways, and we have not seen or communicated with each other for nearly forty years. However, we were somehow reconnected through our professional society activities or student-related connections. We are all reaching our retirement ages. Although our time together was relatively short, our friendships still continued for nearly half a century.

1.2.2 *Looking back*

Looking back to my three classmates at Lehigh during my two years as a graduate student in 1961–1963, Hanson first worked as a researcher at the Portland Cement Association, then joined Wiss, Janney, Elstner Associates (WJE) as a practicing engineer, and he finally returned to the academic world as a distinguished professor of North Carolina State University. He was elected to the National Academy of Engineering in 1992. I was elected to the same academy three years later in 1995.

Badoux returned to the Swiss Federal Institute of Technology, Lausanne, as a faculty and rose steadily to become the President of the Institute (1992–2000). Lew joined the National Bureau of Standards (NBS) in 1968 as a structural research engineer. He successively became Chief of the Structures Division. He was a member of the National Academy of Engineering of Korea. This is, in a nutshell, a brief description of the life journey of three of my classmates, who met nearly 60 years ago at Lehigh University in the steel town called Bethlehem. The Bethlehem Steel Corporation is since long gone; it went bankrupt as the world becomes more flat and global.

1.2.3 *Lynn S Beedle*

Education

Lynn Beedle was born in Orland, California. In 1941, after graduating from the University of California at Berkeley with a BS in civil engineering, he joined the Navy.

Career

In 1947, he joined Lehigh University as an instructor and five years later received his doctorate there in structural engineering. His groundbreaking studies on the properties of steel structures and his creation in 1969 of the Council on Tall Buildings helped the university become a center for civil and structural engineering research.

Honors

He was elected to the National Academy of Engineering in 1972, and in 1999, the Engineering News-Record named him as one of 125 people who made invaluable contributions to the construction industry since the magazine's founding in 1874.

Awards

He was a recipient of a lifetime achievement award from the American Institute of Steel Construction (AISC) in 2000. He also received the Franklin Institute's Frank P Brown Medal, as well as the John Fritz Medal, the Berkeley Engineering

Alumni Society Distinguished Engineering Alumnus Award. He was also named Distinguished Professor of Civil Engineering by Lehigh University.

1.3 Brown University — Solid Mechanics, 1963–1966
(布朗大學－固體力學)

1.3.1 *Turning point in life* (人生轉折點)

Great leaps forward of the development of the modem theory of plasticity are due to two pioneering giants in plasticity at Brown University — William Prager, along with his close associate, Daniel C Drucker, to whom I had the great fortune to be his PhD student (1963–1966) and later to be associated with him professionally until his death in 2001. I received my Doctor of Philosophy degree from Brown University in 1966.

Brown University, Providence, Rhode Island.

The structural limit analysis was a major portion of my work. These methods not only focused on the preliminary design stages but also on checking their ultimate strength stage, which can be accomplished by using the finite element method with high-speed computers.

These limit analysis methods are simple but powerful, emphasizing the physical aspects of the problem while using structural engineering concepts. These methods can be used on very complex problems to provide crude but very useful information that otherwise would be impossible to consider.

NCKU and Lehigh provide a traditional engineering technology education to prepare their graduates in engineering practice. Or simply put, their graduates are trained as professional engineers.

On the other hand, structural engineering and geomechanics are branches of applied mechanics. Brown University (Brown), similar to Harvard University and California Institute of Technology, provides an engineering science education to prepare its graduates in the area of research and development.

In other words, the engineering science education and training are more suitable for the academic world and research jobs. Thus, graduates from Brown are really engineering scientists, rather than professional engineers.

At Brown, I was indeed fortunate to work under Drucker, who I regard as an engineering science guru. One of his most prestigious honors — the National Medal of Science — was presented to him by a US President.

As his doctoral student, I was deeply impressed by his wide range of interests in the properties of materials, structural mechanics, photo-elasticity, material science and soil mechanics. It was no accident that my lifelong career also involves three similar areas of interaction: mechanics, materials and computing.

Looking back on Drucker's teaching, several memorable events come to mind immediately. Let me give you two examples. The first one was that he set a high bar for graduate teaching.

I still remember vividly the opening session of Drucker's beginning class on "Plasticity" (which I took during my first semester at Brown more than 57 years ago), where he pointed out that the materials in technical reports were "at least one year old", the materials in technical papers were "two to three years old", and the textbook materials were "at least five years old". Therefore, there was no point for him to present old material in the classroom since we could all read these ourselves. Instead, he expected us to spend our time reading textbooks and listing all the mistakes and incorrect concepts we found in our class notes. We then had to submit our notebooks to him by the end of the semester as part of our grade. I was shocked since I fully expected to learn the theories from his class presentation, not by reading on my own.

The second example is about technical writing. When I started to write a paper with him, he insisted that I draft an abstract first and then the conclusions from which the table of contents would support the conclusions. His English writing style was very elegant and concise, and he seldom wrote a long reference letter, mostly just a few paragraphs on his observations of your strengths and achievements. This left a deep impression on me. You need to keep your writing concise and to the point. Technical writing is not just about English. It is about how you organize, present and deliver it effectively. This approach has helped me write more than 20 books and 600 papers, and it has never let me down.

1.3.2 *Daniel C Drucker*

TAM NEWSLETTER

Drucker honored .

Education

Daniel Drucker was born in New York City on 3 June 1918 and started his engineering career as a student at Columbia University. His ambition was to design bridges, but as an undergraduate, he met a young instructor named Raymond D Mindlin, who told him that "he would pursue a PhD degree and he would write a thesis on photoelasticity". Drucker complied.

Career

He taught at Cornell University from 1940 to 1943 before joining the Armour Research Foundation. After serving in the US Army Air Corps, he returned to the Illinois Institute of Technology for a short time before he went to Brown University in 1947, where he did much of his pioneering work on plasticity. In 1968, he joined the University of Illinois as Dean of Engineering. During his more than 15 years there, the College of Engineering was consistently ranked among the best five in the nation. The college was known for its insistence on technical excellence. Drucker left Illinois in 1984 to become a graduate research professor at the University of Florida, where he retired in 1994.

Contributions

He is known throughout the world for his contributions to the theory of plasticity and its application to analysis and design in metal structures. He introduced the concept of material stability, now known as Drucker's stability postulate, which provided a unified approach for the derivation of stress-strain relations for the plastic behavior of metals. His theorems led directly to limit design, a technique to predict the load-carrying capacity of engineering structures. He also made lasting contributions to the field of photoelasticity.

Awards

The American Society of Mechanical Engineers (ASME) also honored him with the Timoshenko Medal (1983) and the ASCE presented him the Theodore von Kármán Medal (1966) and the first William Prager Medal (1983) from the Society of Engineering Sciences. He also received the John Fritz Medal (1985) from the Founder Engineering Societies, and the Modesto Panetti and Carlo Ferrari International Prize and Gold Medal (1999) from the Academy of Sciences of Turin.

Honors

In 1988 he received the National Medal of Science. He was a member of the National Academy of Engineering and the American Academy of Arts and Sciences, and he was a foreign member of the Polish Academy of Sciences.

1999 Distinguished Engineering Alumnus Medal, Dean Rod Clifton at Brown University.

Left: With Dean Drucker and my wife, Lily Chen, at the University of Illinois.
Right: With Drucker's son and two daughters in Florida.

The Symposium on Mechanics of Material Behavior (September 1983, University of Illinois, Urbana) marked Dean Drucker's 65th birthday.
Top row: PG Hodge, Jr, W Johnson and JL Sanders, Jr.
Bottom row: LM Kachanov and TH Lin (林同骅).

At the Academia Sinica meeting in Taipei with my former classmates of 1963 from Brown University.

1.4 My Life Journey with Two Giants — Daniel C Drucker and Lynn S Beedle

When I was a graduate student at Lehigh, the University of Illinois Civil Engineering (CE) School under the leadership of N Newmark was considered to be the best civil engineering school in the nation with its top testing facilities and super "who's who" giants on its CE faculty list, including, for example, the soil mechanics giant R Peck who co-authored the first-ever *Soil Mechanics* textbook with the father of soil mechanics K Terzaghi, and the structural guru Hardy Cross who invented the famous classical Moment Distribution Method for us to learn structural engineering. Newmark, on the other hand, helped build the famous Alaska pipeline by overcoming the very difficult permafrost soil foundation problem in order to transport the very hot Alaska oil to mainland US without the supporting foundations sinking. Newmark is now widely considered as one of the founding fathers of Earthquake Engineering.

All engineering faculty at the time were hands-on real-world engineers with deep physical insights. They used little mathematics like differential equations, but they could estimate the right amount of forces in their structures or soil foundations with their deep physical insights and slide rules and simple calculators. Today, we use widely popular limit analysis methods like the simple plastic design methods for steel frames or strut-tie model for reinforced concrete (RC) design, just like how they used their ingenious simple methods.

It was at Brown that I learned for the first time what was known at the time as the "engineering science education" — I was taught the highly mathematical theory of plasticity with tensor notations and to treat every structure as a continuous body known as a continuum. Over the next three years at Brown, we never sketched a single real structure like a beam or a plate or a shell. We drew continuums to represent the structures we had in mind.

After receiving my MS at Lehigh in structural engineering in 1963, I transferred to Brown to continue my graduate study and I wrote my doctoral thesis on soil plasticity and limit analysis under the supervision of Drucker. I was his second-last PhD student at Brown in 1966. Drucker made major contributions to the foundations of plasticity. Among them, the most important impact on me was his discovery and proof of the classical limit theorems on which the limit analysis techniques were developed in the 1950s.

In 1966, I returned to Lehigh to teach. However, there was no opening to teach the mathematical theory of plasticity for structural engineers, and I had to figure out a way to cover the subject that would be useful for civil engineers instead. So limit analysis became my obvious choice. In the teaching process, I developed the course contents tailored to civil engineers' applications in soil mechanics, in particular. In the process, I was able to reproduce almost all Terzaghi's work in his famous soil mechanics book with limit analysis methods. To my surprise, Terzaghi developed most of his solutions based empirically on the large amount of test data he collected during the construction of the New York subway. Simply put, he had all the real-world engineering answers to every geotechnical problem, but without a rigorous mechanics-based theory to guide him.

In 1975, I wrote *Limit Analysis and Soil Plasticity* (below, left), which introduced my limit analysis techniques to geotechnical engineers, in particular, with most examples taken from Terzaghi's book as validation for my techniques

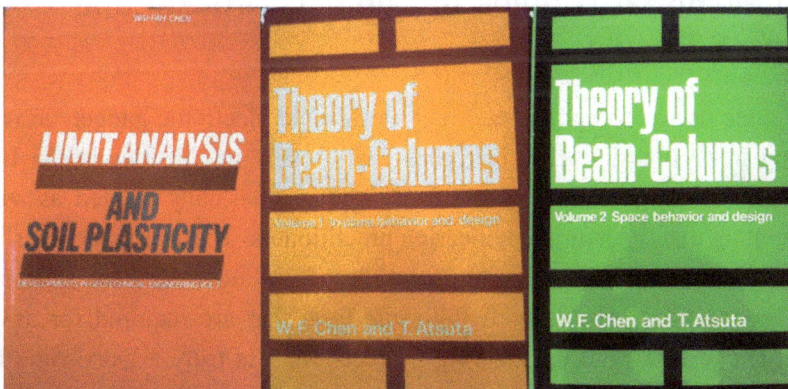

and its solutions for practical applications. This book received an instant and wide acceptance in the geotechnical community and created such excitement around the world. The merit of the limit analysis lies in the fact that engineers can now make practical and safe decisions on the design of complex load-bearing components based on relatively simple calculations.

Only in the last two decades were the lower bound limit analysis methods successfully implemented in the design of reinforced concrete structures, especially for deep beams, web opening and torsion, and beam to column connections. This is now known as *strut-tie model* in the terminology of the RC engineering field.

In the 1970s, our computing power changed drastically with mainframe computing. The Finite Element Methods were well developed and widely used in structural engineering. We were able to apply the theory of stability and the theory of plasticity to simulate the actual behaviors of structural members and frames with great confidence. It was the first time we were able to replace the costly full-scale tests with computer simulation. As a result, the limit state approach to design was advanced, and new specifications were issued.

It was at Lehigh that I worked closely with Beedle and made the major part of my contributions to the theory of plasticity and its application to the design of bi-axially loaded columns in engineering structures with the use of the newly developed plastic theory. As a result, I published a two-volume work in 1976–1977 on the *Theory of Beam-Columns* (above, center and right) with my doctoral student, Toshio Atsuta. This landmark work became the first ever in this field to discuss systematically the complete theory of in-plane and space beam-columns, and it shows how these inelastic theories are applied for the solution of practical design in steel frames beyond their elastic limits in contrast to the well-known elasticity-based books written by Stephen Timoshenko, the father of engineering mechanics in the 1950s.

1.5 My Memorial Tributes to Daniel C Drucker and Lynn S Beedle

With the wide use of computers and real-world field electronic gadgets in real-time monitoring in civil engineering in recent years, many universities in the US have now started to hire new faculty members who are basically trained as structural scientists rather than structural engineering practitioners. These engineering scientists are trusted more on their computer solutions since most of them lack physical insights and have experience with full-scale tests that are required for any large-scale innovative new civil engineering projects. This is truly a worrisome fact for us. Unlike the older generations of structural and geotechnical engineers with their strong gut feelings, the structural and geotechnical scientists of today mostly talk about the applications of AI technology and the use of accumulated large databases seeking their solutions.

I, like my former mentors Beedle and Drucker, still believe that structural and geotechnical engineers must still learn some quick and short methods to re-affirm their sophisticated computer solutions on the back of an envelope, so as to avoid potential disasters like the recent Boeing 737 Max incident that will surely happen again one of these days, as predicted by Murphy's law — "Anything that can go wrong will go wrong". Right!

For almost half a century, advances in the theory of plasticity and the work of Drucker, and the implementation of the plastic theory to steel structures and the leadership of Beedle were intimately linked in the US. It has been a wonderful experience and a rewarding career for me to have the privilege to work under and alongside them. As they both liked to emphasize, the true fulfillment of engineering research and education is "a place in practice". Their impacts on engineering practice and education have been tremendous. They remain role models for us all.

CHAPTER 2
LEHIGH UNIVERSITY, 1966–1976,
BEGINNING TO SHOW
初露頭角

2.1 The Beginning

During my 40-year academic career in the United States (US), ten years were spent at Lehigh University as a faculty member (1966–1976). In addition, I spent two years at Lehigh's Fritz Engineering Laboratory (Fritz Lab) as a graduate research assistant and received my MS degree in 1963. Well, looking back now from the outside, I can provide a better perspective on my university and education. My profession has always made me a constant Lehigh watcher, and I feel that Lehigh's way of teaching and research during Lynn S Beedle's period has been a success — it produced advanced and experienced students who later became leaders in education, the industry, and the armed services and government. Let me share some of my experiences and observations with my own students and their success stories during my time at Lehigh.

2.2 The Leader — Lynn S Beedle

When I was appointed as an assistant professor at Lehigh in 1966, Beedle was director of the Fritz Lab, and he directed research on the plastic design of steel structures. His research laid the groundwork for designing structures on the basis of their load-carrying capacity rather than their allowable stress. His students have gone on to become leaders in limit-state design, load and resistance factor design, and auto-stress design. Most of the experimental and theoretical work was performed at the Fritz Lab and formed the basis of a series of new American Institute of Steel Construction (AISC) specifications for structural steel building design from the 1950s to the present time. In particular, Beedle was a force behind the 1959 AISC Plastic Design in Steel and the 1986 AISC LRFD Specifications, which changed the whole design of steel structures.

Many of Lehigh's faculty members at the time were recruited by him. They came to Lehigh because of him and worked under him. He was a frequent international traveler and was a friend to everyone he came in contact with. The group of faculty members at the Fritz Lab were of an international mix, including, for example, Ted Galambos from Hungary, Alex Ostapenko from Russia, LW Lu (呂立武) and Ben Yen (顏本正) from Taiwan, Lambert Tall from East Europe, and John Fisher and George Driscoll from the US, among others. A good number of his former students were elected to the National Academy of Engineering in later years. Most of them are now leaders of steel research in Europe, Australian, Japan, as well as North America.

I was initially recruited by Lu to work with him on the research project "Columns under Bi-axial Loadings". Later, I was asked to work on the "Beam-to-Column Connections" project under Beedle. He was truly dedicated to the Fritz Lab and put all his effort into it. He pushed his students and junior colleagues to do the best they could. He had a passion for promoting Lehigh's reputation and maintained contacts and relationships with his former colleagues and students over many years. He even remembered my birthday and sent me a congratulation card regularly. He helped almost everyone improve their presentations and showed them how to prepare better slides.

Beedle devoted a tremendous amount of his time serving on several professional societies. He headed the Structural Stability Research Council (SSRC) for almost 25 years. The council is credited with influencing most of the stability research and steel design around the world. He also founded the multidisciplinary, international Council on Tall Buildings and Urban Habitat (CTBUH) in 1969. In this Council, he brought together many different groups of professionals besides structural engineers to look at various aspects of tall buildings from architectural, environmental, and social political viewpoints. He invited and encouraged us to get involved in these organizations and contribute our expertise. He was a firm believer that a true fulfillment of engineering research and education is "a place in practice". This conviction has been deeply rooted for most graduates from the Fritz Lab at Lehigh.

As a student, I used his book *Plastic Design of Steel Frames* to learn the plastic design methods. It was an easy to understand and concise book tailored squarely for practitioners. As an assistant professor, I worked alongside him and contributed to the writing of the ASCE's Manual 41 (Plastic Design in Steel). The Manual became a standard reference book for teaching plastic design methods in the country.

As a colleague, I contributed chapters to his first edited book, *Structural Stability: A World View*, and later, *Tall Buildings and Urban Environment Series*, which were published by McGraw-Hill. These books were truly international in their contents and authorships. He was indeed a giant in the world of structural steel design and a great mentor and my lifelong friend.

He passed away on 30 October 2003 of pancreatic cancer. He was 85.

Together with the Lehigh structural group and friends at one of the Structural Stability Research Council gatherings in the 1970s.

Fritz Engineering Laboratory, Lehigh University, Bethlehem, Pennsylvania.

2.3 Plasticity, Limit Analysis and Structural Design

2.3.1 *Limit analysis and soil plasticity*

When I started my teaching career at Lehigh in 1966, I started to apply the limit analysis techniques to soil mechanics and developed simple methods for obtaining practical geotechnical engineering solutions. Perhaps the most striking feature of the limit analysis method is that no matter how complex the geometry of a problem or loading condition is, it is always possible to obtain a realistic value of the collapse load without the use of a computer. Motivated by the success of this method, I wrote *Limit Analysis and Soil Plasticity* in 1975. This book soon became very popular because geotechnical engineers could now reproduce almost all the existing solutions in *Theoretical Soil Mechanics*, the well-known classical textbook by K Terzaghi, who is regarded as the father of soil mechanics.

I was truly flattered by an email I received many years ago from a graduate student in Brazil, who had been so excited to find *Limit Analysis and Soil Plasticity* at a bookstore in a remote area — the book had been out of print for many years. It was quite moving to read his letter. The graduate student expressed his and his professor's view on my work: "Next Wednesday, I will show this book to my professor and I am sure he will wish for a copy of this book, and I will give him this copy. He is very fond of you, him and all the classmates. In his classes, he always worships you as God."

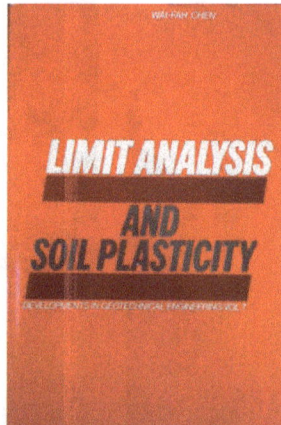

1975
Limit Analysis and Soil Plasticity. Amsterdam, Elsevier Scientific, Inc.
WF Chen

2.3.2 *Charles Scawthorn, Kyoto University, Japan*

Charles Scawthorn is a graduate of the Cooper Union, holds an MSCE degree from Lehigh University and received his Doctoral Engineering degree from Kyoto University, Japan. He was my first MS student in structural engineering at Lehigh, and he wrote a special topic report on "Limit analysis and limit equilibrium solutions

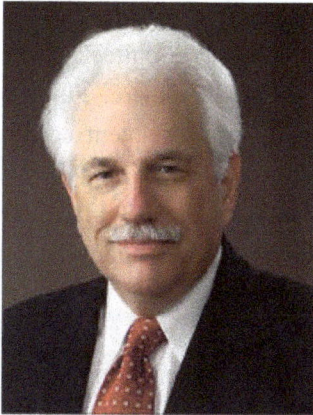

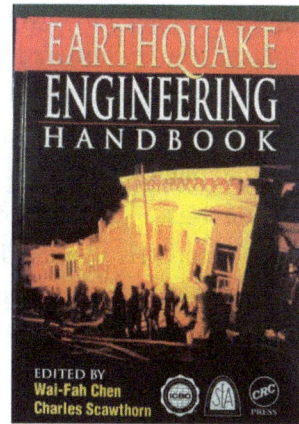

Charles Scawthorn, who co-authored *Limit Analysis in Soil Mechanics* and *Earthquake Engineering Handbook* with me.

in soil mechanics" with me when I started my teaching career. He joined EQE International in 1987 and rose to be one of the senior leaders in the organization. In 2003, the Civil Engineering Department at Kyoto University in Japan invited him to apply for a Professorship, which was a great honor for him and also a wonderful opportunity for him to pursue his professional interests and development.

In more than 40 years of practice, Scawthorn has designed and analyzed buildings and industrial structures and engaged in planning projects and research in the US and internationally. He established himself as an internationally recognized authority for the analysis and mitigation of natural and technological hazards. In 2002, we edited *The Earthquake Engineering Handbook*. He retired in 2008 as Professor and Head of the Earthquake Disaster Prevention Systems Laboratory, Kyoto University.

2.3.3 *Toshio Atsuta, New Industry Research Organization, Kobe, Japan*

Toshio Atsuta was the kind of "super" student all professors dreamed to have. With his top-notch academic credentials and real-world engineering work experiences, he was sent with full company financial support by a major Japanese corporation to Lehigh to study the then-hot topic "plastic design in steels". Atsuta had survived the dropping of the atomic bomb on Hiroshima on 6 August 1945. He was five at the time. Fortunately, his house was located safely away from the center of Hiroshima and so he was free from the contamination of radioactivity. In the spring of 1959, he passed an examination for the medical school but had been disappointed by the choice of medical courses on offered — there were none that incorporated mathematics. In the meantime he was fascinated by a lecture on "*Plastic Design*" offered by Y Fujita of the Department of Naval Architects, who was just returned from Lehigh University after finishing his PhD there. Fujita's lecture had a profound and longlasting impact on him and his subsequent career changes and growth.

He was a graduate of the University of Tokyo in 1963 with a BS in Naval Architecture. He worked there until 1965 as a research associate in civil engineering. Then, he went to Kawasaki Heavy Industries Ltd and worked for the technical institute until 2003, where he retired as director of the company.

In 1968, he heard from H Nishino, who had just returned from Lehigh with his PhD, that Lehigh was looking for his (Nishino) successor. In 1969, Kawasaki gave him a one-year leave and sent him to Lehigh University to get an MS degree. On 3 May that same year, he married the daughter of a close family friend. A few months later in July, they came together to Lehigh to commence his new plastic design in steel adventure and dream education. There, he received his MS and PhD degrees in 1970 and 1972, respectively. Both his degrees were in Structural Engineering. His doctoral dissertation work with me at Lehigh covered almost the entire subject of beam-column analysis, i.e., behavior of cross-sections and beam-columns in elastic-plastic regimes under in-plane or biaxial loading conditions.

For his MS degree, he developed the Column-Curvature-Curve (CCC) method for in-plane beam-column analysis under my direction. The CCC method is essentially a simplification of the then popular Column-Deflection-Curve method. With my helpful request letter to Kawasaki asking for a one-year extension of his leave period to complete his PhD degree (which was approved quickly), Atsuta completed his dissertation entitled "Analysis of inelastic beam-columns" in record speed. Both his MS thesis and PhD dissertation were hand-typed by his wife, Chika, a very challenging task because mathematical equations, Greek letters and mathematics symbols had to be incorporated into the manuscripts — there were no word processors nor personal computers at the time. All equations had to be typed symbol by symbol with a hand insert. It was very hard work for her. She had also been pregnant at the time. His thesis and dissertation work formed the basis of our joint two-volume treatise entitled *Theory of Beam-Columns — Volume 1 In-Plane Behavior and Design* and *Theory of Beam-Columns — Volume 2 Space Behavior and Design* (McGraw-Hill Book Company, 1976; 1977).

Atsuta had been Vice President of the New Industry Research Organization, Kobe, Japan, since its inception in 1997 until his retirement in 2003. The organization had been established to revitalize the regional industry damaged by the big earthquake in Kobe in 1995. Since his retirement in 2003, he applied for 350 patents for Kawasaki, of which 108 have been registered by the Japanese Patent Office.

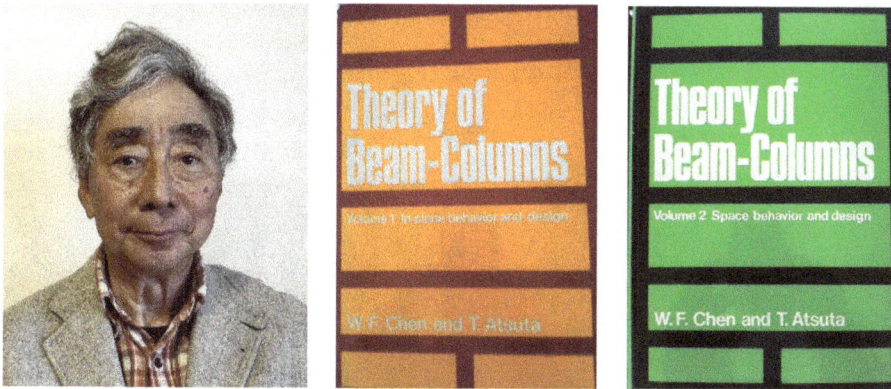

Toshio Atsuta and the *Theory of Beam-Columns* publications.

2.4 A Final Farewell to Bethlehem Steel

If you were a Lehigh graduate during my time, you will remember Bethlehem Steel Corporation as being a giant in the American steel industry, second only to US Steel. Lehigh's Fritz Lab was intimately related to Bethlehem Steel, and we also shared a great first-rate research facility near the Lehigh campus at Lehigh Hill. For us Lehigh Fritz Lab alumni, it is hard to say farewell to this great steel company.

Similarly, it is also hard for us to say farewell to the classical structural analysis methods. The following methods come to my mind immediately: the moment distribution, slope deflection, conjugated beams, column analogy, and Newmark's successive approximations, among others.

To this end, the following structures and mechanics giants come to mind too. This includes William Prager, the father of Applied Mechanics and Theory of Plasticity, S Timoshenko, the father of Engineering Mechanics, Nathan Newmark, the father of Structural Engineering, TY Lin (林同棪), the father of Prestressed Concrete, K Terzaghi, the father of Soil Mechanics, Daniel C Drucker on the Drucker Postulate and Material Stability, and Lynn S Beedle on Plastic Design, among others, but these people had the biggest impact on my career during my time.

We bid a final farewell to Bethlehem Steel on 19 May 2019, as its former headquarters, Martin Tower, which had been the tallest building in the greater Lehigh Valley after it was constructed in 1969, three years after I arrived at Lehigh, but would now implode in Bethlehem, Pennsylvania. Martin Tower opened at the height of Bethlehem Steel's success of being the second-largest steelmaker in the US, employing more than 120,000 people at the time. In a matter of seconds, this 21-story tower came tumbling to the ground, leaving behind only memories of what used to be Bethlehem Steel.

Martin Tower, Bethlehem Steel Headquarter, Bethlehem, Pennsylvania.

2.5 Some Large Scale Structural Steel Tests at a Glance

These tests had been conducted in the Fritz Lab, which was under Beedle, who was director of the laboratory.

2.5.1 *Biaxially loaded steel columns and beam-to-column connections*

The study of steel beam-to-column flange and web connections and biaxially loaded steel columns were sponsored jointly by the American Iron and Steel Institute and the Welding Research Council. The beam-to-column connections program consisted of 12 full-size specimens with various welded and/or bolted symmetrically loaded moment-resisting beam-to-column connections. These connections resulted in two publications — *Ultimate strength of biaxially loaded steel H-columns* (1973) and *Tests of bolted beam-to-column flange moment connections* (1975) — that are of importance in the design and construction of steel multi-story frames

1973
Ultimate strength of biaxially loaded steel H-columns
WF Chen, T Atsuta

1975
Tests of bolted beam-to-column flange moment connections
KF Standig, GP Rentschler, WF Chen

LEHIGH UNIVERSITY

Beam-to-Column Connections

TESTS OF
BOLTED BEAM-TO-COLUMN
FLANGE MOMENT CONNECTIONS

by
Kenneth F. Stanelg
Glenn P. Rentschler
Wai-Fah Chen

Fritz Engineering Laboratory Report No. 333.31A

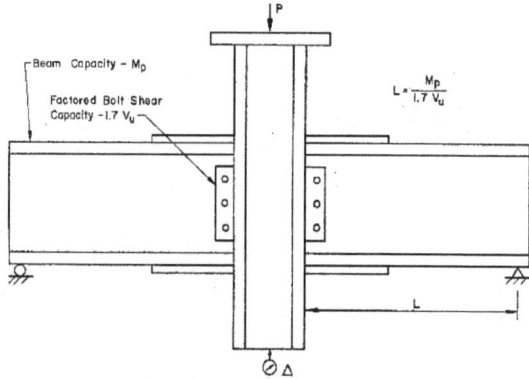

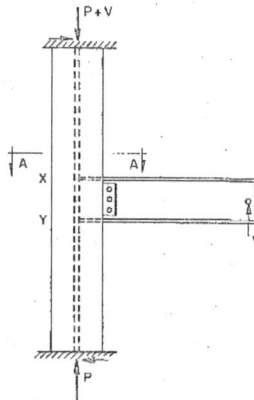

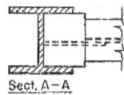

Fig. 1 Specimen Design and Test Setup

1978

Tests of beam-to-column web connection details in beam-to-column building connections: State of the art. In *Beam-to-Column Flange Moment Connections*, ASCE Preprint 80-179, April 1980 (80-9).

GP Rentschler, WF Chen, GC Driscoll

Fig. 1 Web Connection Assemblage

Beam-to-column web moment connections.

Those larger-scale testing results are also documented in a special edition of the *Journal of Constructional Steel Research* in 1988.

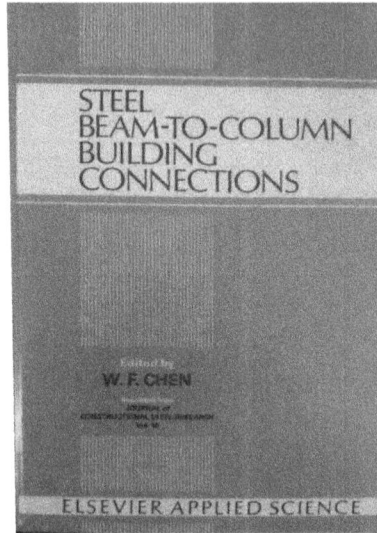

Chen, WF, Editor, *Steel Beam-To-Column Building Connections*, Journal of Constructional Steel Research, Special Issue, Vol. 10 (1988).

2.5.2 *Offshore platform fabricated tubular columns*

The study of fabricated tubular columns used in offshore platforms was sponsored by the American Petroleum Institute (API). The experimental investigation involved the study of the behavior of 10 tubular column specimens in the Fritz Lab's 5 million pound testing machine. These specimens ranged in length from 18 to 36 ft and were of diameters varying between 15 and 22 inches. The figure below showed a medium-length member being prepared for testing. Even for these members, it was clear that the length was greater than the two-story height of the Fritz Lab.

It is interesting to point out that these columns are traditionally manufactured by rolling lengths of a plate into cylinders, which are then welded longitudinally into "cans". Two or more "cans" are usually welded together end-to-end to form a column.

1976

Tests of Fabricated Tubular Columns, January.

DA Ross and WF Chen

Fabricated tubular columns.

CHAPTER 3
PURDUE UNIVERSITY, 1976–1999, HIGHLY PRODUCTIVE YEARS
得心應手

3.1 Stand Alone and Lead (獨當一面)

Purdue University, West Lafayette, Indiana.

I joined Purdue as Professor of Structural Engineering in 1976 after the retirement of Professor John E Goldberg in 1975. Goldberg was an internationally known expert in structural engineering. He served as the Head of the Structures Area in the School of Civil Engineering. He worked at Purdue from 1950 to 1975 and passed away in 1995. He was a legend in structural engineering at Purdue; his stature was similar to Nathan Newmark to the University of Illinois, Lynn S Beedle to Lehigh University, and George Winter to Cornell University. Most of our structural engineering faculty members at Purdue at the time were his students. Their teaching styles and contents were very much influenced by him, and they followed his footsteps. He was the role model for most of them.

Goldberg's impact on classical structural mechanics and structural engineering was profound. His analysis and design techniques were practical and elegant. Even today, engineers apply his knowledge concerning the behavior and design of building structures. After I became Head of Structures Area in 1980, we started to build and expand the core group of faculty members in engineering and applied mechanics. We started several frontier areas of research, including construction safety assessment, domain-specific software development environment, and advanced analysis for steel design, among others. In 1992, I was appointed the first George E Goodwin Distinguished Professor of Civil Engineering.

Left: 1992 Distinguished Professor in the School of Civil Engineering.
Right: Elected to the US National Academy of Engineering in 1995.

Left: The Area Head office of Structural Engineering at the School of Civil Engineering at Purdue in West Lafayette, Indiana, 1980–1999.
Right: My son Arnold and granddaughter Chloe moved to West Lafayette from Sunnyvale, California, in the summer of 2015. Here, they are paying a visit to the Structural Engineering Head Office.

對李遠哲的肯定多過批評。

Elected to the Academia Sinica (Taiwan's National Academy of Science) (中央研究院) in 1998.

3.2 Teaching and Mentoring (言教身教並重)

Over 23 years at Purdue, I taught mainly graduate courses including "Advanced Mechanics", "Structural Stability" and "Structural Plasticity". I only taught a few undergraduate courses. I produced 45 doctoral students but few Master students. The average graduation rate was two per year. This was exactly twice the average production rate of doctoral students per faculty per year at Stanford University. Most of my students were research assistants; they usually came as a cluster, or group, from different regions of the world depending on a specific period of time.

Toshio Atsuta had been my first Japanese student when I began my teaching career at Lehigh. When he returned to Japan and through his recommendations, I accepted two of his colleagues as my research assistants: first, Shouji Toma, and later, Hironori Sugimoto. Both of them worked as design engineers at the Kawasaki Heavy Industries, Ltd. Their experiences expanded the breadth of my technical interests, and the timing was just right for a new research direction in offshore structures. Three more Japanese students came later. They were Eiji Mizuno, Yasuhiro Ohtani and Eiki Yamaguchi. They were recruited through their advisors at the Nagoya, Kobe and Tokyo Universities. All of them had a strong engineering science background. They were all very eager to learn and do hard work.

32   *Plasticity, Limit Analysis, Stability and Structural Design*

Left: At the rooftop of Toshio Atsuta's house: (from left) Shosuke Morino, Arnold Chen, Hironori Sugimoto, me, Toshio, Chika Atsuta and Shouji Toma.
Right: A gathering of Japanese doctoral students at the Kawasaki Industry guesthouse organized by Toshio.

The 2014 Sinica meeting with a lunch gathering with former students and friends at Howard Plaza Hotel:
Seating, left to right: YB Yang (楊永斌), me and FH Wu (吳福祥).
Standing, left to right: WS King (金文森), Lee (李騰芳), JL Peng (彭瑞麟), TK Huang (黃添坤), YL Huang (黃玉麟), YF Li (李有豐), YK Wang (王永康) and K Hwa (華根).

In the 1980s, the Taiwan economy started to take off, and a good number of students were sent abroad by its government for graduate studies. They came from two sources: Academic institutions and the military community. Unlike the University of California at Berkeley, the campus at Purdue was considered a conservative environment suitable for students with a military background. Nine students came from academic institutions: CJ Chang, MF Chang (張明芳), SS Hsieh (謝錫興),

TK Huang (黄添坤), JL Peng (彭瑞麟), YL Huang (黄玉麟), CH Lai (賴志弘), and IH Chen (陳亦宏); most worked as my research assistants. The three students from military institutions were FH Wu (吳福祥), WS King (金文森) and HL Cheng (成曉琳). They were a highly selected group of Army officers.

In 1981, I was invited by the Nanjing Institute of Technology (NIT) (南京工學院) to give a lecture series in Nanjing, China, for two weeks. It was the first time that I returned to China since I escaped to Taiwan with my family in 1950. After I returned to Purdue, I started to accept a few Chinese students. The first two doctoral students were Dajian Han (韓大建) and Xila Liu (劉西拉). Han had been recommended to me by her department chair at the South China Institute of Technology (華南工學院) in Guangzhou (廣州). I met Liu at NIT and knew him in Nanjing. They were characterized by their strong desire to catch up on the missing years lost during the Culture Revolution. They were mature and appreciative of the opportunity to learn. They were very hard workers and quite independent in carrying out research activities.

My lecture series at the Nanjing Institute of Technology (NIT) 南京工學院 in August 1981.

One of my lectures at the Chongqing Architecture University (重慶建築大學), Summer, 1987.
First row, from left, fourth: XL Liu (劉西拉), sixth: me, seventh: DJ Han (韓大建), eighth: SP Zhou (周綏平).

2014 Fremont California reunion with former doctoral students couples: Alexander Chen, Weihong Yang, Mingzhu Duan, Hong Zhang, me and my wife, Lily, and Xila Liu and Lian Duan.

In the years that followed, four more Chinese students were offered research assistantships under my guidance. They were L Duan (段煉), H Zhang (張宏), MZ Duan (段明珠) and WH Yang (楊衛紅). L Duan came from Taiyuan University of Technology directly, while Zhang was recruited from Peking University for his expertise in computer software development. MZ Duan and Yang transferred to Purdue from other universities in the US.

In the 1980s, many Arab students from the Middle East were encouraged and financially supported by their governments to come to the US for graduate studies. They furnished the highlights of my academic career and expanded the breadth of my cultural diversity and technical interests. Their activities were characterized by their desire to return to their countries to teach and to establish new businesses. They wrote well and spoke fluent English. They were generally well prepared in engineering fundamentals. Three students — SI Al Noury, F Al Mashary and KH Mosallam — were from Saudi Arabia, while six — AF Saleeb, SE El Metwally, MA Barakat, MM El Shiekh, MM Abdel Ghaffar and AM El Shahhat — were from Egypt and Jordan. Most of them were supported by their governments through the Peace Scholarship program for graduate studies. I even had one student — M Aboussalah — from the University of Morocco!

The students from Korea seemed to be more interested in steel than in concrete structures. During my last few years at Purdue, I had the opportunity to supervise four Korea graduate students. They were SE Kim, CB Joh, YS Kim and CS Doo. They all worked on steel frames with rigid or semi-rigid connections. Their work ethics and devotion to hard work truly reflected the rise of Korea's economic power after the devastating Korean War.

3.3 Offshore Structures and Plasticity (近海結構和塑性力學)

3.3.1 *Shouji Toma, Hokkai-Gakuen University, Japan*

A brief history of my relationship with my Japanese students and researchers

In the 1960s and 1970s, many Japanese professors and researchers, including Fumio Nishino, Yushi Fukumoto and Toshio Atsuta from leading universities and industries in Japan, visited and attended Lehigh to study steel structural engineering, so we became classmates and good friends. This opened opportunities later for me to recruit many Japanese young structural scholars and engineers to Purdue to study with me.

My first Japanese PhD student at Purdue was Shouji Toma in 1977, followed by Hironori Sugimoto. Both were co-workers of Atsuta at Kawasaki Heavy Industries in Kobe. Then, Eiji Mizuno, who was a student of Fukumoto, came from Nagoya University. In 1983, Yasuhiro Ohtani, who was another of Fukumoto's students, came from Kobe University, and in 1984, Eiki Yamaguchi, who was Nishino's student, came from Tokyo University. Later, Yoshiaki Goto came as a visiting professor from the Nagoya Institute of Technology and stayed with us for one year during his sabbatical year. He was another of Nishino's students too.

In the 1960s, Sumio Nomachi, who was a professor at the Muroran Institute of Technology (later moved to Hokkaido University), had an interest in the theory of folded plates and visited Goldberg at Purdue for his sabbatical leave. This led to more of his students coming to Purdue, including Yuya Honda from the Hokkai-Gakuen University, Kenichi Matsuoka and Norimitsu Kishi, who were both from Muroran Institute of Technology, and Tomoyuki Sawada from the Engineering College of Tomakomai.

In July 1995, I had the opportunity to visit Japan under the US-Japan Exchange Program sponsored by the Japanese government. During this visit, my son Brian and I had the opportunity to travel all the way from the southern to northern parts of Japan, including Kyushu, Hiroshima, Kobe, Nagoya, Tokyo and Sapporo. We were hosted by my former students. We visited almost everywhere. It turned out to be a special period of time because the Great Kobe Earthquake that had occurred a few months before and we had the opportunity to examine almost all types of structural damages that were caused by the earthquake.

Some highlights of my interactions with my Japanese students

Shouji Toma was a graduate of Kobe University with a BSCE degree in 1967. He came to Purdue upon the recommendation of Toshio Atsuta and received his advanced degrees in MSCE in 1978 and his PhD in 1980. He returned to Japan and started his teaching and research career at the Hokkai-Gakuen University, Sapporo, from 1982 to 2014. We published two books: *Analysis and Software of Cylindrical Members* (1996) and *Advanced Analysis of Steel Frames, Theory, Software and Applications* (1994).

We also contributed two chapters ("Bridge structures" and "Statistics of steel weight of highway bridges") in the *Handbook of Structural Engineering* (2005) and *Handbook of Bridge Engineering* (2000), respectively.

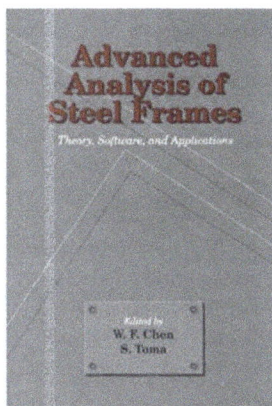

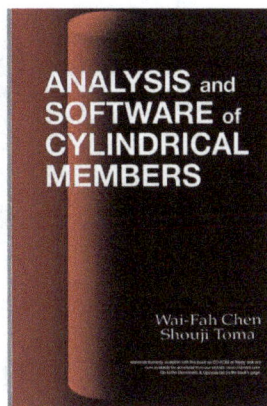

Left: Shouji Toma, Hokkai-Gakuen University, Japan.
Center: *Advanced Analysis of Steel Frames, Theory, Software and Applications.* CRC Press, 1994.
Right: *Analysis and Software of Cylindrical Members.* CRC Press, 1996.

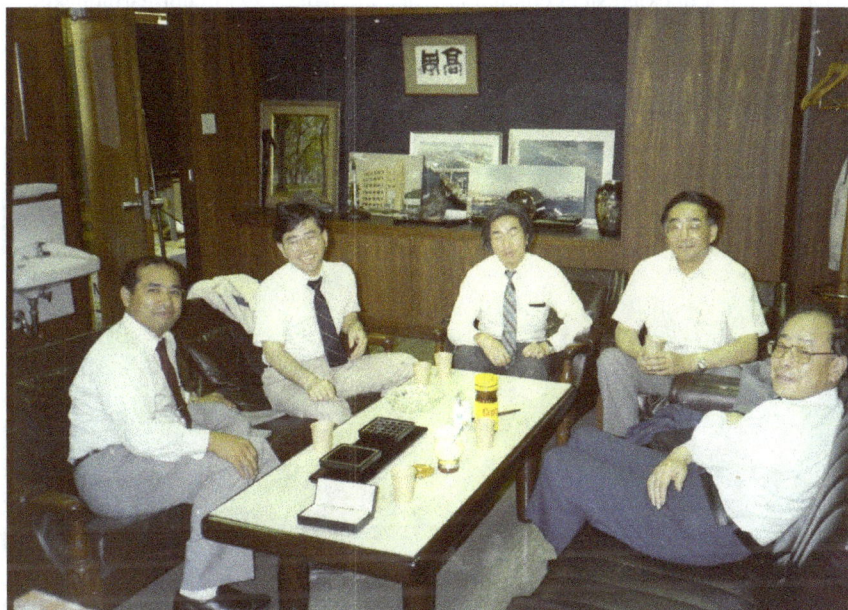

Gathering at the Hokkai-Gakuen University: (from left) Norimitsu Kishi, Yoshihiro Takahashi, Kenichi Matsuoka and Yuya Honda. All have visited Purdue University.

Observing a buckled steel column at Kobe after the 1995 major earthquake.

Visiting the construction site of the Akashi Strait Bridge, the world longest bridge.
From left: Hironori Sugimoto, Shosuke Morino, and me and my son, Brian, in 1995.

Touring the Hokkaido Shrine in 1995.

3.3.2 *Eiji Mizuno, Chubu University, Japan*

Eiji Mizuno was my doctoral student from 1978 to 1981 at Purdue University. He obtained his PhD degree with a dissertation on the subject of landslide problems. After his return to Japan, he worked on research topics of the cyclic behavior of steel and concrete in the structural division at the Department of Civil Engineering, Nagoya University, as a Research Associate (1981–1991) and then Associate Professor (1991–1998), and proposed a modified two-surface model for steel with yield plateau. This constitutive model is currently implemented as one of the material models in the general-purposed finite element analysis program DIANA.

During his career at Nagoya University, we published a classical comprehensive book on *Nonlinear Analysis in Soil Mechanics* (Elsevier, 1990).

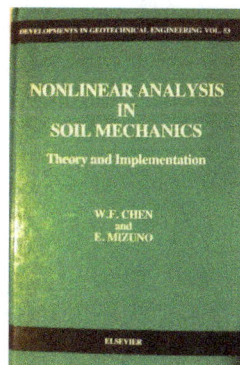

Left: Eiji Mizuno
Right: *Nonlinear Analysis in Soil Mechanics*, Elsevier, Amsterdam, 1990.

Eiji Mizuno at his laboratory in Japan (left); he also came with his students to visit me at the University of Hawaii at the Dean's office (right).

After moving to Chubu University in 1998, Mizuno has engaged in the development of techniques to enhance the seismic resistant-performance of reinforced concrete (RC) columns in the post-peak range. His current topics of research interests include:

– An experimental and analytical study on the cyclic behavior of RC columns in which the boundaries between rebar and steel-fiber reinforced concrete (SFRC) are un-bounded in the plastic-hinged zone.
– An experimental and analytical study on the improvement of seismic resistant-performance of repaired RC columns after heavy damage.

3.3.3 *Atef F Saleeb, University of Akron*

Atef Saleeb is currently a Distinguished Professor in the Civil Engineering department at the University of Akron, Ohio. He was a Cario University graduate with a BSCE degree in 1974. He received both his advanced degrees at Purdue with his MS in 1979 and PhD in 1981 while under me. His MS thesis "Elastic-plastic large displacement analysis of pipes" was published in the 1981 *ASCE Structural Journal*, while his PhD dissertation "Constitutive models for soils in landslides" was published as a two-part article in an ASCE Special publication entitled "Limit equilibrium, plasticity, and generalized stress-strain in geotechnical engineering" in 1981.

During his doctoral dissertation study, Saleeb also completed the draft of the first volume of *Constitutive Equations for Engineering Materials: Elasticity and Modeling* (John Wiley, 1982) while the second volume of *Constitutive Equations for Engineering Materials: Plasticity and Modeling* was published much later in 1994 in

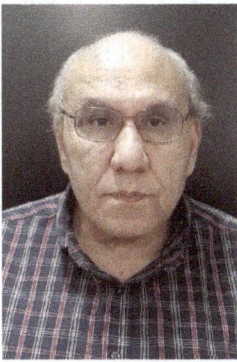

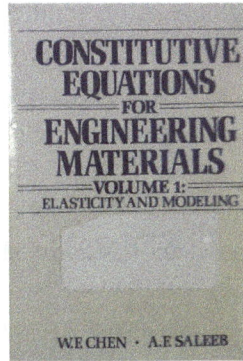

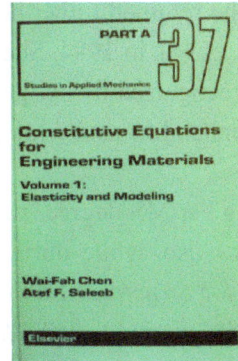

Left: Atef F Saleeb, University of Akron.
Right: *Constitutive Equations for Engineering Materials, Volume 1: Elasticity and Modeling.* Wiley Inter-science, 1982.

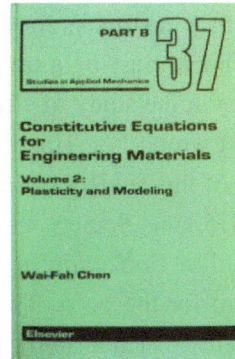

Collaborators WO McCarron (left) and E Yamaguchi (center) on *Constitutive Equations for Engineering Materials, Volume 2: Plasticity and Modeling.* Amsterdam, Elsevier, 1994 (right).

collaboration with two of my former doctoral students, WO McCarron of AMOCO in Tulsa, Oklahoma, and E Yamaguchi of the University of Tokyo, Japan.

As a Distinguished Professor at the University of Akron, Saleeb has been working closely with the Boeing Company and established himself as a world leader in the modeling of nonlinear materials such as Shape Memory Alloys and has become an expert in the field.

3.3.4 *Dajian Han* (韓大建), *South China University of Technology* (華南理工大學), *China*

Dajian Han graduated with a major in solid mechanics from Peking University in 1963. She received her MSCE and PhD in 1982 and 1984, respectively, from

Purdue. She returned to her home university, South China University of Technology (華南理工大學), in 1984. She was promoted quickly to a full professor in 1986. She was well recognized by her peers as a truly productive scholar and role model with a deep understanding and wide knowledge in the field of structural analysis and solid mechanics.

During her MS degree, she completed the draft of *Tubular Members in Offshore Structures*. Also, while during her doctoral study, she also completed the draft of *Plasticity for Structural Engineers*.

In 1988, we published *Plasticity for Structural Engineers* to provide engineers and students in structural engineering with the background needed to make the transition from fundamental theory to computer implementation and engineering practice since the mathematical theory of plasticity was beyond most of their reach. The book was well received by the civil and mechanical engineers and graduate students worldwide. It was also quickly adopted by a good number of universities as a graduate textbook. As the 1989 *Applied Mechanics Review* (Vol. 42) put it: "… This book must be a treasure in the bookshelf of those who seriously commit themselves in the study of structural plasticity .…". The 1989 *Technische Mechanik* (DDR, 10) wrote: "… This book will contribute to reducing the level of 'chaos' in treatment of this subject typical of other books".

In 2015, the Society of Civil and Architecture of the Guangdong Province published *The Collection of Professor DJ Han Papers*, and I was asked to write a Foreword for the book. This was what I wrote in my Foreword, "Wow ,what an achievement in such a short time from coming to Purdue University on February 14, 1980, at the age of 39, right after the Great Culture Revolution which deprived most

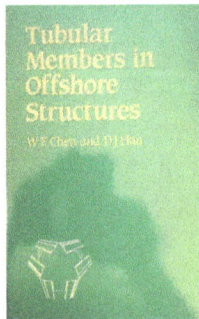

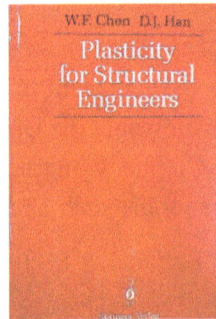

Left: Dajian Han (韓大建), South China University of Technology (華南理工大學), China.
Center, left: *Tubular Members in Offshore Structures*. London, Pitman, 1985.
Center, right: *Plasticity for Structural Engineers*. New York, Springer-Verlag, 1988.
Right: *The Collection of Professor DJ Han Papers*, Society of Civil and Architecture of Guangdong Province, 2015.

young scholars in her generation to study for new knowledge to her present status as an eminent national expert and well-respected scholar in the field of structural engineering." In just about four and a half years at Purdue, she had transformed herself completely and became a very productive scholar.

3.4 Structural Stability (結構穩定)

3.4.1 *Eric Lui* (呂汶), *Syracuse University*

In the 1980s, the Purdue Structures group became the Mecca for a number of young bright graduate students from other competitive schools in the US as well as from some of the top Asian universities. For example, Eric Lui transferred to Purdue from the University of Wisconsin in Madison in 1981 as a teaching assistant. He later became a research assistant in 1983 and finally a post-doctoral research associate in 1985. Lui came from Hong Kong to the University of Wisconsin before transferring to Purdue. At the time he had only planned for an MS degree, but I quickly discovered his talents and made sure that he would continue to pursue his career as a researcher and professor. As a result, and over a period of about a decade, we have already published five books and more than a dozen of technical papers together. In particular, our 1983 paper entitled "End restraint and column design using LRFD" in *Engineering Journal* stood out as an outstanding contribution to steel construction and we were awarded the 1985 TR Higgins Lectureship Award by the American Institute of Steel Construction. Lui has been the Meredith Professor of Civil Engineering at Syracuse University in New York since 2008.

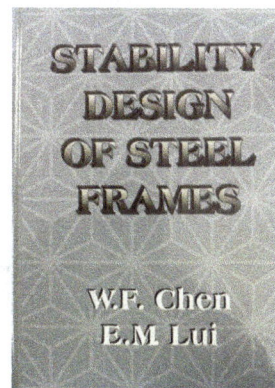

Left: TR Higgins Lectureship Award, AISC, 1985.
Center: *Structural Stability — Theory and Implementation*. New York, Elsevier, 1987.
Right: *Stability Design of Steel Frames*. FL, CRC Press, 1991.

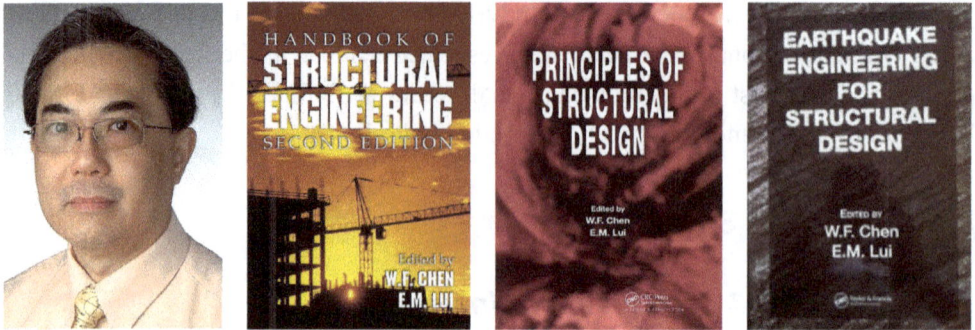

Left: Eric Lui (呂汶), Syracuse University, Syracuse, New York.
Center, left: *Handbook of Structural Engineering*. FL, CRC Press, 1991.
Center, right: *Principles of Structural Design*. FL, CRC Press, 2005.
Right: *Earthquake Engineering for Structural Design*. FL, CRC Press, 2006.

3.4.2 *Lian Duan* (段煉)*, California Department of Transportation*

When Lian Duan was an MS graduate student at the Taiyuan University of Technology (太原理工大學), China, he read and used a reprint of my book *Theory of Beam-Columns*. He had been impressed with the new developments of extending the famous Timoshenko's elastic solutions of beam-columns to the newly developed plastic theory for more realistic inelastic solutions. After he had enrolled at Purdue as my doctoral student in 1986, he came to my office to show me that reprinted copy of my book. I was happy to learn that my books were used in China. So he let me have his reprints as my souvenir. In those early years, Duan frequently showed up at the Civil Engineering Building's mailroom to check his mails from China.

Duan graduated in 1990 with his dissertation "Stability analysis and design of steel structures" split into three parts: "Effective length factor of framed columns", "Design interaction equations for beam-columns", and "Behavior and strength of dented tubular member", based on 10 published papers.

Duan is currently a senior bridge engineer at the California Department of Transportation (Caltrans). He joined Caltrans in 1991. His contribution to Caltrans is threefold: solving highly technical problems for the Seismic Retrofit Program, managing, editing and reissuing Caltrans' Bridge Design Practice Manual, and providing tireless support for steel bridge-related innovations. He has proposed and guided research leading to time and material savings in seismic retrofit and cost-effective seismic-resistant new steel bridge designs.

In the early 1990s, I served as a series editor for the CRC Press in civil and structural engineering areas. To this end, I produced my first *Civil Engineering Handbook* in 1995, followed by the *Structural Engineering Handbook* in 1997.

Left: Copies of *Theory of Beam-Columns.*
Center: Duan in my office at Purdue, 1998.
Right: Duan in the mailroom at Purdue.

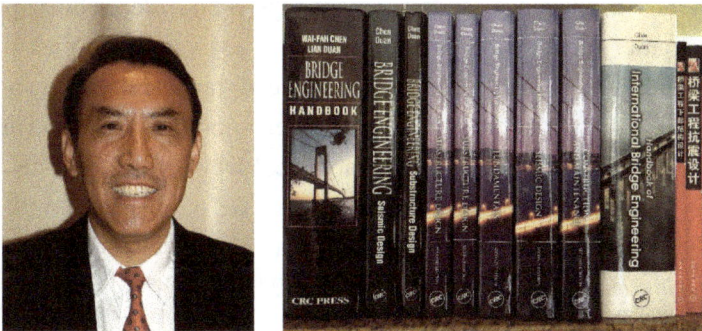

Lian Duan (段煉), California Department of Transportation.
Right, from left to right:
Bridge Engineering Handbook, CRC Press, Boca Raton, FL, 2000.
Bridge Engineering: Seismic Design, CRC Press, Boca Raton, FL, 2003.
Bridge Engineering: Substructure Design, CRC Press, Boca Raton, FL, 2003.
Bridge Engineering: Construction and Maintenance, CRC Press, Boca Raton, FL, 2003.
Handbook of International Bridge Engineering, CRC Press, Boca Raton, FL, 2014.
Bridge Engineering Handbook, Second Edition: Fundamentals, Superstructure Design, Substructure Design, Seismic Design, Construction and Maintenance, CRC Press, Boca Raton, FL, 2014.
Bridge Engineering: Substructure Design, China Machine Press, 2008.
Bridge Engineering: Seismic Design, China Machine Press, 2008.

In 1999, I moved to the University of Hawaii as Dean of Engineering, and it was thanks to Duan's participation that enabled us to continue our series production in the handbooks in *Bridge Engineering* in 2000 and *Earthquake Engineering* (with Scawthorn) in 2002. In 2014, we published the *Handbook of International Bridge Engineering* and also completed the most comprehensive and up-to-date five-volume reference and resource book: *Bridge Engineering Handbook: Fundamentals, Superstructure Design, Substructure Design, Seismic Design and Construction and Maintenance*, 2nd edition (2014).

3.5 The Center of Excellence: Advanced Analysis
(高等分析成就卓越)

3.5.1 *Richard Liew* (劉德源)*, National University of Singapore*

As another example, Richard Liew obtained his Bachelor and Masters of Engineering from the National University of Singapore in 1986 and 1988, respectively, which was followed by a PhD in structural engineering from Purdue in 1992, on a full scholarship provided by the Singapore government that was focused squarely on the emerging topic of advanced analysis in steel construction. He completed his work quickly and produced a landmark paper "Second-order refined plastic-hinge analysis for frame design" that made advanced analysis easy and practical for engineering practitioners, and this has influenced recent steel structural design codes and specifications. His method was further improved by the visiting scholar Yoshiaki Goto from Japan, who was on sabbatical leave from the Nagoya Institute of Technology to Purdue. With their intense effort, we published *Stability Design of Semi-Rigid Frames* in 1995 and several papers as well. Liew and I also edited *The Civil Engineering Handbook*, 2nd Edition, in 2002.

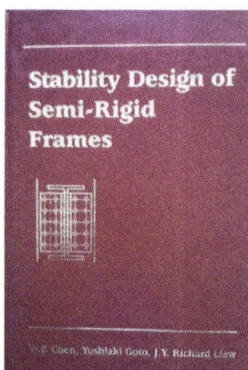

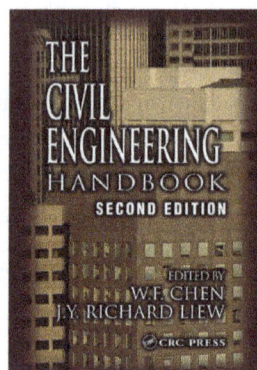

Left: Richard Liew (劉德源), National University of Singapore.
Center: *Stability Design of Semi-Rigid Frame*. New York, Wiley & Sons, 1996.
Right: *The Civil Engineering Handbook*, 2nd Edition. CRC Press, 2002.

Liew is currently a professor at the National University of Singapore and a former President of the Singapore Structural Steel Society. During his career, he has been a member of the Institution of Structural Engineers and the London-based Fire Engineering Task Group. He is also the International Structural Steel Research Advisor of the American Institute of Steel Construction. Liew was behind the construction of One Raffles Place and both the Flower Dome and Cloud Forest at the Gardens by the Bay. He is the key person responsible for developing Singapore's national annexes for Eurocodes 3 and 4, the latest codes of practice for design steel-concrete composite structures in Singapore.

3.5.2 *Norimitsu Kishi, Muroran Institute of Technology, Japan*

Purdue's Semi-Rigid Connection research (now known as Partially Restrained (PR) construction or PR connection) was advanced rapidly with the arrival of another scholar from Japan — Norimitsu Kishi, who came from the Muroran Institute of Technology in Hokkaido. In 1989, he built up a database on semi-rigid steel beam-to-column connections at Purdue University using a large amount of collected experimental test data. To build up this database systematically, he developed the Steel Connection Data Bank Program (SCDB). Using the system of tabulation and plotting in the SCDB program, the moment-rotation characteristics of each connection type were made available, and the appropriate connection models were also given in the program. This database enabled us to develop a more realistic analysis for semi-rigid frame design in North America. To this end, I edited a book entitled *Practical Analysis for Semi-Rigid Frame Design* to cover both the general methods as well as the specific criteria defined by the most recent American Institute of Steel Construction's *Load and Resistance Factor Design Specification*. This publication encompassed codes, databases, modeling, classification, analysis/design, and design tables and aids. Several of my graduate students at Purdue like YS Kim, IH Chen and K Wongkaew, did most of the works reported in this book.

Sometime after I moved to the University of Hawaii as Dean of Engineering, Kishi sent his doctoral student Masato Komuro to continue his work with me on

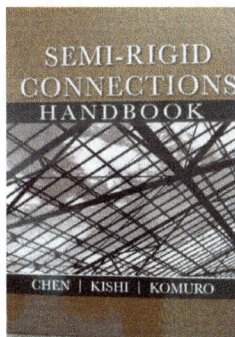

Left: Norimitsu Kishi, Muroran Institute of Technology, Japan.
Center: *Semi-Rigid Connections Handbook.* FL, J. Ross Publishing, Fort Lauderdale, 2018.
Right: Norimitsu Kishi and Eiki Yamaguchi of the Kyushu Institute of Technology, Japan.

the semi-rigid connection database. We used those collected connections database, developed our practical mathematical models for computer implementation, and also provided some case studies on some frames, including composite construction. The three of us compiled whatever information we had into the *Semi-Rigid Connections Handbook* in 2018 to be made readily available for practitioners to use.

3.5.3 *Iqbal S Sohal, Rutgers University*

Another book that grew out of my lecture notes that I used to teach my structural engineering graduate students at Purdue was by my former doctoral student, Iqbal S Sohal. The primary purpose of this textbook, entitled *Plastic Design and Second-Order Analysis of Steel Frames*, was to present the basic concept and methods of analysis of plastic theory and show how to use the theory in practical frame design. It also discussed how the practical design rules in the AISC-LRFD specifications were related to theoretical considerations. This was because the advent of personal computers, particularly in the computing and graphics performance of engineering workstations, had resulted in more sophisticated methods of analysis feasible in design practice. While the use of first-order analysis for elastic or plastic design is still the norm for engineering practice, a new generation of codes has emerged that recommends the second-order theory as the preferred method of analysis.

For the first time, the book came with two diskettes containing two computer programs that were capable of tracing every plastic hinge formation throughout

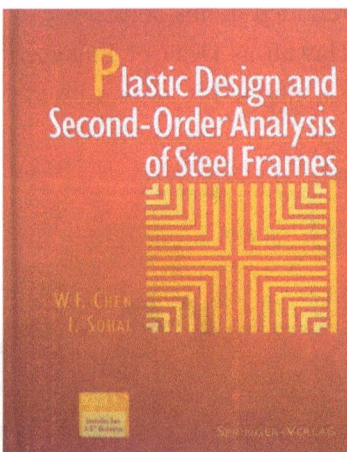

Left: *Plastic Design and Second-Order Analysis of Steel Frames*. New York, Springer-Verlag, 1995.
Right: SL Chan (Hong Kong), R Liew (Singapore), E Yamaguchi (Japan), JL Peng (Taiwan) and SE Kim (Korea) met at an International Conference held in Hong Kong.

the entire range of loading up to plastic collapse or stability failure. The computer programs had been developed by M Abdel-Ghaffar and JYR Liew as part of their PhD thesis work at Purdue for the research project entitled "Second-order inelastic analysis for frame design", sponsored by the National Science Foundation. Sohal graduated in 1986 with his dissertation "Local buckling in the analysis of cylindrical members".

3.5.4 *Seung-Eock Kim, Sejong University, Seoul, Korea*

Though advanced analysis was capable to capture the limit state strength and stability of a structural system along with its each member directly, the use of elastic analysis was still the norm in engineering practice at the time. However, with a new generation of codes being under development in anticipation of the use of the advanced analysis methodology in the near future, the advanced analysis book with my former doctoral student Seung-Eock Kim, entitled *LRFD Steel Design Using Advanced Analysis*, was published in 1997 to address squarely both analysis and design, emphasizing the direct use of the methods in engineering practice. This was a timely and great introduction to an exciting new trend in structural engineering!

Kim is a professor in the Department of Civil and Environmental Engineering and currently serves as the Vice President of Sejong University, Seoul, South Korea. His current research includes the reliability study of semi-rigid frames using advanced analysis, among others. In a recent email from Kim, he said to me: "Sometimes I think of my life in Purdue, and I also think of going to Hawaii and seeing a professor. I am convinced that studying as your student at Purdue has been the basis for my relatively successful achievement as a scholar in Korea."

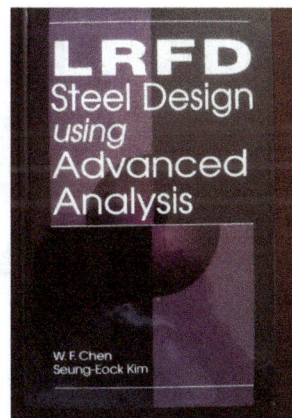

Left: Seung-Eock Kim, Sejong University, Seoul, Korea.
Right: *LRFD Steel Design Using Advanced Analysis*. FL, CRC Press, 1997.

3.5.5 *Fu-Hsiang Wu* (吳福祥), *High-Speed Railway Bureau, Ministry of Transportation and Communications* (交通部高速铁路工程局)*, Taiwan*

After Fu-Hsiang Wu returned to Taiwan from Purdue in 1988, he immediately participated in the construction of the Taipei underground railway project. He was also responsible for the construction of the Taiwan High Speed-Railway project. He first served as the Chief of Engineering Department at the Taipei City Underground Railway Engineering Bureau, and later he served as Director of the Bureau of High Speed Railway (高铁局), Ministry of Transportation and Communications and was fully responsible for its timely completion of the construction over a period of 16 years. In January 2007, on the eve of the opening of the Taiwan High-Speed Rail, he retired from the Bureau and transferred to a teaching position as the Dean of the School of Architecture and Planning of the National Chung Cheng University (國立中正大學).

With his rich practical experiences in railway engineering, he helped the University to establish the first-ever "National Construction Project Management Research and Development Center". Wu graduated from the National Chung-Cheng Institute of Technology (國立中正理工學院) and obtained his MS degree from Ohio State University in Columbus, Ohio. He later transferred to Purdue to earn his PhD under me in 1988 with his dissertation entitled "Semi-rigid connections in steel frames". He had a very wide experience in technical theories, engineering practices and major construction project management skills, especially interpersonal skills in dealing with people on the very challenging task of compulsory land acquisition for the railway construction.

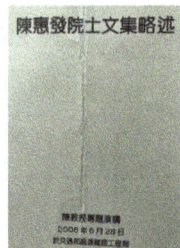

Left: Fu-Hsiang Wu (吳福祥), High-Speed Railway Bureau, Ministry of Transportation and Communications (交通部高速铁路工程局), Taiwan.
Center, left: "The academic achievements and contributions of Academician Chen", "我的学术成就与贡献".
Center, right: In Fu-Hsiang Wu's office.
Right, left to right: FH Wu (吳福祥), me, TK Huang (黃添坤) and JL Peng (彭瑞麟).

Wu invited me to deliver a lecture at the High Speed Railway Bureau in Taipei on 28 June 2006. His assigned lecture topic for me was:

"The academic achievements and contributions of Academician Chen" ("我的学术成就与贡献").

3.5.6 *Won-Sun King* (金文森), *Chaoyang University of Technology* (朝陽科技大學), *Taiwan*

Won-Sun King (金文森) was Professor of Construction Engineering at Chaoyang University of Technology (朝陽科技大學), Taiwan. He retired in 2020. He earned his PhD from Purdue under me in 1990. He had a passion in engineering and technology education for integrity to reduce the level of corruption in the construction industry. He wrote a textbook on *Engineering Ethics* (in Chinese) and taught Engineering Ethics at his University. He is a firm believer that construction corruption could be reduced by ethics education.

He has been active in implementing several of his practical methods for second-order inelastic semi-rigid frames analysis for engineering practice. These methods were published in a series of nine papers over a period of two years (1992–1994), mostly in ASCE, ACI and AISC journals. We published a Chinese textbook entitled *Design of Steel Structures* (鋼結構設計) in Taiwan in 2008.

Left: Won-Sun King (金文森), Chaoyang University of Technology (朝陽科技大學), Taiwan. Right: *Design of Steel Structures* (鋼結構設計). Science and Technology Press, 2008.

A glance at King's articles on Ethic Education

– WS King, L Duan, WF Chen, CL Pan. (2008) "Education improvement in construction ethics", *Journal of Professional Issues in Engineering Education and Practice*, Vol. 134, No. 1.

– WS King, L Duan. (2018) "Engineering ethics and social responsibilities education in civil engineering", *International Journal of Civil Engineering and Construction Science*.
– WS King, L Duan. (2016) "Education effects on engineering ethics", *Journal of Architecture*, Number 95.

3.6 Pioneering Work (開創新的領域)

As mentioned earlier, after I returned from my month-long lecture tour in China, I started to accept a few Chinese students. My first two doctoral students from China were Dajian Han (韓大建) and Xila Liu (劉西拉). They were both graduates of China's most prestigious universities: Han from Peking University (北京大學), who majored in solid mechanics, and Liu from Tsinghua University (清華大學), who majored in structural engineering. This is equivalent to the US's Harvard University and Massachusetts Institute of Technology, respectively. They were the cream of the crop — the best graduates one could dream of, and I was lucky to have both. Though they suffered from China's infamous Cultural Revolution at the time, they were nonetheless very eager to learn and contribute.

To this end, I started two pioneering projects for them to try: one explored the construction load analysis for concrete structures for safe construction and the other was to develop a constitutive model of concrete materials for use in the popular finite element analysis (because our computing power was expanding explosively from the old mainframe computing to the more modern convenient workstation computing at Purdue during this period). Liu initially had some reservations about the construction load analysis project because he was unsure whether this new field would have any relevance for him in China. Han took on the constitutive modeling project happily because her background in Peking University was in plasticity, and I also personally knew of her thesis advisor's reputation in plasticity.

They both finished their doctoral degrees in three and four years, respectively. They also both achieved their milestones in their respective fields admirably well. As a result, we published a series of papers on each subject in high-quality professional society journals. Their pioneering work opened the route map for a group of new students from different universities in different countries. The new students working in the new construction load analysis project were: MM El-Shiekh (1989) from Egypt, KH Mosallam (1991) from Saudi Arabia, AM El-Shahhat (1993) from Egypt, JL Peng (1994) from Taiwan, and MZ Duan (1996) from China. The new students working on the constitutive modeling project were WO McCarron (1985) from the US, and Y Ohtani (1987) and E Yamaguchi (1987), who were both from Japan.

3.6.1 *Xila Liu* (劉西拉), *Tsinghua University* (清華大學), *China*

Xila Liu was a graduate from Tsinghua University (清華大學) in structural engineering in 1963. He received both his MS and PhD in 1984 and 1985, respectively, from Purdue. He returned to China with his wife Chen Chen (陳陳) in 1986. Chen Chen was also a doctoral student at Purdue's Electrical Engineering Department. Xila took a faculty position at Tsinghua University (清華大學) in 1986 while Chen Chen took on an engineering job in power distribution in her native city of Shanghai. Xila was quickly promoted to a full professor in 1987. They were the first married couple with advanced degrees returning to China immediately after their graduation in the US. Their return created such a sensation back home and made instant headlines. They were both received by the highest government officials in China, even including Deng Xiaoping (鄧小平)! Liu is currently active as a popular and admired teacher at the Shanghai Jiaogtong University.

Liu's MSCE thesis was entitled "Reinforced concrete centrifugal pipe columns", in which our theoretical work was verified by the vast test data he collected and brought with him to Purdue on the pipe column tests he did during his early years' assignments on construction sites in the western city of Chongqing (重慶市). This pipe-column paper was published in a two-part article in the *Journal of Structural Division of American Society of Civil Engineers* (ASCE) in 1984. The papers were considered to be the most valuable contribution for 1984, and we received the prestigious 1985 Raymond C Reese Research Prize by ASCE.

He helped me to complete the draft of *Limit Analysis in Soil Mechanics.*

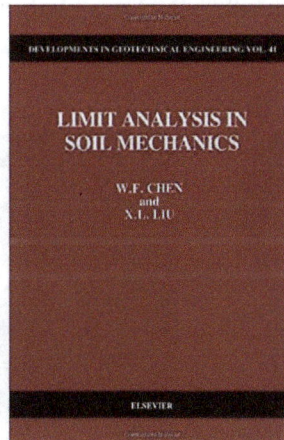

Left: Xila Liu (劉西拉), Tsinghua University (清華大學), China.
Right: *Limit Analysis in Soil Mechanics.* Amsterdam, Elsevier, 1990.

Xila was a very talented individual with a wide interest ranging from participation in symphony orchestra performances to various leadership roles in professional society activities. He won several best teacher awards, and he led a team to draft China's national strategic development white paper on engineering technology in general and structural engineering in particular. As a result he was frequently asked by the Chinese government to represent China in various international society activities. He bought great honor to Purdue and made us all very proud of him.

3.6.2 *Jui-Lin Peng* (彭瑞麟)*, National Yunlin University of Science & Technology* (國立雲林科技大學)*, Taiwan*

Jui-Lin Peng was a 1994 Purdue graduate with his dissertation entitled "Analysis models and design guidelines for high-clearance scaffold systems". He continued his research work in Taiwan but expanded greatly to investigate the actual failure mechanisms in various actual cases and tried to find out and document all possible causes of actual collapse mechanisms of temporary structures failures that were commonly used in construction in Taiwan. In the meantime, he also carried out several large-scale field tests on failure mechanisms of shores, re-shores and scaffolds that were commonly used in Taiwan construction.

As the result of his extensive research and accumulated field investigation experiences, he established himself as the leading authority on construction safety issues in Taiwan. He was frequently invited by government regulatory agencies to serve important roles for monitoring and setup safety regulations for the safe construction of temporary structures. These activities included, for example, to serve as a judge of the 8th "National Industrial Safety Award" of the Executive Yuan

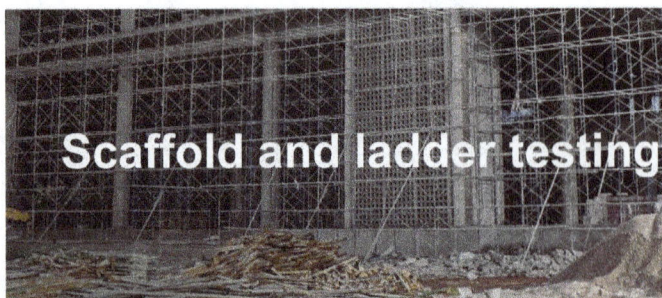

Left: Jui-Lin Peng (彭瑞麟), National Yunlin University of Science & Technology (國立雲林科技大學), Taiwan.
Right: Scaffold systems.

(行政院), a jury member and convener of the 3rd "National Work Safety Award" of the Executive Yuan (行政院), a member of the preliminary evaluation committee in the Ministry of Labor's (勞工部) "Promoting the selection of excellent public works and personnel for labor safety and sanitation", and the cadre of the academic test bank of the level C technicians of the "Steel pipe construction frame" of the Ministry of Labor (勞工部), among others.

Peng was a National Taiwan University (NTU) (臺灣大學) graduate and received the "Outstanding Alumni of Civil Engineering Department of NTU" in 2015. He currently is a distinguished professor with the Department of Civil and Construction Engineering (營建工程系), National Yunlin University of Science & Technology (國立雲林科技大學).

3.7 Outreach and Sharing the Knowledge (推動知識共享)

3.7.1 *Tian Qing Yu* (余天慶)*, Hubei University of Technology* (湖北工業大學)*, China*

In 1996, Professor TQ Yu made a short visit to Purdue and created an unexpected long lasting relationship between us after he started to translate several of my books into Chinese. He and Dr Xunwen Wang (王勳文) and Dr Zaihua Liu (劉再華) started the most challenging one, the two-volume treatise *"Constitutive Equations for Engineering Materials"* and worked over a period of three years and published their work in 2001 by Huazhong University of Science & Technology (華中科技大學). Their books were an instant success and widely popular in China. A big success for them for sure! As a result, Professor Yu was so encouraged by his success and I was happy too. With my blessing, he and his team reorganized it into two separate textbooks: *"Elasticity and Plasticity"* and *"Constitutive Equations for Concrete and Soil."* With more than three years of tireless efforts, their books were published by China Architecture & Building Press in 2004. Professor Yu believed that these books helped bring China's civil engineering technology to the forefront quickly and improved significantly the quality of graduate education in China. The first volume *"Elasticity and Plasticity"* were reprinted 11 times with 14,186 copies from 2004 to 2018, and the second volume *"Constitutive Equations for Concrete and Soil"* were reprinted 4 times with 5288 copies from 2004 to 2019.

In 2006, after I stepped down as the Dean of Engineering at the University of Hawaii, I wrote my biography *"My Life's Journey: Reflections of an Academic"*. When Professor Yu had the opportunity to read it, he was so moved by my life story.

Left: Tian Qing Yu (余天慶), Hubei University of Technology (湖北工業大學), China.
Center: *Constitutive Equations for Engineering Materials*, Huazhong University of Science & Technology (華中科技大學), 2001.
Right: *"Elasticity and Plasticity"* and *"Constitutive Equations for Concrete and Soil"*, China Architecture & Building Press (中國建築工業出版社), 2004.

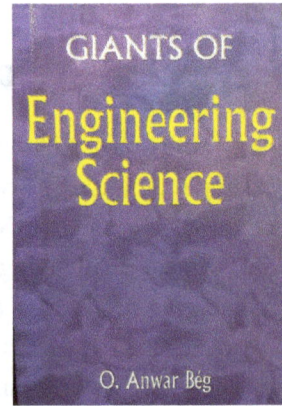

Left: *My Life's Journey, Reflections of an Academic*, World Scientific Publishing Company, Singapore, 2006.
Right: *Giants of Engineering Science, Matador*, UK, 2006.

Later, in the same year, he bought a copy of the book by O Anwar Beg entitled *"Giants of Engineering Science"* published in London in which he was fascinated by the special chapter focused on my academic achievements and contributions. At the end he and Dr Junfeng Shi (石峻峰) decided to translate my bio book for the greater Chinese community around the world.

Prof, Yu, Dr Xiang Ding (丁祥), Dr Jungfeng Shi (石峻峰) and his student team have just completed their translation of my book *"Plasticity in Reinforced Concrete"* published by McGraw-Hill in 1982.

I was so pleased and admired that Prof, Yu and his outstanding team including Dr Xunwen Wang (王勳文), Dr Junfeng Shi Shi (石峻峰), and Dr Xiang Ding

(丁祥), translated several of my books to Chinese so successfully over the last 25 years.

3.7.2 *Suiping Zhou* (周綏平), *Chongqing Architecture University* (重慶建築大學), *China*

Suiping Zhou was a graduate from Tianjin University (天津大學) and a Professor at Chongqing Architecture University (重慶建築大學), China.

Zhou came to Purdue as a visiting scholar and worked with me on the subject "Analysis and design of biaxially loaded columns" from 1983 to 1985. She worked very hard and fared amazingly well during her stay. Her academic achievements were also most impressive. We published a paper on "Design criteria for box-columns under biaxial loading" in the *ASCE Structural Journal* in 1985.

In 1994, based on her extensive and rigorous analysis results on various beam-columns solutions, we contributed a chapter "Plastic-zone analysis of beam-columns and portal frames" to my 1994 CRC book *Advanced Analysis of Steel Frames*, which was co-edited with S Toma.

Zhou (周綏平), XL Liu (劉西拉), DJ Han (韓大建), Q Lu (陸楸), HB Wu (吳惠弼), and WS King (金文森) worked together to translate my 1976–1977 two-volume book *Theory of Beam-Columns* into Chinese. The translated book, which also came in two volumes, was published in Taipei and Beijing in 1995 and 1996, respectively.

In 1996, she translated into Chinese my 1991 book *Stability Design of Steel Frames*, which was co-authored with E Lui. The book was published in China.

Theory of Beam-Columns (2 volumes, in Chinese).
Left: The traditional Chinese version by Science Technology Book Co. (科技圖書股份有限公司), Taipei (臺北市), 1995.
Right: The simplified Chinese version by People Communication House (人民交通出版社), Beijing (北京市), 1996.

Left: Suiping Zhou (周綏平), Chongqing Architecture University (重慶建築大學), China.
Right: *Stability of Steel Frame Design*, World Book Publishing Company, Beijing (北京市), China, 1999.

In 1999, Zhou, King and Duan published *Design of Steel Struc*tures (鋼結構 設計), introducing the newly developed LRFD Method for Chinese readers. She also published a teaching software for the *Structural Matrix Analysis* textbook.

She investigated reinforced concrete bird-shaper roof structures extensively and made a great contribution to its application in practice in China. She won several research and teaching excellence awards from the Sichuan Province (四川省).

With Suiping Zhou (周綏平) in Chongqing (重慶市), China, and my family (Lily, me and my son Arnold) in the summer of 1987.

3.7.3 *Some related translation of our books*

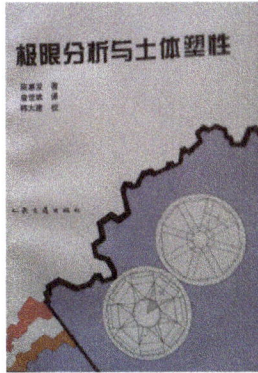

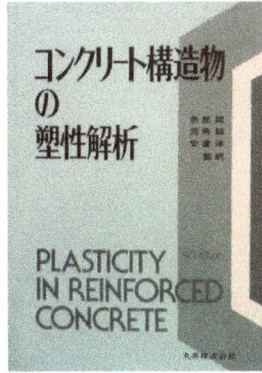

Left: *Limit Analysis and Soil Plasticity.*
Translated by People Communication Publishing House (人民交通出版社), Beijing (北京市), China, 1995 (詹世斌).
Right: *Plasticity in Reinforced Concrete*, 丸善珠式會社, Japan, 1986 (色部誠, 河角誠, 安達洋, 監訳).

CHAPTER 4
UNIVERSITY OF HAWAII, 1999–2006, THE PURSUIT OF EXCELLENCE
追求卓越

4.1 From Faculty to Dean (學而優則仕)

I was appointed as Dean of Engineering at the University of Hawaii (UH) at Manoa, effective 1 September 1999. I was a "hardcore" academic guy managing the College of Engineering to focus squarely on enhancing its reputation. I definitely had big ideas for the College. My vision for the College was simple and ambitious. It would begin with the scholarly faculty: if better research were conducted, quality students would be attracted. This would lead to better graduates for the marketplace, which would, in turn, raise the College's reputation. This would increase stakeholders' support and lead to improved facilities and infrastructure for the College. This was the logic I had for "How to Build a First-Rate Engineering College" in my vision statement, which I prepared when I was appointed Dean.

I was fortunate to receive the full support of the Hawaiian Governor Benjamin Cayetano, in that he wanted to develop a hi-tech industry in Hawaii, and he firmly believed that a world-class engineering school was needed. As a start, I proposed to build the Hawaii Wireless Communication Center because of the College's current strength and Hawaii's geographical location. Ben was impressed with my academic credentials and my timely proposal, and he decided to augment the College's base budget with an additional annual US$1 million to build a center.

Time flies. Over the seven years of my tenure, I focused on building up the college into a top tier engineering school. I did my best to fulfill this goal. My efforts were successful that reflected truly my own interest and expertise in building academic excellence for the University, in particular, the College of Engineering.

4.2 Building a First-Rate Engineering College (急起直追)

Introduction

There are three ingredients to a great engineering college: it must have a first-rate faculty, a highly-selected student body, and very good facilities. A first-rate faculty would attract quality students, which, in turn, would attract quality faculty. Each is dependent on the other. Together they will bring excellence and prestige to the University.

I arrived at the UH campus on 1 September 1999, and I noted that our operating budget's purchasing power was reduced by about 30%, our total number of faculty positions was reduced by about 20%, and our total enrollment has dropped by more than 30% over the last five years. Some of our critical facilities had not been upgraded for many years. To build up the College, these three elements must be excellent.

Renewal

Our highest priority was to strengthen the faculty as their quality would attract excellent students and help to enhance our facilities. Without a strong faculty, there would be no strong college of engineering. During the next few years, the College would have to recruit more than 30% of its current faculty strength, which amounted to more than 15 positions. And we would have to do this at a time when PhDs were in short supply in some critical disciplines. Competition for the best talent would be fierce. To this end, we had to secure funds for start-up packages and use them to recruit the best and brightest and then integrate them with our current faculty in our focus areas of research, so that they could grow and flourish with the College.

Crossing boundaries

Because our resources were limited and the size of our College was relatively small, we could not constantly add new faculty positions in emerging areas of technology and compete at all national levels for excellence. When we recruited, we had to look for individuals with multiple interests who could teach and do research by crossing the specialty boundaries in the same department or even crossing the boundaries with different departments. We had to concentrate our effort to strengthen the focus areas of research in our College, rather than just building one department at a time. We asked and encouraged our faculty to work as a team, integrating their specialties in the creation of focus areas of research in the College, and they are starting to do so. Our Electrical Engineering faculty focused on the area of Advanced Wireless Communication, our Civil Engineering faculty cooperated in the area of Modeling and Simulation of Civil Infrastructure System, while our Mechanical Engineering faculty concentrated their efforts in the area of Advanced Mechanical System and Materials.

Measuring success

How would we know that we have succeeded in renewing our faculty and strengthening our research base and academic programs? One important indicator would be the number of dollars our faculty is able to secure via research grants from federal and state governments as well as from the private sector. Another indicator would be the support we will receive not only from UH alumni but also from national corporations and foundations.

The last indicator would be whether the College is able to recruit and attract excellent promising faculty members as well as top-notch students. All the major Colleges of Engineering in the nation are in a marathon race to maintain their excellent faculty and academic programs and attract high caliber students. We want to join this "club" of excellence. This would be our goal.

Conclusions

To join this "club" of excellence, we would need to have faculty members with national or international reputations. To keep and recruit quality faculty members, we would need attractive salaries, strong colleagues, access to equipment and resources, and opportunities to work with top-notch students. Nothing should be taken for granted. We would have to continue to pay attention to our current faculty while working to attract new ones. Recognition would have to be a critical part of the process by demonstrating how much we value our colleagues' work, achievements and successes. We would highlight our faculty's work in these five areas: Teaching, Research, Scholarship, Mentoring, and Service, and recognize their successes with support and rewards.

Excellence is not something to achieve and then go on to do something else. It is a daily struggle, an annual struggle, and an endless struggle. The struggle should not just be for dollars. Dollars are necessary, but we must have the vision. We should use the resources and potential opportunities to create an environment that would allow us to recruit talented faculty and retain our good faculty, whose reputations would attract quality students. Recruiting top faculty and keeping them happy should not

A graduation cermony at the College of Engineering, Honolulu campus.

solely be the responsibility of myself when I was Dean but also that of the University and the state of Hawaii.

4.3 Hawaii Business: Engineering My Exit (功成身退)

Hawaii Business, December 2005

"Dean Chen's dream of turning the College into one of the best schools in the nation is finally within the arm's reach." So why is he suddenly leaving?

By Jacy L Youn

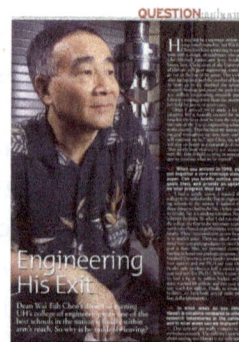

He may not be a superstar athlete or a top rated comedian but Wai-Fah Chen does have something in common with a couple of celebrities who are. Like Michael Jordan and Jerry Seinfeld before him, Chen, Dean of the University of Hawaii's College of Engineering, wants to go out at the top of his game. That is why, after having increased the number of faculty from 45 to 65, doubled the school's research funding and paved the path for a US$25 million engineering research park, Chen is stepping down from the position he has held for just over seven years.

"Since I arrived, we've made a lot of progress. We have basically created the road map for the next dean to turn the school into one of the Top 50 engineering schools in the country. That has been my unwavering goal throughout my time here, and it will continue to be," says Chen, 69, who will stay on board as a research professor. "But surely now that we've got momentum, the time is right to bring in someone new to continue what we've started."

HB: When you arrived in 1999, you put together a very thorough vision paper. Can you briefly outline your goals then, and provide an update on your progress thus far?
A: I had a simple vision. I wanted the college to be ranked in the Top 50 engineering schools in the nation. I outlined the three things we had to do: No. 1 is hire quality faculty. No. 2 is student retention, No. 3 is better facilities. So what I did was make the college the first in the university to give merit raises based on performance to retain faculty. Then I recruited over 20 new faculty at market price. Then we raised enrollment from 500 undergraduates to 760. Our goal is 800. Also, we increased external funding to boost our graduate programs. At Stanford University, every faculty produces four masters and one PhD per year. Our faculty only produces a half a master per year and very few PhDs. When I came in, we had a US$4 or US$5 million budget, we have since reached US$8 million, and this year we may reach US$10 million. Finally, in terms of facilities, we have built several multi-million-dollar laboratories.

HB: In what areas do you think Hawaii is excelling compared to other research laboratories in the nation, and in what areas can we improve?
A: One area we can really compete in is corrosion research. The military's worried about rusting, and Hawaii is the only state that can simulate any climate weather condition on one island, so we have received good funding to do corrosion testing. Wireless communication is another one. All this optical fiber from the Mainland to the Far East goes through Hawaii. A lot of satellites are stationed on Hawaii. And then biomedical research.

HB: What plans does the engineering school have to work with the medical school in developing biomedical research?

A: On the research side, we have some overlapping projects with the National Institute of Health. On the academic side, the first step is to produce a biomedical certificate program — a graduate program. Then we could grow to have an accredited undergraduate biomedical engineering program. Then eventually spin it off as a department.

HB: Given Hawaii's massive infrastructure needs, will we be, or are we currently experiencing an increased demand for an engineering work force?

A: We are increasing enrollment, but not fast enough to meet the local need. Construction is very important part of the Hawaii economy, and yet we are the only accredited school in the state of Hawaii. So we need more civil engineers. The only limitation is facilities. Civil engineering structures are big, but we do not have the space in this building. Also, from the military's point of view, we need more engineers. According to former Commander of the US Pacific Command Admiral Thomas Fargo, the military should be managed by engineers, not by MBAs, because the work is very high-tech.

HB: Once students join the work force, how are we doing at retaining them locally?

A: Right now, mechanical engineers mostly work with Pearl Harbor shipyard. There is going to be a lot of expansion of military in Hawaii, so for those high-tech military jobs, there is good pay and job security. The second part, civil engineer construction, is very booming. They have no trouble finding people to stay here. In fact, 75 percent of all our construction company engineers are UH graduates. The trouble is, we want to produce more. But we do not have the facilities.

HB: What will be the new dean's biggest challenges as he attempts to fill your shoes?

A: The goal is very simple — try to be ranked in the Top 50. He has to improve our graduate program by hiring more doctoral students. That is one of my goals upon leaving. I want to stay onboard and help mentor the junior faculty to grow and be successful, and then I want to help recruit doctoral students. This is a very good time for the school. That is why I decided to step down. I came in at the worst time, and I am leaving at the best time.

4.4 Advising and Mentoring (學術殿堂)

4.4.1 *Doctor and post-doctoral students*

My last doctoral student was Ken Hwa (華根) from Taiwan. He was my MS student 25 years ago at Purdue and then went back to Taiwan to teach at a technical school. He came to UH again to study for his doctoral degree under my supervision by extending the advanced analysis of steel buildings under fire condition in 2003.

At UH as Dean of Engineering, I accepted several visiting scholars and post-doctoral students from several overseas universities including one from Shanghai Jiao Tong University (上海交通大學), one from the Muroran Institute of Technology in Japan, two from Hong Kong Polytechnics University, and one each from Cairo University and Mansoura University, which were both in Egypt. In addition, we had two young UH Civil and Environmental Engineering faculty members joining us for our weekly mentoring seminar, and we had close one-to-one interaction with in-depth discussions on various topics during our group and individual discussions.

This following photos captures one of the seminar gatherings at the UH CEE building:

Left: With my post doctoral students (left most four with one seated) and two faculty teams (right most two with one seated) in the Department of Civil and Environmental Engineering at the University of Hawaii, Honolulu.
Right: With all my post doctoral students.

4.4.2 *Salah El-Metwally, Mansoura University, Egypt*

Salah El-Din E El-Metwally received his MS degree from George Washington University and his PhD under me at Purdue in 1986. His academic positions in Egypt included Head of the Structural Engineering departments at Mansoura University and Tanta University. He has been an active member for many years in

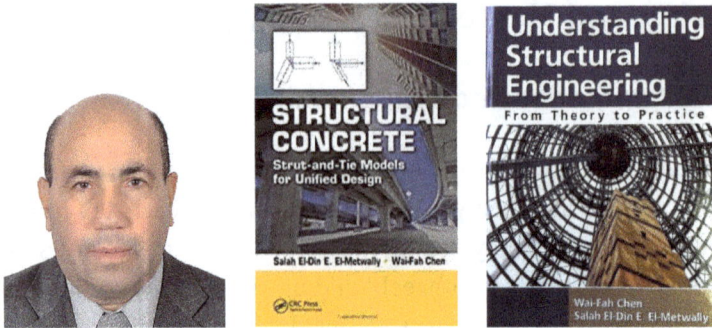

Left: Salah El-Metwally, Mansoura University, Egypt.
Center: *Structural Concrete: Strut-and-Tie Models for Unified Design*, CRC Press, 2017.
Right: *Understanding Structural Engineering: From Theory to Practice*, CRC Press, 2011.

the Standing Committee and Subcommittees of the Egyptian Code for the Design and Construction of Concrete Structures.

He has broad experience in the structural design of many educational and industrial structures, bridges, and large-scale roof structures. He is currently a professor of concrete structures, Mansoura University. He came to visit me at UH several times over the last few years with visiting periods that extended from weeks to months as part of either the Egyptian government program for their faculty sabbatical leave or under the US Fulbright scholarship. As a result, we published two timely and important books: *Understanding Structural Engineering: From Theory to Practice* and *Structural Concrete: Strut-and-Tie Models for Unified Design.*

4.4.3 *Three cardinal rules for a good structural engineer*

1. "Ductility" can be easily forgiven for one's mistake. So watch out, when we deal with materials with less ductility due to size effect, temperature change, or damage due to cold work. For example, steel is considered to be a homogeneous, isotropic and ductile material, but none of these is true at the joint with very thick plates.
2. "Connection Detailing" is everything. Most failures are the result of some detailing problems in joints. Members seldom fail but joints can fail in a brittle manner. We must pay attention to detailing in connections.
3. "Redundancies" are the second-best defense for unexpected failure. Good engineers need to look ahead about any unexpected possible failures due to accidents or quality control. So, to be a good structural engineer, we need to have a good knowledge of materials, especially materials subject to various environmental and stress conditions.

4.5 Structural Engineering: A Place in Practice, Keep Things Simple: Sharing My Wisdom
(分享經驗智慧)

4.5.1 *From theory to practice*

In a career spanning over the last 40 years, I have built a reputation for developing simpler solutions for engineering practice by bridging the gap between theory and practice.

This includes, for example, from the simple plastic analysis theory to practical steel frame design, to simple rules to design beam-to-column connections, to the use of limit theory to solve geotechnical engineering problems, up to my recent work on the direct use of the advanced second-order theory to design a structural system. I have operated on one simple principle — "keep things simple".

In structural engineering, true innovation is more than just applying advanced theories with computing technology. Generally speaking, exotic theories and daring applications frequently complicate things, and design simplifications are always required for their practical implementation.

In the 1950s, the linear elastic theory, coupled with the moment distribution method, made linear analysis practical in designing high-rise building frames with a slide rule. The allowable stress design code was the hallmark of engineering practice.

In the 1960s, structural engineers implemented this complex mathematical theory to engineering practice by introducing the simple plastic hinge concept along with the powerful upper and lower bound limit theorems. As a result, it became a daily engineering tool for designing steel structures in the plastic range in a very simple manner and was adopted in the steel specifications around the world quickly.

In the 1970s, the simple limit analysis method of perfect plasticity made the simple plastic theory realistic in designing multistory steel building frames with calculators. As a result, a plastic design code was issued to provide an alternative design for steel building frames.

In the 1980s, mainframe computing made the second-order inelastic analysis of structural members possible. Simplifications lead to efficiency and saving in materials, labor and time in the development of the new Load and Resistance Factor Design Specifications.

With the current almost free computing environment with personal computers (PC), we now deal routinely with second-order inelastic analysis (commonly called advanced analysis) of complicated structural systems that have hundreds of thousands of degrees of freedom. The modeling of all types of structural systems of high-rise buildings can now be handled quickly and efficiently on relatively inexpensive computers.

In the 1990s, the theory was extended to reinforced concrete (RC) structures with the techniques now known as the Strut-Tie Model. This new lower bound approach can now handle the most difficult tasks of joint detailing, deep beams with square hole cutoffs and slab design with easy, among others. It can also handle the most difficult topics of shear and torsion design in RC with rigorous ease. The S&T Model is now adopted by Concrete Codes around the world.

In the 2000s, research works on advanced analysis of steel structures with PR construction are now in full swing to develop nonlinear procedures and software for practical use in design office. The theory and approaches for advanced analysis of plane frames composed of members of compact sections, fully braced out-of-plane, have been well developed, verified and coded by the American Institute of Steel Construction in the 2005 AISC Specifications as well as others around the world.

These practical applications based on very "sophisticated" theories are the testimony of the power structural engineers, with their great physical intuition resulting in such drastic "simplifications" in order for their use in the real-world engineering practice.

Most recently, the theory has been further extended to *Second-order Inelastic Analysis* (SL Chan), *Toward Performance-based Steel Design* (K Hwa), *Seismic Design With Structural Fuse* (IH Chen), *Strut-Tie Model for Unified RC Design* (S El-Metwally), among others.

Here, as we can state now as one of our structural engineering cardinal rules, "*Simplicity*" is sure of the ultimate achievements of "*sophistication*" through us, the powerful engineering intuition, to reach the practical level of real-world applications. This is why all our engineering researches should always aim to achieve "a place in practice". Otherwise, 不能落實, 一切都是空談.

"Simplicity is the ultimate achievement of sophistication" and "sophistication is a necessary path to get the simplicity" to guide our engineering research to achieve "a place in practice".

4.5.2 *Notable research*

Unlike the older second-order analysis used in seismic and static design that is used as supplemental checking only, the concept of advanced analysis can be used as the primary design and member size framing. Recent collapses of several shallow roof domes in Russia, Germany and Poland show clearly the deficiency of the current practice. The advanced analysis can be easily applied to the design of these special or other conventional structures. Our research team starting from Lehigh University to Purdue University and then to the University of Hawaii has helped us, structural engineering professionals, to be at the forefront of applying the modern science-based theory to our daily engineering practice.

The older guards meet the younger generations in Honolulu, Hawaii

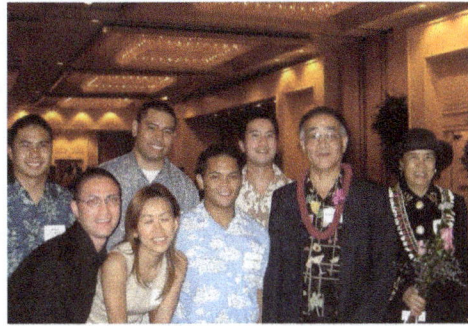

Fund raising banquet at Hawaiian Hilton Village.
Left: With Tang Man-Chung (鄧文中) CEO, TY Lin and Al Yee, CEO Precast Design.
Right: With UH students.

The thirty meter telescope project at UH

President Jean-Lou Chameau of California Institute of Technology (CalTech) and Chancellor Henry Yang (楊祖佑) of the University of California, Santa Barbara (UCSB) came to Honolulu frequently as the co-chairs of the Thirty Meter Telescope project, which has become controversial due to its planned location on Mauna Kea on the island of Hawaii, land which is sacred to Native Hawaiian culture and religion. This telescope would become the largest visible-light telescope on Mauna Kea.

A regular biannual gathering with my former doctoral students at the Howard Plaza Hotel in Taipei whenever we attend the biennial Academia Sinica meeting.

From left to right: (Standing) JL Peng (彭瑞麟), Lee (李騰芳), YK Wang (王永康), H Ken (華根), YL Huang (黃玉麟), FH Wu (吳福祥), YF Li (李有豐) and HL Cheng (成曉琳). (Seated) Me and my wife, Lily.

4.6 A Lecture at Taiwan Bureau of High Speed Railway

10:00 am, Wednesday, 28 June 2006

Lecture delivered at the Taiwan Bureau of High Speed Railway, Taipei, Taiwan

Title:

我的学术成就与贡献

Academic Achievements and Contributions of Academician Chen

Introduction

Thank you very much, Director Wu.

I am honored to be here today to deliver my lecture. Before I go any further, I want to assure you that I fully understand that the purpose of my lecture is inspiration and not to promote myself.

About a few months ago, I was contacted by Director Wu to speak to you today. At that time, he said that I could choose the topic and to make it easy on myself.

No pressure. Then on Thursday, June 15, he sent me an email in which he said and I quote:

"The topic will be: The academic achievements and contributions of Academician Chen. The lecture will last for two hours in a conference hall for nearly two hundred.... I wonder if this arrangement is OK for you."

Wow!!! "...My academic achievements and contributions to engineering ..." ... for two hours!

I recall that a famous Chinese scholar, Mr Lin Yu-Tung, once said, "A good speech is like a girl's skirt, the shorter the better". I promise that I will make my lecture short, to the point, and enjoyable; but not too short beyond what Director Wu would allow me to do.

I am not quite sure how much of an impact my work may have on our civil engineering profession in general and engineering education in particular. However, I will provide you with a few of my personal thoughts, experiences, and perspectives. I hope some of these may help you in your professional career. I will leave it to each of you to determine what sort of impact my comments may have on you.

And... just so that you can keep track of what are my key points and where I am in my presentation, there are four lessons that I have learned in my career and life. I would like to share these lessons with you.

These lessons are:

1. If you do what you love to do, you will be successful.
2. You must be a lifelong learner in this information technology age.
3. You must be flexible in your career choice to be successful.
4. Keep things simple for your engineering work to be useful.

These lessons are interwoven into the story that I am going to tell you. I will point them out along the way rather than address each lesson individually.

My research basically involved three topics of interaction — mechanics, materials and computing. In my research and teaching career over the last 40 years, I have witnessed the tremendous growth and interaction in these three topics. Each played a key role at different periods in time. Together, they have made major contributions to structural and geotechnical engineering applications and practice. This is the story I want to share with you today.

Information technology age

My first thoughts are about what we refer to as the high technology age. Some also refer to this time as the information age. I think, perhaps, it should be called the

Information Technology Age. The foundation of this information technology age is the rapid development of computers, personal computers, in particular.

We can simply refer to this as "computing" power in the terminology of engineering science. Computing power, in my opinion, is one of the greatest accomplishments thus far in the Information Technology Age and it has greatly affected my research.

I think that it is important to reflect on all of the dramatic changes that we have experienced in the Information Technology Age and how this has influenced me and will surely affect you in your future.

Let me use as examples a few glimpses of what a typical engineering office and my research environment looked like 40 years ago and what some of the tools were at that time. Let us compare them very briefly with what we use today. I will show you in this process how these technology changes affected my research and our engineering practice over the years.

In the 1960s

- I was allowed to use FORTRAN language as one of the two foreign language requirements for my PhD degree. Can you imagine that? A computer language as a substitute for French, German or other important scientific languages.
- I taught the first FORTRAN course for our Civil Engineering students at Lehigh University in 1966. That was an experience I would not forget. I recalled that I told my department chairman at the time that I never took the course before but he simply said: "You are young, right, you can learn."
- I used the slide rule to do my homework and HP calculator for my research work. How many of you even know what a slide rule is and what it looks like?
- Specification formulas were simple. They were limited to no more than taking a square root.
- The materials model that was used exclusively was linear elasticity.
- Mathematical solutions were limited to series expansions with closed form.
- Timoshenko's books were the bible of the time. Do any of you know of these books?
- Method of superposition was used extensively.
- Allowable stress design was used exclusively.
- We had a beautiful mathematical theory of plasticity but few solutions.
- Newmark's numerical methods with calculators were popular.
- Hardy Cross's moment distribution method was widely used in engineering practice for building design.

These were the tools and theories from the 1960s. Many of these are probably no longer used and you may have not heard of them. They were, however, the basis that I used at the beginning of my career.

In the 1970s

- We used main-frame computing and waited in computer centers for solutions. This was the time of punch cards and key-punch machines.
- The finite element method was used extensively for numerical solutions.
- The material model used was plasticity with no work-hardening.
- Simple plastic theory was used to design steel frames.
- Ultimate strength method was used for reinforced concrete design.
- We had IBM typewriters to produce manuscripts with equations.
- I wrote my first structural engineering books on the *Theory of Beam-Columns*, extending Timoshenko's work to plastic behavior and design.
- I wrote my first soil mechanics book on the *Limit Analysis and Soil Plasticity* to provide theoretical justifications for the famous Terzaghi solutions. Terzaghi, in case you do not know, was the father of soil mechanics.
- We had our first energy crisis and long lines at gas stations.
- Offshore structural engineering emerged as a new area of research and development.
- I wrote the first book on *Plasticity in Reinforced Concrete*, extending the plasticity theory to concrete materials for offshore structural applications in particular.

In the 1980s

- Along came the 1980s, the third decade of my research. We had computers on our desks with possibly 2MB of disk space.
- Probably laptops were just beginning to be developed. The few that existed weighed 20 lbs.
- Cell phones were slightly larger than the size of a brick and weighed just as much.
- Fax machines were just being developed and used with thermal paper where the ink faded over time.
- Communications were done via hardcopy, US Mail, overnight delivery, fax, and phones.
- We have analog phones.

- Cameras for instant photos were Polaroid where you had to pull out the film hold it for 30 to 60 seconds and then peel them open. That was "instant" at that time.
- Drawings were just that — they were drawn manually. CAD just started.
- Offices included drafting machines or drawing tables, computers with 12" screens, fax machines, hand-held calculators, drawing files and racks, hardcopy libraries, file cabinets, etc.
- We had the Load Resistance Factor Design Code for Steel Structures. It is a maximum strength-based code for member design that uses plastic theory.
- I wrote my first book on *Constitutive Equations of Engineering Materials* to meet the need of finite element analysis. This was the first time that the capability of computing power and finite element analysis method exceeded the advancement of material modeling.
- I wrote the first book on *Soil Plasticity: Theory and Implementation*, which extended the plasticity theory to geotechnical materials for offshore foundation applications, in particular.
- I wrote the popular textbook on *Plasticity for Structural Engineers* to help civil engineers understand the highly mathematical theory of plasticity. Plasticity theory became a basic course for civil engineers.
- I wrote the first book on *Tubular Members in Offshore Structures*, extending the well established wide-flange section members for building design to circular-section tubular members for offshore structural applications.

Present day

- Today, I challenge you to find more than 1 or 2 typewriters in your office.
- We have desktop computers with several MB RAM and disk space in Gigabytes.
- Laptops weight under 3 pounds.
- Cell phones are smaller than your fist, and they can take photos, send emails, give you your calendars, surf the Internet, and so on.
- PDAs are everywhere.
- FTP sites, project websites, emails and IM are the preferred mode of communication.
- Digital cameras provide instant photos.
- Drawings are electronic.
- Offices are mostly desks with computers. We work on projects thousands of miles away but we communicate and transmit data in real-time between people and offices.
- Your personal emails come to your PDA, even when you travel to most destinations in the US, Europe or China. When you are traveling, you use your IP phone on

your laptop. Everyone you talk to thinks that you are in your office and talking to them.

- I work with my secretary next door via the Internet and find most information via Google searches.
- We are connected 24/7.
- I wrote a series of books on *Advanced Analysis for Steel Design*, moving away from traditional member design toward system design leading to performance-based codes for structural engineering.
- Material models include fracture mechanics, composite FRP, and durability, from time-independent to time-dependent behavior.
- Modeling (physics), simulation (computing), visualization (virtual reality) and verification (experimentation) are the rapidly developing fields.
- I edited a series of Handbooks on Civil Engineering, Structural Engineering, Bridge Engineering and on Earthquake Engineering via Internet communication among contributors, publishers, reviewers and editors.

Lifelong learning

Changes over the 40 years of my career have been dramatic and exciting. Most of you have seen only part of it because you grew up during the last 20 years. Many of these tools that I consider "new" have always been there for all of you.

So, what does this all mean? Why am I talking about this Information Technology Age that we are a part of?

Well, what it means to me is that all of you will find, over the next 20 or 30 years, the tools that you will be using, the environment that you will be working in, and the processes that you will be using will change, perhaps dramatically. Things will be very different from what they are today. This will surely affect the way we design and construct structures.

Change is inevitable and more than ever, it is quite rapid. So, in my opinion, you need to hone your ability to change, you need to embed continuous change in your thinking and work processes. You will need to embrace change as your way of life. This change will not only affect your working environment, it will surely affect your personal and family life. Try to make the best out of it.

Focus on the fact that with your education and the experiences that you have been accumulating, each of you has the potential to continuously change, to continuously learn. Simply put, lifelong learning must be an integral part of your life.

This is indeed an exciting time to be engineers in the Information Technology Age.

Flexibility in your career choice

You must be flexible in your career planning and choice. Life is more than just goal-setting as it will have a lot of surprises. This is what happened to me.

I never planned to be a professor but ended up as an academician. I never planned to take an administrative job in a university but ended up as a dean in a major public university. Life is unpredictable but the lessons I have learned — to be a lifelong learner and to be flexible — have helped me greatly in achieving success.

Forty years ago my dreamed universities were Cornell and Stanford, but somehow I ended up at Lehigh University in Bethlehem, PA. At the time, Lehigh's structural group was focusing on the development of a new steel design method known as "Plastic Design" to replace the century-old "allowable stress design". It was a revolutionary concept pioneered at Cambridge University during World War II.

Lehigh's Fritz Laboratory had the world largest testing machine and I participated in the full-scale testing of steel structures. The method was later adopted by the American specifications for steel construction and quickly became a worldwide standard. The steel group became world-famous and I received my MS degree and became part of this group.

As my studies at Lehigh came to an end, I received an offer from Cornell University to continue my steel research and earn my PhD degree. It was a dream-come-true but for some reason, I was attracted to Brown University in Providence, RI, for its highly theoretical work in solid mechanics in general and mathematical theory of plasticity in particular. As a result, I enrolled at Brown and earned my PhD there.

While doing my studies, I found that highly mathematical theories and their applications frequently complicated matters. Design simplifications were always necessary for practical implementation. I came to the conclusion that nothing can be more practical than a simple theory, which is one of the four lessons that I would like you to remember.

Simple limit analysis methods were developed and widely used in steel structures. I focused on the application of this new method, not on steel, but on soil and concrete materials. This was the first time that this type of application was done. It became instantly popular and made me a celebrity of sorts. During the period, I wrote a series of books on this and related subjects.

Concluding remarks

I was surprised and honored to have been selected as one of the ten giants in engineering science in America during the last forty years in the 2003 book by Dr Beg in the UK. The book was entitled *Giants of Engineering Science*, which included well-known figures like YC Fung, CL Tien and TY Wu.

My specialty and contributions were identified in the area of Structural and Geotechnical Engineering Science. This was a special honor as it was the validation of a lifetime of work that I have greatly enjoyed.

Now I realize that everything in a lifetime is a journey. So I say to each of you, your professional career that is ahead of you is a journey. There is no end destination. You need to continue to learn, to grow, to be flexible. And with that, I believe that you will find your career exciting, invigorating, challenging, and everything that you want it to be.

Your career and your life will get stale and old and tired only if you allow it to happen, only if you do not enjoy what you are doing. Do what you love to do and continue your journey; keep things simple, balance your career advancement and family life. It will be a fulfilling life.

Summary

Let us summarize the four lessons we learned today:

1. If you do what you love to do, you will be successful.
2. You must be a lifelong learner in this information technology age.
3. You must be flexible in your career choice to be successful.
4. Keep things simple for your engineering work to be useful.

I believe these lessons that I have learned will help you along your life's journey.
I leave you with one quote by Oprah:
"You don't become what you want, you become what you believe."
Have a great journey.

Acknowledgment

Mr Michael Matsumoto, a CEO and a well-known structural engineer, was the keynote speaker of this year's College of Engineering Convocation at the University of Hawaii. I fully agree with his viewpoint and have intertwined what he said in his speech with my thoughts in this discussion about the Information Technology Age.

4.7 My Life's Journey: Reflections of an Academic
(我的生涯與省思)

4.7.1 *Motivation*

After I stepped down as Dean of Engineering at the University, some of my colleagues, friends and family members, especially my three sons, suggested that I should write a book on my career — engineering — and higher education based on my 40 years of experience as a researcher, educator and engineer. I have been thinking about that suggestion from time to time. So, what is my motivation to write this book? The desire to write *My Life's Journey: Reflections of an Academic* to reflect on my career and life is not the whole story.

To me (and many of my friends and relatives in my generation), the documentation of my struggle as a foreign student, an American immigrant, and an eyewitness to the rise of China represents one of the many real-life stories of my generation. The factors of personal interest and satisfaction in sharing these stories with others and our next generation, in particular, are very significant.

As the famous Nobel Prize winner Pearl S Buck said: "*If you want to understand today, you have to search yesterday.*" To this end, I quote a Chinese saying: "*Throw a brick to attract jades* (抛磚引玉)". My autobiography, just like an ordinary "brick" in the market, will surely attract more and better authors of my generation of similar backgrounds to write much better "*jade-level*" autobiographies for us. The net result is that the sum total of our stories yesterday will help us understand our changes today as a whole.

My granddaughter Chloe was attracted by the description of my escape story from Mainland China to Taiwan in 1949 during China's Civil War. She visited my old Purdue office and was surprised to find a photo of me hanging on the wall of the School of Civil Engineering Building.

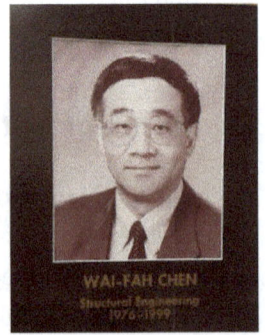

4.7.2 *An interesting interaction with two NCKU graduates*

Bo-Shiuan Wang, 1990 NCKU graduate

Dear Professor Chen:

My name is Bo-Shiuan Wang, graduated from NCKU 90 grades, Master of Science in structural engineering. I am applying for the PhD program at Purdue University and have just completed the application process.

I have read your new book, *My Life's Journey: Reflections of an Academic.* It is such a splendid book and I couldn't stop reading it until I reached the last page. Your four lessons for students are essential and have motivated me to think in a broader and in-depth sense. Your quest of an America dream and journey to be an outstanding scholar really impressed me and inspired my career pursuits.

Thinking back five years ago, the morale of civil engineering has been in a steady decline for years. Most of my peers chose to enter the government sector. Although I was tempted to choose this career path, I decided to join Motorola Electronics and worked on the THSR project as a structural engineer.

Fortunately, this experience has inspired my insight and helped me realize that structural engineering can be applied for many different industries. I have contributed a great deal in collaborating with Motorola in the completion of the THSR project. The company has provided her first successful total solution in the global marketing arena.

In addition, I have gained tremendous knowledge in the field of telecommunications Working on [the] THSR project in designing the high mast tower and tunnel facilities has inspired my interest in structures affected by wind vibration. However, recent design methods may not provide enough evaluation in wind force and time-dependent fatigue on steel structure directly and have puzzled me for a long time. My professional experiences have inspired my pursuit in the advanced study of structural engineering.

In your book, you mentioned that [the] structure-fluid interaction problem and time-dependent behavior of [the] constitutive equation are current trends in structural engineering. This has reinforced my confidence to further my research and study on this subject matter.

When I discussed my planned research subject with Professor **Deh-Shiu Hsu** at NCKU, he suggested I write to you. I am hoping that you may be able to provide some pointers regarding state-of-the-art researches in structure-fluid interaction problem and steel fatigue problem, and recommend professors [who are] experts at this field at Purdue University.

I am bold to write this letter to you. Thank you for your time and understanding. Wish you a happy Chinese New Year!

Bo-Shiuan Wang (Graduate student, NCKU)

This was my reply to him.

Dear Mr Wang:

I am happy to hear that My Life Journey book has brought some impact to your career planning. You made my day.

As I mentioned in my book, most of structural problems are associated with materials. For example, steel is considered to be an isotropic, homogeneous and ductile material in our usual design process. But this is not true at the connecting joints with thick plates. This frequently creates problems in fracture type of brittle failure as witnessed in the MN bridge collapse last year.

Material properties do change with time, cold work and the way it was made. As for the fatigue and fracture problems associated with bridge structures, Professor John Fisher at Lehigh University is a guru on this subject. He has a lot of experiences on the real world of bridge assessment and repair.

Professor CT Sun at Purdue's AA department is an expert on composite materials as used in aerospace. In civil engineering, we begin to use widely the FRP composite materials for bridge repair and rehabilitation. It is a very hot subject right now. The key direction of future research in civil engineering is how to convert the huge amount of data we collected in bridge and building monitoring system through sensors into useful knowledge so that practical actions can be taken for solutions.

Here, like the cancer research in medical field, the aging and corrosion of materials are our problem. We can do all kinds of measurements but just cannot translate this data into useful and actionable knowledge.

This requires a fundamental understanding of materials as a time-dependent dynamic process. It is not a static process as we have been doing for a long time.

I hope this thought may be of some help to your thinking.

Best wishes and Happy Lunar New Year!

WF Chen

This was his reply to mine:

Dear Professor Chen:

Thank you for your quick response. You really are a very kind scholar as I felt from reading your book. Your advices provided me a clearer idea on the overall picture of structural engineering and gave me more confidence to explore this field.

I had read several papers by Professor Fisher and Professor Connor. Their in-depth research on fatigue of connection joint, welded connection, and retrofit for fatigue cracking on steel bridge had been very helpful to me when I designed the connection fringe of steel structure and tried to solve fracture problems. With your advices in mind, I will look into the programs at Lehigh and Purdue and contact Professors Fisher and Connor in person.

Thank you for your kindness again; and hope one day I can say thanks and ask for your guidance personally.

Happy Chinese New Year, The Year of the Rat.

Bo-Shiuan Wang

William Hui-Kang Chu, 1955 Tainan College of Engineering Graduate

Dear Dr Chen:

My name is William Hui-Kang Chu. The last name in Chinese means saving not the common name for red color. I am [from the] class of '36. I attended the Mechanical Engineering Department of Tainan College of Engineering in 1955 as my first choice of field as well as school; even my test score was not good enough for the NTU. I think I must be the first foreigner (from mainland) to play rugby and (probably) also the first "fifth-year-senior" captain of the rugby team there.

I am writing to you after I read your book *My Life's Journey*. In particular, I was moved by your student, Professor Xila Liu's Chapter 15 "My Adviser and I". What a true tribute not only to *his* father but to "A great educationalist in China". (I think the translation would read a little bit more compact as — *Come with a whole heart, Leave without half a blade of grass.*)

I sincerely hope you are in the position to recommend what our adopted country, the US should do immediately, to prevent another bridge collapse like the I-35W bridge in Minnesota. I really think it is not the undersized gusset problem. I saw the picture after the collapse and it looks like the whole section sitting on the (may have been [sunk] a little) bridge abutment [was] sheared off like a giant scissor has just cut it.

It seems to me there is no splice that can compensate the thermal expansion/contraction, to tie these heavily traveled horizontal beams together. If the "rigidity in original geometric angle between the intersection members" is violated, plus the additional vibration and cross-wind loads from the high-level trust to all these joint areas, it is difficult to figure out theoretically or through [a] static test, where is the weakest link. (I wouldn't know any way, just a guess.)

No wonder the press said there are so many other bridges [that] are under the same critical structurally deficient category. I sincerely hope you would get off your retirement/reflection mode and come up with a short-term fix first. I think only you can overcome the bureaucratic political resistance to do something soon. Therefore, if you are already involved, or one way or another, just a brief note that you have received this letter will make my day.

Thank you for your time and your book.

William HK Chu
Hickory, NC

This was my reply to him:

Dear Mr Chu:

It was nice to hear from you on *My Life's Journey* book as well as your comments on the possible cause of the MN bridge collapse.

As I mentioned in my book in Chapter 8 on my three cardinal rules of good structural engineering:

Cardinal Rule One: Ductility can be forgiving of one's mistakes.
Cardinal Rule Two: Connection detailing is everything.
Cardinal Rule Three: Redundancies are the best defense against unexpected failures.

All three rules were violated in the MN bridge design:

1. There is little ductility in the gusset plate because of aging, corrosion, cold work, and low temperature.
2. The joint or gusset is under-designed while there are stress concentration and sharp notch around the rivet holes in the gusset plate.
3. They are no redundancies in the system for the second defense against potential failure.

The real problem right now is that Federal and State Highway Departments have collected a lot of data via various sensors and inspections, but they do not know how to convert these large amounts [of] data into useful information and knowledge so that specific actions can be taken. This is a huge area of research similar to cancer research in [the] medical profession. It is a very challenging one because the root of the problem is to understand how the material ages with time and environment, and how that affects the mechanical properties of the material.

Best wishes!

WF Chen

In response, he wrote back:

Dear Dr Chen:

Thank you very much for your letter. I mean it very sincerely.

I wish you [could] chair a blue-ribbon group of experts to review the data and come up with a priority [on] which bridges need to be fixed first. Congress should reduce the budget increase submitted from President Bush today in defense and homeland security, and move a big chink of that increase to fix the bridges. We ARE the enemy in this regard. We have to fight that enemy here, not there.

That said, I think all these bridge designs violated your other principle about "On Structural Connection" that assume[s] "the beam-to-column connections have sufficient rigidity to maintain its original geometric angle between intersection members until the member's maximum or plastic strength is reached". I feel [that] when the whole span is sheared off and dropped from the bridge abutment, it must mean that one or two uneven sinking abutments prior to that day has contributed the extra stress to the gusset plate, especially around the sharp notch and rivet holes.

Future inspection must measure this shift from the normal resting place. To me, this is a simple measurement to determine which bridge deserves the highest priority to have a short beam splice installed, that will allow the thermal expansion between the two long spans, but to prevent another disaster like the I35 Bridge dropping off from the abutment. In short, the design is not forgiving enough for allowing an imperfect manufacturing process. I sincerely hope this simple hypothesis make sense to you.

I learnt this from a past SAE president many years ago. He said something like this: Engineer is not supposed to keep collecting the data and looking for a perfect solution. Engineer must make the decision sometimes with all the data he has at hand. I was fortunate enough to work for NHTSA from April '68 to December '74 and wrote two Federal Motor Vehicle Safety Standards, FMVSS #214 Side Door Strength and #216 Roof Crush Resistance, *before* the "Industry government political complex" is fully established in NHTSA. I feel you are in a unique position to save these bridges today, honestly.

Thank you for your time again.

Bill Chu 02/04/08
Hickory, NC

4.7.3 *A movie-like life of war, love, academe — tucked into 464 pages*
An interview of a Professor in College of Engineering

By Wenchen Tu, Ka Leo Contributing Reporter, June 2008

An Interview of a Professor in College of Engineering
Wai-Fah Chen has lived a life of war, love and academies his new autobiography

He is sitting behind his desk with a cup of Starbucks coffee. Like many other UH scholars, he flashes an affable smile. But unlike many others, the white hair on his temples signals a life journey that has experienced two wars and three countries, love and four decades as a scholar. And Professor Wai-Fah Chen of UH's College of Engineering has just put his dramatic life story to paper to add to his already long publication list of books about his specialty in the field of structural engineering, stability, and steel design over the past half-century.

His life began in Nanking when he and his twin brother were born in 1936. A year later, when the Sino-Japanese War broke out, his father fled Nanking with the Nationalist Chinese government, and his mother took her babies to her parents' home in a remote mountain village. After the defeat of Japanese forces in 1945, the family was engulfed in China's Civil War. His family moved to Taiwan since the communist forces seized power in Wenchow, where Chen's family had spent one year. In Taiwan, Chen started his formal education that continued at Purdue University in West Lafayette, Indiana, where he later taught engineering for 23 years. He came to UH Manoa in 1999 as a professor and Dean, a position from which he retired last year.

His recently published autobiography titled *My Life's Journey: Reflection of an Academic* is available in English in Singapore and Japan, and soon will be available in America and Europe through World Scientific Publishing Co.

When he stepped down as Dean of Engineering at University of Hawaii in 2006, Chen told Ka Leo in an interview, "Some of my colleagues and friends suggested that I should write a book on my career, engineering, and higher education based on my 40 years of experience as a researcher, educator, and engineer."

Because of his father's career as a general in the Nationalist Air Forces, Chen's family was constantly forced to move from one place to another to escape the hostilities.

During this time, Chen wrote in his 464-page book, "It is difficult nowadays for a foreigner to truly appreciate the depth of patriotism and passion of Chinese people felt toward their country."

Chen wrote on the dangers he and his two brothers faced as they were trying to escape from Mainland China to Taiwan in 1949, explaining, "*It seemed more like a good movie script than real life.*"

As they escaped to Taiwan, the Chen brothers were left on a small island alone with a few people, all of them suffering from hunger and storms. Chen recounted in his newly published book, "We were shocked to learn this fact afterward: they simply abandoned us."

His first time to study without war disturbance was in Taipei in Junior High School and his education then continued in the National Cheng-Kung University. Taiwan's "Americanized Education Model" gave the chance to students like Chen to learn more about Western culture and knowledge.

Purdue University had been selected as the counterpart by the US State Department for National Cheng-Kung University (NCKU). Many of NCKU's faculty members were sent to Purdue, and some Purdue faculty members came to NCKU. Purdue served as the main role model in the school. The "Purdue role model" became an important springboard that encouraged Chen to achieve his American dream.

"As a result of this relationship, I had a deep impression about Purdue and knew about its great engineering school in the US since my undergraduate year," Chen said, adding, this was also the reason he became a Purdue faculty member 20 years later.

Like Chen's success on his professional field, his marriage with Linlin — his lifelong companion — represents a typical happy marriage, as described in his book. Their romance started at the wedding of his twin brother in the summer of 1962, but continued much longer than Chen expected, thinking in traditional Chinese style, in which, he said, "Dating means ready to marry."

But Linlin wanted an American-style romance that meant a long process of dating, going steady, getting pinned, engaged and then married. Four years later, in 1966, after Chen's graduation ceremony, Chen and Linlin got married.

Over 40 years' matrimony did not diminish Linlin's importance in the family. "She is the one who provided the loving encouragement for me, which I sorely missed as a child, which allowed me to advance in my career," Chen wrote in his book, adding, "She is indeed a true central figure in our Chen family and her Hsuan family."

When contrasting his marriage with what he called his nephew's "Internet dating marriage", Chen explained in the interview, "It is okay for today's generation to quickly get married through different ways, but in our time, a marriage is very serious. Marriage equals life."

Besides his career as a prolific academic on engineering subjects, Chen is also an eyewitness to China's rise. From 1981 to present, Chen has been a guest or visiting professor between China and America.

After attending a Business Outlook Forum several years ago, Chen wrote, "I was really surprised to hear that, according to Curtis, the 20th century was the rising of American power replacing European economy, and the 21st century is the rising of China power replacing the American economy." The speaker that impressed him was Ken Curtis, vice chairman of Goldman Sachs Asia.

Chen also experienced the June 4th student unrest in Tiananmen Square in 1989 when the communist government of Deng Xiaoping used troops and tanks to suppress demonstrators. Chen shared with Ka Leo his eyewitnessing of the June 4th movement, saying, "I was standing on the student's side when things just happened, but since time passed, I realized the motive of Deng Xiaoping is probably right, China has [a] different social system and culture background. ... It is clear now that China is rising."

My Life's Journey reflects Chen's career, life and struggle, which go across the century and different cultures. Computer technology allowed him to write and edit the photos to show his remarkable academic life to the public.

Chen quoted Gow Huey Ling, an editor of this book in Singapore as saying, "You should proud you did everything by yourself."

When Chen was named Dean of the College of Engineering in 1999, then UH President Kenneth P Mortimer said, "Dr Chen is seen as a visionary leader in his profession. He will bring a lot of energy to the College of Engineering."

After 7 years working as Dean, Chen accomplished higher enrollment, more research funding and donations that confirmed his contribution to UH as an academic leader.

4.7.4 *A lecture for UH audience*

"*My Life's Journey: Reflections of an Academic*"
By Wai-Fah Chen
Thursday, January 31, 2008 • 12:00 noon to 1:15 p.m.
Center for Biographical Research • Henke Hall 325, 1800 East-West Road

Born in China, and educated there and in the United States, Dr Wai-Fah Chen is a widely recognized structural engineering specialist in the field of mechanics, materials, and computing. *My Life's Journey* provides Dr Chen with the opportunity to reflect on his personal history and his cultural passions: on his role as eyewitness to the rise of China, his struggles to adapt and succeed in the United States, and his career as a researcher, engineer, and educator.

While including accounts of his groundbreaking work in mathematical theory and civil engineering, Dr Chen's memoir covers not just his scientific achievements over the last four decades, but his perspective on the rapid and extensive changes in science, education, and life in China and the US.

Wai-Fah Chen, former Dean of the College of Engineering at the University of Hawaii at Mänoa, is a Researcher in Civil and Environmental Engineering. Previously Head of the Department of Structural Engineering at Purdue University, Dr Chen's honors include the 1990 Shortridge Hardesty Award from the American Society of Civil Engineers, and the 2003 Lifetime Achievement Award from the American Institute of Steel Construction.

My Life's Journey: Reflections of an Academic was published in Fall 2007 by World Scientific Books.

Contact the Center for Biographical Research at 956-3774 or biograph@hawaii.edu for more information.

Dear Dr Chen:

I think our audience would be interested in both, as the subjects are hard to separate. Because some of the audience members will not have read the work, it would be good to discuss the content — and perhaps even read some favorite short sections — but we would also be interested in why you decided, at this time, to write your memoir, and how you went about it — questions such as how did you decide what to include and what to leave out?

Did you mostly rely on your memory, talk to other people, and conduct research? How did writing this book compare to your work as a structural engineer, and how did being an engineer affect the way you conceived and carried out this work? How has your book been received by your colleagues and family?

Our audience members mostly are current or retired faculty and staff, with a few students and members of the general public (whose attendance is often limited by the difficulties of parking on campus).

Because many people associate memoirs with literary or political figures, I think they would be particularly interested in your different perspective as a scientist and educator whose life and career has truly been multicultural.

Stanley Schab (Professor, University of Hawaii)
Center for Biographical Research

4.7.5 *Summary of some comments* (讀後感言)

HF Lee (High school classmate)

"Thanks for the wonderful Christmas gift. My children are going to love it. The book is going to tell them a lot of things from our era that I did not have the chance to tell them. It also gives them a glimpse on how to be an upright person, regardless the success or fame."

KC Chao (Purdue colleague)

"One thing I skipped mentioning in the message I just sent is The American Dream. It is mentioned a number of times in your book. It is your struggle. It is what you struggle for, at least in part. It suddenly dawned on me how true what you say. Though I read it a lot, here and there, and everywhere. Millions of immigrants: businessmen, students, farmworkers, artists, musicians, Nobel Prize winners, etc. All shared that dream! But not me!

Somehow I just never thought of it. I was running away from something, from a nightmare. I just ran and ran. I was scared. I was not happy, because I lost my dreams! But I really had turned around after many years. Now it dawned on me, I really have realized a dream. And you made me see I really had a dream too, I am holding on it, without quite going for it! Thank you! Thank you!"

TV Galambos (*Professor, University of Minnesota*)

"The book is not only enjoyable because it describes your personal and professional history, but also because it contains the history of structural mechanics and structural engineering in the past half-century…"

DJ Han (*Professor and WF Chen's doctoral student*)

"I've received and read your book, *My Life's Journey*. A very good book indeed! I learned more from the book. Thank you very much!"

WF Chen on DJ Han's comments

"For those who grew up around my time like you and Xila Liu, it gives a personal description and experiences of our life struggle during the war time.

For those in [the] structural engineering community, it contains a rich history of structural mechanics and structural engineering and steel design evolution in USA.

For that fortunate younger generation in the rise of China, it provides a perspective of the history of the inevitable changes of China during the last 50 years."

"So, the book is not just my personal life story, but rather it reflects the three areas: An American-Chinese life story for an immigrant, the structural engineering history for [the] engineering community, and the rise of [the] China story for our generation of Chinese."

Helena Yuen (*younger sister*)

"Your book is so full of wisdom and useful knowledge that we are taking more time to read it in detail so that we can draw more enlightenment from it. Your idea to put complex issues into simple solutions is especially endearing to me. In fact, I have now in-corporate[d] this principle into my everyday work activities."

Jaja Hsuan (niece)

"I particularly enjoyed reading about your childhood, your journey to America, your courting with Aunt Lily, and also the anecdotes and fatherly impressions of Eric, Arnold and Brian. We hardly ever hear about pirates these days, so to read about your encounter with pirates was quite fascinating (and scary)."

Subramanian (Professor, India)

"You are one of my idols, and I have read several of your papers and books and [am] fascinated by the quality and quantity of work done by you. I recently enjoyed reading your recent book, in which you have explained about your life and your work."

Jurgen Bathe (Professor, MIT)

"After all the classics you have written… this is another success for you. Congratulations!"

Gen-hua Shi (Professor and friend)

"Actually many of my friends in China talked about your book. I am always your fan."

Peter Wang (Junior and senior high school classmate)

"Your book is not only a careful documentation of your life's journey but [you] have shown us a way of life that we should all tell our children that they should emulate whenever possible. You have set higher standards for the world to follow, which I am very proud of you. We were classmates in junior high class 27 in addition to c33. Indeed it was my privilege to share four years with you in high school days during our formative years.

You have been a great teacher, simply look at the list of your students with their accomplishments. You have taught them well to be a good engineer with knowledge but you also taught them to be a good person and indeed you have succeeded. My friend, I salute you."

Ben Chang (*classmate, NCKU*)

"You are absolutely right. Life is short. Life should be fun, and what fun is life if you don't have good health. I am still working long hours at work. It's no good to my health. I also enjoy reading your article on 'Changing Jobs — Lessons Learned from My Career Path'".

Deh-Shiu Hsu (*Professor, NCKU*)

"As I read the book, seems every one of you have exciting stories I don't know it before. Of course I enjoyed it a lot.

Many times as I read the book, I shared the contents with my wife, Lois. One thing she emphasized: 'Prof. Chen's wife, Lin-Lin, is such a beautiful and lovely girl.' In fact, this is not the first time she gave the comment. She gave the same comment 17 years ago when she met Mrs Chen [for the] first time at Purdue.

Of course, in addition, it is such a surprise to me to know that Lin-Lin is such a competent good wife. So, you are [in] luck! Prof. Chen."

Ta-liang (Leon) Teng (*Professor and high school classmate*)

"…[R]ead through it with great interest and admiration. You really have done an enormous amount of publications and you deserve all those honors bestowed on you.

It particularly reminds me [of] our humble beginning, in an old cargo ship coming across the Pacific. It was a good experience. I missed, or my memory [has] became vague about, some passages. And I believe that you forgot that our first port of call was Kobe in Japan, where we got on shore and played around for about five days while the ship [was] unloading and loading cargoes and waiting for more business.

Even after our second five-day port of call at Yokohama, the ship was only partially loaded and she was so light that part of the propeller got above the water while the ship was pitching in the waves, causing quite a vibration of the whole ship. We were going at about 8 knots. There were three days we were heading to America and the captain told us not move a bit as there was a typhoon blowing against us.

I remember that the ship was sailing for Longview, Oregon. But the captain has to let us get off at Seattle. The crabbing with sailors at both Vancouver and Seattle harbors was great fun, I took my family to Vancouver pier to repeat the crabbing about ten years later using the same technique we learned from the sailors and got two great big Dungeons for a picnic dinner.

Looking back, it was fun and quite a life experience. Now at 70, I just [lie] back and am happy looking back the road we have come from."

Chung-Ling Mao (High school classmate)

"Although we knew most of us come with a very similar background but I still could not believe that we had an almost identical 'childhood' life. I was also born in the city of Nanking in 1936. My father joined the Chinese Army in 1937, the year of Sino-Japanese War. My grandmother, my pregnant mother and I also returned to the native countryside, a very remote farm village near the hill of a small mountain in the county of Ching-Tien Hsein.

The family lost contact with my father. He was injured several times and was sent to VA hospital in Chongqing where he later attended the school of 'Military-Supply' of Chinese Air Force and finally made the contact with the family. After [the] Sino-Japanese war, our family reunited in Shanghai in 1946. I was already ten years old and my sister was eight. Since we never attended any schools we had to play catch-up. We spoke only the Ching-Tien dialect and had to learn the spoken language again.

In 1949 we moved from Shanghai to Fu-Chow and then to Taiwan. The rest is now history! We really appreciate the book and thank you again for sharing your wisdom with us."

Hsiu-Pu Lee (Purdue colleague)

"Your book recounts a vivid history of not just you, but the group of people of our generation whose parents took them to Taiwan to avoid wars in China during the late 1940s.

Growing in a much peaceful and encouraging environment, these kids were able to enjoy better education and good ethical/spiritual training. They have grown up be an achieved group and a group of good human being.

This is a unique generation very different from other generations of the modern China. You are a good example of our generation. I'm so proud of you!"

Andrie Chen 陳啟宗 (Engineer, Exxon)

"I've always considered it a privilege, honor, and pleasure to be one of your students. You've made a significant impact on my life not only when I was studying at Lehigh. It was also because I was one of your students that I was able to retire [at the] Exxon

Production Research Company. Of course, you know that I had a good career with Exxon and retired after 25 years of service."

Einar Dahl-Jorgensen, Norway (Lehigh, MS)

"Your book just arrived today and I have glanced through it. Thank you so much for the gift and for including me. It is greatly appreciated.

I have ordered an additional 5 copies as personal gifts; two will be given to my children for Christmas."

4.8 50th Wedding Anniversary of Wai-Fah and Lily (金婚)

Our 50th Golden Anniversary Celebration was held in Las Costa, San Diego, on 15 June 2016. Our lifelong story is wrapped up in a single drawing shown in the photo below: It starts from our dating days to engagement, marriage and the birth of three boys (see top of drawing), followed by our education and career routing at Brown, Lehigh, Purdue, Stanford and the University of Hawaii (see middle of drawing), and shows the cities we lived in — Boston, Providence, Bethlehem, West Lafayette, Fremont, and lastly to Honolulu — with the different cars we owned, from the Desoto to the GM Station Wagon and then to the Big Ford Van, as sketched at the bottom of the drawing. I thought an ideal opportunity to wrap up my lifelong story and career would be through this simple drawing of a modern 清明上河圖 created by my three sons, Eric, Arnold and Brian.

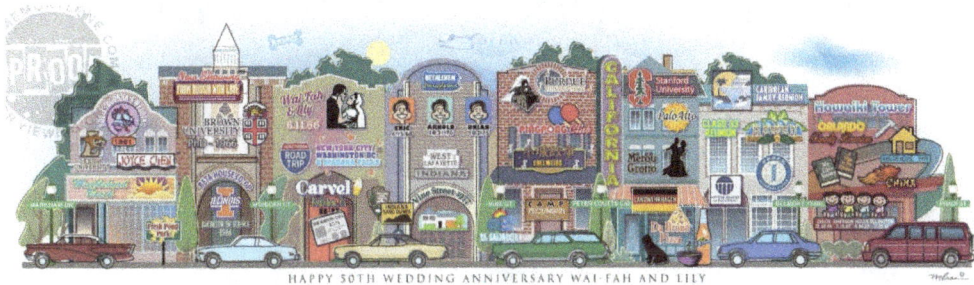

HAPPY 50TH WEDDING ANNIVERSARY WAI-FAH AND LILY

Yuying Chu

兄弟們準備的驚喜禮物, 第一次看過這麼棒的點子, 小小的一幅畫裡面, 有每棟父母住過的房子、每個街牌、每隻寵物、每台車、每個重大事件...好多小細節我以後也要一張!!!

Brothers for the surprise gift, for the first time ever seen such a great idea, little one painting inside, every building parents live through the house, every street, every pet, every car, every major incident...so many small details I will also need a picture!!!

Yuying Chu

孩子們製作了父母從年輕到現在每個階段的幻燈片.

Kids made parents from young to now every stage of the slides.

Yuying Chu
孫子們模仿爺爺奶奶的口頭禪很有趣 XD.
Grandkids imitate, grandpa and grandma mantra fun XD.

Front: Miles (漢敏), Emerson (安祥), Chloe (秀敏) and Ellington (安安)
Seated: Lin (黃慧琳), Wai-Fah (惠發), Lily (宣玲玲) and Christine De Asis
Standing: Arnold (中毅), Eric (中傑), Brian (中宇)

Yuying Chu
可愛的美媽帥爸.
Lovely beautiful mom and handsome dad.

Chapter 5
Selected Academic Bulletins
學術論文通報

This Chapter presents 23 academic article bulletins in their original formats. Those selected articles follow my career path and highlight my 10 major research areas including Limit Analysis, Beam-Columns, Semirigid Construction, Offshore Structures, Concrete Plasticity, Soil Plasticity, Structural Stability, Concrete Building Construction, Advanced Analysis, and Structural Engineering.

5.1 General (總論)

5.1.1

PERGAMON

International Journal of Solids and Structures 37 (2000) 81–92

INTERNATIONAL JOURNAL OF
SOLIDS and STRUCTURES

www.elsevier.com/locate/ijsolstr

Plasticity, limit analysis and structural design

W.F. Chen*

School of Civil Engineering, Purdue University, 1284 Civil Engineering Building, West Lafayette, IN 47907-1284, USA

Abstract

It is important for any discipline from time to time to take stock of its past achievements and to help put the current progress and future directions into perspective. In the case of research and education in plasticity and its impact on structural engineering, these assessments will help shape the directions of our future focus on research opportunities for the next decade. This is described in the present paper. © 1999 Elsevier Science Ltd. All rights reserved.

1. Introduction

What measuring stick should be used to assess the accomplishments of plasticity research for structural engineering profession in the last forty years? Should it be the volume of papers presented, or the number of journal articles published, or the number of Ph.D. theses produced or the number of new courses in the universities offered? I believe that the 'bottom line' for plasticity research should be the amount of research which finds its way into engineering practice. In the following, I shall first summarize briefly in tabular form the 'major advances' of structural engineering that can be attributed to the 'breakthrough' of solid mechanics in general, and plasticity in particular, in the last 40 years. These *'success stories'* fall into a number of broad categories. Within each category are several specific examples where new knowledge has been implemented in structural engineering and, in some measure, the structural engineering practice has been fundamentally changed. I will present some of these success stories from my own experience in the later part of this paper. Since much of my own research over the past 35 years has involved the interaction of *Mechanics*, *Materials* and *Computing*, my view on these developments is therefore strongly influenced by this background. I believe the major advances in solid mechanics and structural engineering in the last forty years are closely related to the interaction of these three areas as:

---
* Tel.: + 1-765-494-2254; fax: + 1-765-496-1105.
   E-mail address: chenwf@ecn.purdue.edu (W.F. Chen)

0020-7683/00/$ - see front matter © 1999 Elsevier Science Ltd. All rights reserved.
PII: S0020-7683(99)00079-7

82 *W.F. Chen / International Journal of Solids and Structures 37 (2000) 81–92*

Solid Mechanics and Structural Engineering: (*Mechanics, Materials, Computing*)

- Elasticity
- Plasticity
- Finite Element
- Fracture Mechanics
- Reliability
- Composite Materials
- Integrated Lifecycle Simulation (*Modeling and Simulation*)

More detailed discussion on this subject can be found in the state-of-the-art paper by Chen (1998).

2. A brief historical sketch on elasticity and plasticity in structural design

2.1. Elasticity and slide-rule computing in the 1950's

This was the period of slide-rule computing in engineering practice, and structural design was based on elastic theory which assumes that structures display a linear response throughout their loading history, ignoring the post-yielding stage of behavior. In this design practice, the elastic analysis is used to compute forces and moments in the members of a structural system, while empirical mumbo-jumbo expressions based on full-scale tests of structural members or elements are used to proportion cross-sections of moment and axial load of members. The most important advances in structural analysis in this decade may be represented by the works of Hardy Cross and N. Newmark of the University of Illinois, and of S.P. Timoshenko of Stanford University. The *Moment Distribution Method*, developed by Hardy Cross in the 1930's, was a very efficient and practical method for the design of high-rise buildings during that period. The subsequent developments of *Numerical Procedures* by N. Newmark in the 1940's and the *Approximate Series* solutions by Timoshenko in the 1950's for structural members, frames, plates and shells form the foundations of modern theory of structural analysis and design.

2.2. Plasticity and computers in their infancy in the 1960's

In this period, the computer was in its infancy while the development of the classical theory of plasticity under the leadership of William Prager of Brown University was in its golden time. We had a rigorous theory but few practical solutions. The problem of determining the stresses in an elastic–plastic structural system requires the prior knowledge of the existing permanent deformations at that particular instant. Since, these, in turn, depend on the previous load history, in practice, most of the loading paths on a structure under consideration are somewhat arbitrary in nature; the question of obtaining the state of stress of a structure under a particular loading path is not only very difficulty but also rather meaningless. For engineering practice, we must develop a simple theory that is realistic and practical. There is nothing more practical than the simple *limit analysis* theorems and techniques developed by Drucker et al. of Brown University in the 1950's (Drucker and Prager, 1952). As a result, the *Plastic Design Methods* for steel structures were developed, refined and implemented into the AISC specifications in the 1960's, under the leadership of Lynn Beedle of Lehigh University in the United States, among many others.

The subsequent extension and expansion of the limit analysis based design applied to reinforced concrete structures as well as to stability problems in geotechnical engineering problems are far-reaching and most impressive. Some of these developments, as well as the directions of current efforts and future trends, will be described in the forthcoming.

W.F. Chen | International Journal of Solids and Structures 37 (2000) 81–92   83

3. A brief historical sketch on limit analysis in civil engineering

In the early 1960's, the computer was in its infancy while the theory of plasticity was in its golden time. As a result, the limit analysis methods were developed and widely applied to steel structures. This was the time I began my teaching career at Lehigh University almost 35 years ago. At Lehigh, I participated in full scale testing of steel structures for the practical development of *Plastic Design Methods in Steels*. The plastic design method was adopted officially by the American Institute of Steel Construction as the new design code in 1963. Only recently has the lower bound stress field method been adopted by the ACI specification for the design of structural members and joints in reinforced concrete, known as the truss model. These developments are briefly summarized in the following:

3.1. Limit analysis applied to steel structures

Building Frames — Plastic Design (1960's)
- 1963 AISC Plastic Design Specification was adopted.
- First-Order plastic-hinge analysis was used for frame design.
- Auto Stress Design was adopted in the 1980's for plastic design of bridges.

3.2. Limit analysis applied to reinforced concrete structures

Concrete is a very old construction material in civil engineering but we were not able to write down its stress–strain relation or the so-called constitutive relations under various combined stress and environmental conditions. This relationship for characterizing concrete's material properties must first be developed before any finite element analysis can be carried out for its structural computer simulation. As a result, the current design practice for reinforced concrete structures is a curious blend of elastic analysis to compute internal forces and moments in a structural system, then plasticity theory is used to size up the members with empirical expressions for member strength based on full scale tests. As pointed out in an article in *Concrete International* by J.G. MacGregor (1984):

> One of the most important advances in reinforced concrete design in the next decade will be the extension of plasticity based design procedures to shear, torsion, bearing stresses, and the design of structural discontinuities such as joints and corners. These will have the advantage of allowing a designer to follow the forces through a structure.

As illustrated in a recent book by Muttoni et al. (1997), the limit theory of perfect plasticity provides a consistent scientific basis, from which simple and, above all, clear models may be derived to determine the statical strength of reinforced concrete structures.

A.A. Gvozdev (1960) appears to be the first to develop the modern concept of limit analysis as it applied to reinforced concrete structures; while K.W. Johansen (1930) used the upper bound techniques of limit analysis, and developed the *Yield-Line Theory for Slab Design* for engineering practice. The application of stress fields to reinforced concrete beam design, based on the concept of lower bound theorem of limit analysis, was first proposed by Drucker (1961). This approach is now being extended and expanded to the design of concrete structures by Muttoni et al. (1997) in a very practical manner. Similarly, upper and lower bound concepts of limit analysis were used by J. Heyman (1966) to explain the modern view of the traditional design and construction of stone skeletons. A comprehensive summary on the applications of limit analysis techniques to reinforced concrete structures is given in a recent book by Nielsen (1998). A general overview of the developments of theory and applications of concrete plasticity can be found in the book by Chen (1982).

84   *W.F. Chen / International Journal of Solids and Structures 37 (2000) 81–92*

3.3. Limit analysis in soil mechanics

The solutions of classical stability problems in soil mechanics have been generally obtained by the well-known *limit equilibrium methods* of Terzaghi (1943) using Coulomb failure criterion. Similar results can be obtained rigorously through the applications of limit analysis of soil plasticity. The limit analysis results were summarized in the books by Chen (1975) and by Chen and Liu (1990). Modern developments of soil plasticity were centered at Cambridge University in the 1960's under the leadership of Professor Roscoe. His research team firmly established the *Critical State Soil Mechanics* (Schofield and Wroth, 1968). Many useful and practical solutions based on the critical state formulation have been obtained by the powerful computer-based finite element method. These results can be found in a recent book by Chen and Mizuno (1990). A brief discussion of the impact of the finite element method and theory of plasticity on structural engineering practice will be given in the forthcoming.

4. A brief historical sketch on the finite element method in structural engineering

In the 1970's, our computing power changed drastically with mainframe computing. The *Finite Element Methods* were well developed and widely used in structural engineering. We were able to apply the theory of stability and the theory of plasticity to simulate the actual behavior of structural members and frames with great confidence. It was the first time we were able to replace the costly full-scale tests with computer simulation. As a result, the limit state approach to design was advanced and new specifications were issues.

In the subsequent years, we had an energy crisis and offshore structural engineering and technology development became the central focus. In the meantime, our computing environment changed drastically and the cost of computing becomes almost insignificant as we entered into the PC and workstation era. This was also the period that we were able to solve almost any kind of structural engineering problems with computer simulation. But now, for the first time, the physical theory is lagging behind the computing power. We need to develop a more refined theory of constitutive equations for engineering materials for finite element types of applications. This marks the beginning of the modern development of *concrete plasticity* in the 1970's. The details of this development were described in the book by Chen (1982). A brief outline of this development will be described in the next section.

Three basic conditions for a valid solution (Ray Clough — U.C. Berkeley)

- Equilibrium Condition (Newton's Law or Physics)
  The *virtual work equation* is used exclusively to establish the relationship between the stress in an element to the *generalized stresses* at nodal points.

- Kinematic Condition (Continuity or Logic)
  The *shape function* is introduced to establish the relationship between the strain in an element to the *generalized strains* at the nodal points.

- Constitutive Relations (Material or Experiment)
  The *theory of plasticity or viscosity* is used to relate the generalized stresses to generalized strains through the use of constitutive equations of engineering materials. The two-volume treatise by Chen (1994a, 1994b), covers most of these developments, among others (Desai and Siriwardane, 1983; Chen and Baladi, 1985).

W.F. Chen / International Journal of Solids and Structures 37 (2000) 81–92   85

The following is a brief summary in tabular form of the impacts of the applications of finite element methods with plasticity theory on structural engineering practice.

4.1. F.E. applied to steel structures

1970s: Development of Member Strength Equations

- Beam strength equation — Beam Design Curve
- Column strength equation — Column Design Curve
- Beam–Column strength equation — Beam–Column Interaction Design Curve
- Biaxially loaded column strength equation for plastic design in steel building frames

These developments were summarized in the two-volume treatise by Chen and Atsuta (1976, 1977).

1980s: Limit State Approach to Design

- Development of reliability-based codes.
- The publication of the 1986 AISC/LRFD Specification.
- The introduction of the second-order elastic analysis to the design codes.
- The explicit consideration of semi-rigid connections in frame design (PR Construction).

These developments were summarized in the book by Chen and Lui (1992).

1990s: Structural System Approach to Design

- Second-Order inelastic analysis for frame design was under intense development.
- The theory of stability is combined with the theory of plasticity for direct frame design.
- The advanced analysis considers explicitly the influence of structural joints in analysis/design process.

These developments were summarized in the book by Chen and Toma (1994).

4.2. F.E. method applied to offshore structures (an illustrative example)

4.2.1. Research behind the success of the offshore structures

The offshore concrete structures constructed in the 1970's for the North Sea oil development were analyzed extensively with the finite element method. Some of the highlights of the analysis process are summarized in the following:

- Solve 100,000 simultaneous equations.
- Designed for a 30 m wave with the platform located in a 300 or 1000 ft deep water.
- Consider 25,000 load combinations.
- Use supercomputer for computing.
- Assume the material to be linearly elastic.

86   *W.F. Chen / International Journal of Solids and Structures 37 (2000) 81–92*

- Cost $7M to develop the computer program.
- Require 250 engineers to input the data.

4.2.2. Failure experience, the problem

The structure failed during the installation process. The reasons for the failure of the analysis are due to:

- Did not consider that the concrete will crack after overloading and redistribute the stresses.
- The anchorage of the reinforced bars was found in the tension zone after the crack of concrete.
- Costs of the failure exceed $0.5B for the structure and $1B for the overall economy.
- The computed shear force by F.E. is about 60% of simple beam hand calculations.

4.2.3. Lessons learned

The subsequent analysis and design for a successful construction of the platform consider the following improvements:

- Improve the modeling on strength and deformation of R.C. element under all possible load combinations and torsion.
- Carry out large-scale element tests for both strength and fatigue.
- Conduct biaxial compression/tension tests:
  - Strength increased by lateral compression, 30%.
  - Strength decreased by every cycle of tension.

- Put much more steels in the shells.
- Use concrete with slump of 260 mm (10 in) instead of 120 mm (4.7 in) to get through.
- Shells are too heavy for installation. Use light-weight concrete to reduce weight.

4.2.4. Conclusions remarks

- Engineers need to develop a good material model for a heavily reinforced concrete plate element.
- Need to do simple hand calculations to check the computer solutions.
- Need experienced engineers to do hand calculation check.
- Need to consider partial failure analysis, like cracks to see possible redistribution.

5. Concrete plasticity: recent developments

5.1. Failure criteria as a start

The early effort to develop a plasticity model for concrete materials has been centered in the search for a suitable failure surface. A failure criterion of Coulomb type with a tension cutoff has been used widely in engineering practice. Based on the knowledge concerning the failure surface, a variety of failure criteria have been proposed in the past 20 years. Most of these criteria are discussed in the book by Chen (1982), where they are classified by the number of material constants appearing in the expressions as one-parameter through five-parameter models. All include the strong influence of the

W.F. Chen / International Journal of Solids and Structures 37 (2000) 81–92 87

normal stress on the shear required in the plane of sliding. All assume the isotropy of the material and convexity of the failure surface.

5.2. Work-hardening as a next step

Once a mathematically and physically attractive failure criterion has been established, the next step is to use the work-hardening theory to establish the stress–strain relation in the plastic range. The relatively sophisticated model developed by Han and Chen (1985) — the model of non-uniform hardening plasticity — illustrates this step. This non-uniform hardening plasticity model:

1. Adopts the most sophisticated five-parameter failure surface of Willam–Warnke as the bounding surface,
2. Assumes an initial yield surface with a shape that is different from the failure surface,
3. Proposes a non-uniform hardening rule for the subsequent loading surfaces with a hydrostatic pressure and Lode-angle dependent plasticity modulus, and
4. Utilizes a non-associated flow rule for a general formulation.

The important features of the inelastic behavior of concrete including: brittle failure in tension, ductile behavior in compression, hydrostatic pressure sensitivity and volumetric dilation under compressive loading, can all be represented by this refined constitutive model.

5.3. Strain-softening as a recent progress

Engineering materials such as concrete, rock and soil exhibit a strong strain-softening behavior in the post-peak stress range, showing a significant elastic–plastic coupling for the degradation of elastic modulus with increasing plastic deformation. Stress-space formulation of plasticity based on Drucker's stability postulate for these materials encounters difficulties in modeling the softening/elastic–plastic coupling behavior. Strain-space formulation is therefore necessary for further progress. As pointed out by Casey and Naghdi (1984) for some years, any arbitrary path in strain space can be specified independently by whether the material work hardens, is perfectly plastic, or strain softens. In this type of formulation the difference in material behavior can be easily described, and it permits a continuous description from one type of behavior to the other with ease. This is in contrast with the conventional stress space formulation for which the work-hardening and strain-softening behaviors must be treated differently. Although the representation of stress–strain behavior in either space can be translated into the other, the use of strain space formulation is more convenient for materials exhibiting the strain-softening behavior. On the other hand, the stress space formulation is often called for when we need a better physical understanding of the material behavior in terms of the applied stress and stress increments, that are normally used in our physical description of material behavior.

As an illustrative example, Han and Chen (1986) presented a consistent form of the constitutive relation for an elastic–plastic material with stiffness degradation in the range of work-hardening as well as strain-softening. Features of this approach include:

1. A relaxation surface is defined in strain space which serves as a criterion for further yielding and fracturing,
2. The dissipated energy due to plastic-fracturing is used as the parameter to record the material history and define both the evolution of the relaxation surface and the elastic degradation,
3. The weak stability postulate of Il'yushin's is used to obtain a relaxation rule, and
4. The consistency condition is used in establishing the constitutive relationships.

Details of the development can be found in the book by Chen and Han (1982).

5.4. Plasticity on the miniscale as the current focus

Strain-softening behavior may not be a material property and any formulation based on continuum mechanics may be misleading (Bažant and Belytschko, 1987). It has been shown that the extension of bond cracks is responsible for the nonlinear behavior of concrete and the growth of these cracks in the form of continuous cracks leads to failure in the case of uniaxial compression. This crack growth may be attributed to the sliding movement at the crack surfaces and to the sideway movement of aggregates. These mechanisms may result in irreversible (plastic) deformation and inelastic volume dilation (Yamaguchi and Chen, 1991). Further, the developments of these cracks could take place around many of the large aggregates. This is probably the reason why the overall stress–strain relationship in pre-peak stress regime provides an adequate representation of material properties in an average sense. However, bond cracks alone cannot cause failure since they are separated from one other. Failure occurs only when there are sufficient bond cracks interconnected with mortar cracks. The development of continuous crack patterns does not lead to immediate loss of load-carrying capacity because concrete at this stage behaves as a highly redundant structure. As successive load paths become inoperative through bond cracking, alternative load paths (either entirely through mortar, or partly through mortar and partly through aggregate) continue to be created and become available for carrying additional load. As the number of paths decreases, the intensity of stress and hence the magnitude of strain on the remaining paths increase at a faster rate than the external load. When the continuous crack pattern is developed extensively, the load-carrying paths are reduced considerably, resulting in a decrease of load-carrying capacity, and the descending branch of the stress–strain curve begins to form. This crack extension introduces various mechanisms including strain localization that govern the failure process of a specimen.

Microscopic observations can serve to reason out the stress–strain response and to capture its fundamental characteristics. Some fascinating applications of mechanics to concrete materials on the mini-scale have been reported in recent years. Details of this development can be found in the ASME state-of-the-art paper by Chen (1994c).

6. Design of steel structures with advanced analysis

When I first began to work in structural stability over 35 years ago, evaluation of the first-order response of a structural system was a significant problem. This includes linear elastic analysis and simple plastic analysis, and the progress made to the present state-of-the-art, which deals routinely with second-order inelastic analysis (commonly called *advanced analysis*) of complicated structural system having hundreds of thousands of degree of freedom, is miraculous. The modeling of all types of structural systems of high-rise buildings can now be handled quickly and efficiently on relatively inexpensive computers. The primary limitation is a sufficient understanding of the response of some secondary structural elements such as concrete floor slab, composite joints and walls that make up the system to develop simple but realistic models that can be incorporated into the analysis programs.

Research works are currently in full swing to develop nonlinear methods and software for practical use in the design office. The theory and approaches for advanced analysis of plane frames composed of members with compact sections, fully braced out-of-plane, have been well developed and verified by tests (Chen and Toma, 1994). Thus, it is feasible to model inelastic member and frame stability directly in a single analysis of planar frames. In fact the Australian Standard (AS4100, 1990) explicitly permits the checking of in-plane member and frame stability solely on the basis of advanced analysis.

Numerical tools for 3-D second-order inelastic analysis methods also have been proposed for analyzing and designing large-scale space frames. However, further work remains before the same can

W.F. Chen / International Journal of Solids and Structures 37 (2000) 81–92   89

be said of modeling the more complex aspects of 3-D member behavior that involves inelastic lateral–torsional effects. At the present time, most practical approaches for stability design still require the separation of in-plane frame and member behavior from out-of-plane member stability checks. The members' slenderness ratios must be checked to ensure that lateral–torsional buckling does not occur. Furthermore, additional checks are required to ensure that inelastic rotation of plastic cross sections (or compact section) must have adequate rotational capacity to allow inelastic redistribution of forces between the frame members.

For advanced analysis to achieve its full potential as a tool for the design of steel frame structures, it must therefore have sufficient generality to cope with both effects, the local buckling of cross section and lateral–torsional buckling of member. At present, little attempt has been made to incorporate the effects of these two effects into a second-order inelastic analysis for practical frame design. Research needs to be accelerated on developing the advanced analysis tools for frame structures, where local buckling and inelastic lateral–torsional buckling become the limit states.

The analytical capability of tracing the performance of an in-plane frame structure into the nonlinear range including seismic loads is currently available. Advanced analysis combines the theory of stability with the theory of plasticity, and traces the gradual plastification of members with rigid or flexible joints in a steel frame. These developments can be found in the books by Chen and Toma (1994) and Chen and Sohal (1995). The power of this new development is the following: with the ability to predict the actual moment distribution at load levels which require members to sustain their plastic moment capacity, 'breaks' can be strategically located throughout the structure. These structural 'fuses' can be designed to fail themselves without risk of the building as a whole falling down, while leaving the majority of the connections in satisfactory condition. This would not only limit the amount of post-quake repair necessary, but also would indicate where the failed connections were and thus greatly reduce the expense of 'exploratory procedures'.

Direct second-order inelastic analysis for steel frame design consistent with the current steel design specifications can be found in the books by Chen and Kim (1997) and Chen et al. (1996), where the necessary software is also provided.

7. The finite block analysis for tension-weak materials

The finite element method is suitable for continuous materials that exhibit equally high tensile and compressive strength. Concrete and soils are strong in compression, weak in tension and become discontinuous materials when tensile cracks develop. The compatibility or continuity implied at the nodal points of the finite element approach requires significant modifications when applied to concrete materials.

The *Finite Block Method* deals with the equilibrium and kinematic of discontinuous block system separated through the existing or assumed crack surfaces in a structural analysis similar to a failure mechanism assumed in the classical upper limit analysis of perfect plasticity. Once a block system is assumed, a system of equilibrium equations for the block assemblage is derived through the minimization of the total potential energy. These equations are then solved by iteration in time increment under specific loading condition until the constraining requirements of no tension and no penetration between blocks are fulfilled. A complete kinematic theory of this type was developed by Shi in his Ph.D. thesis (Shi, 1988). Similar, but much earlier works on the subject area can be found in Cundall (1971) and Kawai (1977), among others.

The finite block method contains characteristic from both the upper and lower bound techniques of limit analysis of perfect plasticity. The method is similar to the upper bound technique in that a failure mode must be predetermined; block boundaries must be defined by the user; and the solution is based

90 *W.F. Chen / International Journal of Solids and Structures 37 (2000) 81–92*

on energy equilibrium. The calculation of frictional dissipation requires that the normal force on the plane of sliding must be known before hand, thus, the finite block method provides an equilibrium stress field for all blocks at every stage of loading. Since Coulomb's friction law is enforced at the interface between blocks, the stress field so obtained is somewhat similar to the statically admissible stress field in the application of the lower bound theorem of limit analysis.

Thus, the finite block method for tension-weak materials may be considered as an extension of the computer-based limit analysis of perfect plasticity. The original version of finite block method is limited to constant-strain state in each block for simplification in the construction of equilibrium system. This may not be suitable for problems with large blocks or stress concentration. A higher order displacement function was therefore assumed and combined with the finite element technique for solving complex engineering problems. Practical applications of the finite block method include stability analysis and support designs for tunnels, slopes, retaining walls, dam abutments and foundations, etc. More detailed description and applications of this method may be found in the state-of-the-art paper by Chen and Huang (1994).

8. Concluding remarks

I believe the future direction of research and education in solid mechanics and structural engineering is in the area of *'modeling and simulation'*. In the current high-performance computing environment, the primary goal of this upcoming focused effort should be in the advancement of the state-of-the-art in a very complex scientific modeling and simulation and its validation for civil engineering applications. The major challenges are the integration of material science, structural engineering and computation and then demonstrating that the results are reliable. In the following, I shall list some of the challenging problems of structural engineering issues involving the interaction of mechanics, materials, and computing for the 21st century:

The Challenging Problems in Structural Engineering

- From Structural System Approach to Life-Cycle Analysis of Structures:
  - Construction: Sequence analysis
  - Service: Performance analysis
  - Degradation: Deterioration science
- From Finite Element Modeling for Continuous Media to Finite Block Analysis for Tension-Weak Materials:
  - Structures with changing geometry and topology
  - Materials with changing properties with time
- From Material Science Research to Structural Engineering Applications:
  - Micro-mechanics level
  - Continuum mechanics level
  - Structural engineering level

The Integrated Lifecycle Simulation of Civil Infrastructure
- Design
- Construction
- Service

W.F. Chen / International Journal of Solids and Structures 37 (2000) 81–92   91

- Deterioration
- Rehabilitation
- Demolition

The '*bottom line*' of my message on the future direction of structural engineering research and education is the following. We must de-emphasize the traditional narrow disciplinary approaches and increase integrative aspects of engineering involving *mechanics, materials, and computing*. We must emphasize that a true fulfillment of engineering research and education is '*a place in practice*'.

References

AS 4100 Standards Australia, 1990. Steel Structures. Standards Australia, Sydney, Australia.

Bažant, Z.P., Belytschko, T., 1987. Strain softening continuum damage: Localization and size effects. In: Desai, C.S., et al. (Eds.), Constitutive Laws for Engineering Materials. Elsevier, New York.

Casey, J., Naghdi, P.M., 1984. Strain-hardening response of elastic–plastic materials. In: Desai, C.S., Gallagher, R.H. (Eds.), Mechanics of Engineering Materials. Wiley, Chichester.

Chen, W.F., 1975. Limit Analysis and Soil Plasticity. Elsevier, Amsterdam.

Chen, W.F., 1982. Plasticity in Reinforced Concrete. McGraw–Hill, New York.

Chen, W.F., 1994a. Elasticity and Modeling, Constitutive Equations for Engineering Materials, vol. 1. Elsevier, Amsterdam.

Chen, W.F., 1994b. Plasticity and Modeling, Constitutive Equations for Engineering Materials, vol. 2. Elsevier, Amsterdam.

Chen, W.F., 1994c. Concrete plasticity: Recent developments. In: Kobayashi, A.S. (Ed.), Mechanics USA 1994 Applied Mechanics Review 47(6), Part 2, June.

Chen, W.F., Atsuta, T., 1976. In-Plane Behavior and Design, Theory of Beam–Columns, vol. 1. McGraw–Hill, New York.

Chen, W.F., Atsuta, T., 1977. Space Behavior and Design, Theory of Beam–Columns, vol. 2. McGraw–Hill, New York.

Chen, W.F., 1998. Structural engineering; past, present, and future. In: Proceedings of the National Structural Engineering Conference, Taipei, Taiwan.

Chen, W.F., Han, D.J., 1982. Plasticity for Structural Engineers. Springer–Verlag, New York.

Chen, W.F., Baladi, G.Y., 1985. Soil Plasticity: Theory and Implementation. Elsevier, Amsterdam.

Chen, W.F., Huang, T.K., 1994. Plasticity analysis in geotechnical engineering: from theory to practice. In: Balasubramaniam, et al. (Eds.), Developments in Geotechnical Engineering. Balkema, Rotterdam.

Chen, W.F., Liu, X.L., 1990. Limit Analysis in Soil Mechanics. Elsevier, Amsterdam.

Chen, W.F., Lui, E.M., 1992. Stability Design of Steel Frames. CRC Press.

Chen, W.F., Mizuno, E., 1990. Nonlinear Analysis in Soil Mechanics. Elsevier, Amsterdam.

Chen, W.F., Toma, S., 1994. Advanced Analysis for Steel Frames: Theory, Software and Applications. CRC Press, Boca Raton, Florida.

Chen, W.F., Goto, Y., Liew, J.Y.R., 1996. Stability Design of Semi-Rigid Frames. John Wiley and Sons, New York.

Chen, W.F., Sohal, I., 1995. Plastic Design and Second-Order Analysis of Steel Frames. Springer–Verlag, New York.

Chen, W.F., Kim, S.E., 1997. LRFD Steel Design Using Advanced Analysis. CRC Press, Boca Raton, Florida.

Cundall, P.A., 1971. A computer model for blocky rock systems. In: Proceedings of the International Symposium on Rock Fracture, France.

Desai, C.S., Siriwardane, H.J., 1983. Constitutive Laws for Engineering Materials. Prentice–Hall, Englewood, N.J.

Drucker, D.C., Prager, W., 1952. Soil mechanics and plastic analysis or limit design. Q. Applied Math. 10 (2), 157–165.

Drucker, D.C., 1961. On Structural Concrete and the Theorems of Limit Analysis. International Association for Bridge and Structural Engineering, Zurich, Abhandlungen, p. 21.

Gvozdev, A.A., 1960. The determination of the value of the collapse load for statically indeterminate system undergoing plastic deformation. Int. J. Mech. Sci. 1, 322–335 (Proc. Conf. Plastic Deformation, 1936, p. 19, translated from Russian by R.M. Haythornthwaite).

Han, D.J., Chen, W.F., 1985. A non-uniform hardening plasticity model for concrete materials. Mechanics of Materials 4, 283–302.

Han, D.J., Chen, W.F., 1986. Strain-space plasticity formulation for hardening–softening materials with elastic–plastic coupling. Solids and Structures 22 (8), 935–950.

Heyman, J., 1996. The stone skeleton. Int. J. Solids and Structures 2, 249–279.

Johansen, K.W., 1930. Styrekeforholden i Stobeskel i Beton (The strength of joints in concrete). Bygningsstat. Medd. 2, 67–68.

Kawai, T., 1977. New discrete structural models and generalization of the method of limit analysis. In: Finite Elements in Nonlinear Mechanics. NIT, Tronheim, pp. 885–906.

MacGregor, J.G.,1984. Challenges and changes in the design of concrete structures. Concrete International.

Muttoni, A., Schwartz, J., Thurlimann, B., 1997. Design of Concrete Structures with Stress Fields. Birkhauser, Berlin.

Nielsen, M.P., 1998. Limit Analysis and Concrete Plasticity, second ed. CRC Press, Boca Raton, FL.

Schofield, M.A., Wroth, C.P., 1968. Critical State Soil Mechanics. McGraw–Hill, London.

Shi, G.H., 1988 Discontinuous deformation analysis: A new numerical model for the static and dynamics of block systems. Ph.D. Thesis, University of California at Berkeley.

Terzaghi, K., 1943. Theoretical Soil Mechanics. Wiley, New York.

Yamaguchi, E., Chen, W.F., 1991. Microcrack propagation study of concrete under compression. Journal of Engineering Mechanics, ASCE 117 (3), 653–673.

5.2 Limit Analysis (極限分析)

5.2.1

Reprinted from *Engineering Plasticity*
edited by J. Heyman and F. A. Leckie
Cambridge University Press, 1968

ON THE USE OF SIMPLE DISCONTINUOUS FIELDS TO BOUND LIMIT LOADS

D. C. DRUCKER AND W. F. CHEN

Abstract

An attempt is made to exhibit some of the connections between the formal approach of the classical theory of perfect plasticity and the intuitive approach of the design engineer. The appropriate choice of a pin-connected truss to support loads is shown to provide an excellent technique for obtaining lower bounds on plastic limit loads. As an aside, the slip-line field under a punch is recovered as a limiting case of infinitely many supporting bars. Similarly, the choice of regions with simple homogeneous fields of deformation, in addition to rigid regions separated by surfaces of discontinuity, can provide good upper bounds. The expected degree of insensitivity to boundary irregularity and material inhomogeneity then is obvious. A result of some interest is obtained for the strip squeezed between two wide or narrow punches.

Introduction

There is an unfortunate tendency on the part of many engineers in practice to feel that the mathematical theories of the various branches of mechanics of solids are too esoteric and complex to be a useful tool for them. Conversely, too many theorists feel that the often intuitive and relatively crude approach of the design engineer is successful only by chance and should be replaced completely by formal mathematical techniques. The need to bring the two together was recognized in the early work of Professor J. F. Baker and in his leadership of the University of Cambridge team in their pioneering and continuing activity in the analysis and the design of structures [1].

The discussion here will be restricted to perfectly plastic solutions and plastic limit analysis and design of continua based on a shear stress criterion of yield. Perfect plasticity was an academic research area only 15 years ago. Today it is a working tool for some engineers in a wide variety of fields from metal deformation processing to the design of turbine rotors and pressure vessels. Despite this dramatic progress and acceptance, however, all is not well. Often the user employs well-known texts, such as Hill [2] or Prager-Hodge [3], or the more recent technical

D. C. DRUCKER AND W. F. CHEN

literature as a magic handbook and tries to fit his problem to the particular solutions he finds. Intuition and innovation seem discouraged by unfamiliarity and apparent complexity. Although the discontinuous fields of stress which have been drawn are simpler to visualize, they too are not often employed in an original manner by the design engineer. Yet, in fact, the concepts are familiar to the structural engineer in his terms and can be utilized by the designer as a working tool. Conversely, the interpretation and extension of slip-line solutions with simple discontinuous fields sheds quite a bit of light on their meaning and validity.

Discontinuous fields of stress viewed as pin-connected trusses

The lower bound theorem of limit analysis [4] states that if an equilibrium state of stress below yield can be found which satisfies the stress boundary conditions, then the loads imposed can be carried without collapse by a stable body composed of elastic—perfectly plastic material. Any such field of stress thus gives a safe or lower bound on the collapse or limit load. Discontinuous fields of stress are found to be especially useful. If the stress fields are chosen for convenience to be at yield in some regions rather than below, the load may be the collapse load itself.

Stress discontinuities in two dimensions have been discussed thoroughly by Winzer and Carrier [5] and by Prager [6] in terms of the jump conditions across the surfaces of discontinuity. Discontinuous fields of stress were employed years ago to give lower bounds to basic problems of interest by Bishop [7], Shield, Brady, and Drucker [8, 9]. Nevertheless, the essential simplicity of the technique is not generally recognized. Perhaps this is a result of an over-identification with exact solutions to perfectly plastic problems and the customary use of stress fields 'at yield'.

An alternative but not basically new point of view is proposed here. Plane strain problems, such as the one illustrated in fig. 1, provide simple and instructive examples. The dimension perpendicular to the plane of the paper will be taken as unity, but all motion is supposed in the plane. The applied force P is carried through the rectangular block of fig. 1a in a very elaborate pattern from the smooth (or rough) punch to the smooth (or rough) supporting plane. Details as well as broad principles are discussed in the original approach of Bishop [7] and in subsequent work by Bishop, Green and Hill [10]. Suppose instead that a pin-connected truss is imagined to carry the load inside the body,

DISCONTINUOUS FIELDS TO BOUND LIMIT LOADS

fig. 1*b*. The forces in the members of the truss are determined directly by summation of forces at each pin as

$$F = \frac{P}{2\cos\beta}, \quad T = F\sin\beta = \frac{P}{2}\tan\beta. \tag{1}$$

In the usual structural design, the cross-sectional area (the width in this plane problem) of each member is taken large enough to give a safe or permissible axial stress. Here the stress must be chosen at or below $2k$, where k is the yield stress in shear, if a lower bound on the limit load is to be found or if the safety of applying P to the block is to be determined. Once definite widths are chosen, the fields of uniaxial tensile or com-

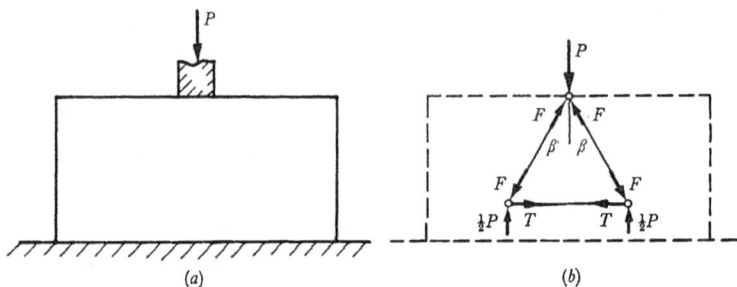

Fig. 1

pressive stress will be seen to overlap. A connection problem arises somewhat as in a real truss when the members all lie in a single plane and cannot pass by each other.

Figure 2 shows a somewhat more general case than fig. 1 to bring out both the approach and the problem more clearly. In fig. 2*a*, two compressive force fields Q, S of arbitrarily chosen widths e, f overlap in ABC. They are balanced by a third compressive force R whose strip width g is determined if no additional lines of discontinuity are to be introduced. No *internal* equilibrium problems arise if Q, R, and S are in equilibrium as indicated by the force triangle in fig. 2. The stress in region ABC is just the sum of a uniaxial compression Q/e and a uniaxial compression S/f at the angles pictured. The different jumps across each discontinuity, BC, AC, AB in the component of normal stress parallel to the line of discontinuity can be computed, but there is no need to do so. Normal and shear tractions across each discontinuity will be continuous because overall equilibrium is satisfied and all regions of space

D. C. DRUCKER AND W. F. CHEN

have been accounted for. If Q/e, S/f, and R/g are each at or below yield ($2k$), the only question which arises is whether Q/e and S/f sum to a state of stress below yield. Note that there are four lines of discontinuity [5] meeting at each point (e.g. BB', BC, BA, BB'') but the jump conditions are not those for fully plastic fields, unless by chance or choice the adjacent fields are 'at yield'.

In fig. 2b, the force S has the same magnitude and direction as in fig. 2a, but now it is tensile instead of compressive. It overlaps with R

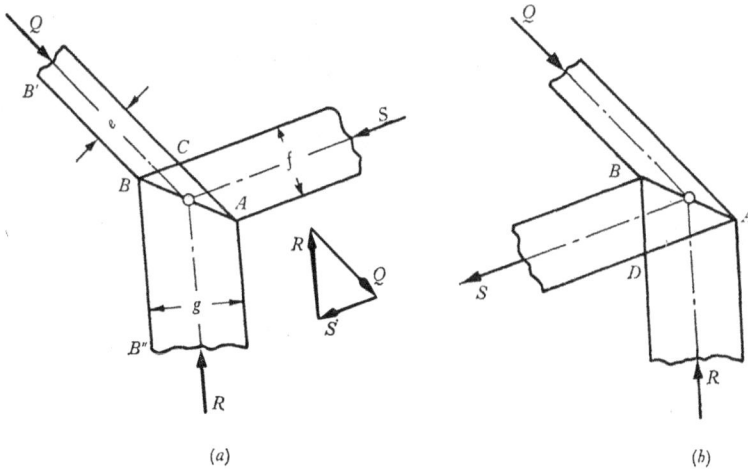

(a)   (b)

Fig. 2. 'Truss' connections.

in the region ABD before it encounters Q. However, the discontinuity line AB is the same line in (b) as in (a). Of course, the combination of a tensile stress S/f and a compressive stress R/g is likely to give a much higher maximum shear stress in ABD than is produced by the combination of the two compressive stresses in ABC.

Figure 3 is a return to fig. 1 with the arbitrary choice of $\beta = 30°$ and a minimum width of the inclined legs so as to give the yield value of $2k$ for the compressive stress in each

$$F = 2kb \cos 30°, \quad P = 2F \cos 30° = 4kb \cos^2 30° = 3kb. \quad (2)$$

This is a proper lower bound for the limit value of P provided the state of stress in the overlap region ABC is not above yield, and further that the forces at the lower pins can be carried without violating yield. The sum of two compressive fields of $2k$ each at an angle of $60°$ has been used much before because it does have the permissible compressive

132

DISCONTINUOUS FIELDS TO BOUND LIMIT LOADS

principal stresses of k and $3k$. (The general result for the sum of two stress fields in the plane is that the mean stresses add algebraically and the maximum shear stresses add as vectors with an angle between them equal to twice their angle in the physical plane.) As far as T and $\frac{1}{2}P$ are concerned, no problem arises either. The picture bears an obvious

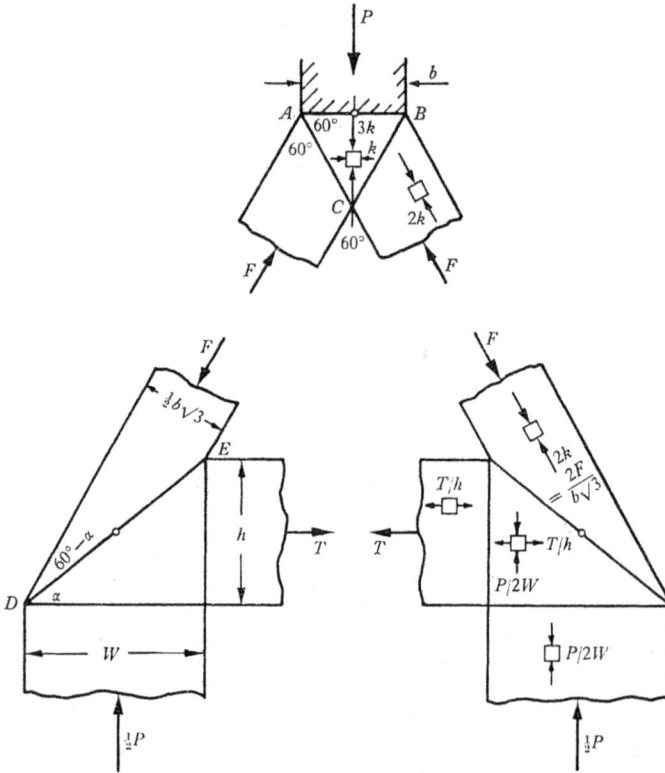

Fig. 3. Details of the truss of fig. 1, $\beta = 30°$.

resemblance to the more general fig. 2b. It is simpler because T and $\frac{1}{2}P$ are at right angles so that the overlap condition is

$$\frac{P}{2W} + \frac{T}{h} \leqslant 2k. \qquad (3)$$

This condition is met for $\alpha \geqslant 30°$. As before, there is no need to look at the jump conditions across the lines of stress discontinuity AC, DE because they are satisfied automatically.

If α is chosen as $30°$ and the picture is drawn in its most compact

D. C. DRUCKER AND W. F. CHEN

configuration, the familiar trapezoidal discontinuous pattern emerges, fig. 4. Its very compactness, however, does tend to hide its simple meaning which is exhibited far better in fig. 3.

The truss picture of fig. 1 *b* clearly does not provide the maximum possible support for the load *P*. A vertical leg, at least, is a reasonable

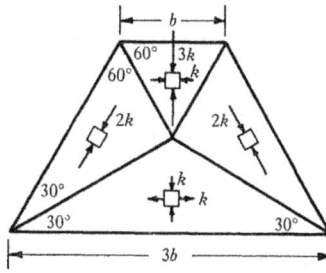

Fig. 4. Compact discontinuous stress fields at yield.

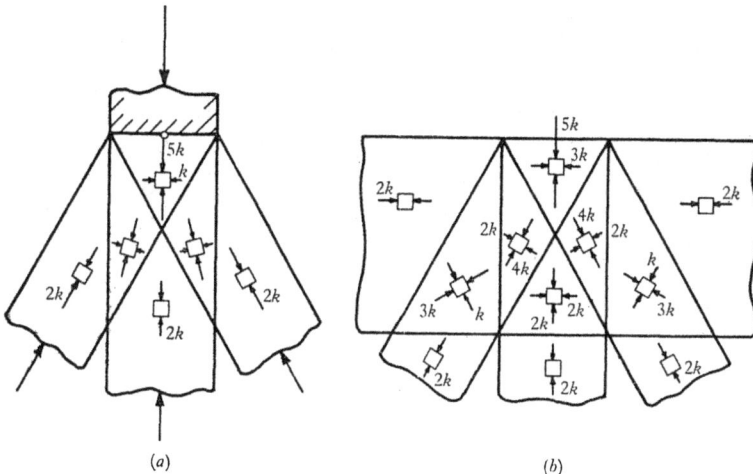

(a)   (b)

Fig. 5. Addition of a vertical leg.

addition if more load is to be carried than $3kb$. Figure 5 *a* shows the addition of the vertical leg to the 30° legs. Again, for simplicity, all legs are taken at yield; the compressive stress in each is $2k$. Now, however, the overlap regions are above yield. Satisfaction of the intuitive feeling that load should be carried both vertically down and along inclined directions thus requires either a reduction of the stresses in the members of the truss, or the addition of a horizontal thrust of $2k$ as shown in

DISCONTINUOUS FIELDS TO BOUND LIMIT LOADS

fig. 5 b to balance the vertical 2k. If an estimate of the maximum load carrying capacity is sought, the horizontal thrust is needed. Often, the problem is to decide on the safety at a given load and fig. 1 b alone may suffice.

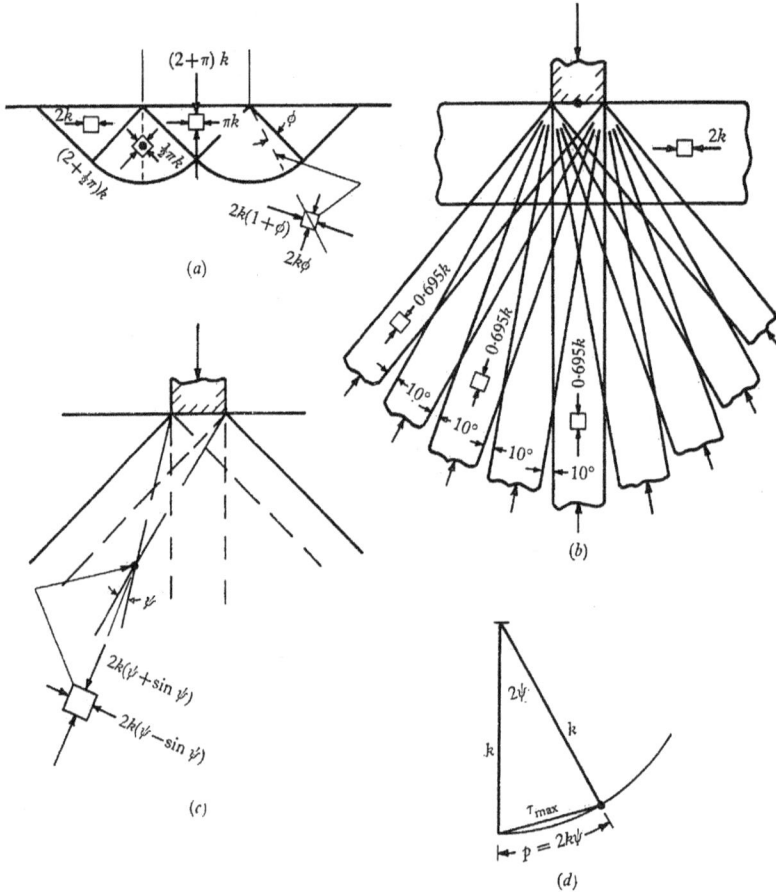

Fig. 6. Supporting legs give slip-line field in limit: (a) slip-line field; (b) nine legs at 10°; (c) limit of infinitely many legs in region $\psi \leqslant$ 30°; (d) addition of overlap fields.

The resemblance to the Prandtl slip-line field of classic plasticity, fig. 6 a, is evident and becomes closer and closer the more the number of supporting legs chosen. A picture for 9 vertical and inclined legs at 10° to each other is shown in fig. 6 b. As the number of legs grows, the stress in each decreases. In the limit, the slip-line field beneath the punch is recovered exactly, while the stress state away from this region

D. C. DRUCKER AND W. F. CHEN

is given by the overlap angle ψ, fig. 6c. The stress of 0·695k in the legs of fig 6b and of 2k in those of fig. 5b are special cases of 4k sin 2θ in fig. 7. When as in fig. 6b, 2θ = 10°, sin 2θ = 0·1737; when as in fig. 5b, 2θ = 30°, sin 2θ = 0·500.

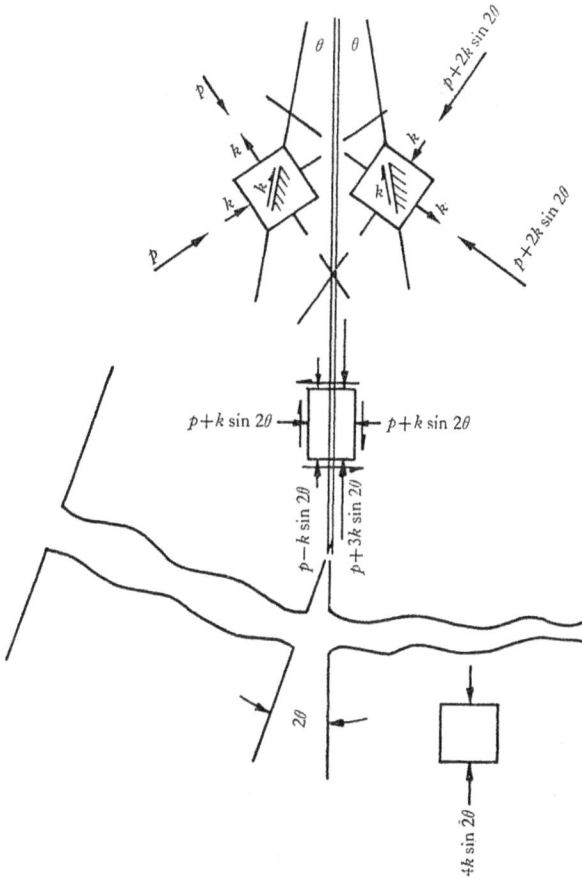

Fig. 7. Addition of compression leg. Stress fields at yield
on both sides of discontinuity.

Addition of the horizontal stress, 2k, although necessary, does not contribute to P in the local sum of vertical forces. If the body is real, and therefore finite in extent, the force in the horizontal truss members must be turned around in the space available without violating yield. This extension of the plastic stress field under the punch was shown to be possible by Bishop [7] who also obtained an explicit solution in a total

DISCONTINUOUS FIELDS TO BOUND LIMIT LOADS

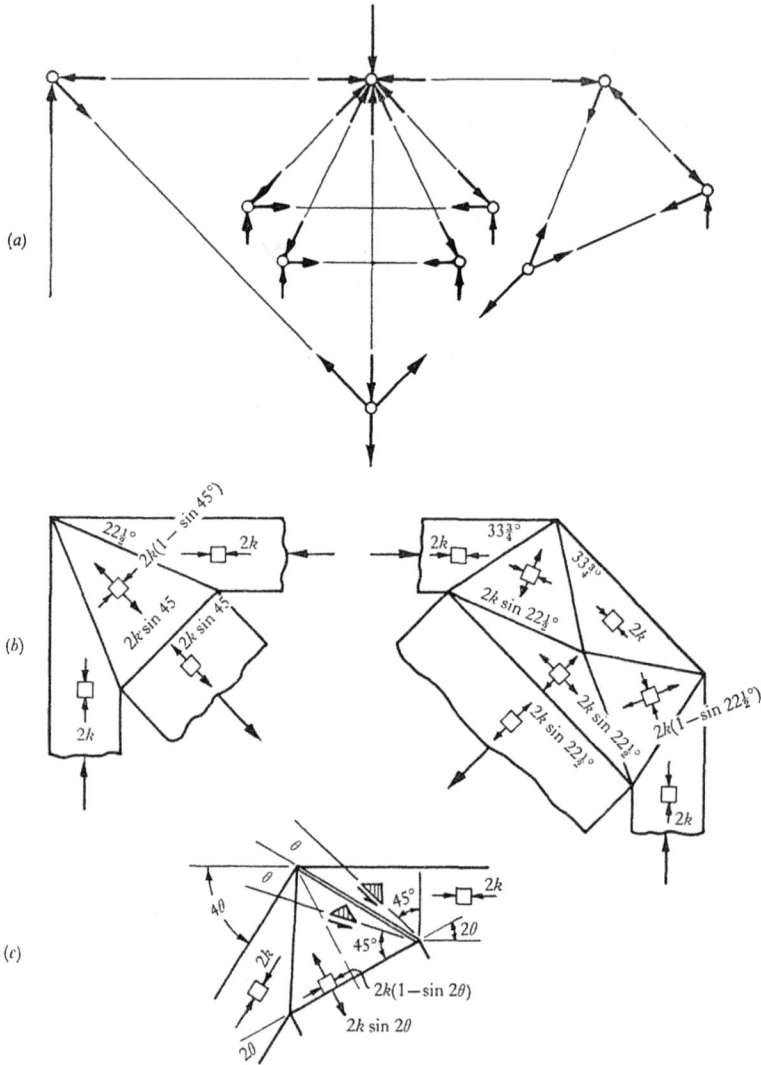

Fig. 8. Change of direction.

width of $8 \cdot 7b$. Infinitely many choices exist when the width available is ample as indicated schematically by the truss members of fig. 8a and the stress patterns of fig. 8b. The more sharply the truss member directions are turned, the larger the force or stress which must be carried in tension. Figure 2b may be thought of as the turning of a compressive force Q to give a compressive force R. The picture for half a turn if the entire

137

D. C. DRUCKER AND W. F. CHEN

picture is symmetric is represented by the lower portion of fig. 3 where the half turn angle is 30°. Figures 2 and 3 show that there is no symmetry requirement or need for fields of stress at yield. The geometric relations for the special choice of stress fields at yield are well known and indicated in fig. 8c. As in fig. 7, the discontinuity line bisects corresponding directions of principal shear stress and so is at an angle θ to the 45° directions, where the total angle of turn is 4θ.

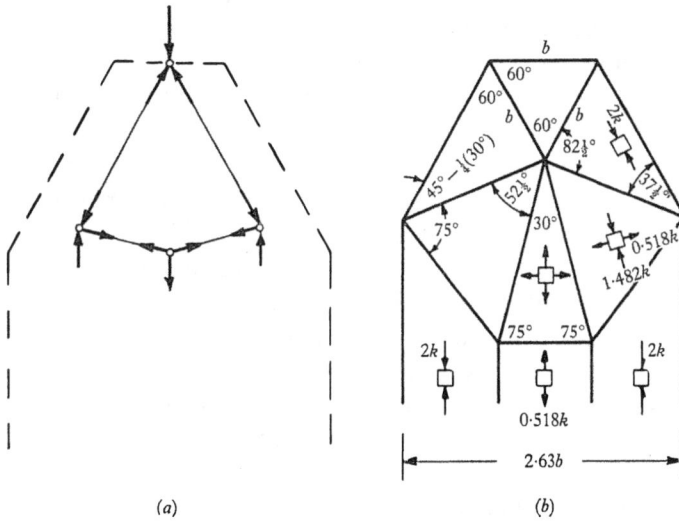

(a)   (b)

Fig. 9

As a last example of the plane truss approach, consider the 30° legs in a trapezoidal or wedge region, figs. 3, 4, 9a. The slip-line field solution gives the total force $P = (2 + \frac{1}{3}\pi)kb$ which is very little higher than the $3kb$ of the discontinuous field, figs. 3 and 4. Suppose the question is asked whether a given geometric configuration will support $3kb$. For example, is the width of $3b$ in fig. 4 needed? It is sufficient, but will a smaller width be enough? This is a very relevant question for the notched bar in tension which is mathematically equivalent to a wedge in compression.

The truss picture of fig. 9a in its most compact form, fig. 9b, provides the information that $3kb$ can be carried in a total width of $2\cdot63b$. This reduction from $3b$ with so little effort seems a worth while accomplishment, but there is no reason to suppose that the minimum value has been found. Discontinuous fields rarely are real. The truss approach is an

138

DISCONTINUOUS FIELDS TO BOUND LIMIT LOADS

engineering tool for obtaining essential but relatively crude information quickly.

Extension of the truss concept to three dimensions is obvious and there does not compete with known solutions and established techniques. Earlier work on the punch problem for a semi-infinite domain [9] provides an indication of its possible usefulness.

Discontinuous velocity fields and strain-rate fields

The upper bound theorem of limit analysis [4] states that the imposed loads cannot be carried if for any assumed fully plastic velocity field the rate of work done by the external forces exceeds the internal rate of dissipation. Equating of external to internal rate of work for any such kinematically admissible field thus gives an unsafe or upper bound on the collapse or limit load. Discontinuous velocity fields not only prove convenient but often are contained in the actual collapse mode. This is in marked contrast to the stress situation where discontinuity is useful and permissible but rarely the actual state.

Figure 10 shows a number of familiar examples of rigid block sliding separated by velocity discontinuities. The upper bounds, $P^U = 2\pi kb$, $4\sqrt{3}\ kb$, and $6kb$, are not extremely close to the actual answer

$$P = (2+\pi)kb.$$

However, they are obtained quickly and easily. Were it not for the fact that the punch problem is so well known, any one of these upper bounds would be considered as providing useful information. Each can prove even more valuable when the material being indented is inhomogeneous and so not easily amenable to exact solution. An infinite variety of such fields can be drawn for this problem or for any other in accord with the intuitive feeling of the designer or analyst for the appropriate mode of failure.

Also indicated in fig. 10 are regions of homogeneous deformation denoted by the symbol $\dot{\epsilon}$ for a field of simple vertical compression and lateral expansion and by $\dot{\gamma}$ for simple shear. Block sliding is not always appealing nor is it always the best or most convenient choice. Intermixing of simple deforming regions and rigid block sliding offers far more scope. Of course, the actual field is more complex because it includes a centred fan, fig. 6a, and a corresponding inhomogeneous distribution of deformation in that region. The addition of just one more line of discontinuity to approximate the fan would improve the

D. C. DRUCKER AND W. F. CHEN

upper bounds considerably. Such diagrams are not drawn because they would obscure the main point. In real problems with complicated shapes and loading, the essential step is to get any reasonable velocity pattern. At present, rigid block sliding is the first choice. Homogeneous deformation in addition is proposed here as an aid in bringing into play the feeling of the analyst for the likely deformation of the body he is studying.

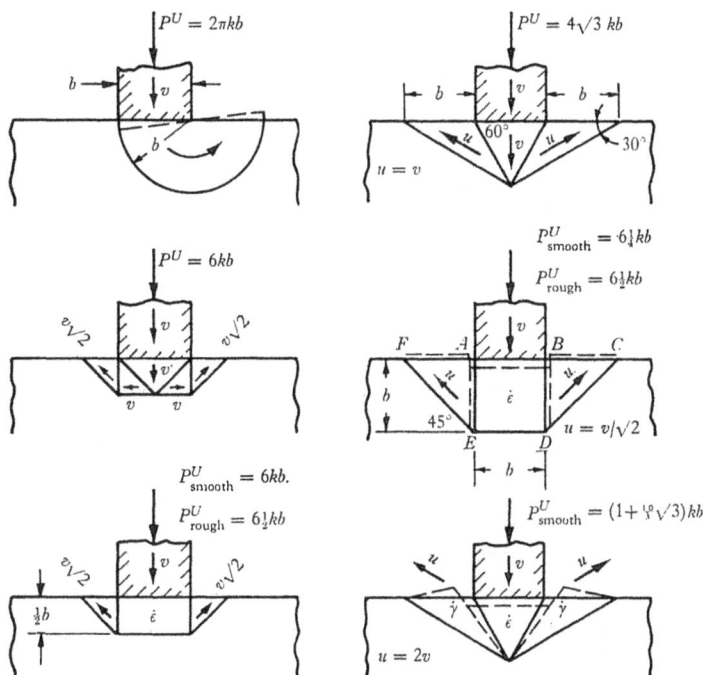

Fig. 10. Discontinuous velocity fields differing greatly from the slip-line field, fig. 6(a), $P = (2+\pi)kb$, and from each other.

Fig. 11b, d illustrates the use of a field of homogeneous deformation along with rigid block sliding in the well-studied problem of a strip squeezed between rigid punches. Certainly the field of homogeneous compression with accompanying lateral expansion in the region $ABB'A'$, designated by $\dot{\epsilon}$, makes for a reasonable start. Volume is conserved in the plastic deformation; the lateral strain is equal and opposite to the vertical strain. Therefore,

$$u = v\frac{W}{h} \quad \text{or} \quad v\frac{b}{H}. \tag{4}$$

DISCONTINUOUS FIELDS TO BOUND LIMIT LOADS

Energy is dissipated throughout the volume at the rate of the yield stress in compression ($2k$) times the strain rate in compression ($2v/h$ or $2v/H$) multiplied by the volume Wh or bH per unit dimension perpendicular to the paper:

$$2k2vW \quad \text{or} \quad 2k2vb. \tag{5}$$

The rate of energy dissipation along the surfaces of discontinuity AA' and BB' is the yield stress in shear (k) multiplied by the relative

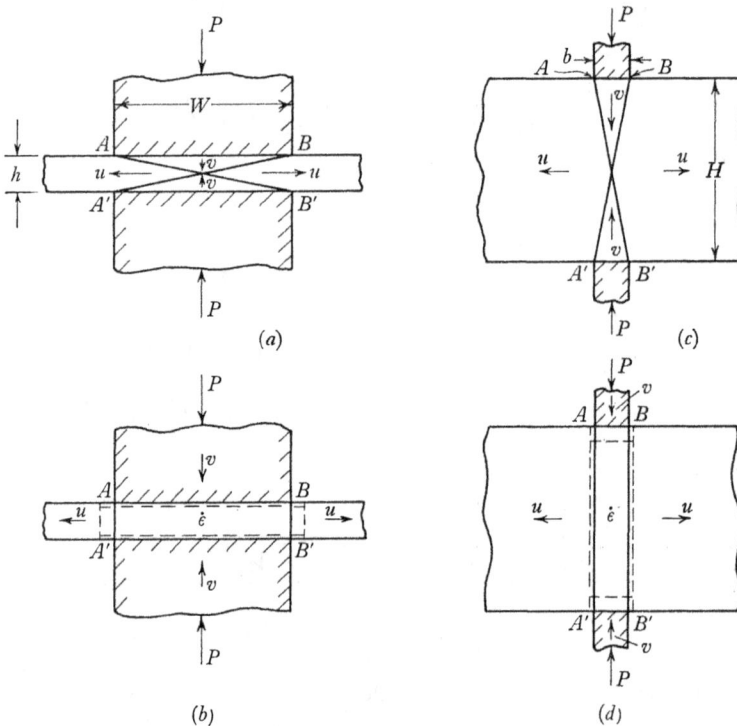

Fig. 11. Wide and narrow rigid punches.

velocity integrated over the area of the surface or the length of the line of discontinuity

$$2k\tfrac{1}{2}vh \quad \text{or} \quad 2k\tfrac{1}{2}vH. \tag{6}$$

The relative velocity appears as $\tfrac{1}{2}v$, the average between the maximum value of v at the top and bottom of the line of discontinuity and the value of zero at the mid-height.

If the punches are smooth, there will be no dissipation at the interfaces. If the punches are rough, the additional rate of dissipation is given

D. C. DRUCKER AND W. F. CHEN

by the average value of the relative velocity ($\tfrac{1}{2}u$) multiplied by the yield stress k in shear and by the total length of contact $2W$ or $2b$.

$$2k\tfrac{1}{2}uW = kvW^2/h \quad \text{or} \quad 2k\tfrac{1}{2}ub = kvb^2/h. \tag{7}$$

For a smooth punch, the upper bounds are obtained by equating $2Pv$, the rate of work done by the external pair of forces P, to the total rate of dissipation

$$P^U = 2kW + kh/2 \qquad \text{or} \quad 2kb + kH/2$$

$$= 2kW\left(1 + \frac{h}{4W}\right) \quad = 2kb\left(1 + \frac{H}{4b}\right). \tag{8}$$

For a rough punch

$$P^U = 2kW\left(1 + \frac{h}{4W} + \frac{W}{4h}\right) \quad \text{or} \quad 2kb\left(1 + \frac{H}{4b} + \frac{b}{4H}\right). \tag{9}$$

It is no accident that the applied *stresses* $P/2kW$ or $P/2kb$ turn out to be equal for the rough punch when $W = H$ and $h = b$ as drawn. All-around hydrostatic pressure in the rectangular region $ABB'A'$ has no effect on yield of a material governed by shear alone. Furthermore, expansion in the long direction and contraction in the short direction requires the same forces and dissipation as contraction in the long direction and an expansion in the short direction.

The rigid block sliding pattern of fig. 11*a* is similarly related to that of fig. 11*c*. There is no relative motion across AB, $A'B'$, AA' or BB' so that no distinction need be made between rough and smooth punches. The result

$$P^U = 2kW\left(\frac{W}{2H} + \frac{h}{2W}\right) \quad \text{or} \quad 2kb\left(\frac{H}{2b} + \frac{b}{2H}\right) \tag{10}$$

requires the computation of the relative velocity across the lines of discontinuity

$$\frac{v}{h}\sqrt{(W^2 + h^2)} \quad \text{or} \quad \frac{v}{H}\sqrt{(H^2 + b^2)} \tag{11}$$

with the relation between u and v necessarily the same as before (4) because there is no volume change.

Each of these simple fields of velocity with their simple discontinuities requires very little computational effort, but neither represents the actual field. In particular, the symmetry of behaviour for the wide and the narrow rough punch is artificial. The region to the left of AA' and to the right of BB' in fig 11*d* will not remain rigid and so does not really

DISCONTINUOUS FIELDS TO BOUND LIMIT LOADS

correspond to the rigid punch regions above AB and below $A'B'$ in fig. 11b. As H/b becomes large, the punches will not interact; a local solution, the slip-line field of fig. 6a will take over. These important questions have been discussed fully by Hill [2], Green [11] and Johnson [12]. The only point to be made here is that the field of homogeneous deformation is very convenient and physically satisfying as a first approximation. Interestingly enough, it does provide a rather good bound for the narrow punch as shown by the cross-hatching in fig. 12.

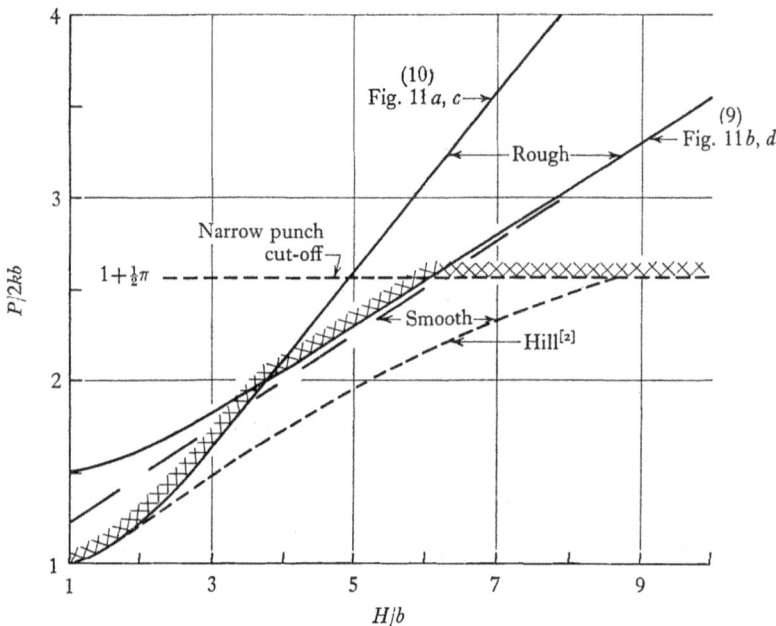

Fig. 12. Upper bounds for narrow punch. *Note:* When punches are rough $P/2kb$ vs. H/b may be interpreted as $P/2kW$ vs. W/h (solid lines only).

Once again it is worth pointing out that inhomogeneity and irregularity often dominate the picture, fig. 13. This is especially true when continuum plasticity is applied on the microscale [13]. Crude bounds obtained from block sliding on a single plane, or combinations of rigid body sliding and homogeneous fields of strain rate as illustrated are useful and give a reasonably clear idea of the influence of irregularities.

Conclusion

Standard and well-understood examples were chosen to demonstrate the role that physical insight and the methods of the structural engineer

D. C. DRUCKER AND W. F. CHEN

can play in the determination of upper and lower bounds to limit loads for perfectly plastic continua. Pin-connected trusses and slip-line fields of plane plasticity have been shown to be intimately related. The truss concept is more generally useful because it is applicable in three dimensions where complete solutions of the equations of plasticity are all but impossible except for the most elementary of problems. Combination of rigid block sliding and fields of homogeneous deformation similarly are usable in three as well as two dimensions.

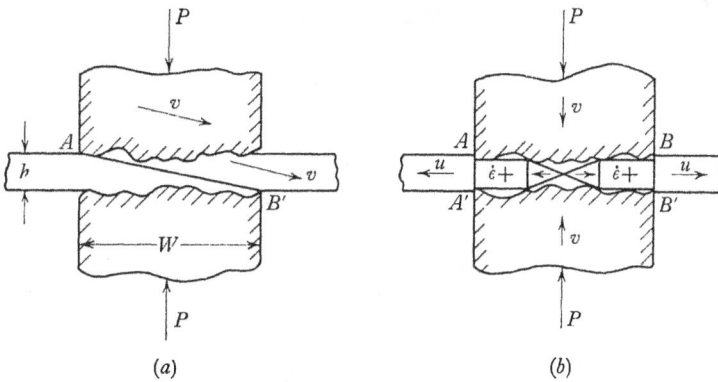

(a) (b)

Fig. 13. Influence of irregularities in contact surface.

References

[1] J. F. Baker. *The Steel Skeleton*, vol. I, Cambridge, 1954; vol. II (with M. R. Horne and J. Heyman), 1956.
[2] R. Hill. *The Mathematical Theory of Plasticity*. Oxford, Clarendon Press, 1950.
[3] W. Prager and P. G. Hodge, Jr. *Theory of Perfectly Plastic Solids*. Wiley, New York, 1951.
[4] D. C. Drucker, W. Prager and H. J. Greenberg. Extended limit design theorems for continuous media. *Q. appl. Math.* 9, 381–389, 1952.
[5] A. Winzer and G. F. Carrier. The interaction of discontinuity surfaces in plastic fields of stress. *J. appl. Mech.* 15; *Trans. Am. Soc. mech. Engrs*, 70, 261–264, 1948.
[6] W. Prager. Discontinuous fields of plastic stress and flow. *Proceedings 2nd U.S. National Congress of Applied Mechanics, Am. Soc. mech. Engrs*, pp. 21–32, 1955.
[7] J. F. W. Bishop. On the complete solution to problems of deformation of a plastic-rigid material. *J. Mech. Phys. Solids*, 2, 43–53, 1953.
[8] W. G. Brady and D. C. Drucker. Investigation and limit analysis of net area in tension. *Proceedings ASCE Separate 296, October 1953, Transactions ASCE* 120, 1133–1154, 1955.

DISCONTINUOUS FIELDS TO BOUND LIMIT LOADS

[9] R. T. Shield and D. C. Drucker. The application of limit analysis to punch indentation problems. *J. appl. mech.* **20**, *Trans. Am. Soc. mech. Engrs*, **75**, 453–460, 1953.

[10] J. F. W. Bishop, A. P. Green and R. Hill. A note on the deformable region in a rigid-plastic body. *J. Mech. Phys. Solids*, **4**, 256–258, 1956.

[11] A. P. Green. Two dimensional problems, Chapter 50 of W. Flügge *Handbook of Engineering Mechanics*. McGraw-Hill, New York, 1962.

[12] W. Johnson and P. B. Mellor. *Plasticity for Mechanical Engineers*, Chapter 12. D. van Nostrand Company, London, New York, 1962.

[13] D. C. Drucker. The continuum theory of plasticity on the macroscale and the microscale, 1966 Marburg Lecture. *J. Mater. ASTM*, **1**, 873–910, 1966.

5.2.2

6450   March, 1969   SM 2

Journal of the

SOIL MECHANICS AND FOUNDATIONS DIVISION

Proceedings of the American Society of Civil Engineers

SOIL MECHANICS AND THEOREMS OF LIMIT ANALYSIS

By W. F. Chen,[1] A. M. ASCE

INTRODUCTION

Stability problems (24)[2] deal with the conditions of ultimate failure of a mass of soil with cohesion c and internal friction ϕ. Problems of earth pressure, bearing capacity, and stability of slopes most often are considered in the realm of plasticity. The most important feature of such problems is the determination of the loads which will cause failure of the mass. A complete solution of stress and strain in a mass of soil under arbitrary loading as the load is increased to failure is almost always impractical and far beyond present understanding. The plastic limit theorems of Drucker, Prager, and Greenberg (4) permit lower and upper bounds on the failure load to be found conveniently when the soil mass is idealized as perfectly plastic.

A review of the theorems along with their applications in soil mechanics has been given recently by Finn (11).

However, the essential simplicity of the approach is not generally recognized. Many engineers have felt that the application of limit theorems requires a comprehensive knowledge of theory of plasticity. Moreover, they may feel that the construction of stress fields as required for the application of the lower bound theorem is unrelated to physical intuition. Without physical insight there is trouble in finding effective ways to alter the stress fields when they do not give a close bound on the collapse or limit load. Therefore, a technique is developed which leads to a clear physical picture of the stress fields.

Furthermore, expressions for the computation of the rate of internal energy dissipation for an assumed velocity field as required for the application of the

Note.—Discussion open until August 1, 1969. To extend the closing date one month, a written request must be filed with the Executive Secretary, ASCE. This paper is part of the copyrighted Journal of the Soil Mechanics and Foundations Division, Proceedings of the American Society of Civil Engineers, Vol. 95, No. SM2, March, 1969. Manuscript was submitted for review for possible publication on May 8, 1968.
[1] Asst. Prof., Dept. of Civ. Engrg., Lehigh Univ., Bethlehem, Pa.
[2] Numerals in parentheses refer to corresponding items in Appendix II.—References.

494   March, 1969   SM 2

upper bound theorem are often obtained from the existing results. They too are not often derived in an original manner by the engineer. The interpretation of a continuous velocity region as the limiting case of infinitely many rigid blocks separated by surfaces of discontinuity provide an excellent technique for obtaining energy dissipation. As a side, this derivation sheds quite a bit of light on the meaning and validity of certain well-known continuous fields, such as the logarithmic spiral zone. However, the main purpose herein is to consider the Plastic Limit Theorems in terms more familiar to the soil engineer and will be restricted to perfectly plastic soils based on the Coulomb Yield Criterion and its associated flow rule.

LIMIT THEOREMS

The two main limit theorems (4) for any body or assemblage of bodies of elastic-perfectly plastic material are:

Theorem 1 (lower bound).—If an equilibrium distribution of stress can be found which balances the applied load and nowhere violates the yield criterion which includes c and ϕ, the soil mass will not fail or will be just at the point of failure.

Theorem 2 (upper bound).—The soil mass will collapse if there is any compatible pattern of plastic deformation for which the rate of work of the external loads exceeds the rate of internal dissipation.

According to the statement of the theorems, it is necessary to find a compatible failure mechanism (velocity field) for the upper bound and a stress field satisfying all conditions in Theorem 1 for the lower bound. If the upper and lower bounds provided by the velocity field and stress field coincide, the exact value of the collapse or limit load is determined.

Some classical methods of soil mechanics are within the basic philosophy of limit analysis. The method of slices, e.g., uses some of the same intuitive ideas about collapse which were used in the development of the upper bound theorem. The method of a slip-line network (23) is another example of an upper bound solution since the extension of the plastic stress field into the remaining regions without violating the yield condition is not insured.

Two points should be clarified. The first is that the changes in geometry are assumed to have a negligible effect on the conditions of equilibrium. Incipient failure only is considered. The second is that the theorems are based on the physical assumption that soil is a perfectly plastic material of the Coulomb type. The Coulomb idealization is a rather drastic one because the associated stress-strain relation or flow rule requires that plastic deformation must always be accompanied by an increase in volume (5). This dilation is larger than that found in practice (8, 9). However, other predictions based on this idealization are remarkably good (6). The theorems strictly are not applicable in general to any process in which energy is dissipated by friction (7), while all real soils are frictional systems to a certain extent. Considerable care then is required in attempting to correlate the theoretical results with the experimental data. Nevertheless, the behavior of a real soil is related to the behavior of a plastic material with the same c and ϕ. A limit analysis approach based upon Coulomb's law of failure in soils does provide useful information, if not the full answer.

It is worth mentioning here that A. C. Palmer (15) in his recent work has

LIMIT ANALYSIS   495

tentatively concluded that the Coulomb yield criterion represents a lower yield condition for real soils.

THE COULOMB YIELD CRITERION AND RATE
OF DISSIPATION OF ENERGY

In deriving the solution of a two-dimensional (plane strain) problem in soil mechanics, it is generally assumed that soil fails by shear as soon as the shearing stress τ on any section satisfies Coulomb's equation (24)

$$\tau = c - \sigma \tan\phi \dots\dots\dots\dots\dots\dots\dots\dots\dots\dots\dots\dots \quad (1)$$

in which σ (here taken to be positive in tension) = the normal stress on the failure section, c = the cohesion and ϕ = the angle of internal friction. In Fig. 1, Eq. 1 is represented by the two straight lines M_oM and M_oM_1, in a plot of τ versus σ. They intersect the horizontal axis at an angle ϕ and the vertical axis at a distance c from the origin. The circles with radius R and R_o are Mohr's

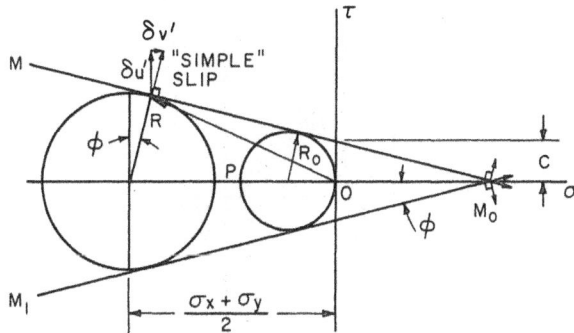

FIG. 1.—MOHR'S STRESS DIAGRAM AND COULOMB YIELD CRITERION

stress circles at failure. The geometric relations shown in the diagram demonstrate that failure occurs as soon as the radius R satisfies the equation

$$R = c \cos\phi - \frac{(\sigma_x + \sigma_y) \sin \phi}{2} \dots\dots\dots\dots\dots\dots\dots\dots \quad (2)$$

The circle with radius R_o represents a uniaxial state of stress for a compression of amount P.

On the basis of Eq. 1, Shield (22), following upon related work by Drucker (6), extended Coulomb's law of failure in two-dimensional problems to a unique yield surface appropriate for the general treatment of three-dimensional problems. In principal stress space this yield surface is a right hexagonal pyramid equally inclined to the σ_1, σ_2, σ_3, axes, and with its vertex at the point $\sigma_1 = \sigma_2 = \sigma_3 = c \cot \phi$.

As stated in the upper bound theorems it is necessary to compare the rate of internal dissipation of energy D per unit volume due to a plastic strain rate with the rate of work of external force. It can be shown in general that the dissipation has the simple form (1)

496   March, 1969   SM 2

$$D = 2 c \tan\left(\frac{\pi}{4} - \frac{\phi}{2}\right) \Sigma \dot{\epsilon}_t \quad \dots \dots \dots \dots \dots \dots \dots \dots \dots \dots (3)$$

in which $\dot{\epsilon}_t$ = a positive principal component of the plastic strain rate tensor.
For the particular case of plane strain, the Eq. 3 reduces to

$$D = c \cos\phi \; \dot{\gamma}_{max} \quad \dots \dots \dots \dots \dots \dots \dots \dots \dots \dots \dots (4)$$

in which $\dot{\gamma}_{max} = [(\dot{\epsilon}_x - \dot{\epsilon}_y)^2 + \dot{\gamma}_{xy}^2]^{1/2}$ is the maximum rate of engineering shear strain.

Eq. 3 for the special case of the Prandtl-Reuss material, for which $\phi = 0$, was obtained previously by Drucker and Shield (19).

An alternative derivation in terms more familiar to the engineer but not basically a new point of view is proposed here. The discussion here will be restricted to the plane strain case. The familiar radial shear zone when $\tau = c$, or more generally the log spiral zone for $c - \phi$ soils, is treated as illustrative examples. The results are adequate in connection with later application.

Energy dissipation in a narrow transition zone.—For the purpose of calculation, it is convenient to have a failure mechanism containing a transition

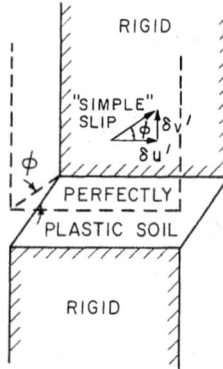

FIG. 2.—"SIMPLE" SLIP ACCOMPANIED BY A SEPARATION FOR $\phi \neq 0$

layer as in Fig. 2 to be a simple discontinuity. The rate of dissipation of energy D_A per unit area along such a surface can easily be obtained by applying the concept of perfect plasticity (3). According to the concept, if the velocity coordinates are superimposed on the stress coordinates as in Fig. 1, the vector representing slip velocity across the failure surface having discontinuous tangential component $\delta u'$ and discontinuous normal separation component $\delta v'$ to the surface is normal to the two failure envelopes M_oM, but some freedom exists at corner M_o (See point M_o, Fig. 1). The dissipation D_A may be interpreted as the dot product of a stress vector (σ, τ) with a velocity vector $(\delta v', \delta u')$, and the geometrical relations reduce the product to the simple form

$$D_A = (\sigma, \tau) \cdot (\delta v', \delta u') = (c \cot\phi, 0) \cdot (\delta u' \tan\phi, \delta u') = c \, \delta u' \quad \dots (5)$$

since the value of this product is the same for all stress points on the envelope. From the same figure it can be seen that

$$\delta v' = \delta u' \tan\phi \quad \dots \dots \dots \dots \dots \dots \dots \dots \dots \dots \dots \dots (6)$$

which states that a simple slip $\delta u'$ must always be accompanied by a separa-tion $\delta v'$ for $\phi \neq 0$ (See Fig. 2). This separation behavior is extremely impor-tant since it makes the ideal soil fundamentally different from that of Coulomb friction sliding for which the limit theorems, proved previously for assem-blages of perfectly plastic bodies, do not always apply.

It is important to mention here that the plane surface and the logarithmic spiral surface of angle ϕ are the only two surfaces of discontinuity which per-mit rigid body motions relative to a fixed surface.

Energy dissipation in a zone of radial shear when $\tau = c$.—An approximation to this zone is given in Fig. 3(a) where a picture for six rigid triangles at an equal central angle $\Delta\theta$ to each other is shown. Energy dissipation takes place along the radial lines 0-A, 0-B, 0-C, etc. due to the discontinuity in velocity between the triangles. Energy also is dissipated on the discontinuous surface D-A-B-C-E-F-G since the material below this surface is considered at rest. Since the material must remain in contact with the surface D-A-B-C-E-F-G the triangles must move parallel to the arc surfaces. Also the rigid triangles

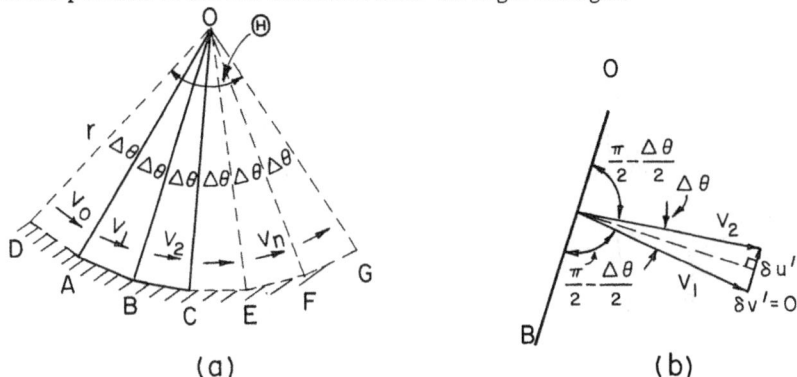

(a)   (b)

FIG. 3.—RIGID TRIANGLES GIVE RADIAL SHEAR ZONE IN LIMIT $V_n = V_2 = V_1$ $= V_0 = V$

must remain in contact with each other so that the compatible velocity diagram of Fig. 3(b) shows that each triangle of the mechanism must have the same speed.

With Eq. 5, the rate of dissipation of energy can easily be calculated. The energy dissipation along the radial line 0-B, for example, is the cohesion c mul-tiplied by the relative velocity, $\delta u'$, and the length of the line of discontinuity:

$$c \, r \left(2V \sin \frac{\Delta\theta}{2} \right) \quad \dots\dots\dots\dots\dots\dots\dots\dots\dots\dots\dots\dots\dots \quad (7)$$

in which the relative velocity $\delta u'$ appears as $(2V) \sin \Delta\theta/2$. Similarly, the energy dissipation along the discontinuous surface A-B is

$$c \left(2r \sin \frac{\Delta\theta}{2} \right) V \quad \dots\dots\dots\dots\dots\dots\dots\dots\dots\dots\dots\dots \quad (8)$$

where the length of A-B is $(2r \sin \Delta\theta/2)$ and $\delta u' = V$. Since the energy dis-

498   March, 1969   SM 2

sipation along the radial line 0-B is the same as along the arc surface A-B, it is natural to expect that the total energy dissipation in the zone of radial shear, D-0-G, with a central angle Θ will be identical with the energy dissipated along the arc D-G. This is evident since Fig. 3(a) becomes closer and closer to the zone of radial shear as the number of n grows. In the limit when n approaches infinity, the zone of radial shear is recovered. The total energy dissipated in the zone of radial shear is the sum of the energy dissipated along each radial line when the number n approaches infinity

$$\lim_{n \to \infty} n \left(2\,c\,r\,V \sin \frac{\Theta}{2n} \right) = 2\,c\,r\,V \lim_{n \to \infty} n \sin \frac{\Theta}{2n} = c\,V\,(r\Theta) \ \ldots \ldots \ (9)$$

where $\quad \Delta\theta = \dfrac{\Theta}{n}$

Energy dissipation in a log spiral zone of c - ϕ soils. —The extension of the above to include the more general case of a log spiral zone for c - ϕ soils is

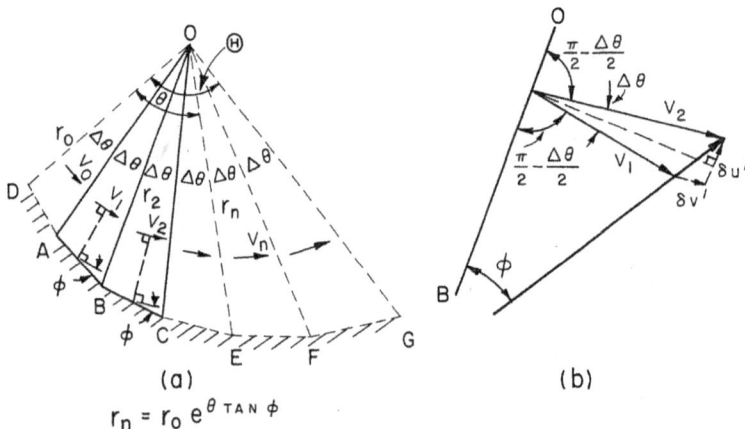

(a)

$r_n = r_0\,e^{\theta \, \text{TAN} \, \phi}$

(b)

FIG. 4.—RIGID TRIANGLES GIVE LOG SPIRAL ZONE IN LIMIT $V_2 = V_1\,(1 + \Delta\theta \, \tan \phi)$

evident. Now a simple slip $\delta u'$ must always be accompanied by a separation $\delta v'$ as required by the Eq. 6 while there is no need for such a separation when the shear strength of a soil is due only to the cohesion. A picture of six rigid triangles at an equal angle $\Delta\theta$ to each other is shown in Fig. 4(a) and the corresponding compatible velocity diagram for the two typical triangles A-0-B and B-0-C is examined [Fig. 4(b)]. If the central angle $\Delta\theta$ is sufficiently small, one may write

$$
\left.
\begin{aligned}
V_1 &= V_0\,(1 + \Delta\theta \, \tan\phi) \\
V_2 &= V_1\,(1 + \Delta\theta \, \tan\phi) \\
V_n &= V_{n-1}\,(1 + \Delta\theta \, \tan\phi)
\end{aligned}
\right\} \quad \ldots\ldots\ldots\ldots\ldots\ldots\ldots\ldots \ (10a)
$$

and from these relations, the velocity in the nth triangle 0-E-F is

SM 2   LIMIT ANALYSIS   499

$$V_n = V_0 (1 + \Delta\theta \tan\phi)^n \quad \ldots\ldots\ldots\ldots \quad (10b)$$

where V_0 is the initial velocity.

Clearly, the log spiral zone is recovered as a limiting case when the number of the rigid triangles grows to infinity. Then, in the limit as $n\to\infty$, Eq. 10b becomes

$$V_0 (1 + \Delta\theta \tan\phi)^n = V_0 \left(1 + \frac{\theta \tan\phi}{n}\right)^n \to V_0 e^{\theta \tan\phi} \quad \ldots\ldots\ldots \quad (11)$$

or $V = V_0 e^{\theta \tan\phi}$

in which V = the velocity at any angular location, θ, along the spiral and agreeing with the value obtained by Shield (20).

With Eq. 5, the rate of energy dissipation along the radial line, say, 0-B, is

$$c \, r_2 (V_1 \, \Delta\theta) \quad \ldots\ldots\ldots\ldots\ldots\ldots\ldots\ldots\ldots\ldots \quad (12)$$

in which $\delta u'$ appears as $V_1 \Delta\theta$. Similarly, the dissipation along the spiral surface A-B is

$$c \left(\frac{r_2\Delta\theta}{\cos\phi}\right) (V_1 \cos\phi) \quad \ldots\ldots\ldots\ldots\ldots\ldots\ldots\ldots \quad (13)$$

in which the length of A-B = $[(r_2 \, \Delta\theta)/\cos\phi]$ and $\delta u' = V_1 \cos\phi$. Again, the dissipation along a radial line is the same as along the spiral surface segment provided that the central angle $\Delta\theta$ is small. Thus, the expression for energy dissipation in the log spiral zone will be identical with the expression along the spiral surface which can easily be obtained by integrating Eq. 13 along the spiral surface $r = r_0 e^{\theta \tan\phi}$

$$\frac{1}{2} c \, V_0 \, r_0 \cot\phi \, (e^{2\Theta\tan\phi} - 1) \quad \ldots\ldots\ldots\ldots\ldots\ldots\ldots \quad (14)$$

agreeing with the value obtained by Haythornthwaite (13).

STRESS FIELD OF LOADED TRUNCATED
WEDGE AS AN ILLUSTRATION

This section, will consider the techniques of constructing the stress field of a loaded truncated wedge. The purpose is mainly illustrative, but the results are also of interest in connection with later application. A somewhat similar approach has been used by Drucker and Chen (10) for a perfectly plastic material for which, $\phi = 0$.

For soils, just as for metals, discontinuous fields of stress are found to be especially useful in the determination of a safe or lower bound on the collapse or limit load. If the stress fields are chosen for convenience to be at yield in some regions rather than below, the load may be the collapse load itself.

Plane strain problems, such as the one illustrated in Fig. 5, provide simple and instructive examples. The dimension perpendicular to the plane of the paper will be taken as unity, but all motion is supposed in the plane. The weight of the soil mass is neglected. The applied pressure Q is carried through the

500   March, 1969   SM 2

trapezoid region of Fig. 5(a) in a very elaborate pattern from the smooth (or rough) footing to the smooth (or rough) supporting rock base. Suppose instead that a pin-connected truss is imagined to carry the load inside the body, Fig. 5(b). The truss action of the truncated wedge would then indicate a stress

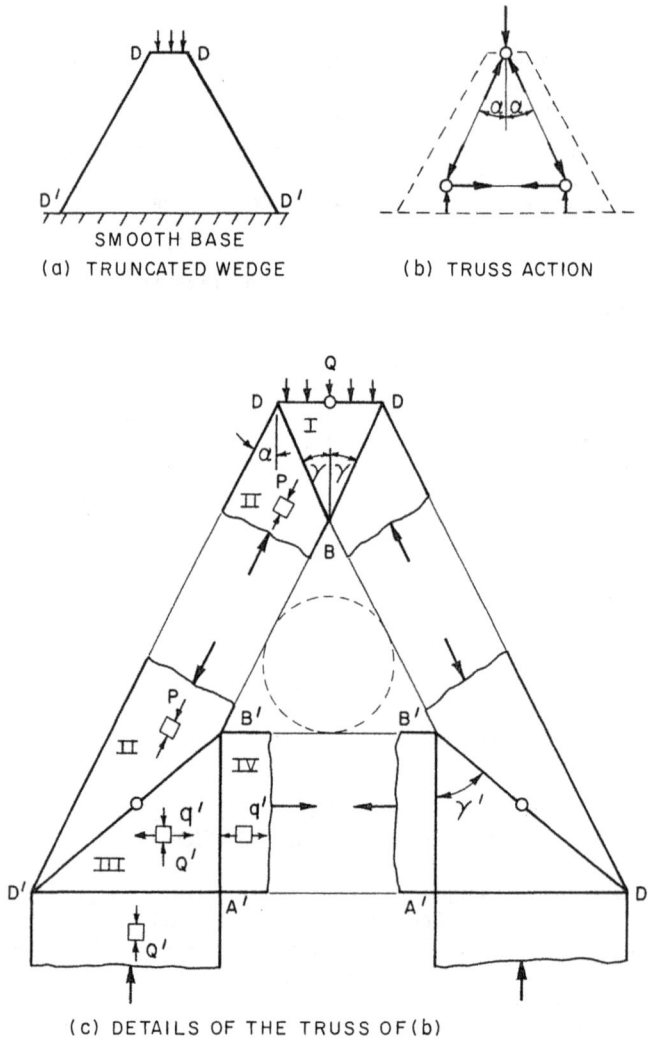

(a) TRUNCATED WEDGE

(b) TRUSS ACTION

(c) DETAILS OF THE TRUSS OF (b)

FIG. 5.—LOADED TRUNCATED WEDGE

field pattern shown in Fig. 5(c) where the three triangles marked I, III are under a biaxial state of stress and the regions marked II, IV are under uni-axial compression P and uniaxial tension q' respectively. Four elementary stress fields are shown in Fig. 6(b), 6(c), 6(b'), 6(c'). The steps are self-

SM 2   LIMIT ANALYSIS   501

evident in Fig. 6. As the field is symmetrical about the axis of applied pressure and the field of (b') or (c') is actually the same as in (b) or (c) except that
α is negative, one only needs to discuss the representative one of Fig. 6(b). To
obtain the highest lower bound of applied pressure Q which is supported by the
horizontal stress q and inclined stress P with the inclination angle α to the
vertical [(Fig. 6(b)], it is clear that one will take the largest permissible value
for the stress P. Since the state of stress in region II [Fig. 5(c)] is uniaxial
compression, the Mohr's circle for this region shown in Fig. 1 then gives
$(P = 2\,R_0)$

$$P - \frac{2\,c\,\cos\phi}{1 - \sin\phi} = 2\,c\,\tan\left(\frac{\pi}{4} + \frac{\phi}{2}\right) \quad \dots\dots\dots\dots\dots\dots\dots\dots\dots \quad (15)$$

The value Q, q, and the inclination angle, γ, of line BD to the vertical are
determined in terms of the known quantities ϕ and α from the following con-

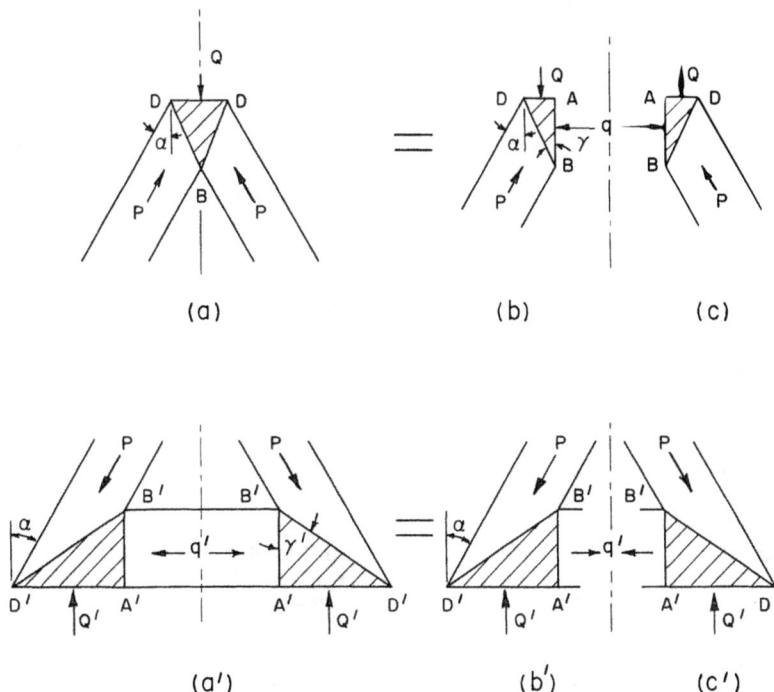

(a)   (b)   (c)

(a')   (b')   (c')

FIG. 6.—TRUNCATED WEDGE STRESS FIELDS

ditions: (1) The over-all equilibrium of forces acting on the triangular element
ABD in vertical and horizontal directions; (2) the assumption that the material
in the triangular element ABD is plastic. It is found that

$$Q = c\,\cot\phi\left[\tan^2\left(\frac{\pi}{4} + \frac{\phi}{2}\right)\frac{\sin\,(\nu + \alpha)}{\sin\,(\nu - \alpha)} - 1\right] \quad \dots\dots\dots\dots\dots \quad (16)$$

502   March, 1969   SM 2

where denoting $\cos \nu = \sin\phi \cos\alpha$. The details are left in Appendix I.

As for the pressure Q', it is simply by substituting $-\alpha$ for α into Eq. 16, that gives

$$Q' = c \cot\phi \left[\tan^2 \left(\frac{\pi}{4} + \frac{\phi}{2} \right) \frac{\sin (\nu - \alpha)}{\sin (\nu + \alpha)} - 1 \right] \dots \dots \dots \dots \dots (17)$$

Fig. 7 shows the most compact configuration of the stress fields in Fig. 5(c) when the triangular area BB'B approaches zero. Here the uniaxial tension region IV (note: q' may not yet reach its yield value) vanishes, therefore, the whole trapezoid region DDD'D' is plastic. Its very compactness, however, does tend to hide its simple meaning which is exhibited far better in Fig. 5(c).

It is interesting to note that the stress field of Fig. 5(c) is also applicable for the case where a hole (as when a flexible pipe is embedded) is present in the triangular region as shown by dotted lines in the figure.

It is important to note that an over-all equilibrium of forces acting on the triangular element ABD (Fig. 8) demands only that the stress component σ_n

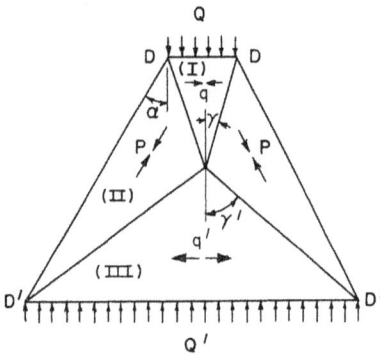

FIG. 7.—LIMITING CASE OF THE STRESS FIELD SHOWN IN FIG. 5(c)

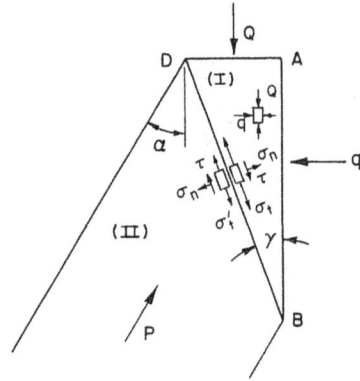

FIG. 8.—STRESS DISCONTINUITY, $\sigma_t \neq \sigma_t'$

normal to BD and the shear stress τ parallel to BD should be the same on both sides, but the components σ_t and σ_t' acting parallel to BD may be different. Clearly the line BD is a line of stress discontinuity. The amount σ_t and σ_t' are restricted by the condition that the material on both sides of the line is nowhere violating the yield-criterion. Prager (16) established the so called "jump condition" at a line of stress discontinuity with both sides in plastic yield for a Prandtl-Reuss material, for which $\phi = 0$, and Shield (21) extended this condition to the soils for the Coulomb type. The stress fields in Fig. 5(c) or Fig. 7 are the discontinuous stress field; the lines DB, D'B', B'A', are lines or stress discontinuities. The results obtained by applying the jump condition for each discontinuous line of the fields in Figs. 5(c) and 7 are met.

Simple variations on the above example are shown in Fig. 9 and Fig. 10, the cases referred to are as follows:

1. Case 1. A strip foundation is supported by a semi-infinite mass of soil

of finite thickness resting on a rough rock base [Fig. 9(a)]. Physical intuition for this problem strongly suggests that the portion of the material under the foundation will behave as a truncated wedge [Fig. 9(b)]. Therefore, a discontinuous stress field [Fig. 9(a)] is constructed by applying the idea of truss action. The foundation is supported by two uniaxial compression "legs" which rest on the base and the shear component of the traction between the "legs" and the base may have any value, since the base is rigid and is perfectly rough. Here the value of the lower bound can be further increased through utilizing the stressfree material more fully, that is, superimposing a vertical compression vertically below the strip foundation and a horizontal compression throughout the material [See Fig. 9(c)]. The states of stress for different re-

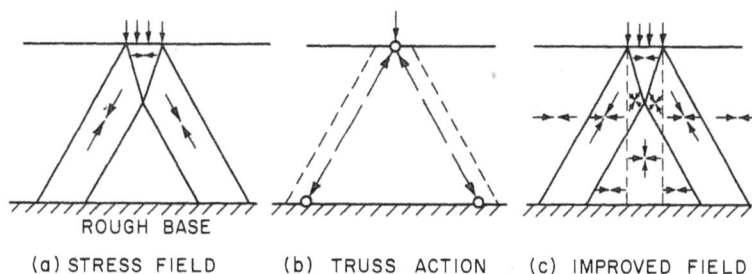

(a) STRESS FIELD   (b) TRUSS ACTION   (c) IMPROVED FIELD

FIG. 9.—STRIP FOUNDATION ON A SEMI-INFINITE BODY RESTING ON A PERFECTLY ROUGH BASE

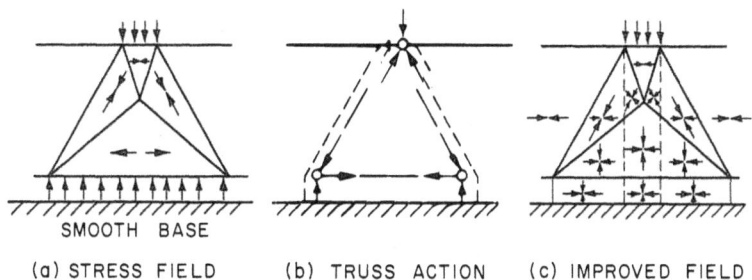

(a) STRESS FIELD   (b) TRUSS ACTION   (c) IMPROVED FIELD

FIG. 10.—STRIP FOUNDATION ON A SEMI-INFINITE BODY RESTING ON A SMOOTH BASE

gions are represented by the small arrows in the figure. Since this stress field is only a special case of the field considered in a following section where a general bearing capacity problem for strip foundation is solved, the details in Fig. 9(c) are omitted here for brevity.

2. Case 2. In Fig. 10, the problem is the same as in Fig. 9 except that the base is smooth. A stress field suggested by the truss action [Fig. 10(b)] is shown in Fig. 10(a), and its improvement by using the same idea as in case 1, is also shown [Fig. 10(c)].

It should perhaps be seen now that the technique of constructing stress fields

504   March, 1969   SM 2

aided only by ordinary mechanics of materials appeals to and makes use of the physical intuition developed by engineers. With the limit theorems, the results so obtained do provide the engineer with a quantitative feeling for his problem although the stress field need bear no resemblance to the actual state of stress according to the theorems.

WEDGE UNDER UNILATERAL PRESSURE

The computation of the critical pressure of the problem of wedges with uniform pressure on one face will serve as an illustration of both the upper and lower bound technique of limit design theorems in soil mechanics.

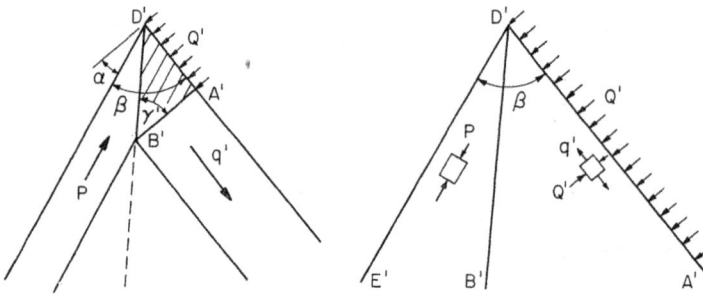

(a) STRESS FIELD OF FIG. 6(c')   (b) EXTENSION OF (a)

FIG. 11.—ACUTE WEDGE $\beta \leq \pi/2$

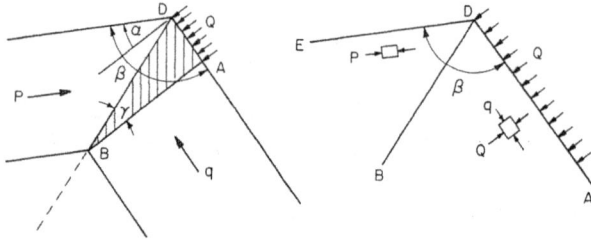

(a) STRESS FIELD OF FIG. 6(b)   (b) EXTENSION OF (a)

FIG. 12.—OBTUSE WEDGE $\beta \geq \pi/2$

The discontinuous stress fields in connection with Figs. 6(c') and 6(b) can now be extended for an obtuse wedge as well as for an acute one. By rotating the stress field of Fig. 6(c') counterclockwise about 90°, one obtains the field of Fig. 11(a) and the extension of the two sides D'A', D'B' of the triangular element D'A'B' to infinity gives a stress field for the acute wedge [Fig. 11(b)]. In the figure, the wedge A'D'E' with a central angle β is loaded by a uniform pressure Q' along A'D' and the line D'B' is the line of stress discontinuities separating the two constant stress regions A'D'E' and B'D'E'. The states of stress for different regions are represented by the small arrows in the figure and Mohr's representation of the stresses for these regions are shown in Appendix I. Since the pressure, Q' obtained previously is a function of the quantities c, ϕ, and α only it can be concluded that here the value of Q' is still given by Eq. 17. In the same way it can be seen that [Fig. 12(b)] gives a stress field for the obtuse wedge and again the value of Q is given by Eq. 16. By the limit theorem 1 stated in the second section, Q or Q' is therefore, a lower bound for the uniform critical pressure of the wedges.

As for an upper bound of the pressure, a failure mechanism is usually indicated by physical intuition and the method for constructing failure line fields in plane strain are of great help in general in obtaining a possible least upper bound. For example, for the case of an acute wedge, the simplest pattern of

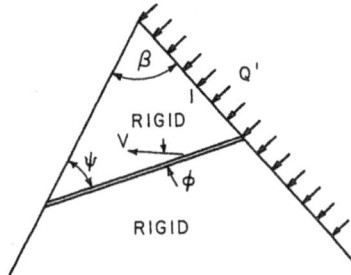

FIG. 13.—RIGID BODY "SLIDE" MOTION FOR ACUTE WEDGE

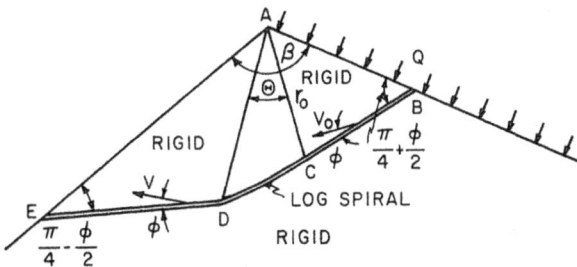

FIG. 14.—FAILURE MECHANISM FOR OBTUSE WEDGE $V = V_0\, e^{\theta \tan\phi}$

failure mechanism may be found by taking a plane slide as shown in Fig. 13. Equating external rate of work to the dissipation Eq. 5 gives

$$Q' \, 1 \, V\cos\left(\phi + \psi + \beta - \frac{\pi}{2}\right) = c \, V\cos\phi \, \frac{\sin\beta}{\sin\psi} \quad\quad\quad (18a)$$

so that
$$Q' = \frac{c \, \cos\phi \, \sin\beta}{\sin\psi \, \sin\,(\psi + \beta + \phi)} \quad\quad\quad\quad\quad (18b)$$

minimizing the right hand side gives

$$\psi = \frac{\pi}{2} - \frac{(\beta + \phi)}{2} \quad\quad\quad\quad\quad\quad\quad\quad (19a)$$

506   March, 1969   SM 2

and $\quad Q_c' \leq Q' = \dfrac{2\,c\,\cos\phi\,\sin\beta}{1 + \cos\,(\beta + \phi)}$ .... (19b)

For the case of the obtuse wedge, the selected failure mechanism is shown in Fig. 14 which consists of two triangular regions ABC, ADE moving as a rigid body and a logarithmic spiral zone ACD. The material below the failure line B-C-D-E remains at rest so that B-C-D-E is a line of velocity discontinuity. Thus the velocity along the line must be inclined at an angle ϕ to the line. In the spiral zone ACD, the velocity increases exponentially from the initial velocity V_0 along the radial line AC to the velocity $V = V_0\,e^{\Theta\tan\phi}$ along the radial line AD as discussed in the third section. Energy is dissipated through-

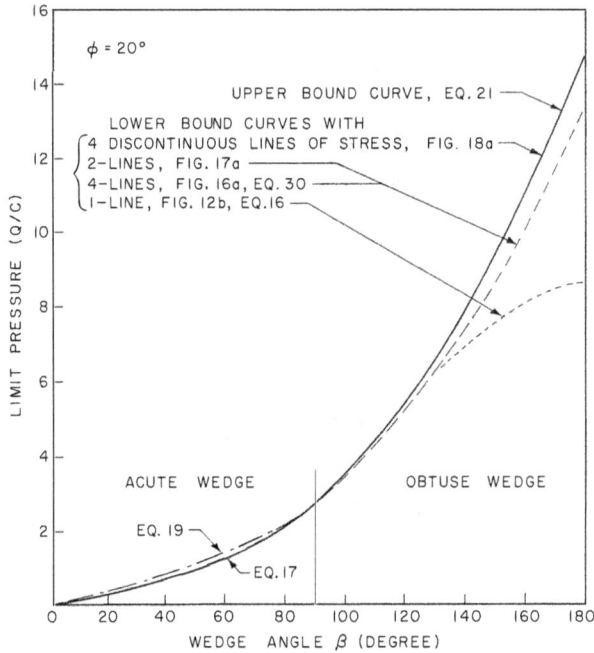

FIG. 15.—LOWER AND UPPER BOUNDS FOR A WEDGE UNDER UNILATERAL PRESSURE, $\phi = 20°$

out the log spiral zone as well as along the spiral surface at an equal rate as given by Eq. 14. Energy is also dissipated along the surface of discontinuity BC and DE. With Eq. 5, the rate of dissipation of energy can easily be calculated. Take the length AC as r_0 so that the length AD is $r_0\,e^{\Theta\tan\phi}$. The energy dissipation along the line segments BC and DE then is given by

$$\overset{\text{BC}}{c\,r_0\,(V_0\,\cos\phi)} + \overset{\text{DE}}{c\,(r_0\,e^{\Theta\tan\phi})(V_0\,e^{\Theta\tan\phi})} \quad .... (20)$$

The upper bound is obtained by equating $2\,Q\,r_0\,\cos\,(\pi/4 + \phi/2)\,V_0\,\sin\,(\pi/4 - \phi/2)$, the rate of work done by the external pressure Q, to the total rate of dissipation

$$Q_c \leq Q^u = c \cot\phi \left[e^{2\Theta \tan\phi} \tan^2 \left(\frac{\pi}{4} + \frac{\phi}{2} \right) - 1 \right] \ldots\ldots\ldots\ldots (21)$$

agreeing with the value obtained by Prandtl (18). By putting $\beta = \pi/2 + \Theta$, Eq. 21 can be written

$$Q_c \leq Q^u = c \cot\phi \left[e^{(2\beta - \pi)\tan\phi} \tan^2 \left(\frac{\pi}{4} + \frac{\phi}{2} \right) - 1 \right] \ldots\ldots\ldots\ldots (22)$$

These lower and upper bounds for the critical pressure of wedges are plotted against wedge angle β in Fig. 15 for $\phi = 20°$.

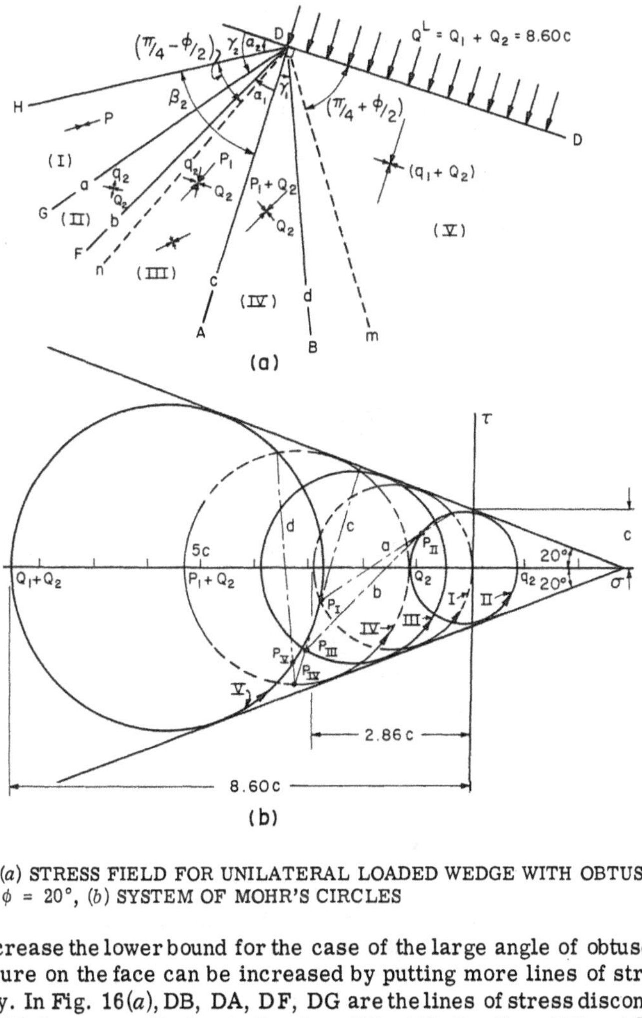

FIG. 16.—(a) STRESS FIELD FOR UNILATERAL LOADED WEDGE WITH OBTUSE ANGLE $\beta = 150°$, $\phi = 20°$, (b) SYSTEM OF MOHR'S CIRCLES

To increase the lower bound for the case of the large angle of obtuse wedge, the pressure on the face can be increased by putting more lines of stress discontinuity. In Fig. 16(a), DB, DA, DF, DG are the lines of stress discontinuities and line DA is perpendicular to the face DD while the lines DF and DB make angles α_1 and γ_1 clockwise and counterclockwise from DA, respectively. The state of stresses for the constant regions D-D-B, B-D-A, A-D-F, F-D-G, and

508   March, 1969   SM 2

G-D-H are represented by the small arrows in the figure. Actually this field is the limiting case of the field given in Fig. 19 for the problem of strip foundation when the loaded face DD goes to infinity along one direction. The critical pressure is still given by Eq. 30. The details will be omitted but the value of Eq. 30 is again plotted in Fig. 15. The question may arise whether the arrangement of four such discontinuous lines a,b,c,d (the same as DB, DA, DF, DG) of Fig. 16(a) are very effective in producing a high pressure along the surface DD. The answer is "no" because of the presence of the tensile stress q_2 in the

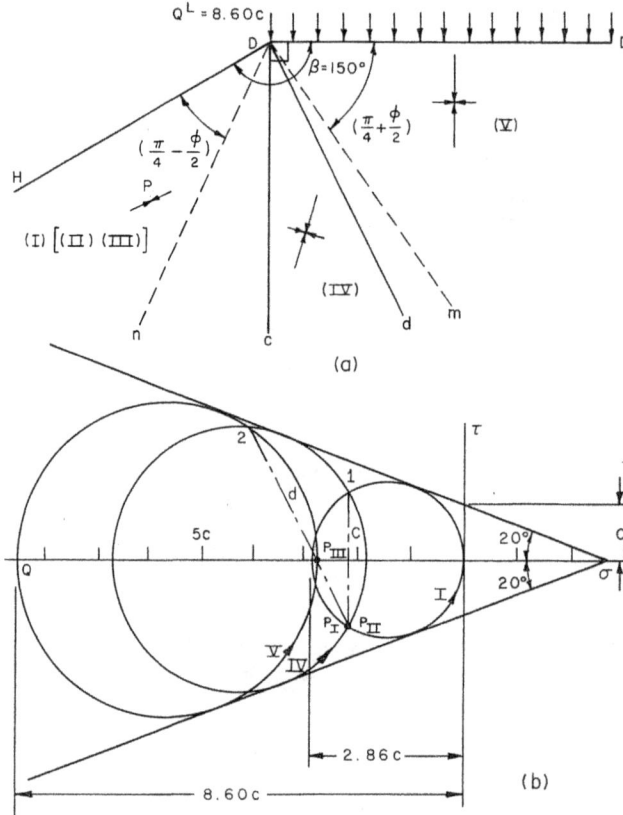

FIG. 17.—(a) STRESS FIELD WITH TWO LINES OF DISCONTINUITY FOR UNILATERAL LOADED WEDGE WITH OBTUSE ANGLE $\beta = 150°$, $\phi = 20°$, (b) GRAPHICAL SOLUTION

region II which moves the Mohr's circle of II to the right instead of to the left as shown in the Fig. 16(b). If one takes away the two discontinuous lines a, b entirely and adjusts the discontinuous line d such that the principal directions in region V are consistent with the boundary surface traction, one can arrive at the same high pressure as those of four discontinuous lines Fig. 16(a). Fig. 17(a) shows the stress system of the case and the regions marked I, IV, V in Fig. 17(a) shown in Fig. 17(b). All the constructions are obvious and do not require further explanation. The points labeled P_I, P_{II}, P_{III} are the poles of the corresponding Mohr's circles. Line P_I-1 is parallel to line c in Fig. 17(a) and P_V-2 is parallel to d. The essential feature is that the discontinuous lines must be located inside the region bounded by the two dotted lines "*om*" and and "*on*" in order to produce a high pressure along DD effectively. The lines "*om*" and "*on*" make angles of $(\pi/4 + \phi/2)$ and $(\pi/4 - \phi/2)$ with the boundary surfaces DD and DH respectively. The region nDm actually is the corre-

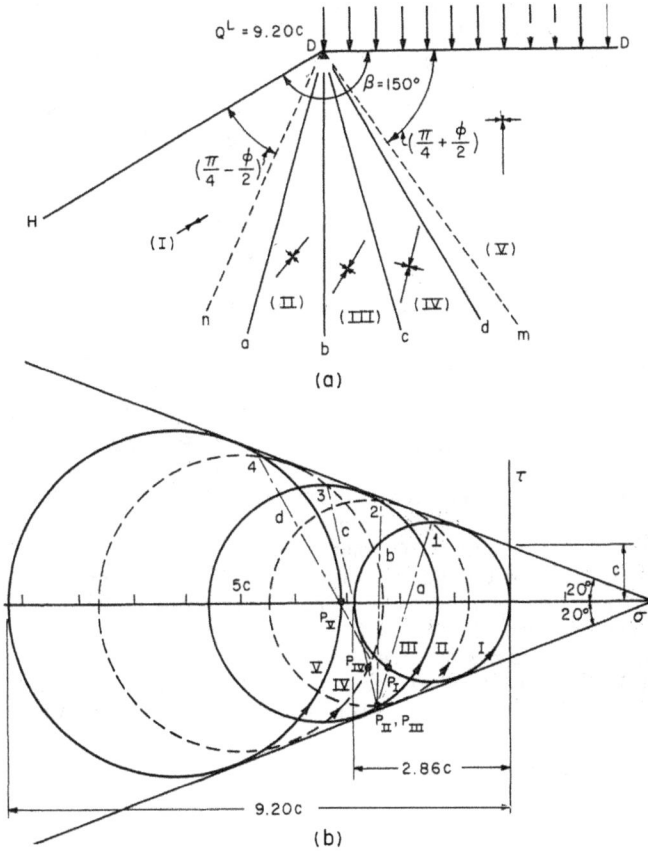

FIG. 18.—(a) STRESS FIELD WITH FOUR LINES OF DISCONTINUITY TO APPROXIMATE THE LOGARITHMIC SPIRAL ZONE OF PRANDTL SOLUTION, $\phi = 20°$, (b) GRAPHICAL SOLUTION

sponding region of logarithmic spiral ACD of Fig. 14. It is interesting to note that the stress field shown in Fig. 17 may be interpreted as an approximated extension of the Prandtl field of Fig. 14 where the continuous changing of stresses in spiral zone ACD is approximated by a few discontinuous lines. Fig. 18 shows four discontinuous lines to approximate the logarithmic spiral zone where the discontinuous lines a, b, c are arbitrarily chosen and line d is

510   March, 1969   SM 2

selected to fit the boundary condition of DD and a pressure of about the amount of the upper bound value is obtained. Any further increase in the pressure by introducing more discontinuous lines for this case will be of no real help as the bounds are already very close. Therefore, it may be concluded that the Prandtl (18) solution gives the correct values since it may be considered as the limiting case when the numbers of discontinuous lines tend to be infinity.

Now it is clear physically from the stress field of Fig. 18(a) that the pressure Q is effectively supported by the constant stress regions marked I, II, III, IV, V where the principal stresses are shown by the small arrows.

It should be mentioned here that the lower bound for the case of the acute wedge is actually the correct answer since a velocity field can be associated with the discontinuous stress field of Fig. 11(b). The method of determining the associated velocity field in the regions for a prescribed velocity on the loaded surface is very similar to that used by Prager (16) for the perfectly plastic material. A detailed discussion is omitted here for the sake of brevity.

STRIP FOUNDATION

A further example of a discontinuous stress field is shown in Fig. 19. The lower bound applies to a rough as well as a smooth punch which is normally and centrally loaded.

The stress field is a combination of the stress fields used in Section 4 and 5 for the loaded trapezoid and the wedge under unilateral pressure. The field is built up in the following way. A vertical pressure Q_1 is supported by the trapezoid [See Fig. 5(c)] stress field D-F-E-B-E-F-D (solid lines) which has an uniaxial compression of amount P_1 in the two "leg" regions D-F-E-B producing a horizontal compression q_1 in the region D-D-B. To improve the lower bound, a vertical and horizontal compression of amount Q_2 is superimposed in the region vertically below DD, that is the region D-C-C-D. The unilateral pressure Q_2 acting on the face DC can be transmitted through the wedge stress field C-D-H (dotted lines) where the line DG is the line of stress discontinuity. The largest permissible value of Q_2 is still given by Eqs. 16 or 17. Stresses acting in the various regions are shown in Fig. 19. The stress P_1 and its inclination α_1 to the vertical are chosen so that regions D-A-B, D-F-E-A are both plastic. Yield in region D-A-B requires that P_1 has the value

$$P_1 = 2 c \tan^3 \left(\frac{\pi}{4} + \frac{\phi}{2} \right) \frac{\sin (\beta_2 - \mu_2)}{\sin (\beta_2 + \mu_2)} \quad \ldots\ldots\ldots\ldots\ldots\ldots \quad (23)$$

where $\sin \mu_2 = \sin \phi \sin \beta_2 \quad 0 \leqslant \mu_2 \leqslant \frac{\pi}{2}$

The value of P_1 is independent of α_1. The yield in region D-F-E-A also requires that P_1 has the value

$$P_1 = \frac{2(Q_2 - q_2)(\sin \phi + \cos 2\alpha_1)}{\cos^2 \phi} \quad \ldots\ldots\ldots\ldots\ldots\ldots \quad (24)$$

since also $\quad Q_2 - q_2 = \left(\frac{2 c \cos \phi}{1 - \sin \phi} \right) \frac{\sin (\beta_2 - \mu_2)}{\sin (\beta_2 + \mu_2)} \quad \ldots\ldots\ldots\ldots\ldots \quad (25)$

Here, the yield condition and Eqs. 16 and 17 have been used. Substituting Eq. 25 into Eq. 24, it follows that

$$P_1 = \left[\frac{4\,c\,(\sin\phi + \cos 2\alpha_1)}{\cos\phi\,(1 - \sin\phi)}\right]\frac{\sin(\beta_2 - \mu_2)}{\sin(\beta_2 + \mu_2)} \quad\dots\dots\dots\dots (26)$$

Equating the values of Eqs. 23 and 26 for P_1 shows that α_1 is given by

$$\sin\alpha_1 = \frac{\cos\phi}{2} \quad\dots\dots\dots\dots\dots\dots\dots\dots (27)$$

The compression values Q_1, q_1 and the inclination γ_1 of the stress discontinuities DB to the vertical can now be determined by the equilibrium of forces in vertical and horizontal directions and also by the condition that the material

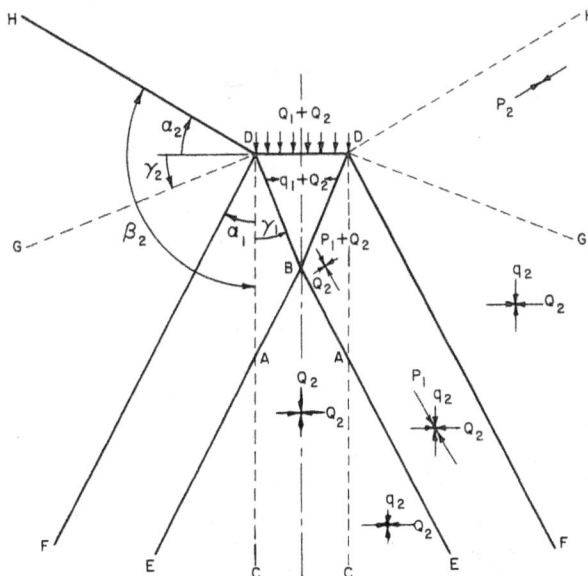

FIG. 19.—STRESS FIELD FOR THE STRIP FOUNDATION—A COMBINATION OF THE STRESS FIELDS, THE LOADED TRAPEZOID OF FIG. 5(c) (SOLID LINES) AND THE WEDGES OF FIGS. 11, 12 (DASHED LINES)

in the triangular element DDB is plastic. As discussed in the fourth section, the pressure is found that

$$Q_1 = \frac{1 - \sin\phi + \sin\phi\sin^2\alpha_1 + \sin\alpha_1 + \sin\alpha_1(1 - \sin^2\phi\cos^2\alpha_1)^{1/2}}{1 - \sin\phi}\,P_1 \quad (28)$$

Substituting the values P_1 of Eq. 23 and α_1 of Eq. 27 into the above expression for Q_1, gives

$$Q_1 = \frac{1}{2}\,c\,\tan^3\left(\frac{\pi}{4} + \frac{\phi}{2}\right)\frac{\sin(\beta_2 - \mu_2)}{\sin(\beta_2 + \mu_2)}\,[4 + \sin\phi + \sin^2\phi$$

$$+ (1 + \sin\phi)(4 + \sin^2\phi)^{1/2}] \quad\dots\dots\dots\dots (29)$$

It follows then that the lower bound of the critical pressure for a strip foundation is given by

512   March, 1969   SM 2

$$Q_1 + Q_2 = c \cot\phi \left[\tan^2 \left(\frac{\pi}{4} + \frac{\phi}{2} \right) \frac{\sin (\beta_2 - \mu_2)}{\sin (\beta_2 + \mu_2)} - 1 \right]$$

$$+ \frac{1}{2} c \tan^3 \left(\frac{\pi}{4} + \frac{\phi}{2} \right) \frac{\sin (\beta_2 - \mu_2)}{\sin (\beta_2 + \mu_2)} \, [4 + \sin\phi$$

$$+ \sin^2\phi + (1 + \sin\phi)(4 + \sin^2\phi)^{1/2}] \quad \cdots\cdots\cdots\cdots \quad (30)$$

The lower bound (Eq. 30) is plotted against the angle of friction for varying inclination angle β in Fig. 20. Expression (30) for the special case $\beta = 0$ was obtained previously by Shield (22).

To find upper bounds two failure mechanisms in an analogous manner to that of wedge problem in Fig. 14 are selected. One by Hill (12), Fig. 21, as-

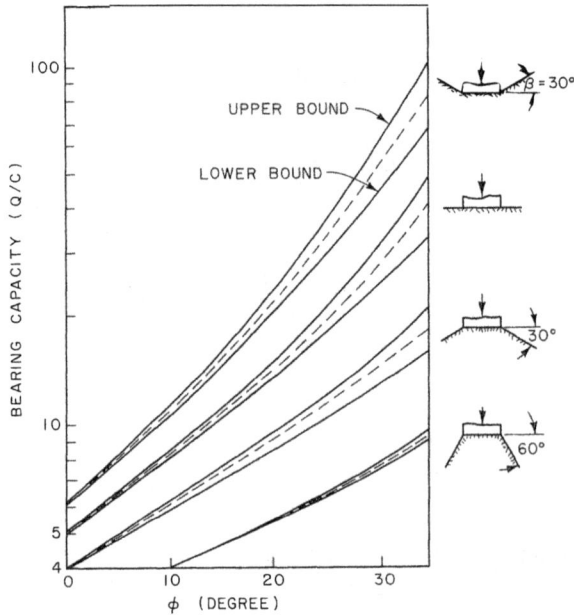

FIG. 20.—SEMI-LOG PLOT OF THE VARIATION OF LOWER AND UPPER BOUNDS OF THE LIMIT PRESSURE WITH ANGLE OF INTERNAL FRICTION FOR VARIOUS INCLI-NATION ANGLES β

sumes zero friction, and appreciable slip does take place. The other by Prandtl (17), Fig. 22, contains a rigid region which acts as an extension of the punch; there is no relative motion between the footing and the contacted soil. In these figures, only the left half failure mechanism is shown since it is symmetrical about the axis of the strip foundation. Both solutions give the same upper bounds of the limit pressure as given by Eq. 21. For comparison, the upper bound values are plotted against ϕ in Fig. 20. They are seen to be rather close.

It is worth mentioning here that Meyerhof's solution (14) is actually based on an upper bound approach and his results are identical to these upper bound calculations.

Effect of Foundation Friction.—The upper and lower bounds obtained Eqs. 21 and 30 are also applicable for all possible finite friction between the footing

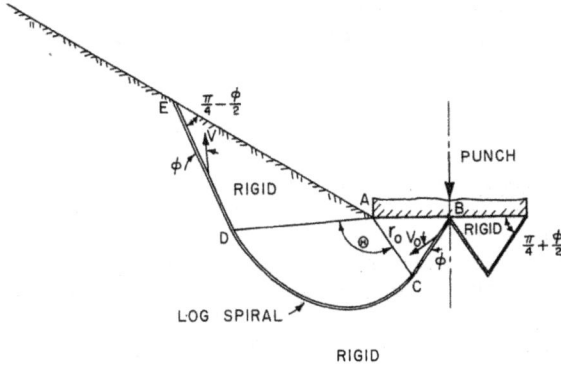

FIG. 21.—THE HILL FAILURE MECHANISM FOR SMOOTH STRIP FOUNDATION IN $C - \phi$ SOIL $V = V_0 \, e^{\theta \tan\phi}$

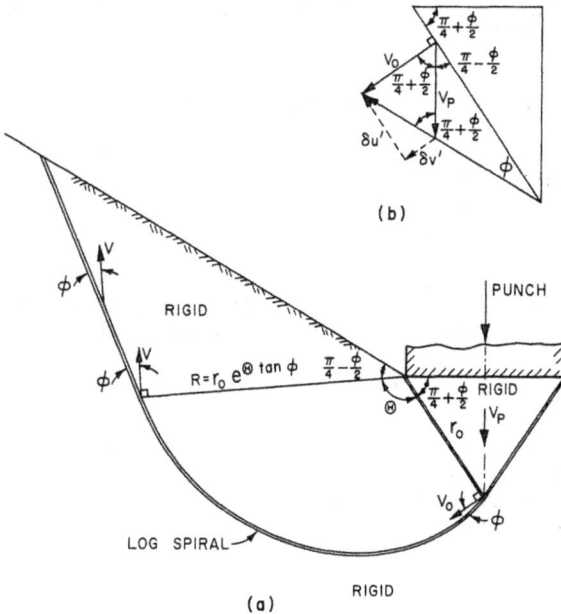

FIG. 22.—PRANDTL FAILURE MECHANISM FOR SMOOTH AS WELL AS FOR ROUGH STRIP FOUNDATION IN $C - \phi$ SOIL $V = V_0 \, e^{\theta \tan\phi}$

and the contacted soil. This follows directly from the intuitively obvious facts that the load which produces collapse for infinite friction (no relative motion) at the frictional interface as given by Prantdl mechanism (Fig. 22) will pro-

514   March, 1969   SM 2

duce collapse for the case of finite friction and the load which is safe for zero friction at the frictional interface as given by the previous stress field (Fig. 19) will be safe with any values of friction. The formal proof of these facts was recognized and stated as frictional theorems of limit analysis by Drucker (7). Occasionally they enable the limit load to be computed precisely for finite non-zero friction (2).

CONCLUSION

The formal approach of the soil plasticity theory has been presented in terms of methods more familiar to the civil engineer. Pin-connected truss and discontinuous stress fields are seen to be closely related. The truss concept provides the design engineer with a rather useful tool for obtaining essential, but relatively crude, information quickly. The extension of the truss concept to three-dimensions is obvious and will be proved to be more useful in providing lower bound solution on plastic limit load of the problems for which the complete solutions of plasticity are all but impossible. The interpretation of the log spiral zone as the limiting case of infinity many rigid triangles sliding does shed quite a bit of light on the nature and meaning of the field. In particular, when soil weight is important, the choice of two or three rigid blocks to approximate the continuous deformation region makes the upper bound calculations convenient.

ACKNOWLEDGMENT

The research reported here was supported by the National Science Foundation under Grant GK 1013 to Brown University. Grateful acknowledgement also is made to D. C. Drucker for his aid and guidance.

APPENDIX I.—LOWER BOUND CALCULATION OF THE TRIANGULAR STRESS FIELD

In the Fig. 23(a) the triangular element ABD is assumed to be in a plastic state of stress due to normal pressure Q, q on the two perpendicular sides AD, AB and the inclined pressure P on the side BD.

If the length of AD is taken as unity, equilibrium of forces [See Fig. 23(b)] gives

Σ in vertical direction = 0

$$Q = \frac{P \sin(\alpha + \gamma)}{\sin\gamma} \cos\alpha \text{ or } \tan\gamma = \frac{P \sin\alpha \cos\alpha}{Q - P \cos^2\alpha} \quad \ldots\ldots\ldots\ldots (31)$$

Σ in horizontal direction = 0

$$q = \frac{P \sin(\alpha + \gamma) \sin\alpha \tan\gamma}{\sin\gamma} = \frac{Q P \sin^2\alpha}{Q - P \cos^2\alpha} \quad \ldots\ldots\ldots\ldots\ldots (32)$$

SM 2   LIMIT ANALYSIS   515

independent of material properties. The yield criterion in Eq. 2 requires

$$(q + Q) \sin\phi + (q - Q) + 2c \cos\phi = 0 \dots\dots\dots\dots\dots\dots \quad (33)$$

The Mohr's circles for the regions marked I, II in Fig. 23(a) are shown in Fig. 23(c) where the poles of the corresponding Mohr's circles are also indi-

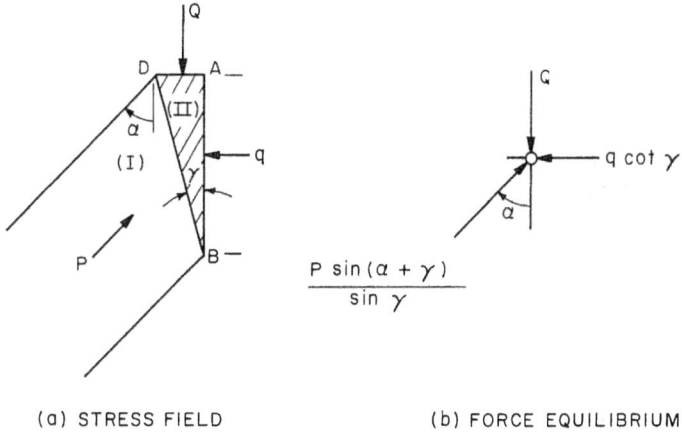

(a) STRESS FIELD   (b) FORCE EQUILIBRIUM

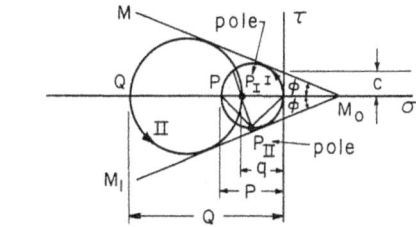

(c) SYSTEM OF MOHR'S CIRCLES FOR (a)

FIG. 23.—THE TRIANGULAR STRESS FIELD

cated. Substituting the value of q in Eq. 32 into Eq. 33, making use of Eq. 15, a quadratic expression for Q in then obtained

$$Q^2 - \frac{2 - 2\sin\phi + 2\sin\phi \sin^2\alpha}{1 - \sin\phi} PQ + P^2 \cos^2\alpha = 0 \dots\dots\dots \quad (34)$$

Taking the larger value of the relevant roots of the quadratic (i.e., positive sign)

$$Q = \frac{2c \cos\phi \left[1 - \sin\phi + \sin\phi \sin^2\alpha + \sin\alpha(1 - \sin^2\phi \cos^2\alpha)^{1/2}\right]}{(1 - \sin\phi)^2} \quad (35)$$

152 *Plasticity, Limit Analysis, Stability and Structural Design*

516 March, 1969 SM 2

When $\phi = 0$ (Prandtl-Reuss material) Eq. 35 takes the form

$$Q = 2c\,(1 + \sin\alpha) \quad\dots\dots\dots\dots\dots\dots\dots\dots\dots\dots \quad (36)$$

agreeing with the value obtained by Prager (16). To put Eq. 35 in a more compact form, denote

$$\cos\nu = \sin\phi\,\cos\alpha \quad 0 \leqslant \nu \leqslant \frac{\pi}{2} \quad\dots\dots\dots\dots\dots\dots\dots \quad (37)$$

After some reduction, Eq. 35 reduces to

$$Q = c\,\cot\phi\left[\tan^2\left(\frac{\pi}{4} + \frac{\phi}{2}\right)\frac{\sin(\nu + \alpha)}{\sin(\nu - \alpha)} - 1\right] \quad\dots\dots\dots\dots\dots \quad (38)$$

agreeing with the value obtained by Shield (21) through the application of the general "jump condition". With this value of Q, the Eqs. 31 and 32 give

$$\gamma = \frac{(\nu - \alpha)}{2} \quad\dots\dots\dots\dots\dots\dots\dots\dots\dots\dots\dots\dots\dots\dots \quad (39a)$$

and $\quad q = c\,\cot\phi\,\tan\alpha\,\tan\frac{1}{2}(\nu - \alpha)\left[\tan^2\left(\frac{\pi}{4} + \frac{\phi}{2}\right)\frac{\sin(\nu + \alpha)}{\sin(\nu - \alpha)} - 1\right] \quad (39b)$

As for the stress field in Fig. 6(b'), the value of Q', q' and γ' are exactly the same as those of Q, q, and γ except for substitution of $-\alpha$ for α in the corresponding equations.

APPENDIX II.--REFERENCES

1. Chen, W. F., "I. On Solving Plasticity Problems of Relevance to Soil Mechanics; II. The Bearing Capacity of Concrete," thesis, Engineering, presented to Brown University, at Providence, R.I., in June, 1966, in partial fulfillment of the requirements for the degree of Doctor of Philosophy.
2. Chen, W. F., discussion on the "Applications of Limit Plasticity in Soil Mechanics," by W. D. Liam Finn, *Journal of the Soil Mechanics and Foundations Division*, ASCE, Vol. 94, No. SM2, Proc. Paper 5424, March, 1968, pp. 608–613.
3. Drucker, D. C., "A More Fundamental Approach to Stress-Strain Relations," *Proceedings*, of the First U.S. National Congress of Applied Mechanics, American Society of Mechanical Engineers, p. 487–491, 1951.
4. Drucker, D. C., Prager, W., and Greenberg, J. J., "Extended Limit Design Theorems for Continuous Media," *Quarterly of Applied Mathematics*, Vol. 9, January, 1952, p. 381–389.
5. Drucker, D. C., and Prager, W., "Soil Mechanics and Plastic Analysis or Limit Design," *Quarterly of Applied Mathematics*, Vol. 10, 1952, p. 157–165.
6. Drucker, D. C., "Limit Analysis of Two and Three Dimensional Soil Mechanics Problems," *Journal of the Mechanics and Physics of Soilds*, Vol. 1, 1953, p. 217–226.
7. Drucker, D. C., "Coulomb Friction, Plasticity and Limit Loads," *Journal of Applied Mechanics*, Vol. 21, 1954, p. 71–74.
8. Drucker, D. C., Gibson, R. E., and Henkel, D. J., "Soil Mechanics and Work-hardening theories of Plasticity," *Proceedings*, ASCE, Vol. 81, 1955.
9. Drucker, D. C., "On Stress-Strain Relations for Soils and Load Carrying Capacity," *Proceedings*, of the First International Conference on the Mechanics of Soil-Vehicle System, held at Turin, Italy, in June, 1961, p. 15–23, Edizioni Minerva Tecnica.

SM 2   LIMIT ANALYSIS   **517**

10. Drucker, D. C., and Chen, W. F., "On the Use of Simple Discontinuous Fields to Bound Limit Loads," Brown University Report, March, 1967, Engineering plasticity edited by J. Heyman and F. A. Leckie, Cambridge University Press, March, 1968, pp. 129–145.

11. Finn, W. D. L., "Applications of Limit Plasticity in Soil Mechanics," *Journal of the Soil Mechanics and Foundations Division*, ASCE, Vol. 93, No. SM5, Proc. Paper 5424, Sept., 1967, pp. 101–120.

12. Hill, R., "The Plastic Yielding of Notched Bars Under Tension," *Quarterly Journal of Mechanics and Applied Mathematics*, Vol. 2, 1949, pp. 40–52.

13. Haythornthwaite, R. M., "Methods of Plasticity in Land Locomotion Studies," *Proceedings*, 1st International Conference on the Mechanics of Soil-Vehicle Systems, Torino-Saint Vincent, Edizion, Minerva Tecnica, held at Turin, Italy, in 1961, pp. 3–19.

14. Meyerhof, G. G., "The Ultimate Bearing Capacity of Foundations," *Geotechnique*, London, England, Vol. 2, No. 4, December, 1951, p. 301–332.

15. Palmer, A. C., "A Limit Theorem for Materials with Non-Associated Flow Laws," Journal De Mechanique, Vol. 5, No. 2, June, 1966.

16. Prager, W., and Hodge, P. G., Jr., "Theory of Perfect Plastic Solids," John Wiley and Sons, New York, 1951, p. 158–163.

17. Prandtl, L., "Uber Die Haerte Plastischer Korper," Nachrichten Von Der Koeniglichen Gesellschaft Der Wissenschaften Zu Goettingen, Mathematisch-physikalische Klasse, 1920, pp. 74–85.

18. Prandtl, L., "Ueber Die Eindringungsfestigkeit (Haerte) Plastischer Baustoffe und die Festigkeit Von Schneiden," Z. angew. Math. Mech., Vol. 1, 1921, p. 15–20.

19. Shield, R. T., and Drucker, D. C., "The Application of Limit Analysis to Punch Indentation Problems," *Journal of Applied Mechanics*, A.S.M.E., Vol..75, p. 453–491, 1951.

20. Shield, R. T., "Mixed Boundary Value Problems in Soil Mechanics," *Quarterly of Applied Mathematics*, Vol. 11, No. 1, 1953, p. 61–75.

21. Shield, R. T., "Stress and Velocity Fields in Soil Mechanics," *Journal of Mathematic Physics*, Vol. 33, No. 2, 1954, p. 144–156.

22. Shield, R. T., "On Coulomb's Law of Failure in Soils," *Journal of the Mechanics and Physics of Solids*, Vol. 4, 1955, p. 10–16.

23. Sokolovskii, V. V., "Statics of Granular Media," 1st ed., Oxford, New York, Pergamon Press, 1965.

24. Terzaghi, K., "Theoretical Soil Mechanics," John Wiley and Sons, New York, 1943, p. 5, p. 22.

APPENDIX III.—NOTATION

The following symbols are used in this paper:

c = cohesion;

D = rate of dissipation of energy per unit volume;

D_A = rate of dissipation of energy per unit area;

P, Q, q = principal stress;

r, R = length parameter defining zone of radial shear;

V = velocity;

$\alpha, \beta, \gamma, \psi$ = angular parameter;

$\dot{\gamma}_{max}$ = maximum rate of engineering shear strain;

$\dot{\epsilon}_t$ = tensile principal component of the plastic strain rate tensor;

θ, Θ = angular parameter defining zone of radial shear;

σ = normal stress;

σ_t, σ'_t = normal stress, see Fig. 8;

τ = shear stress;

ϕ = angle of internal friction;

$\delta u'$ = discontinuous tangential velocity; and

$\delta v'$ = discontinuous normal separation.

5.2.3

7369   June, 1970   EM 3

Journal of the

ENGINEERING MECHANICS DIVISION

Proceedings of the American Society of Civil Engineers

EXTENSIBILITY OF CONCRETE AND
THEOREMS OF LIMIT ANALYSIS

By W. F. Chen,[1] A. M. ASCE

INTRODUCTION

In a recent paper dealing with the bearing capacity of concrete blocks or rock (2), the generalized limit theorems of perfect plasticity (5) were applied to obtain bearing capacity in two dimensions (strip loading) and in three dimensions (circular and square punches). The approach is essentially based on the assumption that the local deformability of concrete or rock in tension and in compression is sufficient to permit the application of limit analysis. The implications of this basic assumption for these classes of materials are far-reaching and when applied, often provide good predictions of bearing capacity (2).

It was considered advisable to check the validity of the limit analysis approach experimentally, and to obtain more data for the deformation of the concrete or rock subjected to biaxial stress states. The experiments described herein were primarily designed to observe the order of the incipient plastic flow strain field during indentation of a circular punch on a circular concrete block. The effect of friction on the observed strength and on the strain field is also analyzed, as is the applicability of limit analysis to brittle materials such as concrete or rock. As a further application of the limit analysis approach, upper and lower bounds are obtained for the bearing capacity of the standardized indirect tensile test for concrete specimens of various shapes. The ana-

Note.—Discussion open until November 1, 1970. To extend the closing date one month, a written request must be filed with the Executive Secretary, ASCE. This paper is part of the copyrighted Journal of the Engineering Mechanics Division, Proceedings of the American Society of Civil Engineers, Vol. 96, No. EM3, June, 1970. Manuscript was submitted for review for possible publication on December 9, 1968.
[1] Assist. Prof. of Civ. Engrg., Fritz Engrg. Lab., Lehigh Univ., Bethlehem, Pa.

342   June, 1970   EM 3

lytically derived results are compared with the results derived from elasticity theory, and good agreement is observed.

PREVIOUS WORK

Although the limit design concept of plasticity originated from, and has been accepted in, the field of reinforced concrete for many years (9,18), the use of the generalized limit theorems of perfect plasticity in two-and three-dimensional concrete media and its extension to rock mechanics, has just begun. A number of interesting solutions in this area have been published (2,7, 10). A brief description of the various approaches and idealizations is therefore helpful and will be analyzed herein.

If the tensile strength of concrete or rock is assumed to be zero for dimensions of engineering interest, and not just a small fraction of the compressive strength, then the limit theorems of perfect plasticity will hold rigorously for

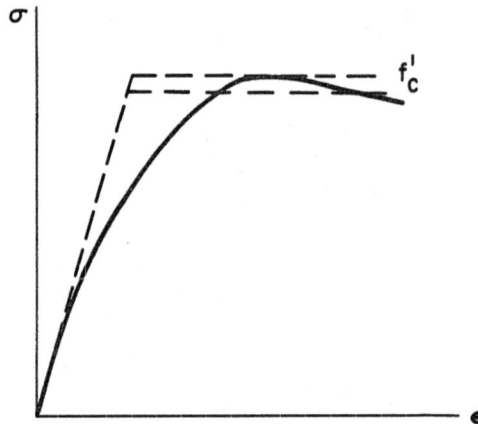

FIG. 1.—STRESS-STRAIN CURVE FOR CONCRETE AND ELASTIC-PERFECTLY PLASTIC IDEALIZATIONS

this idealization. They can be applied easily when a tensile crack is introduced in the failure mechanism. The early application of the limit theorems approach to the analysis and design of Voussoir arches is a good example. Kooharian (10) made the extreme but not unreasonable assumption that concrete is unable to take tension and behaves as a rigid, infinitely strong material in compression. This idealization was selected because it was simple and provided a first approximation to the behavior of a real Voussoir arch.

If concrete and rock are assumed to have zero tensile strength and a finite compressive strength, they may almost be considered real materials. The assumption of infinite ductility in tension at zero strength is rigorous and conservative for the application of limit analysis (2,6,10), but the assumption of infinite ductility in compression at selected compressive strength is more questionable. Brittleness in compression and a falling stress-strain curve after a maximum strength is usually observed, as indicated in Fig. 1. This is not compatible with the limit theorems. If the compressive strain of concrete is small and is not repeated often, then the deformability of concrete in compression prior to an appreciable fall-off of stress may be sufficient to permit the consideration of limit theorems with the concrete idealized as perfectly plastic at a yield stress in compression that approximates the ultimate strength f'_c. However, with the safe assumption that concrete or rock is unable to resist any tension, Chen and Drucker (2) have recently shown that the bearing capacity is just the unconfined compression strength of the column of material directly under the load.

Obviously, the idealization of zero tensile strength is not true. When a small but significant tensile strength is assumed, and if the Mohr-Coulomb surface for failure in compression is taken to represent a perfectly plastic yield surface, the predicted capacity is found to be in good agreement with published test results (2). These results convincingly show that tensile strength and local extensibility of concrete or rock are essential to the ability of concrete blocks and rock to carry the load.

Even the more modern point of view expressed by Winter and his co-workers (17), Robinson (15), Oladapo (13), and many others, which indicates a completely brittle but initially stable fracturing of concrete in tension, does not rule out the application of the assumption of perfectly plastic failure in tension. It is only necessary for the local tensile strength to be maintained for the required range of tensile strain in order for the limit theorems to apply. Furthermore, it is worth mentioning that a substantial increase in the tensile strength and thus tensile strain of concrete may be achieved with short lengths of wire in random orientation, but nearly uniform spacing throughout the concrete (14). This may provide even better correlation between the theoretical predictions of limit analysis and test results than was obtained by a rigid punch on a plain concrete block, as reported in Ref. 2.

EXPERIMENTAL STUDY OF STRAIN FIELD

Localized tensile strains of the order of four times those obtained in a simple tension test have been reported in a concrete cylinder loaded by a concentric punch (1). Therefore, tests were carried out to ascertain whether the strain field developed in the concrete block at the instant of collapse is sufficient to allow complete plasticity.

Mortar cylinders at different sizes (6 in.-diam by 12 in.-height, and 6 in.-diam by 6 in.-height) were tested. Three different base conditions were used. A 7 in. by 7 in. by 3/8 in. steel plate was intended to provide high base friction. For some 6 in. by 6 in. cylinders, a thin sheet of teflon was inserted between the cylinder and the supporting base to minimize friction effects. To obtain a smooth condition at the base, the 6 in. by 12 in. cylinders were tested by a double punch (short steel cylinders, 2 in. in diameter) in which the load was applied to a cylinder at two opposite faces (Fig. 2). This condition ensured zero shear stress over the midheight section of the cylinder.

Specimens and Testing Procedure.—Specimens were cased in timber molds. Each specimen contained a center hole 5/8 in. in diameter along the axis of the cylinder. Strain gauges were attached to a typical cylinder at the positions indicated in Fig. 2. The surfaces required for the attachment of the gages were scrubbed clean and smooth. The exact positions of the gages were marked

344   June, 1970   EM 3

out, and the interior gages were pressed to the inside cylinder wall by applying internal pressure to an inserted rubber hose. Preliminary experiments
had shown that this procedure gave good bond and resulted in stable readings
within a few minutes of attachment.

The mortar used for the tests contained a 1:3 mix by weight of normal
Portland cement, to sand having particle dimensions in the range from No. 4
to No. 100 U.S. Sieve. The water-cement ratio was 0.5. The cylinders were
tested at an age of 32 days to 37 days in a Baldwin Hydraulic Testing Machine.
All cylinders were moisture-cured at $\pm75°$ F for 6 days to minimize drying
out effects.

A spherical seat was placed on top of the short steel cylinder to insure
even load distribution. Loads were applied to the specimens continuously until
failure occurred at a rate not exceeding 1,000 lb per sec. The control tests
were carried out on standard concrete cylinders. Both compression tests, and
splitting-tension (indirect tensile) tests were performed on cylinders having
the same age as the concentric punch tests.

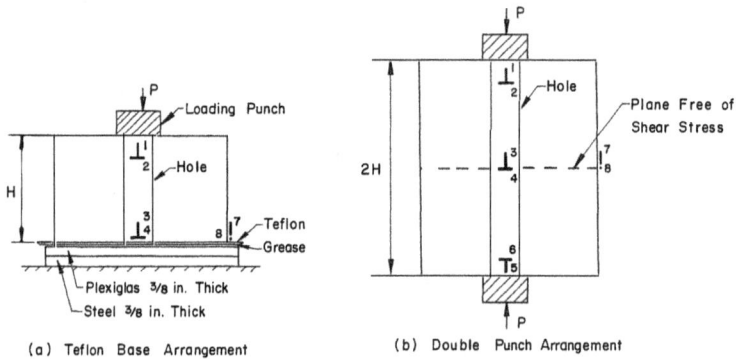

(a) Teflon Base Arrangement   (b) Double Punch Arrangement

FIG. 2.—LOAD TESTS OF CONCRETE CYLINDERS

At the end of each test, the attachment of each gage to the concrete was
examined and it proved satisfactory in all cases. The details of the experiment are described in Ref. 3.

Results.—The vertical and horizontal strain distributions along the axis of
the cylinder at the instant of collapse are shown in Fig. 3. Each curve plotted
is the result of three separate tests on similar specimens. The variation in
strain reading on similar specimens was found to be small in most cases,
which suggests that the gage application technique is reliable.

The strain curves shown in Fig. 3 differ in the numerical values, but show
the same general distribution of strain with depth in the cylinder. A marked
increase was observed in vertical compressive strain of the order of at least
twice that obtained in a simple compression test (300×10^{-5} in. per in.) near
the region directly below the punch. The horizontal tensile strains are seen to
be distributed rather uniformly along the axis of a cylinder, but reverse to a
compressive strain near the bottom, for blocks with teflon and steel bases.

An important point in the strain distribution curves is that the average tensile strains along the axis of the cylinder near ultimate failure in the punch

EM 3   LIMIT ANALYSIS   345

tests are greater than those in the flexure and direct tension tests. On the average, the tensile strains for smooth punch tests (teflon base or double punch) are about four times those in the flexure tests (13) (10 × 10^{-5} in. per in.), and five times as large as the average strain in simple tension tests (11). Although the average tensile strains in rough punch tests (steel base) are much smaller than those in smooth cases, the average values reached in the specimen are still as large as the average strain in flexure tests.

The strain readings for the gages on the surface of a block (gage 8 in Fig. 2) indicate that an average horizontal tensile strain of up to 80 % or more of the average tensile strain in flexure tests is reached in the cylinder just prior to collapse for all base conditions. This suggests that the tensile plastification tends to relieve the more highly-stressed parts in the center portion of the specimen, and to throw the tensile stress onto those parts near the surface of the specimen where the stress is lower.

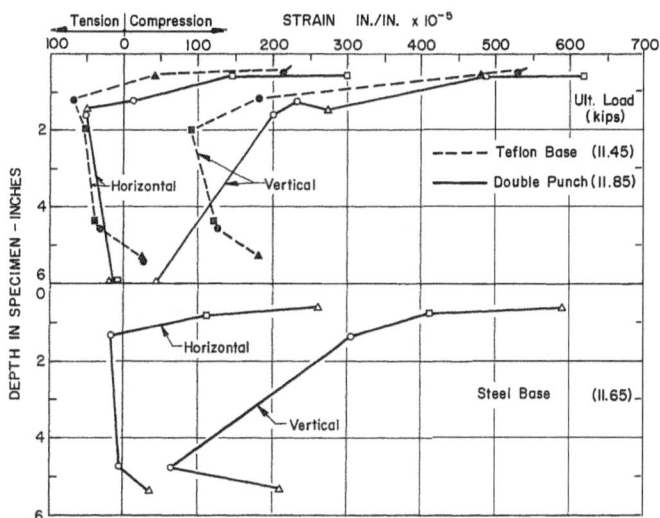

FIG. 3.—HORIZONTAL AND VERTICAL STRAIN DISTRIBUTION AT FAILURE IN PUNCH TESTS. SPECIMEN TESTED AT ABOUT 34-DAYS AGE

For the case of a double punch, the vertical strain distribution on the middle horizontal section of the specimen is found to be almost uniform (gage 7 in Fig. 2.) However, the vertical strain distribution over the teflon or steel base is not uniform, and it varies from about zero at the edge of a cylinder to a compressive strain of order 200 × 10^{-5} in. per in. or more at the center. This seems to indicate a possible lower cone developed, immediately above the base, for the teflon and steel base specimens. The double punch specimen, therefore, has a greater region of horizontal tensile strain than do the teflon and steel base specimens (Fig. 3). This indication of greater distribution of tensile strain in double punch specimens suggests that the load carrying capacity for a double punch specimen may be higher than those of similar specimens with steel or teflon supporting bases. Indeed, this was found to be the case for most of the specimens tested.

346   June, 1970   EM 3

The strain distribution diagrams (Fig. 3) thus suggests that a high compressive stress more comparable with a triaxial test is developed in the region directly beneath the punch. The plane passing through the axis of the cylinder exhibits an almost uniform tensile strain over that plane (except the region near the supporting base), and supports the previous work of Chen and Drucker, who assume that the concrete can strain sufficiently to develop complete plasticity throughout the material so that the limit analysis technique can be applied.

As mentioned previously, the large strains observed may be caused by the internal bond cracks of concrete instead of the so-called plastification. The fact is that these initial bond cracks do not cause structural deterioration, so that the plastification of concrete may be interpreted in this sense.

LIMIT ANALYSIS OF INDIRECT TENSILE TESTS

The bearing capacity of a concrete block is closely related to the behavior of an indirect tensile test (splitting), in which a compressive load is applied to

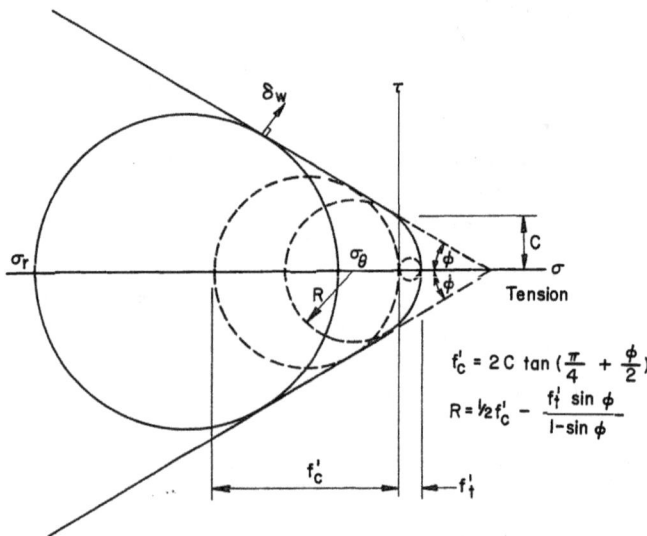

$$f'_c = 2C \tan\left(\frac{\pi}{4} + \frac{\phi}{2}\right)$$

$$R = \tfrac{1}{2} f'_c - \frac{f'_t \sin\phi}{1-\sin\phi}$$

FIG. 4.—MODIFIED MOHR-COULOMB CRITERION

a cylinder along two opposite generators [Fig. 5(b)]. The distribution of stress and hence the relevant formula for computing the tensile strength of the indirect tensile test have been analyzed by the theory of elasticity 16. A plasticity treatment of this problem is given herein. The analysis is based on the fundamental assumption that the local tensile strain of concrete under multiaxial stress conditions is sufficient to permit the application of limit analysis. In addition, it is assumed, as in Ref. 2, that the concrete may be idealized as perfectly plastic with a Mohr-Coulomb failure surface as the yield surface in compression, and a small but nonzero tension cutoff (Fig. 4). In Fig.

4, f_c' and f_t' denote the simple compression and simple tensile strength respectively, and c is cohesion and ϕ is the angle of internal friction of the concrete. The vector δw representing slip velocity across a failure surface is normal to the yield curve.

The Upper Bound Theorem of limit analysis (5) states that the concrete cylinder will collapse, if, for any assumed failure mechanism, the rate of work done by the applied loads exceeds the internal rate of dissipation. Equating external and internal energies for any such mechanism thus gives an upper bound on the collapse load.

Fig. 5(b) shows a failure mechanism consisting of two rigid wedge regions ABC, and a simple tension crack CC connecting these two wedges. The wedges move toward each other as a rigid body, and displace the surrounding material horizontally sideways. The relative velocity vector δw at each point along the lines of discontinuity AC and BC is inclined at an angle ϕ to these lines (6).

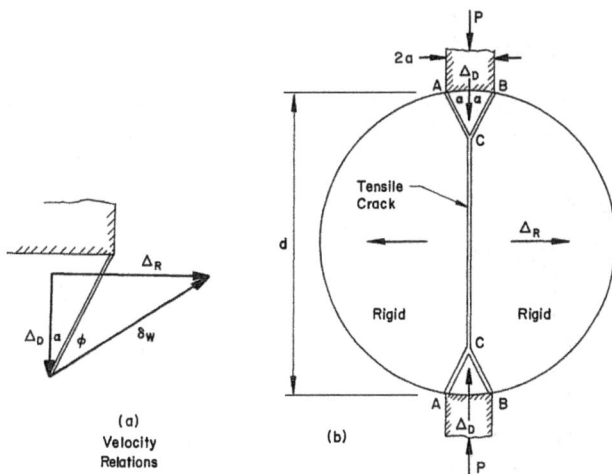

FIG. 5.—BEARING CAPACITY OF AN INDIRECT TENSILE TEST

The compatible velocity relations are shown in Fig. 5(a). The rate of dissipation of energy along the wedge surfaces may be found by multiplying the area of these discontinuous surfaces by f_c' (1 - sin ϕ)/2 times the discontinuity in velocity δw across the surfaces (2). Similarly, the rate of dissipation of energy along the separation surface CC is found by multiplying the area of separation by f_t' times the relative separation velocity $2\Delta_R$ across the separation surface. Equating the external rate of work to the total rate of internal dissipation yields

$$P^u = \left(\frac{a}{\sin \alpha}\right) \left[\frac{f_c'\, l\, (1 - \sin \phi)}{\cos (\alpha + \phi)} - 2 f_t'\, l\, \cos \alpha \, \tan (\alpha + \phi)\right]$$

$$+ f_t'\, l\, d\, \tan (\alpha + \phi) \quad \dots\dots\dots\dots\dots\dots\dots\dots\dots\dots\dots (1)$$

in which d = the diameter; and l = the length of the cylinder.

The upper bound has a minimum value when α satisfies the condition $\partial P^u / \partial \alpha$ = 0, which is

$$\cot \alpha \quad \tan \phi + \sec \phi \left[1 + \frac{\frac{d}{2a} \cos \phi}{\left(\frac{f'_c}{f'_t}\right)\left(\frac{1 - \sin \phi}{2}\right) \sin \phi} \right]^{1/2} \quad \dots \dots (2)$$

Eq. 1 is, in fact, a simple modification of the two-dimensional solution obtained in Ref. 2. For the dimensions used in the standardized indirect tension test (ASTM-C649-62T): $2a = 1/2$ in. and $d = 6$ in., and the average values for concrete: $f'_c/f'_t = 10$ and $\phi = 30°$. (ASTM specifies that the width of the plywood strip placed between the punch and the concrete cylinder is 1 in., however, the load is actually distributed through the plywood to the concrete cylinder over a band of appreciable less width (say 1/2 in.) The upper bound has a minimum value at the point $\alpha = 16.1°$, and

$$P \le P^u = 1.83 \, l \, d \, f'_t \dots \dots \dots \dots \dots \dots \dots \dots \dots \dots \dots (3)$$

Therefore $f'_t \ge 0.548 \dfrac{P}{ld} \dots \dots \dots \dots \dots \dots \dots \dots \dots (4)$

The Lower Bound Theorem of limit analysis (5) states that if an equilibrium distribution of stress can be found in the concrete cylinder which no-

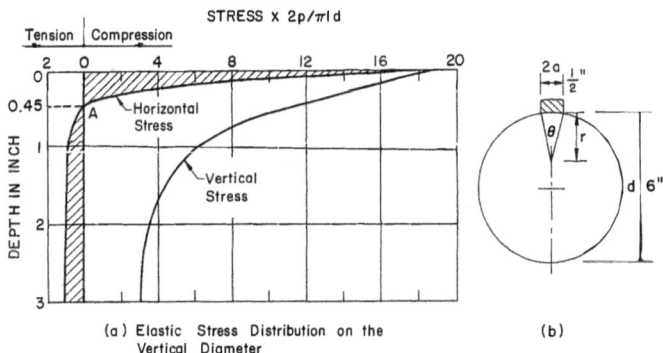

(a) Elastic Stress Distribution on the Vertical Diameter   (b)

FIG. 6.—STRESS DISTRIBUTIONS IN CYLINDER LOADED OVER A WIDTH OF 1/2 IN.

where exceeds the Mohr-Coulomb yield criterion in Fig. 4, then the loads imposed can be carried without collapse, or will just be at the point of collapse. Clearly, any stress distribution obtained through the theory of elasticity will give a safe or lower bound on the collapse load, provided that the chosen stress field nowhere violates the yield criterion.

Stress distribution in a plane disc (plane stress) subjected to loads perpendicular to the axis of a disc [Fig. 5(b)] has been analyzed thoroughly by Frocht (8) and Timoshenko (16). The stress distribution for the disc will be a statically admissible stress field if the magnitude of the maximum shearing stress on any section through the concrete in the disc is not greater than an amount which depends linearly upon the hydrostatic pressure (Fig. 4). If the load is assumed to be uniformly distributed over the width $2a$ (say 1/2 in.), it can be shown that if $2a < d/10$, the stresses on the vertical diameter may be adequately approximated by:

Vertical Stress $\sigma_r = \dfrac{-2P}{\pi l d} \left[\dfrac{d}{4a} (\theta + \sin\theta) + \dfrac{d}{d-r} - 1 \right]$ .... (5)

Horizontal Stress $\sigma_\theta = \dfrac{2P}{\pi l d} \left[1 - \dfrac{d}{4a} (\theta - \sin\theta) \right]$ .... (6)

Term θ is the angle subtended by the loaded area at the point considered [Fig. 6(b)], and tensile stress is taken as positive. The stress distribution along the vertical diameter, calculated for $2a = d/12$, is shown in Fig. 6(a).

It can be shown that the critical points which decide the maximum value of P are those points along the vertical diameter jointing the applied forces (the stresses along the vertical diameter are principal stresses). The vertical stress at point A [Fig. 6(a), $r = 0.45$ in.] must not be greater than f_c' in order that the yield condition not be violated, as the horizontal stress is zero at this point. Because the cylinder is long, it approximates closely to plane strain condition. The condition above Point A is thus comparable with a triaxial test. Yielding does not, therefore, occur here. The points vertically below A ($r >$ 0.45 in.) are under a biaxial state of compression-tension with a tension of amount near $2P/\pi l d$. It is found that the critical point along the vertical diam-

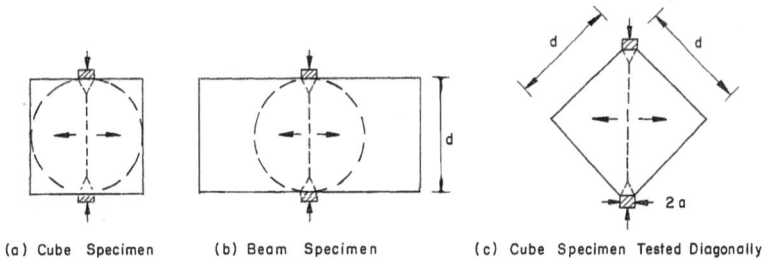

(a) Cube Specimen   (b) Beam Specimen   (c) Cube Specimen Tested Diagonally

FIG. 7.—THREE POSSIBLE SPLITTING TEST ARRANGEMENTS

eter plane, which first reaches the yield condition, is the point $r = 0.5$ in. $[\sigma_r = -10.49\ (2P/\pi l d),\ \sigma_\theta = 0.31\ (2P/\pi l d)]$. The modified Mohr-Coulomb yield condition can be written (Fig. 4)

$\sigma_r = \sigma_\theta \tan^2 \left(\dfrac{\pi}{4} + \dfrac{\phi}{2} \right) - f_c'$ .... (7)

For $\phi = 30°$ and $f_c' = 10\,f_t'$, Eq. (7) reduces to

$\sigma_r = 3\,\sigma_\theta - 10\,f_t'$ .... (8)

Substituting the values σ_r and σ_θ at $r = 0.5$ in. into Eq. 8, a lower bound on the collapse load of the indirect tensile test is thus obtained:

$P \geq P^l = 1.37\ l\ d\ f_t'$ .... (9)

Therefore $f_t' \leq 0.728\ \dfrac{P}{ld}$ .... (10)

Therefore, the stress field of Fig. 6 and the velocity field of Fig. 5 show that the tensile strength of concrete for the indirect tensile test lies within ±14 % of the value $0.638\ P/ld = 2P/\pi l d$. It is interesting to note that the aver-

350   June, 1970   EM 3

age value of the upper and lower bounds solutions given previously, is identical to that derived from elasticity theory.

The upper and lower bounds obtained in Eqs. 4 and 10 are also applicable for splitting tests on concrete specimens of other shapes such as, for example, the cases shown in Fig. 7(a) (cube specimen) and Fig. 7(b) (beam specimen). This follows directly from the facts that the assumed failure mechanism in the upper bound calculations for a cylinder specimen is also applicable to a cube specimen or a beam specimen [see dashed lines in Figs. 7(a) and 7(b)], and the admissibility of the elastic stress field used to obtain the lower bound for the cylinder specimen [or directly from the theorem that the addition to a body of weightless material cannot result in a lower collapse load (5)]. This then indicates that the formula $f_t^{\,\prime} = 2P/\pi l d$ may be considered valid for the splitting tests on cube specimen or beam specimen. This conclusion is identical with that of Davies and Bose (4) using a finite element method with linear elastic idealization for concrete.

The extension of the preceding information to obtain a relevant formula for computing the tensile strength of a cube specimen tested diagonally is evident. An identical failure mechanism of the problem is shown in Fig. 7(c). For the dimensions: $2a = 1/2$ in., as an example, and $d = 6$ in., the upper bound Eq. 1 has a minimum value at the point $\alpha = 14.6°$, and

$$P \le P^u = 2.16\ l\ d\ f_t^{\,\prime} \dots\dots\dots\dots\dots\dots\dots\dots\dots\dots\dots \quad (11)$$

(NOTE: The appropriate value d in Eqs. 1 and 2 is $\sqrt{2}\ d - 2a = 8$ in.) Therefore

$$f_t^{\,\prime} \ge 0.463\ \frac{P}{ld} \dots\dots\dots\dots\dots\dots\dots\dots\dots\dots\dots\dots \quad (12)$$

The elastic solution of Davies and Bose will give a safe or lower bound on the collapse load, provided that the stress field nowhere violates the modified Mohr-Coulomb yield condition. (The solution is valid for the case of point load, however, as pointed out by Davies and Bose, the pattern of stress distribution for the point load and distributed load is similar except near the loading zones. Thus, it will not effect the analyses herein.) It is found that the critical point, which decides the maximum value of P, is the point at the center of the diagonal cube specimen. It can then be shown that when the maximum tensile stress at this point $[\sigma_\theta = 0.77\ (2P/\pi ld)]$ reaches $f_t^{\,\prime}$, a maximum lower bound load of the diagonal cube specimen is obtained:

$$P \ge P^l = 2.04\ l\ d\ f_t^{\,\prime} \dots\dots\dots\dots\dots\dots\dots\dots\dots\dots\dots \quad (13)$$

Therefore $f_t^{\,\prime} \le 0.49\ \dfrac{P}{ld} \dots\dots\dots\dots\dots\dots\dots\dots\dots\dots\dots\dots \quad (14)$

Thus, the tensile strength of a diagonal cube specimen lies within $+3$ per-cent of the value $0.476\ P/ld = 0.75\ (2P/\pi ld)$. It is interesting to note that the diagonal cube specimen has been used in the Soviet Union and the relevant formula for computing the tensile strength is assumed to be about 80 % of the value $2P/\pi ld$. This agrees quite well with the present analysis.

CONCLUSION

Local extensibility of concrete is seen to be sufficient to develop almost complete plasticity throughout a concrete cylinder loaded by a concentric flat-

EM 3 LIMIT ANALYSIS 351

ended punch. The application of limit theorems to concrete, therefore, appears reasonably justified. The results of plasticity analysis being similar with that of elasticity analysis in the case of the indirect tensile tests should prove both interesting and useful. It should be kept in mind, however, that justification of the perfect plasticity idealization for concrete may be proper for some selected cases but which may have no relevance for other cases. Caution should be exercised in order not to overestimate the significance of the selected cases in appraising the overall validity of the approach.

It seems clear, at this stage, that more problems of theoretical significance and practical importance must be investigated so that the implications of plasticity to this class of material may be better understood. Thus, the present theory should be considered as only a first step in this particular application of limit analysis, and it should be extended to practical problems to be useful.

ACKNOWLEDGMENTS

The research reported herein was supported by the National Science Foundation under Grant GK-3245 to Lehigh University. The comments from the reviewers are also gratefully acknowledged.

APPENDIX I.—REFERENCES

1. Campbell-Allen, D., Discussion of "Bearing Capacity of Concrete," by William Shelson, Proceedings, *Journal of the American Concrete Institute,* Vol. 54, June, 1958, pp. 1185–1187.
2. Chen, W. F., and Drucker, D. C., "Bearing Capacity of Concrete Blocks or Rock," *Journal of the Engineering Mechanics Division,* ASCE, Vol. 95, No. EM4, August, 1969, pp. 955–978.
3. Chen, W. F., and Hyland, M. W., "Bearing Capacity of Concrete Blocks," *Journal of the American Concrete Institute,* Vol. 67, March, 1970, pp. 228–236.
4. Davies, J. D., and Bose, D. K., "Stress Distribution in Splitting Tests, Proceedings, *Journal of the American Concrete Institute,* Vol. 65, No. 8, August, 1968, pp. 662–669.
5. Drucker, D. C., Prager, W., and Greenberg, H. J., "Extended Limit Design Theorems for Continuous Media," *Quarterly of Applied Mathematics,* Vol. 9, 1952, pp. 381–389.
6. Drucker, D. C., and Prager, W., "Soil Mechanics and Plastic Limit Analysis or Limit Design," *Quarterly of Applied Mathematics,* Vol. 10, 1952, pp. 157–165.
7. Drucker, D. C., "On Structural Concrete and the Theorems of Limit Analysis," *Engineering,* Vol. 21, Zurich, 1961, pp. 49–59.
8. Frocht, M. M., *Photoelasticity,* Vol. 2, John Wiley and Sons, Inc., New York, 1948, pp. 121–129.
9. Gvozdev, A. A., "The Determination of the Value of the Collapse Load for Statically Indeterminate Systems Undergoing Plastic Deformation," Proceedings of the Conference on Plastic Deformations, December, 1936, Akademiia Nauk SSSR, Moscow-Leningrad, 1938, p. 19. Translated from the Russian by R. M. Haythornthwaite, *International Journal of Mechanical Science,* Pergamon Press Ltd., Vol. 1, 1960, pp. 322–335.
10. Kooharian, A., "Limit Analysis of Voussoir (Segmental) and Concrete Arches," *Journal of the American Concrete Institute,* Vol. 24, 1952, pp. 317–328.
11. Kaplan, M. F., "Strains and Stresses of Concrete at Initiation of Cracking and Near Failure," *Journal of the American Concrete Institute,* Vol. 60, July, 1963, pp. 853–879.

352   June, 1970   EM 3

12. Nadai, A., *Theory of Flow and Fracture of Solids,* Vol. 1, McGraw-Hill Book Co., New York, 1950, p. 207.
13. Oladapo, I. O., "Extensibility and Modulus of Rupture of Concrete," *Bulletin No. 18,* Structural Research Laboratory, Technical University of Denmark, 1964.
14. Romualdi, J. P., and Mandel, J. A., "Tensile Strength of Concrete Affected by Uniformly Distributed and Closely Spaced Short Lengths of Wire Reinforcement," Proceedings, *Journal of the American Concrete Institute,* Vol. 61, No. 6, June, 1964, pp. 657–670.
15. Robinson, G. S., "Behavior of Concrete in Biaxial Compression," *Journal of the Structural Division,* ASCE, Vol. 93, No. ST1, Proc. Paper 5090, February, 1967, pp. 71–86.
16. Timoshenko, S., *Theory of Elasticity,* McGraw-Hill Book Co., Inc., New York, 1934, pp. 104–108.
17. Winter, G., et al., "Properties of Steel and Concrete and the Behavior of Structures," *Journal of the Structural Division,* ASCE, Vol. 86, No. ST2, Proc. Paper 2384, February, 1960, pp. 33–62. Also, *Journal of the American Concrete Institute.*
18. Yu, C. W., and Hognestad, Eivind, "Review of Limit Design for Structural Concrete," *Journal of the Structural Division,* ASCE, Vol. 84, No. ST8, Proc. Paper 1878, December, 1958.

APPENDIX II.—NOTATION

The following symbols are used in this paper:

a = punch width;

c = cohesion;

d = cylinder diameter;

f'_c = simple compression strength;

f'_t = simple tensile strength;

l = cylinder length;

P, P^u, P^l = collapse load, upper bound, lower bound;

p = ultimate bearing pressure;

R = Mohr circle defined in Fig. 5;

r = distance defined in Fig. 6(b);

α = angle defined in Fig. 4(b);

Δ_D = downward velocity, Fig. 4(b);

Δ_R = horizontal velocity, Fig. 4(b);

δw = relative velocity;

ϵ = axial strain;

θ = angle defined in Fig. 6(b);

σ = normal stress;

σ_θ = horizontal stress, Eq. 6;

σ_r = vertical stress, Eq. 5;

τ = shearing stress; and

ϕ = angle of friction.

5.3 Beam-Columns (樑柱理論)

5.3.1

7482   August, 1970   EM 4

Journal of the

ENGINEERING MECHANICS DIVISION

Proceedings of the American Society of Civil Engineers

GENERAL SOLUTION OF INELASTIC BEAM-COLUMN PROBLEM[a]

By W. F. Chen,[1] A. M. ASCE

INTRODUCTION

The elastic-plastic behavior of an eccentrically loaded beam column under a concentrated load at the midspan, as shown in Fig. 1, is an important technical problem with frequent engineering applications. The obvious example is a compression member in building frames; bridge trusses are another. It is also of great significance in the theory of structures as it is one of the simplest beam-column problems involving only simple loadings and boundary conditions. It is natural, therefore, to expect that this problem should possess a long history of study. Analytical solutions that describe the elastic behavior of beam columns with various end conditions comprise the most highly developed aspect of beam-column research. Analyses of the solutions and theories are described admirably by Timoshenko and Gere (12). Unfortunately, past attempts have not been successful in obtaining analytical solutions to beam-column problems composed of common structural sections when the beam-columns are stressed beyond the elastic limit. In part, this difficulty was caused by the inability to obtain a relatively simple moment-curvature-thrust relationship for commonly used structural sections. In addition, it is pointed out (2) that direct solutions to the differential equations of deflection are generally impossible to obtain because of the high nonlinearity of the basic equations.

A relatively easy alternative is to find a more appropriate variable than deflection for the elastic-plastic stability analysis, with a view to obtaining a relatively simple differential equation. Curvature was demonstrated to be a more appropriate variable than deflection for the elastic-plastic analysis of eccentrically loaded columns (2). This approach is herein adopted.

Note.—Discussion open until January 1, 1971. To extend the closing date one month, a written request must be filed with the Executive Director, ASCE. This paper is part of the copyrighted Journal of the Engineering Mechanics Division, Proceedings of the American Society of Civil Engineers, Vol. 96, No. EM4, August, 1970. Manuscript was submitted for review for possible publication on December 16, 1969.

[a]Presented at the June 15-19, 1970, Sixth U.S. National Congress of Applied Mechanics, held at Cambridge, Mass.

[1]Asst. Prof., Fritz Engrg. Lab., Dept. of Civ. Engrg., Lehigh Univ., Bethlehem, Pa.

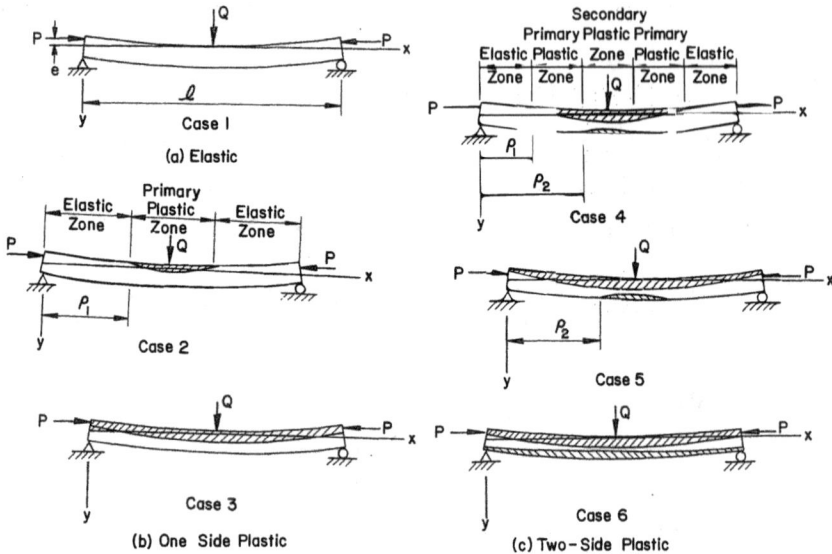

FIG. 1.—ECCENTRICALLY LOADED BEAM COLUMN

Herein the writer essentially presents a continuation of the work by Chen and Santathadaporn (2). Here, the moment-curvature-thrust relationship derived in Ref. 2 for a rectangular section is generalized to represent any complex shape of structural section to a high degree of approximation. The rectangular section is, of course, a special case of the generalized moment-curvature-thrust relationship. In addition, the generalized relationship also includes the idealized wide flange and box sections of Hauck and Lee (6) (they assume the flange elements are very thin) as a special case with even better accuracy, so that it may be considered a generalization of their formulation.

PREVIOUS WORK

The elastic-plastic behavior of laterally loaded beam-columns has not been studied thoroughly, but a number of numerical solutions have been obtained. Wright developed an approximate formula for the case of a beam-column loaded by a concentrated load at the midspan (Fig. 1 set e = 0), see Ref. 14. The same case was also studied by Ketter, and interaction curves for predicting the strength of wide-flange beam columns were developed (9). Horne and Merchant (7) recently proposed an empirical method for estimating the strength of a beam column. However, the validity of their method has not been verified by either numerical solutions or laboratory tests. More recently, Lu and Kamalvand (10), following the related work on eccentrically loaded columns of von Karman (13), Chwalla (4), and Ojalvo (11), extended the numerical integration procedure further to include the effect of the additional lateral load. The moment-curvature-thrust relationship used in their computations was represented in graphical form and programmed on a computer.

Interaction curves for the ultimate lateral load carrying capacity of a variety of beam-columns were presented (for the case, $e = 0$).

STATEMENT OF BEAM-COLUMN PROBLEM

The beam-column problem under consideration is shown in Fig. 1. Because of symmetry, only one half of the beam-column need be considered. It is therefore convenient to take the origin of the x-y axes at the left support of the beam column, the x axis being directed along the member, and the y axis being perpendicular to the x axis, positive downwards.

It is assumed that lateral-torsional buckling of the beam-column is effectively prevented so that failure is always caused by excessive bending in the plane of the applied loads. The axial force P at the ends of the beam column is assumed to be applied first and maintained at a constant value as the lateral load Q is continuously increased from zero to its maximum and then drops off steadily beyond the maximum point. This assumption on the loading condition may be unnecessarily restrictive, but it does agree with the requirement that unloading of material stressed into the plastic range does not take place for an initially stress-free perfectly plastic beam column. Furthermore, it will be seen later that such a loading condition can reduce greatly the work on the numerical evaluation of the solution for this problem. Other loading paths are, of course, permissible provided that the degree of unloading is small. For example, a previous investigation (1) has indicated that the loads (axial force and lateral load) that are increased proportionately from zero do produce unloading, but the effect appears to be small. The predicted collapse loads were found to be conservative. In the present analysis irreversibility of plastic deformation in the beam column is not attempted because it makes the solution highly complicated.

GENERALIZED MOMENT-CURVATURE-THRUST RELATIONSHIPS

It is necessary to introduce nondimensional quantities so that the moment-curvature-thrust relationship, and therefore the basic differential equations, may be written in a form more appropriate for computation.

The initial yield quantities for moment, M; axial force, P; and curvature, Φ of a section are defined by

$$M_y = \sigma_y S, \ P_y = \sigma_y A, \ \Phi_y = \frac{2\epsilon_y}{h} \quad \dots \dots \dots \dots \dots \dots \dots \dots (1)$$

in which σ_y = the yield stress in tension or compression; and ϵ_y = the corresponding strain. Other quantities used are: area, A; elastic section modulus, S; and height, h. One further defines the nondimensional variables by

$$m = \frac{M}{M_y}, \ p = \frac{P}{P_y}, \ \phi = \frac{\Phi}{\Phi_y} \quad \dots \dots \dots \dots \dots \dots \dots \dots \dots (2)$$

A general m-ϕ-p curve of a common structural section with or without residual stress usually has the shape shown diagrammatically in Fig. 2. The curve may be divided into three regimes: elastic, primary plastic, and secondary plastic. They are separated by the points (m_1, ϕ_1) and (m_2, ϕ_2) as shown in the figure. The general characteristic of such a curve is that the

424   August, 1970   EM 4

rate of curvature-hardening falls steadily and the curve bends over more and more. For a perfectly plastic material the moment will asymptotically approach the value m_{pc} as ϕ tends to infinity but will not attain it for any finite curvature. In most metals, however, it is certainly realistic to assume that the limit value, m_{pc}, can actually be attained and even exceeded at a finite curvature because of strain hardening.

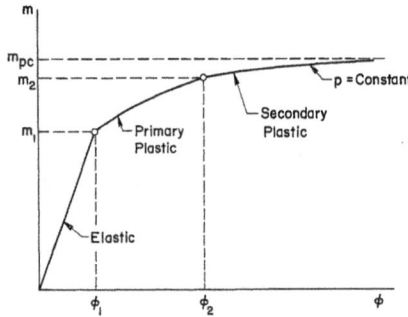

FIG. 2.—MOMENT-CURVATURE-THRUST RELATIONSHIP FOR COMMON STRUC-TURAL SECTION

It appears reasonably to assume that the general m-ϕ-p curve could be fitted by the following three equations. These equations are the direct generalization of the expressions derived previously for the rectangular section (2).

Elastic Regime.

$$m = a\,\phi \quad \dots \dots \dots \dots \dots \dots \dots \dots \dots \dots \dots \dots \quad (3)$$

valid for $0 \le \phi \le \phi_1$ . . . . (4)

Primary Plastic Regime.

$$m = b - \frac{c}{\phi^{1/2}} \quad \dots \dots \dots \dots \dots \dots \dots \dots \dots \dots \dots \quad (5)$$

valid for $\phi_1 \le \phi \le \phi_2$ . . . . (6)

Secondary Plastic Regime.

$$m = m_{pc} - \frac{f}{\phi^2} \quad \dots \dots \dots \dots \dots \dots \dots \dots \dots \dots \dots \quad (7)$$

valid for $\phi_2 \le \phi$ . . . . (8)

in which a, b, c, f = arbitrary constants. These constants can be evaluated easily by solving simultaneous equations which will arise if the particular values ϕ_1, ϕ_2, and ϕ_∞ are inserted and the moments equated to the appropriate moments m_1, m_2, and m_{pc}. As an example, this had been done for idealized wide-flange and box sections with very thin flange elements (6). If R denotes the ratio of the sum of the flange areas to the sum of the web areas, for

$$p \ge \frac{1}{1 + R} \quad \dots \dots \dots \dots \dots \dots \dots \dots \dots \dots \dots \dots \dots \dots \dots \quad (9)$$

these constants are $a = 1$ . . . . (10)

$$b = \frac{3\,(1 + R)}{1 + 3R}\,(1 - p) \cdot \cdot \cdot \cdot \cdot \cdot \cdot \cdot \cdot \cdot \cdot \cdot \cdot \cdot \cdot \cdot \cdot \cdot \cdot \cdot \cdot \cdot \cdot \cdot \cdot \cdot \quad (11)$$

$$c = \frac{2}{1 + 3R}\,(1 - p)^{3/2} \cdot \cdot \cdot \cdot \cdot \cdot \cdot \cdot \cdot \cdot \cdot \cdot \cdot \cdot \cdot \cdot \cdot \cdot \cdot \cdot \cdot \quad (12)$$

valid for $\phi_1 = 1 - p$ . . . . (13)

and $\phi_2 = \infty$ . . . . (14)

Here, secondary plastic m-ϕ-p relationships are absent, while the primary

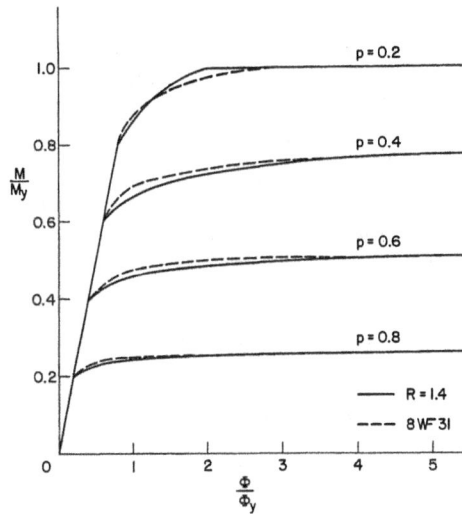

FIG. 3.—MOMENT-CURVATURE-THRUST RELATIONSHIPS (ACTUAL CURVES SHOWN DASHED

plastic m-ϕ-p relationships are extended from the elastic limit, $\phi_1 = 1 - p$, to infinity.

For $p \leq \dfrac{1}{1 + R}$ . . . . (15)

these constants are $a = 1$ . . . . (16)

$$b = 1 - p + \frac{2p\left[1 + 2R - (1 + R)^2\,p\right]}{(1 + 3R)\{1 - (1 - p)^{1/2}\left[1 - (1 + R)\,p\right]^{1/2}\}} \quad \cdot \cdot \cdot \cdot \cdot \cdot \quad (17)$$

$$c = \frac{2p\,(1 - p)^{1/2}\left[1 + 2R - (1 + R)^2\,p\right]}{(1 + 3R)\{1 - (1 - p)^{1/2}\left[1 - (1 + R)\,p\right]^{1/2}\}} \quad \cdot \cdot \cdot \cdot \cdot \cdot \cdot \cdot \cdot \cdot \cdot \quad (18)$$

$$f = \frac{1}{2\,(1 + 3R)} \quad \cdot \cdot \cdot \cdot \cdot \cdot \cdot \cdot \cdot \cdot \cdot \cdot \cdot \cdot \cdot \cdot \cdot \cdot \cdot \cdot \cdot \cdot \cdot \quad (19)$$

$$m_{pc} = \frac{3}{2}\,\frac{1 + 2R - (1 + R)^2\,p^2}{1 + 3R} \quad \cdot \cdot \cdot \cdot \cdot \cdot \cdot \cdot \cdot \cdot \cdot \cdot \cdot \cdot \cdot \cdot \cdot \cdot \cdot \quad (20)$$

426   August, 1970   EM 4

valid for $\phi_1 = 1 - p$ .... (21)

and $\phi_2 = \dfrac{1}{1 - (1 + R) p}$ .... (22)

Evidently, the area ratio, R, assumes the value of zero for a rectangular cross section. As Eqs. 9 to 22 are derived for idealized wide-flange and box sections, R may be treated as an arbitrary curve-fitting parameter to fit the actual m-ϕ-p curves. For example, if the value of R is assumed to be 1.4, the actual m-ϕ-p curves for the 8 WF 31 section can be represented closely by the expressions described herein. Fig. 3 shows the plot of m versus ϕ for several values of p. Comparison of these curves with those obtained for the actual shape indicates that the actual curves are generally too large by approximately 4 %.

The present curves (Fig. 3) are seen to provide a better approximation than the results given by Hauck and Lee (6) for the overall range of curvature. For the particular 8 WF 31 section cited, their results are generally too large by approximately 10 %.

The m-ϕ-p relationships for an arbitrary shape of structural section including the influence of residual stress can also be fitted by the generalized m-ϕ-p relationships proposed herein. Of course, Eqs. 9 to 22 in their present forms do not apply to wide-flange sections with residual stress. Equations applicable to such sections are reported in Ref. 3.

DIFFERENTIAL EQUATIONS AND SOLUTIONS

The equation of equilibrium for bending of the beam column shown in Fig. 1 is, of course, independent of the mechanical behavior of the member. In the usual notation (12)

$$\frac{d^2m}{dx^2} + k^2 \phi = 0 \quad .... (23)$$

in which k is given by $k^2 = \dfrac{P \, \Phi_y}{M_y} = \dfrac{P}{EI}$ .... (24)

The quantity EI represents the flexural rigidity of the beam column in the plane of bending. Combining the statical Eq. 23 with the generalized m-ϕ-p relationships (Eqs. 3, 5, and 7), one can express the differential equations of the axis of the beam column in the following forms.

Elastic Zone.

$$\frac{d^2\phi}{dx^2} + \frac{k^2}{a} \phi = 0 \quad .... (25)$$

Primary Plastic Zone.

$$\phi \frac{d^2\phi}{dx^2} - \frac{3}{2} \left(\frac{d\phi}{dx}\right)^2 + \frac{2k^2}{c} \phi^{7/2} = 0 \quad .... (26)$$

Secondary Plastic Zone.

$$\phi \frac{d^2\phi}{dx^2} - 3 \left(\frac{d\phi}{dx}\right)^2 + \frac{k^2}{2f} \phi^5 = 0 \quad .... (27)$$

Eqs. 25, 26, and 27 are the basic differential equations for bending of beam columns. If the constants a, c, and $f = 1$, $2(1 - p)^{3/2}$ and $1/2$ respectively, these equations reduce to the equations developed previously for bending of a rectangular section (2).

Eqs. 26 and 27 are exact differential equations with ϕ^{-4} and ϕ^{-7} as the integral factors respectively. Six integration constants are to be expected in a general solution of Eqs. 25, 26, and 27. Proceeding with the solution, Eqs. 26 and 27 may be integrated once to give the following.

Primary Plastic Zone.

$$\frac{d\phi}{dx} = \frac{2\sqrt{2}k}{\sqrt{c}} \, (D - \phi^{1/2})^{1/2} \, \phi^{3/2} \quad \dots \dots \dots \dots \dots \dots \dots \dots \dots \quad (28)$$

Secondary Plastic Range.

$$\frac{d\phi}{dx} = \frac{k}{\sqrt{f}} \, (1 + G\phi)^{1/2} \, \phi^{5/2} \quad \dots \dots \dots \dots \dots \dots \dots \dots \dots \quad (29)$$

To solve for the curvature, one has to integrate Eqs. 28 and 29 for the solution. It is difficult to express ϕ explicity in terms of x. Alternatively, it is more convenient to express the curvature implicitly in the form $x = x(\phi)$. The general solutions of the Eqs. 25, 28, and 29 are given in the following.

Elastic Zone.

$$\phi = A_1 \cos\frac{kx}{\sqrt{a}} + B \sin\frac{kx}{\sqrt{a}} \quad \dots \dots \dots \dots \dots \dots \dots \dots \dots \quad (30)$$

Primary Plastic Zone.

$$x - x_p = -\left(\frac{\sqrt{c}}{\sqrt{2}k}\right)\frac{1}{D}\left[\frac{(D - \phi^{1/2})^{1/2}}{\phi^{1/2}} + \frac{1}{D^{1/2}}\tanh^{-1}\frac{(D - \phi^{1/2})^{1/2}}{D^{1/2}}\right] \quad (31)$$

Secondary Plastic Zone.

$$x - x_s = \frac{2}{3}\frac{\sqrt{f}}{k}\left(G + \frac{1}{\phi}\right)^{1/2}\left(2G - \frac{1}{\phi}\right) \quad \dots \dots \dots \dots \dots \dots \quad (32)$$

Thus, if the constants of integration A_1, B, D, G, x_p, and x_s in Eqs. 30, 31, and 32 are known from the boundary conditions of the beam column and the continuity conditions of the curvature curves (or jump conditions to be analyzed in the following section), x becomes a known function of ϕ. In general, there are six different distributions of stress zone possible as shown in Fig. 1. Therefore, the constants corresponding to each case have to be considered separately. They will be analyzed in some detail when the solution for each case is presented.

DISCONTINUITY IN DERIVATIVE OF CURVATURE CURVE

The analysis of the differential equations, developed generally in the preceeding section developed later in the following three sections according to various individual cases) implicitly assumes that the curvature curve is a continuous differentiable function of x. However, consideration of the gen-

428   August, 1970   EM 4

eralized m-ϕ-p relationship (Fig. 2) shows that one should not expect a great deal in the way of differentiability on the curvature curve. At the point $\phi = \phi_1$ or $\phi = \phi_2$, two $dm/d\phi$ values are associated with the curvature. Any such discontinuity in the m-ϕ-p relationship will show up where it is applicable in the study of the solutions of the differential equations in general. In addition, discontinuity in the derivative of the curvature curve can also arise in the case when a concentrated load is present. In such a case, there is a discontinuity in the derivative of the curvature curve at the section where the concentrated load is applied. In all these circumstances, the "jump condition" (or discontinuity condition) instead of the continuity condition for the derivative of the curvature curve must be used locally for the determination of the

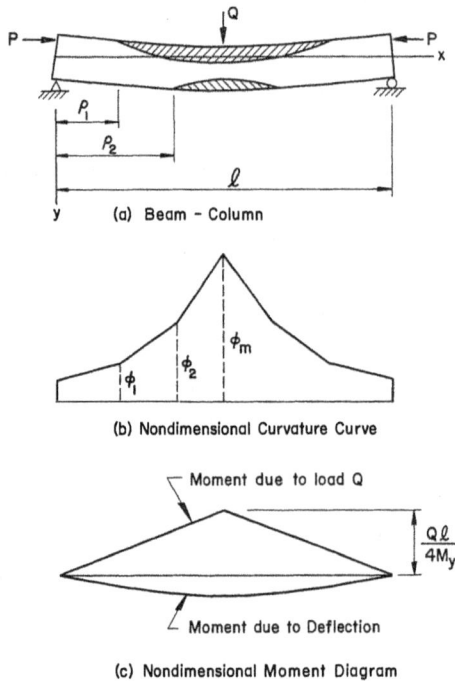

(a) Beam – Column

(b) Nondimensional Curvature Curve

(c) Nondimensional Moment Diagram

FIG. 4.—DISCONTINUITIES IN DERIVATIVE OF CURVATURE CURVE

integration constants. The purpose of the present analysis is to show how the permissible jump conditions are determined.

A beam column, a moment diagram, and a curvature curve are sketched in Fig. 4. The jump in $d\phi/dx$ will occur either across the boundaries ρ_1, ρ_2, or under the concentrated load Q. There will be two values of $d\phi/dx$ depending upon whether x is allowed to approach the jump from the right side or from the left side.

Jumps in $d\phi/dx$ to Jumps in m-ϕ-p Curve.

Jump Condition at Elastic-Primary Plastic Boundary, $x = \rho_1$.—As the value dm/dx at $x = \rho_1$ is continuous (see the sketched moment diagram in Fig.

EM 4 INELASTIC BEAM-COLUMN 429

4), it follows that

$$\left(\frac{dm}{d\phi}\right)_{\phi_1^-}\left(\frac{d\phi}{dx}\right)_{\rho_1^-} = \left(\frac{dm}{d\phi}\right)_{\phi_1^+}\left(\frac{d\phi}{dx}\right)_{\rho_1^+} \quad\dots\dots\dots\dots\dots\dots \quad (33)$$

(ρ_1^- denotes $x \to \rho_1$ from left, i.e., x is always $< \rho_1$) or, using the m-ϕ-p relations (Eqs. 3 and 5), Eq. 33 reduces to

$$\frac{2a}{c}\,\phi_1^{3/2}\left(\frac{d\phi}{dx}\right)_{\rho_1^-} = \left(\frac{d\phi}{dx}\right)_{\rho_1^+} \quad\dots\dots\dots\dots\dots\dots \quad (34)$$

Jump Condition at Primary-Secondary Plastic Boundary, $x = \rho_2$.—Similarly, if ϕ_1 and ρ_1 in Eq. 33 are substituted by ϕ_2 and ρ_2, one obtains

$$\frac{c}{4f}\,\phi_2^{3/2}\left(\frac{d\phi}{dx}\right)_{\rho_2^-} = \left(\frac{d\phi}{dx}\right)_{\rho_2^+} \quad\dots\dots\dots\dots\dots\dots \quad (35)$$

Jump Condition at Elastic-Secondary Plastic Boundary, $x = \rho_1 = \rho_2$.—For the special case, $p = 0$ (beam problem), $\phi_1 = \phi_2$, one obtains

$$\frac{a}{2f}\,\phi_2^3\left(\frac{d\phi}{dx}\right)_{\rho_2^-} = \left(\frac{d\phi}{dx}\right)_{\rho_2^+} \quad\dots\dots\dots\dots\dots\dots \quad (36)$$

Jumps in $d\phi/dx$ Due to Concentrated Load.—Since

$$\frac{d\phi}{dx} = \frac{\dfrac{dm}{dx}}{\dfrac{dm}{d\phi}} \quad\dots\dots\dots\dots\dots\dots \quad (37)$$

It follows that the jump conditions under the concentrated load Q which is applied at the midsection of the beam column are given in the following (Fig. 4).

Elastic Regime.

$$\left.\frac{d\phi}{dx}\right|_{x=\frac{l-}{2}} = \frac{\dfrac{Q}{2M_y}}{a} = \frac{Q}{2aM_y} \quad\dots\dots\dots\dots\dots\dots \quad (38)$$

Primary Plastic Regime.

$$\left.\frac{d\phi}{dx}\right|_{x=\frac{l-}{2}} = \frac{\dfrac{Q}{2M_y}}{\dfrac{c}{2\phi_m^{3/2}}} = \frac{Q}{cM_y}\,\phi_m^{3/2} \quad\dots\dots\dots\dots\dots\dots \quad (39)$$

Secondary Plastic Regime.

$$\left.\frac{d\phi}{dx}\right|_{x=\frac{l-}{2}} = \frac{\dfrac{Q}{2M_y}}{\dfrac{2f}{\phi_m^3}} = \frac{Q}{4fM_y}\,\phi_m^3 \quad\dots\dots\dots\dots\dots\dots \quad (40)$$

in which ϕ_m denotes the curvature at the central cross section of the beam column.

SOLUTION OF ELASTIC BEAM-COLUMN—CASE 1

Fig. 1, Case 1, Case 2, and Case 4 are found to be of major importance in the practical application of beam-column solutions. For this reason, these three cases will be considered in detail in the following, and the solutions for the other cases will be given in Ref. 8. The general solution of the elastic beam column [Fig. 1(a)] is given by Eq. 30. As the beam column and the loading are symmetrical about the midsection of the beam column, it is necessary to consider only the portion to the left of the load.

The constants of integration A_1 and B are determined from the conditions at the end of the beam column and at the point of application of the load Q. As the curvature at the end of the beam column is $\phi_o = m_o/a$ and the derivative of the curvature at the point of application of the load Q must satisfy the jump condition as given by Eq. 38, one concludes that

$$A_1 = \frac{m_o}{a} \quad \dots \dots \dots \dots \dots \dots \dots \dots \dots \dots \dots \dots \dots \dots \dots (41)$$

$$B = \frac{Q}{2\sqrt{a}\,kM_y} \sec \frac{kl}{2\sqrt{a}} + \frac{m_o}{a} \tan \frac{kl}{2\sqrt{a}} \quad \dots \dots \dots \dots (42)$$

in which $m_o = \frac{1}{2} p \left(\frac{e}{r}\right)\left(\frac{h}{r}\right) \quad \dots \dots \dots \dots \dots \dots \dots \dots \dots \dots (43)$

in which r = radius of gyration of the section about the axis of bending and e = eccentricity of the axial force P.

Substituting into Eq. 30 the values of constants from Eqs. 41 and 42, one obtains the equation for the curvature curve

$$\phi = \left[\frac{m_o}{a} \cos \frac{k(l-2x)}{2\sqrt{a}} + \frac{2}{\sqrt{a}} \left(\frac{\alpha q}{kl}\right) \sin \frac{kx}{\sqrt{a}}\right] \sec \frac{kl}{2\sqrt{a}} \quad \dots \dots (44)$$

valid for $0 \leq x \leq l/2$ in which q = the nondimensional lateral load and is given by

$$q = \frac{Q}{Q_p} = \frac{Ql}{4\alpha M_y} \quad \dots \dots \dots \dots \dots \dots \dots \dots \dots \dots \dots \dots \dots \dots (45)$$

in which Q_p = the plastic limit load of the beam column according to simple plastic theory and α = the shape factor of the section about the axis of bending.

The maximum curvature occurs at the center, which is

$$\phi_m = \phi_{x=l/2} = \frac{m_o}{a} \sec \frac{kl}{2\sqrt{a}} + \frac{2}{\sqrt{a}} \left(\frac{\alpha q}{kl}\right) \tan \frac{kl}{2\sqrt{a}} \quad \dots \dots \dots (46)$$

valid for $0 \leq \phi_m \leq \phi_1$. Eq. 44 or Eq. 46 shows that for a given longitudinal force the curvatures of an elastic beam column are proportional to the lateral load q.

SOLUTION OF ELASTIC-PLASTIC BEAM-COLUMN YIELDED ON ONE-SIDE—CASE 2

Referring to Fig. 1(b), Case 2, the primary plastic zone begins at a distance, ρ_1, from the ends. The curvature at the center of the beam column is denoted by ϕ_m. The elastic end portions of the beam column will be investigated first. The general solution of this portion is given by Eq. 30. The constants of integration A_1 and B are now determined from the conditions at the end and at the elastic-primary boundary. Since the curvature at the end of the beam column is $\phi_o = m_o/a$ and, at the elastic-primary plastic boundary $(x = \rho_1) = \phi_1$, it follows that the equation for the curvature curve in the elastic zone is

$$\phi = \left[\frac{m_o}{a} \sin \frac{k(\rho_1 - x)}{\sqrt{a}} + \phi_1 \sin \frac{kx}{\sqrt{a}} \right] \csc \frac{k\rho_1}{\sqrt{a}} \quad \dots\dots\dots\dots \quad (47)$$

valid for $0 \le x \le \rho_1$. The unknown distance ρ_1 will be determined from the solution for the primary plastic zone.

The derivative of the curvature curve for the primary plastic zone is given by Eq. 28. The constant D and the unknown quantity, ϕ_m, can be determined from the jump conditions at the central cross section and at the section $x = \rho_1$. As the derivative of the curvature curve at these two sections must satisfy the jump conditions, as given by Eqs. 39 and 34, respectively, it can be concluded that

$$D = \phi_1^{1/2} \eta \quad \dots\dots\dots\dots\dots\dots\dots\dots\dots\dots\dots\dots \quad (48)$$

$$\phi_m = \left[\phi_1^{1/2} \eta - \frac{2}{c} \left(\frac{aq}{kl} \right)^2 \right]^2 \quad \dots\dots\dots\dots \quad (49)$$

valid for $\phi_1 \le \phi_m \le \phi_2$ in which η is defined as

$$\eta = 1 + \frac{a}{2c} \phi_1^{3/2} \left[\cot \frac{k\rho_1}{\sqrt{a}} - \left(\frac{m_o}{a\phi_1} \right) \csc \frac{k\rho_1}{\sqrt{a}} \right]^2 \quad \dots\dots\dots\dots \quad (50)$$

The general solution of the curvature curve for the primary plastic zone is given by Eq. 31. The constant x_p and the value ρ_1 are found from the conditions that ϕ in Eq. 31 must be $= \phi_m$ and ϕ_1 for $x = l/2$ and $x = \rho_1$, respectively. Using Eqs. 48 and 49, one obtains

$$\frac{x_p}{l} = \frac{1}{2} + \frac{\sqrt{c}}{\sqrt{2}} \frac{1}{\phi_1^{3/4} \eta^{3/2}} \frac{1}{kl} \left\{ \frac{\sqrt{2} \sqrt{c} \; \phi_1^{1/4} \eta^{1/2} \left(\frac{aq}{kl} \right)}{c \phi_1^{1/2} \eta - 2 \left(\frac{aq}{kl} \right)^2} \right.$$

$$\left. + \tanh^{-1} \left[\frac{\sqrt{2}}{\sqrt{c}} \frac{1}{\phi_1^{1/4} \eta^{1/2}} \left(\frac{aq}{kl} \right) \right] \right\} \quad \dots\dots\dots\dots \quad (51)$$

and $\dfrac{\rho_1}{l} = \dfrac{x_p}{l} - \dfrac{\sqrt{c}}{\sqrt{2}}\, \dfrac{1}{\phi_1^{3/4}\, \eta^{3/2}}\, \dfrac{1}{kl}\left[\eta^{1/2}\,(\eta - 1)^{1/2} + \tanh^{-1}\left(1 - \dfrac{1}{\eta}\right)^{1/2}\right]$ (52)

For a given beam column and for the known values of the constants which define the generalized m-ϕ-p relationship, Eq. 52 represents the desired condition for the determination of the elastic-primary plastic boundary, ρ_1, which can be obtained by trial and error.

Values of ρ_1 found from this equation lie between the limits $l/2$ and ρ_1^l (as yet undefined, $\rho_1^l \geq 0$). The value $l/2$ corresponds to $\phi_m = \phi_1$, which means that the value of the load q (loaded from Case 1 to Case 2) is that value for which the elastic limit is reached at the central cross section. The load q is obtained by substituting $\phi_m = \phi_1$ in Eq. 46 or Eq. 49, which gives

$$q_{\phi_m=\phi_1} = \dfrac{\sqrt{c}}{\sqrt{2}}\, \phi_1^{1/4}\, \dfrac{kl}{\alpha}\left(\eta_{\rho_1=l/2} - 1\right)^{1/2} \quad \dots\dots\dots\dots\dots\dots \quad (53)$$

When the central cross section curvature, as given by Eq. 49 reaches ϕ_2, the value of q is

$$q_{\phi_m=\phi_2} = \dfrac{\sqrt{c}}{\sqrt{2}}\, \phi_1^{1/4}\, \dfrac{kl}{\alpha}\left[\eta - \dfrac{\phi_2^{1/2}}{\phi_1^{1/2}}\right]^{1/2} \quad \dots\dots\dots\dots\dots\dots \quad (54)$$

and the corresponding lower limit value, ρ_1^l, of ρ_1 can be found from Eq. 52 by trial and error.

Assume that the end loadings (p and m_o) are small enough to keep the whole beam column in the elastic range. If the lateral load q is now gradually increased to its maximum and then drops off steadily beyond the maximum point, the curvature at the central cross section of the beam column goes through the successive stages of ϕ_1 and ϕ_2 and therefore through Case 1, Case 2, and Case 4. For $q \leq q_{\phi_m=\phi_1}$ the beam column will behave elastically throughout, and the relation between q and ϕ_m is linear (Eq. 46). For $q > q_{\phi_m=\phi_1}$ there will be a region of the beam column which will behave plastically. As ρ_1 decreases from $l/2$ to ρ_1^l, this region will spread through an increasing portion of the beam column. At any instant in this interval one can assume a value of ρ_1 and obtain the value of x_p from Eq. 52. With this value of x_p the corresponding lateral load q can be computed from Eq. 51 by trial and error. Finally, for $\rho_1 < \rho_1^l$ yielding on the lower portion of the beam column at the center also develops and Case 4 takes over.

Suppose that the end loadings are large enough to cause part of the beam column to become plastic on one side before the lateral load q is applied. Let us assume that this initial plastic zone begins at a distance, ρ_1^o, say, from the ends. Since the value ρ_1^o corresponds to $q = 0$, the constant x_p in Eq. 51 is $l/2$, and thus the value ρ_1^o can be obtained readily from Eq. 52 by trial and error. Again, it is convenient to obtain numerical results by first assuming a value of ρ_1 ($\rho_1^l \leq \rho_1 \leq \rho_1^o$) and obtaining the corresponding value of x_p from Eq. 52 and, therefore, the lateral load q from Eq. 51 by trial and error.

To obtain the curvature curve over the central portion of the beam column where the cross sections are primary plastic, one simply substitutes the values of constants from Eqs. 48 and 51 into Eq. 31, and the resulting equation is the curvature curve for the primary plastic zone which is expressed implicitly as a function of x

$$\frac{x}{l} = \frac{1}{2} - \frac{\sqrt{c}}{\sqrt{2}} \frac{1}{\phi_1^{3/4} \eta^{3/2}} \frac{1}{kl} \left\{ \eta^{1/2} \left[\frac{\phi_1}{\phi} \eta - \left(\frac{\phi_1}{\phi}\right)^{1/2} \right]^{1/2} \right.$$

$$- \frac{\sqrt{2} \sqrt{c} \; \phi_1^{1/4} \; \eta^{1/2} \left(\frac{\alpha q}{kl}\right)}{c \, \phi_1^{1/2} \, \eta - 2\left(\frac{\alpha q}{kl}\right)^2} + \tanh^{-1} \left[1 - \left(\frac{\phi}{\phi_1}\right)^{1/2} \frac{1}{\eta} \right]^{1/2}$$

$$\left. - \tanh^{-1} \left[\frac{\sqrt{2}}{\sqrt{c}} \frac{1}{\phi_1^{1/4} \eta^{1/2}} \left(\frac{\alpha q}{kl}\right) \right] \right\} \dots\dots\dots\dots\dots\dots\dots (55)$$

valid for $\rho_1 \le x \le l/2$.

SOLUTION OF ELASTIC-PLASTIC BEAM-COLUMN
YIELDED ON TWO SIDES—CASE 4

Referring to Fig. 1(c), Case 4, in addition to the primary plastic zone, the secondary plastic zone extends a distance ρ_2 from the ends. The procedure for solving the problem is similar to that used before. The curvature curve (Eq. 47) for the wholly elastic end portions of the beam column is still applicable if the appropriate value of ρ_1 is used.

As for the primary plastic zone for which, $\rho_1 \le x \le \rho_2$, the derivative of the curvature function is given by Eq. 28. The constant of integration D in this equation can be determined from the jump condition, as given by Eq. 34, for the section where the elastic and primary plastic zones meet. Using the fact that $\phi_{x=\rho_1} = \phi_1$, it can be concluded that

$$D = \phi_1^{1/2} \eta \dots\dots\dots\dots\dots\dots\dots\dots\dots\dots\dots (56)$$

The derivative of the curvature function for the secondary plastic zone, for which, $\rho_2 \le x \le l/2$, is given by Eq. 29. The constant G and the unknown curvature, ϕ_m, at the center of the beam column can now be determined from the jump condition, as given by Eq. 40, for the section where the concentrated load is applied and the jump condition, as given by Eq. 35, for the section where the primary plastic zone and secondary plastic zone meet. These two conditions together with $\phi_{x=l/2} = \phi_m$ and $\phi_{x=\rho_2} = \phi_2$ give

$$G = \frac{c}{2f} (\phi_1^{1/2} \eta - \phi_2^{1/2}) - \frac{1}{\phi_2} \dots\dots\dots\dots\dots\dots\dots\dots (57)$$

$$\phi_m = \frac{f}{\left(\frac{\alpha q}{kl}\right)^2 + \frac{f}{\phi_2} - \frac{c}{2} \left(\phi_1^{1/2} \eta - \phi_2^{1/2}\right)} \dots\dots\dots\dots\dots\dots (58)$$

valid for $\phi_m \ge \phi_2$.

To determine the curvature curve in the primary plastic zone as well as in the secondary plastic zone, the constants of integration x_p and x_s in Eqs. 31 and 32 must be determined. The constants are found from the conditions that ϕ, as given by Eqs. 31 and 32, must = ϕ_1 for $x = \rho_1$ and must = ϕ_m for $x = l/2$, respectively.

Substituting the values of x_p and x_s thus obtained into Eqs. 31 and 32, respectively, yields the curvature curve expressed implicitly as a function of x

434					August, 1970					EM 4

for these two zones. Using the relations given in Eqs. 57 and 58, it follows that

$$\frac{x}{l} = \frac{\rho_1}{l} + \frac{\sqrt{c}}{\sqrt{2}} \frac{1}{\eta^{3/2} \phi_1^{3/4}} \frac{1}{kl} \left[\eta^{1/2} (\eta - 1)^{1/2} - \left(\frac{\eta^2 \phi_1}{\phi} - \frac{\eta \phi_1^{1/2}}{\phi^{1/2}} \right)^{1/2} \right.$$

$$\left. + \tanh^{-1} \left(1 - \frac{1}{\eta} \right)^{1/2} - \tanh^{-1} \left(1 - \frac{\phi^{1/2}}{\eta \phi_1^{1/2}} \right)^{1/2} \right] \quad \dots \dots \dots \dots \quad (59)$$

valid for $\rho_1 \leq x \leq \rho_2$. In addition

$$\frac{x}{l} = \frac{1}{2} - \frac{2}{3} \frac{\sqrt{f}}{kl} \left\{ \left[\frac{c}{2f} (\phi_1^{1/2} \eta - \phi_2^{1/2}) - \frac{1}{\phi_2} + \frac{1}{\phi} \right]^{1/2} \left[\frac{2}{\phi_2} + \frac{1}{\phi} \right. \right.$$

$$\left. - \frac{c}{f} (\phi_1^{1/2} \eta - \phi_2^{1/2}) \right] - \frac{1}{\sqrt{f}} \left(\frac{\alpha q}{kl} \right) \cdot \left[\frac{1}{f} \left(\frac{\alpha q}{kl} \right)^2 + \frac{3}{\phi_2} \right.$$

$$\left. \left. - \frac{3c}{2f} (\phi_1^{1/2} \eta - \phi_2^{1/2}) \right] \right\} \quad \dots \dots \dots \dots \dots \dots \quad (60)$$

valid for $\rho_2 \leq x \leq l/2$ in which η = function of the as yet unknown distance ρ_1, as defined in Eq. 50. The unknown distances, ρ_1 and ρ_2, which denoted elastic-primary plastic and primary-secondary plastic boundaries, respectively, can now be determined from the conditions at the section $x = \rho_2$. At this section, the two portions of the curvature curve, as given by Eqs. 59 and 60, have a common value and must = ϕ_2. After some algebraic manipulation, one obtains

$$\frac{\rho_2}{l} = \frac{\rho_1}{l} + \frac{\sqrt{c}}{\sqrt{2}} \frac{1}{\eta^{3/2} \phi_1^{3/4}} \frac{1}{kl} \left[\eta^{1/2} (\eta - 1)^{1/2} - \left(\frac{\phi_1}{\phi_2} \eta^2 - \frac{\phi_1^{1/2}}{\phi_2^{1/2}} \eta \right)^{1/2} \right.$$

$$\left. + \tanh^{-1} \left(1 - \frac{1}{\eta} \right)^{1/2} - \tanh^{-1} \left(1 - \frac{\phi_2^{1/2}}{\phi_1^{1/2}} \frac{1}{\eta} \right)^{1/2} \right] \quad \dots \dots \dots \dots \quad (61)$$

and also the desired equation for the elastic-primary plastic boundary ρ_1:

$$\left(\frac{\alpha q}{kl} \right)^3 + 3 \left[\frac{f}{\phi_2} - \frac{c}{2} (\phi_1^{1/2} \eta - \phi_2^{1/2}) \right] \left(\frac{\alpha q}{kl} \right) + 3f \left\{ \frac{kl}{4} - \frac{k\rho_2}{2} \right.$$

$$+ \frac{\sqrt{c}}{3\sqrt{2}} (\phi_1^{1/2} \eta - \phi_2^{1/2})^{1/2} \left[\frac{c}{f} (\phi_1^{1/2} \eta - \phi_2^{1/2}) - \frac{3}{\phi_2} \right] \right\} = 0 \quad \dots \dots \quad (62)$$

in which ρ_2 is also a function of ρ_1.

It is seen that this equation is a cubic equation in q (or $\alpha q/kl$) and the analytical solutions of this equation can be found directly. At the same time the solution for the unknown distance ρ_1 is much more complicated since this quantity enters into hyperbolic functions containing ρ_1 and a method of trial and error must be utilized in order to obtain a solution. Therefore, it is convenient to obtain numerical results by first solving this equation for q expressed as a function of ρ_1. Then, by assuming a value of ρ_1, the corresponding value of q can be obtained. Once this is done the corresponding values of the physical characteristics of the beam column can be computed in a rather straightforward manner.

For simplification the two coefficients of this cubic equation are denoted by s and t, respectively, and then Eq. 62 becomes

$$\left(\frac{\alpha q}{kl}\right)^3 + s \left(\frac{\alpha q}{kl}\right) + t = 0 \quad \dots\dots\dots\dots\dots\dots\dots\dots\dots \quad (63)$$

Solving Eq. 63 for q, one obtains

$$q = \frac{kl}{\alpha} \left\{ \left[-\frac{t}{2} + \left(\frac{t^2}{4} + \frac{s^3}{27}\right)^{1/2} \right]^{1/3} + \left[-\frac{t}{2} - \left(\frac{t^2}{4} + \frac{s^3}{27}\right)^{1/2} \right]^{1/3} \right\} \quad \dots \quad (64)$$

if $[(t^2/4) + (s^3/27)] > 0$. In addition

$$\left. \begin{array}{l} q_1 = 2 \dfrac{kl}{\alpha} \left(-\dfrac{t}{2}\right)^{1/3} \\[3mm] q_2 = -\dfrac{kl}{\alpha} \left(-\dfrac{t}{2}\right)^{1/3} \end{array} \right\} \quad \dots\dots\dots\dots\dots\dots\dots\dots\dots\dots\dots\dots \quad (65)$$

if $[(t^2/4) + (s^3/27)] = 0$, and

$$\left. \begin{array}{l} q_1 = 2 \dfrac{kl}{\alpha} \left(-\dfrac{s}{3}\right)^{1/2} \cos \dfrac{\psi}{3} \\[3mm] q_2 = 2 \dfrac{kl}{\alpha} \left(-\dfrac{s}{3}\right)^{1/2} \cos \left(\dfrac{\psi}{3} + 120°\right) \\[3mm] q_3 = 2 \dfrac{kl}{\alpha} \left(-\dfrac{s}{3}\right)^{1/2} \cos \left(\dfrac{\psi}{3} + 240°\right) \end{array} \right\} \quad \dots\dots\dots\dots\dots\dots\dots \quad (66)$$

if $[(t^2/4) + (s^3/27)] < 0$ in which ψ is defined as

$$\cos \psi = - \frac{\dfrac{t}{2}}{\left(-\dfrac{s^3}{27}\right)^{1/2}} \quad \dots\dots\dots\dots\dots\dots\dots\dots\dots\dots \quad (67)$$

in which s and t = functions of the parameter ρ_1.

For a given value of ρ_1 there will exist one (Eq. 64), two (Eq. 65), or three (Eq. 66) real roots for the lateral load q, possibly. Each root of q will give a value of ϕ_m. The values of q and ϕ_m so obtained can either be positive or negative. The negative sign of the value q indicates that q acts opposite to the direction assumed in Fig. 1, but can produce either positive or negative curvature at the central portion of the beam-column. However, if one considers only the particular case of positive q and positive ϕ_m, then the solution for q, satisfying such conditions, will be unique, and the unique value of q will be given by q_1.

Values of ρ_2 must lie between the limits $l/2$ and ρ_2^l (as yet undefined, $\rho_2^l \geq 0$). The value $l/2$ corresponds to $\phi_m = \phi_2$, the value of q is given by Eq. 54, and the corresponding upper limit value of ρ_1 is ρ_1^u, which is, of course, equal to the lower limit value of ρ_1^l already obtained in Case 2. When the curvature ϕ_m, as given by Eq. 58, approaches infinity the plastic regions in the upper and lower portions of the beam column meet at the center and a so-called plastic hinge forms, and the value of q is

$$q\,\phi_{m\rightarrow\infty} = \frac{kl}{\alpha} \left[\frac{c}{2} (\phi_1^{1/2} \eta - \phi_2^{1/2}) - \frac{f}{\phi_2} \right]^{1/2} \quad \dots\dots\dots\dots\dots\dots \quad (68)$$

and the corresponding lower limit value, ρ_1^l, of ρ_1, and therefore the lower limit value, ρ_2^l, can readily be found from Eq. 62 by trial and error.

Again, as analyzed in the preceding section, it is possible that the end

436   August, 1970   EM 4

loadings may be large enough to cause part of the beam column to become plastic on two sides before the lateral load q is applied. In such circumstances the initial upper limit value of ρ_2 will be ρ_2^0 instead of the value $l/2$. The value of ρ_1^0, which corresponds to $q = 0$ (or $\rho_2 = \rho_2^0$), can be obtained from Eq. 62 by equating q to 0 and by using a method of trial and error. Values of ρ_1 must now lie between the limits ρ_1^0 and ρ_1^l and the values of ρ_2 must lie between ρ_2^0 and ρ_2^l. The lateral load q corresponding to $\rho_1 = \rho_1^0 = 0$. The load q corresponding to $\rho_1 = \rho_1^l$ is given by Eq. 68. For other values of q Eqs. 64, 65, and 66 can be used readily by assuming various values of ρ_1.

NUMERICAL RESULTS

Interaction Curves.—A complete numerical evaluation of this solution for the beam-column problem has been performed on a CDC 6400 Digital Com-

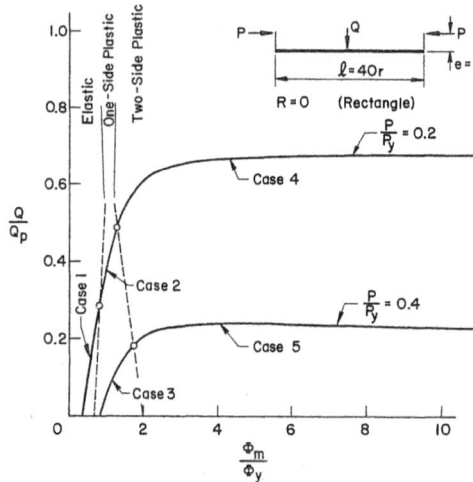

FIG. 5.—LATERAL LOAD VERSUS MID-SECTION CURVATURE CURVES $E = 30{,}000$ KSI, $\sigma_y = 34$ KSI

puter. Details of the method of computation are reported in Ref. 8. The results of calculations made with various individual cases are calculated for structural steel with $E = 30{,}000$ ksi and $\sigma_y = 34$ ksi.

The desired (q, ϕ_m) curve can be obtained directly from various individual cases. A typical plot of q versus ϕ_m for a beam column of given eccentricity ratio e/r and slenderness ratio l/r, subjected to the constant load p, is shown in Fig. 5 in which values of $e/r = 1$, $l/r = 40$, and $p = 0.2, 0.4$ have been used. The small circles and the dotted lines in the figure indicate the different stages of plastification within the beam column. It is known that for any given axial force p it is necessary to increase the lateral load q at the beginning in order to produce an increase of mid-section curvature, while beyond a certain limit, given by the maximum point of the curve, further curvature may proceed with a diminishing of the load. Although the diminish-

EM 4   INELASTIC BEAM-COLUMN   437

ing of the load with respect to an increase in curvature is not seen for the
curve $p = 0.2$ (within the range of $\phi < 10$) and is not rapid for the curve $p = 0.4$, as shown in Fig. 5. It is observed, however, that when the slenderness
ratio of the beam column or the axial compressive force, or both, are in-
creased, there is a rapid diminishing portion of the curve beyond the maximum point. Thus the maximum points of the curves in Fig. 5 represent, for

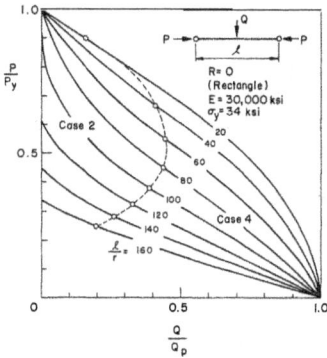

FIG. 6.—INTERACTION CURVES
FOR   RECTANGULAR   CROSS
SECTION

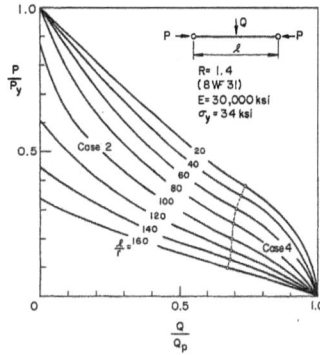

FIG. 7.—INTERACTION CURVES
FOR   WIDE-FLANGE   CROSS
SECTION

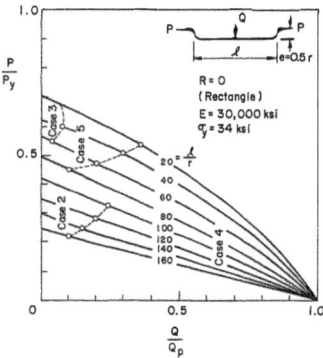

FIG. 8.—INTERACTION CURVES
FOR   RECTANGULAR   CROSS
SECTION WITH CONSTANT EC-
CENTRICITY

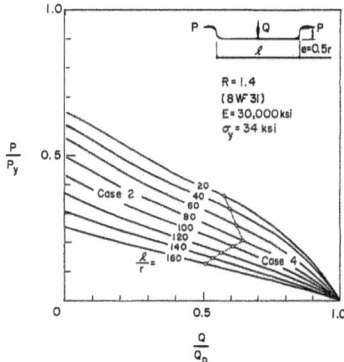

FIG. 9.—INTERACTION   CURVES
FOR WIDE-FLANGE CROSS SEC-
TION WITH CONSTANT ECCEN-
TRICITY

the given slenderness ratio and assumed eccentricity, the maximum strength
of the beam column.

The maximum loads obtained in this way for various values of axial force
p, slenderness ratio l/r, and assumed eccentricity ratio e/r are represented
by the interaction curves in Figs. 6 to 9. The curves are drawn for values of
l/r from 20 to 160 and each graph is plotted for a value of e/r and a particular cross section. For each value of e/r (0 and 0.5), two sets of such curves

438   August, 1970   EM 4

are shown, one for $R = 0$, valid for a solid rectangular cross section, and the other for $R = 1.4$, valid for a wide-flange cross section (8 W 31). The dotted-open circle curves in Fig. 6 to Fig. 9 represent the dividing lines between the range of applicability of various individual cases. Each curve is for a particular beam column and gives the combinations of axial force and lateral load that can be safely supported by the beam column. Since the interaction curves are nondimensionalized, they can be directly used in analysis and design computations.

Assuming that the axial force p is maintained at a constant value, it can be seen from the solutions of the preceding sections that the functional relationship between the lateral load q and the mid-section curvature ϕ_m only involves two independent parameters e/r (or m_o) and kl. The parameter kl is of importance in the analysis and design of beam column, because it enables the interaction curves prepared for one yield stress level to be used for different yield stress levels. Since the axial force p is assumed to remain con-

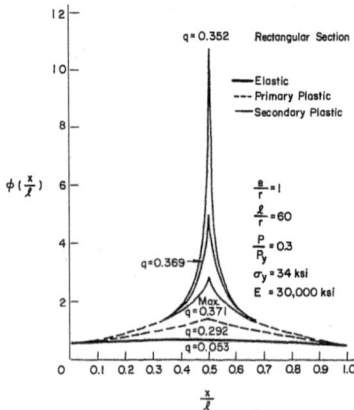

FIG. 10.—CURVATURES ALONG LENGTH OF BEAM COLUMN

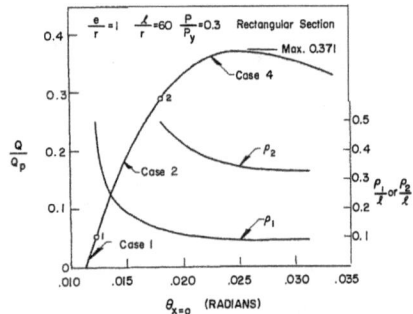

FIG. 11.—LATERAL LOAD VERSUS END SLOPE CURVE AND CORRESPONDING VALUES FOR ρ_1 AND ρ_2

stant for a particular beam-column problem, the parameter kl becomes a function of the two ratios, σ_y/E and l/r:

$$kl = \sqrt{\frac{p\,\sigma_y}{E}}\left(\frac{l}{r}\right) \quad\quad\quad\quad\quad\quad\quad\quad\quad\quad\quad (69)$$

As long as the value for the parameter kl remains the same, the (q, ϕ_m) curves corresponding to different combinations of σ_y/E and l/r will be identical for any assumed eccentricity ratio. Thus the interaction curves, prepared here for steel with a yield stress of 34 ksi, can be applied to steels of other yield stress levels by simply substituting an equivalent slenderness ratio:

$$\left(\frac{l}{r}\right)_{equ} = \sqrt{\frac{34}{\sigma_y}}\left(\frac{l}{r}\right) \quad\quad\quad\quad\quad\quad\quad\quad\quad\quad\quad (70)$$

Here E is assumed to be the same for all different yield stress steels. This conclusion is identical with that of Ref. 5.

Curvature Curves.—The curvature curves obtained in the previous sections may be plotted in a single diagram for various values of q. Fig. 10 shows a family of such curves with $e/r = 1$, $l/r = 60$, and $p = 0.3$. The extent of the elastic, primary plastic and secondary plastic zones of the beam column corresponding to different stages of loading, q, are denoted by heavy solid lines, dotted lines, and light solid lines, respectively, in the figure. These curves were computed for beam column with a rectangular cross section and again with material properties $\sigma_y = 34$ ksi and $E = 30,000$ ksi.

Knowing the values of curvature along the elastic-plastic beam column, the lateral load versus end slope or the lateral load versus mid-span deflection can be readily obtained through the use of the following two expressions:

$$\theta_{x=0} = \frac{2\left(\frac{\sigma_y}{E}\right)\left(\frac{l}{r}\right)}{\left(\frac{h}{r}\right)} \int_0^{0.5} \frac{\Phi}{\Phi_y}\, d\left(\frac{x}{l}\right) \quad \dots\dots\dots\dots\dots\dots\dots\dots \quad (71)$$

$$\frac{y_{x=l/2}}{h} = \frac{2\left(\frac{\sigma_y}{E}\right)\left(\frac{l}{r}\right)^2}{\left(\frac{h}{r}\right)^2} \int_0^{0.5} \left(\frac{\Phi}{\Phi_y}\right)\left(\frac{x}{l}\right) d\left(\frac{x}{l}\right) \quad \dots\dots\dots\dots\dots\dots \quad (72)$$

As an example, taking the curvature curves as shown in Fig. 10 as the basis for computations, the corresponding lateral load versus end slope curve is shown in Fig. 11 in which point 1 corresponds to the value of load $q = 0.053$ at which the yield point is just reached in the most compressed fiber of the mid section of the beam column; point 2 corresponds to the value of $q = 0.292$ for which the fiber of maximum tensile stress of the midsection on the convex side of the beam column has also reached its yield point, and the maximum value of the curve for q is 0.371 which defines the maximum strength of the beam-column.

To see the rate of expansion of the plastic zones as the end slope increases, the distances ρ_1 and ρ_2 which specify the elastic-primary plastic boundary as well as the primary-secondary plastic boundary, respectively, are also plotted against the end slope in Fig. 11. It is seen that the initial rate of expansion of the plastic zone on the compressive side of the beam column (see curve marked ρ_1) is very rapid and this rate falls steadily, and the curve for ρ_1 bends over more and more as the end slope increases. When the end slope reaches the value approximately 0.025 radians, the value of ρ_1 becomes practically a constant and equals the value 0.093 l. The curve for ρ_2 is very similar to the curve of ρ_1, but the initial rate of plastic expansion on the tensile side is less rapid and the value of ρ_2 practically approaches the constant 0.328 l when the end slope reaches the value approximately 0.0275 rad.

From this analysis it can be concluded that when the load q is gradually increased to its maximum and then drops off steadily beyond the maximum point, the plastic zones spread first toward the ends of the beam column with a high initial rate of expansion with respect to the rate of expansion toward the axis of the beam column. Beyond a certain value of q, further end rotation may proceed with mainly expanding the plastic zones toward the axis of the beam column while the expansion toward the ends has practically ceased. In the limit the plastic regions in the upper and lower portions of the beam col-umn will meet at the center and a so-called plastic hinge forms.

CONCLUSIONS

An analytical solution that describes the elastic-plastic behavior of an ec-centrically loaded beam column under a concentrated load at the midspan was obtained. In the analysis the moment-curvature-thrust relationships derived previously for a rectangular section are generalized to represent any com-plex shape of structural section to a high degree of approximation. Further-more, curvature instead of deflection was utilized to simplify greatly the mathematical aspect of the problem. It is shown how the load-curvature curves of beam columns can be obtained directly from the solution. The in-teraction curves for the applied lateral load versus axial force with various values of slenderness ratio and eccentricity ratio of the beam column at the instant of collapse was obtained from the load-curvature curves.

APPENDIX I.—REFERENCES

1. Baker, J. F., Horne, M. R., and Roderick, J. W., "The Behavior of Continuous Stanchions," Proceedings of the Royal Society, Serial A, Vol. 198, pp. 493, 1949.
2. Chen, W. F., and Santathadaporn, S., "Curvature and the Solution of Eccentrically Loaded Columns," *Journal of the Engineering Mechanics Division,* ASCE, Vol. 95, No. EM1, Proc. Paper 6382, Feb., 1969, pp. 21–39.
3. Chen, W. F., "Further Studies of an Inelastic Beam-column Problem," *Fritz Engineering Laboratory Report No. 331.6,* Lehigh University, Bethlehem, Penna., Jan., 1970.
4. Chwalla, E., "Aussermittig Gedruckte Baustahlstabe Mit Elastisch Eingespannten Enden und Verschieden Grossen Angriffshebeln," *Der Stahlbau,* Vol. 15, Nos. 7 and 8, 1937, pp. 49 and 57.
5. Column Research Council, "Guide to Design Criteria for Metal Compression Members," B. J. Johnson, ed., John Wiley and Sons, Inc., New York, N.Y., 1966, Chapter 6.
6. Hauck, G. F., and Lee, S. L., "Stability of Elasto-Plastic Wide-Flange Columns, *Journal of the Structural Division,* ASCE, Vol. 89, No. ST6, Proc. Paper 3738, Dec., 1963, pp. 297–324.
7. Horne, M. R., and Merchant, W., "The Stability of Frames," Pergamon Press, Inc., New York, N.Y., 1965.
8. Iyengar, S. N., and Chen, W. F., "Computer Program for an Inelastic Beam-Column Prob-lem," *Fritz Engineering Laboratory Report No. 331.7,* Lehigh University, Bethelehem, Penna., May, 1970.
9. Ketter, R. L., "Further Studies on the Strength of Beam-Columns, *Journal of the Structural Division,* ASCE, Vol. 87, No. ST6, Proc. Paper 2910, Aug., 1961, pp. 135.
10. Lu, L. W., and Kamalvand, H., "Ultimate Strength of Laterally Loaded Columns," *Journal of the Structural Division,* ASCE, Vol. 94, No. ST6, Proc. Paper, June, 1968, pp. 1505–1523.
11. Ojalvo, M., "Restrained Columns," ASCE, Vol. 86, No. EM5, Proc. Paper 2615, Oct., 1960, pp. 1–11.
12. Timoshenko, S. P., and Gere, J. M., *Theory of Elastic Stability,* 2nd ed., McGraw-Hill Book Co., Inc., New York, 1961.
13. Von Karman, T., "Untersuchungen Uber Knickfestigkeit," Mitteilungen Uber Forschungsar-beiten, Herausgegeben Vom Verein Deutscher Ingenieure, No. 81, Berlin, Germany, 1910.
14. Wright, D. T., "The Design of Compressed Beams," *The Engineering Journal,* Canada, Vol. 39, Feb., 1956, pp. 127.

APPENDIX II.--NOTATION

The following symbols are used in this paper:

A = area of section;

A_1, B, D, G, x_p, x_s = constants of integration;

$a, b, c, f, m_1, m_2, m_{pc}, \phi_1, \phi_2$ = arbitrary constants define generalized m-ϕ-p curve;

E = modulus of elasticity;

e = eccentricity;

h = depth of section;

I = moment of inertia of section about axis of bending;

$k = P/EI$;

l = length of beam column;

M = bending moment;

M_o = applied moment at end of beam column;

M_y = moment which causes first yielding in section;

$m = M/M_y$;

$m_o = M_o/M_y$;

P = thrust;

P_y = axial yield load;

$p = P/P_y$;

Q = concentrated lateral load;

Q_p = plastic limit load according to simple plastic theory;

$q = Q/Q_p$;

q_1, q_2, q_3 = roots of cubic Eq. 62;

R = section variable;

r = radius of gyration about axis of bending;

S = elastic section modulus;

x, y = coordinate axes;

α = shape factor of section about axis of bending;

ϵ_y = strain at yield point;

ρ_1 = distance from end to primary plastic zone of beam column;

ρ_2 = distance from end to secondary plastic zone of beam column;

η = function defined in Eq. 50;

σ_y = yield stress;

Φ = curvature;

Φ_m = midspan curvature;

Φ_y = curvature at initial yielding for pure bending moment;

$\phi = \Phi/\Phi_y$; and

$\phi_m = \Phi_m/\Phi_y$.

5.3.2

9613   March, 1973   ST 3

Journal of the

STRUCTURAL DIVISION

Proceedings of the American Society of Civil Engineers

ULTIMATE STRENGTH OF BIAXIALLY LOADED STEEL H-COLUMNS

By Wai F. Chen,[1] M. ASCE and Toshio Atsuta,[2] A. M. ASCE

INTRODUCTION

The maximum strengths of biaxially loaded columns simply supported at their ends are frequently used in engineering structures, but design data to assess their maximum values are very scant. This lack of information is largely due to the very difficult procedures in analysis encountered and the labor involved in computations when conventional methods are used. In particular, the biaxially loaded column under unsymmetric loading condition poses special problems. And even with drastically simplified analysis such as assuming elastic-perfectly plastic moment curvature relationships for the cross sectional and sinusoidal, or parabolic forms for the deflections and equilibrium conditions satisfied only at selected sections, etc., an analytical formula is still not possible.

Recently, an alternate but extremely simple approximate analysis was developed and simple interaction equations of in-plane beam-columns were presented for various loading conditions (3). In that work, the moment-curvature-thrust relationship for a uniaxially loaded cross section was idealized to be elastic-perfectly plastic. The approximate average flow moment, M_{mc}, was assumed to lie between the initial yield moment, M_{yc}, and the plastic limit moment, M_{pc}.

Extension of this two-dimensional approximate approach for in-plane beam-columns to three-dimensional analysis for biaxially loaded columns is attempted herein. The biaxial moment-curvature-thrust relationship is assumed to be elastic-perfectly plastic. The approximate averaged flow surface is assumed to lie between the initial yield surface and the fully plastic limit surface. Once the proper values for the averaged flow surface are selected,

Note.—Discussion open until August 1, 1973. To extend the closing date one month, a written request must be filed with the Editor of Technical Publications, ASCE. This paper is part of the copyrighted Journal of the Structural Division, Proceedings of the American Society of Civil Engineers, Vol. 99, No. ST3, March, 1973. Manuscript was submitted for review for possible publication on May 5, 1972.

[1] Assoc. Prof. of Civ. Engrg., Fritz Engrg. Lab., Lehigh Univ., Bethlehem, Pa.

[2] Engr., Ocean Engrg. Dept., Kobe Ship Building Div., Kawasaki Heavy Ind., Ltd., Japan; Formerly Grad. Student, Dept. of Civ. Engrg., Lehigh Univ., Bethlehem, Pa.

the maximum load carrying capacity of the biaxially loaded column can be computed in a rather simple manner by an elastic analysis.

ELASTIC ANALYSIS OF COLUMN

Governing Equations.—Consider a column of length L in a space (X, Y, Z) as shown in Fig. 1. The location of a cross section is defined by the Z-coordinate, on which the local coordinates (ξ, η, ζ) are placed. All the ex-

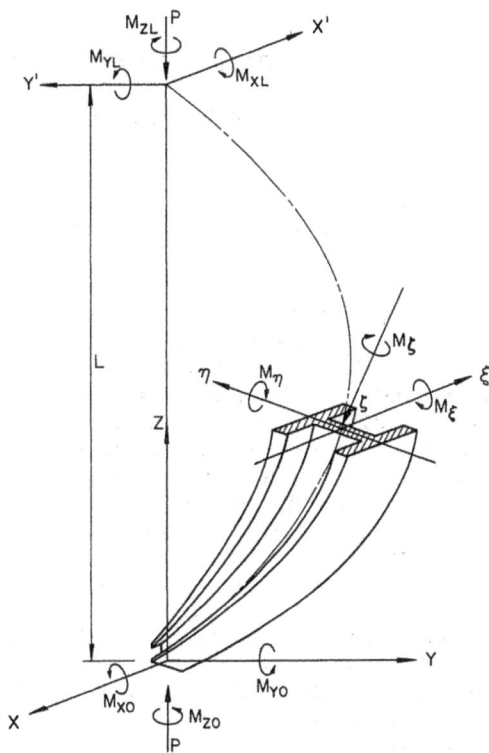

FIG. 1.—INTERNAL FORCES AND EXTERNAL FORCES

ternal and internal moments and axial force shown in Fig. 1 are in the positive direction.

The centroidal displacements of the column are chosen to be: U = displacement in X direction; V = displacement in Y direction; and Θ = rotation about Z axis.

Neglecting the second order terms, the internal moments are approximately related to the displacements by (6)

$$M_\xi = E I_\xi (V'' - \Theta U'') \approx E I_\xi V'' \quad\dotsfill (1a)$$

$$M_\eta = - E I_\eta (U'' + \Theta V'') \approx - E I_\eta U'' \quad\dotsfill (1b)$$

$$M_\xi = E I_\omega \Theta''' - (GK_T + \overline{K}) \Theta' \quad \dots\dots\dots\dots\dots\dots \quad (1c)$$

in which E, G = moduli of elasticity; I_ξ, I_η = moment of inertia; I_ω = warping moment of inertia; K_T = St. Venant torsion constant; and $\overline{K} = \int_A \sigma (\xi^2 + \eta^2) \, d\xi \, d\eta$.

The external forces and moments acting at each end of the column ($Z = 0$ and $Z = L$) are: P = axial force; M_{XO}, M_{XL} = bending moments about X axis; M_{YO}, M_{YL} = bending moments about Y axis; and M_{ZO}, M_{ZL} = twisting moments about Z axis in which the axial force, P, and the bending moments, M_x and M_y, are the applied forces and the twisting moments, M_z, are considered to be the reactions associated with the rotation constraints at the ends.

The equilibrium equations to be satisfied by the external forces and the internal forces are

$$M_\xi = M_X - P V + M_Y \Theta \quad \dots\dots\dots\dots\dots\dots\dots\dots \quad (2a)$$

$$M_\eta = M_Y + P U - M_X \Theta \quad \dots\dots\dots\dots\dots\dots\dots\dots \quad (2b)$$

$$M_\xi = M_Z + M_X U' + M_Y V' - \frac{V}{L} (M_{YL} - M_{YO})$$

$$- \frac{U}{L} (M_{XL} - M_{XO}) \quad \dots\dots\dots\dots\dots\dots\dots \quad (2c)$$

in which $M_X = M_{XO} + \dfrac{Z}{L} (M_{XL} - M_{XO}) \quad \dots\dots\dots\dots\dots\dots \quad (3a)$

$$M_Y = M_{YO} + \frac{Z}{L} (M_{YL} - M_{YO}) \quad \dots\dots\dots\dots\dots\dots \quad (3b)$$

$$M_Z = M_{ZO} = M_{ZL} \quad \dots\dots\dots\dots\dots\dots\dots\dots\dots \quad (3c)$$

The governing differential equations of the biaxially loaded column are obtained from Eqs. 1 and 2:

$$E I_\xi V'' + P V - M_Y \Theta = M_X , \quad \dots\dots\dots\dots\dots\dots\dots \quad (4a)$$

$$E I_\eta U'' + P U - M_X \Theta = - M_Y \quad \dots\dots\dots\dots\dots\dots \quad (4b)$$

$$E I_\omega \Theta''' - (G K_T + \overline{K}) \Theta' - M_X U' - M_Y V'$$

$$+ \frac{V}{L} (M_{YL} - M_{YO}) + \frac{U}{L} (M_{XL} - M_{XO}) = M_Z \quad \dots\dots\dots\dots \quad (4c)$$

Nondimensionalization.--The coordinates and displacements are nondimensionalized by

$$\left. \begin{aligned} x &= \frac{X}{L}; \ y = \frac{Y}{L}; \ z = \frac{Z}{L} \\ u &= \frac{U}{L}; \ v = \frac{V}{L}; \ \theta = \Theta \end{aligned} \right\} \quad \dots\dots\dots\dots\dots\dots\dots\dots \quad (5)$$

and the axial force, bending moments, and twisting moment are nondimensionalized by

$$p = \frac{P}{P_Y}; \ m_x = \frac{M_X}{M_{p\xi}}; \ m_y = \frac{M_y}{M_{p\eta}}; \ m_z = \frac{M_Z}{M_{p\omega}} \quad \dots\dots\dots\dots\dots \quad (6)$$

in which the full plastic states of the cross section are given by the initial

472   March, 1973   ST 3

yield normal stress, σ_y, and the initial yield shear stress, τ_y, and the plastic section moduli, Z_ξ, Z_η, and Z_ω, thus

$$P_y = \sigma_y A; \quad M_{p\xi} = \sigma_y Z_\xi; \quad M_{p\eta} = \sigma_y Z_\eta; \quad M_{p\omega} = \tau_y Z_\omega \cdots\cdots\cdots (7)$$

The plastic section moduli are defined by

$$A = \int_A dA; \quad Z_\xi = \int_A \eta \, dA; \quad Z_\eta = \int_A \xi \, dA;$$

$$Z_\omega = \int_A \sqrt{\xi^2 + \eta^2} \, dA \cdots\cdots\cdots\cdots\cdots\cdots (8)$$

The modulus, Z_ω, is the fully plastic section modulus due to warping torsion only for sections doubly symmetric. In the case of symmetric section, the value, $\bar{K}$, as given in Appendix II can be expressed in the simple form

$$\bar{K} = -\frac{P}{A}(I_\xi + I_\eta) \cdots\cdots\cdots\cdots\cdots\cdots (9)$$

Eqs. 4 can now be rewritten in the nondimensional form as

$$v'' + p_\xi v - m_{y\xi} \theta = m_{x\xi} \cdots\cdots\cdots\cdots\cdots\cdots (10a)$$

$$u'' + p_\eta u - m_{x\eta} \theta = - m_{y\eta} \cdots\cdots\cdots\cdots\cdots\cdots (10b)$$

$$\theta''' - p_\omega \theta' - m_{x\omega} u' - m_{y\omega} v' + v\,(m_{y\omega L} - m_{y\omega 0})$$

$$+ u\,(m_{x\omega L} - m_{x\omega 0}) = m_{z\omega} \cdots\cdots\cdots\cdots\cdots\cdots (10c)$$

in which $p_\xi = \dfrac{\sigma_y}{E} \dfrac{AL^2}{I_\xi} p; \quad p_\eta = \dfrac{\sigma_y}{E} \dfrac{AL^2}{I_\eta} p$

$\left. \qquad\qquad\qquad\qquad\qquad\qquad\qquad\right\} \cdots\cdots\cdots\cdots\cdots\cdots (11)$

$p_\omega = \dfrac{G}{E} \dfrac{K_T L^2}{I_\omega} + \dfrac{\sigma_y}{E} \dfrac{L^2}{I_\omega} p\,(I_\xi + I_\eta)$

and $m_{x\xi} = \dfrac{\sigma_y}{E} \dfrac{Z_\xi L}{I_\xi} m_x; \quad m_{y\xi} = \dfrac{\sigma_y}{E} \dfrac{Z_\eta L}{I_\xi} m_y$

$m_{x\eta} = \dfrac{\sigma_y}{E} \dfrac{Z_\xi L}{I_\eta} m_x; \quad m_{y\eta} = \dfrac{\sigma_y}{E} \dfrac{Z_\eta L}{I_\eta} m_y$

$\left. \qquad\qquad\qquad\qquad\qquad\qquad\qquad\right\} \cdots\cdots\cdots\cdots\cdots (12)$

$m_{x\omega} = \dfrac{\sigma_y}{E} \dfrac{Z_\xi L^3}{I_\omega} m_x; \quad m_{y\omega} = \dfrac{\sigma_y}{E} \dfrac{Z_\eta L^3}{I_\omega} m_y$

$m_{z\omega} = \dfrac{\tau_y}{E} \dfrac{Z_\omega L^3}{I_\omega} m_z$

Deflection Functions. — Assuming the deflected shapes of the column as

$$u = A_0 \sin \pi z + z(1 - z)(A_1 + A_2 z) \cdots\cdots\cdots\cdots\cdots\cdots (13a)$$

$$v = B_0 \sin \pi z + z(1 - z)(B_1 + B_2 z) \cdots\cdots\cdots\cdots\cdots\cdots (13b)$$

$$\theta = C_0 \sin \pi z + z(1 - z)(C_1 + C_2 z) \cdots\cdots\cdots\cdots\cdots\cdots (13c)$$

in which A_0, A_1, A_2, B_0, B_1, B_2, C_0, C_1, and C_2 are the constants to be determined from the governing equations and the boundary conditions.

These deflection functions satisfy the simply supported boundary conditions except for the warping deformations

$$u = 0; \quad v = 0; \quad \theta = 0 \text{ at } z = 0 \text{ and } 1 \quad \dots\dots\dots\dots\dots\dots \quad (14)$$

Warping Restraint.— The warping restraint factors, $\gamma_{\omega 0}$ and $\gamma_{\omega L}$, at the ends of the column are introduced by

$$\left. \begin{aligned} C_1 &= -\gamma_{\omega 0}\, \pi\, C_0 + (1 - \gamma_{\omega 0})\, C_2 \\ C_1 &= -\gamma_{\omega L}\, \pi\, C_0 - (2 - \gamma_{\omega L})\, C_2 \end{aligned} \right\} \quad \dots\dots\dots\dots\dots\dots \quad (15)$$

So that the warping boundary conditions may be satisfied at both ends: $\gamma_\omega = 0$ implies $\theta'' = 0$, warping permitted; and $\gamma_\omega = 1$ implies $\theta' = 0$, warping restrained.

From Eq. 15 the constants, C_1 and C_2, are solved in terms of C_0, thus

$$\left. \begin{aligned} C_1 &= -\pi\, \gamma_1\, C_0 \\ C_2 &= \pi\, \gamma_2\, C_0 \end{aligned} \right\} \quad \dots\dots\dots\dots\dots\dots\dots\dots\dots \quad (16)$$

in which
$$\left. \begin{aligned} \gamma_1 &= \frac{2\,\gamma_{\omega 0} + \gamma_{\omega L} - 2\,\gamma_{\omega 0}\,\gamma_{\omega L}}{3 - \gamma_{\omega 0} - \gamma_{\omega L}} \\ \gamma_2 &= \frac{\gamma_{\omega 0} - \gamma_{\omega L}}{3 - \gamma_{\omega 0} - \gamma_{\omega L}} \end{aligned} \right\} \quad \dots\dots\dots\dots\dots\dots \quad (17)$$

Method of Solution.— Substituting the deflection functions, Eqs. 13, into the governing differential equations, Eqs. 10, results in

$$\pi^2 B_0 \sin \pi z - 2(B_2 - B_1) + 6 B_2 z - p_\xi [B_0 \sin \pi z$$
$$+ z(1 - z)(B_1 + B_2 z)] + m_{y\xi}[C_0 \sin \pi z + z(1 - z)(C_1$$
$$+ C_2 z)] = -m_{x\xi} \quad \dots\dots\dots\dots\dots\dots\dots \quad (18)$$

$$\pi^2 A_0 \sin \pi z - 2(A_2 - A_1) + 6 A_2 z - p_\eta [A_0 \sin \pi z$$
$$+ z(1 - z)(A_1 + A_2 z)] + m_{x\eta}[C_0 \sin \pi z + z(1 - z)(C_1$$
$$+ C_2 z)] = m_{y\eta} \quad \dots\dots\dots\dots\dots\dots\dots \quad (19)$$

$$\pi^3 C_0 \cos \pi z + 6 C_2 + p_\omega [\pi C_0 \cos \pi z + C_1$$
$$+ 2(C_2 - C_1) z - 3 C_2 z^2] + m_{x\omega}[\pi A_0 \cos \pi z + A_1$$
$$+ 2(A_2 - A_1) z - 3 A_2 z^2] + m_{y\omega}[\pi B_0 \cos \pi z + B_1$$
$$+ 2(B_2 - B_1) z - 3 B_2 z^2] - (m_{y\omega L} - m_{y\omega 0})[B_0 \sin \pi z$$
$$+ z(1 - z)(B_1 + B_2 z)] - (m_{x\omega L} - m_{x\omega 0})[A_0 \sin \pi z$$
$$+ z(1 - z)(A_1 + A_2 z)] = m_{z\omega} \quad \dots\dots\dots\dots\dots \quad (20)$$

Evaluating Eqs. 18, 19, and 20 at the end, $z = 0$ and $z = 1$, the following are obtained:

$$2(B_2 - B_1) = m_{x\xi 0} \quad \dots\dots\dots\dots\dots\dots\dots\dots \quad (21a)$$

$$2(A_2 - A_1) = - m_{y\eta 0} \quad \dots \dots \dots \dots \dots \dots \dots \dots \dots \dots \dots \quad (21b)$$

$$\pi^3 C_0 + 6 C_2 + p_\omega (\pi C_0 + C_1) + m_{x\omega 0} (\pi A_0 + A_1)$$

$$+ m_{y\omega 0} (\pi B_0 + B_1) = m_{z\omega 0} \quad \dots \dots \dots \dots \dots \dots \dots \dots \quad (21c)$$

and $\quad 2(B_1 + 2 B_2) = - m_{x\xi L} \quad \dots \dots \dots \dots \dots \dots \dots \dots \dots \dots \quad (22a)$

$$2(A_1 + 2 A_2) = m_{y\eta L} \quad \dots \dots \dots \dots \dots \dots \dots \dots \dots \dots \dots \quad (22b)$$

$$\pi^3 C_0 - 6 C_2 + p_\omega (\pi C_0 + C_1 + C_2) + m_{x\omega L} (\pi A_0$$

$$+ A_1 + A_2) + m_{y\omega L} (\pi B_0 + B_1 + B_2) = - m_{z\omega L} \quad \dots \dots \dots \dots \quad (22c)$$

in which the moments with subscript 0 and L are the moments (Eqs. 12) evaluated at the ends $Z = 0$ or $Z = L$, respectively.

From the first two equations in Eqs. 21 and Eqs. 22, the four constants, A_1, A_2, B_1, and B_2, are obtained as

$$A_1 = \frac{1}{6} (m_{y\eta L} + 2 m_{y\eta 0}) \quad \dots \dots \dots \dots \dots \dots \dots \dots \dots \dots \quad (23a)$$

$$A_2 = \frac{1}{6} (m_{y\eta L} - m_{y\eta 0}) \quad \dots \dots \dots \dots \dots \dots \dots \dots \dots \dots \quad (23b)$$

$$B_1 = - \frac{1}{6} (m_{x\xi L} + 2 m_{x\xi 0}) \quad \dots \dots \dots \dots \dots \dots \dots \dots \dots \quad (23c)$$

$$B_2 = - \frac{1}{6} (m_{x\xi L} - m_{x\xi 0}) \quad \dots \dots \dots \dots \dots \dots \dots \dots \dots \quad (23d)$$

From the last equation in Eqs. 21 and 22, the following is obtained:

$$2\pi (\pi^2 + p_\omega) C_0 + 2 p_\omega C_1 + p_\omega C_2 = - m_{x\omega 0} (\pi A_0 + A_1)$$

$$- m_{x\omega L} (\pi A_0 + A_1 + A_2) - m_{y\omega 0} (\pi B_0 + B_1)$$

$$- m_{y\omega L} (\pi B_0 + B_1 + B_2) \quad \dots \dots \dots \dots \dots \dots \dots \dots \quad (24)$$

The constants, A_0, B_0, C_0, and $m_{z\omega}$, in Eqs. 18, 19, and 20 can be determined from the condition at the midspan of the column ($z = 1/2$) and Eq. 24 as

$$(\pi^2 B_0 + 2 B_1 + B_2) - \frac{1}{8} p_\xi (8 B_0 + 2 B_1 + B_2) + \frac{1}{16} (m_{y\xi 0}$$

$$+ m_{y\xi L})(8 C_0 + 2 C_1 + C_2) = - \frac{1}{2} (m_{x\xi 0} + m_{x\xi L}) \quad \dots \dots \dots \dots \quad (25)$$

$$(\pi^2 A_0 + 2 A_1 + A_2) - \frac{1}{8} p_\eta (8 A_0 + 2 A_1 + A_2)$$

$$+ \frac{1}{16} (m_{x\eta 0} + m_{x\eta L})(8 C_0 + 2 C_1 + C_2) = \frac{1}{2} (m_{y\eta 0} + m_{y\eta L}) \quad \dots \quad (26)$$

$$\frac{1}{8} (m_{x\omega 0} + m_{x\omega L}) A_2 + \frac{1}{8} (m_{y\omega 0} + m_{y\omega L}) B_2 - \frac{1}{8} (m_{y\omega L}$$

$$- m_{y\omega 0})(8 B_0 + 2 B_1 + B_2) - \frac{1}{8} (m_{x\omega L} - m_{x\omega 0})(8 A_0 +$$

$$2 A_1 + A_2) + 6 C_2 + \frac{1}{4} p_\omega C_2 = m_{z\omega} \quad \dots \dots \dots \dots \dots \dots \dots \quad (27)$$

Solving A_0, B_0, C_0, and $m_{z\omega}$ from Eqs. 24, 25, 26, and 27, using Eq. 16 and Eq. 23, the following four equations are obtained:

$$\frac{1}{2}\,(m_{x\omega 0} + m_{x\omega L})\,A_0 + \frac{1}{2}\,(m_{y\omega 0} + m_{y\omega L})\,B_0 + \left[\pi^2 + p_\omega\left(1 - \gamma_1\right.\right.$$

$$\left.\left. + \frac{1}{2}\,\gamma_2\right)\right] C_0 = \frac{1}{12\,\pi}\,[- m_{x\omega 0}\,(m_{y\eta L} + 2\,m_{y\eta 0}) + m_{y\omega 0}\,(m_{x\xi L}$$

$$+ 2\,m_{x\xi 0}) - m_{x\omega L}\,(2\,m_{y\eta L} + m_{y\eta 0}) + m_{y\omega L}\,(2\,m_{x\xi L} + m_{x\xi 0})] \qquad (28)$$

$$(\pi^2 - p_\xi)\,B_0 + \frac{1}{16}\,(m_{y\xi 0} + m_{y\xi L})(8 - 2\,\pi\,\gamma_1 + \pi\,\gamma_2)\,C_0$$

$$= \frac{-1}{16}\,p_\xi\,(m_{x\xi 0} + m_{x\xi L}) \quad\dotfill \qquad (29)$$

$$(\pi^2 - p_\eta)\,A_0 + \frac{1}{16}\,(m_{x\eta 0} + m_{x\eta L})(8 - 2\,\pi\,\gamma_1 + \pi\,\gamma_2)\,C_0$$

$$= \frac{1}{16}\,p_\eta\,(m_{y\eta 0} + m_{y\eta L}) \quad\dotfill \qquad (30)$$

$$m_{z\omega} = -\,(m_{x\omega L} - m_{x\omega 0})\,A_0 - (m_{y\omega L} - m_{y\omega 0})\,B_0$$

$$+ \left(6 + \frac{1}{4}\,p_\omega\right)\gamma_2\,C_0 - \frac{1}{24}\,m_{x\omega 0}\,(m_{y\eta L} + 2\,m_{y\eta 0})$$

$$+ \frac{1}{24}\,m_{y\omega 0}\,(m_{x\xi L} + 2\,m_{x\xi 0}) + \frac{1}{24}\,m_{x\omega L}\,(2\,m_{y\eta L} + m_{y\eta 0})$$

$$- \frac{1}{24}\,m_{y\omega L}\,(2\,m_{x\xi L} + m_{x\xi 0}) \quad\dotfill \qquad (31)$$

Stability Considerations.—Eqs. 28, 29, and 30 are the three simultaneous equations in terms of the three unknowns, A_0, B_0, and C_0, or in the matrix form

$$[K]\,[D] = [F] \quad\dotfill \qquad (32)$$

in which

$$[K] = \begin{bmatrix} \left[\pi^2 - p_\eta\right] & 0 & \left[\frac{1}{16}\,(m_{x\eta 0} + m_{x\eta L})(8 - 2\pi\,\gamma_1 + \pi\,\gamma_2)\right] \\[2ex] 0 & \left[\pi^2 - p_\xi\right] & \left[\frac{1}{16}\,(m_{y\xi 0} + m_{y\xi L})(8 - 2\pi\,\gamma_1 + \pi\,\gamma_2)\right] \\[2ex] \left[\frac{1}{2}\,(m_{x\omega 0} + m_{x\omega L})\right] & \left[\frac{1}{2}\,(m_{y\omega 0} + m_{y\omega L})\right] & \left[\pi^2 + p_\omega\left(1 - \gamma_1 + \frac{1}{2}\,\gamma_2\right)\right] \end{bmatrix} \quad (33)$$

$$[D] = \begin{bmatrix} A_0 \\ B_0 \\ C_0 \end{bmatrix} \quad\dotfill \qquad (34)$$

$$\text{and } [F] = \begin{bmatrix} \left[\frac{1}{16} p_\eta \left(m_{y\eta 0} + m_{y\eta L}\right)\right] \\ \left[-\frac{1}{16} p_\xi \left(m_{x\xi 0} + m_{x\xi L}\right)\right] \\ \frac{1}{12\pi}\left[-m_{x\omega 0}\left(m_{y\eta L} + 2\,m_{y\eta 0}\right) + m_{y\omega 0}\left(m_{x\xi L} + 2\,m_{x\xi 0}\right)\right. \\ \left. -m_{x\omega L}\left(2\,m_{y\eta L} + m_{y\eta 0}\right) + m_{y\omega L}\left(2\,m_{x\xi L} + m_{x\xi 0}\right)\right] \end{bmatrix} \quad (35)$$

In order to investigate the physical meaning of Eq. 32, it will be expressed in terms of dimensional parameters. Use of Eqs. 11 and 12 and some operations of constant multiplication converts Eq. 33, Eq. 34, and Eq. 35 into

$$[K] = \begin{bmatrix} \left[\frac{\pi^2 EI_\eta}{L^2} - P\right] & 0 & \left[\frac{M_{XO} + M_{XL}}{2L}\right] \\ 0 & \left[\frac{\pi^2 EI_\xi}{L^2} - P\right] & \left[\frac{M_{YO} + M_{YL}}{2L}\right] \\ \left[\frac{M_{XO} + M_{XL}}{2L}\right] & \left[\frac{M_{YO} + M_{YL}}{2L}\right] & \frac{\pi^2 EI_\omega}{L^4} - \left(P\frac{I_\xi + I_\eta}{AL^2} - \frac{GK_T}{L^2}\right)\frac{1 - \gamma_1 + \frac{1}{2}\gamma_2}{1 - \frac{\pi}{4}\gamma_1 + \frac{\pi}{8}\gamma_2} \end{bmatrix} \quad (36)$$

$$[D] = \begin{bmatrix} A_0 \\ B_0 \\ C_0\left(1 - \frac{\pi}{4}\gamma_1 + \frac{\pi}{8}\gamma_2\right) \end{bmatrix} \cdots\cdots\cdots\cdots\cdots\cdots\cdots (37)$$

$$[F] = \begin{bmatrix} \left[\frac{PL\left(M_{YO} + M_{YL}\right)}{8\,EI_\eta}\right] \\ \left[-\frac{PL\left(M_{XO} + M_{XL}\right)}{8\,EI_\xi}\right] \\ \frac{L}{12\pi}\left[\frac{M_{YO}\left(2M_{XO} + M_{XL}\right) + M_{YL}\left(M_{XO} + 2M_{XL}\right)}{EI_\xi}\right. \\ \left. -\frac{M_{XO}\left(2M_{YO} + M_{YL}\right) + M_{XL}\left(M_{YO} + 2M_{YL}\right)}{EI_\eta}\right] \end{bmatrix} \cdots (38)$$

In the special case when the loads are symmetric ($M_{XO} = M_{XL}$ and $M_{YO} = M_{YL}$) and when end warpings are free ($\gamma_1 = \gamma_2 = 0$), the matrix $[K]$ is identical to the coefficient matrix derived by Culver in Ref. 4. From Eqs. 36-38, all information about elastic stability of a column is obtained. First, when the column is centrally loaded, $[F] = 0$, the simultaneous equations, Eq. 32, become homogeneous as

$$\begin{bmatrix} \frac{\pi^2 EI_\eta}{L^2} - P & 0 & 0 \end{bmatrix}$$

$$
\begin{bmatrix}
0 & \dfrac{\pi^2 \, EI_\xi}{L^2} - P & 0 \\[3mm]
0 & 0 & \dfrac{\pi^2 \, EI_\omega}{L^4} - \left(P \dfrac{I_\xi + I_\eta}{AL^2} - \dfrac{GK_T}{L^2} \right) \dfrac{1 - \gamma_1 + \frac{1}{2}\gamma_2}{1 - \frac{\pi}{4}\gamma_1 + \frac{\pi}{8}\gamma_2}
\end{bmatrix}
$$

$$
\begin{bmatrix}
A_0 \\[2mm]
B_0 \\[2mm]
C_0\left(1 - \dfrac{\pi}{4}\gamma_1 + \dfrac{\pi}{8}\gamma_2\right)
\end{bmatrix} =
\begin{bmatrix}
0 \\ 0 \\ 0
\end{bmatrix}
\quad \dots\dots\dots\dots\dots \quad (39)
$$

In order that a nontrivial solution for A_0, B_0, and C_0 exists, det $[K] = 0$, from which three stability limit loads are solved:

$$
P = \frac{\pi^2 \, EI_\eta}{L^2} \quad \dots\dots\dots\dots\dots\dots\dots\dots\dots\dots\dots\dots \quad (40a)
$$

$$
P = \frac{\pi^2 \, EI_\xi}{L^2} \quad \dots\dots\dots\dots\dots\dots\dots\dots\dots\dots\dots\dots \quad (40b)
$$

$$
P = \frac{A}{I_\xi + I_\eta}\left(\frac{\pi^2 \, EI_\omega}{L^2}\, \frac{1 - \frac{\pi}{4}\gamma_1 + \frac{\pi}{8}\gamma_2}{1 - \gamma_1 + \frac{1}{2}\gamma_2} + GK_T \right) \quad \dots\dots\dots\dots \quad (40c)
$$

Eqs. 40*a* and 40*b* are flexure buckling loads about the weak axis and the strong axis of the cross section, respectively. Eq. 40*c* is torsion buckling load. If end warpings are not restrained ($\gamma_1 = \gamma_2 = 0$), then this torsion buckling load coincides with the well-known torsion-buckling expression such as Timoshenko's solution (8).

    Lateral torsional buckling can also be examined. Assuming in-plane bending moment about strong axis only (M_{XO}, M_{XL}), Eq. 32 yields

$$
\begin{bmatrix}
\dfrac{\pi^2 \, EI_\eta}{L^2} - P & 0 & \dfrac{M_{XO} + M_{XL}}{2L} \\[3mm]
0 & \dfrac{\pi^2 \, EI_\xi}{L^2} - P & 0 \\[3mm]
\dfrac{M_{XO} + M_{XL}}{2L} & 0 & \dfrac{\pi^2 \, EI_\omega}{L^4} - \left(P \dfrac{I_\xi + I_\eta}{AL^2} - \dfrac{GK_T}{L^2} \right) \dfrac{1 - \gamma_1 + \frac{1}{2}\gamma_2}{1 - \frac{\pi}{4}\gamma_1 + \frac{\pi}{8}\gamma_2}
\end{bmatrix}
$$

$$
\times
\begin{bmatrix}
A_0 \\[2mm]
B_0 \\[2mm]
C_0\left(1 - \dfrac{\pi}{4}\gamma_1 + \dfrac{\pi}{8}\gamma_2\right)
\end{bmatrix} =
\begin{bmatrix}
0 \\[2mm]
-\dfrac{PL\,(M_{XO} + M_{XL})}{8\,EI_\xi} \\[2mm]
0
\end{bmatrix}
\quad \dots\dots\dots\dots\dots \quad (41)
$$

$$\text{or } B_0 = - \frac{\dfrac{PL\,(M_{XO} + M_{XL})}{8\,EI_\xi}}{\left(\dfrac{\pi^2\,EI_\xi}{L^2} - P\right)} \qquad\qquad\qquad (42)$$

and

$$
\begin{bmatrix}
\left[\dfrac{\pi^2\,EI_\eta}{L^3} - P\right] & \left[\dfrac{M_{XO} + M_{XL}}{2L}\right] \\[3mm]
\left[\dfrac{M_{XO} + M_{XL}}{2L}\right] & \dfrac{\pi^2\,EI_\omega}{L^4} - \left(P\,\dfrac{I_\xi + I_\eta}{AL^2} - \dfrac{GK_T}{L^2}\right)\dfrac{1 - \gamma_1 + \frac{1}{2}\gamma_2}{1 - \frac{\pi}{4}\gamma_1 + \frac{\pi}{8}\gamma_2}
\end{bmatrix}
\begin{bmatrix}
A_0 \\[3mm]
C_0\left(1 - \frac{\pi}{4}\gamma_1 + \frac{\pi}{8}\gamma_2\right)
\end{bmatrix}
=
\begin{bmatrix}
0 \\[3mm]
0
\end{bmatrix}
\quad (43)
$$

Equating the determinant of the coefficient matrix (Eq. 43) to zero, the lateral torsional buckling condition is obtained as

$$\left(\frac{\pi^2\,EI_\eta}{L^2} - P\right)\left[\frac{\pi^2\,EI_\omega}{L^4} - \left(P\,\frac{I_\xi + I_\eta}{AL^2} - \frac{GK_T}{L^2}\right)\frac{1 - \gamma_1 + \frac{1}{2}\gamma_2}{1 - \frac{\pi}{4}\gamma_1 + \frac{\pi}{8}\gamma_2}\right]$$

$$= \left(\frac{M_{XO} + M_{XL}}{2L}\right)^2 \qquad\qquad\qquad (44)$$

When the column is subjected to biaxial bending as well as axial thrust, Eqs. 32 are not homogeneous; the column must be treated as a load deflection problem. The solutions for this problem exist only when

$$\det [K] \geq 0 \qquad\qquad\qquad (45)$$

This criterion can be checked conveniently by defining the stability function

$$F_{st} = \frac{1}{32}\left[\frac{8 - 2\pi\,\gamma_1 + \pi\,\gamma_2}{\pi^2 + p_\omega\left(1 - \gamma_1 + \frac{1}{2}\gamma_2\right)}\right]\left[\frac{(m_{x\omega 0} + m_{x\omega L})(m_{x\eta 0} + m_{x\eta L})}{\pi^2 - p_\eta}\right.$$

$$+ \frac{(m_{y\omega 0} + m_{y\omega L})(m_{y\xi 0} + m_{y\xi L})}{\pi^2 - p_\xi}\right] \qquad\qquad (46)$$

in which $F_{st} < 1$ implies $\det [K] > 0$, stable; $F_{st} = 1$ implies $\det [K] = 0$, stability limit; and $F_{st} > 1$ implies $\det [K] < 0$, unstable.

Solutions to Deflection Coefficients.—When the column is stable ($F_{st} < 1$), Eq. 32 can be solved for A_0, B_0, and C_0 as

$$C_0 = \left\{\frac{8}{3\pi}\;\frac{[m_{y\omega 0}\,(m_{x\xi L} + 2\,m_{x\xi 0}) + m_{y\omega L}\,(2\,m_{x\xi L} + m_{x\xi 0}) - m_{x\omega 0}\,(m_{y\eta L} + 2\,m_{y\eta 0}) - m_{x\omega L}\,(2\,m_{y\eta L}}{32\left[\pi^2 + p_\omega\left(1 - \gamma_1 + \frac{1}{2}\gamma_2\right)\right](1 - F_{st})}\right.$$

$$\left. + m_{y\eta 0})] - \frac{p_\eta}{\pi^2 - p_\eta}\,(m_{x\omega 0} + m_{x\omega L})(m_{y\eta 0} + m_{y\eta L}) + \frac{p_\xi}{\pi^2 - p_\xi}\,(m_{y\omega 0} + m_{y\omega L})(m_{x\xi 0} + m_{x\xi L})\right\} \quad (47)$$

$$\text{and } A_0 = -\left[\frac{8 - 2\pi\,\gamma_1 + \pi\,\gamma_2}{16}\right]\left[\frac{m_{x\eta 0} + m_{x\eta L}}{\pi^2 - p_\eta}\right]C_0$$

$$+ \frac{p_\eta}{16} \left[\frac{m_{y\eta_0} + m_{y\eta}L}{\pi^2 - p_\eta} \right]$$

$$B_0 = - \left[\frac{8 - 2\pi\,\gamma_1 + \pi\,\gamma_2}{16} \right] \left[\frac{m_{y\xi_0} + m_{y\xi}L}{\pi^2 - p_\xi} \right] C_0 \qquad \left. \right\} \quad \ldots\ldots\ldots (48)$$

$$- \frac{p_\xi}{16} \left[\frac{m_{x\xi_0} + m_{x\xi}L}{\pi^2 - p_\xi} \right]$$

Since all the deflection functions have been determined, the elastic column has been solved under biaxial loading. The moments at an arbitrary section can now be computed from Eqs. 1 and 13, as

$$m_\xi = - \frac{E}{\sigma_y} \frac{I_\xi}{Z_\xi L} \left[\pi^2 B_0 \sin \pi z + 2(B_1 - B_2) + 6 B_2 z \right] \quad \ldots\ldots \quad (49a)$$

$$m_\eta = \frac{E}{\sigma_y} \frac{I_\eta}{Z_\eta L} \left[\pi^2 A_0 \sin \pi z + 2(A_1 - A_2) + 6 A_2 z \right] \quad \ldots\ldots \quad (49b)$$

$$m_\zeta = - \frac{E}{\tau_y} \frac{I_\omega}{Z_\omega L^3} \left[\pi(\pi + p_\omega) C_0 \cos \pi z + (1 - 2z) p_\omega C_1 \right.$$

$$\left. + (6 + 2 p_\omega z - 3 p_\omega z^2) C_2 \right] \quad \ldots\ldots\ldots\ldots\ldots\ldots\ldots \quad (49c)$$

Critical Section.—The critical section is the section where the yield condition

$$F_y [p, m_\xi, m_\eta, m_\zeta) - 1 = 0 \quad \ldots\ldots\ldots\ldots\ldots\ldots\ldots\ldots \quad (50)$$

is reached first. Since in the elastic regime, $F_y < 1$, the location of the critical section $(z = z^*)$ is determined by

$$\left. \frac{\partial F_y}{\partial z} \right|_{z = z^*} = 0 \quad \ldots\ldots\ldots\ldots\ldots\ldots\ldots\ldots\ldots\ldots\ldots \quad (51)$$

As will be seen later, the expression for the yield function, F_y, is not simple enough such that Eq. 51 can be solved analytically. Herein, the simple approximate function

$$F_y = |p| + |m_\xi| + |m_\eta| \quad \ldots\ldots\ldots\ldots\ldots\ldots\ldots\ldots\ldots \quad (52)$$

is used for the determination of the critical section. Substituting Eqs. 49 into Eq. 52, then

$$F_y = \left\{ \pi^2 \left(\frac{I_\eta}{Z_\eta L} A_0 - \frac{I_\xi}{Z_\xi L} B_0 \right) \sin \pi z + 6 \left(\frac{I_\eta}{Z_\eta L} A_2 \right. \right.$$

$$\left. - \frac{I_\xi}{Z_\xi L} B_2 \right) z + 2 \left[\frac{I_\eta}{Z_\eta L} (A_1 - A_2) - \frac{I_\xi}{Z_\xi L} (B_1 - B_2) \right] \right\} \frac{E}{\sigma_y} + p \qquad (53)$$

From the condition, $\partial F_y/\partial z = 0$ at $z = z^*$, the location of the critical section is

$$z^* = \frac{1}{\pi} \cos^{-1} \left(- \frac{6}{\pi^3} \frac{\dfrac{I_\xi}{Z_\xi L} B_2 - \dfrac{I_\eta}{Z_\eta L} A_2}{\dfrac{I_\xi}{Z_\xi L} B_0 - \dfrac{I_\eta}{Z_\eta L} A_0} \right) \quad \ldots\ldots\ldots\ldots\ldots \quad (54)$$

480   March, 1973   ST 3

Thus the moments at the critical section are obtained from Eqs. 49 as

$$m_\xi^* = - \frac{E}{\sigma_y} \frac{I_\xi}{Z_\xi L} [\pi^2 B_0 \sin \pi z^* - 2(B_2 - B_1) + 6 B_2 z^*] \quad \ldots \ldots \quad (55a)$$

$$m_\eta^* = \frac{E}{\sigma_y} \frac{I_\eta}{Z_\eta L} [\pi^2 A_0 \sin \pi z^* - 2(A_2 - A_1) + 6 A_2 z^*] \quad \ldots \ldots \quad (55b)$$

$$m_\xi^* = - \frac{E}{\tau_y} \frac{I_\omega}{Z_\omega L^3} [\pi(\pi^2 + p_\omega) C_0 \cos \pi z^*$$

$$+ (1 - 2 z^*) p_\omega C_1 + (6 + 2 p_\omega z^* - 3 p_\omega z^{*2}) C_2] \quad \ldots \ldots \ldots \quad (55c)$$

The biaxially loaded column reaches its ultimate strength when these moments on the critical section satisfy an interaction relationship for the cross section as given by Eq. 50. This interaction relationship will be examined further in the following section.

For the special case of a symmetric loading $(\gamma_{\omega 0} = \gamma_{\omega L}, m_{x0} = m_{xL}, m_{y0} = m_{yL})$, the critical section is at the midheight $(z^* = 1/2)$, and Eq. 55 reduces to

$$m_\xi^* = \frac{E}{\sigma_y} \frac{I_\xi}{Z_\xi L} \left[\frac{\pi}{2} \frac{m_{y\xi}}{\pi^2 - p_\xi} \frac{\left(1 - \frac{\pi}{4} \frac{p_\xi}{\pi^2 - p_\xi}\right) m_{y\omega} m_{x\xi} - \left(1 - \frac{\pi}{4} \frac{p_\eta}{\pi^2 - p_\eta}\right) m_{x\omega} m_{y\eta}}{\frac{\pi^2 - p_\omega (1 - \gamma_\omega)}{1 - \frac{\pi}{4} \gamma_\omega} - \frac{1}{32} \left(\frac{m_{x\omega} m_{yn}}{\pi^2 - p_\eta} + \frac{m_{y\omega} m_{y\xi}}{\pi^2 - p_\xi}\right)} \right.$$

$$\left. + \left(1 + \frac{\pi^2}{8} \frac{p_\xi}{\pi^2 - p_\xi}\right) m_{x\xi} \right] \quad \ldots \ldots \ldots \ldots \ldots \ldots \ldots \ldots \ldots \ldots \quad (56a)$$

$$m_\eta^* = \frac{E}{\sigma_y} \frac{I_\eta}{Z_\eta L} \left[-\frac{\pi}{2} \frac{m_{xn}}{\pi^2 - p_\eta} \frac{\left(1 - \frac{\pi}{4} \frac{p_\xi}{\pi^2 - p_\xi}\right) m_{y\omega} m_{x\xi} - \left(1 - \frac{\pi}{4} \frac{p_\eta}{\pi^2 - p_\eta}\right) m_{x\omega} m_{y\eta}}{\frac{\pi^2 - p_\omega (1 - \gamma_\omega)}{1 - \frac{\pi}{4} \gamma_\omega} - \frac{1}{32} \left(\frac{m_{x\omega} m_{yn}}{\pi^2 - p_\eta} + \frac{m_{y\omega} m_{y\xi}}{\pi^2 - p_\xi}\right)} \right.$$

$$\left. + \left(1 + \frac{\pi^2}{8} \frac{p_\eta}{\pi^2 - p_\eta}\right) m_{y\eta} \right] \quad \ldots \ldots \ldots \ldots \ldots \ldots \ldots \ldots \ldots \quad (56b)$$

$$m_\xi^* = 0 \quad \ldots \ldots \ldots \ldots \ldots \ldots \ldots \ldots \ldots \ldots \ldots \ldots \quad (56c)$$

Further, for an uniaxial bending case (either $m_x = 0$ or $m_y = 0$), only the last term will be nonzero.

AVERAGED INTERACTION EQUATIONS FOR BIAXIALLY LOADED CROSS SECTION

Following the previous work on in-plane beam-columns (3), a certain averaged interaction relationship $(F_A = 0)$ for a biaxially loaded cross section between initial yielding and the fully plastic state of the cross section

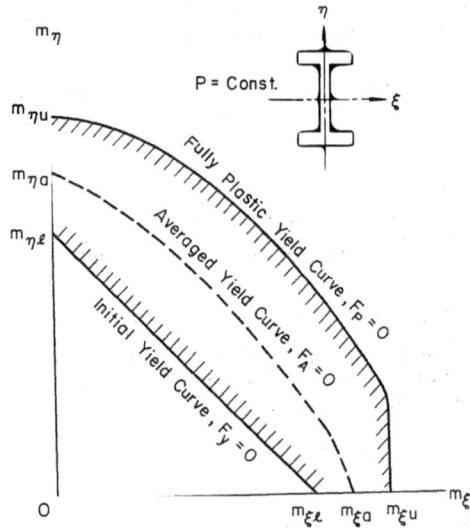

FIG. 2.—AVERAGED INTERACTION CURVES OF SECTION

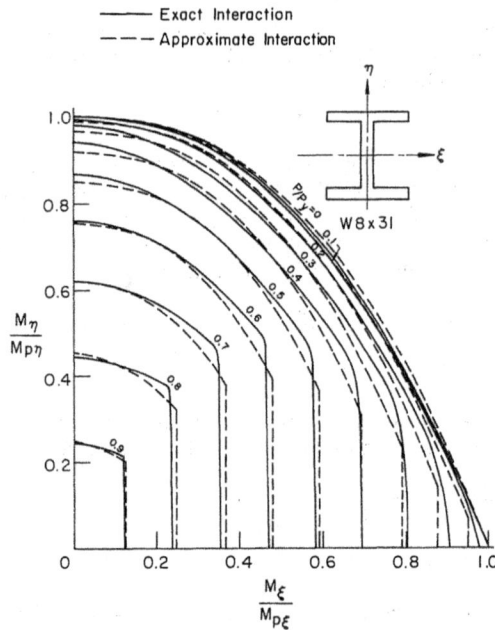

FIG. 3.—INTERACTION CURVES OF WIDE-FLANGE SECTION

can be found among the axial force (p) and the biaxial bending moments $(m_\xi,$ $m_\eta)$ at the critical section when the ultimate state of the column is reached. The contribution of the twisting moment (m_ζ) to the yield condition is neglected herein.

This averaged interaction relationship $(F_A = 0)$ is bounded by the initial yield condition $(F_y = 0)$ and the fully plastic limiting yield condition $(F_p = 0)$ as seen in Fig. 2:

$$F_y \leq F_A \leq F_p \dots \dots \dots \dots \dots \dots \dots \dots \dots \dots \dots \dots \quad (57)$$

In the following section, the wide flange shape of cross section is selected as an example for illustration. The initial yield function for a wide flange shape of cross section is

$$F_y = f_{s\xi} m_\xi + f_{s\eta} m_\eta + p - 1 = 0 \dots \dots \dots \dots \dots \dots \dots \dots \quad (58)$$

in which $f_{s\xi}$, $f_{s\eta}$ = shape factors of cross section.

The fully plastic yield function of a wide flange cross section has been derived approximately by Chen and Atsuta (2) as

$$\left. \begin{array}{l} F_p = \dfrac{m_\xi^a}{1 - p^b} + m_\eta + p^c - 1; \ (m_\eta \geq \overline{m}_\eta) \\[2ex] F_p = m_\xi + p^d - 1; \ (m_\eta \leq \overline{m}_\eta) \end{array} \right\} \dots \dots \dots \dots \dots \dots \quad (59)$$

in which $\overline{m}_\eta = 1 - p^c - \dfrac{(1 - p^d)a}{1 - p^b}$ $\dots \dots \dots \dots \dots \dots \dots \dots \dots \dots \quad (60)$

The four constants, a, b, c, and d, are determined according to the size of a particular wide flange cross section. For the cross section W8×31, e.g., these constants are found to be (2)

$$a = 2.453; \ b = 1.209; \ c = 2.714; \ d = 1.987 \dots \dots \dots \dots \dots \dots \quad (61)$$

The approximate fully plastic interaction curves $(F_p = 0)$ calculated from Eq. 59 using the foregoing constants are compared in Fig. 3 with the exact curves reported in Ref. 2. Good agreement is observed.

In order to obtain the averaged interaction relationship, the averaging factors, α and β, are introduced as shown in Fig. 2:

$$\left. \begin{array}{l} m_{\xi a} = m_{\xi u} - \alpha(m_{\xi u} - m_{\xi l}) \\[2ex] m_{\eta a} = m_{\eta u} - \beta(m_{\eta u} - m_{\eta l}) \end{array} \right\} \dots \dots \dots \dots \dots \dots \dots \dots \quad (62)$$

in which $m_{\xi l}$, $m_{\eta l}$ = initial yield moments in uniaxial bending; and $m_{\xi u}$, $m_{\eta u}$ = fully plastic moments in uniaxial bending. From Eqs. 58 and 59

$$\left. \begin{array}{l} m_{\xi l} = \dfrac{1 - p}{f_{s\xi}}; \ m_{\eta l} = \dfrac{1 - p}{f_{s\eta}} \\[2ex] m_{\xi u} = 1 - p^d; \ m_{\eta u} = 1 - p^c \end{array} \right\} \dots \dots \dots \dots \dots \dots \dots \dots \quad (63)$$

For simplicity, the averaged yield curve, F_A, is assumed to have the same form as that of the fully plastic yield curve, Eq. 59, but the biaxial bending moments, m_ξ and m_η, are proportionally reduced by the factors, $m_{\xi a}/m_{\xi u}$ and $m_{\eta a}/m_{\eta u}$, respectively. Using Eq. 62, then

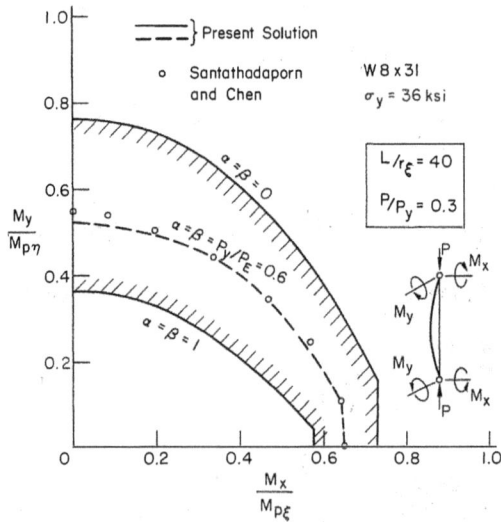

FIG. 4.—BOUNDED INTERACTION CURVE OF COLUMN

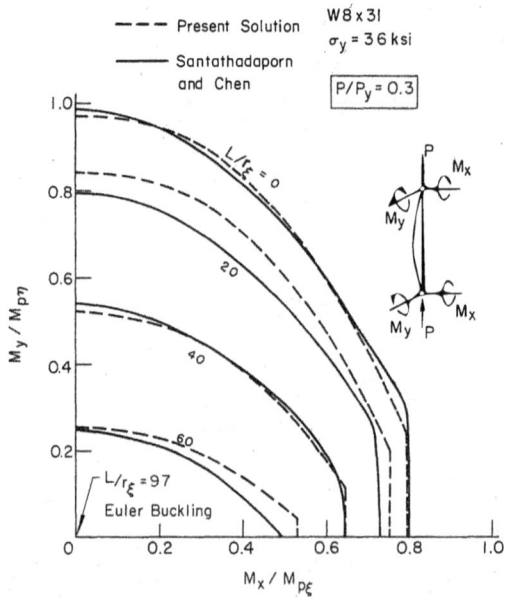

FIG. 5.—COMPARISON ON SYMMETRIC LOADING CASES

$$F_A = \frac{1}{1 - p^b}\left[\frac{m_\xi}{1 - \alpha + \dfrac{\alpha\, m_{\xi l}}{m_{\xi u}}}\right]^a + \frac{m_\eta}{1 - \beta + \dfrac{\beta\, m_{\eta l}}{m_{\eta u}}}$$

$$+ p^c - 1; \quad (m_\eta \geq \overline{\overline{m}}_\eta) \quad \dotfill \quad (64a)$$

$$F_A = \frac{m_\xi}{1 - \alpha + \dfrac{\alpha\, m_{\xi l}}{m_{\xi u}}} + p^d - 1; \quad (m_\eta \leq \overline{\overline{m}}_\eta) \quad \dotfill \quad (64b)$$

in which $\overline{\overline{m}}_\eta = \left(1 - \beta + \dfrac{\beta\, m_{\eta l}}{m_{\eta u}}\right)\left[1 - p^c - \dfrac{(1 - p^d)a}{1 - p^b}\right] \dotfill (65)$

When $\alpha = \beta = 0$, the averaged yield curve, $F_A = 0$, becomes identical to the

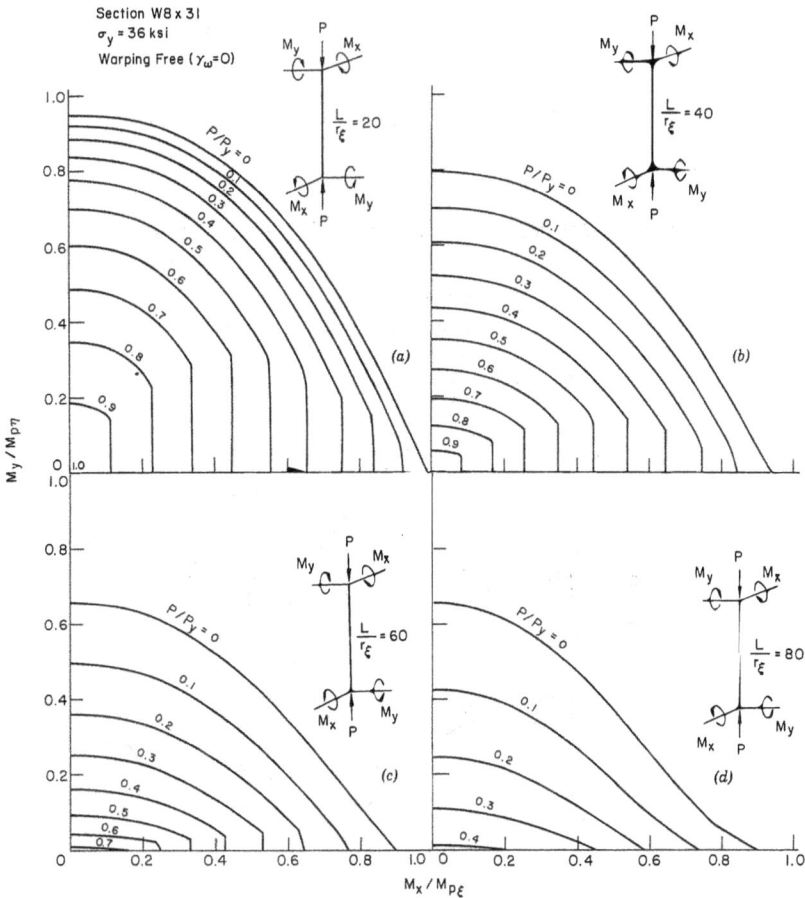

FIG. 6.—INTERACTION CURVES OF SYMMETRIC COLUMN

fully plastic yield curve, $F_p = 0$ (upper bound), and when $\alpha = \beta = 1$, the curve passes through the two points which represent the initial yield condition, $F_y = 0$, in the uniaxial bending cases ($m_\xi = 0$ or $m_\eta = 0$), but the two curves are not identical. For practical purposes, the values of α and β are assumed to be bounded by

$$0 \le \alpha \le 1; \ 0 \le \beta \le 1 \ \dots\dots\dots\dots\dots\dots\dots\dots\dots\dots \ (66)$$

The choice of a proper value for α and β for a biaxially loaded column at its ultimate state will be considered in the following.

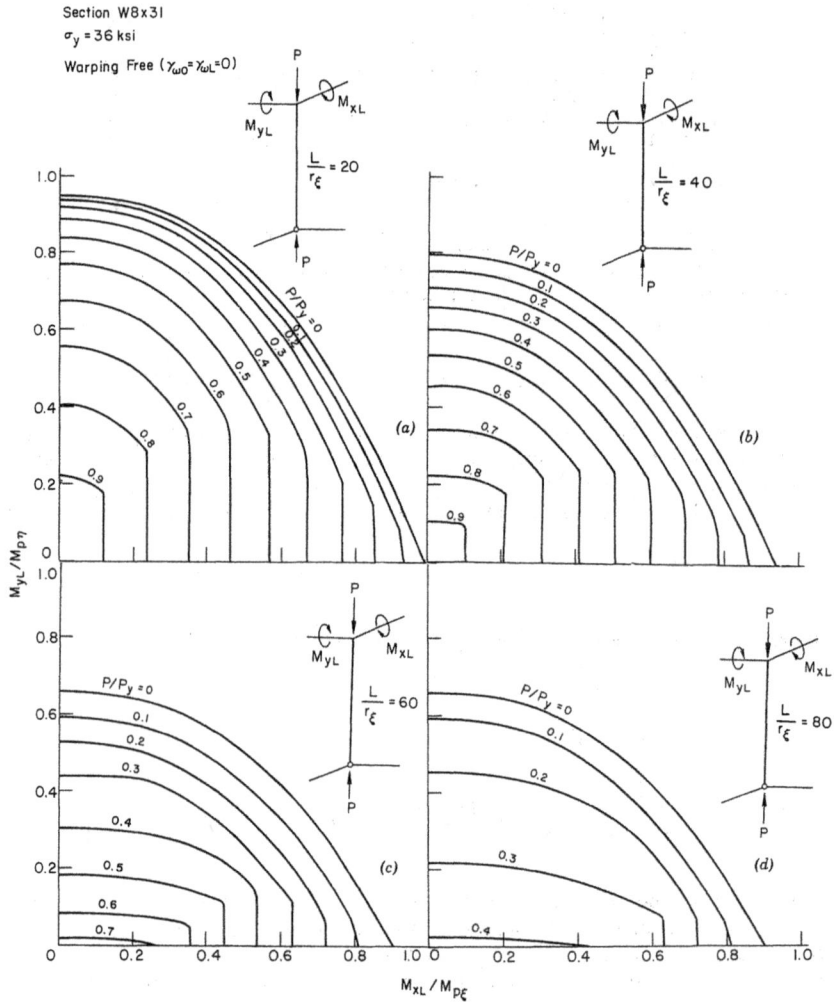

FIG. 7.—INTERACTION CURVES OF UNSYMMETRICALLY LOADED BIAXIAL COLUMN

TABLE 1.—ULTIMATE

Number	Size	Interaction Constants of Sections (Eq. 59)				Yield stress, σ_y
		a	b	c	d	
(1)	(2)	(3)	(4)	(5)	(6)	(7)
1	H6×6	2.339	1.553	2.738	1.365	33
2	W5×18.5	2.372	1.539	2.660	1.308	36
3	W5×18.5	2.365	1.528	2.641	1.297	36
4	W6×25	2.432	1.331	2.767	1.329	36
5	H5×5	2.327	1.516	2.716	1.353	36
6	H5×5	2.327	1.516	2.716	1.353	36
7	H5×6	2.427	1.212	2.879	1.373	36
8	H5×6	2.427	1.212	2.879	1.373	36
10	14×8	2.610	1.035	3.466	1.566	36
12	W5×18.5	2.365	1.491	2.667	1.314	36
13	W4×13	2.399	1.343	2.810	1.366	65
14	W4×13	2.399	1.343	2.810	1.366	65
B1	W12×79	2.427	1.313	2.689	1.282	33
H1	W14×43	2.566	1.070	3.128	1.392	33

Note: Material in parentheses indicates location of critical section from the bottom.
a Experimental results are due to Birnstiel (1).

ULTIMATE STRENGTH OF BIAXIALLY LOADED COLUMN

Using the averaged yield condition, Eq. 64, and the bending moments at the critical section of the column, Eq. 55, the ultimate strength of the column can be calculated as the combination of the applied end forces (p, m_{xo}, m_{yo}, m_{xL}, m_{yL}). This is the maximum strength of the biaxially loaded column

$$F_c = (p, m_{xo}, m_{yo}, m_{xL}, m_{yL}) = 0 \qquad \cdots\cdots\cdots\cdots (67)$$

in which $F_c = \dfrac{1}{1 - p^b} \left[\dfrac{m_\xi^*}{1 - \alpha + \dfrac{\alpha}{f_{s\xi}}} \right]^a + \dfrac{m^*}{1 - \beta + \dfrac{\beta}{f_{sn}}}$

$$+ p^c - 1; \ (m_\eta^* \ge m_\eta) \qquad (68a)$$

$$\dfrac{}{1 - \alpha + \dfrac{\alpha}{f_{s\xi}}} + p^d - 1; \ (m_\eta^* \le m_\eta) \ \cdots\cdots\cdots\cdots (68b)$$

Fig. 4 shows the interaction curves for a column (W8×31) under the symmetric loading condition. The open circles are the results reported in Ref. 6. These results are bounded by the two limiting curves (solid lines) and the averaging factors of $\alpha = \beta = 0.6$ are seen to give a good approximation to the more accurate results (6) as shown by the dotted line.

After a number of trial calculations, it is found that the following formula gives a good estimation of the averaging flow factors

STRENGTH OF COLUMNS

Length, L	Eccentricities		Ultimate Load		
	e_{xo}, e_{xL}	e_{yo}, e_{yL}	Experiment[a]	Syal and Sharma (7)	Present solution
(8)	(9)	(10)	(11)	(12)	(13)
96	1.61	2.78	92.8	93.1	102.7
96	1.60	3.21	54.1	50.75	49.4
120	0.80	2.63	62.7	60.84	55.3
96	1.66	2.95	86.3	84.35	84.9
96	2.36	3.17	49.6	50.20	46.2
120	2.38	2.51	47.9	47.76	39.6
96	0.89	2.82	76.6	80.16	75.7
96	0.34	1.87	109.4	110.10	110.6
96	0.19	2.60	85.0	78.70	79.5
120	0.77	2.78	51.0	56.23	53.3
120	0.42	2.72	46.1	47.24	45.6
120	0.83	2.35	38.7	40.12	44.1
180	6.00	12.00	—	150.95	155.9
264.6	0.50 (0.00)	5.00 (0.00)	152.0	144.19 (0.34)	143.8 (0.37)

$$\left. \begin{array}{l} \alpha = \beta = \dfrac{P_y}{P_E}; \quad \text{if } P_E \geq P_y \\[2ex] \alpha = \beta = 1.0; \quad \text{if } P_E \leq P_y \end{array} \right\} \quad \dots \dots \dots \dots \dots \dots \dots \quad (69)$$

in which P_E = the Euler's buckling load given by

$$P_E = \frac{\pi^2 E I_\eta}{L^2} \quad \dots \dots \dots \dots \dots \dots \dots \dots \dots \dots \dots \dots \dots \dots \quad (70)$$

This is a reasonable estimation because when $P_E \ll P_y$, elastic buckling will occur first, thus the initial yield condition will govern ($\alpha = \beta = 1$). If $P_E \gg P_y$, yielding of section will take place first and the ultimate state will be much closer to the fully plastic yield condition ($\alpha = \beta = 0$).

Using the formula (Eq. 69), the interaction curves for the biaxially loaded column (W8×31) with different slenderness ratios were calculated where the axial force was kept constant ($p = 0.3$) as shown in Fig. 5. A good agreement is observed with the results reported in Ref. 6.

NUMERICAL EXAMPLES

Fig. 6 shows the numerical results for a symmetrically loaded column of W8×31 cross section. The averaging factors, α and β are determined according to Eq. 69. Since the loading is symmetric, the critical section is at the midspan ($z^* = 1/2$). In such a case, a good agreement with more accurate results was usually observed (Fig. 5).

Fig. 7 shows results for an unsymmetrically loaded column with W8×31 section. The bending moments are applied only at the top end of the column $(z = 1)$ and the critical section is located between the midspan and the top end $(0.5 < z^* < 1.0)$.

Harstead and Birnstiel reported experimental and theoretical analyses of biaxially loaded H-Columns (1,5). Using the same specimens, Syal and Sharma calculated their ultimate strength (7). Details of the H-sections are listed in Table 1. The results of the ultimate strength are also compared in the table with the present approximate solutions. In Table 1, example No. H1 is the only unsymmetric loading case. The present solution is seen to give good approximation comparing with the results reported by Syal and Sharma (7).

CONCLUSIONS

A simple method to obtain the maximum strength of a biaxially loaded column is presented using an averaged yield condition of a cross section. Although it is an approximate solution, its validity has been shown by comparison with other reported results in symmetric as well as in an unsymmetric loading case. The interaction curves presented for the ultimate strength of a biaxially loaded column will be useful in the design of such columns for the various loading conditions specified.

ACKNOWLEDGMENTS

The junior writer is grateful to Kawasaki Heavy Industries, Ltd., Japan, for providing him with the opportunity to study at Lehigh University. The preparation of this paper was sponsored by the National Science Foundation, under Grant GK35886 to Lehigh University.

APPENDIX I.—REFERENCES

1. Birnstiel, C., "Experiments on *H*-Columns under Biaxial Bending," *Journal of the Structural Division*, ASCE, Vol. 94, No. ST10, Proc. Paper 6186, Oct., 1968, pp. 2429-2449.

2. Chen, W. F., and Atsuta, T., "Interaction Equations for Biaxially Loaded Sections," *Journal of the Structural Division*, ASCE, Vol. 98, No. ST5, Proc. Paper 8902, May, 1972, pp. 1035-1052.

3. Chen, W. F., and Atsuta, T., "Simple Interaction Equations for Beam-Columns," *Journal of the Structural Division*, ASCE, Vol. 98, No. ST7, Proc. Paper 9020, July, 1972, pp. 1413-1426.

4. Culver, C., "Exact Solution of the Biaxial Bending Equations," *Journal of the Structural Division*, ASCE, Vol. 92, No. ST2, Proc. Paper 4772, Apr., 1966, pp. 63-83.

5. Harstead, G. A., Birnsteil, C., and Leu, K. C., "Inelastic Behavior of *H*-Columns Under Biaxial Bending," *Journal of the Structural Division*, ASCE, Vol. 94, No. ST10, Proc. Paper 6173, Oct., 1968, pp. 2371-2398.

6. Santathadaporn, S., and Chen, W. F., "Analysis of Biaxially Loaded Steel *H*-Columns," *Journal of the Structural Division*, ASCE, Vol. 99, No. ST3, Mar., 1973, pp. 491-509.

7. Syal, I. C., and Sharma, S. S., "Biaxially Loaded Beam-Column Analysis," *Journal of the Structural Division*, ASCE, Vol. 97, No. ST9, Proc. Paper 8384, Sept., 1971, pp. 2245-2259.

8. Timoshenko, S. P., and Gera, J. M., *Theory of Elastic Stability*, 2nd ed., McGraw-Hill Book Co., New York, N.Y., 1961.

APPENDIX II.—NOTATION

The following symbols are used in this paper:

A = area of section;

A_i, B_i, C_i = coefficients of deflection functions (i = 0, 1, 2);

a, b, c, d = constants in interaction of wide flange sections, Eq. 59;

$[D]$ = deflection coefficient vector;

E, G = moduli of elasticity;

$[F]$ = external force vector;

F_c = interaction function of column;

F_{st} = stability function;

F_y, F_A, F_p = initial, averaged, and fully plastic yield functions for cross section;

$f_{s\xi}, f_{s\eta}$ = shape factors of section;

I_ξ, I_η = bending moments of inertia of section;

I_ω = warping moment of inertia of section;

$\overline{K}$ = $\int_A \sigma \, (\xi^2 + \eta^2) \, d\xi \, d\eta$;

$[K]$ = stiffness matrix;

K_T = St. Venant torsion constant;

L = length of column;

M = moment;

$M_{p\xi}, M_{p\eta}, M_{p\omega}$ = full-plastic moments of section;

$M_{xo}, M_{xL}, M_{yo}, M_{yL},$
M_{zo}, M_{zL} = end moments;

M_ξ, M_η, M_ζ = moments acting on section;

m = moment nondimensionalized by $M_{p\xi}, M_{p\eta}, M_{p\omega}$;

$m_\xi^*, m_\eta^*, m_\zeta^*$ = moments on critical section;

$m_{\xi a}, m_{\eta a}$ = averaged moments;

$m_{\xi l}, m_{\eta l}$ = initial yield moments in uniaxial bending;

$m_{\xi u}, m_{\eta u}$ = full plastic moments in uniaxial bending;

P_E = Euler's buckling load;

P_y = axial force at yielding;

P, p = axial force (p = P/P_y);

U, V, u, v = deflection of column (u = U/L, v = V/L);

X, Y, Z, x, y, z = global coordinates (x = X/L, y = Y/L, z = Z/L);

α, β = averaging factors in interaction of section;

$\gamma_{\omega o}, \gamma_{\omega L}$ = warping restraint factors;

γ_1, γ_2 = warping restraint constants—Eq. 17;

Θ, θ = rotation angle of column;

ξ, η, ζ = local coordinates on section;

σ = normal stress; and

σ_y, τ_y = yield stresses.

5.4 Semi-rigid Construction (柔性節點結構)

5.4.1

11854   JANUARY 1976   ST1

JOURNAL OF THE
STRUCTURAL DIVISION

TESTS OF WELDED STEEL BEAM-TO-COLUMN
MOMENT CONNECTIONS

By John Parfitt, Jr.[1] and Wai-F. Chen,[2] M. ASCE

INTRODUCTION

With the many types of connections available for construction, one of the decisive factors in the choice of a particular type of connection is economy, especially in high-rise steel buildings. One of the most common types of connections used in high-rise steel buildings is the moment-resisting beam-to-column connection.

In 1971 a test program was initiated at Lehigh University to study the behavior, and to develop the design method of moment-resisting beam-to-column connections. These are connections in which the beams are framed into the column flanges with the beams causing bending of the column about the major axis. The test program, consisting of 12 full-size beam-to-column connections, was under the guidance of the Welding Research Council (WRC) Task Group on Beam-to-Column Connections. The details of this test program are described elsewhere (3).

Presently, all 12 tests of the series of 12 specimens have been completed. Ref. 8 presents the result of a complete analysis of a fully welded connection. This specimen (test C12 of Ref. 3) serves as a control specimen for the purpose of evaluating the performance of several other connections of different joint design in the series. Refs. 1 and 4 summarize the theoretical and experimental results on the phase of the flange-welded web-bolted connections (C1, C2, and C3) in the series. The test results on fully-bolted connections (web-bolted and flange-bolted through moment plate, C6, C7, C8, and C9) are currently under preparation (9).

This paper presents the test results on three interrelated welded connections in this series (C4, C5, and C12). Fig. 1 shows the geometry of these three

Note.—Discussion open until June 1, 1976. To extend the closing date one month, a written request must be filed with the Editor of Technical Publications, ASCE. This paper is part of the copyrighted Journal of the Structural Division, Proceedings of the American Society of Civil Engineers, Vol. 102, No. ST1, January, 1976. Manuscript was submitted for review for possible publication on January 16, 1975.

[1] Engr., Pennsylvania Dept. of Transportation, Allentown, Pa.
[2] Prof. of Civ. Engrg., Fritz Engrg. Lab., Lehigh Univ., Bethlehem, Pa.

connections. All three connections, composed of the same size columns (W14 × 176) and beams (W27 × 94), are made of American Society for Testing and Materials (ASTM) A572 grade 55 steel. The primary difference among the three is the method by which shear is carried. The fully welded connection

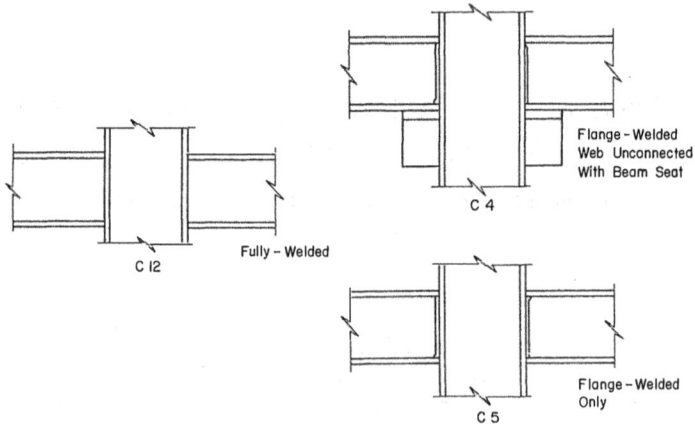

Flange–Welded
Web Unconnected
With Beam Seat

C 4

C 12　　　Fully – Welded

Flange – Welded
Only

C 5

FIG. 1.—Connection Geometries

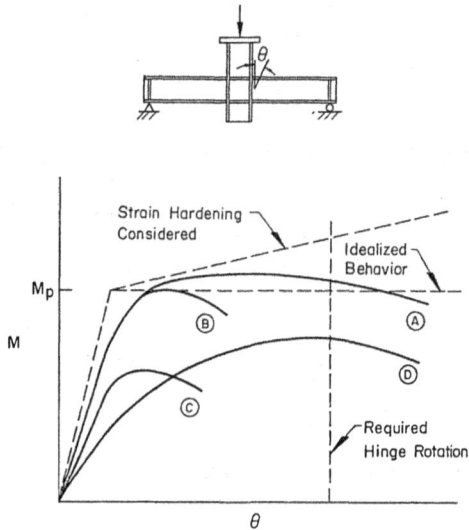

θ

Strain Hardening
Considered

Idealized
Behavior

Mp

M

Ⓑ

Ⓐ

Ⓓ

Ⓒ

Required
Hinge Rotation

θ

FIG. 2.—Moment Rotation Curves

(C12, the control specimen) utilizes the beam web to carry shear; the second connection, the flange-welded, web unconnected with a beam seat (C4) carries shear by means of a beam seat; and finally the third connection, which is only flange-welded (C5) carries both moment and shear in the beam flanges.

DESIGN CONCEPTS AND CRITERIA

The three connections examined herein represent actual interior beam-to-column moment connections and are designed according to plastic analysis procedures. Connections C4 and C12 comply with American Institute of Steel Construction (AISC) specifications (5). The schematic moment-rotation curve in Fig. 2 shows the behavior of a beam-to-column connection under symmetric loading. By properly designing the joint and preventing possible premature failure, the connection will be able to carry the plastic moment of the beam with sufficient rotation capacity and overall stiffness, as indicated by Curve A. However, if the design is unsatisfactory, the connection behavior will not be adequate. This is depicted by Curves B, C, and D. Connections C4 and C12 are proportioned so that Curve A can be obtained.

As will be seen later, C4 (the connection with the beam seat) failed prematurely with the beam web buckling. This connection was then redesigned to include a beam web stiffener, labeled C4R. In the redesign, the web was considered to act as a column with one end hinged and one end fixed.

The connections, along with all the others in the series, were designed so that the plastic moment of the beam section would be obtained. The column section chosen was that which had the least size permitted without requiring horizontal stiffeners. The connection members were proportioned such that at the beam-to-column junction the plastic moment and factored shear capacity would be achieved simultaneously.

The shear capacity used [374 kips (1,660 kN)] was the shear capacity obtained for the bolts in the design of the shear plate for test C2 (see Ref. 3). The three connections were then designed using a 374-kip (1,660-kN) shear capacity (94.7% of shear force that produces full yielding of beam web), and the beam span was then calculated as the ratio of moment to shear. The theoretical plastic limit loads modified to include the effect of shear force in beam flanges for test C5 and beam web for test C12 are 370 kips (1,650 kN) and 590 kips (2,630 kN) respectively. In the calculations, Von Mises yield condition for normal stress and shear stress in the beam section is utilized. The plastic limit load for test C4 using simple plastic theory, not including the effect of shear force on the beam flanges, is 748 kips (3,330 kN). The detailed design procedure of these specimens is given elsewhere (3,6).

The specimens were welded according to the American Welding Society (AWS) Building Code (2) with E70TG electrodes. the electrodes for fillet welds were E7028. Type NR311 filler metal was used for beam flange groove welds; and NR202 filler metal was used for beam web groove welds. All groove welds were inspected by ultrasonic testing and fillet welds by magnetic particle with respect to the AWS code, and all welds met the requirements for building structures.

TEST PROGRAM

Description of Connections.—The joint detail of C12 is shown in Fig. 3; the beam flanges and beam web are connected by groove welds to the column flanges. To simulate field practice, an erection plate is tack-welded to the column flange, and A307 bolts are used temporarily to attach the beam to the column.

192   JANUARY 1976   ST1

The joint detail of C4 is shown in Fig. 4. Vertical shear is resisted by a two-plate welded stiffener seat, which is designed according to Table VIII of the AISC manual. The beam flanges are groove-welded to the column flanges, and the seat plates and stiffener plates are fillet welded to the column flanges. To simulate field conditions for this connection, the seat plates and stiffener plates are attached to the column at the fabrication shop first, and then the beam is held in place by the A307 bolts during the welding of the flanges.

As seen in Fig. 5, only the groove welds of the flanges connects the beam to the column flanges for test C5. It has neither an erection seat nor an erection clip. The strength of this connection should be weaker than that of test C4. A detailed description of these test specimens is given in Ref. 3.

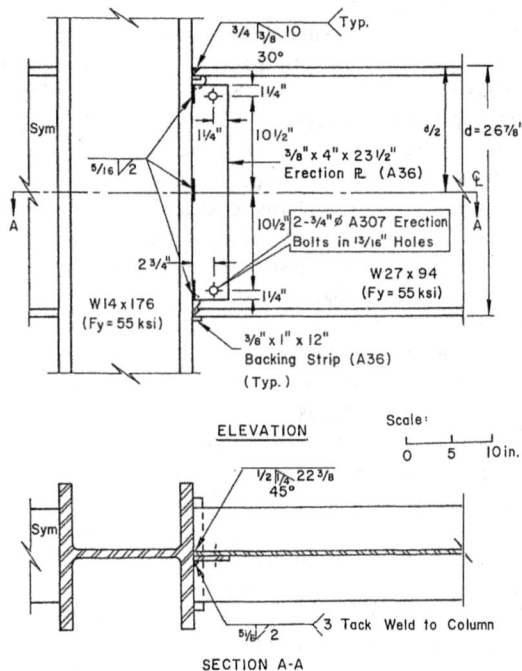

FIG. 3.—Test C12 Detail (1 in. = 25.4 mm)

Material Properties.—Properties used in determining stresses are as follows: Modulus of elasticity $E = 29{,}570$ ksi (204,000 MN/m^2); yield strain $\epsilon_y = 0.001857$ in./in.; yield stress $\sigma_y = 54.9$ (379 MN/m^2); strain at onset of strain hardening $\epsilon_{st} = 0.0150$ in./in.; strain hardening modulus $E_{st} = 581$ ksi. (4,000 MN/m^2). A detailed report of material properties is included in Ref. 7.

Instrumentation.—Electric strain gages were applied on beam flanges to provide checks for possible lateral buckling, and to determine the stress distribution. Strain gages were attached at the upper portion of the column, and were used to aline the connection and testing machine crosshead. Deflection dial gages were located directly under the column for measuring overall deflection and

ST1   WELDED STEEL TESTS   193

in the column web compression region for determining web buckling. Level bars were attached near the beam-to-column juncture to determine the rotation capacity of the joint.

Strain gages were also provided in the beam webs and column web to obtain the general stress distributions and flow throughout the web plates. Strain rosettes in the center of column compression web were placed on opposite sides at the same location. These values were averaged to account for any early web buckling.

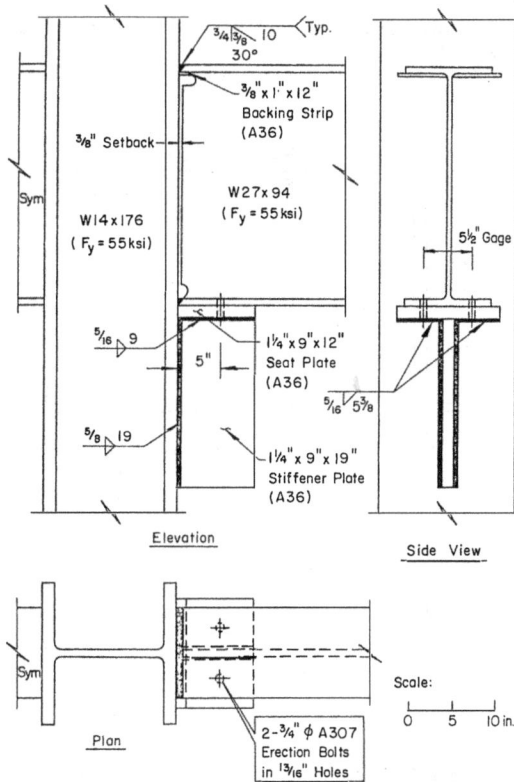

FIG. 4.—Test C4 Detail (1 in. = 25.4 mm)

Test Setup.—The test setup is shown in the inset of Fig. 2. A 5,000,000-lb (2,300,000-kg) capacity hydraulic testing machine was used to apply axial load in the column. The beams were supported by two pedestals resting on the floor. Rollers were used to simulate simply supported end conditions. Because of the size of sections and the short span of the beam used, no lateral bracing was needed to provide stability. Bearing stiffeners were provided over supports to ensure that web crippling would not occur in the beam.

Test Results and Review

As described previously, the connections simulate actual interior symmetrical-ly-loaded beam-to-column moment connections. The test setup as shown in the inset of Fig. 2 is in an inverted position, so that a concentrated load can be utilized.

Test Procedures, General Behavior, and Failure of Connections.—The applied load for the tests was increased continuously until failure, with all the strain and dial gage readings recorded after each load increment. Vertical alinement was checked by transit after each load to ensure that no lateral buckling occurred. The load deflection curve of each test, C12, C4, C4R, and C5 as shown in Fig. 6 was plotted continuously so that general specimen behavior could be observed and compared.

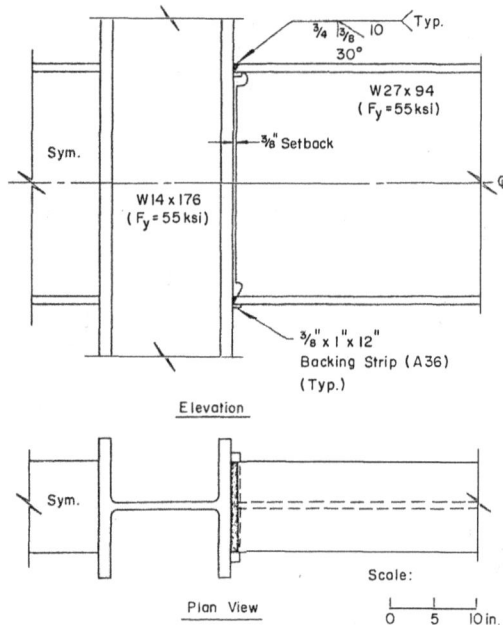

FIG. 5.—Test C5 Detail (1 in. = 25.4 mm; 1 kip = 4.45 kN)

Specimen C12.—Load increments of 25 kips (110 kN) were used initially until 600 kips (2,700 kN) was attained. Then, 20-kip (90-kN) increments were used until 680 kips (3,000 kN) was reached, and the connection was unloaded completing the first loading cycle. On the second loading cycle after reloading to 680 kips (3,000 kN), the load was increased another 20 kips–700 kips (90 kN–3,100 kN). From there on including the third loading cycle, the increments were changed from a load rate to a deflection rate of 0.20 in. (5 mm). After each deflection, the load was allowed to stabilize until there was no further movement of the sensitive crosshead, with the loading valve closed.

The first yield lines began forming in the compression web of the column at an applied load of 475 kips (2,110 kN). Both localized yielding at the toe of the fillet and yielding at the web center were observed. At this point the load deflection curve began to deviate from the linear. At 600 kips (2,700 kN) yielding was observed in the tension region of the column web near the toe of the fillet. Yielding now appeared to extend completely through the web in

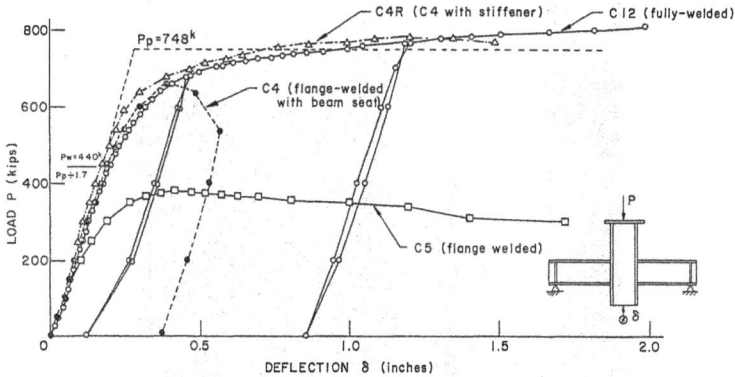

FIG. 6.—Load-Deflection Curves (P_p = Theoretical Plastic Limit Load; P_w = Working Load, $P_w = P_p/1.7$) (1. in. = 25.4 mm; 1 kip = 4.45 kN)

FIG. 7.—Weld Fracture at Tansion Flange of C12

the compression region and in the upper beam web area near the compression flange.

The connection attained a maximum load of 838 kips (3,730 kN) at a deflection of approx 2.7 in. (12 mm). Fig. 7 shows a view of the fracture of the weld at the tension flange. As seen by the picture, the weld did not fail, but pulled out the surrounding column flange material. Fig. 8 shows fraature of weld along the beam-web, which occurred simultaneously. The connection is shown at the end of the test in Fig. 9.

FIG. 8.—Fracture of Weld along Beam Web of C12

FIG. 9.—Connection C12 at End of Test

Specimen C4.—A 50-kip (220-kN) load increment was used until a load of 450 kips (2,000 kN) was attained. Then a deflection increment was used until

FIG. 10.—Yielding at Cope of Test C4

FIG. 11.—C4 at Failure

a load of 660 kips (2,940 kN) was reached. At this stage of loading, the specimen began to unload. Yielding was first observed at the cope of the beam web under the beam seat while loading the specimen from 150 kips–200 kips (670 kN–890 kN), as shown in Fig. 10. Maximum load was attained at 660 kips

FIG. 12.—Connection C4R at End of Test

FIG. 13.—Connection C5 at End of Test

(2,900 kN) with a deflection of 0.386 in. (9.8 mm). At this point, the beam web near the junction began to buckle, and caused the specimen to unload. The beam web then tore at the cope; an overall view of the specimen at failure is shown in Fig. 11.

Specimen C4R.—Because of the premature failure of C4 caused by excessive

beam buckling, it was decided to retest the connection. The buckled beam was replaced, and vertical stiffeners were added on both sides of both beams. The stiffeners, 5-in. × 5-in. × 3/8-in. (130-mm × 130-mm × 9.5-mm) angles attached with five A307 bolts, were designed assuming the beam web near

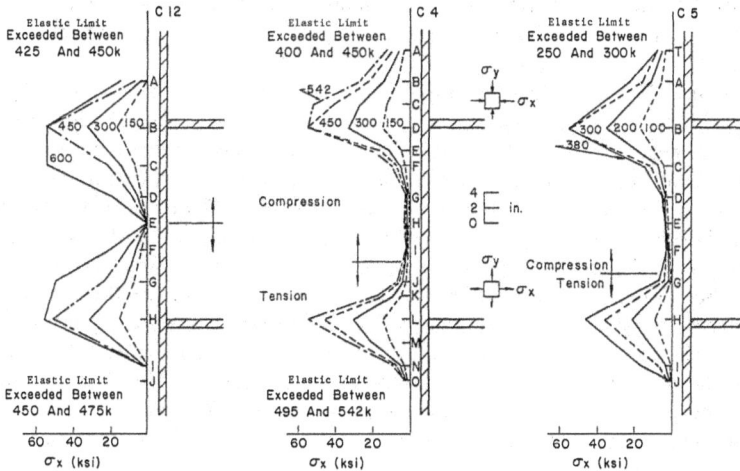

FIG. 14.—Variation of Horizontal Stress σ_x along Column Innerface (k-line) (1 ksi = 6.9 MN/m²), 1 kip = 4.45 kN)

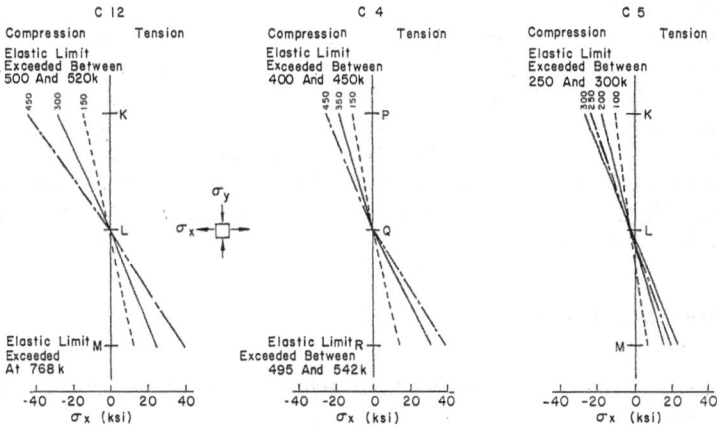

FIG. 15.—Variational Horizontal Stress along Column Center Line (1 ksi = 6.9 MN/m²; 1 kip = 4.45 kN)

the junction to act as a column which was hinged at one end and fixed at the other end. This assumption was based on the observations of the way the beam web plate of specimen C4 buckled. Specimen C4R was then loaded in the same manner as specimen C4. The only data obtained for this retest specimen were for deflection which are plotted in Fig. 6 along with the load-deflection curves of the other tests.

First signs of yielding occurred at a load of 450 kips (2,000 kN) in the beam web directly under the beam seat. Failure due to weld fracture at the heat-affected zone of the tension flange occurred while the specimen was unloading at a load of 768 kips (3,420 kN). The maximum load attained was 776 kips (3,450 kN) with a corresponding deflection of 1.2 in. (30.5 mm). A picture of the specimen at the end of the test is shown in Fig. 12.

Specimen C5.—The load increment for this test was also 50 kips (220 kN) until 350 kips (1,560 kN) was attained, and then deflections were used. Yielding first occurred at a load of 300 kips (1,300 kN) in both beam tension flanges and one compression flange. Yielding was also observed at the toe of the column fillet welds in the compression area. At a load of 375 kips (1,670 kN), local plastic hinges formed. Fig. 13 shows this connection at the end of the test.

Analysis of Results.—Fig. 14 compares the horizontal stress variation, σ_x, along the column innerface (k-line) for the three connections, with compressive stresses occurring in the upper region and tensile stresses occuring in the lower region. The greater distribution of σ_x in test C12 is due to the effective use of the beam web to carry shear; whereas in C4 and C5 the shear is carried by the beam seat and the beam flanges, respectively. The pattern of the stress distribution in the tension and compression zones of column web is seen to be the same for all connections.

The horizontal stress variations along the center line of the column web as shown in Fig. 15 are in close proximity to being linear and approximately equal to 0 at the center line intersection. However, linearity is not maintained after the initial yielding has been attained which occurred in the compressive region of each connection.

As the load-deflection curve in Fig. 6 begins to deviate from linearity, yielding begins to occur either in the beam flanges or in the panl zone adjacent to the column flanges. The stresses then redistribute to the adjacent area with the majority of the stress being distributed over a small distance from the center line of each flange as shown in Fig. 14. Simultaneously, the connections begin to rotate inelastically, with C12 and C4 being fairly equal. From the load deflection curve of Fig. 6, it can be seen that the AISC specifications (5) are adequate for the design of C4R as well as C12.

SUMMARY AND CONCLUSIONS

Herein, test results on three interrelated welded steel beam-to-column moment connections are reported. The size of columns and beams of these three connections is identical. The primary distinction among them is the way the beam shear force is carried. The fully welded connection (C12) utilizes the beam web to carry a significant part of the shear force; the stiffened seated beam connection (C4) carries the shear force through both the beam seat and the beam flanges; and the connection (C5), which is beam flange-welded only, carries the entire shear force and moment capacity through the beam flanges. On the basis of the test results in this study, the following conclusions have been reached.

1. The AISC specification provides adequate rules in design of such fully welded connections as C12 or stiffened seated beam connections as C4. For the latter case, however, the possibility of buckling of the beam web above the stiffened seat must be checked, and beam web stiffeners may be added (C4R). This type of connection can be used in plastic design as the plastic limit load, sufficient rotation capacity, and adequate elastic stiffness are developed (Fig. 6).

2. Although the stiffened seated beam connection (C4) fails by excessive buckling of the beam web, and eventually fractures at the cope hole of the beam web, specimen (C4) does exhibit sufficient stiffness under working load.

3. The fully welded connection (C12), and the stiffened seated beam connection with beam web stiffeners (C4R) are basically identical in their general behavior to the applied loads, and may be used interchangeably (Fig. 6).

4. The flange-welded only connection (C5) attains 51% of its predicted plastic limit load based upon the whole section, and showed substantial deformation and rotation capacity. The unloading was due to a beam-web fracture, which initiated at the cope hole. Testing was terminated when the tearing of the beam web became excessive and the beam web buckled (Fig. 13).

5. The basic patterns of stress distribution in the panel zone of the column are essentially the same for all the connections tested. However, the stress distributions in the beam flanges and web are effected significantly by the amount of shear force transferred to the beam flange.

ACKNOWLEDGMENTS

The project is sponsored jointly by the American Iron and Steel Institute and the Welding Research Council. Research work is carried out under the technical advice of J. A. Gilligan and the Welding Research Council Task Group. Thanks are extended to Joseph Huang, John Regec, and Glenn Rentschler for participating in the design and testing of the specimens.

APPENDIX.—REFERENCES

1. Chen, W. F., Huang, J. S., and Beedle, L. S., "Recent Results on Connection Research at Lehigh," *Proceedings*, Regional Conference on Tall Buildings, Bangkok, Thailand, Jan., 1974, pp. 799–813.
2. "Code for Welding in Building Construction," *AWS D1.0-69*, 9th ed., American Welding Society, 1969.
3. Huang, J. S., and Chen, W. F., "Steel Beam-to-Column Moment Connections," presented at the April 9-12, 1973, ASCE National Structural Engineering Meeting, held at San Francisco, Calif. (Preprint 1020).
4. Huang, J. S., Chen, W. F., and Beedle, L. S., "Behavior and Design of Steel Beam-to-Column Moment Connections," *Welding Research Council Bulletin 188*, Oct., 1973.
5. "Manual of Steel Construction," 7th ed., American Institute of Steel Construction, 1970.
6. Parfitt, J., Jr., and Chen, W. F., "Tests of Welded Steel Beam-to-Column Moment Connections," *Fritz Engineering Laboratory Report No. 333.30*, Lehigh University, Bethlehem, Pa., Dec., 1974.
7. Regec, J. E., Huang, J. S., and Chen, W. F., "Mechanical Properties of C-Series Connections," *Fritz Engineering Laboratory Report No. 333.17*, Lehigh University, Bethlehem, Pa., Apr., 1972.

8. Regec, J. E., Huang, J. S., and Chen, W. F., "Test of a Fully-Welded Beam-to-Column Connections," *Welding Research Council Bulletin 188*, Oct., 1973.
9. Standig, K. F., Rentschler, G. T. and Chen, W. F., "Tests of Bolted Beam-to-Column Flange Moment Connections," *Welding Research Council Bulletin*, 1976 (in press).

5.4.2

15386   MAY 1980   ST5

JOURNAL OF THE STRUCTURAL DIVISION

TESTS OF BEAM-TO-COLUMN WEB MOMENT CONNECTIONS [a]

By Glenn P. Rentschler,[1] A. M. ASCE, Wai F. Chen,[2] and George C. Driscoll,[3] Members, ASCE

INTRODUCTION

Background.—One of the most influential elements in the behavior and cost of multistory steel building frames is the moment resisting beam-to-column connection. A majority of these connections are column flange connections where the beam frames into the column flange. Considerable research work has been done on this type of connection at Lehigh University (5,6,10) as well as many other research institutions, e.g., Ref. 8.

Another type of moment resisting connection commonly found in building frames is the web connection. It is a study of this type of connection that was conducted at Lehigh University. In this connection, the beams are attached to the column perpendicular to the plane of the column web (Fig. 1). The action of the beam bending moment tends to bend the column about its weak axis.

Previous research done in the United States on web connections has been limited to static testing of symmetric web connections (4) and testing of unsymmetrical web connections under repeated and reversed loading (8) with no axial force. Research at Lehigh University centered on a study of unsymmetrical web connections where there was a beam on only one side of the column (Fig. 1) and axial load was applied on the column. This is a more severe type of loading on the beam and column assemblage than the symmetrically loaded connections previously studied. This study was under the guidance of the Welding Research Council Task Group on Beam-to-Column Connections.

Note.—Discussion open until October 1, 1980. To extend the closing date one month, a written request must be filed with the Manager of Technical and Professional Publications, ASCE. This paper is part of the copyrighted Journal of the Structural Division, Proceedings of the American Society of Civil Engineers, Vol. 106, No. ST5, May, 1980. Manuscript was submitted for review for possible publication on September 24, 1979.

[a] Presented at the April 24–28, 1978, ASCE National Spring Convention and Continuing Education Program, held at Pittsburgh, Pa. (Preprint 3202).

[1] Engr., Gilbert Assocs., Reading, Pa.; formerly, Research Asst. and Instr., Fritz Engrg. Lab., Lehigh Univ., Bethlehem, Pa.

[2] Prof., Struct. Engrg., Purdue Univ., West Lafayette, Ind.

[3] Prof., Civ. Engrg., Fritz Engrg. Lab., Lehigh Univ., Bethlehem, Pa.

1006   MAY 1980   ST5

Scope.—The entire study of beam-to-column web connections was divided into two distinct phases of activity. Each phase consisted of both experimental and theoretical investigations.

The first phase of activity was called the pilot test program. In attempting to organize a comprehensive research program of web connections, it was felt

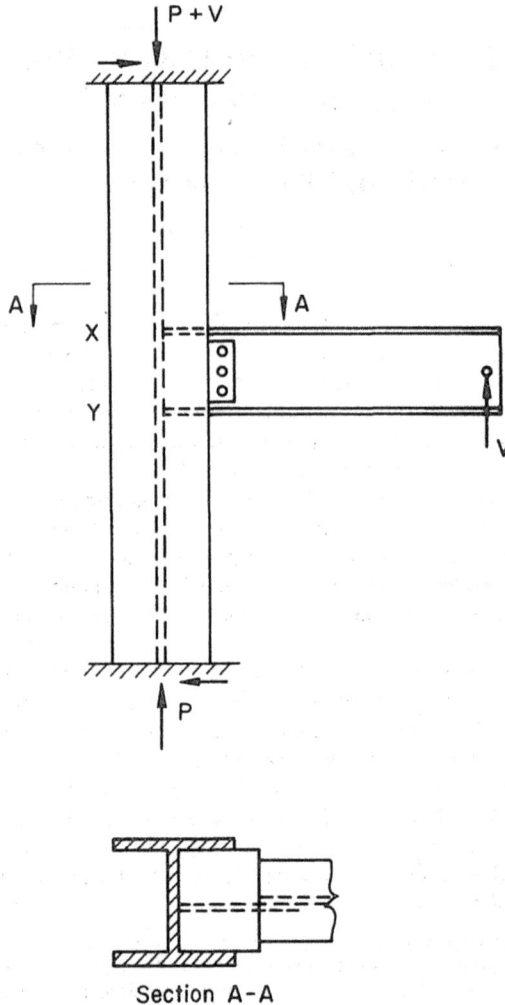

Section A-A

FIG. 1.—Web Connection Assemblage

that, by isolating certain variables, a better insight into different aspects of connection behavior could be obtained. Since the study centered around moment-resisting web connections, the critical variable examined prior to development of full-scale connection assemblages was the effects that concentrated beam bending forces have on a column when applied to simulate a web connection.

For the pilot study, the effects of column axial load and beam shear on the behavior of the web connections were ignored. The main purpose of the pilot tests was to learn how to design the full-scale specimens. These pilot tests were planned to provide answers concerning member sizes, connection geometry and stiffener requirements. Details of the testing program and test results are reported in Ref. 2.

The second phase, and the emphasis of this paper, was the testing of four full-scale web connection assemblages. Each assemblage consisted of an 18-ft (5.5-m) long column and a beam approx 5 ft (1.5 m) long connected at midheight of the column. Four different geometries of welding and bolting the beam to the column were tested. These connections simulated actual building connections

FIG. 2.—Test Setup

with the beam transmitting shear and moment to the column and the column being acted upon by an axial load. Presented herein is a summary of the results of this testing program. Theoretical investigations are currently being undertaken and will be combined with the test results to formulate design recommendations at a future time.

Objective.—For the beam-to-column web connection assemblage shown in Fig. 1, theoretically, the maximum strength of this assemblage is reached when plastic hinges form at sections X and Y in the column or in the beam at the beam-to-column juncture. However, other factors exist that may limit the maximum strength of such connections; e.g., if the beam flange is narrower than the distance between column fillets, and the beam is welded directly to

1008   MAY 1980   ST5

the column web, a yield line mechanism may form in the column web before the formation of any plastic hinges. This depends upon the width of beam flange, depth of beam, and column web thickness. Even if the attachment of the beam to the column is such that the yield line mechanism will not form, the maximum load based on simple plastic theory might not be attained due to high stress concentrations or a lack of connection ductility leading to fracture of the material. If it is necessary to carry a beam load larger than that obtained at the occurrence of the previously described limits to load carrying capacity, then stiffening of the connection must be considered.

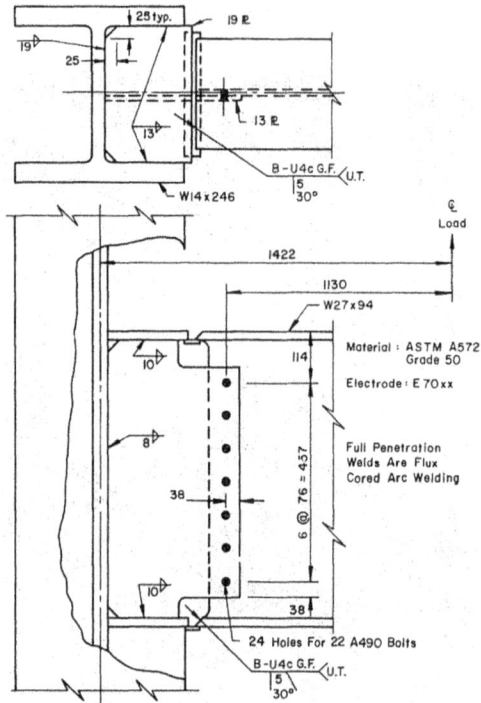

FIG. 3.—Connection 14-1

Since lateral deflection is very important in the design of tall buildings, web connections must also be looked at from a deformation standpoint. If such a connection lacks sufficient stiffness in the working load range, due to large deformations of components or to localized yielding its value as a structural component is suspect—even if it has the ability to achieve a load required to form a plastic hinge. Again, stiffening may have to be examined as a means of increasing the stiffness of the joint.

Thus the overall objective of the study is to examine web connections from the viewpoint of strength, stiffness, and ductility and to consider connection stiffening when required to attain the desired connection load or stiffness. The ultimate objective is the formulation of guidelines for the design of such connections.

Testing Program

General.—The test program consisted of four different connection geometries. The connections were full-scale using realistic beam and column sections, unsymmetrically loaded by an increasing monotonic load to simulate static conditions. A complete description of the testing program is given in Ref. 9.

The specimens were designed according to the 1969 American Institute of Steel Construction (AISC) Specification (1). The connections were proportioned to resist the moment and shear generated by the full factored load. Since the

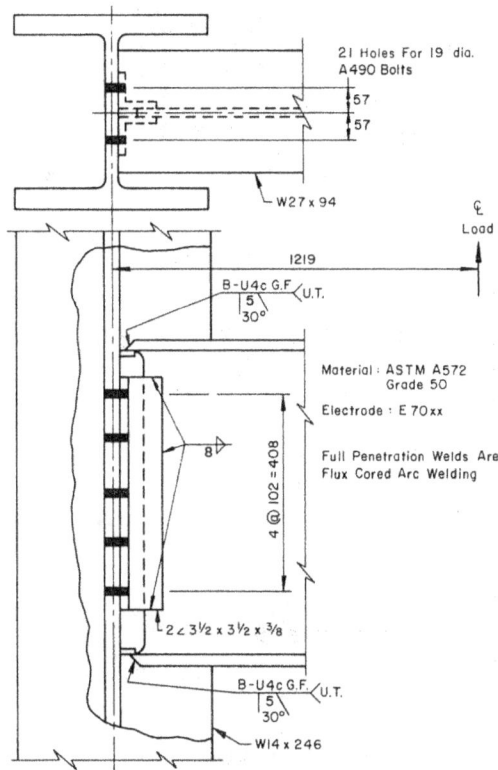

FIG. 4.—Connection 14-2

loading condition resembles gravity-type loading (dead load plus live load) the load factor used was 1.7. The stresses used in proportioning welds, shear plates, and top and bottom moment plates were then equal to 1.7 times those given in Section 1.5 of the AISC Specification. For A490 high-strength bolts in bearing-type connections, the design shear stresses used were equal to 1.7 × 40 ksi (276 MPa), instead of 32 ksi (221 MPa), as suggested in the 1969 Specification. The bearing stresses were 1.35 F_y.

The specimens were fabricated from American Society for Testing and Materials (ASTM) A572 Grade 50 steel. This steel was selected due to its increased use

1010   MAY 1980   ST5

in building design and because there is a narrower margin between yield and ultimate for this high-strength steel than for a lower grade steel. Thus, if the web connections performed well using the high strength steel, similar connections using mild strength steel should also perform well.

The column and beam sizes were the same for all four specimens. The column was a W14 × 246 and the beam was a W27 × 94. These connection components were chosen so that there would be a realistic combination of members to simulate a connection in a multistory frame.

The column section was chosen to avoid a failure in the column outside

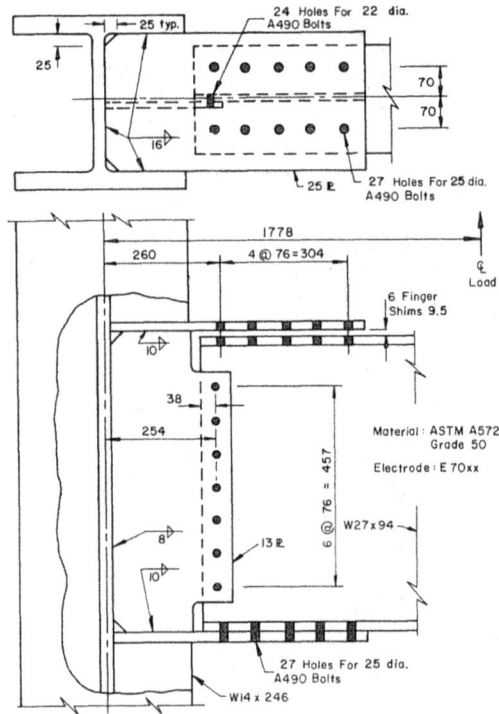

FIG. 5.—Connection 14-3

of the connection region. This section was selected after considering stability and bending of the 18-ft (5.5-m) long column under axial load, and yielding of the column's cross section above and below the beam. Since in the test setup the column was to have fixed ends, the column length of 18 ft (5.5 m) was chosen so that the distance between column inflection points was 12 ft (3.6 m), a figure fairly reflective of buildings currently being designed.

The beam lengths varied for the four specimens. The connections were proportioned such that the beam section at the beam-to-column juncture could resist the beam plastic bending moment M_p and 81% of the beam shear, V_p, required to cause shear yielding of the beam web. This section is called the critical section and is different for the various connections. (The locations of the critical section for the four connections will be given in a later paragraph.) The beam span to the critical section from point of application of the beam load is then simply the ratio of M_p to 81% V_p. For the beam section used, the length of the beam from application of load to the critical section was 48 in. (1.2 m).

In connections where some of the elements were bolted, ASTM A490 bolts were used to assemble the joint. All bolted joints were designed as bearing-type using an allowable bolt shear stress for A490 bolts of 40 ksi (276 MPa). The

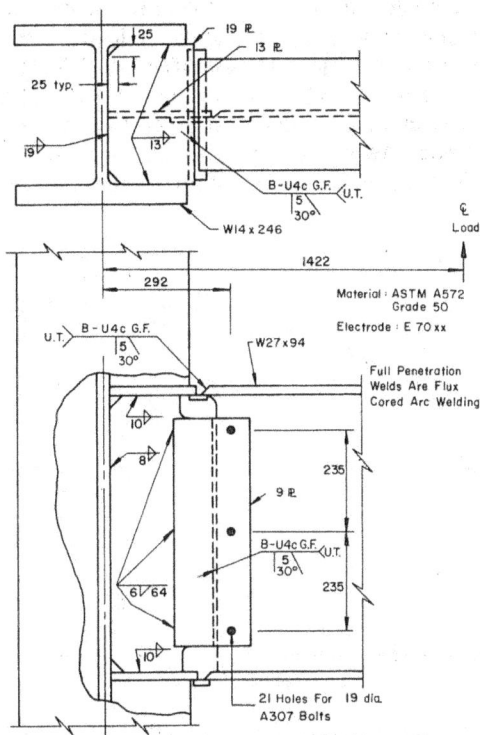

FIG. 6.—Connection 14-4

use of this higher allowable shear stress has been proven satisfactory when used in previous beam-to-column connection studies (5).

Although oversize and slotted holes are desirable to facilitate erection adjustments, the effect of using holes cut in this manner on the performance of beam-to-column connections has already been shown (5). It has been observed that slotted holes do not affect the strength of bearing-type joints. For this reason, it was decided to use round holes 1/16 in. (1.6 mm) larger than the bolt diameter to assess their effects on the behavior of bolted beam-to-column connections. Holes were punched, subpunched and reamed, or drilled, whichever was required for the hole size by the 1969 AISC Specification. All bolts were installed by the turn-of-nut method.

1012   MAY 1980   ST5

The connection specimens were welded according to the American Welding Society (AWS) Building Code (3). For fillet welds, the weld electrodes were E70XX. In determining the size of the fillet weld, the design shear stress on the effective throat was 1.7×21 ksi (145 MPa). The full penetration groove welds were made using the flux-cored arc welding technique. All welds were checked ultrasonically for defects.

Test Setup.—A schematic view of the test setup for the full-scale tests is shown in Fig. 1. Fig. 2 is a photograph of the setup taken during testing. The assemblage was placed in the 5,000,000-lb (22,240-kN) universal testing machine with that machine applying a constant axial load to the column. An axial load was applied to the column to have as realistic a load on the assemblage as possible. It was found that axial load has an effect on yielding and deformation of connections (7). The lower end of the column was bolted to the floor and the upper end was held in a fixed-end condition position by the testing machine head and bracing beam. An upward load to the beam was then applied by a hydraulic jack in increments to simulate static loading.

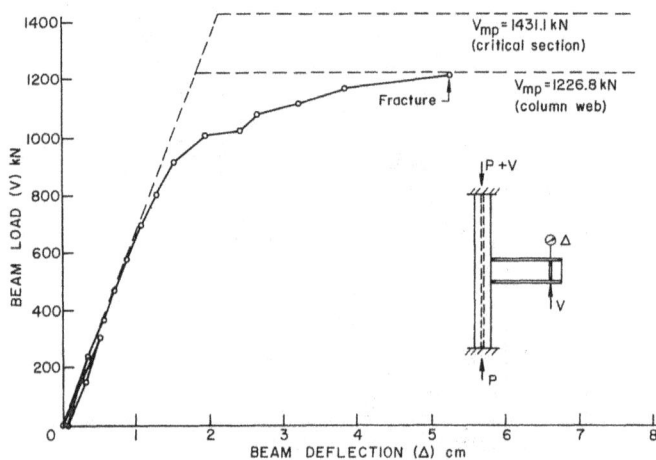

FIG. 7.—Connection 14-1 Load-Deflection Curve

After it was placed in the testing machine and properly alined, the column was loaded in 250-kip (1,112-kN) increments to a load of 1,520 kips (6,761 kN). This is equal to the value of the column axial load P obtained from P/P_y = 0.5 (1,810 kips) (8,051 kN) – 290 kips (1,290 kN). The value of 290 kips (1,290 kN) is the beam load, V, calculated to cause M_p in the beam at the critical section. The value P_y is the axial load required to cause yielding in the column. Both P and V values were calculated using nominal yield stress values.

The beam was then loaded in increments of approx 25 kips (111 kN) until deflections became excessive, at which time deflection increments were applied. The value of the column load as applied by the upper head of the testing machine was adjusted at each increment to read 1,520 kips (6,761 kN) plus the beam load V. Thus, the column in the top half of the assemblage had

an axial load of $P + V$ and the column in the bottom half had a value of
P. Once the theoretical plastic moment of the beam was attained at the critical
section, the value of the axial load in the upper column was equal to the desired
value of $P/P_y = 0.5$. If the value of V exceeded 290 kips (1,290 kN), the
axial load in the column was allowed to increase beyond $P/P_y = 0.5$. Thus,
the test assemblage simulated an inverted assemblage of a building frame where
the load on a particular floor level increases the load on the column below
that level relative to the column above.

Description of Specimens.—Specimen 14-1 shown in Fig. 3 was a flange-welded
web-bolted connection. The beam flanges were groove welded to the flange
moment plates which in turn were fillet welded to the column web and flanges.
A one-sided shear plate bolted with seven 7/8-in. (22-mm) diam A490 high-strength
bolts was used to resist vertical shear. The shear plate was fillet welded to

FIG. 8.—Fracture of Connection 14-1

both the column web and flange moment plates. Round holes 1/16 in. (1.6
mm) greater than the bolt diameter were used in the web plate and beam web.
The flange moment plates were 3/4 in. (19 mm) thick which is the thickness
of the beam flanges, and the web plate was 1/2 in. (13 mm) thick, which
is the beam web thickness. The critical section here was at the column flange
tip approx 8 in. (203 mm) from the center line of the column web. This then
provided a beam span length of 56 in. (1,400 mm) from the center line of
column to the point of application of the beam load.

Specimen 14-2 (Fig. 4) was also a flange-welded web-bolted connection. The
beam flanges were welded directly to the column web by a full penetration
groove weld. The beam web was bolted directly to the column web by a pair
of back-to-back angles to resist shear. The angles were 3-1/2 in. × 3-1/2 in.
× 3/8 in. (89 mm × 89 mm × 9.5 mm) and the bolts were eight 3/4 in.

(19 mm) A490 high-strength bolts. The angles were fillet welded to the web of the beam. Here, the critical section was at the center line of the columns giving a beam span length of 48 in. (1,200 mm).

Shown in Fig. 5 is connection 14-3, a fully bolted connection. Top and bottom moment plates were bolted to the beam flange by 10 1-in. (25-mm) diam A490 high-strength bolts in 1-1/16-in. (27-mm) round holes. These moment plates were fillet welded to the column web and flanges. This connection was designed as a flange bearing connection. The beam web shear attachment was the same as Specimen 14-1. Here, the critical section was taken as the outer row of flange bolts giving a beam span length of 70 in. (1,800 mm).

FIG. 9.—Connection 14-1 after Testing

Specimen 14-4 shown in Fig. 6 was a fully-welded connection and was used as a control test. The connection was similar to 14-1 in that the beam flanges were groove welded to the flange moment plates which in turn were fillet welded to the column web and flanges. However, in this connection, the beam web was groove welded to the shear plate to transfer the beam shear. The web shear plate was again fillet welded to both the column web and flange moment plates. The beam web was welded to the web shear plate after being held in position by three 3/4 in. (19 mm) A307 erection bolts. As in 14-1, the critical section was at the column flange tips with a similar beam span length of 56 in. (1,400 mm).

Test Results

Connection 14-1.—The load-deflection curve plot of beam load V versus beam deflection Δ is given in Fig. 7. The connection initially exhibited a definite linear-elastic V-Δ slope. The effect of yielding of the assemblage was indicated

TABLE 1.—Tensile Specimen Test Results

Test number (1)	Source (2)	Static yield stress, in kips per square inch (megapascals) (3)
170-1	Column flange	45.00 (310)
170-2	Column flange	59.66 (411)
170-3	Column web	48.95 (337)
170-4	Column web	49.19 (339)
180-1	Beam flange	55.30 (381)
180-2	Beam flange	55.57 (383)
180-3	Beam web	62.30 (430)
180-4	Beam web	62.40 (430)
181-1	Column flange	59.20 (408)
181-2	Column flange	59.66 (411)
181-3	Column web	63.70 (439)
181-4	Column web	64.10 (442)
	1/2-inch plate	51.74 (357)
	3/4-inch plate	53.56 (369)
	1-inch plate	47.91 (330)

Note: Connection columns 14-1 and 14-2 from material designated 170, and 14-3 and 14-4 from material designated 181.

by a reduction of stiffness at higher load levels. Failure of this specimen occurred at a beam load of 273 kips (1,214 kN) which is 85% of the beam load V_{mp} calculated to produce the beam plastic moment M_p at the critical section. Since most design and analysis of beams use span lengths from center to center of column, the percentage of computed M_p attained on this basis is much higher. If the center line of the column were to be taken as the critical section, the

1016   MAY 1980   ST5

load level reached was 99% of the load calculated to produce M_p. Beam deflection at the maximum load was 2.07 in. (53 mm).

Failure of this specimen was due to tearing across the entire width of the tension flange connection plate in the region of the transverse groove weld as shown in Fig. 8. Failure was instantaneous with no evidence of tearing prior to the last load increment. The beam load dropped to zero immediately with no opportunity to observe an unloading slope for the connection. Fig. 9 gives a view of the panel zone of the connection showing the extent of yielding at the conclusion of testing.

The elastic theoretical slope shown in the graph in Fig. 7, and later graphs, for comparison, is based upon accounting for beam bending, beam shear deformations, joint rotation, and the effect of small end rotations at the top of the column. It does not include items such as loss of column stiffness due to axial load. The actual test curve is very close to the theoretical stiffness in the elastic range. The upper theoretical horizontal line is the beam shear

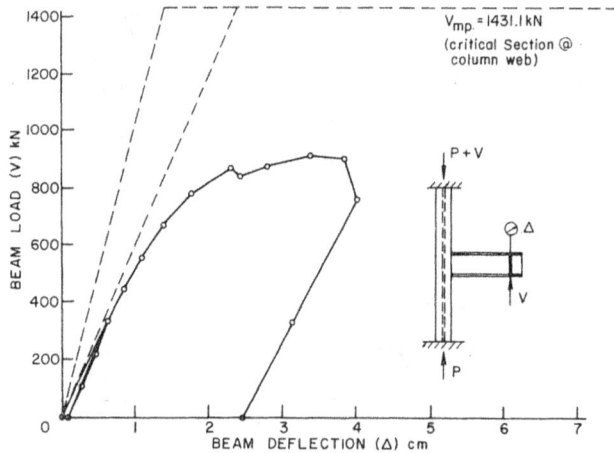

FIG. 10.—Connection 14-2 Load-Deflection Curve

required to cause the plastic moment M_p in the beam at the critical section and is based on nominal yield strength values. For the four connections, the yield strengths of material from all beams were well above nominal (see Table 1).

Connection 14-2.—Fig. 10 shows the beam load versus beam deflection plot for connection 14-2. The curve has a definite linear elastic V-Δ slope up to a load of approx 100 kips (445 kN). The effect of yielding of connection components is indicated by the nonlinear behavior at higher loads. This nonlinear behavior was primarily due to yielding of the column web under the action of the beam flange forces, the column web alone having to resist the beam bending forces because the beam flanges were not attached to the column flanges.

The maximum load in this specimen was 205.4 kips (914 kN) which was 64% of V_{mp} at the critical section, which in this case is taken to be the center line of the column web. The failure of this specimen was indicated by two related events. First, at a beam load of 195 kips (867 kN), the column web fractured on one side of the beam tension flange where the beam was welded

FIG. 11.—Tearing of Connection 14-2 at Tension Flange-Column Web Junction

FIG. 12.—Connection 14-2 Column Web Yielding

to the column web. The fracture did not completely penetrate the column web but caused a redistribution of stress in the beam tension flange. The fracture caused an increase in stress (and related strain) on the portion of the beam still intact with the column web. Ultimate failure then occurred at a load of 201.9 kips (898 kN) when a portion of the weld still connecting the beam flange

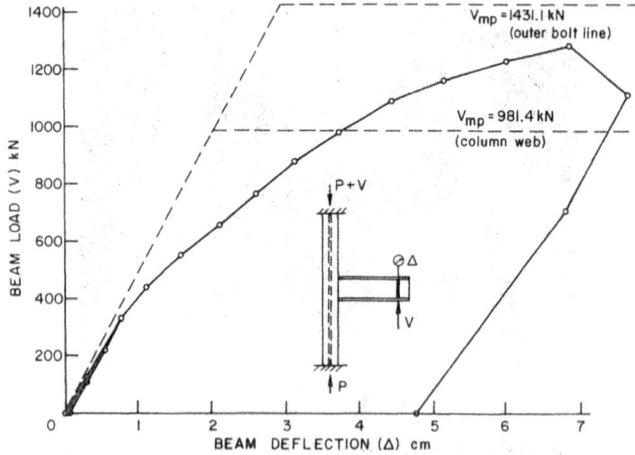

FIG. 13.—Connection 14-3 Load-Deflection Curve

FIG. 14.—Tearing of Connection 14-3 Flange Plate

to the column web fractured. Since the fracture did not proceed across the entire beam flange, the load did not drop off completely, but no further loading was attempted. The maximum beam deflection attained was 1.58 in. (40 mm).

Shown in Fig. 11 is a view of the beam tension flange showing the fracture in the region of the beam flange-to-column web groove weld. The severe deformations to which the column web was subjected are visible in Fig. 12 which is a view of the opposite side of the column. The beam tension flange is in the lower part of the photograph.

The elastic stiffness of the tested connection does not compare well to the theoretical stiffness, calculated as described earlier. The major difference between the two curves is the deformation of the column web. Without stiffening present on the opposite side of the column, the flexible column web deforms significantly out-of-plane due to the action of the beam flange forces.

Connection 14-3.—The beam load V versus beam deflection Δ for connection 14-3 is given in Fig. 13. The plot shows an initial linear elastic slope up to approx 90 kips (400 kN) and then a secondary linear slope up to a load of 200 kips (890 kN). This general type of behavior of two distinct slopes agrees

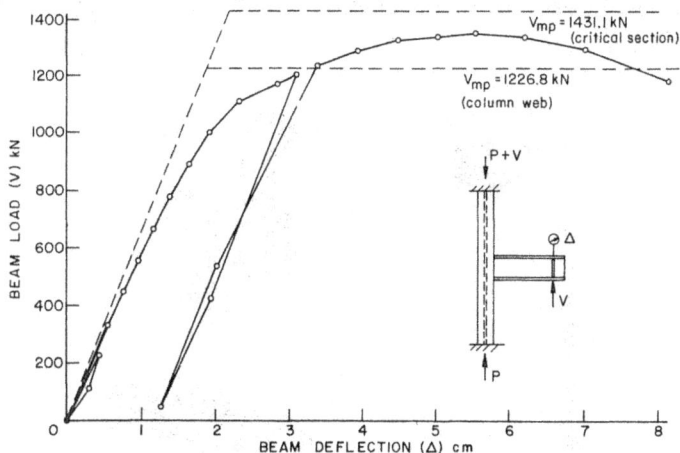

FIG. 15.—Connection 14-4 Load-Deflection Curve

quite favorably with results of tests on bolted connections recently conducted at Lehigh University (10). The second linear slope is due to many minor slips of the bolted flange plates into bearing. There was no one major slip during the test of this connection or during previous beam-to-column connection tests. The load-deflection curve then gradually loses stiffness due to yielding of elements within the assemblage. The maximum load attained on this test was 289 kips (1,286 kN) which is 90% of the beam load required to cause M_p at the critical section (131% if the critical section is taken as the center line of the column web). During the next load interval, a tear developed in the tension flange connection plate as shown in Fig. 14 and the load dropped to 249 kips (1,108 kN). The load reached a value of approx 300 kips (1,334 kN) before the tear occurred. No further loading was attempted and the connection was completely unloaded.

As in 14-1, the initial elastic slope compares favorably with the theoretical

TABLE 2.—Summary of Connection Test Beam Loads

Category (1)	14-1 (2)	14-2 (3)	14-3 (4)	14-4 (5)
Maximum beam load, in kilonewtons	1,214.4	913.7	1,285.5	1,350.0
As a percentage of V_p, at critical section	85	64	90	95
As a percentage of V_p, at center line of column	99	64	131	110
Maximum beam deflection Δ_{max}, in centimeters	5.26	4.01	7.67	8.18
Δ_{max}/Δ_p	2.46	1.68	2.60	3.83
Failure mode	fracture	fracture	fracture	large deformations

Note: Δ_p = theoretical deflection at beam load of V_{mp} (includes experimental deflection due to column web movement for Connection 14-2); V_{mp} = 1427.0 kN (320.8 kips) at critical section.

FIG. 16.—Connection 14-4 after Testing

slope. However, the secondary elastic slope due to beam flange plate bolt slip shows considerable reduction in stiffness. The fact that this stiffness reduction occurs in what would be the working load range is of considerable importance.

Connection 14-4.—A curve depicting the beam load V versus beam deflection

Δ behavior is given in Fig. 15. The connection assemblage exhibits a linear load-deflection slope up to a beam load of approx 150 kips (667 kN) at which time the stiffness is reduced due to local yielding. Again, the elastic slope deviates slightly from the predicted stiffness.

The maximum loading on the specimen was 303.5 kips (1,350 kN) which is 95% of the beam load required to produce plastic moment at the critical section (110% if the critical section were considered to be the center line of the column). The testing was terminated when, due to the large beam deflection and other deformations, no further purpose would have been served by continuing to load. The load started to fall off from its peak value due to out-of-plane deformation of the beam compression flange and the vertical web connection plate. The beam deflection at the end of testing was 3.22 in. (82 mm). A photo of Connection 14-4 at the conclusion of testing is provided in Fig. 16.

Presented in Table 2 is a summary of maximum beam loads and the maximum load as a percentage of the plastic moment load computed at both the critical section and at the center line of column. Also given is the maximum deflection in centimeters and as a ratio of the theoretical deflection, Δ_p, at the start of plastic beam behavior as well as failure mode of the four connections.

CONCLUSIONS

A series of four full-scale beam-to-column moment-resisting web connection assemblages have been tested to observe their behavior under simulated static loading. This testing program in combination with a previous testing program and future theoretical work will provide a thorough understanding of such connections and will lead to recommendations and guidelines for those involved in their design.

The following conclusions can be made regarding the test results of these four assemblages:

1. When considering the maximum beam load evaluated at the column center line, Connections 14-1, 14-3, and 14-4 all achieved load levels beyond the plastic moment load.

2. The maximum load level of Connection 14-2 was only 64% of the plastic moment load.

3. Connections 14-1, 14-2, and 14-4 all exhibited a linear elastic stiffness followed by gradual plastification of subassemblage elements.

4. Connection 14-3 exhibited two linear elastic slopes prior to the start of local connection yielding. The secondary elastic slope was due to minor slips of the bolted flange plates into bearing.

5. The deviation of the actual initial elastic slope from that predicted can be accounted for by the out-of-plane movement of the column web under the action of beam flange forces.

6. The out-of-plane movement of the column web on Connection 14-2 and the resultant reduction in connection stiffness is quite significant and could be an important factor when such connections are used in design.

7. Column stiffening must be examined, if it is desired to limit the out-of-plane deformation of the column web, especially on the type of connection where the beam is attached only to the column web as in 14-2.

1022   MAY 1980   ST5

8. Although bolted connections such as 14-3 exhibit very good strength, the reduction of stiffness in the working load range due to bolt slippage must be considered when such connections are used.

9. The failure of 14-1, 14-2, 14-3 was by fracture of connection material with the fractures in 14-1 and 14-3 occurring after the connection was loaded well into the plastic range. The fractures of all three connections can be related to high-stress concentrations.

10. Use of column stiffening as a means of reducing stress concentrations is a definite possibility and should be examined, especially in connections similar to 14-2.

Acknowledgments

The investigation reported herein was conducted in the Fritz Engineering Laboratory of Lehigh University, Bethlehem, Pa. Lynn S. Beedle is Director of the Laboratory. The study of steel beam-to-column web connections was sponsored jointly by the American Iron and Steel Institute and the Welding Research Council (WRC). Research was carried out under the technical guidance of the WRC Task Group on Beam-to-Column Connections, of which John A. Gilligan is Chairman. The interest, encouragement and guidance of this committee is gratefully acknowledged.

Appendix.—References

1. *American Institute of Steel Construction, Manual of Steel Construction,* 7th Ed., American Institute of Steel Construction, New York, N.Y., 1970.
2. Chen, W. F., and Rentschler, G. P., "Tests and Analysis of Beam-to-Column Web Connections," *Methods of Structural Analysis,* Vol. II, ASCE, 1976, pp. 957–976.
3. "Code for Welding in Building Construction," *A WS D1.0-69,* 9th ed., American Welding Society, Miami, Fla., 1969.
4. Graham, J. D., Sherbourne, A. N., Khabbaz, R. N., and Jensen, C. C., "Welded Interior Beam-to-Column Connections," *AIA File No. 13-C,* American Institute of Steel Construction, New York, N.Y., 1959.
5. Huang, J. S., Chen, W. F., and Beedle, L. S., "Behavior and Design of Steel Beam-to-Column Moment Connections," *Welding Research Council Bulletin No. 188,* New York, N.Y., Oct., 1973.
6. Parfitt, J., Jr., and Chen, W. F., "Tests of Welded Steel Beam-to-Column Moment Connections," *Journal of the Structural Division,* ASCE, Vol. 102, No. ST1, Proc. Paper 11854, Jan., 1976, pp. 189–202.
7. Peters, J. W., and Driscoll, G. C., Jr., "A Study of the Behavior of Beam-to-Column Connections," *Fritz Engineering Laboratory Report No. 333.2,* Lehigh University, Bethlehem, Pa., 1968.
8. Popov, E. P., and Pinkney, R. B., "Cyclic Yield Reversal in Steel Building Connections," *Engineering Journal,* American Institute of Steel Construction, Vol. 8, No. 3, July, 1971.
9. Rentschler, G. P., and Chen, W. F., "Test Program of Moment-Resistant Steel Beam-to-Column Web Connections," *Fritz Engineering Laboratory Report No. 405.4,* Lehigh University, Bethlehem, Pa., May, 1975.
10. Standig, K. F., Rentschler, G. P., and Chen, W. F., "Tests of Bolted Beam-to-Column Flange Moment Connections," *Bulletin No. 218,* Welding Research Council, New York, N.Y., Aug., 1976.

5.4.3

SEMIRIGID STEEL BEAM-TO-COLUMN CONNECTIONS: DATA BASE AND MODELING

By Wai-Fah Chen, [1] Member, ASCE, and N. Kishi[2]

ABSTRACT: The development of a data base on semi-rigid steel beam-to-column connections at Purdue University using collected experimental test data is described. To build-up this data base systematically, the Steel Connection Data Bank Program (SCDB) is developed. Using the system of tabulation and plotting in the SCDB program, the moment-rotation characteristics of each connection type are made available, and the appropriate connection models are also given in the program.

INTRODUCTION

In designing a steel framework, it is customary to represent the actual connection behavior between beam and column by one of two kinds of models: rigid moment connection or flexible pinned connection. While it is recognized that actual beam-to-column connections always provide some rigidity, it is very difficult to evaluate the actual restraint provided by these semi-rigid connections. At the University of Illinois in 1917, Young performed the first experiment to estimate the rigidity of steel beam-to-column connections. Since then, experimental testing has been continued.

The newly published AISC design code, referred to as the Load and Resistance Factor Design (LRFD) Specification (1986), designates two types of constructions in its provisions: Type FR (fully restrained) and Type PR (partially restrained). If Type PR construction is used, the effect of connection flexibility on the behavior and strength of the frame must be taken into account in the analysis and design procedures.

The primary distortion of a steel beam-to-column connection is its rotational deformation, θ_r, caused by the in-plane bending moment, M (Fig. 1). The effect of this connection deformation has a destablizing effect on frame stability since additional drift will occur as a result of the decrease in effective stiffness of the members to which the connections are attached. An increased frame drift will intensify the P-Δ effect and hence the overall stability of the frame will be affected.

Thus, the nonlinear characteristics of beam-to-column connections play a very important role in steel structural design. To this end, the connection behavior must be adequately estimated or predicted as well as the specified methods for design analysis established. In recent years, several researchers have published papers discussing the influences of connection rigidity on steel frame structures for all connection types. These have contributed much to the LRFD specification. For example, Goverdhan (1983) collected much of the available test data on moment-rotation characteristics and tried to for-

[1]Prof. and Head of Struct. Engrg., School of Civ. Engrg., Purdue Univ., West Lafayette, IN 47907.

[2]Assoc. Prof., Civ. Engrg., Muroran Inst. of Tech., Muroran, Japan 050.

Note. Discussion open until June 1, 1989. To extend the closing date one month, a written request must be filed with the ASCE Manager of Journals. The manuscript for this paper was submitted for review and possible publication on September 16, 1987. This paper is part of the *Journal of Structural Engineering*, Vol. 115, No. 1, January, 1989. ©ASCE, ISSN 0733-9445/89/0001-0105/$1.00 + $.15 per page. Paper No. 23104.

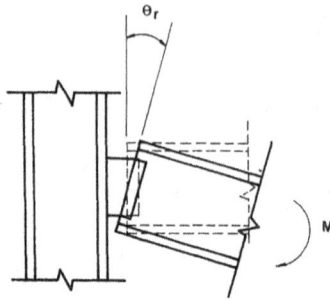

FIG. 1.   Rotational Deformation of a Connection

mulate a prediction equation for each curve. Nethercot (1985) conducted a literature survey for the period 1915–1985. In his paper, he reviewed steel beam-to-column connection test data and the corresponding curve representations.

The aim of this study is to provide moment-rotation characteristics and corresponding parameters of semirigid beam-to-column connections used frequently in steel construction. Based on the collected experimental test data, a data bank on steel beam-to-column connections was developed at the Purdue University Computer Center. To control this data base systematically, the Steel Connection Data Bank Program (SCDB) has been developed. The SCDB program provides seven main functions, including routines for tabulating and plotting the experimental test data and some proposed prediction equations. Three prediction equations, among many others, are chosen and installed in the program.

MODELING OF CONNECTION M-θ_r BEHAVIOR

Several analytical models have been developed to represent connection flexibility. These models are generally either a sophisticated numerical simulation or an approximation based on test data. Early models used the initial stiffness of the connection as the key parameter in a linear M-θ_r model. Although the linear model is very easy to use, it has a serious disadvantage. It is not suitable for a full range of rotation. A closer approximation of true connection behavior can be obtained by using either bilinear and piecewise linear models.

Jones et al. (1980, 1981) proposed a cubic-B-spline model to obtain a more suitable function. However, this model requires a large number of sampling data during the formulation process. Richard et al. (1980) and Ang and Morris (1984) proposed power models for some special connections independently. Chen and Lui (1985) and Lui and Chen (1986) formulated an exponential model which was used in a second-order structural analysis. Chen-Lui's model gives a good representation for an experimentally obtained monotonic M-θ_r curve, similar to that predicted by a cubic-B-spline model. However, the exponential model may not completely represent the M-θ_r curve with a few linear components. Thus, each model has some disadvantages and limitations.

In the data bank program (SCDB) installed at the Purdue University Computer Center, three prediction equations are included. These will be described briefly.

Polynominal Model

A popular model for structural analysis is the polynominal function proposed by Frye and Morris (1975). The Frye-Morris model was developed based on a procedure by Sommer (1969). They used the method of least squares to determine the constants of the polynomial. The primary disadvantage of this model is that the first derivative of this function which indicates connection stiffness, may be discontinuous and/or possibly negative, which is physically impossible. It has the form

$$\theta_r = C_1 \cdot (KM) + C_2 \cdot (KM)^3 + C_3 \cdot (KM)^5 \quad \dots \dots \dots \dots \dots \dots \dots \dots \dots \quad (1)$$

where C_i = curve-fitting constants; and K = the standardization factor (a dimensionless factor whose value depends on the size parameters for the particular connection considered)

Modified Exponential Model

The Chen-Lui exponential model has been refined to accommodate linear components by Kishi and Chen (1986) and is referred to herein as the modified exponential model.

This model is represented by a function of the form

$$M = M_0 + \sum_{j=1}^{m} C_j \left[1 - \exp\left(-\frac{|\theta_r|}{2j\alpha} \right) \right]$$

$$+ \sum_{k=1}^{n} D_k (|\theta_r| - |\theta_k|) H[|\theta_r| - |\theta_k|] \dots \dots \dots \dots \dots \dots \dots \dots \dots \quad (2)$$

where M_0 = initial connection moment; α = scaling factor; C_j, D_k = curve-fitting parameters; θ_k = starting rotation of kth linear component given from experimental M-θ_r curve; and $H[\ ]$ = Heaviside's step function.

$$H[\theta] = 1 \quad \text{for} \quad \theta \geq 0 \dots \dots \dots \dots \dots \dots \dots \dots \dots \dots \dots \quad (3)$$

$$H[\theta] = 0 \quad \text{for} \quad \theta < 0 \dots \dots \dots \dots \dots \dots \dots \dots \dots \dots \dots \quad (4)$$

Using the linear interpolation technique for original M-θ_r data, the weight function for each M-θ_r datum is nearly equal. The constants C_j and D_k for the exponential and linear terms of the function are determined by a matrix. Individual terms of the matrix are obtained by the method of least squares for a given set of moment-rotation data similar to that of the Chen-Lui exponential model.

The instantaneous connection stiffness R_k at an arbitrary rotation $|\theta_r|$ can be evaluated by differentiating Eq. 2 with respect to $|\theta_r|$.

When the connection is loaded, we have

$$R_k = R_{kt} = \frac{dM}{d|\theta_r|}\bigg|_{|\theta_r| = |\theta_r|} = \sum_{j=1}^{m} \frac{C_j}{2j\alpha} \exp\left(-\frac{|\theta_r|}{2j\alpha} \right) + \sum_{k=1}^{n} D_k H[|\theta_r| - |\theta_k|] \quad \dots \dots \dots \quad (5)$$

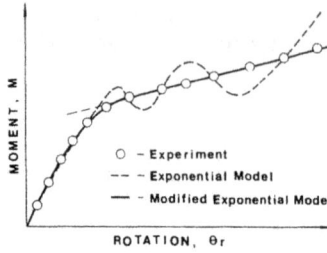

FIG. 2. Comparison between Results by Exponential and Modified Exponential Models for M-θ_r Data Including a Linear Term

When the load is removed, we have

$$R_k = R_{ki} = \left. \frac{dM}{d|\theta_r|} \right|_{|\theta_r|=0} = \sum_{j=1}^{m} \frac{C_j}{2j\alpha} + D_k H[-|\theta_k|]]|_{k=1} \quad\dots\dots\dots\dots\dots\dots\dots (6)$$

This model has the following merits:

    1. The formulation is relatively simple and straightforward.
    2. It can deal with connection loading and unloading for the full range of rotation in a second-order structural analysis.
    3. The abrupt changing of the connection stiffness among the sampling data is only generated from inherent experimental characteristics.

    The curve-fitting and connection stiffness values from the experimental data for each type of connection examined are calculated with $m = 6$ in Eqs. 2 and 5 in the SCDBprogram. The comparison between the Chen-Lui exponential model and the modified exponential model for numerical example test data including a linear component is shown in Fig. 2.

Power Model with Three Parameters
    The modified exponential model mentioned earlier is a curve-fitting equation obtained by using the least mean square technique for the test data. From a different viewpoint, Kishi and Chen (1987) and Kishi and Chen et al. (1987) developed another procedure to predict the moment-rotation characteristics of steel beam-to-column connections. In this procedure, the initial elastic stiffness and ultimate moment capacity of the connection are determined by a simple analytical model. Using those values so obtained, a three-parameter power model given by Richard and Abbott (1975), was adopted to represent the moment-rotation relationship of the connection.
    The generalized form of this model is

$$M = \frac{R_{ki}\ \theta_r}{\left\{ 1 + \left(\dfrac{\theta_r}{\theta_0} \right)^n \right\}^{1/n}} \quad\dots\dots\dots\dots\dots\dots\dots\dots\dots\dots\dots\dots\dots\dots\dots\dots\dots\dots (7)$$

where R_{ki} = initial connection stiffness; M_u = ultimate moment capacity; θ_0 = a reference plastic rotation = M_u/R_{ki}; and n = shape parameter.

TABLE 1. Connection Types

Type number (1)	Connection type (2)
1	Single web-angle connections
2	Double web-angle connections
3	Top- and seat-angle connections with double web-angle
4	Top- and seat-angle connections
5	Extended end-plate connections
6	Flush end-plate connections
7	Header plate connections

GENERAL REMARKS REGARDING COLLECTED DATA

The literature survey encompassed experimental data, published from 1936 (Rathbun) to the present, on riveted, bolted, and welded connections. All of the 303 tests collected so far are classified into seven types, as shown in Table 1. The references and number of experimental curves of each connection type are listed in Table 2.

Single Web-Angle Connections/Single Plate Connections

Single web-angle connections consist of an angle either bolted or welded to both the column and the beam web, as shown in Fig. 3. On the other hand, single plate connections use the plate instead of the angle. This connection type requires less material than a single web-angle connection. Generally, in designing these connections the single web-angle connections have moment rigidity equal to about one-half of the double web-angle connections, and the single plate connection have rigidity equal to or greater than the single web-angle connections.

Double Web-Angle Connections

Double web-angle connections consist of two angles either bolted or riveted to both the column and the beam web as shown in Fig. 4. The earliest tests on double web-angle connections conducted by Rathbun (1936) used rivets as fasteners. In the 1950s, most specifications for the design of steel structures allowed the use of high strength bolts in lieu of rivets. To clarify the effect of high strength bolts on connection behavior when used in conjunction with rivets, Bell et al. (1958) conducted experiments on riveted and bolted beam-to-column connections. Lewitt et al. (1966) tested the same kind of connection to supplement the data obtained from the previous research. Today, double web-angle connections with high strength bolts are used popularly. The connection rigidity of this type of connection is stiffer than those of the single web-angle and single plate connections.

Top- and Seat-Angle Connections with Double Web-Angle

A typical top- and seat-angle connection with double web-angle is shown in Fig. 5. Top- and seat-angle connections with or without double web-angles are semi-rigid connections and belong to Type PR construction of the AISC-LRFD specification. Double web-angles are used to improve the connection restraint characteristics of top- and seat-angle connections, and for

TABLE 2.   Author and Number of Experimental M-θ, Curves for Each Connection Type

References for experimental curves (1)	Number of tests (2)
(a) Single web-angle connections, single plate connections	
S. L. Lipson (1968)	30
L. E. Thompson et al. (1970)	12
S. L. Lipson (1977)	8
R. M. Richard et al. (1982)	4
(b) Double web-angle connections	
J. C. Rathbun (1936)	7
W. C. Bell et al. (1958)	4
C. W. Lewitt et al. (1966)	6
W. H. Sommer (1969)	4
L. E. Thompson et al. (1970)	48
B. Bose (1981)	1
(c) Top- and seat-angle connections with double web-angle	
J. C. Rathbun (1936)	2
A. Azizinamini et al. (1985)	20
(d) Top- and seat-angle connections	
J. C. Rathbun (1936)	3
R. A. Hechtman et al. (1947)	12
S. M. Maxwell et al. (1981)	12
M. J. Marley (1982)	26
(e) Extended end-plate connections	
L. G. Johnson et al. (1960)	1
A. N. Sherbourne (1961)	5
J. R. Bailey (1970)	26
J. O. Surtees et al. (1970)	6
J. A. Packer et al. (1977)	3
S. A. Ioannides (1978)	6
R. J. Dews (1979)	3
P. Grundy et al. (1980)	2
N. D. Johnstone et al. (1981)	8
(f) Flush end-plate connections	
J. R. Ostrander (1970)	24
(g) Header plate connections	
W. H. Sommer (1969)	20

shear transfer. In order to evaluate the inherent ductility offered by the flexural deformation capacities of both the flange and the web angles in the legs attached to the column under static and earthquake loading, experimental studies on this connection type were performed by Altman et al. (1982) and Azizinamini et al. (1985) at the University of South Carolina. The tests program included specimen subject to cyclic loading as well as static loading. Cyclic loading test data are excluded from the present data base. The pre-

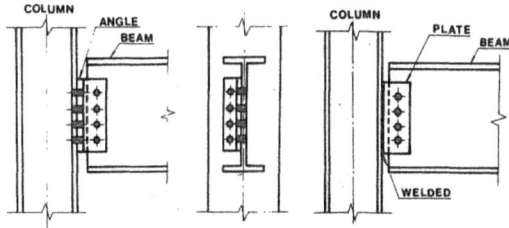

FIG. 3. Typical Single Web-Angle and Single Plate Connections: (a) Single Web-Angle Connection; (b) Single Plate Connection

diction equations for this type of connection based on the empirical model derived by Frye and Morris (1975), are developed by Altman et al. (1982) using these connections M-θ_r data.

Top- and Seat-Angle Connections

A typical top- and seat-angle connection is shown in Fig. 6. The former AISC specification (1978) describes the top- and seat-angle connection as follows: (1) The seat angle is influenced to transfer only vertical reaction and should not give significant restraining moment on the end of the beam; and (2) the top angle is merely for laterally stability and is not considered to carry any gravity loads. However, according to the experimental results, these connections will transfer some end moment from the beam. They are included as Type PR construction of the AISC-LRFD specification. The experiments conducted by Rathbun (1936) and Hectman et al. (1947) on this

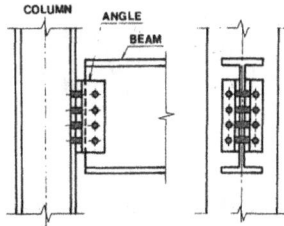

FIG. 4. Typical Double Web-Angle Connection

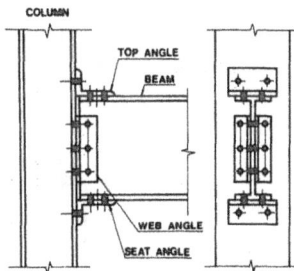

FIG. 5. Typical Top- and Seat-Angle Connection with Double Web-Angle

111

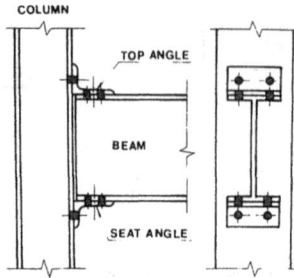

FIG. 6. Typical Top- and Seat-Angle Connection

type of connections used rivets for the fasteners. Today, most connections are fastened by high strength bolts.

Extended End-Plate Connections/Flush End-Plate Connections

In general, end plate connections are welded to the beam end along both the flanges and web in the fabricator's shop and bolted to the column in the field. The end plate connection has been used extensively since the late 1960s. The extended end-plate connections are classified into two types as extended

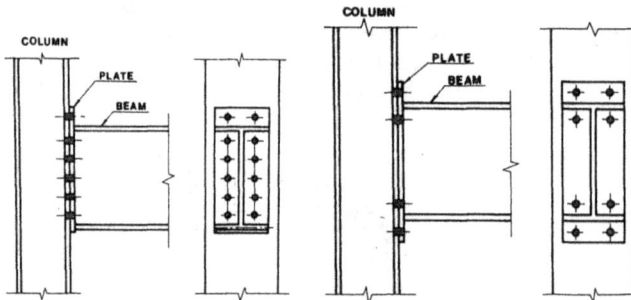

FIG. 7. Typical Extended End-Plate Connection: (a) Extended on Tension Side Only; (b) Extended on Tension and Compression Sides

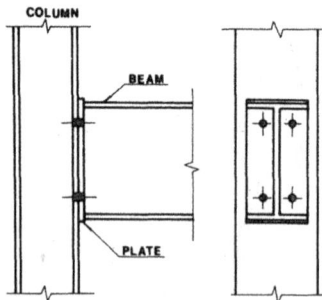

FIG. 8. Typical Flush End-Plate Connection

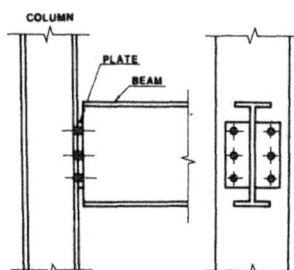

FIG. 9. Typical Header Plate Connections

end-plate either on the tension side only or on both the tension and compression sides as shown in Fig. 7. A typical flush end-plate connection is shown in Fig. 8. Since some end-plate connections are considered as Type FR rather than Type PR connections, they have often been used as means of transferring beam end moment to the column. The extended end-plate connection on the tension side only is commonly used. The extended end-plate on both sides is preferred when the connection is subjected to moment reversal, as during severe earthquake loading. While the flush end-plate connection is weaker than the extended end-plate connection, this connection type is often used in roof details. The behavior of end-plate connections depends on whether the column flange near the connection is stiffened or not. The stiffeners of the column flanges act to prevent flexural deformation of the column flange, thereby influencing the behavior of the plate and fasteners.

Header Plate Connections

A header plate connection consists of an end plate, whose length is less than the depth of the beam, welded to the beam web and bolted to the column, as shown in Fig. 9. The moment-rotation characteristics of these connections are similar to those of double web-angle connections. Accordingly, a header plate connection is used mainly to transfer the reaction of the beam

TABLE 3. Fastening Mode Patterns

Pattern number, N (1)	Fastening mode pattern (2)
1	All riveted
2	All bolted
3	Riveted-to-beam and bolted-to-column
4	Bolted-to-beam and riveted-to-column
5	Riveted-to-beam and welded-to-column
6	Bolted-to-beam and welded-to-column
7	Welded-to-beam and riveted-to-column
8	Welded-to-beam and bolted-to-column
9	All riveted without column stiffeners
10	All riveted with column stiffener
11	All bolted without column stiffener
12	All bolted with column stiffener
20	All modes included

II – 10

```
CONNECTION TYPE : DOUBLE WEB-ANGLE CONNECTIONS
            MODE : RIVETED-TO-BEAM AND BOLTED-TO-COLUMN

TESTED BY  :  W.C.BELL ET AL (1958)          U.S.A.
TEST ID.   :  FK-4C

COLUMN : W12X65                FASTENERS: A325- -3/4"D
BEAM   : W18X50                          13/16" OVERSIZE HOLES
ANGLE  : 6 X 4 X 3/8           MATERIAL : G40.21
                                     FY = 38.10 KSI
                                     FU = 64.50 KSI

                    MAJOR PARAMETERS

       LP = 11.5000"   LU = 3.2500"   LL = 3.2500"   TA = 0.3750"
       GB = 4.5000"    GC = 2.5730"   CU = 1.2500"   CL = 1.2500"
       PB = 3.0000"    PC = 3.0000"   GB = 3.0000"
       NB = 2 X 4      NC = 1 X 4

REMARK  1)  HIGH STRENGTH BOLTS USED
        2)  BOLTS : FY=87.9 KSI, FU=136.6 KSI,  RIVETS : FY=36.5 KSI, FU=63.2KSI
```

NO	MOMENT (K-IN)	ROTATION (RADIANS) X 1/1000		NO	MOMENT (K-IN)	ROTATION (RADIANS) X 1/1000
1	0.0	0.00		26	410.3	44.97
2	26.8	0.57		27	420.4	48.46
3	42.2	1.01		28	430.4	51.94
4	57.5	1.45		29	453.3	56.09
5	71.9	1.89		30	476.1	60.25
6	86.2	2.33		31	481.7	63.56
7	112.9	3.59		32	487.4	66.86
8	127.7	3.93		33	510.8	68.62
9	142.6	4.27		34	525.0	72.70
10	158.5	4.61		35	539.1	76.78
11	174.4	4.95		36	553.3	80.86
12	188.8	5.24		37	567.4	84.94
13	203.2	5.53				
14	227.8	6.79				
15	255.2	9.06				
16	275.6	10.91				
17	290.5	13.98				
18	305.3	17.05				
19	320.1	20.11				
20	328.0	24.03				
21	336.0	27.94				
22	343.9	31.85				
23	351.9	35.77				
24	376.1	38.62				
25	400.3	41.48				

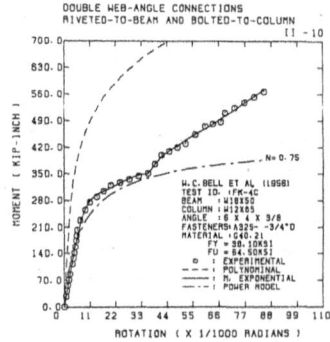

DOUBLE WEB-ANGLE CONNECTIONS
RIVETED-TO-BEAM AND BOLTED-TO-COLUMN
II – 10

```
          H.C.BELL ET AL (1958)
          TEST ID.  : FK-4C
          COLUMN    : W12X65
          BEAM      : W18X50
          ANGLE     : 6 X 4 X 3/8
          FASTENERS: A325- -3/4"D
          MATERIAL : G40.21
                 FY = 38.10KSI
                 FU = 64.50KSI
          o     : EXPERIMENTAL
          --- : POLYNOMIAL
          ─── : M. EXPONENTIAL
          ─·─ : POWER MODEL
```

```
MOMENT-ROTATION PREDICTION EQUATIONS     ( R : X 1/1000 RADIANS )

* FRYE AND MORRIS POLYNOMIAL MODEL :     R = SUM ( AI X ( K*BM )**PI X 10**QI )
      XD = 11.500000"   G = 5.146000"   T = 0.375000"
      A1 = 3.660000    A2 = 1.150000   A3 = 4.570000    K = 0.021481
      P1 =        1    P2 =        3   P3 =        5    Q1 =     -1   Q2 =    -3   Q3 =   -5

MODIFIED EXPONENTIAL MODEL :
      BM = SUM ( AI X ( 1 - EXP( - R/( 2*I*C ) ) ) )  +    SUM ( RKJ X ( R - RJO ) )  +  BMO
      C  =  0.85040917E+00          BMO=  0.00000000E+00     NEXP= 6     NLINER= 2
      A1 =  0.43582680E+03  -0.71415568E+04   0.33242692E+05  -0.66136250E+05   0.60354957E+05  -0.20432854E+05
      RJO =        35.7700         41.4800
      RKJ =  0.76301928E+01  -0.29302073E-01

POWER MODEL :     BM = ( RKI X R ) / ( 1 + ( R/RO )**RN )**( 1/RN )
      RN = 0.750     RKI =  0.83485722E+02          RMU =  0.45369160E+03
```

NO	ROTATION (RADIANS) X 1/1000		MOMENT (KIP-INCH)				CONNECTION STIFFNESS (KIP-INCH) X 1000		
		EXPRI.	POLY.	M.EXPO.	P.MODEL.		POLY.	M.EXPO.	P.MODEL.
1	0.00	0.0	0.0	0.0	0.0		0.1272E+03	0.4551E+02	0.8349E+02
3	1.01	42.2	124.9	39.4	60.5		0.1156E+03	0.3518E+02	0.4666E+02
5	1.89	71.9	214.0	69.5	95.8		0.8605E+02	0.3397E+02	0.3494E+02
7	3.59	112.9	320.1	127.5	144.1		0.4476E+02	0.3350E+02	0.2313E+02
9	4.27	142.6	347.6	149.8	158.8		0.3670E+02	0.3218E+02	0.2026E+02
11	4.95	174.4	370.6	171.1	171.8		0.3099E+02	0.3016E+02	0.1793E+02
13	5.53	203.2	387.4	187.9	181.6		0.2738E+02	0.2804E+02	0.1631E+02
15	9.06	255.2	460.5	260.5	226.9		0.1610E+02	0.1339E+02	0.1015E+02
17	13.98	290.5	523.7	296.2	266.0		0.1043E+02	0.3505E+01	0.6278E+01
19	20.11	320.1	577.2	313.5	296.8		0.7396E+01	0.2958E+01	0.4023E+01
21	27.94	336.0	626.7	337.6	322.1		0.5479E+01	0.2738E+01	0.2612E+01
23	35.77	351.9	665.0	351.8	339.3		0.4398E+01	0.8224E+00	0.1857E+01
25	41.48	400.3	688.6	396.6	348.9		0.3861E+01	0.4411E+01	0.1504E+01
27	48.46	420.4	713.7	424.8	358.2		0.3374E+01	0.3788E+01	0.1200E+01
29	56.09	453.3	737.9	453.2	366.5		0.2974E+01	0.3740E+01	0.9666E+00
31	63.56	481.7	758.9	481.9	373.0		0.2673E+01	0.3952E+01	0.8014E+00
33	68.62	510.8	772.0	502.3	376.9		0.2503E+01	0.4120E+01	0.7134E+00
35	76.78	539.1	791.5	536.9	382.2		0.2276E+01	0.4350E+01	0.6006E+00
37	84.94	567.4	809.2	573.1	386.7		0.2090E+01	0.4504E+01	0.5138E+00

FIG. 10. Connection Information Out of Data Base Obtained by Using SCDB Program (Example 1)

to the column. Almost all moment-rotation tests on header plate connections were performed by Sommer at the University of Toronto in 1969. He tested 20 specimens (see Table 2). On the basis of the results of these tests, a standardized moment-rotation curve has been established which was developed as polynomial model by Frye and Morris (1975).

STEEL CONNECTION DATA BANK PROGRAM (SCDB)

The experimental connection test data collected in the previous section can be used to assess analytically the performance of framed structures with

```
********************************************
*                                          *
*   THIS FILE ( BANK01 ) IS DATA BANK       *
*          FOR SINGLE WEB CLEAT CONNECTIONS  *
*                                          *
********************************************

    MOMENT-ROTATION PREDICTION EQUATIONS     ( R : X 1/1000 RADIANS )
        ***** US UNIT VERSION ( INCH,KIP-INCH ) *****

    FRYE AND MORRIS POLYNOMINAL MODEL :     R = SUM ( AI X ( K*BM )**PI X 10**QI )

    MODIFIED EXPONENTIAL MODEL :
        BM = SUM ( AI X ( 1 - EXP( - R/( 2*I*C ) ) ) ) +       SUM ( RKJ X ( R - RJO ) ) + BMO

    POWER MODEL :     BM = ( RKI X R ) / ( 1 + ( R/RO )**RN )**( 1/RN )


    I - 1
    CONNECTION TYPE : SINGLE WEB-ANGLE CONNECTIONS
             MODE : ALL BOLTED
         TESTED BY : S.L.LIPSON (1968)                  CANADA
         TEST ID. : AA-2/1

    FRYE AND MORRIS POLYNOMINAL MODEL :
         XD = 5.500000"    G = 2.762500"    T = 0.250000"
         A1 = 4.280000    A2 = 1.450000    A3 = 1.510000    K = 0.239361
         P1 =      1    P2 =      3    P3 =      5    Q1 =      0    Q2 =      -6    Q3 =      -13
    MODIFIED EXPONENTIAL MODEL :
         C =   0.75435250E+00              BMO=   0.00000000E+00        NEXP= 6        NLINER= 2
         AI =   0.29149748E+01  -0.29963327E+02   0.73967132E+01   0.37852068E+03  -0.81221008E+03   0.47015517E+03
         RJO =          13.9800                     32.3200
         RKJ =   0.92425535E-02  -0.35940356E-01
    POWER MODEL :
         RN = 0.750        RKI =   0.11095362E+01        RMU =   0.28735590E+02        RO =   0.25898741E+02


    I - 2
    CONNECTION TYPE : SINGLE WEB-ANGLE CONNECTIONS
             MODE : ALL BOLTED
         TESTED BY : S.L.LIPSON (1968)                  CANADA
         TEST ID. : AA-2/2

    FRYE AND MORRIS POLYNOMINAL MODEL :
         XD = 5.500000"    G = 2.762500"    T = 0.250000"
         A1 = 4.280000    A2 = 1.450000    A3 = 1.510000    K = 0.239361
         P1 =      1    P2 =      3    P3 =      5    Q1 =      0    Q2 =      -6    Q3 =      -13
    MODIFIED EXPONENTIAL MODEL :
         C =   0.42240750E+00              BMO=   0.00000000E+00        NEXP= 6        NLINER= 1
         AI =  -0.25526444E+02   0.37619327E+03  -0.18708497E+04   0.41350032E+04  -0.42052390E+04   0.16063543E+04
         RJO =          11.5000
         RKJ =   0.92198431E-01
    POWER MODEL :
         RN = 1.000        RKI =   0.11095362E+01        RMU =   0.28735590E+02        RO =   0.25898741E+02


    I - 3
    CONNECTION TYPE : SINGLE WEB-ANGLE CONNECTIONS
             MODE : ALL BOLTED
```

FIG. 11. File Content Composing from Values of Three Prediction Equations Obtained by Using SCDB Program (Example 2)

semirigid connections. These test data are stored in the computer as the steel beam-to-column connection data base at the Purdue University Computer Center. The items from each test that are registered in the data base are:

1. Connection type and fastening mode.
2. Author, test I.D., and tested country.
3. The material properties and size of fasteners.
4. The material strength of the angles designed for connection elements.
5. All parameters used in beam-to-column connections.
6. Moment-rotation test data.

Numerical values registered in the data base are standardized in U.S. customary units by executing a subprogram as the preprocessor. The connection type is classified as shown in Table 1 and fastening mode as shown in Table 3.

The data base can be controlled by the Steel Connection Data Bank program (SCDB). This program, at the Purdue University Computer Center, consists of the following main routines;

1. Transformation of the outputted unit system from original U.S. customary units to the MKS system.
2. Set and print the selected test data.
3. Set and print general tables of test data concerning connection type and mode.

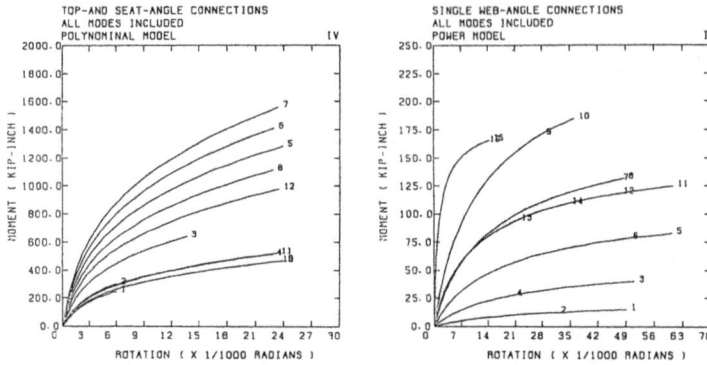

FIG. 12. Comparison Figures among Results from Prediction Equation for Experimental Test Data with Same Connection Type and Fastening Mode, Obtained by SCDB Program (Example 3)

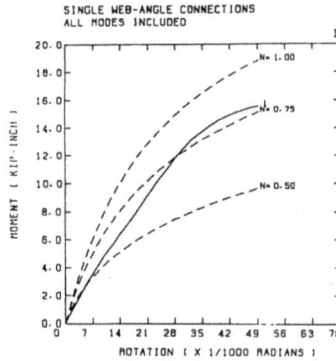

FIG. 13. Comparison Figures to Determine Appropriate Value *n* of Power Model, Obtained by SCDB Program (Example 4)

4. Determine and print the values of three prediction curves for selected test data.

5. Plot experimental curve together with its prediction curves for selected test data.

6. Set a file to be able to transfer it to a second order frame analysis program the values of three prediction equations for given test data.

7. Make a figure composing of the curves obtained by using a selected prediction equation for test data with common fastening mode.

Depending on the users demand, it is possible to execute the SCDB program together with the above-mentioned routines. Adding users special routines to the SCDB program, the prediction equations with/without curve-fitting techniques can be developed. This program enables the user to make comparative studies of the role of different joint parameters on M-θ, behavior.

Figs. 10–13 are sample output obtained by executing the SCDB program.

Summary and Conclusions

The limit states in the LRFD specification of AISC include two conditions:

1. The ultimate limit state concerning structural safety.
2. The serviceability limit state restricting the intended use and occupancy of the structure. This state includes deflection, vibration and permanent deformation.

The LRFD specification considers a second-order nonlinear structural analysis including the effect of secondary moments in frames due to axial loads. In the second-order analysis, the flexible connection (Type PR construction) plays a very important role.

To properly apply the LRFD specification, it is necessary to develop a practical means for modeling the moment-rotation relationship of a semirigid connection (Type PR construction). Also, we must provide the means for designers to analyze complicated steel framed structures quickly and accurately. As a first step, available moment-rotation experimental test data of steel beam-to-column connections have been collected. The data are now stored in the computer. Three prediction equations to evaluate the moment-rotation characteristics of the beam-to-column connections are provided. These are the Frye-Morris polynomial model, the Kishi-Chen modified exponential model based on a curve-fitting technique, and a three-parameter power model with the values derived by a simple analytical procedure (Kishi and Chen et al. 1987). Available test data on steel beam-to-column connections and comparison between experimental and theoretical results calculated using the three prediction equations are available by executing the SCDB program. The Kishi-Chen SCDB program (1986), consisting of multi-functions to operate the data base, can be used easily and effectively for the second-order nonlinear analysis of a frame with semirigid connections.

Appendix. References

Altman, W. G., Jr., et al. (1982). "Moment-rotation characteristics of semi-rigid steel beam-to-column connection." Civ. Engrg. Dept., Univ. of South Carolina, Columbia, S.C.

Ang, K. M., and Morris, G. A. (1984). "Analysis of three-dimensional frame with flexible beam-column connections." Can. J. Civ. Engrg., 11, 245–254.

Azizinamini, A., Bradburn, J. H., and Radziminski, J. B. (1985). "Static and cyclic behavior of semi-rigid steel beam-column connections." Dept. of Civ. Engrg., Univ. of South Carolina, Columbia, S.C.

Bailey, J. R. (1970). "Strength and rigidity of bolted beam-to-column connections," Conf. on Joints in Structures, Univ. of Sheffield, Sheffield, England, Vol.1, Paper 4.

Bell, W. G., Chesson, E., Jr., and Munse, W. H. (1958). "Static tests of standard riveted and bolted beam-to-column connections." Univ. of Illinois Engrg. Experiment Station, Urbana, Ill.

Bose, B. (1981). "Moment-rotation characteristic of semi-rigid joints in steel structures." J. Inst. Engrs. (India), Part CI, Civ. Engrg., Div., 62(2), 128–132.

Chen, W. F., and Lui, E. M. (1985), "Column with end restraint and bending in load and resistance factor design." AISC Journal, Third Quarter, 105–132.

Dews, R. J. (1979). "Experimental test results on experimental end-plate moment connections." Thesis presented to Vanderbilt University, at Nashville, Tenn., in partial fulfillment of the requirements for the degree of Master of Science.

Frye, M. J., and Morris, G. A. (1975). "Analysis of flexibly connected steel frames." *Can. J. Civ. Engrg.*, 2, 280–291.

Goverdhan, A. V. (1983). "A collection of experimental moment-rotation curves and evaluation of prediction equations for semi-rigid connections." Thesis presented to Vanderbilt University, at Nashville, Tenn., in partial fulfillment of the requirements for the degree of Master of Science.

Grundy, P., Thomas, I. R., and Bennetts, I. D. (1980). "Beam-to-column moment connections." *J. Struct. Div.*, ASCE, 106(ST1), 313–330.

Hechtman, R. A., and Johnston, B. G. (1947). "Riveted semi-rigid beam-to-column building connections, progress report no. 1," AISC Research at Lehigh Univ., Bethlehem, Pa.

Ioannides, S. A. (1978). "Flange behavior in bolted end-plate moment connections." Thesis presented to Vanderbilt University, at Nashville, Tenn., in partial fulfillment of the requirements for the degree of Doctor of Philosophy.

Johnson, L. G., and Cannon, J. C., and Spooner, L. A. (1960). "High tensile pre-loaded bolted joints for development of full plastic moments." *British Welding J.*, 7, 560–569.

Johnstone, N. D., and Walpole, W. R. (1981). "Bolted end-plate beam to column connections under earthquake type loading." *Research Report 81-7*, Dept. of Civ. Engrg., Univ. of Canterbury, Christchurch, New-Zealand.

Jones, S. W., Kirby, P. A., and Nethercot, D. A. (1980). "Effect of semi-rigid connections on steel column strength." *J. Const. Steel Res.* 1(1), Sept., England, 38–46.

Jones, S. W., Kirby, P. A., and Nethercot, D. A. (1981). "Modeling of semi-rigid connection behaviour and its influence on steel column behaviour." *Joints in Structural Steelwork*. J. H. Howlett, W. M. Jenkins, and R. Stainsby, eds., Pentech Press, United Kingdom, 5.73–5.78.

Kishi, N., and Chen, W. F. (1986). "Steel connection data bank program." CE-STR-86-18, School of Civ. Engrg., Purdue Univ., W. Lafayette, Ind.

Kishi, N., and Chen, W. F. (1986). "Data base of steel beam-to-column connections." CE-STR-86-26, School of Civ. Engrg., Purdue Univ., W. Lafayette, Ind.

Kishi, N., and Chen, W. F. (1987). "Moment-rotation relation of top- and seat-angle connections." CE-STR-87-4, School of Civ. Engrg., Purdue Univ., W. Lafayette, Ind.

Kishi, N., et al. (1987). "Moment-rotation relation of top- and seat-angle with double web-angle connections." CE-STR-87-16, School of Civ. Engrg., Purdue Univ., W. Lafayette, Ind. (*Proc. of the State-of-the-Art Workshop on Connections and the Behavior, Strength and Design of Steel Structures*, R. Bjorhovde, J. Brozzetti, and A. Colson, eds., Ecole Normale Superieure de Cachan, France, May 25–27, 1987).

Kishi, N., et al. (1987). "Moment-rotation relation of single/double web-angle connections." CE-STR-87-18, School of Civ. Engrg., Purdue Univ., W. Lafayette, Ind. (*Proc. of the State-of-the-Art Workshop on Connections and the Behavior, Strength and Design of Steel Structures*, R. Bjorhovde, J. Brozzetti, and A. Colson, eds., Ecole Normale Superieure de Cachan, France, May 25–27, 1987).

Lewitt, C. W., Chesson, E., Jr., and Munse, W. H. (1966). "Restraint characteristics of flexible riveted and bolted beam-to-column connections." Dept. of Civ. Engrg., Univ. of Illinois, Urbana, Ill.

Lipson, S. L. (1968). "Single-angle and single-plate beam framing connections." *Proc., Canadian Structural Engineering Conf.*, Toronto, Ontario, 141–162.

Lipson, S. L. (1977). "Single angle welded-bolted connections," *J. Struct. Div.*, ASCE, 103(ST3), 559–571.

Load and Resistance Factor Design (LRFD) Specification for Structural Steel Buildings. (1986). AISC, Chicago, Ill.

Lui, E. M., and Chen, W. F. (1986). "Analysis and behavior of flexibly jointed frames." *Engrg. Struct.*, Butterworth, U.K, 8(2), 107–118.

Marley, M. J., and Gerstle, K. H. (1982), "Analysis and tests of flexibly-connected steel frames." Report to AISC under Project 199, AISC, Chicago, Ill.

Maxwell, S. M., et al. (1981). "A realistic approach to the performance and application of semi-rigid joints in steel structures." *Joints in Structural Steelwork,* J. H. Howlett, W. M. Jenkins, and R. Stainsby, eds., Pentech Press, 2.71–2.98.

Nethercot, D. A. (1985). *Steel beam-to-column connections—A review of test data.,* CIRIA, London.

Ostrander, J. R. (1970). "An experimental investigation of end-plate connections." Thesis presented to the University of Saskatchewan, at Saskatoon, Saskatchewan, in partial fulfillment of the requirements for the degree of Master of Science.

Packer, J. A., and Morris, L. J. (1977). "A limit state design method for the tension region of bolted beam-column connections." *The Struct. Eng.,* 55(10), 446–458.

Rathbun, J. C. (1936). "Elastic properties of riveted connections." *Trans. ASCE,* Paper No. 1933, Vol. 101, 524–563.

Richard, R. M., and Abbott, B. J. (1975). "Versatile elastic-plastic stress-strain formula." *J. Engrg. Mech. Div.,* ASCE, 101(EM4), 511–515.

Richard, R. M., et al. (1980). "The analysis and design of single plate framing connections," *AISC Engrg. J.,* 2d Quarter, 38–52.

Richard, R. M., Kriegh, J. D., and Hormby, D. E. (1982). "Design of single plate framing connections with A307 bolts." *AISC Engrg. J.,* 4th Quarter, 209–213.

Sherbourne, A. N. (1961). "Bolted beam-to-column connections," *The Struct. Engr.,* 39, Jun., 203–210.

Sommer, W. H. (1969), Behavior of welded-header-plate connections." Thesis presented to University of Toronto, at Toronto, Canada, in partial fulfillment of the requirements for the degree of Master of Applied Science.

Specification for the design, fabrication and erection of structural steel for buildings. (1978). AISC, Chicago, Ill.

Surtees, J. O., and Mann, A. P. (1970). "End plate connection in plastically designed structures." Conf. on Joints in Structures, 1(5), Univ. of Sheffield, Sheffield, England.

Thompson, L. E., McKee, R. J., and Visintainer, D. A. (1970). "An investigation of rotation characteristic of web shear framed connections using A-36 and A-441 steels." Dept. of Civ. Engrg., Univ. of Missouri-Rolla, Rolla, Mo.

Young, C. R. (1917). *Bulletin No.104, Engineering Experiment Station,* Univ. of Illinois, Urbana, Ill.

5.5 Offshore Structures (近海結構)

5.5.1

12809   MARCH 1977   ST3

JOURNAL OF THE STRUCTURAL DIVISION

TESTS OF FABRICATED TUBULAR COLUMNS[a]

By Wai F. Chen,[1] M. ASCE and David A. Ross[2]

INTRODUCTION

A relatively unstudied development in structural engineering is the use of fabricated tubular steel beams and columns. This trend is growing, particularly in the design of multistory and offshore oil structures.

Designers of such tubular columns face an immediate problem in the lack of a reliable design guide since such columns are usually fabricated in diameters far greater than those for which previous research data are available. This lack of knowledge on the *strength* of these members, suitably based on experimental evidence, hampers the designer in his efforts to design a safe but relatively economic structural member. There is also a more fundamental problem with such structural members arising due to the lack of knowledge of the *behavior* of members fabricated by relatively new fabrication processes. Among problems associated with prediction of member behavior are the effects of two-dimensional residual stresses in members introduced during fabrication and the unknown importance of initial imperfections in fabrication.

This has motivated the present study of fabricated tubular columns subjected to concentric axial load. A research program currently underway at Lehigh University and Purdue University has both theoretical and experimental phases, both of which attempt to provide design assistance to such members. This paper reports on the experimental phase of the investigation. Included in the investigation was the measurements of residual stresses in a typical fabricated cylindrical column, the testing of three stub columns, and the testing of 10 full-scale pin-ended long columns under axial load, with slenderness ratios ranging from 39 to 83 and diameter-to-wall thickness ratios of 48 and 70.

Note.—Discussion open until August 1, 1977. To extend the closing date one month, a written request must be filed with the Editor of Technical Publications, ASCE. This paper is part of the copyrighted Journal of the Structural Division, Proceedings of the American Society of Civil Engineers, Vol. 103, No. ST3, March, 1977. Manuscript was submitted for review for possible publication on April 27, 1976.

[a] Presented at the April 5–8, 1976, ASCE National Water Resources & Ocean Engineering Convention, held at San Diego, Calif., (Preprint 2660).

[1] Prof. of Struct. Engrg., School of Civ. Engrg., Purdue Univ., W. Lafayette, Ind., formerly Lehigh Univ., Bethlehem, Pa.

[2] Research Asst., Fritz Engrg. Lab., Lehigh Univ., Bethlehem, Pa.

SCOPE OF TEST PROGRAM

It is appropriate here to describe briefly the process by which fabricated tubular structural members are commonly made. Usually, the tubular member is formed by several cycles of repeated cold-rolling of a flat plate until opposite edges come together. A cylinder or "can" is then formed by welding down this longitudinal seam. Manufacturing limitations usually limit the length of these

TABLE 1.—List of Specimens

Specimen number (1)	Nominal length, in meters (feet) (2)	Nominal L/r ratio (3)	Effective length factor, K (4)	Outside diameter, D_o, in millimeters (inches) (5)	Diameter to thickness ratio, D_o/t (6)	Central heat lot[a] (7)
1	5.5 (18)	42	0.89–0.95	380 (15)	48	I, II[b]
2	5.5 (18)	42	0.95	380 (15)	48	I, II[b]
3	7.6 (25)	60	0.88–0.92	380 (15)	48	II
4	7.6 (25)	60	0.96	380 (15)	48	II
5	7.6 (25)	39	0.60–0.68	560 (22)	70	II[c]
6	7.6 (25)	39	0.72–0.76	560 (22)	70	II[c]
7	11 (36)	83	0.78–1.0	380 (15)	48	II
8	11 (36)	83	0.61–0.69	380 (15)	48	I, II[b]
9	11 (36)	58	0.75–0.86	560 (22)	70	II[c]
10	11 (36)	58	0.64–0.83	560 (22)	70	II[c]

[a] The yield stress of heat lot II was higher than that for heat lot I.
[b] Circumferential weld near center, different heat lots on each side.
[c] All pipe heat lot II.

cans to about 3 m (10 ft), but any number of cans may be welded together end-to-end to form the desired member. A possibility of longitudinal weld tearing in a completed member when loaded is avoided by staggering the welds between "cans," usually making the weld in one can about 180° out-of-phase to the weld in the next can. [American Petroleum Institute (API) Specifications (12) require at least 90° out-of-phase.]

This forming process clearly introduces significant circumferential residual stresses that vary through the thickness of the plate while the longitudinal welding process introduces significant longitudinal residual stresses. Particular attention of this research has focused on the magnitudes and distributions of these stresses which are a necessary prelude to any analytical investigation of the effects of these stresses on beam-column behavior under load. The measurement was undertaken on a short column of a size similar to that used in the three stub column tests.

The stub column tests were undertaken in order to plot the column buckling strength curve based on the tangent-modulus theory and included the effect of residual stresses. The long columns varied in length from 5.5 m–11 m (18 ft–36 ft) and in diameter from 380 mm–560 mm (15 in.–22 in.). An important feature of these tests was the use of spherical end bearing blocks during column

TABLE 2.—Material Properties[a]

Origin (1)	Yield values σ_y and Young's Modulus, E (2)	Heat lot I (3)	Heat lot II (4)
Mill Report	Dynamic σ_y	318 (46.1)	328 (47.5)
	Static σ_y	—	—
	E	—	—
Lehigh Laboratory	Dynamic σ_y	288 (41.7)	321 (46.5)
	Static σ_y	271 (39.3)	308 (44.6)
Test[b]	E	211,000 (30,600)	212,000 (30,700)
CB&I Laboratory	Dynamic σ_y	293 (42.5)	324 (47.0)
	Static σ_y	271 (39.3)	308 (44.6)
Test[c]	E	214,000 (31,000)	213,000 (30,800)

[a] Values are given in Meganewtons per square meter (kips per square inch).
[b] Maximum strain rate = 0.64 mm/min (0.025 in./min).
[c] Maximum strain rate = 1.28 mm/min (0.05 in./min).
Specimens taken from plate before rolling.

testing. Apart from simulating "column effective length" as closely as possible, this also allowed the column to buckle in its preferred direction.

Table 1 gives the list and dimensions of specimens supplied for testing. The specimens were fabricated in accordance with the requirements of American Petroleum Institute Specifications (12), with welding procedures conforming to American Welding Society (13) requirements. The sections used to form the columns were from American Society for Testing and Materials A36 steel plate in which the original rolling direction was perpendicular to the longitudinal axis of the finished columns. Two heat lots of steel were included in the specimens and the properties of these, as found in various tensile tests, are recorded in Table 2. The wall thickness of all specimens was 7.8 mm (5/16 in.).

622   MARCH 1977   ST3

SUPPLEMENTARY TESTS

The material properties, as determined from tensile testing, are listed in Table 2. The stub column test gives a stress-strain curve showing the effect of residual stresses. The proportional limit, the static yield stress level, and the elastic and the elastic-plastic moduli are the important data furnished by the curve. Data from the stub column test are necessary for the prediction of column buckling strength based on tangent modulus theory.

The stub column specimens were tested in the 5,000,000-lb Baldwin Testing Machine in the Fritz Engineering Laboratory of Lehigh University using the technique reported and recommended in Ref. 14. Details of these tests are summarized in Ref. 8.

The method of "sectioning" was used to obtain the experimental or measured values of longitudinal strains and, consequently, longitudinal residual stresses. A series of 10-in. gage holes were laid out on the specimen and measured with 1/10,000-in. accuracy Whittemore strain gage. The difference in length before and after the sectioning is a measure of residual stress (15). A "hole drilling technique," described in Ref. 7, was used to measure the variation of circumferential residual stresses through the wall thickness. A critical review of the "hole drilling technique" along with other methods is given in Ref. 14. A brief analysis of the results obtained by these testing techniques is presented.

RESIDUAL STRESSES

Longitudinal residual stresses introduced by longitudinal welding of the "cans" were measured by the method of "sectioning" (15). The residual stress distribution thus obtained is shown in Fig. 1. This distribution represents an average through the wall thickness. The solid curve shows a possible approximation of the test points by a curve of the type predicted by Marshall (5) and the dotted lines are a straight-line approximation suggested as a simplified alternative. Note that near the weld the material has effectively yielded in tension.

If x is the distance from the weld, R, the tubular member radius, σ_L, the longitudinal residual stress at a point, and σ_y, the material yield stress, then the following values may be adopted as end points in this straight line approximation:

$$\frac{\sigma_L}{\sigma_y} = 1.0 \quad \text{at} \quad \frac{x}{R} = 0; \quad \frac{\sigma_L}{\sigma_y} = 0 \quad \text{at} \quad \frac{x}{R} = 0.15;$$

$$\frac{\sigma_L}{\sigma_y} = -0.30 \quad \text{at} \quad \frac{x}{R} \approx 0.3; \quad \frac{\sigma_L}{\sigma_y} = 0 \quad \text{at} \quad \frac{x}{R} = 1.0;$$

$$\frac{\sigma_L}{\sigma_y} = 0.1 \quad \text{at} \quad \frac{x}{R} \approx 1.2; \quad \frac{\sigma_L}{\sigma_y} = 0 \quad \text{at} \quad \frac{x}{R} \geq 2.0 \quad \dots\dots\dots\dots\dots (1)$$

Tensile stress is assumed to be positive, and the resultant axial force must be zero to maintain equilibrium. However, in Eq. 1, no attempt has been made to balance bending moment about an axis perpendicular to the weld, as the out-of-balance moment was found to be negligible. It is particularly noteworthy that, except near the weld, longitudinal residual stresses as approximated by Eq. 1 go through zero points at distances from the weld equal to integer multiples of the column radius. Similar results were also observed elsewhere (6). Ref. 6 suggests that the straight line approximation may be adequate for column radii up to a maximum of about 380 mm (15 in.), i.e., for radii in excess of this value, R should be taken as 380 mm (15 in.). This is reasonable when it is considered that a finite amount of heat is added to a "can" in the longitudinal welding process. There are indications that there may be a dependence of the

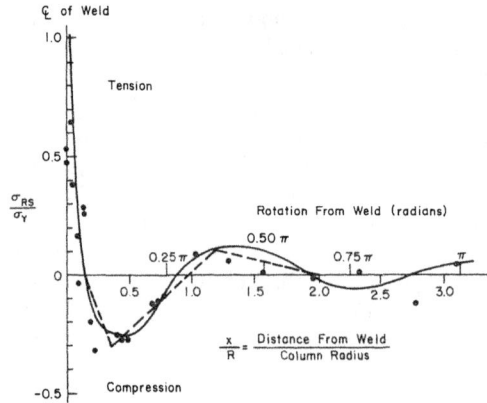

FIG. 1.—Longitudinal Residual Stress Distribution Obtained from Method of Sectioning

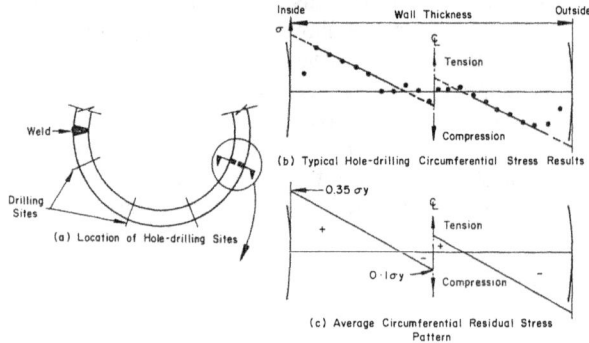

FIG. 2.—Circumferential Residual Stress Pattern

σ_L/σ_y ratio on the yield strength of the material and the welding procedure used.

Circumferential residual stresses were measured by using a hole-drilling technique (7), in which surface strain measurements were taken of the strain released due to drilling at the base of a small diameter hole in the tubular column wall. For a given location, the experiment was conducted both from inside and outside surfaces of the tubular column [Fig. 2(a)]. No significant variation in the distribution of circumferential residual stresses through the thickness of the wall was found at different locations on the same cross section [Fig. 2(a)]. Fig. 2(b) shows a typical experimental result. The hole-drilling technique has a limited range of validity such that the results near the surface, as well as those taken near the center line of the wall, may contain possible inaccuracies. Thus, the straight line approximation is dotted in these areas. Fig. 2(c) shows the average circumferential residual stress pattern obtained.

If the forming of a flat plate into a cylindrical shape by several cycles of repeated cold-rolling can be idealized as the process of pure bending of a beam, a crude estimate of the pattern of circumferential residual stresses distribution

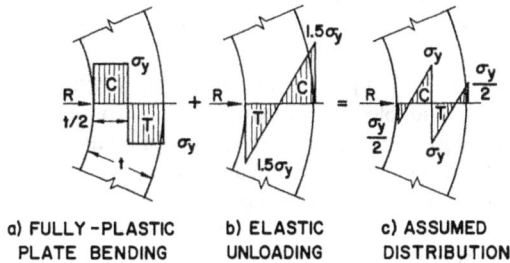

a) FULLY-PLASTIC   b) ELASTIC   c) ASSUMED
   PLATE BENDING   UNLOADING   DISTRIBUTION

FIG. 3.—Estimated Pattern of Circumferential Residual Stress Distribution

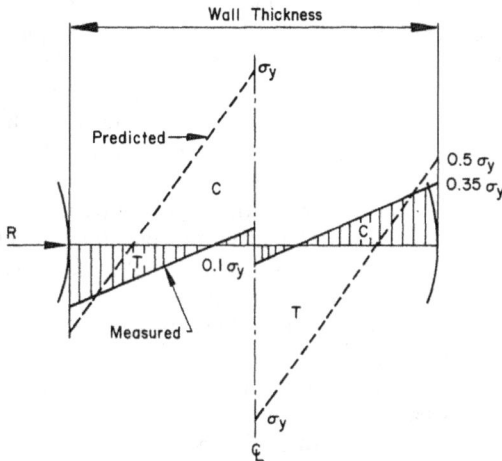

FIG. 4.—Comparison of Estimated and Measured Circumferential Residual Stresses

is then possible. This is shown in Fig. 3. In the cold-rolling process, fully plastic yielding of the plate must occur [Fig. 3(a)]. Before welding, the "can" is allowed to "spring back" [Fig. 3(b)] and it results in the circumferential residual stress distribution shown in Fig. 3(c). Fig. 4 shows that this estimate differs markedly in magnitude from the measured distribution [taken from Fig. 2(c)], but it does show the correct pattern of distribution. Repeated cold-rolling, incomplete spring back and subsequent cooling of the welds, and residual stresses existing in the flat plate before forming can all explain these discrepancies.

LONG COLUMN TESTING

General.—In the 10 full-scale long column tests, the maximum nominal length, L, of the column was fixed by the height of the Baldwin 5,000,000-lb testing machine in Lehigh's Fritz Engineering Laboratory [about 12 m (40 ft)] and the minimum column diameter (and thus the radius of gyration, r) was controlled by the forming equipment of the specimen manufacturers.

A special feature of these tests was the use of spherical bearing heads at each end of the specimen during testing. Unlike a column of I or H section in which the buckling direction is well defined, it was impossible to predict the buckling direction of a fabricated tubular column. The use of spherical bearing heads thus allowed the column to buckle in its preferred, or weakest, direction.

A true pin-ended condition is practically impossible to attain in practice because the normal spherical end blocks will always have some frictional resistance to end rotation. Strain gages therefore were mounted on each specimen at quarter points along the specimen length and near each end. By plotting column curvatures as measured, a good approximation to the true column effective length of each specimen could be found. In the 10 tests conducted, the effective column length factor K ranged from 0.6 to 1.0 (Table 1).

Initial Imperfections.—Measurements of the out-of-roundness of a fabricated specimen were made. It was found that, in general, there was less than 1% difference between two perpendicular diameters at all positions along the column length. This measurement therefore was not made on subsequent specimens.

The American Petroleum Institute has specifications (12) for allowable fabrication imperfections for out-of-straightness. The specifications allow 3 mm (1/8 in.) in 3 m (10 ft) (or approximately one part in 1,000) with the restriction that the out-of-straightness not exceed 9 mm (3/8 in.) in 12 m (40 ft) (or 7.5 parts in 10,000). The test specimens met these tolerances. However, since specimen out-of-straightness could be a critical parameter in determining column performance, in particular by influencing the buckling direction, extensive measurements were made of the initial out-of-straightness. This was done with the specimen in an upright position using a theodolite.

Clearly, there is a problem in establishing diametrical planes on which to take these out-of-straightness measurements. An attempt was made to find an axis of maximum out-of-straightness by rolling the specimen on a flat surface. The longitudinal welds hampered this process.

A typical out-of-straightness pattern is shown in Fig. 5 for an 11-m long and 0.38-m diam specimen. The distribution of heat lots along the length is shown in Fig. 5, which also shows a diagram exploded along line A to show the relative weld positions. Each weld is fixed at between 25 mm and 50 mm (1 in. and 2 in.) from either line A or line C as indicated. Table 3 gives the magnitude of the out-of-straightness, and also the form of the out-of-straightness pattern. In general, the API specified tolerances for out-of-straightness have not been exceeded. It appears that the out-of-straightness on a diametrical plane nearly parallel to the weld locations is greater than that on the perpendicular diametrical plane.

Experimental Procedure.—Lateral deflections at quarter points along the length of each specimen and rotations of the spherical end bearing blocks were measured.

626 MARCH 1977 ST3

Since longitudinal lines had been established on each specimen for out-of-straightness measurements, these lines were also used to establish points on

TABLE 3.—Maximum Column Out-of-Straightness

| Specimen number (1) | Plane A-C (Fig. 5) | | Plane B-D (Fig. 5) | |
	Millimeters (inches) (2)	Form of curvature (3)	Millimeters (inches) (4)	Form of curvature (5)
1	2 (0.08)	Single	8 (0.31)	Single
2	4 (0.16)	Local imperfections	5 (0.20)	Single
3	5.6 (0.22)	Single	8.4 (0.33)	Single
4	6 (0.24)	Single	2 (0.08)	Single
5	3.7 (0.15)	Single	2.6 (0.10)	Double
6	2 (0.08)	Triple	2.6 (0.10)	Triple
7	4.4 (0.17)	Triple	2 (0.08)	Local imperfections
8	5.5 (0.22)	Double	4 (0.16)	Single
9	5.4 (0.21)	Triple	4.5 (0.18)	Double
10	4.2 (0.17)	Triple	2 (0.08)	Single

TABLE 4.—Initial End Eccentricities (Center of Pipe Relative to End Block)

| Specimen number (1) | Top Head | | Bottom Head | |
	δ_A, in millimeters (2)	δ_B, in millimeters (3)	δ_A, in millimeters (4)	δ_B, in millimeters (5)
1	—	—	—	—
2	—	—	0.7	0.7
3	—	—	−1.4	−1.4
4	−2.1	0	—	—
5	5.6	−5.6	0	0
6	1.0	5.7	−1.1	−1.1
7	−10.8	4.0	−3.4	3.4
8	−6.1	−5.0	0	0
9	−0.5	−1.7	4.0	10.8
10	−6.9	−6.9	4.0	−0.7

the circumference for measurement of axial strain and lateral displacement. Rotations in two perpendicular directions of the bottom bearing block could

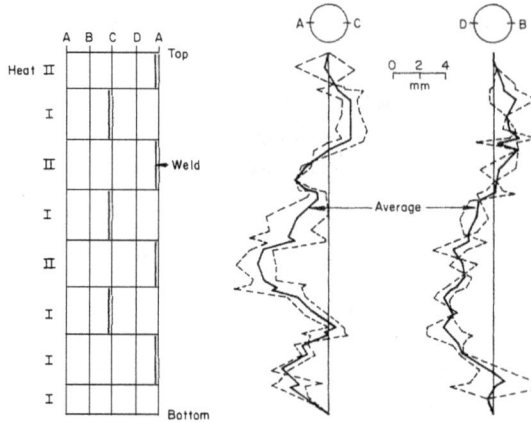

FIG. 5.—Typical Specimen Out-of-Straightness as Measured Prior to Testing

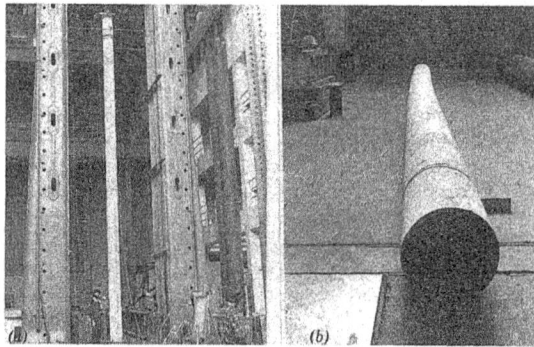

FIG. 6.—Column Testing: (*a*) Prior to Testing (11-m × 0.38-m Diam Specimen); (*b*) Buckled Column After Testing

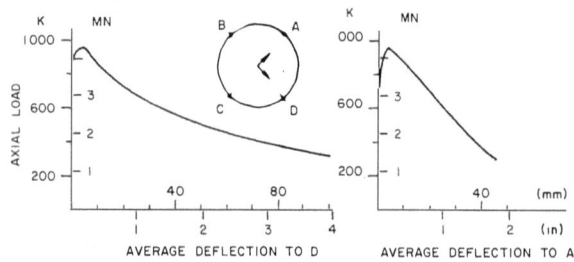

FIG. 7.—Axial Load-Lateral Deflection Curves at Midheight

readily be measured manually with a dial gage and level apparatus. However, this same procedure was difficult for the head rotation measurements at an elevation of up to 12 m (40 ft). A plumb-bob type rotation gage was then used in which the curvature of a sheet metal plumb-bob support was measured with electric-resistance strain gages. The lateral deflections at quarter points were measured by potentiometers (four at each level) and also at midheight by dial gages (also four). For stub column tests, the end alinement was ensured by a process of trial-and-error loads until equal straining was observed at

TABLE 5.—Failure Mechanism and Buckling Direction

Specimen number (1)	Failure mode (2)	Location critical section x/L^a (3)	Direction of buckling with respect to weld (4)	Remarks (5)
1	Plastic instability	0.48	30° to weld	Yield zones in lower yield strength can
2	Plastic instability	0.48	30° to weld	Yield zones in lower yield strength can
3	Plastic instability	0.57	Perpendicular to weld	Yield zones in lower yield strength can
4	Plastic instability	0.62	Parallel to weld	Initial buckling in lower strength can
5	Plastic instability then local buckling	0.38	Perpendicular to weld	Local buckling near but out of central can
6	Local buckling	0.82	60°–70° to weld	Local buckling near but out of central can
7	Plastic instability	0.41	Perpendicular to weld	Yield zones mainly in lower yield strength can
8	Plastic instability	0.64	45° to weld	Yield zones in lower yield strength can
9	Local buckling	0.27	40° to weld	Local buckling in lower end can
10	Local buckling	0.56	Perpendicular to weld	Local buckling in upper can

^{a}Distance x measured from base of specimen.

circumferential points on a section. For the long column tests, the best possible alinement was obtained and the remaining end eccentricity noted. Table 4 gives these measured eccentricities. The end eccentricity is essential in the theoretical analysis in which the column is treated as a biaxially eccentrically loaded member.

The axial load was applied in increments and the static readings of column behavior recorded. Fig. 6(a) shows a typical specimen prior to testing, while Fig. 6(b) shows the same specimen after testing. Maximum lateral deflections measured were of the order of 20 cm (8 in.) but considerable elastic straightening of the specimen was noted as the applied load was released after the test.

Test Results and Observations.—Typical axial load-lateral deflection curves are shown in Fig. 7 for one of the specimens. Each of these curves was plotted as an average deflection of two opposite sides of the specimen. For most specimens, some lateral movement was noted at approx 70%–80% of the recorded maximum load. Furthermore, as is characteristic of column behavior, there was a marked reduction of load carrying capacity beyond maximum load.

Table 5 contains a brief description of the mode of failure of each column. The location of the critical section characterized by either local buckling or plastic instability with an extensive yielded zone occurred frequently at some distance away from the center of the specimen. This was caused by the combined

TABLE 6.—Midheight Deflections and End Rotations of Columns at End of Tests

Specimen number (1)	Deflection, δ_m, in millimeters (inches) (2)	δ_m/L × 10^2 (3)	Top end rotation, θ_T/bottom end rotation, θ_B (4)	Predicted θ_T/θ_B (5)
1	53 (2.07)	0.96	—	—
2	57 (2.24)	1.04	0.70	1.0
3	84 (3.31)	1.10	1.15	1.33
4	96 (3.79)	1.26	1.01	1.63
5	79 (3.10)	1.03	0.67	0.61
6	49 (1.92)[a]	—	4.42	4.56
7	184 (7.26)	1.68	0.57	0.70
8	140 (5.50)	1.27	1.53	1.78
9	106 (4.16)[a]	—	0.32	0.37
10	121 (4.77)	1.10	1.74	1.27

[a] Maximum deflection closer to a quarter-point.

effects of various orientations of longitudinal welds, the lower yield strength of material in some "cans," and the magnitude and orientation of out-of-straightness existing in different "cans" of the same specimen. Note that specimens 6 and 9 failed at exceptionally large distances from the midheight of the specimen.

Table 5 indicates that four specimens are buckled in a diametrical plane that is perpendicular to the diametrical plane containing the longitudinal weld seams. Specimen 4 is the only specimen that buckled in the plane parallel to the plane containing the welds.

The two diameter-to-thickness ratios ($D_o/t = 48$ and 70) used in the specimens encompass the transition from the local buckling type of failure, recognized by the checkerboard pattern of cross-sectional distortion, to the plastic instability type of failure characterized by an extensive yielding of a "can" and relatively little cross-sectional distortion. From Table 5, it appears that a (D_o/t) ratio of about 60 is the critical ratio for which the failure mode of a long column changes from local buckling to the plastic instability type of failure. Columns with $D_o/t = 70$ were found to retain their load-carrying capacity higher than that predicted by tangent modulus theory using the results of stub column tests, even in the presence of pronounced local buckling at failure. This will be discussed further in the following section.

The critical "can" of specimen 10 was not the central can. Examination of the out-of-straightness patterns suggests that this may have been the major influencing factor on the behavior of this column. The critically yielded "can" of specimens 6 and 9 was located very near the ends, despite the fact that the entire specimens were made from steel of the same heat lot. In both cases, the bulge of local buckling occurred only a few inches from a circumferential weld. Examination of the out-of-straightness patterns again shows that this may have been the major factor. In these it was also observed that there was a larger than average "misfit" between the longitudinal axes of consecutive cans at the critical circumferential welds. Factors such as out-of-straightness and "misfit" are crucial in determining the position of critical "can" along the column length, especially for columns with large D_o/t ratios (i.e., $D_o/t > 70$).

Two other aspects of post-buckling behavior are illustrated by the data in Table 6. The maximum lateral deflections measured at the midheight of a specimen were recorded numerically. These deflections are of the order of approx 1.0%–1.7% of the column length. At these deflections, the remaining axial load-carrying capacity of the member is in the order of 40% of the buckling load.

In Table 6, the rotation of the top head, θ_T, is recorded as a fraction of the rotation of the bottom head, θ_B. A simple prediction of this ratio is also made by assuming that the buckled column has all plastic rotation concentrated at the critical location, i.e., that the deflected shape of a buckled column is essentially bilinear. Comparison of observed and predicted values shows good agreement, except for specimens 4 and 10, for which a secondary critical location could be found.

Tables 1 and 5 show that the larger diameter column specimens all failed in an interactive local and overall instability buckling mode and were the only specimens to do so. The failure was initially in the form of plastic instability, followed immediately by local buckling. Large diameter columns cannot sustain finite deformations without developing local cross-sectional distortion. For all the columns tested, however, the column buckling strength is not significantly affected by its diameter-to-thickness ratio. There was a sudden loss of axial load-carrying capacity in specimens with $D_o/t = 70$ which failed by local buckling. In contrast, the loss of load-carrying capacity for specimens with $D_o/t = 48$ was less sudden and a significant post-buckling strength (plastic deformation capacity) was usually observed. The transition from general yielding to local buckling type of failure was observed at D_o/t ratios between 50 and 70 for all the l/r ratios tested.

Comparison with Column Strength Curves.—Fig. 8 defines both "static" axial load P and "dynamic" load P_d, as obtained during testing. The "static" load is essentially the maximum load corresponding to zero strain rate, whereas the "dynamic" load is taken as the maximum load the specimens sustained during loading. The dynamic-to-static load ratios varied within the range P_d/P = 1.02 to 1.07. A summary of the maximum measured loads for the 10 columns and three stub columns tested is given in Table 7.

Fig. 9 shows theoretical strength curves together with the 10 long column

TABLE 7.—Maximum Column Loads

Speci-men (1)	Dynamic			Static		
	P_{ult}[a] (2)	P_y[a] (3)	P_{ult}/P_y (4)	P_{ult}[a] (5)	P_y[a] (6)	P_{ult}/P_y (7)
1	2,581 (580)	2,674 (601)	0.965	2,476 (556)	2,523 (567)	0.981
2	2,648 (595)	2,674 (601)	0.990	2,492 (560)	2,523 (567)	0.988
3	2,403 (540)	2,674 (601)	0.899	2,270 (510)	2,523 (567)	0.900
4	2,403 (540)	2,674 (601)	0.899	2,296 (516)	2,523 (567)	0.910
5	4,370 (982)	4,406 (990)	0.992	4,263 (958)	4,228 (950)	1.01
6	4,361 (980)	4,406 (990)	0.990	4,112 (924)	4,228 (950)	0.973
7	2,270 (510)	2,674 (601)	0.849	2,212 (497)	2,523 (567)	0.877
8	2,465 (554)	2,674 (601)	0.921	2,367 (532)	2,523 (567)	0.938
9	4,272 (960)	4,406 (990)	0.970	4,183 (940)	4,228 (950)	0.99
10	4,228 (950)	4,406 (990)	0.960	4,094 (920)	4,228 (950)	0.968
Stub column						
1	3,596 (808)	2,986 (671)	1.21	3,444 (774)	2,861 (643)	1.20
2	3,373 (758)	2,674 (601)	1.26	3,222 (724)	2,523 (567)	1.28
3	4,583 (1,030)	4,406 (990)	1.04	4,450 (1,000)	4,228 (950)	1.05

[a] Values are given in kilonewtons (kips).

test results. Data from the three stub column tests are used for the derivation of these curves based on tangent modulus theory. Static yield stress values based on tensile coupon tests are used in the calculation of the plotted values. The "barbell" plotted for each test reflects the uncertainty in effective lengths of specimens at buckling. It can be seen that most of the test points lie above the predicted value.

In Fig. 10(a) the basic Column Research Council (CRC) curve is shown together with the results of the 10 long column tests. This column curve is the basis of allowable stresses for columns given by the 1969 Specification of the American Institute of Steel Construction (11). Fig. 10(a) is plotted on a "static" loading basis and Fig. 10(b) is on a "dynamic" loading basis. Note that the CRC column curve was developed mainly on the basis of tests for hot-rolled wide-flange steel shapes of small and medium sizes. The comparison shows that the fabricated tubular members also exhibited a strength close to that implied by the CRC column curve for the case of short columns, but that they had somewhat greater

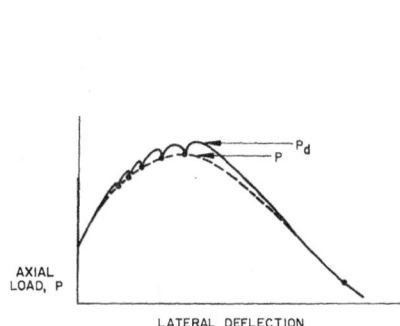

FIG. 8.—Definition of Dynamic and Static Loads

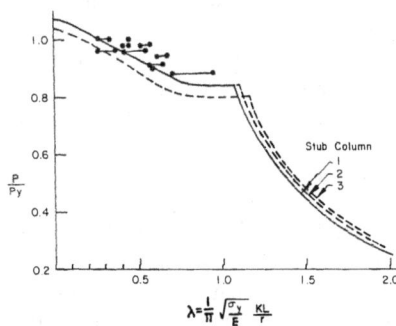

FIG. 9.—Comparison of Test Results with Column Strength Curves Derived from Stub Column Tests Based on Tangent Modulus Theory

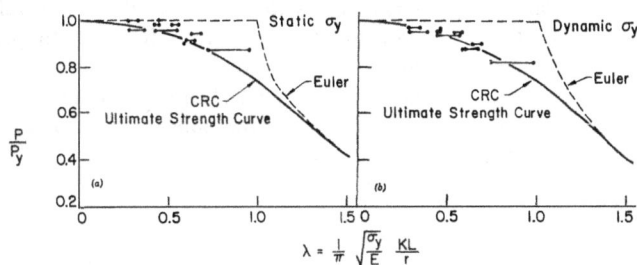

FIG. 10.—Comparison of Test Results with CRC Column Strength Curve Based on: (a) Static Yield Stresses; (b) Dynamic Yield Stresses

strength than that predicted by this curve for the case of intermediate length of columns [Fig. 10(a)].

Conclusions

As described in this paper, the following problems were investigated experimentally—the magnitude and distribution of longitudinal and circumferential residual stresses in fabricated steel tubular columns, the stub column tests, and the strength and behavior of 10 full-scale fabricated cylindrical columns of medium slenderness ratios with diameter-to-thickness ratios of 48 and 70.

Some of the following conclusions are of immediate significance to the designer, and others may be of importance for future research:

1. The longitudinal residual stress distribution has the general shape shown in Fig. 1, and may be represented by straight line approximation (Eq. 1).

2. The variation of circumferential residual stress distribution for different locations along the circumference of a cross section is not appreciable except near the weld. A typical circumferential residual stress distribution through the thickness of the wall is shown in Fig. 2(c).

3. The combined effect of longitudinal and circumferential residual stresses on the fabricated "can" is considerable, as indicated by a recent theoretical study of moment-curvature-thrust relations for such "cans." Details of this study are reported elsewhere (10).

4. The theoretical ultimate load analysis based on the tangent modulus theory of an initially straight column underestimated the strength of fabricated tubular members (Fig. 9).

5. Except for the shortest columns, these fabricated members exhibited a strength higher than that implied by the CRC column curve by amounts varying from 8% to 16% [Fig. 10(a)].

6. It appears that the transition from general plastic yielding to a local buckling type of failure occurs at a D_o/t ratio of about 60 for all Kl/r ratios tested. The maximum column strength does not appear to be affected by the failure mode.

7. The results of this study have indicated that future theoretical work must consider the following factors: (a) The combined effect of longitudinal and circumferential residual stresses on the behavior and strength of fabricated tubular members must be considered—these effects can be reflected in the development of moment-curvature-thrust relationships; and (b) the effects of staggering the longitudinal welds between "cans" as well as the transverse welds connecting two adjacent "cans" on the behavior and strength of fabricated tubular members must be considered. This can be reflected in the consideration of the out-of-straightness along the member.

Acknowledgments

The investigation reported herein was conducted in Fritz Engineering Laboratory, Lehigh University, Bethlehem, Pa. Dr. L. S. Beedle is Director of the Laboratory. The experimental phase of the research was supported by the American Petroleum Institute through the Column Research Council. The theoretical phase of the work is currently supported by a grant from the National Science Foundation under NSF Grant No. ENG 75-10171 to Lehigh University. The interest, encouragement and guidance of the Advisory Committee, of which Mr. L. A. Boston was Chairman, is gratefully acknowledged.

Appendix I.—References

1. Chen, W. F., and Atsuta, T., "Theory of Beam-Columns," *In-Plane Behavior and Design*, Vol. 1, McGraw-Hill Book Co., Inc., New York, N.Y., Dec., 1976.
2. Chen, W. F., and Atsuta, T., "Theory of Beam-Columns," *Space Behavior and Design*, Vol. 2, McGraw-Hill Book Co., Inc., New York, N.Y., Mar., 1977.

634   MARCH 1977   ST3

3. Chen, W. F., and Ross, D. A., "The Axial Strength and Behavior of Cylindrical Columns," *OTC paper No. 2683*, Eighth Annual Offshore Technology Conference, Houston, Tex., May 3-6, 1976, pp. 741-754.
4. Johnston, B. G., ed., *The Stability Research Council Guide to Stability Design Criteria for Metal Structures*, 3rd ed., John Wiley and Sons, Inc., New York, N.Y., Apr., 1976.
5. Marshall, P. W., "Stability Problems in Offshore Structures," presented at the March 25, 1970, Annual Technical Meeting, Column Research Council, held at St. Louis, Mo.
6. Ostapenko, A., and Gunzelman, S. X., "Local Buckling of Tubular Steel Columns," *Proceedings of the National Structural Engineering Conference on Methods of Structural Analysis*, Vol. II, ASCE, August 22-25, 1976, W. E. Saul and A. Hopzyrot, eds., pp. 549-568.
7. Redner, S., "Measurement of Residual Stresses by the Blind Hole Drilling Method," *Bulletin TDG-5*, Photolastic, Inc., Malvern, Pa., 1974.
8. Ross, D. A., and Chen, W. F., "Preliminary Tests of Fabricated Tubular Columns," *Fritz Engineering Laboratory Report No. 393.5*, Lehigh University, Bethlehem, Pa., Jan., 1976.
9. Ross, D. A., and Chen, W. F., "Tests of Fabricated Tubular Columns," presented at the April 5-8, 1976, ASCE National Water Resources and Ocean Engineering Convention, held at San Diego, Calif. (Preprint No. 2660).
10. Ross, D. A., and Chen, W. F., "Behavior of Fabricated Tubular Columns under Biaxial Bending," *Proceedings of the Specialty Conference on Mechanics of Engineering*, ASCE, Waterloo, Ontario, Canada, May 26-28, 1976.
11. "Specification for the Design, Fabrication and Erection of Structural Steel for Buildings," American Institute of Steel Construction, New York, N.Y., 1969.
12. "Specification for Fabricated Structural Steel Pipe," American Petroleum Institute, *API Specification 2B*, Oct., 1972.
13. "Specification for Welded Highway and Railway Bridges," American Welding Society, *AWS D1.1-72*, 1972.
14. Tall, L., "Stub Column Testing Procedure," Document C-282-61, Class C Document, International Institute of Welding, Oslo, Norway, June 1962; also, appendix to *Column Research Council Guide*, 2nd ed., B. J. Johnston, ed., John Wiley and Sons., Inc., New York, N.Y., 1966.
15. Tebedge, N., Alpsten, G., and Tall, L., "Residual Stress Measurement by the Sectioning Method," *Experimental Mechanics*, Vol. 13, No. 2, Feb., 1973, pp. 88-96.

APPENDIX II.—NOTATION

The following symbols are used in this paper:

D_o = nominal outside diameter of column;
E = modulus of elasticity;
K = effective length factor;
L = column length;
P = static buckling load;
P_d = dynamic buckling load;
P_y = yield axial load;
R = column radius;
r = radius of gyration;
t = wall thickness;
θ_B = bottom head rotation;
θ_T = top head rotation;
λ = $(1/\pi)(\sqrt{\sigma_y/E})(L/r)$;
σ_L = longitudinal residual stress; and
σ_y = yield stress.

5.5.2

International Journal of Offshore and Polar Engineering
Vol. 4, No. 2, June 1994 (ISSN 1053-5381)
Copyright © by The International Society of Offshore and Polar Engineers

Ultimate Strength of Damaged Tubular Members

L. Duan
California Department of Transportation, Sacramento, California, USA

W. F. Chen
Purdue University, West Lafayette, Indiana, USA

J. T. Loh
Exxon Production Research Co., Houston, Texas, USA

ABSTRACT

A moment-thrust-curvature-based procedure for calculating the behavior and ultimate strength of damaged tubular members is presented. The new set of moment-thrust-curvature expressions for damaged or undamaged tubular members developed by the authors is used. A computer program, BCDENT, was developed, whose capability includes analysis of single or multident tubular members subjected to axial compression, end moments, and distributed or concentrated lateral loads. In this paper, the validity and accuracy of the moment-thrust-curvature approach for determining the ultimate strength of damaged tubular members are verified by comparing analytical predictions with the available 151 test results.

INTRODUCTION

Owing to their low drag coefficient in comparison to other structural shapes, tubular members are used extensively in offshore structures. These members are generally subjected to gravity, wind, wave and current loads. For members in the wave zone, they often experience localized damage caused mainly by supply workboat collisions or dropped heavy object impacts. In the last two decades, experimental and analytical research on structural tubes has made significant progress in establishing refined criteria for the design of undamaged cylindrical tubular members in offshore platforms (Marshall, 1970; Sherman, 1976; Chen and Ross, 1977; Toma and Chen, 1979; Sherman, 1982; Chen and Han, 1985; Loh, 1990). However, available design specifications (API-RP-2A, 1989; API-RP-2A-LRFD, 1989; AISC-LRFD, 1986; AISC-LRFD, 1989) give no specific information on how localized damage affects the behavior and strength of dented tubular members under field service conditions. To assess the fitness of these offshore structures in service, technical information is needed for these dented members in terms of both their behavior and ultimate strength.

Damaged tubular members were first studied experimentally by Smith, Kirkwood and Swan (1979). During the '80s, a considerable amount of experimental and theoretical research about the effects of damage on the strength and behavior of tubular members has been conducted (Smith and Dow, 1981; Smith, Somerville and Swan, 1981; Ellinas, 1984; Ueda and Rashed, 1985; Richards and Andronieous, 1985; Yao, Taby and Moan, 1986; Taby and Moan, 1985 and 1987). Recently, MacIntyre and Birkemoe (1989) used a nonlinear finite element shell analysis (ABAQUS) to investigate the denting and subsequent axial compression of dented tubular members. This is the most rigorous

Received November 22, 1991: revised manuscript received by the editors January 18, 1994. The original version (prior to the final revised manuscript) was presented at the Second International Offshore and Polar Engineering Conference (ISOPE-92), San Francisco, USA, June 14-19, 1992.
KEY WORDS: Beam-column, column, damaged tubular member, dent, moment-thrust-curvature, stability, strength.

procedure among all existing analyses, but it requires a considerable computing effort.

Research reported in the open literature has, in the past, focused particular attention on dented members subjected to axial compression combined with negative bending (compression at the dent). In actual offshore structures, however, local damages may occur in any orientations relative to applied end moments. Dent locations vary along the member, and lateral loading can often accompany axial compression. Little attention has been paid to dented members subjected to loads of different directions with respect to the dents. It is the purpose of this study to develop a computer model for the analysis of a dented tubular beam-column subjected to biaxial bending with respect to the dents.

Based on the M-P-Φ relationships developed previously by Duan, Loh and Chen (1993), an analytical procedure and the computer program BCDENT were developed to calculate the general behavior of a dented (single or multiple dents) tubular beam-column subjected to loads of different directions and combinations. A brief comparison of BCDENT predictions to test results for 151 dented member tests is also presented. In general, good agreement was obtained, confirming the validity of the M-P-Φ approach in dented member analysis. The program listing of the BCDENT and typical input/output details are given in a recent book by Chen and Toma (1995).

GENERAL DESCRIPTION

The tubular beam-column under BCDENT consideration is treated as an individual member. The initial geometrical imperfections, w_i (out-of-straightness), is considered, and its boundary conditions are pinned. The dents can be multiple in arbitrary orientations and can be located anywhere along the member length. The loading cases include:

1. Constant end axial loads, increasing end bending moments.
2. Constant end axial loads, increasing lateral loads (either linearly distributed or two concentrated loads).
3. Constant end bending moments and/or lateral loads, increasing end axial loads.

The following assumptions were made:

1. Deformations are small. That is, small displacement beam theory was used in the formulation.
2. Shear and torsional deformations were neglected.
3. Strain reversal does not occur in the member, i.e., the member deformations are always monotonically increasing.
4. The member is divided into a number of segments. The properties of these segments are described by their corresponding M-P-Φ and M-P-ε_0 expressions.
5. For an undented section (dd/t < 1.0), the nonlinear M-P-Φ relation is represented by equations reported previously by Duan, Loh and Chen (1993). The M-P-Φ expressions (Fig. 1) include both ascending and descending branches due to local buckling for undented tubular sections with 1% out-of-roundness and with or without residual stresses.
6. For a dented section (dd/t < 1.0), the three-regime M-P-Φ expressions, developed previously by the authors (1993), are used (Fig. 2).
7. A single set of M-P-ε_0 expressions, reported by Duan, Loh and Chen (1993), is used for both dented and undented sections with or without residual stresses, and with or without local buckling.

The coordinate system for a biaxially loaded, dented tubular beam-column is shown in Fig. 3. The right-hand rule is used for the sign convention in the analysis (Fig. 4). Fig. 5 shows a typical beam-column subjected to biaxial loadings.

NUMERICAL PROCEDURE

The analytical procedure used in the computer program BCDENT is based on the incremental deflection method. The deflected shape of the member with a specific deflection at a control station is first assumed. The deflected shape of the member corresponding to this deflection is then computed, followed by computation of the bending moments, taking into account the axial load and the lateral loads. The new deflections are calculated by integrating the curvatures obtained from the known M-P-Φ expres-

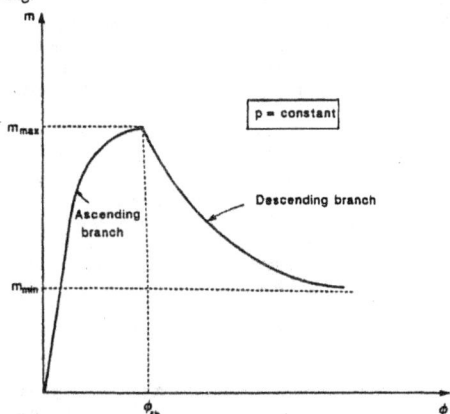

Fig. 1   Moment-thrust-curvature curve for undented tubular sections with local buckling

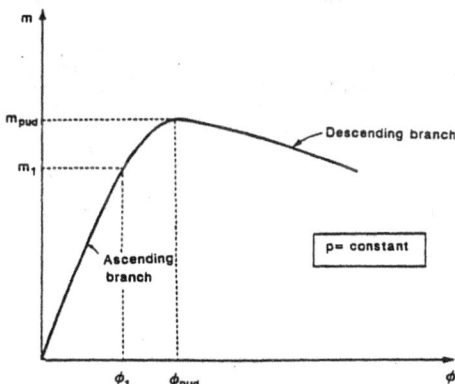

Fig. 2   Moment-thrust-curvature curve for dented tubular section

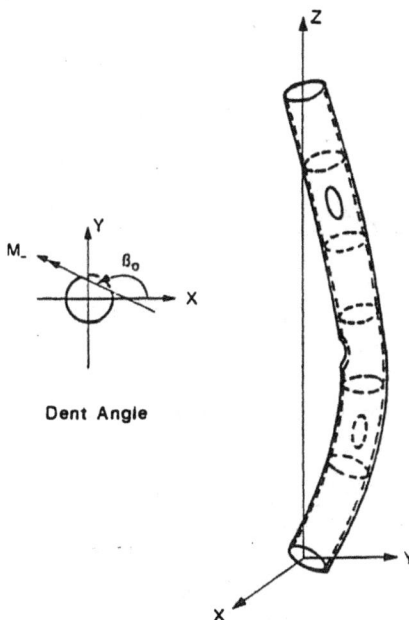

Fig. 3   Coordinate system for biaxially loaded dented tubular beam-column

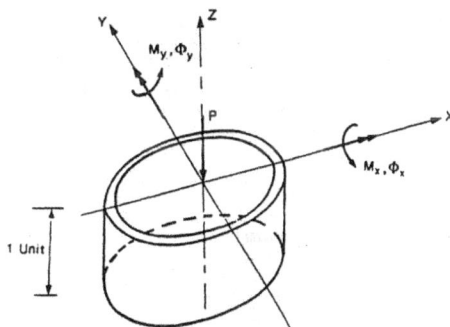

Fig. 4   Moment, curvature and axial force in cross-section

International Journal of Offshore and Polar Engineering   *129*

sions along the length of the member. Comparisons of the computed new deflections with the assumed deflections are then made to check whether convergence has been achieved. Using the updated deflected shape and repeating this procedure for successive deflection increments at the control station eventually lead to the desired full load-deflection relationship for the dented member.

The following steps may be summarized as follows (Fig. 6):

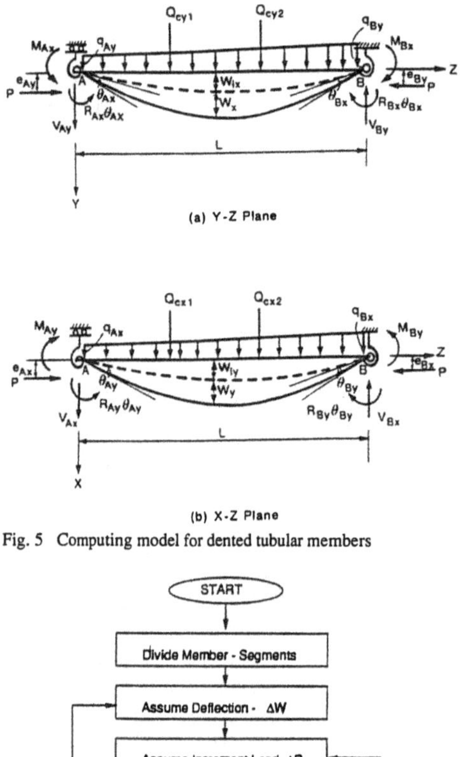

(a) Y-Z Plane

(b) X-Z Plane

Fig. 5   Computing model for dented tubular members

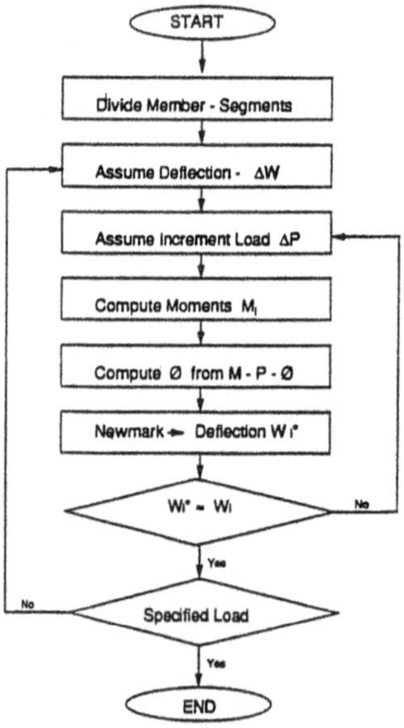

Fig. 6   Flowchart of numerical procedure

1. Divide the member into a sufficient number of segments. (The length of each segment may be different.) The division points are called stations or nodes.

2. Assume initial deflections w_i at the stations along the member.

3. Assume an incremental deflection Δw_i of the same shape as the initial deflection.

4. Assume an increment of axial load, Δp, or an increment of lateral load, Δq. (Increments can be negative or positive).

5. Calculate the *current peak* moments and curvatures for all stations; Φ_{1b} and M_{max} for undented sections; Φ_{pud} and M_{pud} for dented sections.

6. Calculate the bending moment M_i at all stations considering the second-order effect (P-δ) due to the axial load. Update Φ_i at all stations using known M-P-Φ expressions.

(a) If $M_i < M_{pud}$ (or M_{max}) and $\Phi_i < \Phi_{pud}$ (or Φ_{1b}), the curvature is on the ascending branch.

(b) If $M_i < M_{pud}$ (or M_{max}) and $\Phi_i > \Phi_{pud}$ (or Φ_{1b}), the curvature is on the descending branch.

7. Compare M_i with M_{pud} or M_{max}, and the updated Φ_i with Φ_{pud} or Φ_{1b}. If $M_i > M_{pud}$ (or M_{max}), repeat Step 4.

8. Using the Newmark's integration procedure described in the following section, determine the deflections at all stations, w_a^*.

9. Compare w_a^* with $w_a = w_i + \Sigma \ \Phi w$ at the control station. If the maximum difference is within the tolerance (say, $10^{-3} \ w_a$), go to next step. Otherwise, keep the deflection δ at the control station constant (Fig. 7), and scale the current deflection shape to the newly obtained w_a^*. Repeat Step 4.

10. Compare w_a^* with $w_a = w_i + \Sigma \ \Delta w$ at all other stations. If the maximum difference is within the tolerance (say, $10^{-3} \ w_a$), increase the next additional deflection at the control station with the same deflection shape newly obtained in the last comparison. Otherwise, update the current deflection shape to the newly obtained w_a^*. Repeat Step 6.

11. Repeat Steps 4~10 until loads decrease to a specified percentage level of the peak loads.

NEWMARK'S INTEGRATION PROCEDURE

Newmark's integration method (Newmark, 1943; Chen and Atsuta, 1976) is a useful procedure to compute the deflection shape from a given curvature distribution. It is based on the conjugate beam concept. One of the characteristics of Newmark's method is that the curvature distribution on the conjugate beam is replaced by a series of *Equivalent Concentrated Loads*. The magnitude of the Equivalent Concentrated Loads can be expressed in terms of the curvature diagram at the points of concentrated loading and the distance between the concentrated loads. They are obtained from equilibrium consideration. Here, we shall only outline the key steps and derive the Equivalent Concentrated Loads

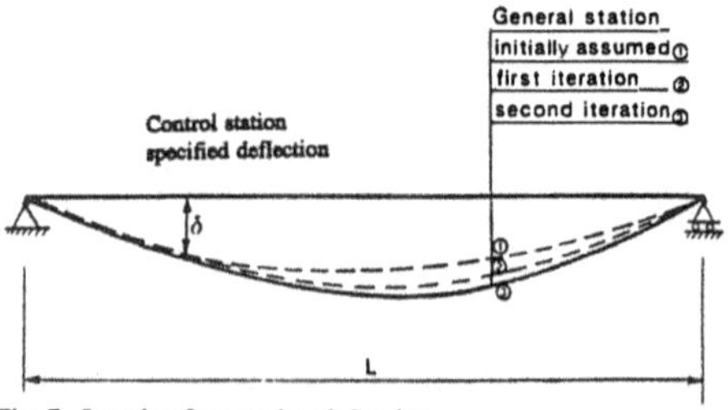

Fig. 7   Iteration for member deflection

for segments with different length. Newmark's procedure includes the following:

1. Assume the distribution of curvature between two stations to be quadratic, and calculate the Equivalent Concentrated Loads acting at these stations.
2. Calculate slopes at all stations by a numerical integration of curvatures.
3. Integrate the slopes along the length of a member to obtain the deflections at all stations.

In Newmark's method, the member is usually divided into several equal-length segments. For dented tubular members, the length of dented segments may be different from that of undented segments. As the formulas of Equivalent Concentrated Loads for unequal-length segments are not available in open literature, the general expressions for the magnitude of the equivalent concentrated loads are derived below.

The representation of the curvature diagram between two stations by a second-order polynomial expression is shown in Fig. 8. The value of the curvature diagram at the stations are denoted as ϕ_1, ϕ_2, and ϕ_3. The lengths of neighbor segments are denoted as h_1 and h_2. The equivalent concentrated loads are denoted as R_{12}, R_{21} and R_2. From the principle of static equivalence, general expressions for equivalent concentrated loads are obtained as follows:

$$R_{12} = \frac{\phi_1 h_1 h_2 (3h_1 + 4h_2) + \phi_2 h_1 (h_1^2 + 3h_1 h_2 + 2h_2^2) - \phi_3 h_1^3}{12 h_2 (h_1 + h_2)} \tag{1}$$

$$R_{21} = \frac{\phi_1 h_1 h_2 (h_1 + 2h_2) + \phi_2 h_1 (h_1^2 + 5h_1 h_2 + 4h_2^2) - \phi_3 h_1^3}{12 h_2 (h_1 + h_2)} \tag{2}$$

$$R_2 = \frac{\phi_1 A + \phi_2 B + \phi_3 C}{12 h_1 h_2 (h_1 + h_2)} \tag{2}$$

where

$$A = h_1^2 h_2 (h_1 + 2h_2) - h_2^4 \tag{4}$$

$$B = \left(h_1^2 + h_2^2\right)^2 + 6h_1^2 h_2^2 + 5h_1^3 h_2 + 5h_1 h_2^3 \tag{5}$$

$$C = h_1 h_2^2 (h_2 + 2h_1) - h_1^4 \tag{6}$$

For equal-length segments, $h_1 = h_2 = h$, the general formulas of equivalent concentrated loads, Eqs. 1~3, reduce to the following conventional formulas, respectively.

$$R_{12} = \frac{h(7\phi_1 + 6\phi_2 - \phi_3)}{24} \tag{7}$$

$$R_{21} = \frac{h(3\phi_1 + 10\phi_2 - \phi_3)}{24} \tag{8}$$

$$R_2 = \frac{h(\phi_1 + 10\phi_2 + \phi_3)}{12} \tag{9}$$

LOAD-SHORTENING RELATIONSHIPS

The total axial-shortening (Δ) of a beam-column consists of two parts: the axial shortening due to axial strain, and the axial shortening due to geometric change of lateral deflections:

$$\Delta = \Delta_s + \Delta_g \tag{10}$$

where Δ_s = axial-shortening due to axial strain, and Δ_g = axial-shortening due to geometric change

The axial-shortening due to axial strain is obtained from:

$$\Delta_s = \sum_{i=1}^{i=N} (\Delta L_i) \varepsilon_{oi} \tag{11}$$

where N = number of segments, ΔL_i = length of segment i, and ε_{oi} = axial strain of segment i determined from the M-P-ε_o expressions proposed by Duan, Loh and Chen (1993).

As soon as the deflection of a member is obtained, the corresponding axial-shortening due to geometric change of deflection is obtained from:

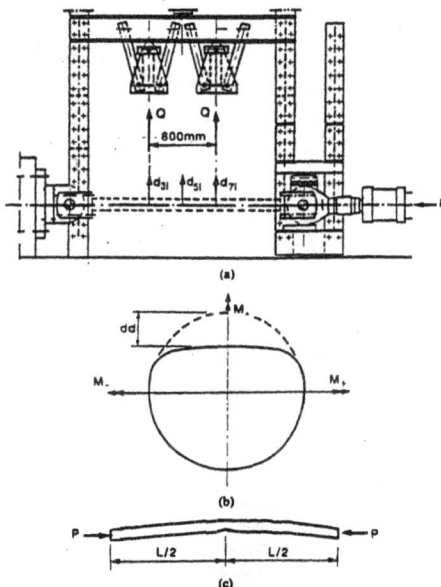

(a)

(b)

(c)

Fig. 9   Dented tubular member test: (a) test problems nos. 3~5
(b) dented section (c) test problems nos. 1-2

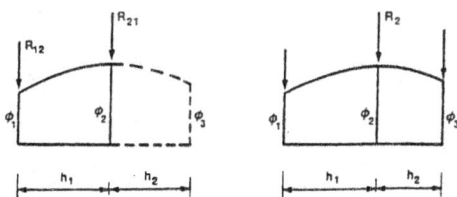

Fig. 8   Equivalent concentrated loads

International Journal of Offshore and Polar Engineering   *131*

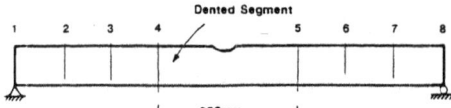

Fig. 10   Divided segment and stations for analysis of Landet and Johnsen's tests (1987)

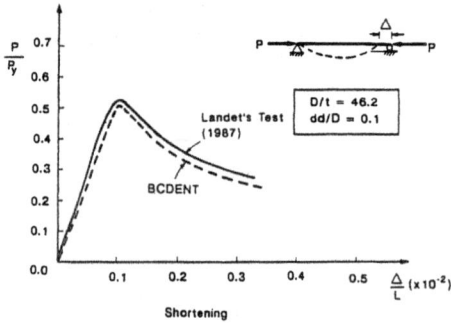

Fig. 11   Axial load-shortening curves for dented member test D1-35

$$\Delta_R = \sum_{i=1}^{i=N} \left\{ \Delta L_i - \sqrt{(\Delta L_i)^2 - (w_{1i} - w_{2i})^2} \right\}$$ (12)

where w_{1i}, w_{2i} = total resultant deflection at two ends of segment i.

DENTED TUBULAR MEMBERS

The accuracy and the validity of the M-P-Φ approach are examined by comparing BCDENT predictions to 5 test results reported by Landet and Johnsen (1987). The geometry and material properties for the 5 dented tubular members are given in Table 1 and Fig. 9.

Pin-Ended Dented Column Tests

In the present numerical analysis of Landet's test results (1987), the member is divided into 7 segments with 8 stations. Because the adopted M-P-Φ expressions for dented tubular sections are based on an average moment and curvature along an 800-mm segment, the length of the *dented* segment is taken here as 800 mm. The dented M-P-Φ curves are adopted at the two end stations of the dented segment, as shown in Fig. 10. The axial load-shortening curves computed by BCDENT are compared with these results for dent depth ratios (dd/D) of 0.1 (D1-35) and 0.2 (D1-36) in Figs. 11 and 12. It is seen that the predicted maximum strength is quite close to the test result, but analytical curves are generally lower than the tested curves. Several factors contribute to this difference. In the analytical studies, the ends of members are assumed to be perfectly pin-ended. In the experiments, some end restraints always exist. Other contributing factors are that the M-P-Φ curves used in the BCDENT are developed on the basis of constant axial load tests, and that the idealized M-P-Φ curves at the pre-maximum region are softer than those tested (Duan, Loh and Chen, 1993), especially for the case of dd/D = 0.2.

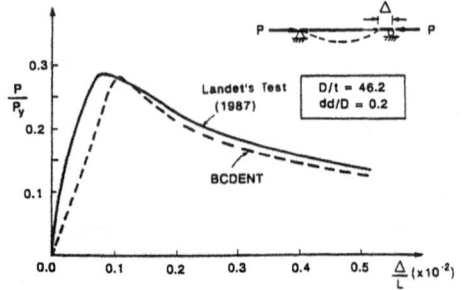

Fig. 12   Axial load-shortening curves for dented member test D1-36

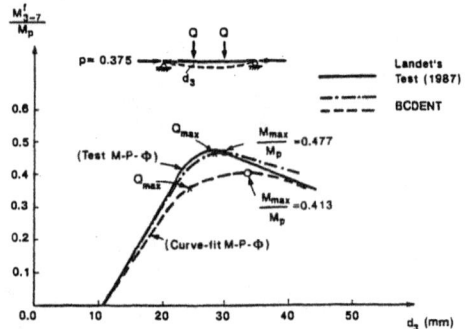

Fig. 13   Moment-deflection curves for dented beam-column test D2-37 (negative bending)

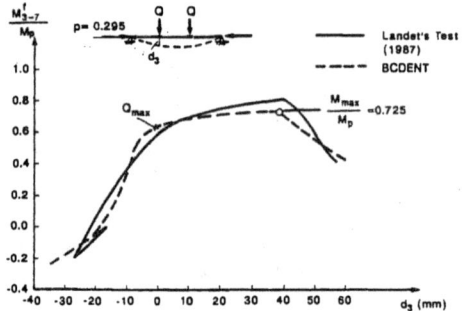

Fig. 14   Moment-deflection curves for dented beam-column test C3-25 (positive bending)

Dented Beam-Column Tests

The load-deflection curves computed by BCDENT are compared with three of Landet's beam-column test results (Landet and Johnsen, 1987) in Figs. 13~15. These three tests had two lateral concentrated loads coupled with a constant axial compression. Fig. 13 is for negative bending, Fig. 14 for positive bending, and Fig. 15 for neutral bending. On the analytical curves, crosses denote the location of maximum lateral concentrated loads Q_{max}; circles represent the maximum bending moments for the critical dent sections M_{max}. It can be seen that the M_{max} point usually

132

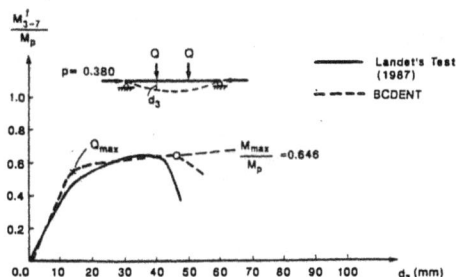

Fig. 15 Moment-deflection curves for dented beam-column test C4-28 (neutral bending)

occurs after the Q_{max} point.

In Fig. 13, the dashed line is the analytical result based on the general curve-fit M-P-Φ expressions, implemented in BCDENT, while the dot-dashed line is based on the actual measured M-P-Φ behavior for the particular individual specimen. From Fig. 13, it may be concluded that the M-P-Φ approach can predict adequately the behavior of dented tubular members. The accuracy of the predictions depends mainly on the accuracy of the M-P-Φ expressions used in the analysis.

Ultimate Strength of Dented Members

To further calibrate BCDENT, this section briefly summarizes a comprehensive study of comparisons of BCDENT's ultimate strength predictions to all of the available 151 test data in the open literature. Details of comparisons and test data are given in an EPR report by Loh (1991). The available test data are made up of 130 tests with $D/t < 80$ and 21 tests with $D/t = 88 - 122$; member length $L = 0.5$ mm — 7.75 m; dent depth to diameter ratio $dd/D = 0 - 0.23$; and dent depth to thickness ratio $dd/t = 0 - 25$. 107 test specimens are axial and eccentric load cases, and 44 test specimens are combined axial load and moment cases.

A distribution of 130 ratios for dented members with $D/t < 80$, representing all load types, is shown in Fig. 16. The mean value of measured ultimate strength/strength predicted by BCDENT is 1.01, and the standard deviation is 0.12. A total of 151 ratios is plotted in Fig. 17 for comparisons, including 21 cases with high D/t ratios ($D/t = 88 - 122$) for which BCDENT is not valid. The mean value of measured ultimate strength/strength predicted by BCDENT is 1.02, and the standard deviation is 0.13. These comparisons indicate that BCDENT gives a good mean prediction of ultimate strength with a small deviation.

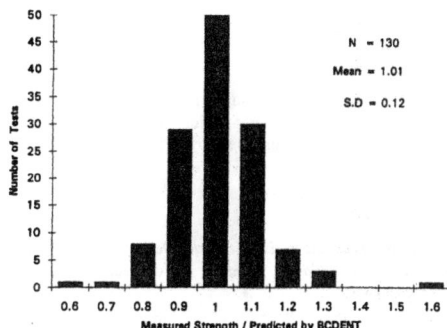

Fig. 16 Comparison of ultimate strength prediction by BCDENT to test data ($D/t<80$)

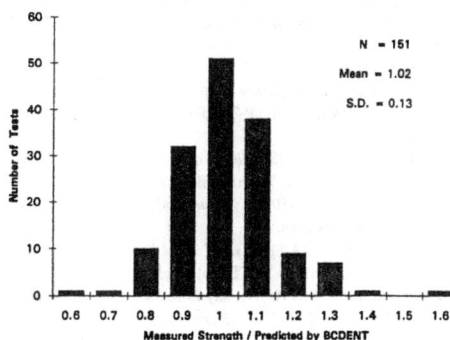

Fig. 17 Comparison of ultimate strength prediction by BCDENT to test data

SUMMARY AND CONCLUSIONS

1. An analytical procedure and a computer program BCDEN have been developed for predicting the general behavior of dented tubular beam-columns based on the M-P-Φ expressions reported previously by Duan, Loh and Chen (1993).

2. The validity of the analysis procedure has been confirmed by comparisons of BCDENT predictions with available test results for dented tubular members subjected to various load combinations. The mean value of the ultimate strength of dented tubular members predicted by BCDENT is in good agreement with the experimental tests. Comparison of BCDENT with the test results

No.	Specimen	D mm	t mm	L mm	K	F_y MPa	dd mm	d_{3i} mm	d_{5i} mm	d_{7i} mm	P kN	M_-	M_+	M^*
1	D1-35	140.0	3.03	2677	0.946	368	14.0	-9.5	-11.0	-9.5	AP			
2	D1-36	140.0	3.02	2677	0.946	400	28.0	-19.5	-24.0	-19.5	AP			
3	D2-37	140.0	3.01	2677	0.982	388	13.9	11.0	11.8	11.4	-188.4	NM		
4	C3-25	140.0	3.02	2677	0.991	388	27.9	-17.1	-22.1	-17.2	-148.5		PM	
5	C4-28	140.0	3.03	2677	0.986	405	14.0	0.0	0.0	0.0	-200.7			CM

Table 1 Geometry and properties of dented member tests (Landet and Johnsen, 1987)

Note: See Fig. 9.

D	= outside diameter	K = measured effective length factor	AP = axial load column test
t	= tube thickness	F_y = measured static yield strength	NM = negative bending beam-column test
L	= member length	dd = initial dent depth	PM = positive bending beam-column test
		d_i = initial deflection	CM = neutral bending beam-column test

International Journal of Offshore and Polar Engineering   *133*

showed reasonably small scatter.

ACKNOWLEDGEMENTS

This study was financially supported by Exxon Production Research Company. The computations were done at Purdue University. The contributions made by EPR Staff are gratefully acknowledged.

REFERENCES

AISC (1986). *Load and Resistance Factor Design Specification for Structural Steel Buildings*, American Inst of Steel Construction, Chicago, Illinois.

AISC (1989). *Specification for Structural Steel Buildings — Allowable Stress and Plastic Design*, American Inst of Steel Construction, Chicago, Illinois.

API (1989). *Recommended Practice for Planning, Designing and Constructing Fixed Offshore Platform*, 18th ed, API-RP-2A, Amer Petroleum Inst, Washington, DC.

API (1989). *Draft Recommended Practice for Planning, Designing and Constructing Fixed Offshore Platforms - Load and Resistance Factor Design*, 1st ed, API-RP-2A-LRFD, Amer Petroleum Inst, Washington, DC.

Chen, WF, and Atsuta, F (1976). *Theory of Beam-Columns, Vol I: In-plane Behavior and Design*, McGraw-Hill, NY.

Chen, WF, and Han, DJ (1985). *Tubular Members in Offshore Structures*, Pitman, London.

Chen, WF, and Ross, DA (1977). "Test of Fabricated Tubular Columns," *J Struct Division*, ASCE, Vol 103, No ST3, pp 619-634.

Chen, WF, and Toma, S (1995). *Analysis and Software of Tubular Members*, CRC Press, Boca Raton, Florida. To appear.

Duan, L, Loh, JT, and Chen, WF (1993). "Moment-Thrust-Curvature Relationships for Dented Tubular Sections," *J Struct Eng*, ASCE, Vol 119, No 3, pp 809-830.

Ellinas, CP (1984). "Ultimate Strength of Damaged Tubular Bracing Members," *J Struct Eng*, ASCE, Vol 110, No 2, pp 245-259.

Landet, E, and Johnsen, RH (1987). "Investigation on Ultimate Strength of Dented Pipes," *Tech Rept* No 87-3278, VERITEC.

Loh, JT (1990). "A Unified Design Procedure for Tubular Member," *Proc Offshore Tech Conf*, OTC 6310, Houston, pp 365-379.

Loh, JT (1991). *Comparison of Dented Member Strength Predictions to Test Data*, Exxon Production Research Company, Houston.

MacIntyre, J, and Birkemoe, PC (1989). *Damage of Steel Tubular Members in Offshore Structures: A Nonlinear Finite Element Analysis*, Dept Civil Eng, Univ of Toronto, Ontario, Canada.

Marshall, PW (1970). "Stability Problems in Offshore Structures," *Proc Annual Tech Sessions, Column Res Council*, St. Louis, Missouri.

Newmark, NM (1943). "Numerical Procedure for Computing Deflection, Moments, and Buckling Loads," *Trans ASCE*, Vol 108, p 1161.

Richards, DM, and Andronicou, A (1985). "Residual Strength of Dented Tubulars: Impact Energy Correlation," *Proc Int Offshore Mech and Arctic Eng Symp*, Dallas, pp 522-527.

Sherman, DR (1976). "Tests of Circular Steel Tubes in Bending," *J Struct Division*, ASCE, Vol 102, No ST11, pp 2181-2195.

Sherman, DR (1982). "Research in North America on the Stability of Circular Tubes," *Proc Annual Tech Sessions, Struct Stability Res Council*, New Orleans.

Smith, CS, Kirkwood, W, and Swan, JW (1979). "Buckling Strength and Post-Collapse Behavior of Tubular Bracing Members Including Damage Effects," *Proc Int Conf on Behavior of Offshore Structures* (BOSS'79), London, pp 303-326.

Smith, CS, and Dow, RS (1981). "Residual Strength of Damaged Steel Ships and Offshore Structures," *J Const Steel Res*, Vol 1, No 4, pp 2-15.

Smith, CS, Somerville, JW, and Swan, JW (1981). "Residual Strength and Stiffness of Damaged Steel Bracing Members," *Proc Offshore Tech Conf*, Houston, pp 273-291.

Taby, J, and Moan, T (1985). "Collapse and Residual Strength of Damaged Tubular Members," *Proc Int Conf Behavior of Offshore Structures*, Delft, pp 395-408.

Taby, J, and Moan, T (1987). "Ultimate Behavior of Circular Tubular Members with Large Initial Imperfections," *Proc Annual Tech Sessions, Struct Stability Res Council*, Houston, pp 79-104.

Toma, S, and Chen, WF (1979). "Analysis of Fabricated Tubular Columns," *J Struct Division*, ASCE, Vol 105, No ST11, pp 2343-2366.

Ueda, Y and Rashed, SMH (1985). "Behavior of Damaged Tubular Structural Members," *Proc Offshore Mech and Arctic Eng Symp*, ASME, Dallas, pp 528-536.

Yao, T, Taby, J and Moan, T (1986). "Ultimate Strength and Post-Ultimate Strength Behavior of Damaged Tubular Members in Offshore Structures," *Proc Offshore Mech and Artic Eng Symp*, ASME, Vol 3, pp 301-308.

5.6 Concrete Plasticity (混凝土塑性力學)

5.6.1

◣ _____ 11

IABSE COLLOQUIUM

Constitutive Equations for Concrete

Equations de base du béton

Stoffgleichungen für Beton

KOPENHAGEN 1979

Plasticity in Reinforced Concrete

W.F. CHEN
Professor of Structural Engineering
Purdue University
West Lafayette, Indiana, USA

REPRINT FROM THE INTRODUCTORY
REPORT EXTRAIT DU RAPPORT
INTRODUCTIF SONDERDRUCK AUS DEM
EINFÜHRUNGSBERICHT

SUMMARY

A summary of the current state-of-the-art in the mathematical modeling of the mechanical behaviour of concrete is presented. A general discussion of some experimental facts is followed by a detailed description of the five basic types of models: 1. uniaxial and equivalent uniaxial models; 2. linear elastic and fracture models; 3. nonlinear elastic and variable moduli models; 4. elastic-perfectly plastic and fracture models; and 5. elastic-strain hardening plastic and fracture models. Their relative merits and limitations are discussed and some of the interrelationship between various models are demonstrated. Directions of further research are indicated.

RESUME

Le rapport présente l'état actuel des connaissances sur les modèles mathématiques exprimant le comportement mécanique du béton. Quelques remarques générales sont faites à partir de quelques faits expérimentaux. Une description détaillée des cinq types fondamentaux de modèle est donnée: 1. modèle uniaxial et quasi-uniaxial; 2. modèle linéaire élastique et de rupture; 3. modèle non-linéaire élastique; modèle élastique-parfaitement plastique et de rupture; 5. modèle linéaire élastique-écrouissable et de rupture. Leur valeur relative et leurs limites sont discutées et quelques relations entre les différents modèles sont données. Des directions sont indiquées pour des recherches futures.

ZUSAMMENFASSUNG

Es wird ein Überblick gegeben über die heutigen Kenntnisse an mathematischen Modellen, die das mechanische Verhalten von Beton beschreiben. Nach einer Erörterung einiger Aspekte des in Versuchen beobachteten Verhaltens werden fünf Arten von Modellen beschrieben: 1. einachsiale und quasieinachsiale Modelle; 2. linear elastisches Verhalten bis zum Bruch; 3. nichtlinear elastisches Verhalten; 4. elastisch-ideal plastisches Verhalten bis zum Bruch; 5. elastisch-verfestigend plastisches Verhalten bis zum Bruch. Die Vorzüge und Grenzen der einzelnen Modelle werden einander gegenübergestellt und einige Beziehungen zwischen den verschiedenen Modellen werden dargestellt. Richtungen für die weitere Forschungstätigkeit werden angedeutet.

1. INTRODUCTION

A complete *progressive failure analysis* of reinforced concrete structures under
static and seismic loading conditions requires the consideration of loading and
ground motion input, generalized material behavior, and analytical procedure.
The terms "loading" and "ground motion" refer to the specific loadings and motions that should be considered in design and analysis of reinforced concrete
structures. This includes the magnitude and direction of loading, the duration
and frequency content of the motion, and the acceleration, velocity and displacement parameters. This is beyond the scope of the present paper.

The term "generalized material behavior" refers to multi-dimensional stress-
strain relations which adequately describe the basic characteristics of reinforced concrete materials subjected to monotonical and cyclic loading. These
are called *constitutive equations*. These are the most fundamental relations required for any analysis of reinforced concrete structures. This paper sets out
those which have been most widely and successfully used in analytical and numerical approaches to reinforced concrete problems. To discuss the mathematical
modeling of non-linear reinforced concrete behavior, three areas must be examined:
the behavior of concrete, the response of steel reinforcement and the bond-slip
phenomenon between steel and concrete.

Since steel reinforcement is comparatively thin, it is generally assumed capable
of transmitting axial force only and thus, a uniaxial stress-strain relationship
is sufficient for general use. The most commonly used plasticity model for steel
reinforcement is the linearly elastic-perfectly plastic type which ignores the
Bauschinger effect but allows elastic unloading. As for concrete, however, a
knowledge of multiaxial stress-strain behavior is required. This is still far
from complete, although a large variety of models have been proposed in recent
years. Herein, a critical review of these constitutive equations for concrete
is given in the forthcoming. For the present, we limit our discussion to the
cases of short-time loading for which the creep effects may be neglected. Although the analysis and design of reinforced concrete structures require not only
each relationship between stresses and strains of steel and concrete but also the
bond-slip relation between steel and concrete, only the constitutive relations
for plain concrete will be reviewed and evaluated here. Once the stress-strain
relation of each material is available and a bond-slip relation is assumed, steel
reinforcements can then be placed in proper positions in concrete elements and
constitutive equations for the composite response of reinforced concrete element
can be readily formulated. The mechanics of bond-slip phenomenon between steel
and concrete will not be considered here. In most practical applications, a
perfect bond is generally assumed.

The term "analytical procedure" refers to the mathematical and numerical aspects
of calculation used to obtain solutions. In recent years there has been a growing interest in the application of the *finite element* procedure to the analysis
of reinforced concrete problems. The present state of development of computer
programs and finite-element method are limited to linear, non-linear, or plasticity material models for two-dimensional analysis of reinforced concrete structures under static and seismic loading conditions. The relative merits and
limitations of different constitutive models for concrete will be reviewed here
with particular emphasis on their use in the numerical analysis of reinforced
concrete structures.

2. EXPERIMENTAL DATA

We begin by examining some typical experimental data for concrete under uniaxial, biaxial and triaxial states of stress. These data, which are essential in the generalized development of mathematical modeling of concrete, serve the following two major purposes: (1) to give a guidance on the proper type of material behavior to be developed in the mathematical modeling; and (2) to provide data for the determination of the various material constants which appear in the mathematical models.

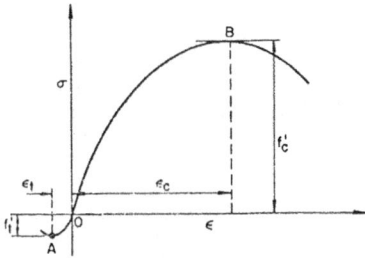

Fig. 1 Typical Stress-Strain Curve for Concrete Under Uniaxial Tension and Compression

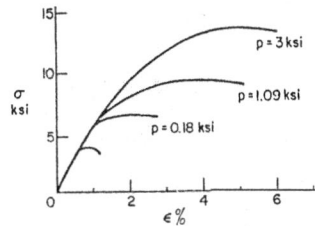

Fig. 2 Typical Stress-Strain Curve for Concrete Under Compression and Lateral Pressure (1 ksi = 6.89 NM/m^2)(Richart et al. 1928, Ref. 1)

Under ordinary experience, concrete is brittle in tension and limitedly deformable in compression. A typical uniaxial stress-strain curve for concrete is shown in Fig. 1. The maximum tensile stress f_t' is about 8-12 times *less* than the maximum compressive stress f_c'. Similar proportions occur between corresponding strains ε_t and ε_c; whereas ε_c takes on values of about 0.25%, ε_t does not usually exceed 0.015%. The crushing of concrete is usually followed by an unstable portion of strain softening with a crushing strain ε_u of about 0.35%. The brittleness of the tensile failure does not necessarily mean that the behavior is linearly elastic up to the fracture, as shown in Fig. 1. A characteristic of tensile type of failure is that the failure itself is abrupt with the development of one major crack, while in the case of compressive type of failure, more major cracks develop. Crack formation at failure occurs in the direction perpendicular to the direction of the greatest principal strain or stress.

Ordinary experience simply is not adequate to predict the behavior of concrete under triaxial compression state of stresses. The stress-strain curves of Fig. 2 illustrate the change in behavior of a concrete as it deforms under increasingly higher confined pressure. Hydrostatic pressure is seen largely to increase both maximum stress and maximum strain during compression, and the unstable strain softening portion gradually vanishes for increasing pressures.

Considerable experimental data are available regarding the strength, deformational characteristics and microcracking behavior of concrete subjected to biaxial state of stresses. Fig. 3 shows typical experimental stress-strain curves for concrete under biaxial compression (Fig. 3a), combined tension and compression (Fig. 3b) and biaxial tension (Fig. 3c).

First, it is seen that the maximum compressive strength increases for the biaxial compression state. A maximum strength increase of approximately 25% is achieved at a stress ratio of $\sigma_2/\sigma_1 = 0.5$ and is reduced to about 16% at an equal biaxial compression state ($\sigma_2/\sigma_1 = 1$). Under biaxial compression-tension, the compressive strength decreases almost linearly as the applied tensile stress is increased. Under biaxial tension, the strength is almost the same as that of uniaxial tensile strength (see Fig. 4).

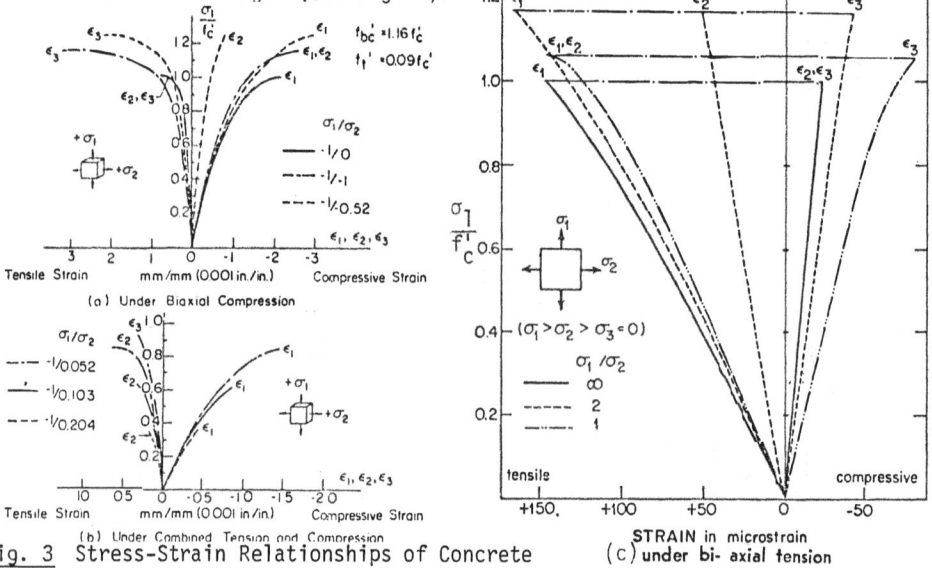

Fig. 3   Stress-Strain Relationships of Concrete

Second, concrete "ductility" under biaxial state of stresses has different values depending on the nature of stress states: compressive type or tensile type. For uniaxial and biaxial compression type (Fig. 3a), the average value of the maximum strain is about 2500 microstrain, and the average value of the maximum tensile strain varies approximately from 1000 to 3000 microstrain. The tensile ductility is greater under biaxial compression state than in uniaxial compression (Fig. 3a). In biaxial compression-tension (Fig. 3b), the magnitude at failure of both the principal compressive strain and the principal tensile strain decrease as the tensile stress increases. In uniaxial, axial and biaxial tension (Fig. 3c), the average value of the maximum principal tensile strain is about 150 microstrain. Note that the existence of a descending branch under biaxial stress states has not generally been observed. However, by using a constant rate of straining, Nelissen [3] has been able to achieve the descending portions of stress strain curves in biaxial loading test.

Third, failure of concrete occurs by tensile splitting with the fractured surface orthogonal to the direction of the maximum tensile strain. Tensile strains are found to be of crucial importance in the failure criterion and failure mechanism of concrete.

Fourth, as the failure point is approached, an increase in volume occurs as the compressive stress continues to increase, as shown in Fig. 5. This inelastic volume increase is called "dilatancy", and is usually attributed to progressive growth of major microcracks of concrete.

There are several approaches for defining this complicated stress-strain behavior of concrete under various stress states. They can be divided in three main groups: (1) representation of given stress-strain curves by using curve-fitting methods, interpolation or mathematical functions, (2) linear and nonlinear elasticity theories, and (3) perfect and work-hardening plasticity theories. We shall describe some of the main constitutive models used in the analysis of reinforced concrete structures.

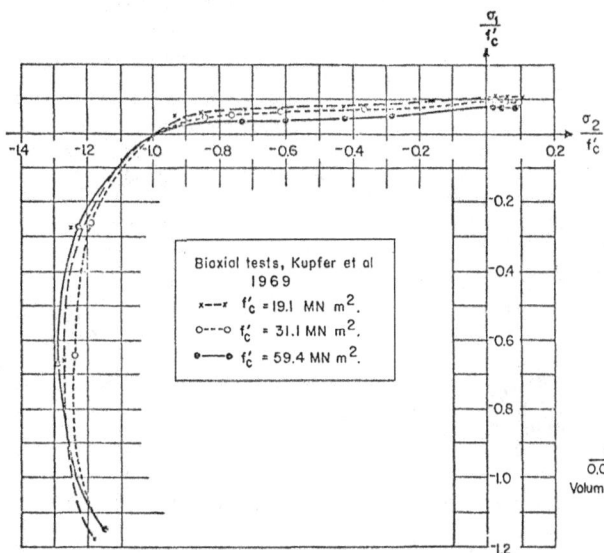

Fig. 4 Biaxial Strength Envelope of Concrete.

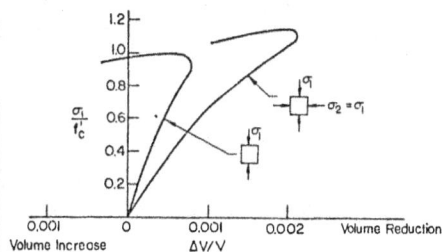

Fig. 5 Typical Stress-Strain Curve for Concrete Volume Change Under Biaxial Compression.

3. UNIAXIAL AND EQUIVALENT UNIAXIAL MODELS

The widely used function for simulation of stress-strain curves of concrete under uniaxial as well as biaxial state of stresses has the form [5]

$$\sigma = \frac{a\varepsilon}{1 + \left(\dfrac{a\varepsilon_p}{\sigma_p} - 2\right)\left(\dfrac{\varepsilon}{\varepsilon_p}\right) + \left(\dfrac{\varepsilon}{\varepsilon_p}\right)^2} \tag{1}$$

where σ, ε = stress and strain in the principal stress direction,
     σ_p, ε_p = experimentally determined values of maximum principal stress and corresponding strain, and
     a = experimentally determined coefficient which represents the initial tangent modulus.

Equation (1) has a horizontal tangent modulus at the point of peak stress and corresponding strain $(\sigma_p, \varepsilon_p)$. For a uniaxial stress state, the points of peak stress are either (f'_c, ε_c) or (f'_t, ε_t) as shown in Fig. 1 and the uniaxial initial elastic modulus is a = E.

For biaxial stress states, the maximum stress point, σ_p, is the value which can be determined from the biaxial strength envelop such as the one shown in Fig. 4, and the corresponding value of maximum strain, ε_p, in the major direction may be fixed at about 2500 micro-compressive-strain for uniaxial and biaxial compression state (Fig. 3a), or at about 150 micro-tensile-strain for uniaxial and biaxial tension state (Fig. 3c).   Since under biaxial compression-tension state, the compressive strength, σ_p, decreases almost linearly as the tensile stress is increased, the corresponding decrease in compressive strain, ε_p, can be estimated by proportioning its value to the tensile stress increase (Fig. 3b). Note that the value of ε_p in the minor direction will vary. Various curve-fitting expressions for the biaxial ultimate strength envelope and corresponding major and minor strains can be found in References [4-7].

The biaxial compressive stress-strain curves of Fig. 3(a) show increasing *initial* stiffnesses for increasing values of lateral compression and this is caused mainly by the effect of Poisson's ratio. Thus the strain measured in the same direction as the stress includes the contribution from lateral direction. Similarly, Poisson's ratio has a decreasing effect on the *initial* stiffnesses of the biaxial tensile stress-strain curves (Fig. 3c). For a linearly elastic, isotropic material, the biaxial stress-strain relation can be expressed as

$$\sigma = \frac{E\varepsilon}{1 - \nu\alpha} \tag{2}$$

where α = ratio of the principal stress in the orthogonal direction to the principal stress in the direction considered,
           E = initial tangent modulus in uniaxial loading, and
           ν = Poisson's ratio in uniaxial loading.

As an approximation, the *effective initial modulus*, $E/(1-\nu\alpha)$, due solely to the Poisson's effect can be used as the initial modulus, $a = E/(1-\nu\alpha)$, in Eq. (1) which describes the nonlinear biaxial stress-strain relationship of concrete. Since this biaxial stress-strain equation passes through the point of peak stress and strain $(\sigma_p, \varepsilon_p)$, which accounts for mainly the microcrack confinement effect in the presence of biaxial stress, the tangent modulus which is the slope at any point of the biaxial stress-strain curve, $E_t = d\sigma/d\varepsilon$, include both the microcrack confinement effect as well as the Poisson's effect.

The basic concept of this model is to treat the biaxial stress-strain behavior of concrete as an *equivalent uniaxial* relation. According to this approach, the strain increment in each principal direction is evaluated solely by the principal stress increment in the same direction and the corresponding tangent stiffness, which is a function of the prinxipal stress ratio, α, accounts for all the biaxial effects. In this, Poisson's ratio is assumed to be a constant which is near 0.2. According to experimental evidence, this is a fairly reasonable approximation up to about 80% of peak stress, but after this point it progressively deviates (see Fig. 5)

The main advantages of this model are that it is simple and that the required data are either readily obtainable from uniaxial tests on the concrete or readily available from various biaxial tests reported in literature. The model is mainly applicable to planar problems such as beams, panels and thin shells where the stress is predominantly biaxial. However, it is immediately apparent from Fig. 5 that there is an abrupt volume increase near peak stress under biaxial compression and from Fig. 2 that there is a marked influence of hydrostatic pressure on the behavior of concrete under triaxial stress states. These behaviors can not be accounted for by the present equivalent uniaxial approach. Thus the model described will have little validity in three-dimensional situations.

4. LINEAR ELASTIC-FRACTURE MODELS

One of the most important characteristics of concrete is its low tensile strength which results in tensile cracking at very low stress compared to compressive stresses. The tensile cracking reduces the stiffness of the concrete and is usually the major contributor to the nonlinear behavior of some reinforced concrete structures such as panels and shells where the stress is predominantly biaxial tension-compression type. For these structures, accurate modeling of cracking behavior of concrete is undoubtedly the most important factor and linear elastic-fracture models have been developed and used by many investigators to study the nonlinear response of reinforced concrete beams, panels and shells [see, for example, Refs. 8 and 9].

In the following, a linear elastic-fracture model under general stress states is developed in three parts: elasticity, fracture criteria and fractured concrete.

4.1 Linear Elasticity

The basic relations between stress and strain, upon which uncracked and cracked concrete are based, are those of *linear elasticity* which is described completely by two elastic constants, Poisson's ratio, ν, and Young's modulus, E. The elastic limit envelope in general stress space is the same as that of the fracture envelope.

For the uncracked isotropic concrete, the constitutive relationships are well known, in the usual notation [10]

$$p = \frac{\sigma_{kk}}{3} = \frac{E}{3(1 - 2\nu)} \, \varepsilon_{kk} = K \, \varepsilon_{kk} \tag{3}$$

$$s_{ij} = \frac{E}{1 + \nu} \, e_{ij} = 2G \, e_{ij} \tag{4}$$

where $K = E/3(1-2\nu)$ and is called the *bulk modulus* and $G = E/2(1+\nu)$ and is the *shear modulus*. Volume change $\varepsilon_{kk} = \varepsilon_{11} + \varepsilon_{22} + \varepsilon_{33} = \varepsilon_x + \varepsilon_y + \varepsilon_z$ is produced by the mean normal stress $p = \sigma_{kk}/3 = (\sigma_{11} + \sigma_{22} + \sigma_{33})/3 = (\sigma_x + \sigma_y + \sigma_z)/3$; distortion or shear deformation $e_{ij} = \varepsilon_{ij} - \varepsilon_{kk}\delta_{ij}/3$ is produced by the shear stress or the stress deviation $s_{ij} = \sigma_{ij} - \sigma_{kk}\delta_{ij}/3$. Each is independent of the other. A generalization of Eqs. (3) and (4) for defining isotropic nonlinear stress-strain behavior of concrete will be given later when the nonlinear elastic and variable moduli models are presented.

For the cracked concrete, the stress-strain relations are still linear elastic but the material stiffness or the tangent elasticity matrix is modified to reflect the fact that the stresses in the direction normal to the crack are zero. Overall this implies that crack propagation is solved by succession of transitions from one instantaneous elastic stiffness to another. However, the entire amount of stresses existing in the crack plane before cracking must be released suddenly and thrown back into the structure for redistribution. This redistribution may introduce additional cracking and results in a considerable complication in numerical analysis. Stress-strain relationships of fractured concrete including the sudden release of fracuring stresses will be given later when the general criteria for defining fracture of concrete is presented in what follows.

4.2 Fracture Criteria

When the state of stress reaches a certain critical value, concrete will fail
by fracturing. Fracture of concrete can occur in two different types: (1) the
"cracking" type of fracture occurs when the principal stresses are either in the
tension-tension state or tension-compression state and their value exceeds the
limit value; and (2) the *"crushing"* type of fracture occurs when the principal
stresses are in the compression-compression state and their value exceeds the
limit value. When concrete cracks the material only loses its tensile strength
normal to the crack direction but retains its strength parallel to the crack
direction. On the other hand, when concrete crushes, the element loses its
strength completely.

To determine the fracture of concrete under multi-axial state of stresses or
strains, a *fracture criterion* which specifies the limit value is needed. Most
of the existing fracture (or failure) criteria for brittle materials are written
in terms of stresses, which, in many cases, are not adequate to predict the
failure characteristics of concrete material. In the following, a dual representation of fracture criterion expressed in terms of both stresses and strains
is described. The dual criterion has the following forms:

(1) stress criterion

$$f(\sigma_{ij}) = 0 \tag{5}$$

The simplest type of fracture criterion in compression states is to assume a
linear relation between octahedral shear stress $\tau_{oct} = \sqrt{2J_2/3}$ and octahedral
normal stress $\sigma_{oct} = I_1/3 = p$,

$$\tau_{oct} = a + b\,\sigma_{oct} \tag{6}$$

or alternatively,

$$\sqrt{J_2} = k - \alpha p \tag{7}$$

where a and b or k and α are material constants to be determined from experimental data and

$$I_1 = \sigma_{ii} = 3p$$
  = first stress invariant which corresponds to the mean normal stress com-
      ponent of the stress state   (8)

$$J_2 = \frac{1}{2} s_{ij} s_{ij} = \frac{1}{2} \sigma_{ij} \sigma_{ij} - \frac{1}{6} \sigma_{ii}^2$$
  = second invariant of deviatoric stresses   (9)

If α is zero, Eq. 7 reduces to the von Mises yield condition for metal. The
criterion of Eq. 7 is called the *extended von Mises criterion*. Drucker and
Prager have shown that the function (7) reduces to the well-known Coulomb
criterion in the case of plane strain condition if [11]:

$$\alpha = \frac{\tan \phi}{\sqrt{(9 + 12 \tan^2\phi)}} \quad , \quad k = \frac{3c}{\sqrt{(9 + 12 \tan^2\phi)}} \tag{10}$$

where the angle ϕ is known as the angle of internal friction and c is the
cohesion of a concrete. The constants c and ϕ can be looked upon simple as
parameters which characterize the total resistance of the concrete to shear.

On the basis of biaxial experimental data, a fracture function for concrete in compression states in the form of

$$3J_2 + f'_c I_1 + I_1^2/5 = f'^2_c/9 \tag{11}$$

has been proposed by Buyukozturk [12].

In tension-compression and tension-tension states, the maximum stress (or strain) criterion is generally adopted in most applications

In a recent development, a fracture function for concrete of the following form has been proposed in [13]:

compression-compression domain $(\sqrt{J_2} \leq -\dfrac{1}{\sqrt{3}} I_1$ and $I_1 \leq 0)$

$$f(\sigma_{ij}) = J_2 + \frac{A_u^c}{3} I_1 = \tau_u^{2c} \tag{12}$$

tension-compression or tension-tension domain $(\sqrt{J_2} > -\dfrac{1}{\sqrt{3}} I_1$ or, $I_1 > 0)$

$$f(\sigma_{ij}) = J_2 - \frac{1}{6} I_1^2 + \frac{A_u^T}{3} I_1 = \tau_u^{2T} \tag{13}$$

where A_u and τ_u are material constants

$$A_u^c = \frac{(f'_{bc}/f'_c) - 1}{2(f'_{bc}/f'_c) - 1} f'_c \quad , \quad A_u^T = \frac{1 - (f'_t/f'_c)}{2} f'_c \tag{14}$$

$$\tau_u^{2c} = \frac{2(f'_{bc}/f'_c) - (f'_{bc}/f'_c)^2}{3[2(f'_{bc}/f'_c) - 1]} f'^2_c \quad , \quad \tau_u^{2T} = \frac{(f'_t/f'_c)}{6} f'^2_c$$

to be determined from the uniaxial tensile strength f'_t, uniaxial compressive strength f'_c, and equal biaxial compression strength f'_{bc}.

Fig. 6 shows the fracture surface of Eqs. (12) and (13) in the general $(I_1, \sqrt{J_2})$ stress space together with the initial discontinuous surface and subsequent loading surfaces. These initial and subsequent surfaces are obtained by simply scaling the fracture surface down to different sizes. For biaxial tests, initial discontinuity is seen to occur at about 75% of the fracture strength in the uniaxial and biaxial compression tests (Fig. 3a); and at about 60% of the fracture strength in tests involving direct tension (Fig. 3c). These surfaces are required in the later part of the paper where the theory of work-hardening plasticity is applied to construct the incremental stress-strain relationships for the elastic-plastic behavior of concrete.

(2) strain criterion

$$g(\varepsilon_{ij}) = 0 \tag{15}$$

The simplest type of fracture criterion in terms of strains in compression states is to assume that it has the same form as that of stress criterion. For instance, in the case of Equation (12) the corresponding strain criterion in compression states with the maximum tensile strain criterion as cut-off in tension states has the forms

$$g(\varepsilon_{ij}) = J_2(\varepsilon_{ij}) + \frac{A_u^c}{3}\left(\frac{\varepsilon_u}{f_c'}\right) I_1(\varepsilon_{ij}) = (\tau_u^c)^2 \left(\frac{\varepsilon_u}{f_c'}\right)^2 \tag{16}$$

$$\text{or maximum of the principal tensile strains} = \varepsilon_t \tag{17}$$

in which

$$I_1(\varepsilon_{ij}) = \varepsilon_{ii} \tag{18}$$

$$J_2(\varepsilon_{ij}) = \frac{1}{2}\varepsilon_{ij}\varepsilon_{ij} - \frac{1}{6}\varepsilon_{ii}^2, \text{ and} \tag{19}$$

A_u^c is defined in Eq. (14) and ε_u and ε_t specify the maximum ductilities of concrete under uniaxial compressive and tensile loading conditions, respectively. The fracture surface defined by Eqs. (16) and (17) in the biaxial principal strain space is shown in Fig. 7.

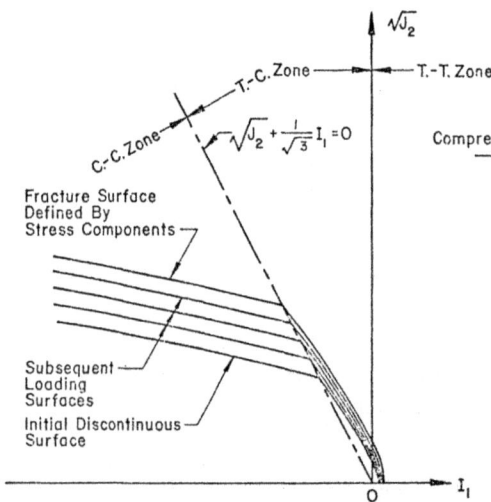

Fig. 6   Loading Surfaces of Concrete in $(I_1, \sqrt{J_2})$ Space.

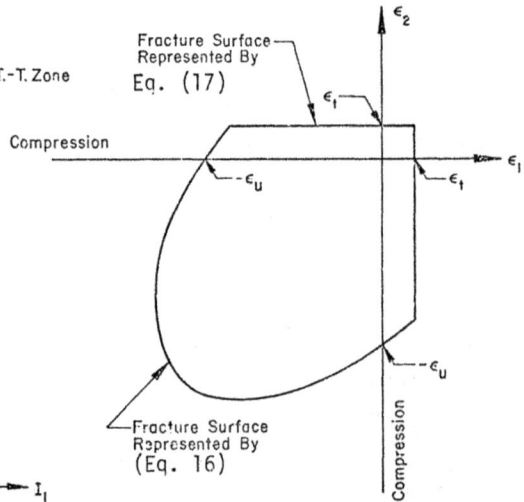

Fig. 7   Fracture Surface Defined by Strain Components in the Biaxial Strain Space.

When the stress state in the concrete satisfies either the stress criterion, Eq. (5), or the strain criterion, Eq. (15), fracture of concrete is assumed to occur.

4.3   Stress-Strain Relations of Fractured Concrete

In the following, we discuss the stress-strain relations of a fractured concrete, using a proper physical model which simulates the kinematics of a fractured concrete: crushing type and cracking type.

The term *"crushing"* is used to indicate the complete rupture and disintegration of the material under compression type of stress states. After crushing, the current stresses drop suddenly to zero and the concrete is assumed to lose its resistance completely against further deformation. The term *"cracking"* is used to indicate a partial collapse of the material across the plane of cracking under tensile type of stress states. An infinite number of parallel fissures are assumed to occur in the direction normal to the offending principal tensile stress or strain. Once a crack has formed, the tensile stress across the crack drops suddenly to zero and the resistance of the material normal to the crack direction is reduced to zero in this direction against further deformation. However, material parallel to the crack is assumed to carry stress according to the uniaxial or biaxial conditions prevailing parallel to the crack. The conditions for further cracks or opening and closing of the existing cracks will be described later when the matrix constitutive relationships for fractured concrete are discussed.

The tensile type of stress states (including tension-compression type) and multi-axial compressive type of stress states can be defined in the following manner:

(1) In terms of stress invariants

For example, in the case of Eqs. (12) and (13), it can be shown that when a stress state satisfies the condition

$$\sqrt{J_2} \leq -\frac{1}{\sqrt{3}} I_1 \quad \text{and} \quad I_1 \leq 0 \tag{20}$$

The stress state is of compression type and crushing type of fracture is assumed to occur. Otherwise, it is of tensile type and cracking type of fracture is assumed to occur.

(2) In terms of principal stresses σ_1, σ_2 and σ_3:

If all the three principal stresses are compressive (negative) or zero, the stress state is of compressive type and "crushing" type of fracture is assumed to occur. Otherwise, the stress state is of tensile type and "cracking" type of fracture is assumed to occur.

Incremental Stress-Strain Relationships

The fracture model shown schematically in Fig. 8 is used here to discuss the required incremental stress-strain relationships of fractured concrete. The slopes of lines 0-1 and 2-3 represent the "material stiffness" before and after the occurence of fracture. The total stresses released are denoted by the stress vector $\{\sigma_0\}$ (line 1-2 in Fig. 8). The released stresses are redistributed to adjacent material of the entire structure. The released stresses are assumed to be generated discontinuously from zero to the specified magnitudes at the instant of fracturing. The incremental stress-strain relationship after fracturing can be represented by the well-known relationship

$$\{d\sigma\} = [D]_c \{d\varepsilon\} \tag{21}$$

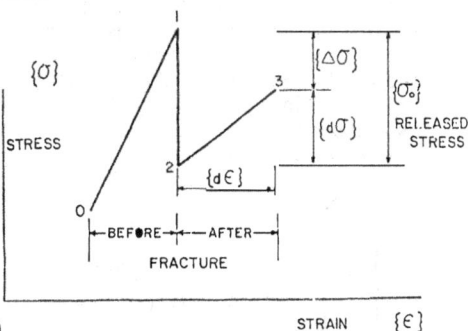

Fig. 8 Stress-Strain Model of a Fractured Concrete.

The total stress change $\{\Delta\sigma\}$ in the fractured material during this process can be written formally as

$$\{\Delta\sigma\} = \{d\sigma\} - \{\sigma_0\} = [D]_c\{d\varepsilon\} - \{\sigma_0\} \qquad (22)$$

in which

$[D]_c$ = material stiffness matrix after fracturing (either the cracking type or the crushing type),

$\{\sigma_0\}$ = the released stress vector during fracturing

Crushing Type of Fracture

It is usually assumed that at the instant of crushing, all stresses at the point just prior to the crushing are released complete and thereafter the concrete is assumed to lose its resistance completely against any type of further deformation. This implies that the stress point 2 in Fig. 8 drops to zero and the slope of line 2-3 is also zero ($[D]_c$ = 0 in Eq. 22 and $\{\sigma_0\}$ = current stress vector at the point just prior to crushing).

Cracking Type of Fracture

A crack is assumed to form in the planes (or surfaces for axisymmetric problems) perpendicular to the maximum principal tensile stress direction if the stress fracture criteria controls (Eq. 5), or perpendicular to the maximum principal tensile strain direction if the strain fracture criteria controls (Eq. 15). In order to avoid the complexities of the problem, further restriction pertaining to the crack formation is usually introduced. Cracks are assumed to form only in the planes perpendicular to the plane for planar problems or only in the axisymmetrical surfaces for axisymmetric problems.

It is further assumed that at the instant of the crack formation, only the normal stress perpendicular to the cracked plane and the shear stress parallel to the cracked direction are released and the other stresses are assumed to remain unchanged. It follows that the stress states of the cracked material are reduced to (1) the uniaxial stress states parallel to the cracked direction for plane stress problems; (2) the biaxial stress states in the cracked and z directions for plane strain problems; or (3) the biaxial stress states in the cracked and circumferential (θ) directions for axi-symmetric problems, respectively.

Assuming the behavior of the sliced material between two adjacent cracked planes is linearly elastic, the incremental stress-strain relationships for the cracked material can be derived. For instance, in the case of plane stress, the stress-strain relationship in the global coordinate system has the form, in the usual notation for stresses and strains,

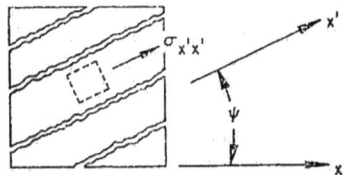

Fig. 9   Pattern of Cracks and Stress Distribution in a Cracked Concrete.

$$\begin{Bmatrix} \Delta\sigma_x \\ \Delta\sigma_y \\ \Delta\tau_{xy} \end{Bmatrix} = \left[E\{b(\Psi)\}\{b(\Psi)\}^T \right] \begin{Bmatrix} d\varepsilon_x \\ d\varepsilon_y \\ d\gamma_{xy} \end{Bmatrix} - \left[[I]-\{b(\Psi)\}\{b'(\Psi)\}^T \right] \begin{Bmatrix} \sigma_x \\ \sigma_y \\ \tau_{xy} \end{Bmatrix} \qquad (23)$$

where σ_x, σ_y and τ_{xy} are the current stress components at the point just prior to the formation of a crack and

$$\{b(\Psi)\} = \begin{Bmatrix} \cos^2\Psi \\ \sin^2\Psi \\ \sin\Psi\ \cos\Psi \end{Bmatrix}, \quad \{b'(\Psi)\} = \begin{Bmatrix} \cos^2\Psi \\ \sin^2\Psi \\ 2\sin\Psi\ \cos\Psi \end{Bmatrix}, \quad [I] = \begin{bmatrix} 1 & 0 & 0 \\ 0 & 1 & 0 \\ 0 & 0 & 1 \end{bmatrix} \quad (24)$$

in which Ψ = the direction of the cracks as shown in Fig. 9.

For the case of plane strain, the stress-strain relationship (23) still holds except that E be replaced by $E/(1-v^2)$. Details of this development together with a plasticity formulation for the nonlinear behavior of concrete for both plane and axisymmetric problems can be found in Ref. 14.

Further Cracking or Crushing of the Cracked Concrete

After the formation of initial cracks, the structure can often deform further without overall collapse. Thus, the possibility of crack closing and opening and the formation of further cracks may arise. These additional transition charac- teristics in the cracked material can be taken into account throughout the analysis. This is described further in what follows.

Formation of the Secondary Cracks

For a cracked concrete, the stress state reduces to the uniaxial one for plane stress problems or to the biaxial ones for plane strain and axisymmetric problems, respectively (see Fig. 9). The dual fracture criteria as defined by Eqs. (5) and (15) are still applicable for such a concrete. Once the reduced stress state satisfies one of the fracture criteria, further fracture (crushing or cracking) is assumed to occur and the direction of the new cracks are allowed to occur perpendicularly to the first cracks. The first cracks and the new cracks are called the *primary* cracks and the *secondary* cracks, respectively.

In the following, the incremental stress-strain relations for the fractured con- crete are summarized:

(1) For the crushed concrete or for the cracked concrete in which more than a single set of cracks are open, the material stiffness is assumed to be zero for further loading and the current state of stress at the point just prior to fracture is assumed to be released completely;

(2) For the cracked concrete in which a single set of cracks are open, the incremental stress-strain relation of the type as given by Eq. (23) can be used;

(3) For the cracked concrete in which all sets of cracks are closed, the concrete is assumed to be completely healed and it behaves as a linear elastic material.

5. NONLINEAR ELASTIC AND VARIABLE MODULI MODELS

The linear relations between the mean response and the deviatoric response of concrete

$$p = K\varepsilon_{kk} \quad (25)$$

$$s_{ij} = 2Ge_{ij} \qquad (26)$$

written in the preceding section are *isotropic* and *reversible*. Clearly, then, a simple extension of these relations with the elastic constants replaced by scalar functions associated with either the stress invariants or strain invariants will have the property of isotropy and reversibility also. Experimental results show that approximately unique relationships existed between hydrostatic stress p and volumetric strain ϵ_{kk} (or octahedral normal stress-strain $\sigma_{oct} = p$, $\epsilon_{oct} = \epsilon_{kk}/3$), and between deviatoric stress and strain (conveniently represented by octahedral shear stress and strain $\tau_{oct} = \sqrt{2J_2/3}$, $\gamma_{oct} = 2\epsilon_{oct} = \sqrt{8J_2^{(\epsilon)}/3}$) until fairly close to peak or ultimate stress conditions [15,16]. The unique octahedral normal stress and strain, and octahedral shear stress and strain relations have been determined experimentally by several investigators [7,15,16,17] from which the tangent bulk and shear moduli as nonlinear functions of the stress and/or strain tensor invariants can be derived. Note that these nonlinear relationships are developed mainly from experimental data under multiaxial compressive stress state.

The simplest type of such invariant relationships is to assume the tangent bulk modulus K_t and the tangent shear modulus G_t in Eqs. (25) and (26) to be functions of the first and second stress invariants I_1 and J_2 respectively, i.e.

$$K_t = K_t(I_1) = K_t(\sigma_{ii}) \qquad (27)$$

$$G_t = G_t(J_2) = G_t(\tfrac{1}{2} s_{ij}s_{ij}) \qquad (28)$$

These relationships can then be introduced into Eqs. (25) and (26) in terms of stress rate and strain rate (incremental) relations

$$\dot{s}_{ij} = 2G_t\dot{e}_{ij} \qquad (29)$$

$$\dot{p} = K_t\dot{\epsilon}_{kk} \qquad (30)$$

values of E_t and ν_t can be obtained by standard expressions from linear elasticity. Once these current tangent moduli are known, the current tangent elasticity matrix, with any modification due to cracking as described in the preceding section, can be determined. Thus the nonlinear deformation response of concrete up to fracture can now be simulated *incrementally* as a piece-wise linear elastic-fracture model, using the updated or current values of the tangent moduli.

Note that there is a neat and logical separation between the mean response and the deviatoric or shear response of concrete, which is exhibited very clearly by Eqs. (29) and (30). However, there will be interaction between the two through the change in the magnitude of the scalar function G_t and K_t with variation in $I_1 = \sigma_{kk}$ and $J_2 = \frac{1}{2} s_{ij}s_{ij}$ if the scalar functions are generalized to be functions of both I_1 and J_2 as $G_t(I_1,J_2)$ and $K_t(I_1,J_2)$. This implies that volume change ϵ_{kk} will not depend solely on the mean normal stress p. Similarly, distortion or shear deformation, e_{ij} will not depend solely on the stress deviation or shear stress, s_{ij}. They will depend on each other, and interact through the variation of both bulk and shear moduli, if these moduli are assumed to be functions of both I_1 and J_2.

Equations (25) and (26) are nonlinear stress-strain relations for isotropic, elastic concrete which reduce to the linear forms when G and K are constants.

They represent elastic (reversible) behavior because the state of strain is determined uniquely by the current state of stress without regard to the history
of the loading. This is obviously not the case for concrete at relatively high
stress levels. These relationships also assume that the principal axes of strain
increment always coincide with the principal axes of stress increment (Eqs. 29
and 30). This is true only at low stress levels. At high stresses, especially
near failure, however, the strain increment axes in concrete probably coincide
more closely with the principal axes of stress (and not of stress increment).
Knowledge regarding this aspect of concrete deformation behavior under triaxial
stress states is still lacking and mostly unknown. Further, the biaxial compression data show clearly that concrete dilates near and after failure or peak
stresses and considerable volume expansion occurs as the compressive stresses increase. This *"dilatancy"* is not predicted by the present nonlinear elastic model.
Some of these limitations on nonlinear elastic modeling of concrete can be overcome by the introduction of the so-called *"variable moduli"* models.

Here, as in nonlinear elastic models, Eqs. (29) and (30), both the bulk and
shear moduli are taken as nonlinear functions of the stress and/or strain tensor
invariants. However, unlike the nonlinear elastic models, different functions
are used in initial loading, unloading, and reloading to reflect the irreversible
characteristics of inelastic deformations of concrete. A model of this type can
fit to a rather complete set of laboratory data for a particular concrete material.
At the present time, however, this approach has not been attempted in the field
of structural concrete due to the lack of available experimental data on concrete
stress-strain behavior subjected to biaxial and triaxial stress reversal and
cyclic loading.

The other problem associated with this type of formulation is that the model may
not satisfy all rigorous theoretical requirements for all stress histories. To
illustrate, for example, when the stresses are near *neutral loading* in shear,
the variable moduli model fails to satisfy the continuity condition along such
stress paths, and the resulting numerical solutions in practical applications
may become questionable. This was found to be the case in a ground shock computation in soil mechanics [18]. Such a concern on continuity condition also
can occur in the analysis of reinforced concrete problems.

All these limitations can be overcome if an elastic-plastic stress-strain relationship is employed. If the difference between the nonlinear elastic or varia-
blemoduli and more rigorous plasticity models to be described in what follows is
not s gnificant when applied to reinforced concrete structures under certain
loading conditions, then the present approach will be very attractive. This is
because the present modeling, unlike the plasticity modeling, contains no explicit yield condition and hence it is computationally simple. It is particularly
well suited for finite element code calculations where local stiffness is required. The variable moduli model, which is similar to the *deformation theory* of
plasticity, is capable of fitting repeated hysteretic data in cyclic loading.
At present, not enough data are available for such applications in reinforced
concrete structures. In the following, we discuss two plasticity models: an
elastic-perfectly plastic-fracture model and an elastic-plastic-strain-hardening-
fracture model.

6. ELASTIC-PERFECTLY PLASTIC-FRACTURE MODELS

It is known that under triaxial compression, concrete can flow like a ductile material on the yield or failure surface before reaching its crushing strains. The assumption made in Sec. 4 that concrete crushes completely once the fracture surface is reached is rather rough, but a fair first approximation. To account for this limited plastic flow ability of concrete before crushing, a perfectly plastic model can be introducted.

A typical elastic-perfectly plastic and fracture model is generally assumed to set linearly elastically below yield or fracture surface, and is perfectly plastic on the surface until crushing or cracking strains are reached. Thus under uniaxial conditions, the compressive stress-strain curve for this model would exhibit a short horizontal stress plateau at the peak stress value, f_c' and this plateau terminates when the crushing strain of $-\varepsilon_u$ is reached (Fig. 10). More generally, a limited tensile ductility can also be assumed for concrete under tensile type of loading.

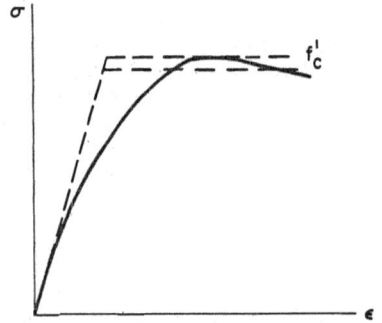

Fig. 10 Stress-Strain Curve for Concrete and Elastic-Perfectly Plastic Idealizations.

6.1 Incremental Stress-Strain Relationships for the Elastic-Plastic Behavior

The complete stress-strain relationships are developed in three parts: (1) before yield; (2) during plastic flow; and (3) after fracture. The linear elastic stress-strain relationships before yield and after fracture have been developed and described in Sec. 4. Only the plastic stress-strain relationship during the plastic flow needs to be added here. To achieve this one must first define (1) the condition of yield; and (2) the strain criterion for fracture. With these boundaries determined, the plastic stress-strain relationship in an incremental form can then be established.

The fracture criterion written in Sec. 4 in terms of stress invariants can be taken as the perfectly plastic yield surface. A considerable amount of numerical work has been done using (1) von Mises criterion, (2) extended von Mises criterion (or Drucker-Prager criterion); and (3) Coulomb or modified Coulomb criterion. With the exception of the von Mises criterion which is used widely in metal plasticity, all yield criteria developed for concrete incorporate a dependence of yield point stress on the mean normal stress $3p = I_1$ in addition to the dependence on the invariant of the "averaged" maximum shear stress J_2.

To construct the stress-strain relation in the plastic range, the *normality* of the plastic deformation rate vector to the yield surface or the so-called *flow rule* is commonly employed. The dependence of yield function on the mean normal stress and the concept of flow rule lead, in general, to a plastic volume increase under pressure. This *"dilatancy"* near failure is usually observed in concrete and rock type of materials. It is of interest to note that this phenomenon is generally not observed in soil type of medium except possibly in densely packed dry sands. Consequently, recent development in soil plasticity attempts to keep this effect as close to zero as possible. Here, as in rock materials, we attempt to allow for dilatancy. The historical development of various types of plasticity models used in soil mechanics as well as in rock-like materials, such as concrete, along with some numerical results using linearly elastic-perfectly plastic models, is given in a recent book by Chen [10].

The normality condition of the plastic deformation vector $d\varepsilon_{ij}^p$ to the yield surface $f(\sigma_{ij})$ can be expressed mathematically in the form:

$$d\varepsilon_{ij}^p = \lambda \frac{\partial f}{\partial \sigma_{ij}} \tag{31}$$

in which $\lambda > 0$ is a scalar proportionality factor. Once a yield function is determined, the plastic stress-strain relationship can be derived in a rather straight-forward manner by performing the operations indicated in the above equation.

Recently, the post-fracture behavior of concrete under biaxial compression states has been reported [16]. The normality flow rule used for plastic flow of concrete before fracture is not strictly observed in the case of *fractured* concrete. This seems to indicate that Eq. (31) is not proper for the development of a stress-strain relationship of a fractured concrete. Instead, this relationship should be developed on the basis of a proper physical model which simulates closely the kinematics of a fractured concrete: crushing type and cracking type. This is described in Sec. 4.

Due to the lack of available experimental data on concrete ultimate deformation capacity under multiaxial stress states, the strain criterion required for the fracture of concrete under compression states is usually developed by simply converting the yield criterion in stresses directly into strains. This is accomplished, for example, by simply replacing f'_c with ε_u, τ_{oct} or J_2 with the octahedral shearing strain ε_{oct} or $J_2^{(\varepsilon)}$, and p with the hydrostatic strain ε_{kk}. The maximum tensile strain criterion is often used as the cut-off plane for this compression surface. The resulting fracture envelope in the biaxial strain space has the shape shown in Fig. 7. Note that the magnitude of the maximum tensile strains at failure is generally not a constant in the tension-compression states, but increases with the degree of compression. From biaxial stress tests, it is found that concrete can sustain significantly higher indirect tensile strains than direct tensile strains. A general strain criterion for fracture of concrete is still incomplete and further research on this is urgently needed.

At the present time, little of the "typical" experimental concrete data under cyclic loading conditions are available. As more experimental data are expected to be available in the future, some modifications of the present perfect plasticity models are expected in order to match better with these new cyclic loading data. These modifications can be grouped into the following three types: (1) the generalization of the yield condition to include *strain softening* beyond the peak flow stress; (2) the use of nonlinear relations of the type described in the preceding section for concrete before yield; and (3) the generalization of the flow rule for plastic concrete by introducing the concept of *plastic potential surface* which is not assumed to be identical to that of the yield surface. These modifications will lead to the development of a series of extended and advanced elastic-perfectly plastic concrete models.

6.2 Ductility Assumptions and Limit Analysis

If the tensile strength of concrete is assumed to be zero for dimensions of engineering interest, and not just a small fraction of the compressive strength, then the *limit theorems* of perfect plasticity will hold rigorously for this idealization [10]. They can be applied easily when a tensile crack is introduced in the failure mechanism. The early application of the limit theorems approach to the analysis and design of Voussoir arches is a good example.

Kooharian [19] made the extreme but not unreasonable assumption that concrete is unable to take tension and behaves as a rigid, infinitely strong material in compression. This idealization was selected because it was simple and provided a first approximation to the behavior of a real Voussoir arch.

If concrete is assumed to have zero tensile strength and a finite compressive strength, it may almost be considered as a real material. The assumption of infinite ductility in tension at zero strength is rigorous and conservative for the application of limit analysis, but the assumption of infinite ductility in compression at selected compressive stregnth is more questionable. Brittleness in compression and a falling stress-strain curve after a maximum strength is usually observed, as indicated in Fig. 10. This is not compatible with the limit theorems. If the compressive strain of concrete is small and is not repeated often, then the deformability of concrete in compression prior to an appreciable fall-off of stress may be sufficient to permit the consideration of limit theorems with the concrete idealized as perfectly plastic at a yield stress in compression that approximates the ultimate strength f_c'. However, with the safe assumption that concrete is unable to resist any tension, Chen and Drucker [20] have shown that the bearing capacity is just the unconfined compression strength of the column of material directly under the load.

Obviously, the idealization of zero tensile strength is not true. When a small but significant tensile strength is assumed, and if the Coulomb surface for failure in compression is taken to represent a perfectly plastic yield, the predicted capacity is found to be in good agreement with published test results [20]. These results convincingly show that tensile strength and local extensibility of concrete are essential to the ability of concrete blocks to carry the load.

In view of the fact that: (1) a complete elastic-plastic and fracture analysis of concrete and reinforced concrete structures by finite element method is always complicated; and (2) failure or collapse load is usually the governing condition in so many problems in reinforced concrete structures, the drastic idealization of concrete as a rigid, perfectly plastic material together with a proper failure criterion, such as the modified Coulomb criterion with a small tension cut-off, will lead to the simple and efficient method of limit analysis from which the collapse load can be calculated directly. Such idealizations are found acceptable in many practical situations such as (1) the load-carrying capacity of a construction joint in plain concrete [21]; (2) shear strength of reinforced concrete [22]; (3) load carrying capacity of concrete blocks [10,20,21]; and (4) plastic collapse analysis of reinforced concrete plates and shells [23].

7. ELASTIC-STRAIN HARDENING PLASTIC AND FRACTURE MODELS

The latest step in the development of concrete constitutive models is the use of the strain-hardening theory of plasticity. A yield surface called *loading surface* which combines both perfect plasticity and strain-hardening is postulated and an associated flow rule is used for the plastic concrete before fracture. This approach can be considered as a generalization of all previous models, and at the same time satisfies rigorously the basic principles of continuum mechanics such as the requirements of *uniqueness* of solution and *continuity* of near neutral loading paths. It also allows a good fitting of material property data. This is given in the forthcoming.

According to this approach, the stresses in a structure under operation condition are expected to be in the initial discontinuous range such that the concrete behavior can be expected to be characterized as linear elastic and formation of microcracks can be minimized, i.e., one may expect to avoid fatigue. This *initial discontinuous surface* is the limiting surface for elastic behavior. The elastic limit is defined here as the initial discontinuous surface similar to the fracture surface but a certain distance within the fracture surface. Fig. 11 shows a trace of these two extreme surfaces in a two-dimensional principal stress space. When the state of stress lies within the initial discontinuous surface, the material is assumed to be linear elastic and the linear elastic constitutive equations can be applied.

When the material is stressed beyond the initial discontinuous or elastic limit surface, a subsequent new discontinuous surface called the *loading surface* is developed. This new surface replaces the initial discontinuous surface. If the material is unloaded from and reloaded within this subsequent loading surface, no additional irrecoverable deformation will occur until this new surface is reached. If straining is continued beyond this surface, further discontinuity and additional irrecoverable deformation results. In other words, at each stage of the history of loading there exists a loading surface in stress space containing all states of stress which can be reached by elastic changes and any straining beyond this surface is accompanied by the irrecoverable plastic deformation. Beyond the elastic limit, the normality condition or the so-called associated flow rule is assumed to govern the post yielding stress-strain relations for concrete. Once the loading surface is defined, constitutive equations based on the concept of flow rule can be derived.

A crush and crack surface called *strain fracture surface* in terms of strains is also postulated to define the complete collapse for the yielded concrete. The *stress fracture surface* in terms of stresses is defined as the outermost extreme of the loading surface similar to that of perfectly plastic yield surface. Once the stress fracture surface is reached, the concrete begins to flow under constant stresses. Finally, the concrete is assumed to crack or crush depending upon the nature of the stress states, when the strain fracture surface is reached. The stress-strain relationship developed in Sec.4 for a fractured concrete can now be applied.

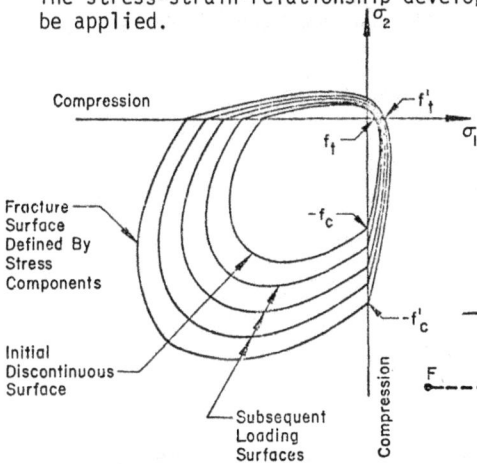

Fig. 11   Loading Surfaces of Concrete in Biaxial Stress Plane.

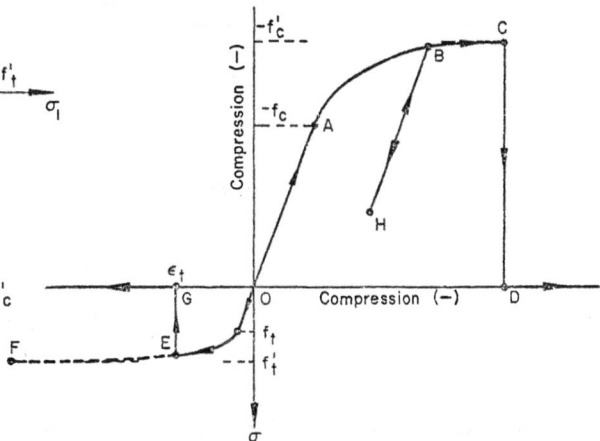

Fig. 12   Idealized Uniaxial Stress-Strain Curve for Concrete.

7.1 Uniaxial Stress-Strain Curve

The stress-strain characteristics of concrete under general multi-axial stress
states described above can best be explained from the uniaxial stress-strain
curve shown in Figure 12. The concrete is considered to be a linear elastic
strain (or work) hardening plastic, and fracture material in compression as well
as in tension. The modulus of elasticity in the elastic range is assumed to be
the same for both tensile and compressive stress states. The transition compressive and tensile stresses from elasticity to plasticity are $-f_c$ and f_t respectively. At stress level $-f_c'$, the concrete becomes perfectly plastic until crushing occurs at a compressive strain of $-\varepsilon_u$. The concrete is assumed to have a
limited tensile strength of f_t' and limited tensile strain of ε_t. When the tensile stress or strain exceeds its limiting value f_t' or ε_t, a crack is assumed to
occur in a plane normal to the direction of the offending stress or strain.

Unloading and reloading (line BH in the figure) in the plastic range follow the
initial elastic modulus E. The transition stresses from strain hardening plasticity to perfectly plastic flow in compression, which will eventually lead to
crushing of concrete, is $-f_c'$ and the transition to brittle fracture in tension
which will lead to cracking is f_t'. These limiting stresses are called fracture
stresses of concrete.

The complete stress-strain relationships under general stress states are developed
in three parts" (1) linear elasticity before yielding; (2) strain hardening plasticity for a post yielded concrete; and (3) linear elasticity after fracture.
Items (1) and (3) have been developed in Sec. 4. Item (2) will be discussed in
what follows.

7.2 Strain-Hardening

For a plastic concrete, a strain-hardening plasticity model as proposed in Ref.
[13] is used here to illustrate the procedure. The loading function adopted in
this model has the following form:

Compression-compression domain $(\sqrt{J_2} \leq \frac{1}{\sqrt{3}} I_1$ and $I_1 \leq 0)$

$$f(\sigma_{ij}) = \frac{J_2 + \frac{\beta}{3} I_1}{1 - \frac{\alpha}{3} I_1} = \tau^2 \tag{33}$$

Tension-compression or tension-tension domain $(\sqrt{J_2} > -\frac{1}{\sqrt{3}} I_1$ or $I_1 > 0)$

$$f(\sigma_{ij}) = \frac{J_2 - \frac{1}{6} I_1^2 + \frac{\beta}{3} I_1}{1 - \frac{\alpha}{3} I_1} = \tau^2 \tag{34}$$

where α and β are material constants, and τ is strain-hardening parameter. The
loading functions, Eqs. (33) and (34), can be depicted as the loading surfaces
shown in Figure 11 in the biaxial principal stress space and in Figure 6 in the
$(I_1, \sqrt{J_2})$ space for the more general case of triaxial stress states. For $\tau = \tau_0$
or $\tau = \tau_u$, the loading function represents the initial discontinuous (yield)
or fracture surface in the stress space, respectively, whereas $\tau_0 < \tau < \tau_u$
represents a subsequent loading surface. The constants τ_0, τ_u, α and β are to
be determined from the following three tests: uniaxial compression f_c', uniaxial
tension f_t' and equal biaxial compression f_{bc}' which may be assumed to be 1.16 f_c'.

Further, all the initial yield values for f_c, f_t and f_{bc} can be taken to be 30-60% of the corresponding maximum strength values.

In constructing the relationship between stress and strain increments in the plastic range for general stress states, one must define (a) the current condition for yield; (b) the general form of the stress-strain relationship; and (c) a criterion for work hardening. In the present formation, the current condition for yielding is determined by the maximum value of $\tau = \tau_{max}$ which the concrete has already experienced in its prior loading history. This value of τ_{max} determines the current loading function which controls the condition for yielding.

As for the incremental stress-strain relationships for the elastic-plastic behavior of concrete, we assume that the "normality" condition as commonly used in the incremental theory of metal plasticity still holds for plastic concrete. This normality condition (flow rule) results in the following incremental plastic stress-strain relationship.

$$de^p_{ij} = \frac{1}{H'} \cdot \frac{\partial f/\partial \sigma_{ij}}{\sqrt{(\partial f/\partial \sigma_{mn})(\partial f/\partial \sigma_{mn})}}\, df \tag{35}$$

in which

$$H' = \frac{dF}{d\varepsilon_p} = \text{work hardening modulus} \tag{36}$$

where

$$f(\sigma_{ij}) = F(\varepsilon_p) = \text{effective stress anc} \tag{37}$$

$$\varepsilon_p = \sqrt{de^p_{ij}\, de^p_{ij}} = \text{effective strain} \tag{38}$$

Matrix constitutive equations based on the concept of subsequent loading surface and associated flow rule have been derived [24]. A comprehensive review of the plasticity theory as applied to concrete mechanics is given in [10].

The effective stress $F = f(\sigma_{ij})$ is defined here to be able to extrapolate from a simple uniaxial compression test into the multi-dimensional situation. The initial yield occurs when the effective stress, F, equals the initial yield stress, f_c, measured in the uniaxial compression test. With loading continued, the subsequent yielding occurs and plastic strains in the uniaxial direction, ε^p_1, as well as in the lateral directions, $\varepsilon^p_2 = \varepsilon^p_3$, are introduced. For each value of effective stress, F = f, the corresponding effective strain,

$\varepsilon_p = \sqrt{\varepsilon^{p^2}_1 + 2\varepsilon^{p^2}_2}$ can be calculated by simply subtracting the elastic contribution from the total strains in the uniaxial compression test. This results in $\varepsilon^p_1 = \varepsilon_1 - \varepsilon^e_1$ and $\varepsilon^p_2 = \varepsilon_2 - v\varepsilon^e_1$. The slope of (F,ε_p) curve, H', is known as work hardening modulus. A crushing type of fracture occurs when the effective stress, F, equals the plastic flow stress, f'_c and the uniaxial compressive strain reaches the limit value, ε_u.

To include the microcrack-confinement effect on the work hardening modulus, H', in the multi-dimensional stress situation, the effective stress-strain curve can also be constructed from a biaxial compression test similar to that of uniaxial compression test. Since a biaxial stress-strain expression, expressed in terms of the principal stress ratio, has been developed in Ref. [5], closed-form expressions for the effective stress-strain curve and hence the effective work hardening modulus, H', under a biaxial stress state may be derived. Similarly, the effective stress-strain curve may also be developed from an almost unique relationship between octahedral stress and strain described in Sec. 5.

Equation (35) which determines the plastic strain increment $d\varepsilon_{ij}^p$, together with the elastic Hooke's law for the elastic strain increment $d\varepsilon_{ij}^e$, give the complete relation between stress increment, $d\sigma_{ij}$, and strain increment

$$d\varepsilon_{ij} = d\varepsilon_{ij}^e + d\varepsilon_{ij}^p \tag{39}$$

for a strain-hardening plastic concrete.

8. SUMMARY AND CONCLUSIONS

A brief review of various mathematical models for concrete that are currently being used in reinforced concrete analysis have been given in this paper. Based on the results and observations described herein, the following conclusions can be made:

1. The equivalent uniaxial model is the simplest model for fitting to biaxial experimental data. While it approximates many (but not all) features of biaxial experimental tests adequately, it can not match stress-strain data for the triaxial test. The model is particularly appealing to planar type of problems such as beams, panels and shells because (1) the stresses in these structures are predominantly biaxial, and (2) the model has its broad data base.

2. The variable moduli model gives a good fit to the full set of tests (uniaxial, biaxial and triaxial compression data under proportional loading) available. Unique relationships have been established between hydrostatic stress and volumetric strain (or octahedral normal stress and strain), and between deviatoric stress and strain (or octahedral shear stress and strain). From these relationships, tangent bulk and shear moduli can be derived. Thus the nonlinear deformational response of concrete is simulated incrementally as a piecewise linear elastic material with variable moduli. The model is therefore computationally simple and it is particuarly well suited for finite element code calculations where local stiffness is required. The model has also the capability of producing repeated hysteretic effects in cyclic loading. However, experimental data under reversed loading are still very scarce. This does not allow for further development. Variable moduli approach does not model accurately the behavior of concrete at high stress levels (near failure). It may have the continuity problem when stress paths are on or near neutral loading.

3. The strain-hardening plasticity model can be considered as a generalization of all the previous models and it satisfies all the basic principles of continuum mechanics. A set of equations has been presented that provide the general framework for further development. Many of the details have been worked out, but others require further investigation. Some of the interrelationships between linear elasticity, work-hardening plasticity, dual fracture criterion, mechanics of fractured concrete, empirical equation for biaxial stress modulus, and the empirical equations for tangent moduli are demonstrated.

4. Limit analysis which is developed on the basis of perfect plasticity model can be used to calculate the collapse load of some plain and reinforced concrete structures such as bearing capacity of concrete blocks, shear in reinforced concrete beams and slabs. In such a situation, the complicated nonlinear irreversible mechanical behavior of concrete and reinforced concrete can be drastically

idealized as a rigid-perfectly plastic material.

REFERENCES

1. RICHART, F. E., BRANDTZAEG, A., and BROWN, R. L.: "A Study of the Failure of Concrete Under Combined Compressive Stresses", Engineering Experiment Station, University of Illinois, Bulletin No. 185, 1928, 104 pp.
2. KUPFER, HILSDORF and RUSCH: "Behavior of Concrete Under Biaxial Stresses", ACI Journal, Proceedings, Vol. 66, No. 8, August, 1969, pp. 656-666.
3. NELISSEN, L. J. M.: "Biaxial Testing of Normal Concrete", Heron (Delft), Vol. 18, No. 1, 1972, 90 pp.
4. TASUJI, M. E., SLATE, F. O., and NILSON, A. H.: "Stress-Strain Response and Fracture of Concrete in Biaxial Loading", ACI Journal, Proceedings, Vol. 75, No. 7, July, 1978, pp. 306-312.
5. LIU, T. C. Y., NILSON, A. H., and SLATE, F. O.: "Biaxial Stress-Strain Relations for Concrete", Journal of the Structural Division, ASCE, Vol. 98, No. ST5, May, 1972, pp. 1025-1034.
6. LIU, T. C. Y., NILSON, A. H., and SLATE, F. O.: "Stress-Strain Response and Fracture of Concrete in Uniaxial and Biaxial Compression", ACI Journal, Proceedings, Vol. 69, No. 5, May 1972, pp. 291-295.
7. KUPFER, H. B, and GERSTLE, K. H.: "Behavior of Concrete Under Biaxial Stresses", Journal of the Engineering Mechanics Division, ASCE, Vol. 99, No. EM4, August, 1973.
8. NGO, D., and SCORDELIS, A. C.: "Finite Element Analysis of Reinforced Concrete Beams", ACI Journal, Vol. 64, No. 3, March 1967.
9. PHILLIPS, D. V., and ZIENKIEWICZ, O. C.: "Finite Element Non-linear Analysis of Concrete Structures", Proceedings, Institute of Civil Engineers, Vol. 61, Part 2, March, 1976, pp. 59-88.
10. CHEN, W. F.: "Limit Analysis and Soil Plasticity", Elsevier Scientific Publishing Co., Amsterdam, The Netherlands, 1975, 638 pp.
11. DRUCKER, D. C., and PRAGER, W.: "Soil Mechanics and Plastic Analysis or Limit Design", Quarterly of Applied Mathematics, Vol. 10, 1952, pp. 157-165.
12. BUYUKOZTURK, O.: "Non-linear Analysis of Reinforced Concrete Structures" Computers & Structures, Vol. 7, 1977, pp. 149-156.
13. CHEN, A. C. T., and CHEN, W. F.: "Constitutive Relations for Concrete", Journal of the Engineering Mechanics Division, ASCE, Vol. 101, No. EM4, August, 1975, pp. 465-481.
14. CHEN, W. F., and SUZUKI, H.: "Constitutive Models for Concrete", Preprint Volume, ASCE Annual Convention, Chicago, IL, October 16-20, 1978.
15. PALANISWAMY, R. and SHAH, S. P.: "Fracture and Stress-Strain Relationship of Concrete Under Triaxial Compression", Journal of the Structural Division, ASCE, Vol. 100, No. ST5, May, 1974, pp. 901-916.
16. ANDENAES, E., GERSTLE, K. and KO, H-Y.: "Response of Mortar and Concrete to Biaxial Compression", Journal of the Engineering Mechanics Division, ASCE, Vol. 103, No. EM4, August, 1977, pp. 515-525.
17. CEDOLIN, L., CRUTZEN, Y. R. J. and POLI, S. D.: "Triaxial Stress-Strain Relationship for Concrete", Journal of the Engineering Mechanics Division, ASCE, Vol. 103, No. EM3, June, 1977, pp. 423-439.
18. NELSON, I. and BALADI, G. Y.: "Outrunning Ground Shock Computed with Different Models", Journal of the Engineering Mechanics Division, ASCE, Vol. 103, No. EM3, June, 1977, pp. 377-393.
19. KOOHARIAN, A.: "Limit Analysis of Voussoir (Segmental) and Concrete Arches", ACI Journal, Vol. 24, 1952, pp. 317-328.
20. CHEN, W. F. and DRUCKER, D. C.: "Bearing Capacity of Concrete Blocks or Rock", Journal of the Engineering Mechanics Division, ASCE, Vol. 95, No. EM4, August, 1969, pp. 955-978.

21.   JENSEN, B. C.: "Some Applications of Plastic Analysis to Plain and Rein-
      forced Concrete", The Institute of Building Design Report No. 123, Techni-
      cal University of Denmark, DK-2800 Lynghby, 1977, 119 pp.
22.   NIELSEN, M. P. and BRAESTRUP, M. W.: "Plastic Shear Strength of Reinforced
      Concrete Beams", Bygningsstatiske Meddelelser, Vol. 46, No. 3, 1975.
23.   SAVE, M. A., and MASSONNET, C. E.: "Plastic Analysis and Design of Plates,
      Shells and Disks", North Holland, Amsterdam, 1972, 478 pp.
24.   CHEN, A. C. T. and CHEN, W. F.: "Constitutive Equations and Punch-Indenta-
      tion of Concrete", Journal of the Engineering Mechanics Division, ASCE, Vol.
      101, No. EM6, December, 1975, pp. 889-906.

5.6.2

Int. J. Mech. Sci. Vol. 35, No. 12, pp. 1097–1109, 1993
Printed in Great Britain.

0020–7403/93 $6.00 + .00
© 1993 Pergamon Press Ltd

CONCRETE PLASTICITY: MACRO- AND MICROAPPROACHES

W. F. CHEN

School of Civil Engineering, Purdue University, West Lafayette, IN 47907, U.S.A.

Abstract—The development and the present state of some key aspects of the theory of concrete plasticity are summarized. Emphasis is placed on the representation and interpretation of the inelastic deformation of concrete materials in the postelastic range. The first part of the paper discusses the development of plasticity-based models in the prepeak stress range. A general formulation of softening behavior from the plasticity theory combined with the fracturing (damage) theory is then described. The difficulties of characterizing concrete deformations in the postpeak stress range led to an extensive study of the relationship between macrospace features and microscopic events in recent years. This is briefly outlined, and some key aspects of the development on strain softening and strain localization are reviewed. Against the background of this information, attempts are then made to develop an elastic–plastic–damage applicative model. Special emphasis is placed here on the practicality of the unified model combining the classical theory of plasticity with the modern theory of continuum damage mechanics. Some thoughts on possible numerical algorithms and program development for this type of model are also discussed. The role of cement–paste–sand interfaces in providing a deeper understanding of the behavior of concrete materials at the microscale are explored. Directions for further research are indicated.

1. INTRODUCTION

Constitutive equations are of central importance to concrete mechanics and the engineering design of reinforced concrete structures. In the elastic range, the range of reversible deformation, the constitutive equations are embodied in Hooke's law. The mechanics of concrete in this range are well understood. Constitutive equations for concrete deformed beyond the elastic range represent an area of great importance, but they are still neither well established nor fully understood, despite the fact that this has been the subject of intensive research for many years. At present, most of the so-called complete stress–strain relations or curves under uniaxial and biaxial stress states, as commonly used in actual structural calculations, are fitted with the experimental facts because of the lack of proper and reliable constitutive equations. For example, at present, data from the so-called complete direct tension stress–strain curve cannot be related to the flexural tension stress–strain curve or splitting tension stress–strain curve.

To calculate the distribution of stress and the progress of the postelastic and postpeak stress deformations in a reinforced concrete structure, constitutive equations for micro-cracked, damaged concrete are needed that will allow one to track the progressive failure evolution of the composite system from first crack and spalling through subsequent crack propagation to the development of localized crack zones, and eventually leading to the crushing of unconfined concrete materials.

In this paper, special emphasis is placed here on the representation and interpretation of the inelastic deformation of concrete materials in the postelastic range. The difficulties of characterizing concrete deformations in the inelastic range in a realistic way result in different constitutive theories. At present, there are two basic approaches to characterize concrete behavior: the classical plasticity theory and the elastic–damage theory.

The stress–strain relationships in the postelastic range derived from the classical plasticity theory are based on the continuum viewpoint, despite the fact that this is precisely the area where concrete exhibits some "discontinuous" nature beyond "yielding", and a clear localized progressive failure process nature beyond the "peak" stress. The classical plasticity theory using a pressure-dependent yield function together with its associated flow rule enables one to go beyond the elastic range in a theoretically consistent way, because of the well-established theory in which uniqueness, stability and continuity can all be assured. Consistency and rigorousness are admirable, but the physical phenomenon of microcracked, damaged concrete that is the necessary evolution to failure for a composite system is the physical reality that cannot be avoided for further progress. The basic physical principles then must be used to obtain the relations between stress and deformation. It is this basic "constitution" of the concrete that determines the actual relations between stress and deformation.

The classical plasticity theory requires the postelastic deformation to proceed at a work-hardening stress level equal to the current yield strength of the concrete in an associated flow rule manner. However, the fact remains that few laboratory experiments have been done that substantiate the assumption of "normality" and the associated flow rule for concrete. In fact, concrete materials are made of the mortar–aggregate composite system with preexisting voids and microcracks, and are probably more fractured and frictional than plastic to a great extent, so that on theoretical grounds alone, one cannot expect much from the classical plasticity theory, other than that it gives a crude delineation of zones stressed in the elastic range, beyond the elastic range, and into the postpeak range resulting in a decrease in strength for further deformation (strain softening and localization).

In the elastic–damage theory, the inelastic deformation is described entirely as elastic damage of the material. The inelastic behavior of concrete materials is reflected only by the stiffness degradation, while there is no permanent deformation in the material after a complete unloading. However, permanent deformation is an important feature of the inelastic deformation of concrete materials, and this has been observed in many experiments. Despite this, the elastic–damage theory provides a useful technique to relate the mechanism of deformation obtained in a micromechanical study and observed in experiments to the macroconstitutive equations in continuum mechanics. A great deal of research has been done in recent years in this area, which firmly establishes the basic framework of the elastic–damage theory. A much better understanding has now been achieved on several fundamental issues, and a better modeling technique is now in order, but much remains to be done.

In this paper, the development of plasticity-based models in the prepeak stress range is first summarized. Emphasis is placed on the fundamental concepts peculiar to concrete materials. A general formulation of softening behavior by the plasticity theory combined with the fracturing (damage) theory is then discussed. Since the phenomenological models may not be adequate for describing the material behavior in the postpeak stress range, an extensive study of the relationship between macrospace features and microscopic events has been made in recent years. This will be briefly outlined and some of the state-of-the-art models for strain softening and strain localization will be reviewed. Against the background of this information, attempts will then be made to develop an elastic–plastic–damage applicative model. Special emphasis will be placed on the practicality of the model. A unified approach will be taken to develop such a model combining the elastic–damage theory with the plasticity theory. Considerations on possible numerical algorithms for an efficient and reliable implementation of the proposed model are also discussed. Directions of further research are indicated.

2. FAILURE CRITERIA AS THE START

There are three basic assumptions used in the development of an incremental theory of plasticity: an initial yield surface, a hardening rule and a flow rule. Since the plasticity theory was originally developed for metals, its application to frictional materials like concrete requires a considerable modification on these three assumptions. The early efforts to develop a plasticity model for concrete materials have been centered in search of a suitable failure surface. Then, the initial and subsequent yield surfaces are assumed in accordance with the shape of the failure surface. Since the failure surface serves as the strength criterion for concrete, it is the key element in the constitutive modeling of concrete materials. A yield or failure criterion of the Coulomb type combined with the theory of elasticity has been used widely in engineering practice to solve many interesting problems.

Experimental investigations performed in the late 1960s and early 1970s have revealed the shape of the failure surface in principal stress space as shown in Fig. 1. The failure

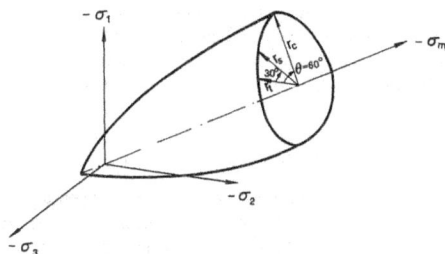

FIG. 1. Failure surface of concrete.

surface has curved meridians, representing the fact that the hydrostatic pressure produces the effect of increasing the shearing capacity of the material. The meridians start from the hydrostatic tension failure point, open in the direction of the negative hydrostatic axis. The tensile meridian, r_t, the compressive meridian, r_c, and the shear meridian, r_s, corresponding to $\theta = 0°$, $30°$, $60°$, respectively, satisfy the condition $r_t < r_s < r_c$. As shown in Fig. 1, the traces of the cross-sections of the failure surface are nearly triangular for tensile and small compressive stresses and become increasing bulged (more circular) for higher compressive stresses. The failure surface is convex.

Based on the knowledge concerning the failure surface, a variety of failure criteria have been proposed. Most of these criteria have been discussed in the book by Chen [1], where they are classified by the number of material constants appearing in the expressions as one-parameter through five-parameter models, and all include the strong influence of the normal stress on the shear stress required on the plane of sliding.

3. WORK HARDENING AS A NEXT STEP

Once a mathematically and physically attractive failure criterion has been established, work hardening was an obvious next step to establish stress–strain relations in the plastic range. To this end, a relatively sophisticated model—the model of nonuniform hardening plasticity of Han and Chen [2] was developed. The nonuniform hardening plasticity model adopts the most sophisticated failure surface of Willam–Warnke as the bounding surface; assumes an initial yield surface with a shape that is different from the failure surface; proposes a nonuniform hardening rule for the subsequent loading surfaces with a hydro-static pressure and Lode-angle-dependent plasticity modulus; and utilizes a nonassociated flow rule for a general formulation.

The work-hardening stress–strain behaviors of concrete based on the nonuniform hardening plasticity model are found to be in good agreement with experimental results involving a wide range of stress states and different types of concrete materials. The important features of inelastic behavior of concrete, including brittle failure in tension, ductile behavior in compression, hydrostatic pressure sensitivity and volumetric dilation under compressive loadings, can all be represented by this improved constitutive model.

Key phenomenological features of the inelastic behavior of concrete materials can be appropriately reflected in the classical theory of plasticity including irreversibility, the small stiffness near failure when compared with its resistance to elastic deformation, and approximate time independence. At present, the basic concepts of concrete plasticity are well understood, and satisfactory models of concrete in the prepeak stress range can be developed for practical use. Since the postulated rules for work-hardening materials have been applied, the number of model parameters can be significantly reduced. However, it should be noticed that the nonassociated flow rule, which has succeeded in controlling the volumetric strain, yields a nonsymmetric stiffness matrix. As a result, much more com-putational effort is usually required when the model is used in numerical analysis of a concrete structure. As pointed out by Drucker [3], the associated flow rule should not simply be viewed as a mathematical convenience; it defines a very comforting type of response in contrast with many of the consequences of nonassociated flow rules including possible nonuniqueness for a boundary value problem. Consequently, the associated flow rule should not be abandoned easily unless great care is taken with nonassociated rule formulations.

4. STRAIN SOFTENING AS RECENT PROGRESS

Engineering materials such as concretes, rocks and soils exhibit a strong strain-softening behavior in the postfailure range, showing a significant elastoplastic coupling for the degradation of elastic modulus with increasing plastic deformation. The stress-space formulation of plasticity based on Drucker's stability postulate for these materials encounters difficulties in modeling the softening/elastoplastic coupling behavior; strain-space formulation is therefore necessary for further progress.

As pointed out by Casey and Naghdi [4] some years ago, any arbitrary path in strain space can be specified independently by whether the material work hardens, is perfectly plastic, or strain softens. In this type of formulation, the difference in material behavior can be easily described, and it permits a continuous description from one type of behavior to the other with ease. This is in contrast with the conventional stress-space formulation for which the work-hardening and strain-softening behaviors must be treated differently. Although the representation of stress–strain behavior in either space can be translated into the representation in the other [5], the use of a strain-space formulation is more convenient for materials exhibiting strain-softening behavior. On the other hand, the stress-space formulation is often called for when we need a better physical understanding of the material behavior in terms of the applied stress and stress increments that are normally used in our physical description of material behavior.

In the axial compression test, the gradual degradation of elastic modulus in subsequent cycles of a repetition loading program is generally observed (Fig. 2). The elastic modulus decreases with increasing straining. This behavior is considered to be attributed to microcracking or fracturing and can be treated elegantly by the theory of progressive fracturing solids of Dougill [7]. The plastic–fracturing theory of Bazant and Kim [8] accounts for both the plastic deformation and the elastic degradation by combining the conventional stress-based plasticity theory with the fracturing theory. This combined approach involves two loading surfaces; one is the yield surface specified in stress space, and the other is the fracturing surface specified in strain space. Difficulties therefore arise in the definition of loading criterion. This is especially serious for the softening regime in which the strains continue to increase with the decrease of stresses. To avoid this problem, a strain-space plasticity approach can be used in formulating the plastic–fracturing theory, as

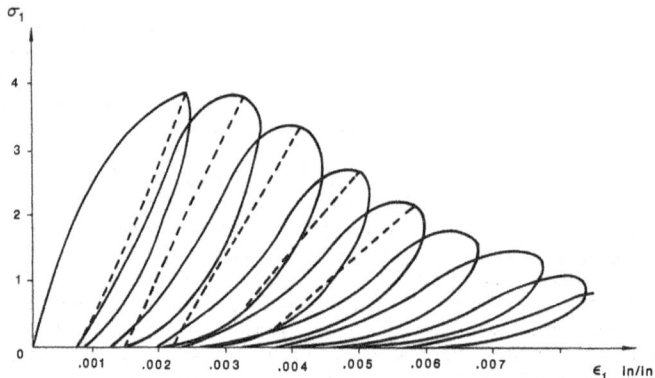

FIG. 2. Stress–strain curve under cyclic uniaxial compressive loading [6].

proposed by Han and Chen [9]. This later approach presents a consistent form of the constitutive relation for an elasto-plastic material with stiffness degradation in the range of work hardening as well as of strain softening. Features of this approach include the following: the relaxation surface is defined in strain space which serves as a criterion for further yielding and fracturing; the dissipated energy due to plastic fracturing is the parameter to record the material history and to define both the evolution of the relaxation surface and the elastic degradation; the weak stability postulate of Il'yushin is used to obtain a relaxation rule; the consistency condition is used in establishing the constitutive relation. Details of this development can be found in the book by Chen and Han [5].

A decrease in elastic modulus with increasing inelastic deformation in concrete materials is caused by the creation and growth of cracks that clearly leave less solid material per unit of total area in position to resist an increase in the average stress. Sooner or later, a critical stage will be reached beyond which there will be a continuing drop of load in a structural element or of average stress in a material with an increase in deformation or average strain. This results in a softening stress–strain curve for the material. At this postpeak stage of loading, it is obviously not possible for the structural element to maintain its macroscopically homogeneous condition, and increasing inhomogeneity of deformation must be developing as the result of the increasing concentration of deformation at the weakest or most overstressed regions. This phenomenon is known as *strain softening and localization*, for which a concentration of deformation develops in a small volume corresponding to an increase in strain with a decrease in stress. This physical process leads Bazant and Belytschko [10] to suggest that a minimum thickness of the zone of cracking be specified to provide not only computational stability but also a better description of the physical behavior of concrete materials.

The plastic–fracturing models proposed by Bazant and Kim [8] and Han and Chen [9] for describing the softening behavior of concrete materials have therefore been criticized in the past, in that the stress–strain relationship in the softening range is merely a nominal property, not a *material* property. In the postpeak stress range, strain localization usually occurs and the descending branch of the load–deformation curve may not be interpreted as the strain softening of the material. The macroscopic model of plastic fracturing may lose its physical basis. However, if the geometrical and structural effects are lumped together and are considered by some means like the model of Frantziskonis and Desai [11], the continuous description of the softening stress–strain relation may be reasonable.

5. PLASTICITY ON THE MICROSCALE AS THE CURRENT FOCUS

As discussed in the preceding section, in the postpeak stress range, concrete will soften and result in a decrease in strength, and the assumption of "normality" and the associated flow rule is questionable. It is well known that Coulomb friction invalidates "normality". Concrete materials are frictional to some extent and cannot be modeled properly by the classical plasticity theory beyond the peak stress range. Here is where microcracks and crack growth calculations interact with continuum calculations to achieve a real understanding of the behavior of concrete materials. Thus, strain softening represents an important subject for concrete mechanics research.

To this end, constitutive equations for microcracked, damaged concrete are needed. It is in this area where the fundamental research in concrete mechanics must be addressed. Subjects of concern include:

1. Micromechanical studies of concrete in the postpeak stress range including mechanisms of deformation, localization, strain softening, and brittle and ductile transition.
2. Micromechanical behavior of concrete pertaining to the microstructure of concrete at the interface of cement–paste–fine (sand) and coarse aggregate particles.

We have discussed so far the use of plasticity theory to describe the macroscopic stress–strain behavior of concrete. We shall now turn to plasticity on the *miniscale* and *microscale*. Microscopic observations can serve to reason out the stress–strain response and to capture its fundamental characteristics. Some fascinating applications of mechanics to concrete materials on the microscale have been reported in recent years. We shall start with a review of the relationship between macroscopic features and microscopic events that were studied recently by Yamaguchi and Chen [12] for the simple cases of uniaxial tension and uniaxial compression. The significance and role of the interfaces of the coarse aggregate with cement paste and sand particles in the stress–strain behavior of concrete are then explored. Directions for further research are indicated at the end.

It has been shown that the extension of bond cracks is responsible for the nonlinear behavior of concrete, and the growth of these cracks in the form of continuous cracks leads to failure in the case of uniaxial compression. This crack growth may be attributed to the sliding movement at the bond crack surfaces and to the sideways movement of aggregates (Fig. 3a). These mechanisms may result in irreversible (plastic) deformations and inelastic volume dilatation. Further, the development of these cracks could take place around many of the large aggregates. This is probably the reason why the overall stress–strain relationship in the prepeak stress regime provides an adequate representation of material properties in an average sense.

However, bond cracks alone cannot cause failure since they are separated from one another. Failure occurs only when there are sufficient bond cracks interconnected with mortar cracks. The development of continuous crack patterns does not lead to the immediate loss of load-carrying capacity because concrete at this stage behaves as a highly redundant structure. As successive load paths become inoperative through bond cracking, alternative load paths (either entirely through mortar, or partly through mortar and partly through aggregate) continue to be created and become available for carrying additional load. As the number of load paths decreases, the intensity of stress and hence the magnitude of strain on the remaining paths increase at a faster rate than the external load.

When the continuous crack pattern is developed extensively, the load-carrying paths are reduced considerably, resulting in a decrease of load-carrying capacity, and the descending branch of the stress–strain curve begins to form. This cracks extension process introduces various mechanisms that govern the failure process of a specimen.

In the uniaxial tension case, the ascending branch of a stress–strain curve is nearly linear up to its peak load. This is because microcracks tend to be arrested much less frequently under tension. Thus, once cracks are initiated, they grow faster and lead to failure sooner than in a compression test. A specimen fails at a single critical section (one major crack) in the transverse direction [14]. The development of this crack across the width of a specimen is achieved soon after failure [15]. Although concrete is considered to be a brittle material in general, it exhibits a substantial postpeak resistance, probably due to the bridging of the cracked surface by aggregates and fibrous crystals.

Postpeak stress deformation is mostly the result of widening of a single crack across the specimen. The unloading behavior of the zones away from the cracked region is observed while the overall deformation advances [15]. This localized behavior is confirmed by another observation that the area under a load–displacement diagram, which may be regarded as strain energy dissipation during the crack formation process, is hardly affected by halving the height of a tension specimen [14]. This localization presents a striking contrast to the prepeak deformation, which is almost uniform over the entire specimen.

As a consequence of these preliminary studies, we have come a long way towards understanding the significance of cement–paste–sand particle interfaces in the stress–strain behavior of concrete materials in compression. In order to achieve a deeper understanding, a new approach to study the micromechanical behavior of concrete by linking its material science to structural engineering aspects has been under way at Purdue University. This approach requires considerations from a microscopic viewpoint in order to develop the basic understanding of the underlying mechanisms. With this understanding, microstructure study can be used to quantify the influence of interfaces at a larger scale with respect to the overall structural engineering behavior. This is described briefly later. Further discussions on this subject can be found in the paper by Chen and Cohen [13].

In studying the response of concrete to mechanical loading, researchers have traditionally considered concrete to be composed of three basic phases: (1) mortar; (2) coarse aggregate; and (3) mortar–coarse aggregate interface (or transition zone). Mortar is defined as cement

Concrete plasticity   1103

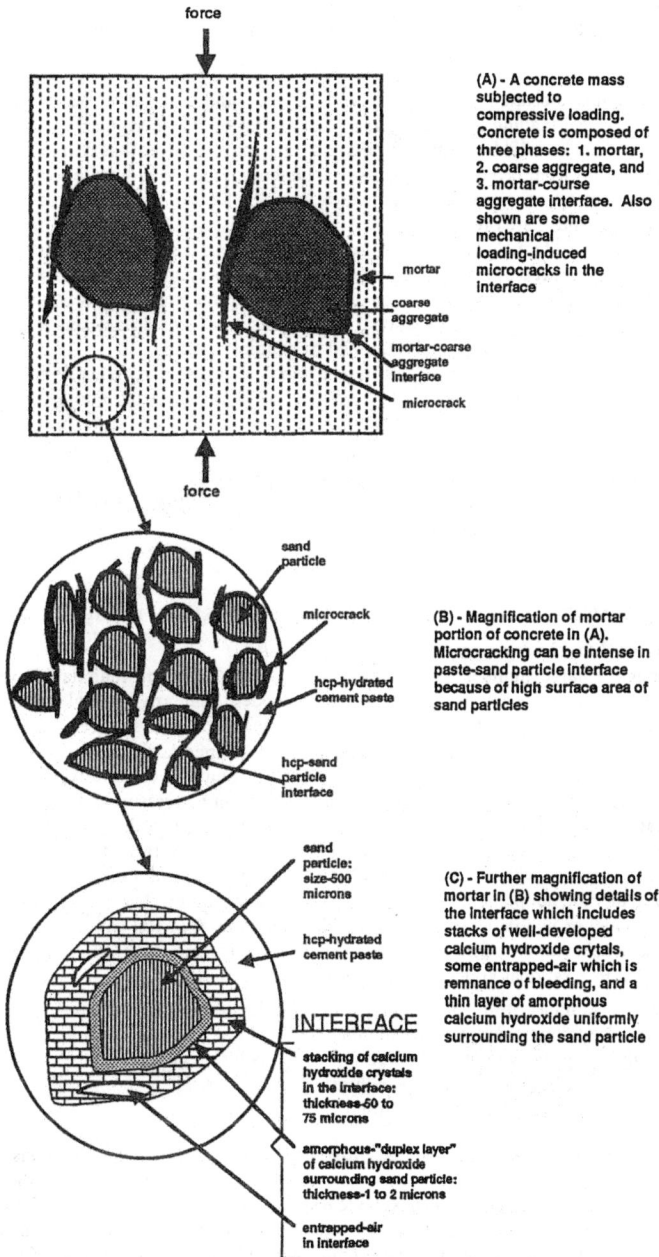

(A) - A concrete mass subjected to compressive loading. Concrete is composed of three phases: 1. mortar, 2. coarse aggregate, and 3. mortar-course aggregate interface. Also shown are some mechanical loading-induced microcracks in the interface

(B) - Magnification of mortar portion of concrete in (A). Microcracking can be intense in paste-sand particle interface because of high surface area of sand particles

(C) - Further magnification of mortar in (B) showing details of the interface which includes stacks of well-developed calcium hydroxide crytals, some entrapped-air which is remnance of bleeding, and a thin layer of amorphous calcium hydroxide uniformly surrounding the sand particle

FIG. 3. Details of concrete mass and magnification of its mortar portion for (A) concrete, (B) mortar portion of concrete, and (C) further magnification of mortar [13].

paste plus sand particles. This is illustrated in Fig. 3a which shows a concrete mass and its three phases. Also shown are some mortar–coarse aggregate interface cracks. Generally, external loading- and shrinkage-induced microcracks are initiated in these interfaces because they are most often the weaker links or flaws in conventional concrete. Earlier researches have demonstrated that in compression tests the behavior of concrete is very similar to that of mortar. This suggests that the role of sand particles in concrete behavior under mechanical loading is more significant than that of coarse aggregate particles and confirm that, unlike what has been assumed to date in many models, that the mortar matrix in concrete is far from behaving as a linear-elastic, homogeneous and continuum material.

Recent independent microstructural investigations by Bentur *et al.* [16] have confirmed these findings and therefore demonstrated that the phases to be considered in concrete should be modified to include: (1) cement paste; (2) inclusions (air voids, sand particles), and coarse aggregate particles; and (3) the cement paste–inclusion interface. Here, the mechanical properties and concentration of each phase are important in an analytical work. This is a significant departure from the traditional phases considered by mechanicians in analytical modeling of concrete behavior under applied mechanical loading.

Figure 3b is a magnification of the mortar portion of concrete. Whereas mortar has traditionally been regarded by mechanicians to be a linear-elastic, homogeneous and continuum phase, recent microstructural investigations discussed above have provided strong evidence that mortar is strongly discontinuous and heterogeneous, and that the role of sand particles and cement paste–sand particle interfaces (Fig. 3c) should not be ignored, at least in favor of mortar–coarse aggregate interfaces, as suggested earlier. This is the subject of current investigation at Purdue University [13].

6. ELASTIC–PLASTIC–DAMAGE THEORY AS THE NEXT LOGICAL STEP

6.1. *The need for the development of a realistic model*

Although a large number of constitutive models for concrete materials have been developed in recent years, realistic applicative models that are rational, reliable and practical and can be directly implemented into a general-purpose finite element analysis code with relative ease are still very limited. A rational concrete model should be able to describe adequately the main characteristics of the complete constitutive behavior of concrete materials, ranging from a tension and lower confining pressure state to a very high confining pressure state, as well as from the prefailure regime to the postfailure regime. Furthermore, the reliability of an applicative model is closely related to the numerical stability of a model, which in turn depends on the formulation of a model, as well as the numerical method adopted for its computer code implementation. The practicality requires that the model should be as simple as possible, as long as the main characteristics of the constitutive behavior of concrete materials are captured. The main characteristics that must be included in a rational applicative model of concrete materials are:

1. The high nonlinearity of the stress–strain behavior in the prefailure regime, i.e. the progressive failure characteristics of concrete.
2. The unrecoverable volume dilatation in the postfailure regime.
3. The softening behavior in the postfailure regime.
4. The elastic stiffness degradation caused by microcracking.

Based on the micromechanical studies together with our experience on constitutive modeling accumulated in recent years, as well as recent advances in the theory of continuum-damage mechanics, the next logical step is to develop a combined and practical model that can incorporate all these new advances and developments into an unified elastic–plastic–damage model for concrete materials. This is given elsewhere [16]. In this combined effort, we attempt to develop a framework for this comprehensive three-dimensional elastic–plastic–damage stress–strain relation for concrete materials, as well as for its possible implementation into a general-purpose nonlinear finite element analysis program. It is hoped that some simpler versions of this program currently under development at Purdue University can be modified for use on a microcomputer with disk storage.

6.2. *Essential aspects of elastic–plastic–damage modeling*

Herein, we shall review some recent achievements in the elastic–plastic–damage modeling of concrete materials. No attempt will be made here to review comprehensively the vast literature in this area. Comprehensive state-of-the-art reviews on the plasticity-based models can be found in Eberhardsteine *et al.* [17], and on the continuum-damage–mechanics-based models in Lemaitre [18] and in Chaboche [19].

As mentioned previously, the failure of concrete materials is attributed to the progressive propagation and concentration of microcracks in the material. This is defined as the *damage* of a material. The *continuum–damage mechanics* [20–22] based on thermodynamics can be used to model this phenomenon. In this approach, the stiffness degradation behavior is assumed to be caused by material damage. There are two types of damage variable usually employed to represent the damage. One is the isotropic or scalar damage which is related to the collapse of the microporous structure of the material. The other is the anisotropic or tensorial damage which is related to the creation of surfaces in the material due to decohesion. If a material is virgin isotropic elastic, then the scalar damage preserves its property, while the tensorial damage induces anisotropy for the elastic behavior of the material. This is called deformation-induced anisotropy. The damage growth in the material can be described by either prescribing a *damage evolution law* [23] or using the concept of damage surface [22].

Two types of damage models, the elastic–damage model and the elastic–plastic–damage model have been suggested for concrete materials. In the elastic–damage models, the inelastic behavior of concrete materials is reflected only by the stiffness degradation, and there is no permanent deformation in the material after a complete unloading. Dougill's work belongs to this catalogue. He first suggested the progressive fracturing theory in which the elastic stiffness is assumed to be a function of dissipated energy in a deformation process. A fracturing loading surface was introduced in strain space to define all states that can be reached without further fracturing. This surface was also employed as a potential surface to specify the direction of the fracture stress decrement.

To describe the coupling between the plasticity of concrete materials and its elastic stiffness degradation, the elastic–plastic–damage models have been suggested. Ortiz and Popov [24], among others, derived an elastic–plastic coupling constitutive relation using the law of thermodynamics and the theory of interacting continuum. The plastic–fracturing theory of Bazant and Kim [8] combines the conventional plasticity theory with the fracturing theory.

Frantziskonis and Desai [11] recently developed a strain-softening constitutive model combining plasticity theory with damage theory. The behavior of the material element is assumed to consist of two parts: a topical part and a damaged part. The damaged part obeys a rigid–perfectly plastic constitutive relation with zero yield strength, while the topical part follows an elasto-plastic constitutive relation without softening. Although both parts are nonsoftening, the overall response obtained by the theory of mixture turns out to be a hardening–softening type with an elastic stiffness degradation. This idea is instructive and reveals that a structural change in the material may result in a strain–softening behavior in an average sense even though the material itself does not soften.

To prevent the mesh-sensitiveness associated with strain softening in the finite element analysis, Bazant *et al.* [25] adopted a modified nonlocal continuum approach, in which the nonlocal averaging was only applied to the variables that control strain softening or damage. Combining this approach with the elastic–damage theory, an efficient algorithm to overcome the mesh-sensitivity has been obtained.

In summary, the elastic–plastic–damage model that combines the conventional plasticity theory with the continuum damage theory is very promising and should provide a reasonable modeling technique for describing the behavior of concrete materials. At present, it is still under active development.

7. SOME THOUGHTS ON NUMERICAL ALGORITHMS AND PROGRAM DEVELOPMENT

In the preceding section, we have discussed some existing elastic–plastic–damage models to highlight the particular difficulties associated with the constitutive modeling of concrete materials. A brief description of the basic concept of an elastic–plastic–damage model has been given. Some thoughts on the possible numerical algorithms used to implement this type of model in a finite element code together with the plan of possible program development will be presented in the following.

7.1. *Numerical algorithms*

The finite element incremental formulation of a structure with a nonlinear material property results in a system of nonlinear simultaneous equations. Solving the system of nonlinear simultaneous equations is an essential part of the finite element nonlinear material analysis. To perform the analysis efficiently and accurately, the algorithm involved in solving the nonlinear equation system must be carefully designed. Two types of algorithms are involved in this procedure. One is the algorithm used to solve the system of nonlinear simultaneous equations, and the other is the algorithm used to determine the current stress state, corresponding to a strain increment, for a given stress state, damage state, and plastic deformation history.

The quasi-Newton method [26] can be used in the present work for solving the system of nonlinear simultaneous equations. An important advantage of the quasi-Newton method is that its convergent rate is not sensitive to the stiffness matrix actually used in the iteration procedure [27]. We can even use the initial elastic stiffness matrix of the structure throughout the incremental steps without losing much of its efficiency. As a result, this method is most suitable for analyzing a structure made of material that exhibits strain softening and stiffness damage.

The incremental constitutive equation derived from an elastic–plastic–damage model relates an infinitesimal stress increment to an infinitesimal strain increment, which must be integrated in a finite element analysis. The particular algorithm used to perform this integration is very important. It not only affects the accuracy of the stress computed, but also affects the convergence of the iteration in a solution of the system of nonlinear simultaneous equations. The Runge–Kutta algorithm with a carefully designed procedure, as developed previously at Peking University by Yin and Zhang [27] in their program NOLM83, provides a good example for such an integration algorithm for practical use.

To prevent the mesh-sensitivity associated with the strain softening and stiffness damage, the nonlocal approach suggested by Bazant *et al.* [25] may be adopted to modify the algorithms mentioned above for the local approach. In the nonlocal approach, spatial averaging can be applied to the damage-related variables.

7.2. *Program development*

One of the current trends in the development of finite element engineering application programs is to implement new material models and new numerical methods into existing general-purpose finite element codes. New material models and new numerical methods reported in the open literature will certainly make the current analysis programs more reliable and more efficient. However, because of a lack of good organization for most well-established engineering software systems, they are generally not suitable for a direct updating to include these new features as required for such developments. New architectures should be used in the design of such software systems.

To this end, research jointly between the Purdue Schools of Civil Engineering and Computer Science has been in progress for the past 3 years and is ongoing in the development of a Structural Engineering Software Development Environment (SESDE) [28]. The overall goal of this work is to improve the quality and productivity in structural engineering research and instructional software development through creation of a domain-specific programming environment consisting of reusable software components and Computer-Aided Software Engineering (CASE) tools which support software reuse. The SESDE is an object-oriented programming environment based on C^{++} which will help alleviate many of the difficulties commonly encountered in maintaining the integrity of existing software components during the development of a new research capabilities.

Accompanying the present study, we will attempt to implement such an elastic–plastic–damage model for concrete materials as a module for the library. It is hoped that the creation of such a platform for constitutive model development will facilitate the sharing and transfer of information between researchers and practitioners in structural engineering modeling and computing. The saving from the rapid incorporation of research results to practicing engineering could be significant, if the theory and practice are properly considered in the developments.

8. CONCLUDING REMARKS

To calculate the distribution of stress and the progress of postelastic deformations in a reinforced concrete structure, constitutive equations are of central importance. The difficulties of characterizing concrete deformations in the inelastic range in a realistic way result in different constitutive theories. The classical plasticity theory and the elastic–damage theory are the two basic approaches currently available for the characterization of the inelastic constitutive behavior of concrete materials.

However, the fact is that concrete materials exhibit both extremes of behavior: that is, ductile flow and brittle fracture. Under certain conditions, a transition between brittle and ductile behavior occurs, yet no distinction between the two modes has been made in any of the present failure criteria and related theories. Ideally, constitutive equations should reflect both the modes of deformation and the possibility of a brittle–ductile transition.

In this regard, the elastic–plastic–damage theory is found useful to provide the basic framework for describing such characteristics of abrupt properties change as brittle fracture, ductile flow, softening, hardening and crack healing. In this approach, the mechanisms of deformation obtained from micromechanics studies can be related and represented in macroconstitutive equations in continuum mechanics. Herein, we have proposed a unified elastic–plastic–damage theory based on our recent fundamental understanding of concrete micromechanics. The combined theory of plasticity and damage mechanics is within the present state-of-the-art, and it in turn requires the development of a proper damage evolution law for concrete materials. This is a key element in the combined theory.

Particular features to be considered in this continuing development include: (1) postelastic deformation of brittle cracked concrete; (2) constitutive description admitting transition between brittle and ductile regimes; (3) the response of a concrete composite system in a nonuniform or nonhomogeneous stress field; and (4) postpeak stress deformation of fractured concrete, mode of deformation (localized), strain softening, crack healing, and ductile fracture.

Great progress has been made in the application of elasticity, plasticity and damage mechanics to concrete materials. A unified treatment of various existing mathematical models of concrete has been attempted from which a comprehensive three-dimensional elastic–plastic–damage stress–strain relationship for concrete can be formulated. It is hoped that this unified approach will lead to the development of a rational, reliable and applicative model for practical use. The purpose of this type of formulation is for the development of a general-purpose three-dimensional concrete structural analysis program, where the concrete is modeled by constitutive equations that reflect the essential features of concrete behavior, that result from the microcracks of mortar–aggregate interaction.

Good progress has also been made in the application of mechanics on the microscale to concrete materials. The connection between the mortar–coarse aggregate interface and the stress–strain behavior does appear to be rather well understood for simple compression and tension types of loading. However, the connection is not clear for concrete subjected to combined stress states, even at simple biaxial stress states. Much remains to be learned about the role of cement-paste–sand particle interfaces in the stress–strain behavior of concrete that is subject to, say, even an simple compression force.

A further improvement of our modeling techniques for material depends on our deeper understanding of the behavior of the material on the microscale. The constitutive modeling of concrete in the postpeak stress range is an important issue in the analysis of concrete structures. Much more remains to be done. In fact, a broad topic of great interest and importance is the design of tailor-made concrete materials themselves on the microscale to achieve desired macroscopic mechanical properties (strengths, fracture toughness, ductility, durability, etc.). It would be a dream for engineers to design the materials to match their requirements at an affordable cost. This dream cannot be realized until the governing mechanisms of softening behavior from a microscale forming mechanics viewpoint is fully understood.

REFERENCES

1. W. F. CHEN, *Plasticity in Reinforced Concrete*. McGraw-Hill, New York (1982).
2. D. J. HAN and W. F. CHEN, A nonuniform hardening plasticity model for concrete materials. *Mech. Mater.* **4**, 283–302 (1985).
3. D. C. DRUCKER, Conventional and unconventional plastic response and representation. *Appl. Mech. Rev.* **41**, 151–167 (1988).
4. J. CASEY and P. M. NAGHDI, Strain-hardening response of elastic–plastic materials, in *Mechanics of Engineering Materials* (edited by C. S. DESAI and R. H. GALLAGHER), pp. 61–89. Wiley, Chichester (1984).
5. W. F. CHEN and D. J. HAN, *Plasticity for Structural Engineers*. Springer, New York (1988).
6. B. P. SINHA, K. H. GERSTLE and L. G. TULIN, Stress–strain relations for concrete under cyclic loading. *J. Am. Concrete Inst.* **61**, 195–211 (1964).
7. J. W. DOUGILL, On stable progressively fracturing solids. *ZAMP* **27**, 423–437 (1976).
8. Z. P. BAZANT and S. KIM, Plastic-fracturing theory for concrete. *J. Engng Mech. Div., ASCE* **105**, 407–428 (1979).
9. D. J. HAN and W. F. CHEN, Strain-space plasticity formulation for hardening–softening materials with elasto-plastic coupling. *Int. J. Solids Structures* **22**, 935–950 (1986).
10. Z. P. BAZANT and T. BELYTSCHKO, Strain softening continuum damage: localization and size effect, in *Constitutive Laws for Engineering Materials* (edited by C. S. DESAI et al.), pp. 11–33. Elsevier, New York (1987).
11. G. FRANTZISKONIS and C. S. DESAI, Analysis of a strain softening constitutive model. *Int. J. Solids Structures* **23**, 751–767 (1987); Constitutive model with strain softening. *Int. J. Solids Structures* **23**, 733–750 (1987); Elastoplastic model with damage for strain softening geomaterials. *Acta Mechanica* **68**, 151–170 (1987).
12. E. YAMAGUCHI and W. F. CHEN, Microcrack propagation study of concrete under compression. *J. Engng Mech. ASCE* **117**, 653–673 (1991).
13. W. F. CHEN and M. D. COHEN, Micromechanical considerations in concrete constitutive modeling. *Proc. 10th ASCE Structures Congr.*, San Antonio, TX, pp. 270–273 (1992).
14. K. W. B. HURLBUT and S. STURE, Experimental constitutive and computational aspects of concrete failure. *Proc. U.S.–Japan Seminar on Finite Element Analysis of Reinforced Concrete Structures*, Tokyo, Vol. 1, pp. 149–171 (1985).
15. V. S. GOPALARATNAM and S. P. SHAH, Softening response of plain concrete in direct tension. *ACI J.* **82**, 310–323 (1985).
16. A. BENTUR and M. D. COHEN, Effect of condense silica fume on the microstructure of the interfacial zone in Portland cement mortars. *J. Am. Ceramic Soc.* **70**, 738–743 (1987); W. F. CHEN, Concrete plasticity: macro and micro approaches. Structural Engineering Report No. CE-STR-92-25, School of Civil Engineering, Purdue Univeristy, West Lafayette, IN (1992).
17. J. EBERHARDSTEINE, G. MESCHKE and H. MANG, Triaxial constitutive models for concrete. Report from Institut fur Festigkeitslehre, Technische Universitat, Vienna (1987).
18. J. LEMAITRE, Local approach of fracture, *Engng Fracture Mech.* **25**, 523–537 (1986).
19. J. L. CHABOCHE, Continuum damage mechanics: part I—general concepts. *J. Appl. Mech.* **55**, 59–64 (1988); Continuum damage mechanics: part II—damage growth, crack initiation and crack growth. *J. Appl. Mech.* **55**, 65–72 (1988).
20. L. M. KACHANOV, *Introduction to Continuum Damage Mechanics*. Martinus Nijhoff, The Netherlands (1986).
21. L. M. KACHANOV, Continuum model of medium with cracks. *J. Engng Mech., ASCE* **106**(EM5), 1039–1051 (1980).
22. D. KRAJCINOVIC and G. U. FONSEKA, The continuous damage theory of brittle materials. *J. Appl. Mech., ASME* **48**, 809–824 (1981).
23. J. MAZARS, A description of micro- and macroscale damage of concrete structures. *Engng Fracture Mech.* **25**, 729–737 (1986); A model of a unilateral elastic damageable material and its application to concrete, in *Fracture Toughness and Fracture Energy of Concrete* (edited by F. H. WITTMANN), pp. 61–71. Elsevier, New York (1986).
24. M. ORTIZ and E. P. POPOV, A physical model for the inelasticity of concrete. *Proc. R. Soc. Lond.* **A383**, 101–125 (1982); Plain concrete as a composite material. *Mech. Mater.* **1**, 139–150 (1982).
25. Z. P. BAZANT and F. B. LIN, Non-local yield limit degradation. *J. Numer. Methods Engng* **26**, 1805–1823 (1988); Nonlocal smeared cracking model for concrete fracture. *J. Structural Engng ASCE* **114**, 2493–2510 (1988); Nonlocal continuum damage, localization instability and convergence. *J. Appl. Mech.* **55**, 287–293 (1988).
26. K. J. BATHE, *Finite Element Procedure in Engineering Analysis*. Prentice-Hall, New York (1982).

Concrete plasticity 1109

27. Y. YIN and H. ZHANG, NOLM83—a finite element program for elasto-plastic stress-deformation and stability analysis of rock–soil system. Technical Report, Department of Geology, Peking University, China (1986).

28. W. F. CHEN, H. E. DUNSMORE, D. W. WHITE and H. ZHANG, SESDE—an envisioned software development environment for structural engineering. Structural Engineering Report CE-STR-90-6, School of Civil Engineering, Purdue University, West Lafayette, IN (1990); Preparing structural engineering research and education for the 21st Century. *J. Chinese Inst. Civil Engng and Hydraulic Engng* **2**, 95–106 (1990); Domain-specific object-oriented environment for parallel computing, steel structures. *J. Singapore Struct. Steel Soc.* **3**, 47–60 (1992).

5.7 Soil Plasticity (土壤塑性力學)

5.7.1

Recent Developments in Laboratory and Field Tests and Analysis of Geotechnical Problems / Bangkok / 1983

Plasticity modeling and its application to geomechanics

E.MIZUNO
Nagoya University, Japan

W.F.CHEN
Purdue University, West Lafayette, Ind., USA

ABSTRACT: The general background for nonlinear stress analysis in geotechnical engineering problems and mathematical modeling of soil behavior are presented within the framework of theory of continuum mechanics. Recently developed advanced plasticity models, as well as classical models, are reviewed and discussed with respect to their advantages and limitations for application to geotechnical engineering problems. Incremental constitutive matrices of several types of material models are then presented in a form that is suitable for direct numerical analysis, under both associated and non-associated flow rule assumptions. For a large-deformation analysis of soil response with advanced plasticity models, general incremental finite element equilibrium equations are developed. Further, step-by-step computer implementation procedures for Drucker-Prager model and strain-hardening cap models are presented in a process where the incremental equilibrium equations can be numerically integrated. In the analytical study, these plasticity models are applied to slope stability problems under seismic loading conditions. The finite element solutions corresponding to different material models are compared to each other. They are also evaluated from the viewpoint of the modern limit analysis method. Finally, the applicability and limitation of the plasticity models to geotechnical engineering problems in general, and earthquake induced landslide problems in particular, are discussed, based on the results from present large deformation finite element analyses.

1 INTRODUCTION

To parallel the application of advanced computational procedures into the nonlinear stress analyses on geotechnical engineering problems, further research related to the material models, or the characteristics of material strength, is urgently needed to provide more precise predictions of the complicated behavior of geological media. However, since the mechanical behavior of soils is influenced by many factors, such as density, residual stresses, stress history, existence of discontinuity, void ratio, temperature, time, and pore water pressure, etc., the problem is how to represent these factors with mathematical models and to incorporate these models into the numerical analysis. Consequently, the problem of modeling the mechanical behavior of geotechnical materials for use in analytical studies still remains one of the most difficult challenges in geotechnical engineering.

Although the application of limit equilibrium method and modern limit analysis method to geotechnical problems has been in the area of soil statics, and more

recently in the area of soil dynamics, the progressive failure behavior prior to collapse has not yet been analyzed. But a valid assessment of geotechnical engineering problems requires a complete progressive failure analysis under dynamic as well as static loading conditions. In turn, this requires a better understanding of the mechanical behavior of soil, the development of more advanced material models and the application of such material models into the finite element analysis. Also, this analysis is essential for the verification of the limit analysis method.

In the present paper, the following items are discussed:

1. The general background for nonlinear stress analysis in geotechnical engineering is presented.

2. Existing plasticity-based material models are reviewed and assessed with respect to their advantages and limitations as they apply to the numerical analyses in geotechnical engineering problems.

3. The theoretical development of plasticity models and their finite element implementation are presented within the context of

large and small deformation analyses.

4. A computational study of the progressive failure behavior of selected soil problems is conducted by employing existing plasticity material models.

5. Comparison of plasticity models with an available analytical technique of the limit analysis method is made by performing these analyses on the same selected benchmark problems.

6. Verification and critical appraisal of the "pseudo-static" method of limit analysis are made from the finite element analysis on seismic landslide problems.

7. Applicability and limitations of the plasticity models for solving geotechnical engineering problems in general, and earthquake induced landslide problems in particular, are established, based on the results from the finite element analysis.

The study will be limited to the case of short-term loading, for which creep effects can be neglected. In particular, the strain softening behavior of soils is not considered in the numerical analyses.

2 GENERAL BACKGROUND

Since soils are multi-phase materials comprised of mineral grains, air voids and water, the mathematical characterization of their behavior should ideally be based on a consideration of the behavior of the individual constituent elements and their interaction. However, such an approach can be rather complex and would not be particularly fruitful in geotechnical engineering applications. For the problems treated in geotechnical engineering, the scale of practical interest is in the range of tens to thousands of feet. Thus, these "microscopic" effects may be averaged and the soil can be idealized as a continuum. Soil mechanics, which is treated within the framework of such idealization, is therefore a branch of mechanics of solids. The word "mechanics" implies a mathematical formulation of the problem and of the basic equations to be used in its solution. In the continuum theory of soil mechanics that includes the mathematical theories of elasticity, plasticity, and viscosity, three basic sets of equations such as equations of equilibrium, equations of compatibility and constitutive equations must be applied. The essential set of equations that differentiates the soil from other solids is the relations between stress and strain. These relations may be simple or can be extremely complex, depending on the conditions to which soil material has been subjected (Chen and Saleeb, 1982, 1983).

Once soil constitutive equations are estab-lished, the general formulation for the solution of a solid mechanics problem can be completed. With the present state of development of finite element computer programs, a range of solutions in solid mechanics is not limited to linear problems but can be extended to include problems of various kinds involving material and geometric nonlinearities (Chen, 1982a; Mizuno and Chen, 1983a and 1983b).

2.1 Characteristics of soil behavior

Some typical stress-strain curves for soils in the triaxial tests are schematically shown in Fig.1. As can be seen from Fig.1a, the axial strain-deviatoric stress relation for normally consolidated clay in a drained test is characterized by a highly nonlinear response curve which rises to a maximum value and then remains at that level as straining continues. Here, the further straining is always associated with an increase in stress. This phenomenon is known as strain-hardening. The stress-strain curves for normally consolidated and over-consolidated clays in an undrained test exhibit the same overall behavior as that of normally consolidated clay in a drained test; only the values of initial modulus and/or peak stress at failure are different. However, overconsolidated clay in a drained test behaves differently from these three types of clay. The stress-strain curve has a clearly defined peak occuring at a low strain level. An element of this clay strained beyond the strain corresponding to the peak stress point, becomes weaker than it was at that point. This phenomenon is known as strain-softening.

Similar conclusions can be made from Fig. 1b for the behavior of sand. Dense sand in an undrained test and loose sand in a drained test show highly nonlinear responses that include strain-hardening. Dense sand in a drained test shows an initially linear portion and peak stress followed by strain-softening to a residual stress. The behavior of dense sand in a drained test is similar to that of overconsolidated clay in a drained test except that the magnitudes of peak stress and residual stress are substantially different. The curve for loose sand in an undrained test reaches its peak point in a region of low stress and relatively high strain, and shows strain-softening with a mild slope.

2.2 Idealizations of soil behavior

The stress-strain behavior of soils, as presented in the previous section, is not

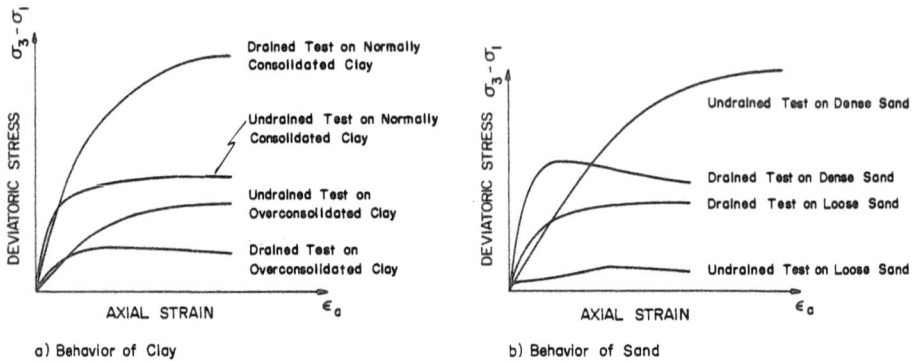

Fig.1 Typical soil stress-strain curves

linearly elastic for the entire range of loading of practical interest. In fact, actual behavior of soils is very complicated and they show a great variety of behavior when subjected to different conditions. Drastic idealizations for modeling soil behavior are therefore essential in order to develop simple mathematical models for practical applications. No one mathematical model completely describes the complex behavior of real soils under all conditions. Each soil model is aimed at a certain class of phenomena, captures their essential features, and disregards what is considered to be of minor importance in that class of applications. Thus, this constitutive model meets its limits of applicability where a disregarded influence becomes important. Although Hooke's law has been used successfully in soil mechanics to describe the general behavior of soil media under short-term working load conditions, it fails to predict the behavior and strength of a soil-structure interaction problem near ultimate strength conditions under which plastic deformation attains a dominating influence, while elastic deformation becomes of minor importance.

Under short-term loading, soil behavior may be idealized as time-independent, where the effects of time can be neglected. This time-independent idealization of soils can be further idealized as elastic behavior (a reversible one) and plastic behavior (an irreversible one). For an elastic material there exists a one-to-one coordination between stress and strain. Thus of a body that consists of this idealized material returns to its original shape whenever all stresses are reduced to zero. This reversibility is not the case for a plastic material. As a first step in constitutive modeling of

soils, it is therefore logical to utilize and define the classical theories of elasticity and plasticity as developed for an idealized material. However, there are in many cases considerable differences between the properties of soils and those of the idealized bodies. These differences may have a significant influence on the solution of some boundary value problems in soil mechanics. In such cases, the classical theories must be modified and extended so that the special properties of soils in certain practical applications are taken into consideration.

2.3 Elasticity and modeling

A linear elastic model is the simplest one, and it can be classified as isotropic, transversely isotropic, orthotropic or anisotropic, depending on the materials assumed in the analysis. Although the behavior of soil or rock shows in general the nonlinearities mentioned previously, an analysis with this model may give a reasonable solution, even under the assumption of a linear stress-strain relationship, provided the elastic parameters, such as Young's modulus and shear modulus, are properly determined. To overcome the material nonlinearity as well as the geometric nonlinearity, a piecewise linear model is considered. This model assumes that the stress-strain curve can be represented by piecewise linear relations. Consequently, the linear elastic model is used with different material parameters for each linear interval.

A nonlinear elastic model is regarded as an extension of the piecewise linear elastic model using infinitesimal linear intervals. The material constants are assumed to be a function of stress, strain, or their invariants. The elastic moduli at an arbitrary stress (or strain) level can be determined from experimental data with an interpolation method and curve fitting method (Desai, et.al. 1979). However, several shortcomings are associated with this model.

1. Since the model predicts only the elastic strain, the direction of the strain increment coincides with that of the stress increment. As the stress state at an element approaches the yielding or failure condition, however, the direction of the strain increment observed in experimental work becomes in general coincident with that of the total stress.

2. Dilatancy cannot be taken into consideration because there is no change of plastic volume caused by the change of the shear stress.

3. The behavior under cyclic loading cannot be treated.

Further, advanced nonlinear elastic models, such as the hyperelastic and hypoelastic models, have been developed in order to overcome some of the shortcomings listed above. The behavior predicted by hyperelastic models is independent of the stress path because the stresses are expressed only in terms of the strains. On the other hand, hypoelastic models, which include terms of the stress increments and strain increments as well as the stresses and strains, can predict behavior with stress dependence. Though these advanced nonlinear elastic models can more accurately predict the behavior of soils or rocks, the number of material constants in the models increases, and calibration of the models with the experimental data becomes complicated.

As in the nonlinear elastic models, variable moduli models contain no explicit yielding condition, but the bulk and shear moduli are both taken as nonlinear functions of the stress and/or strain tensor invariants. Thus, unlike the nonlinear elastic models, different functions are used for initial loading, unloading, and reloading to reflect the irreversible characteristics of inelastic deformations. A model of this type can be fitted to a rather complete set of laboratory data for a particular soil. However, a major problem with this type of formulation is that the model cannot satisfy all the rigorous theoretical requirements for all stress histories. For example, when the stresses are near neutral loading in shear, the variable moduli model fails to satisfy the continuity condition along such stress paths, and the resulting numerical solutions in practical applications become highly questionable. Thus the model is restricted to practical applications in which no substantial neutral loading, or near neutral loading occurs. For proportional loading, the model is theoretically correct.

2.4 Plasticity and modeling

All the limitations discussed above can be overcome if an elastic-plastic stress-strain relationship based on incremental theory of plasticity is employed. In the most fundamental sense, it is reasonable to assume that the soil behaves as an elastic-plastic material where the initial yield condition, which defines the occurrence of irreversible plastic strains and the subsequent yield condition, which defines the post yielding response beyond the initial yield condition, are taken into consideration.

Such drastic idealizations are thus valuable not only for the ease of treatment of practical engineering problems, but also conceptually for a clear physical understanding of the essential features of the complex behavior of a material under certain conditions. Therefore, for soils, as for metals, perfect plasticity is still an excellent design assumption, while very complex stress-strain relations of soil which require an ever-increasing elaboration in detail of a mathematical description may be approximated crudely by the simple isotropic, kinematic, or mixed hardening models, as will be explained in a later section. Recently proposed and developed plasticity models are all within the realm of this simplification.

The plasticity based formulations give a reasonably good fit of material property data from laboratory tests. Existing plasticity models are defined by a few material parameters which can be determined from standard tests. In general, these models can represent such important material characteristics as dilatancy, dependency of strength on stress or strain history, nonlinear hysteretic behavior under cyclic loading, and coincidence of strain increment and stress increment axes at low stress levels, with transition to coincidence of strain increment and stress axes at high stress levels. Further, these models rigorously satisfy the basic requirements of continuum mechanics such as uniqueness, stability, and continuity.

2.5 Basic concepts of incremental theory of plasticity

In further development of the elastic-plastic stress-strain relations for geotechnical materials, the following three concepts of incremental theory of plasticity such as yield function, flow rule, and hardening rule are required:

1. In general a yield function can be visualized as a yield surface in multi-dimensional stress space. This function defines the stress conditions at which plastic deformation will occur for a material element. A yield function f has the general form

$$f(\sigma_{ij}) \quad f_c \tag{1}$$

where σ_{ij} is a stress tensor and f_c is a constant for perfectly plastic material, but becomes a variable for strain-hardening material.

2. The flow rule defines the relationship between the next increment of plastic strain $d\varepsilon_{ij}^p$, and the present state of stress σ_{ij} for a yielded element subjected to further loading. This relationship is established using the concept of plastic potential function g, which is defined in terms of ε_{ij}^p and σ_{ij} and has the general form

$$g = g(\sigma_{ij}, \varepsilon_{ij}^p) \tag{2}$$

The plastic strain increments are thus written as

$$d\varepsilon_{ij}^p = d\lambda \frac{\partial g}{\partial \sigma_{ij}} \tag{3}$$

where $d\lambda$ is a positive scalar function. As can be understood from Eq. (3), the direction of the plastic strain incremental vector is defined to be normal to the plastic potential surface at the current stress point. As a result, the direction of the principal axes for incremental strain is in general not coincident with that for incremental stress.

3. According to experimental evidence, it is known that during the process of incremental plastic deformation the yield surface is changing continuously in size and shape. A law governing this aspect of the problem, i.e. of one which defines the manner of constructing the subsequent yield surface, is called a hardening rule. The variable f_c in Eq. (1) is usually written as

$$f_c = f_c(\sigma_{ij}, \varepsilon_{ij}^p) \tag{4}$$

There are several hardening rules that have been proposed to describe the growth of subsequent yield surfaces for strain-hardening (softening) materials. The choice of a specific rule depends on the ease with which it can be applied and its ability to represent the hardening behavior of a particular material. In general, three types of hardening rules have been commonly utilized

(Chen 1982a). These are isotropic hardening, kinematic hardening, and mixed hardening.

In an isotropic hardening rule, the initial yield surface is assumed to expand (or contract) uniformly without distortion as plastic flow continues. On the other hand, the kinematic hardening rule assumes that during plastic deformations, the loading surface translates without rotation as a rigid body in stress space, maintaining the size and shape of the initial yield surface. This rule provides a means of accounting for the Bauschinger effect. Therefore, kinematic hardening models are particularly suitable for materials with a pronounced Bauschinger effect, such as soils under cyclic and reversed types of loadings.

A combination of isotropic and kinematic hardening rules leads to a more general one, and therefore provides for more flexibility in describing the hardening behavior of the material. For such a mixed (combined) hardening model, the loading surface experiences translation as well as expansion (contraction) in all directions, and different degrees of Bauschinger effect may be simulated. Kinematic and mixed types of hardening rules are generally known as anisotropic hardening models.

In the last few years, several plasticity models with more complex hardening rules, combining the concepts of kinematic and isotropic hardening, have been developed and applied to describe the behavior of soils under cyclic loading. Some of these plasticity models will be discussed in a later section.

2.6 Nonlinear stress analyses in geotechnical engineering

By utilizing the finite element method, which had previously been employed mainly in structural engineering problems, the application of geological material models into soil and rock mechanics has rapidly advanced and become more powerful for nonlinear stress analyses that include both material and geometric nonlinearities. As a result, iterative calculations based on the linearly incremental formulation can be treated with comparative ease within the framework of finite element software. Consequently, practical problems, which previously could not have been solved analytically, can now be analyzed, and more reasonable data for design can be obtained from finite element calculations.

Since geotechnical engineering problems are so complex, idealizations must be performed with great care to obtain reliable solutions by finite element analysis. In recent years, the development of more sophisticated material models based on the concept of continuum mechanics has rapidly advanced with the aid of finite element computational work. Consequently, the range of application for finite element method has been extended not only to nonlinear stress analyses of foundation, slope, embankment and excavation problems, but also to discontinuity problems and interaction problems between structure and soil under dynamic as well as static loading conditions. Although the finite element method has had a profound effect on the analysis of nonlinear stress-strain behavior of geotechnical materials, there are still some analytically and theoretically difficult problems such as:

1. Selection or evaluation of material models,
2. Selection or evaluation of material constants,
3. Idealization of ground conditions,
4. Assessment of analytical results, and
5. Evaluation of analytical procedures in the finite element method.

When an elaborate idealization of the problem is made, the finite element nonlinear stress analysis becomes meaningful and powerful. Considering all the elaborate calculations involved, however, it is easy to forget that the numerical results are only a response to idealized conditions and may not be taken as the real behavior. Such obtained results must be assessed by experimental data and test data in situ. Then, the applicability of the material models and their numerical implementation in the nonlinear stress analysis can be re-evaluated.

3 PLASTICITY MODELS IN GEOTECHNICAL ENGINEERING

In this section, plasticity models of geotechnical materials are presented within the framework of a historical review (Chen 1980; Mizuno and Chen 1982b and Chen 1982b). It will be noted that the continuum mechanics sign convention (tension positive) is utilized to represent the plasticity models.

3.1 Classical perfect plasticity models

Perfect plasticity is an appropriate idealization for a structural metal because it captures the essential features of its behavior. However, perfect plasticity is not at all appropriate for soils. Some of the troubles, and the justification for adopting this idealization for practical use, were discussed in the paper "Concept of Path Independence and Material Stability

for Soils" by Drucker (1966).

For the most part, the concept of perfect plasticity has been used extensively in the past in conventional soil mechanics in assessing the collapse load in stability problems.

The Tresca and Von Mises models are well-known yield criteria for metals. Tresca's criterion is known as the maximum shear stress criterion. Von Mises's criterion is called the J_2-theory, the octahedral shear stress criterion, or the distortional energy theory. Von Mises criterion considers the effect of intermediate principal stress on the failure or yielding strength. Since these criteria were developed primarily for metals whose yield strengths are insensitive to hydrostatic pressure, these criteria are not suitable for application in soil mechanics. Then, taking into account the significant effect of hydrostatic pressure on soil strength, extended Tresca and extended von Mises criteria have been proposed.

The Coulomb model is certainly the best known failure criterion in soil mechanics. This criterion, which was proposed for geotechnical materials much earlier than that of Tresca's and Von Mises's yield criteria for metals, is the first type of failure criterion that takes into consideration the effect of hydrostatic pressure. This criterion states that failure occurs when the shear stress τ and the normal stress σ acting on any element in the material satisfy the linear equation

$$\tau + \sigma \tan\phi - c = 0 \qquad (5)$$

where c and ϕ denote the cohesion and the angle of internal friction respectively.

As shown by Shield (1955), Coulomb's failure surface is an irregular hexagonal pyramid in principal stress space. For some problems in geomechanics, Chen and Drucker (1975) proposed a criterion combining the Coulomb criterion with a small tensile strength cut-off. Even though the Coulomb criterion is simple in graphical form, the Coulomb surface exhibits corners or singularities in the three dimensional generalization. The resulting general yield or failure function with singularities gives rise to some difficulties in numerical analysis.

In addition, the Coulomb criterion neglects the influence of intermediate principal stress on shear strength. Nevertheless, for the most part, this criterion has in the past been used through necessity and for simplicity to obtain reasonable solutions to important, practical problems in geotechnical engineering.

For practical purposes, a smooth

surface is often used to approximate the yield surface, with singularities in elastic-plastic finite element analyses under more general stress conditions. The Drucker-Prager perfectly plastic model (Drucker and Prager, 1952) can be considered as a first attempt to approximate the well-known Coulomb criterion by a simple smooth function. This criterion is expressed as a simple stress invariant function of the first invariant of the stress tensor, I_1, and the second invariant of the deviatoric stress tensor, J_2, together with two material constants α and k. It has the simple form

$$\alpha I_1 + \sqrt{J_2} = k \qquad (6)$$

where α and k are material constants. The Drucker-Prager material constants, α and k, may be selected in several ways by matching this model with the well-known Coulomb criterion under plane strain, plane stress, and three dimensional stress space. An assessment of various possible matching schemes is given by chen and Mizuno (1979).

If α is zero, Eq. (6) reduces to the well-known Von Mises yield condition for metal. When ϕ is zero, the Coulomb criterion reduces to the Tresca criterion for metal. In a sense, the Von Mises criterion may be considered an approximate version of the Tresca yield criterion.

The Drucker-Prager model cannot predict plastic volumetric strain during hydrostatic loading. To improve this, an extended Von Mises model with a convex end cap was proposed by Drucker, Gibson, and Henkel (1957). Details will be given in the next section.

3.2 Strain-hardening plasticity cap models

Recent research has focused on the application of the strain-hardening theory of plasticity to soil media. From a theoretical point of view, strain-hardening models are very attractive because they are capable of treating the conditions of unloading, stress path dependency, dilatancy, and the effects of intermediate principal stress.

Drucker, Gibson, and Henkel (1957) were the first to suggest that soil might be modeled as an elastic-plastic strain-hardening material. They introduced a spherical end cap to the Drucker-Prager model, as shown in Fig.2, in order to control the plastic volumetric change of soil, or dilatancy. As the soil strain-hardens, both the cone and cap expand. In the development of their model, two important innovations are included. The first is the introduction of the idea of a spherical end cap fitted to the cone. The second is the use of current soil density (specific volume or

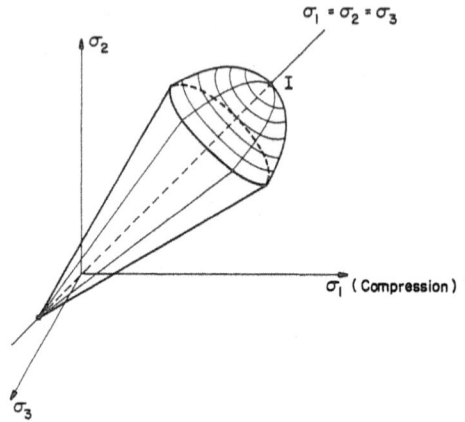

Fig.2   Drucker-Prager type of strain-hardening cap model

void ratio) as the strain-hardening parameter to determine the successive loading surfaces, such as the surface marked I in Fig.2 for a particular value of soil density. There will be a succession of such surfaces, all geometrically similar, but of different sizes, for different densities. This strain- or work-hardening model was a major step forward toward a more realistic representation of soil behavior. Two specific isotropic hardening models such as Cambridge model (concept of critical state), and generalized cap models, will now be discussed.

A few researchers have investigated the possibility of modeling soils as a strain-hardening material. The Cambridge group in the U.K., under the leadership of Prof. Roscoe, extended the basic concept of Drucker, Gibson, and Henkel, and developed several plasticity-based soil models based on experimental data from triaxial tests. Roscoe, Schofield, and Wroth (1958) published a paper that contained the basis for a number of subsequent strain-hardening models for soil. Later, Roscoe, Schofield, and Thurairajah (1963) utilized the strain-hardening theory of plasticity to formulate a complete stress-strain model for normally consolidated or lightly overconsolidated clay in a triaxial test. These results are presented together in a book entitled "Critical State Soil Mechanics" (Schofield and Wroth, 1968) in which the models for sand and for normally consolidated or lightly overconsolidated clay are named, respectively, Granta-gravel and Cam-clay models. In the former model, volumetric strain and shear strain are assumed to be

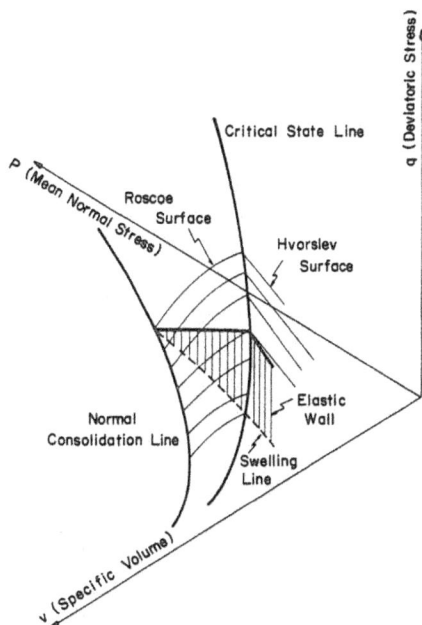

Fig.3 State boundary surface and elastic wall

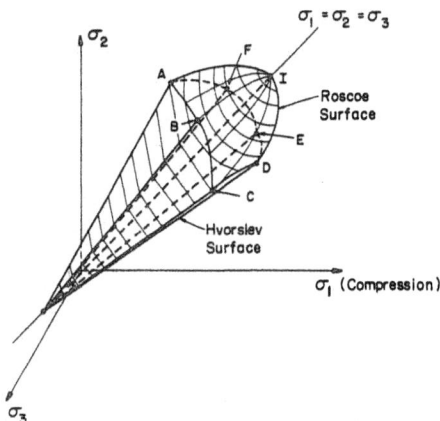

Fig.4 Complete state boundary surface in principal stress space: Cambridge type of strain-hardening cap model

irrecoverable. In the latter model, volumetric strain and shear strain are assumed to be, respectively, partially recoverable and irrecoverable. Important concepts such as Roscoe and Hvorslev surfaces, critical state line, and elastic wall in the mean normal stress-deviatoric stress-specific volume space are advanced, based on the work of Roscoe, Schofield, and Wroth (1958), and Parry (1958). The Roscoe and Hvorslev surfaces, which meet at the critical state line as shown in Fig. 3, determine respectively the plastic behavior of normally consolidated and overconsolidated soils. The complete failure or yield surface of Roscoe and Hvorslev surfaces is termed a state boundary surface (e.g. Atkinson and Bransby, 1978; and Roscoe and Burland, 1968).

The critical state line is the locus of failure points for all shear tests under both drained and undrained conditions. This line has the crucial property that failure of normally consolidated soils occurs once the stress state reaches the line, irrespective of the stress path to the critical state line. At this critical state, large shear distortions occur with any change in stress. The elastic wall is the plane in

which elastic behavior of overconsolidated soils occurs.

A modified version of the Cam-clay model for a general three dimensional stress state was proposed by Roscoe and Burland (1968). In this model the Roscoe surface and the Hvorslev surface are extended into three dimensional stress space. The three dimensional view of the model is shown in Fig. 4. The Hvorslev surface may be fitted to Coulomb's surface since soil is governed by the Coulomb criterion at failure (Bishop, 1966). The failure or yield surface corresponding to Coulomb criterion in principal stress space is an irregular hexagon pyramid whose apex lie on the space diagonal as shown in Fig. 4. The three dimensional cap of the Roscoe surface, marked "I" in the figure, corresponds to a specific volume. The Roscoe surface intersects with the Hvorslev surface in a line ABCDEFA. This line is called the critical state locus, and it is simply a line which separates the two surfaces corresponding to one fixed value of the specific volume (or void ratio). There will be similar surfaces of different size corresponding to different values of specific volume. In this model, the volumetric strain and shear strain are assumed to be, respectively, partially recoverable and irrecoverable. For certain stress histories, this model strain-softens rather than strain-hardens.

Since the development of Cambridge models, many attempts have been made to use various versions of these models in the numerical solutions of boundary value problems and

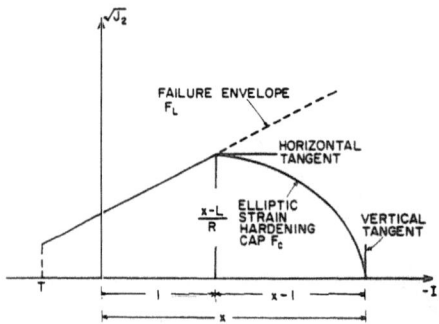

Fig.5 Elliptic cap model

prediction of soil behavior in the field (Palmer,1973). Also, extensive reviews of various types of Cambridge models have been given in two symposia (Palmer,1973; Parry, 1972)

In recent years, the concept of state boundary surface, particularly the Roscoe surface, has been further modified and refined by DiMaggio and Sandler (1971), and Baladi and Rohani (1979), among others.

The yield function they used consists of a perfectly-plastic (failure) portion fitted to a strain-hardening elliptic cap in $I_1-\sqrt{J_2}$ space, as shown in Fig.5. Associated flow rule was employed for the failure and cap functions. This model is in general called a cap model.

In this model, the functional form for both the perfectly-plastic and strain-hardening portions may be quite general and would allow for fitting of a wide range of material properties.

The movement of the cap is controlled by the increase or decrease of the plastic volumetric strain. Strain-hardening in this model can therefore be reversed. It is this mechanism that leads to an effective control on dilatancy, which can be kept quite small (effectively zero) as required for many soils. Their model has also been adopted for rocks (Sandler, DiMaggio and Baladi,1976) by allowing only expansion of the cap (i.e. hardening). In this variation of the model, the cap movement is assumed to depend only on the maximum previous value of the plastic volumetric strain, and consequently the cap is not reversible. In this way, the model allows representation of a relatively large amount of dilatancy, which is often observed during the failure of rocks at low pressures.

These generalized cap models have also been expanded to include rate effects, and also anisotropic behavior within the yield sur-

face and viscoplastic behavior during yielding (Nelson, 1978; Sandler and Baron, 1979). Many of these variations of the generalized cap models are now widely used in ground shock computations (Nelson, 1978; Nelson and Baladi, 1977; Sandler and Baron, 1979).

3.3 Advanced plasticity models

Since cap type of strain-hardening models which take into consideration the soil density (or void ratio) as a hardening parameter has been proposed by Drucker, many modified versions of plasticity models have been developed and presented, introducing internal variables controlling soil behavior.

Prevost and Höeg (1975) suggested a double-hardening model consisting of a shear loading surface, determined by the plastic shear strain, and a volumetric loading surface determined by the plastic volumetric strain. A similar double-hardening elastoplastic model was established by Vermeer (1978) for the initial loading, unloading and reloading of sand. Further, the modified Cam-clay model was extended by Van Eekelen and Potts (1978) for Drammen clay under cyclic loading. In this development, the prefailure behavior of clay is governed by a shear loading surface and a cap type of volumetric loading surface, while at failure the clay is governed by the Coulomb criterion.

In what follows it is worth mentioning recently advanced anisotropic plasticity models such as the nested yield surface models and the bounding surface models for cyclic strain-hardening or softening materials. Although the isotropic strain- or work-hardening models described above have been so far used in soil mechanics because of their simplicity for direct computational applications, these models are not adequate for the prediction of the behavior of soil which has undergone unloading-loading reversals. The isotropic model predicts elastic behavior only until the stress is fully reversed. However, it has been observed from experiments that upon unloading both elastic and plastic deformations occur well before the stress is fully reversed. Hence, an alternative approach to the isotropic hardening model is provided by the use of a kinematic type of strain-hardening rules, or a combination of isotropic and kinematic types of strain-hardening rules. Recently, this approach has been employed by several researchers to provide a more realistic representation of soil behavior under reversed, and particularly cyclic, loading conditions.

Iwan (1967), following the related work of Masing (1926), proposed one dimensional plasticity models which consist of a collection of perfectly elastic and rigid-plastic or slip elements arranged in either a series-parallel or a parallel-series combination. The model can contain a very large number of elements, and the properties of these elements can be distributed such that they can match the particular form of hysteretic behavior of a certain type of soil. Such models are known as the overlay or mechanical sublayer models (Zienkiewicz et. al. 1977).

In order to extend the one dimensional model to three dimensional situations, an extended formulation of the classical incremental theory of plasticity has been proposed by Iwan (1967). Instead of using a single yield surface in stress space, he postulated a family (nest) of yield surfaces with each surface translating independently in a pure kinematic manner, or individually obeying a linear work-hardening model. Their combined action, in general, gives rise to a nonlinear work-hardening behavior for the material as a whole. The approach leads to a realistic Bauschinger effect of a type that could not be obtained by using a single yield surface and a nonlinear work-hardening rule even with kinematic hardening. The same concept of using a field of nested yield surfaces was also proposed independently by Mroz (1967). These models have been used recently for soils and are usually known as multi-surface plasticity models. In all the proposed models of this type, the associated flow rule has been utilized.

Recently, Prevost (1977, 1978a) has extended the Iwan/Mroz model for the undrained behavior of clays under monotonic and cyclic loading conditions, utilizing the von Mises types of nested yield surfaces.

The rule used to govern the translation of the nested yield surfaces during plastic loading was that suggested by Mroz (1967). According to this rule, when the stress point P (Fig.6) reaches the yield surface $f^{(m)}$, then, upon further loading (i.e. when the stress increment σ_{ij} is applied), the instantaneous translation of $f^{(m)}$ towards the next yield surface $f^{(m+1)}$ will be along $\overline{PR}$, where R (known as the conjugate point) is the point on $f^{(m+1)}$ with the outward normal in the same direction as the normal to $f^{(m)}$ at P (i.e., $\overline{n}^{(m+1)} = \overline{n}^{(m)}$ in Fig.6). Thus, in this model development, the basic strain-hardening rule is still of a kinematic type, but a simultaneous isotropic hardening (or softening) is allowed.

This model has been adopted recently by Prevost et. al.(1981) in the finite element analyses of soil-structure interaction of centrifugal models under both monotonic and cyclic loadings simulating the situation encountered in the analysis of off-

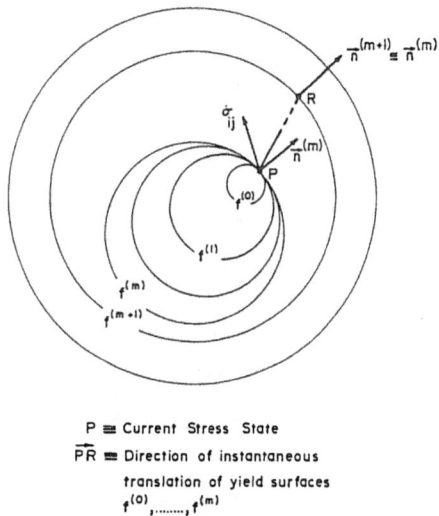

P ≡ Current Stress State

$\overline{PR}$ ≡ Direction of instantaneous translation of yield surfaces $f^{(0)}, \dots\dots, f^{(m)}$

Fig.6 Nested Yield surface model

shore gravity structure foundations under wave forces. The results obtained from the analysis agree quite well with those measured experimentally in the centrifuge. This study has demonstrated the ability of the multi-surface model to provide realistic representation of soil behavior under complex loadings.

Prevost (1978b) has also extended the idea of a nested yield surface model for undrained conditions to one for drained conditions, by taking into account the effect of a hydrostatic pressure on the yielding of soils. It is assumed in this model that the hardening modulus and the current size of the yield surface are functions of plastic volumetric strain or plastic shear distortions, or both, in order to use a combination of isotropic and new kinematic strain-hardening rules. A non-associated flow rule is employed for the inner yield surfaces, while an associated flow rule is taken for the outermost yield surface. In addition, a potential surface to determine the directions of plastic strains is defined so that the plastic deviatoric strain increment vector is normal to the projection of the yield surface on the deviatoric plane (π-plane).

Since the plastic modulus varies along the yield surface due to the requirement of consistency condition, experimental observations such as "consolidation", "dilatant", and "critical state", can be explained, particularly on the outermost surface, according to the conditions of plastic hardening

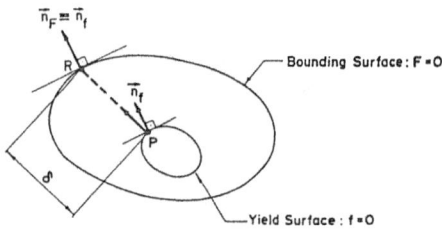

Fig.7 Bounding surface model

modulus H' such as H'>0, H'<0, and H'=0, respectively.

A theoretical interpretation of this model is given in detail by Prevost (1978b) for cases of undrained axial soil tests and consolidation tests.

A nested yield surface model may ·be capable of representing the anisotropic behavior of soils which cannot be described by classical plasticity models. However, the necessity to define, renew and keep in memory such nested yield surfaces in the model may make a computational implementation of the model intricate and expensive.

Much simpler alternatives of anisotropic strain-hardening plasticity models have been recently employed for soils based on the bounding (consolidation or limiting) surface concept introduced earlier for metals (Dafalias and Popov, 1975; Krieg, 1975). A two-surface model of this type was proposed by Mroz, Norris and Zienkiewicz (1978, 1979) for clays. A bounding surface, F=0, representing the consolidation history of the soil, and a yield surface, f=0, defining the elastic domain within the bounding surface (Fig.7) were employed in the model.

The bounding surface was assumed to expand or contract isotropically, but the yield surface was allowed to translate, expand or contract within the domain enclosed by the bounding surface. The translation of the yield surface is governed by the same rule as the multi-surface models described earlier (i.e., f will translate towards the bounding surface along $\overline{PR}$ in Fig.7).

The hardening modulus on the yield surface is assumed to vary depending on the relative configuration of the yield and bounding surfaces. It was this last assumption that distinguished the present formulation from the previous multi-surface models. Here, instead of using a field of nested yield surfaces with associated hardening moduli, an interpolation rule is utilized to define the variation of the hardening moduli between the yield and bounding surfaces. In the interpolation rule used by Mroz et. ·al.

(1978 and 1979), the hardening modulus, H', was taken as a function of the distance δ (Fig.7) between the current stress point, P, on the yield surface and its conjugate point, R, on the bounding surface. Very detailed discussions of the present model and its application in representing the behavior of clays under monotonic and cyclic triaxial test conditions have been given by Mroz et. al. (1979).

Mroz et. al. (1981) further introduced the concept of the accumulated strain fatigue parameter which determines the degree of soil stiffness degradation, in order to account for the rearrangement of soil structure in the course of cyclically plastic deformation and inelastic behavior such as progressive densification/dilatation and/or corresponding pore water changes. The two-surface anisotropic hardening model is extended to a modified one which consists of three surfaces; yield surface, hardening surface (consolidation surface), and intermediate surface which divides the domain within the consolidation surface into two sub-domains. In the extended two-surface model, the progressive material degradation is simulated by an intermediate surface. Further, an alternative description is provided by the new version of the hardening model (Mroz et. al, 1981) with an infinite number of loading surfaces, accounting for the cyclic degradation effect into the plastic hardening modulus. The development of models of a similar type can be found in recent papers by Pietruszczak and Mroz (1983) and Mroz and Pietruszczak (1983).

Dafalias and Herrmann (1980) have also used the bounding surface formulation to describe the behavior of clay under cyclic loadings. However, no explicit yield surface was postulated within the bounding surface. An associated flow rule was utilized for the bounding surface. In this model, the variation of the plastic (hardening) modulus within the bounding surface is defined based on a very simple radial mapping rule. For each actual stress point within (or on) the bounding surface, a corresponding "image" point on the surface is specified as the intersection of the surface with the straight (radial) line connecting the origin with the current stress point (the origin was assumed to be always within the bounding surface). The actual hardening modulus is then assumed to be a function of the hardening modulus on the bounding surface, at the "image" point, and the distance between the actual stress point and its "image".

In order to take into consideration the difference in the behavior of the soil between compression and tension, the bounding ·surface theory has been presented, introducing all

three stress invariants for initially isotropic materials, and then the qualitative
behavior of the model has been assessed in
detail under conditions of monotonic,
cyclic, drained, and undrained loadings
(Dafalias and Herrmann, 1982).

4 THEORETICAL DEVELOPMENT OF PLASTICITY MODELS

At the beginning of yielding and during
subsequent loadings, soils undergo both
elastic and plastic deformations. In order
to estimate quantitatively elastic and
plastic components in the total deformations,
the concepts of the theory of elasticity
and plasticity based on continuum mechanics
are to be employed in this section. In the
development of the stress increment-strain increment relationship, the total strain increment $d\varepsilon_{ij}$ is assumed to be the sum of the
elastic strain increment $d\varepsilon_{ij}^{e}$ and the plastic strain increment $d\varepsilon_{ij}^{p}$, i.e.:

$$d\varepsilon_{ij} = d\varepsilon_{ij}^{e} + d\varepsilon_{ij}^{p} \tag{7}$$

The elastic strain increment in Eq. (7) is
assumed to be completely described within
the framework of incremental Hooke's law,
where two material constants such as bulk
modulus, K, and shear modulus, G, are constants or functions of stress invariant
and/or strain invariant. On the other hand,
as for the estimate of plastic strain increment, the previously mentioned basic
concepts of incremental theory of plasticity are required. Utilizing such concepts of
plasticity as well as elasticity theory, the
complete stress increment-strain increment
relationship can be developed for any kind
of soil material, such as perfect plastic,
isotropic hardening, kinematic hardening,
and mixed hardening types.

4.1 Elastic-plastic constitutive develop- ment

For the development of elastic-plastic constitutive equations, the estimate of plastic strain increment associated with the
stress increment is of an important concern.
The plastic strain increment $d\varepsilon_{ij}^{p}$, as given
by the flow rule, is in general

$$d\varepsilon_{ij}^{p} = d\lambda \frac{\partial g}{\partial \sigma_{ij}} \tag{8}$$

where g is the plastic potential function
and $d\lambda$ is a positive scalar function. The
parameter $d\lambda$ is further written as

$$d\lambda = \frac{1}{H'} (\partial f / \partial \sigma_{ij} \cdot d\sigma_{ij}) \tag{9}$$

where H' is a plastic hardening modulus and
f is a yield or failure function. Therefore,
Eq. (8) may take the following form

$$d\varepsilon_{ij}^{p} = \frac{1}{H'} (\frac{\partial f}{\partial \sigma_{mn}} d\sigma_{mn}) \frac{\partial g}{\partial \sigma_{ij}} \tag{10}$$

If the yield or failure function f is taken
as the plastic potential function of g, the
normality condition is satisfied. This is
known as a material with an associated flow
rule. If the plastic potential function, g,
is different from the yield or failure function, f, the material is one with a non-
associated flow rule. Although the normality
condition is not always satisfied in soil,
rock and concrete, the associated flow rule
has generally been used in the past for simplicity. According to the incremental theory of plasticity as well as the theory of
elasticity, an incremental elastic-plastic
constitutive equation with a non-associated
flow rule can be written in the general
form

$$d\sigma_{ij} = \left[D_{ijkl}^{e} - \frac{\frac{\partial g}{\partial \sigma_{rs}} D_{ijrs}^{e} D_{mnkl}^{e} \frac{\partial f}{\partial \sigma_{mn}}}{H' + \frac{\partial f}{\partial \sigma_{mn}} D_{mnrs}^{e} \frac{\partial g}{\partial \sigma_{rs}}} \right] d\varepsilon_{kl} \tag{11}$$

where $d\sigma_{ij}$, $d\varepsilon_{ij}$, D_{ijkl}^{e} are respectively
incremental stress tensor, incremental
strain tensor, and elastic stiffness tensor.
Eq. (11) is the general constitutive form
for a perfect plastic, isotropic strain-hardening, and anisotropic strain-hardening
materials. The plastic hardening modulus H'
in Eq. (11) plays an important role in predicting the behavior of a material, and it
has different forms for various types of
material. In what follows, the plastic hardening moduli H' are represented for perfect
plastic, isotropic hardening, and anisotropic hardening materials.

4.2 Plastic hardening modulus of perfect plastic material

The model contains a "failure envelope" or
"yield surface" fixed in the stress invariant space. Its function is in general expressed by

$$f = f(\sigma_{ij}) \tag{12}$$

During plastic flow, the yield function f
should satisfy the consistency condition
df = 0. That is,

$$df = \frac{\partial f}{\partial \sigma_{ij}} d\sigma_{ij} \tag{13}$$

Since $(\partial f / \partial \sigma_{ij}) d\sigma_{ij}$ is zero, H' should also be zero from Eq. (10). In other words, the perfect plastic material is the one whose plastic hardening modulus H' is equally zero along the fixed surface.

4.3 Plastic hardening modulus of isotropic strain/work hardening model

The general function presenting an isotropic strain-hardening model can be expressed by

$$f = f[\sigma_{ij}, x(\varepsilon_{ij}^P), \kappa(\varepsilon_p)] \qquad (14)$$

in which $x(\varepsilon_{ij}^P)$, and $\kappa(\varepsilon_p)$ are respectively hardening parameter which is a function of plastic strain tensor ε_{ij}^P, and material parameter which is a function of the effective plastic strain, $\varepsilon_p = \int C\sqrt{(d\varepsilon_{ij}^P . d\varepsilon_{ij}^P)}$ where the parameter C is constant depending on the yield function. From the consistency condition, df = 0, we obtain

$$df = \frac{\partial f}{\partial \sigma_{ij}} d\sigma_{ij} + \frac{\partial f}{\partial x}\frac{\partial x}{\partial \varepsilon_{ij}^P} d\varepsilon_{ij}^P +$$
$$\frac{\partial f}{\partial \kappa}\frac{d\kappa}{d\varepsilon_p} C\sqrt{d\varepsilon_{ij}^P \, d\varepsilon_{ij}^P} = 0 \qquad (15)$$

Substituting Eq. (10) into Eq. (15), the solution for the plastic hardening modulus H' is expressed by

$$H' = -\frac{\partial f}{\partial x}\frac{\partial x}{\partial \varepsilon_{ij}^P}\frac{\partial g}{\partial \sigma_{ij}} - C\frac{\partial f}{\partial \kappa}\frac{d\kappa}{d\varepsilon_p}\sqrt{\frac{\partial g}{\partial \sigma_{ij}}\frac{\partial g}{\partial \sigma_{ij}}} \qquad (16)$$

Further, consider the work-hardening model whose function is expressed by

$$f = f[\sigma_{ij}, \kappa(W_p)] \qquad (17)$$

where κ is a function of plastic work W_p which is defined as $W_p = \int \sigma_{ij} \, d\varepsilon_{ij}^P$. In this case, the consistency condition is

$$df = \frac{\partial f}{\partial \sigma_{ij}} d\sigma_{ij} + \frac{\partial f}{\partial \kappa}\frac{d\kappa}{dW_p}\sigma_{ij} \, d\varepsilon_{ij}^P = 0 \qquad (18)$$

Thus, the substitution of $d\varepsilon_{ij}^P$ in Eq. (10) leads to the solution of the plastic hardening modulus H' expressed by

$$H' = -\frac{\partial f}{\partial \kappa}\frac{d\kappa}{dW_p}\sigma_{ij}\frac{\partial g}{\partial \sigma_{ij}} \qquad (19)$$

The plastic hardening modulus, H', in Eq. (16) is the one for generalized cap models as well as the isotropic strain-hardening type of Coulomb, Tresca and von Mises criteria, while H' in Eq. (19) is for models such as the Lade and Duncan model (Lade and Duncan, 1975), and the Lade model (Lade, 1977 and 1979) where the hardening parameter is a function of the plastic work W_p.

4.4 Plastic hardening modulus of an anisotropic strain-hardening model

The general loading function for an anisotropic material is simply expanded from the isotropic material function by introducing a kinematic hardening rule as well as an isotropic hardening rule. The form of function is written as

$$f = f[\sigma_{ij} - \alpha_{ij}, x(\varepsilon_{ij}^P), \kappa(\varepsilon_p)] \qquad (20)$$

where α_{ij} denotes the translation of the center of the yield surface in stress space, and is usually assumed to be a function of, the plastic strain tensor ε_{ij}^P. Similarly, the consistency condition implies

$$df = \frac{\partial f}{\partial \sigma_{ij}} d\sigma_{ij} - \frac{\partial f}{\partial \sigma_{ij}}\frac{\partial \alpha_{ij}}{\partial \varepsilon_{kl}^P} d\varepsilon_{kl}^P +$$
$$\frac{\partial f}{\partial x}\frac{\partial x}{\partial \varepsilon_{ij}^P} d\varepsilon_{ij}^P + \frac{\partial f}{\partial \kappa}\frac{d\kappa}{d\varepsilon_p} C\sqrt{d\varepsilon_{ij}^P d\varepsilon_{ij}^P} \qquad (21)$$

The plastic hardening modulus H' is thus written as

$$H' = \frac{\partial f}{\partial \sigma_{ij}}\frac{\partial \alpha_{ij}}{\partial \varepsilon_{kl}^P}\frac{\partial g}{\partial \sigma_{kl}} - \frac{\partial f}{\partial x}\frac{\partial x}{\partial \varepsilon_{ij}^P}\frac{\partial g}{\partial \sigma_{ij}} -$$
$$\frac{\partial f}{\partial \kappa}\frac{d\kappa}{d\varepsilon_p} C\sqrt{\frac{\partial g}{\partial \sigma_{ij}}\frac{\partial g}{\partial \sigma_{ij}}} \qquad (22)$$

As a special case, let us consider Prager's and Ziegler's pure kinematic hardening rules. The yield function followed by a pure kinematic hardening rule is written as

$$f = f(\sigma_{ij} - \alpha_{ij}) \qquad (23)$$

Therefore, H' in Eq. (22) becomes

$$H' = \frac{\partial f}{\partial \sigma_{ij}}\frac{\partial \alpha_{ij}}{\partial \varepsilon_{kl}^P}\frac{\partial g}{\partial \sigma_{kl}} \qquad (24)$$

According to Prager's hardening rule, the translation of the center, $d\alpha_{ij}$, is in general expressed by

$$d\alpha_{ij} = c \, d\varepsilon_{ij}^P \qquad (25)$$

where c is the work-hardening constant, which is characteristic for a given material. Substitution of the relationship in Eq. (25) into Eq. (24) yields

$$H' = c\frac{\partial f}{\partial \sigma_{ij}}\frac{\partial g}{\partial \sigma_{ij}} \qquad (26)$$

On the other hand, Ziegler's hardening rule assumes the rate of translation to take place in the direction of the reduced-stress vector $\bar{\sigma} = \sigma_{ij} - \alpha_{ij}$ in the form

$$d\alpha_{ij} = d\mu\bar{\sigma}_{ij} = d\mu(\sigma_{ij} - \alpha_{ij}) \qquad (27)$$

where $d\mu$ is a positive proportionality factor which depends on the history of the deformation. For simplicity, the factor $d\mu$ can be assumed in the form

$$d\mu = a \, d\varepsilon_p \qquad (28)$$

where a is a positive constant which is characteristic for a given material. The plastic hardening modulus H' for this material thus takes the form

$$H' = aC \frac{\partial f}{\partial\sigma_{ij}} \sqrt{\frac{\partial g}{\partial\sigma_{mn}} \frac{\partial g}{\partial\sigma_{mn}}} (\sigma_{ij} - \alpha_{ij}) \qquad (29)$$

Furthermore, in engineering applications for soil materials, the concept of combined isotropic and kinematic hardening, called mixed hardening, is attractive; i.e. the yield surface is allowed both to expand and to translate.

In combined isotropic and kinematic hardening, the yield function can in general be written as

$$f = f[\sigma_{ij} - \alpha_{ij}, \kappa(\bar{\varepsilon}_p)] \qquad (30)$$

where $\bar{\varepsilon}_p$ is the reduced-effective plastic strain. $\bar{\varepsilon}_p$ is similarly defined as

$$\bar{\varepsilon}_p = \int d\bar{\varepsilon}_p = C\int \sqrt{(d\bar{\varepsilon}^p_{ij} d\bar{\varepsilon}^p_{ij})} \qquad (31)$$

which governs the process of isotropic hardening; the increment of reduced-plastic strain $d\bar{\varepsilon}^p_{ij}$ will be defined in what follows. The increment of plastic strain is now simply split into two co-linear components

$$d\varepsilon^p_{ij} = d\varepsilon^{p(i)}_{ij} + d\varepsilon^{p(k)}_{ij} \qquad (32)$$

where $d\varepsilon^{p(i)}_{ij}$ is associated with the expansion of the yield surface and $d\varepsilon^{p(k)}_{ij}$ is associated with the translation of the yield surface. These two plastic strain components may be written as

$$d\varepsilon^{p(i)}_{ij} = M \, d\varepsilon^p_{ij} \qquad (33)$$

$$d\varepsilon^{p(k)}_{ij} = (1-M) \, d\varepsilon^p_{ij} \qquad (34)$$

where M is the material parameter in a range between the values of -1 and 1, defining the share rate of isotropic hardening in the total amount of hardening. M is called the parameter of mixed hardening. Since the parameter M can also be given negative values, isotropic softening can be considered. This implies that during the translation the yield surface is assumed either to expand or to contract. Such a contraction is sometimes observed in experimental studies.

The isotropic share part of plastic strain increment $d\varepsilon^{p(i)}_{ij}$ associated with the expansion of the yield surface is now used to define the reduced-plastic strain increment $d\bar{\varepsilon}^p_{ij}$, that is,

$$d\bar{\varepsilon}^p_{ij} = d\varepsilon^{p(i)}_{ij} = M \, d\varepsilon^p_{ij} \qquad (35)$$

On the other hand, the kinematic share part of the plastic strain increment $d\varepsilon^{p(k)}_{ij}$ is used to define the rate of translation of the yield surface.

For the case of Prager's hardening rule the relationship between $d\alpha_{ij}$ and $d\varepsilon^{p(k)}_{ij}$ is given by

$$d\alpha_{ij} = c \, d\varepsilon^{p(k)}_{ij} = c(1-M) \, d\varepsilon^p_{ij} \qquad (36)$$

Thus the consistency condition for this type of mixed hardening material is

$$df = \frac{\partial f}{\partial\sigma_{ij}} d\sigma_{ij} - \frac{\partial f}{\partial\sigma_{ij}} d\alpha_{ij} + \frac{\partial f}{\partial\kappa} \frac{d\kappa}{d\bar{\varepsilon}_p} d\bar{\varepsilon}_p = 0 \qquad (37)$$

Substituting $d\alpha_{ij}$ in Eq. (36) and $d\bar{\varepsilon}_p$ in Eq. (31) into Eq. (37), we obtain

$$df = \frac{\partial f}{\partial\sigma_{ij}} d\sigma_{ij} - c(1-M) \frac{\partial f}{\partial\sigma_{ij}} d\varepsilon^p_{ij} +$$
$$CM \frac{\partial f}{\partial\kappa} \frac{d\kappa}{d\bar{\varepsilon}_p} \sqrt{d\varepsilon^p_{ij} d\varepsilon^p_{ij}} = 0 \qquad (38)$$

Further, substituting Eq. (10) and solving for plastic hardening modulus H', we find

$$H' = c(1-M) \frac{\partial f}{\partial\sigma_{ij}} \frac{\partial g}{\partial\sigma_{ij}} -$$
$$CM \frac{\partial f}{\partial\kappa} \frac{d\kappa}{d\bar{\varepsilon}_p} \sqrt{\frac{\partial g}{\partial\sigma_{ij}} \frac{\partial g}{\partial\sigma_{ij}}} \qquad (39)$$

For the case of introducing Ziegler's hardening rule, the relationship between $d\alpha_{ij}$ and $d\varepsilon^{p(k)}_{ij}$ is given by

$$d\alpha_{ij} = aC \sqrt{d\varepsilon^{p(k)}_{mn} d\varepsilon^{p(k)}_{mn}} (\sigma_{ij} - \alpha_{ij})$$
$$= aC(1-M)\sqrt{d\varepsilon^p_{mn} d\varepsilon^p_{mn}} (\sigma_{ij} - \alpha_{ij}) \qquad (40)$$

Therefore, H' becomes

$$H' = aC(1-M) \frac{\partial f}{\partial\sigma_{ij}} \sqrt{\frac{\partial g}{\partial\sigma_{mn}} \frac{\partial g}{\partial\sigma_{mn}}} (\sigma_{ij} - \alpha_{ij}) -$$
$$CM \frac{\partial f}{\partial\kappa} \frac{d\kappa}{d\bar{\varepsilon}_p} \sqrt{\frac{\partial g}{\partial\sigma_{ij}} \frac{\partial g}{\partial\sigma_{ij}}} \qquad (41)$$

Since the translation rate of the center of the yield surface $d\alpha_{ij}$ has been so far described explicitly in terms of the plastic strain increment $d\varepsilon^p_{ij}$, the plastic hardening modulus H' can be easily obtained. The rate of translation $d\alpha_{ij}$ may in general take an implicit form with respect to the plastic strain increment $d\varepsilon_{ij}^p$. For instance, in nested yield surface models by Mroz (1967) and Prevost (1978a and 1978b), the plastic hardening modulus H' is first assumed to be constant or a function of the effective plastic strain ε_p, and then the increment of translation, $d\alpha_{ij}^p$, is determined to satisfy the consistency condition.

5 STIFFNESS FORMULATION OF SOIL PLASTICITY MODELS

In this section, Eq. (11) will be transformed into a form readily applicable to finite element analysis. To this end, a general description of Eq. (11) is given for several plasticity models.

In general, $\partial f / \partial \sigma_{ij}$ and $\partial g / \partial \sigma_{ij}$ can be written respectively as

$$\frac{\partial f}{\partial \sigma_{ij}} = \frac{\partial f}{\partial I_1}\delta_{ij} + \frac{\partial f}{\partial J_2}s_{ij} + \frac{\partial f}{\partial J_3}t_{ij} \qquad (42)$$

and

$$\frac{\partial g}{\partial \sigma_{ij}} = \frac{\partial g}{\partial I_1}\delta_{ij} + \frac{\partial g}{\partial J_2}s_{ij} + \frac{\partial g}{\partial J_3}t_{ij} \qquad (43)$$

where $t_{ij} = \dfrac{\partial J_3}{\partial \sigma_{ij}} = s_{ik}s_{kj} - \dfrac{2}{3}J_2\delta_{ij}$

Substituting Eqs. (42) and (43) into Eq. (11), we obtain, after some simplifications, the following final form:

$$d\sigma_{ij} = \left[D_{ijkl}^e - \frac{H_{ij}^* H_{kl}}{H} \right] d\varepsilon_{kl} \qquad (44)$$

where

$H = 3AL(3\lambda + 2\mu) + 2B\mu(2MJ_2 + 3NJ_3) +$

$\qquad 2C\mu(3MJ_3 + Ns_{ik}s_{kj}s_{il}s_{lj} - \frac{4}{3}NJ_2^2) + H'$,

$H_{ii} = A(3\lambda + 2\mu) + 2\mu Bs_{ii} + 2\mu Ct_{ii}$ (no sum)

$H_{ii}^* = L(3\lambda + 2\mu) + 2\mu Ms_{ii} + 2\mu Nt_{ii}$ (no sum)

H_{ij} $(i \neq j) = 2\mu Bs_{ij} + 2\mu Ct_{ij}$,

H_{ij}^* $(i \neq j) = 2\mu Ms_{ij} + 2\mu Nt_{ij}$,

$A = \dfrac{\partial f}{\partial I_1}, B = \dfrac{\partial f}{\partial J_2}, C = \dfrac{\partial f}{\partial J_3}, L = \dfrac{\partial g}{\partial I_1}$,

$M = \dfrac{\partial g}{\partial J_2}, N = \dfrac{\partial g}{\partial J_3}$, and

H' = plastic hardening modulus, λ and μ are Lame constants.
Therefore, the elastic-plastic constituent tensor D_{ijkl}^{ep} is given by

$$D_{ijkl}^{ep} = D_{ijkl}^e - \frac{H_{ij}^* H_{kl}}{H} \qquad (45)$$

As can be seen from Eq. (45), the stiffness tensor is not symmetric for non-associated flow rule case but becomes a symmetric one for the associated flow rule case.

The plastic hardening modulus H' and the equations for A, B, C, L, M, N, etc., are derived below for several plasticity models, assuming an, associated flow rule as well as a non-associated flow rule.

5.1 Stiffness coefficients of classical plasticity models

Herein, perfect plasticity models such as Coulomb, Tresca and Drucker-Prager models are presented for the case of an associated flow rule. For these perfectly plastic materials, H' is now zero.

The Coulomb criterion in terms of stress invariants is given by Zienkiewicz et. al. (1975) as

$$f = g = I_1\sin\phi + \frac{3(1-\sin\phi)\sin\theta + \sqrt{3}(3+\sin\phi)\cos\theta}{2}\sqrt{J_2}$$

$$-3c\cos\phi = 0 \qquad (46)$$

where $\theta = \dfrac{1}{3}\cos^{-1}(\dfrac{3}{2}\sqrt{3}\,\dfrac{J_3}{J_2\sqrt{J_2}})$, ϕ = angle of internal friction, and c = cohesion. Taking derivatives of Eq. (46) with respect to I_1, J_2, and J_3, we obtain

$$\frac{\partial f}{\partial I_1} = A = L = \sin\phi, \qquad (47)$$

$$\frac{\partial f}{\partial J_2} = B = M = \frac{3(1-\sin\phi)\sin\theta + \sqrt{3}(3+\sin\phi)\cos\theta}{4\sqrt{J_2}} +$$

$$\frac{3\sqrt{3}J_3[3(1-\sin\phi)\cos\theta - \sqrt{3}(3+\sin\phi)\sin\theta]}{8J_2^2\sin3\theta}, \qquad (48)$$

$$\frac{\partial f}{\partial J_3} = C = N =$$

$$\frac{\sqrt{3}[3(1-\sin\phi)\cos\theta - \sqrt{3}(3+\sin\phi)\sin\theta]}{4J_2\sin3\theta} \qquad (49)$$

The Tresca criterion in terms of stress invariants is derived easily by substituting $\phi = 0$ into Eq. (46) of the Coulomb criterion. Thus, we obtain

$$f = g = \frac{3\sin\theta + 3\sqrt{3}\cos\theta}{2}\sqrt{J_2} - 3c = 0 \qquad (50)$$

Similarly, we obtain from Eqs. (47), (48) and (49) with $\phi = 0$

$$\frac{\partial f}{\partial I_1} = A = L = 0, \qquad (51)$$

$$\frac{\partial f}{\partial J_2} = B = M = \frac{3\sin\theta + 3\sqrt{3}\cos\theta}{4\sqrt{J_2}} +$$

$$\frac{3\sqrt{3}J_3(3\cos\theta - 3\sqrt{3}\sin\theta)}{8J_2^2 \sin3\theta} \ , \tag{52}$$

$$\frac{\partial f}{\partial J_3} = C = N = \frac{\sqrt{3}(3\cos\theta - 3\sqrt{3}\sin\theta)}{4J_2\sin3\theta} \tag{53}$$

The equation of the Drucker-Prager criterion is again written as

$$f = g = \alpha I_1 + \sqrt{J_2} - k = 0 \tag{54}$$

Therefore, we find

$$\frac{\partial f}{\partial I_1} = A = L = \alpha, \tag{55}$$

$$\frac{\partial f}{\partial J_2} = B = M = \frac{1}{2\sqrt{J_2}} \ , \quad \text{and} \quad \frac{\partial f}{\partial J_3} = C = N = 0 \tag{56}$$

In the case of $\alpha = 0$, stiffness coefficients A, B, C, L, M, and N become identical to those of the von Mises criterion.

In the non-associated flow rule case, the failure or hardening function f is not identical to the plastic potential function g. As an example, for analysis with the Coulomb criterion, a different type of Coulomb function or Drucker-Prager function may be taken as the plastic potential function g. By combining the stiffness coefficients described above, the appropriate stiffness matrix can be derived by users.

5.2 Stiffness coefficients of a modified Cam-clay model

A modified Cam-clay elliptic yield surface can be expressed in terms of stress invariants.

$$f = I_1^2 - I_{ol}\,I_1 + \frac{9J_2}{M^2} = 0 \tag{57}$$

where M is a material constant and I_{ol}, hardening parameter, is the value of I_1 at the intersection of an elliptic yield cap with the I_1-axis. Further, the relationship between the hardening parameter change, dI_{ol}, and the change in plastic volumetric strain, $d\varepsilon_{kk}^p$, is written as

$$dI_{ol} = -\frac{(1+e_v)\,I_{ol}}{\Lambda - \eta}\,d\varepsilon_{kk}^p \tag{58}$$

in which e_v, Λ, and η are respectively void ratio, the slope of virgin consolidation line, and the slope of rebound-reloading line in void ratio e_v and $\ln(-I_1/3)$ space, which is obtained from the response of soil to hydrostatic pressure.

The critical state line, which controls the failure of material, intersects the elliptic yield surface at its maximum point and is defined by

$$\sqrt{J_2} = -\,Mp \tag{59}$$

that is, the critical stress state is defined by an extended Von Mises type expression.

The stress-strain relation of a Cam-clay model is now completely defined since we have specified the failure and yield surfaces and the relationship between the strain-hardening parameter and the plastic volumetric strain. Thus we obtain stiffness coefficients such as A, B, C, L, M, N, etc., for the elliptic yield surface, that is,

$$\frac{\partial f}{\partial I_1} = A = L = 2I_1 - I_{ol}, \tag{60}$$

$$\frac{\partial f}{\partial J_2} = B = M = \frac{9}{M^2} \ , \quad \text{and} \quad \frac{\partial f}{\partial J_3} = C = N = 0 \tag{61}$$

Further, the plastic hardening modulus H' in Eq. (16) takes the following form.

$$H' = -\frac{\partial f}{\partial I_{ol}}\frac{dI_{ol}}{d\varepsilon_{kk}^p}\frac{\partial \varepsilon_{kk}^p}{\partial \varepsilon_{ij}^p}\frac{\partial g}{\partial \sigma_{ij}} \tag{62}$$

Utilizing f in Eq. (57) into Eq. (62), we find

$$H' = -\frac{3(1+e_v)\,I_{ol}\,I_1}{\Lambda - \eta}\frac{\partial g}{\partial I_1} \tag{63}$$

In order to overcome the numerical difficulty in treatment of a strain-softening region such as $I_1 > I_{ol}/2$, we could introduce a perfectly plastic idealization of the critical state line as a perfect plastic yield surface. In this case, the extended von Mises type expression in Eq. (59) can be similarly utilized to obtain the stiffness coefficients.

5.3 Stiffness coefficients of generalized cap models

Cap models consist of the Drucker-Prager type of failure (or yield) surface and a plane or elliptic cap hardening surface. As for the stiffness coefficients of the Drucker-Prager type expression, these have been presented in Section 5.1. In order to prevent an excessive amount of dilatancy on the failure surface, the modified version of the Drucker-Prager surface has been recently utilized (Baladi 1979). The function of the modified surface is written as

$$f = \sqrt{J_2} - [a - c\,\text{Exp}(bI_1)] = 0 \tag{64}$$

where a, b, and c are material constants. For the associated flow rule case, the stiffness coefficients are expressed by the following forms:

$$\frac{\partial f}{\partial I_1} = A = L = bc\,\text{Exp}(bI_1) \tag{65}$$

$$\frac{\partial f}{\partial J_2} = B = M = \frac{1}{2\sqrt{J_2}}, \quad \text{and} \quad \frac{\partial f}{\partial J_3} = C = N = 0 \qquad (66)$$

Desai, et.al. (1982) proposed the following form of a failure surface.

$$f = \alpha I_1 + \beta \, \mathrm{Exp}(\gamma I_1) + \sqrt{J_2} - \xi = 0 \qquad (67)$$

where α, β, γ, and ξ are material constants determined from the conventional triaxial tests. For this case, the stiffness coefficients are expressed under the assumption of an associated flow rule, by

$$\frac{\partial f}{\partial I_1} = A = L = \alpha + \beta\gamma \, \mathrm{Exp}(\gamma I_1) \qquad (68)$$

$$\frac{\partial f}{\partial J_2} = B = M = \frac{1}{2\sqrt{J_2}}, \quad \text{and} \quad \frac{\partial f}{\partial J_3} = C = N = 0 \qquad (69)$$

On the other hand, the stiffness coefficients corresponding to the plane cap and elliptic cap hardening surfaces are as follows.

The function of the plane cap surface is

$$f = I_1 - x(\varepsilon_{kk}^P) = 0 \qquad (70)$$

where x is the hardening function which causes the plastic volumetric change $d\varepsilon_{kk}^P$. The location of the cap x is related to the plastic volumetric strain ε_{kk}^P according to the following equation:

$$\varepsilon_{kk}^P = W[\mathrm{Exp}(Dx) - 1] \qquad (71)$$

in which D and W are material constants. Therefore, stiffness coefficients and the hardening modulus for a plane cap surface are

$$A = L = 1, \quad B = M = 0, \quad C = N = 0 \qquad (72)$$

and

$$H' = \frac{3L}{D(\varepsilon_v^P + W)} \qquad (73)$$

On the other hand, the elliptic cap function has the form of a quarter of an ellipse such as

$$f = (I_1 - \ell)^2 + R^2 J_2 - (x - \ell)^2 = 0 \qquad (74)$$

in which ℓ is the value of I_1 at the center of the elliptic cap, and R is the ratio of the major to minor axis lengths for the elliptic cap, which may be a function of ℓ or constant. Similarly, we obtain

$$A = L = 2(I_1 - \ell), \quad B = M = R^2, \quad C = N = 0 \qquad (75)$$

and

$$H' = 2(x - \ell) \frac{3L}{D(\varepsilon_v^P + W)} \qquad (76)$$

6 LARGE-DEFORMATION FINITE-ELEMENT IMPLEMENTATION FOR SOIL PLASTICITY MODELS

In this section, general incremental finite element equilibrium equations are developed for a large-deformation analysis of soil response with advanced soil plasticity models. To formulate these incremental equilibrium equations, which reflect the change, in geometry and material properties of soil mass, a mixed incremental formulation using the current configuration of the body as a known material reference state is utilized, and this reference state is updated at each incremental step (Mizuno and Chen, 1982a).

6.1 Mixed incremental formulation

Two different configurations of the body, which are an initial configuration and a subsequent configuration, are used to formulate the incremental equilibrium equations governing an increment of deformation. The stresses, strains and displacements in the initial configuration are assumed to be known and have been determined through a sequence of incremental steps.

Because a Lagrangian formulation is employed here, the equilibrium equations for the subsequent configuration are written in terms of the geometry of the initial configuration. Note that the initial stresses are referred to a global reference frame, and that the initial geometry is a deformed geometry determined through a sequence of previous increments.

Herein, two different strain tensors are introduced. The incremental Green's strain tensor is defined as

$$d\overset{*}{\varepsilon}_{ij} = \frac{1}{2}(du_{i,j} + du_{j,i} + du_{k,i} \, du_{k,j}) \qquad (77)$$

and the infinitesimal strain increment tensor is

$$d\varepsilon_{ij} = \frac{1}{2}(du_{i,j} + du_{j,i}) \qquad (78)$$

where du_i is the incremental displacement vector from the initial configuration of the body, and a comma followed by a subscript indicates a partial derivative with respect to current material coordinates (x_1, x_2, x_3).

Consider the equilibrium conditions of the initial state and the subsequent state by using the virtual work equation. First, the equilibrium equation of the body at the initial state is expressed by

$$\int_s T_i \delta(du_i) \, ds + \int_v \rho_0 F_i(x) \, \delta(du_i) \, dv$$
$$= \int_v \sigma_{ij} \delta(d\varepsilon_{ij}) \, dv \qquad (79)$$

where T_i is the boundary traction vector per unit of initial area, and $F_i(x)$ is the body force vector per unit mass and ρ_0 is the mass density of the initial state. It is assumed that the body force is purely a function of position in the fixed reference frame.

Second, consider the equilibrium of the body in the subsequent configuration. Recalling that the equation of equilibrium for the subsequent configuration is to be written in terms of the geometry of the initial configuration, it should be noted that the internal virtual work per unit of initial volume is the product of Kirchhoff's stress tensor S_{ij} and the variation of the incremental Green's strain $d\varepsilon_{ij}^*$ tensor. Thus, equilibrium of the subsequent configuration is implied by the virtual work equation.

$$\int_s (T_i + dT_i)\,\delta(du_i)\,ds + \int_v \rho_0 \{F_i(x) + dF_i\}\delta(du_i)\,dv$$

$$= \int_v S_{ij}\,\delta(d\varepsilon_{ij}^*)\,dv \qquad (80)$$

where dT_i is the incremental surface traction vector per unit of initial area and dF_i is the incremental body force vector per unit mass of initial configuration.

After an increment of deformation, S_{ij} can be expressed by

$$S_{ij} = \sigma_{ij} + dS_{ij} \qquad (81)$$

where dS_{ij} is the incremental Kirchhoff's stress tensor. Using Eq. (77), the variation of the incremental Green's strain tensor, $\delta(d\varepsilon_{ij}^*)$, can be given as

$$\delta(d\varepsilon_{ij}^*) = \delta(d\varepsilon_{ij}) + \frac{1}{2}\{du_{k,i}\,\delta(du_k)_{,j} +$$

$$du_{k,j}\,\delta(du_k)_{,i}\} \qquad (82)$$

Substituting Eqs. (81) and (82) into Eq. (80), we obtain

$$\int_s dT_i \delta(du_i)\,ds + \int_v \rho_0 dF_i \delta(du_i)\,dv +$$

$$\{\int_s T_i \delta(du_i)\,ds + \int_v \rho_0 F_i(x)\delta(du_i)\,dv -$$

$$\int_v \sigma_{ij}\delta(d\varepsilon_{ij})\,dv\} = \int_v \frac{1}{2}\sigma_{ij}\{du_{k,i}\,\delta(du_k)_{,j} +$$

$$du_{k,j}\,\delta(du_k)_{,i}\}dv + \int_v dS_{ij}\delta(d\varepsilon_{ij}^*)\,dv \qquad (83)$$

It should be noted that the expression in brackets on the left-hand side of Eq. (83) would be identically zero if the initial stress distribution satisfied the equation of equilibrium (Eq. 79). If the incremental stress and displacement distributions satis-

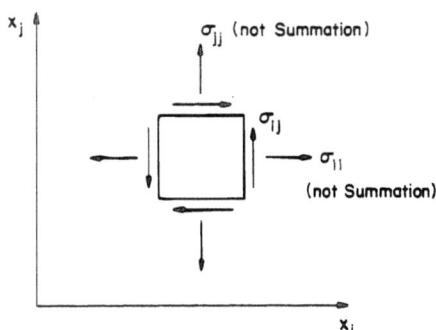

Fig.8 Stress state in initial configuration

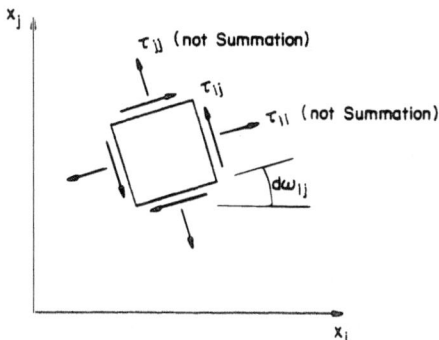

Fig.9 Stress state in subsequent configuration

fying Eq. (83) could be found for all kinematically admissible virtual displacement increments, the subsequent configuration would be in an equilibrium state.

6.2 Linearized incremental equation (large strain and large rotation)

Equation (83) represents an exact condition of equilibrium for the subsequent configuration. Here, Eq. (83) will be linearized under conditions of a large strain and a large rotation without making any assumptions concerning the relative magnitudes of strains and rotations.

Consider now the Eulerian stress tensor σ_{ij} for a three dimensional body as shown in Fig. 8. During an increment of deformation, the neighborhood of a generic point translates, rotates and deforms. The rotation increment to the first order is defined by

$$d\omega_{ij} = \frac{1}{2}\left[\frac{\partial(du_j)}{\partial x_i} - \frac{\partial(du_i)}{\partial x_j}\right] \qquad (84)$$

Physical stresses τ_{ij} in the subsequent configuration, referred to a local coordinate system rotated an amount $d\omega_{ij}$ from the fixed reference frame, are sketched in Fig. 9. The relation between σ_{ij} and τ_{ij} is written as

$$\tau_{ij} = \sigma_{ij} + d\tau_{ij} \tag{85}$$

where $d\tau_{ij}$ is the incremental Cartesian stress tensor for subsequent configuration referred to a local coordinate. $d\tau_{ij}$ is obtained from the following general form (Davidson and Chen, 1974)

$$d\tau_{ij} = D^{ep}_{ijkl} \, d\varepsilon_{kl} \tag{86}$$

where D^{ep}_{ijkl} is the elastic-plastic stiffness described in sections 4 and 5.

It is now necessary to express the incremental Kirchhoff's stress dS_{ij} in Eq. (83) in terms of $d\tau_{ij}$, σ_{ij} and $d\varepsilon_{ij}$ so that Eq. (83) can be linearized to the equation for large strain and large rotation. dS_{ij} related to $d\tau_{ij}$ to the first order is generally written as

$$dS_{ij} = d\tau_{ij} + \psi_{ijkl} \, d\varepsilon_{kl} \tag{87}$$

where ψ_{ijkl} is, in general, expressed in terms of the current stress components, σ_{ij}.

The incremental Kirchhoff's stress tensor dS_{ij} for the plane strain condition, for example, is given as

$$
\begin{Bmatrix} dS_{xx} \\ dS_{yy} \\ dS_{xy} \end{Bmatrix} =
\begin{Bmatrix} d\tau_{xx} \\ d\tau_{yy} \\ d\tau_{xy} \end{Bmatrix} +
$$

$$
\begin{bmatrix} -\sigma_{xx}, & \sigma_{xx}, & -\sigma_{xy} \\ \sigma_{yy}, & -\sigma_{yy}, & -\sigma_{xy} \\ 0, & 0, & -(\sigma_{xx}+\sigma_{yy})/2 \end{bmatrix}
\begin{Bmatrix} d\varepsilon_{xx} \\ d\varepsilon_{yy} \\ d\gamma_{xy} \end{Bmatrix} \tag{88}
$$

The cases for plane stress and axisymmetric conditions are similarly derived.

Substituting the general form of Eq. (87) into Eq. (83) and rearranging some terms leads to the following linear incremental equilibrium equation.

$$\int_s (T_i + dT_i)\,\delta(du_i)\,ds + \int_v \rho_0 \{F_i(x) + dF_i\}\delta(du_i)\,dv$$

$$-\int_v \sigma_{ij}\,\delta(d\varepsilon_{ij})\,dv = \int_v \sigma_{ij}\,du_{k,i}\,\delta(du_k)_{,j}\,dv$$

$$+\int_v (d\tau_{ij} + \psi_{ijkl}\,d\varepsilon_{kl})\,\delta(d\varepsilon^*_{ij})\,dv \tag{89}$$

Here note that the $[\psi]$ matrix is in general asymmetric. If a finite element procedure is employed in conjunction with Eq. (89), the asymmetry of the $[\psi]$ matrix will cause the set of discrete equilibrium equations to be asymmetric, too.

To linearize the incremental equilibrium equation (89), it should now be recalled that the incremental deformations and rotations are "small". This means that the relative elongations and shears as defined by Novozhilov (1953) are given by the incremental Green's strain tensor (Eq. 77). Utilizing the following identity

$$du_{i,j} = d\varepsilon_{ij} - d\omega_{ij} \tag{90}$$

into Eq. (89), and eliminating the higher-order products of incremental quantities, we have

$$\int_s (T_i + dT_i)\,\delta(du_i)\,ds + \int_v \rho_0 \{F_i(x) + dF_i\}\delta(du_i)\,dv$$

$$-\int_v \sigma_{ij}\,\delta(d\varepsilon_{ij})\,dv = \int_v \sigma_{ij}\,[d\varepsilon_{ki}\,\delta(d\varepsilon_{kj}) -$$

$$d\varepsilon_{ki}\,\delta(d\omega_{kj}) - d\omega_{ki}\,\delta(d\varepsilon_{kj}) + d\omega_{ki}\,\delta(d\omega_{kj})]\,dv +$$

$$\int_v (d\tau_{ij} + \psi_{ijkl}\,d\varepsilon_{kl})\,\delta(d\varepsilon_{ij})\,dv \tag{91}$$

Although Eq. (91) is an approximate equation governing the incremental response of a soil mass, it can describe the response of a large deformation and a large rotation.

6.3 Linearized incremental equation (small strain and large rotation)

Herein, the incremental equilibrium equation, Eq. (83), will be linearized by assuming that the order of magnitude of the incremental strains is less than that of the incremental rotations. If it is further assumed that the incremental Green's strains, $d\varepsilon^*_{ij}$, are an order of magnitude less than the incremental rotations, $d\omega_{ij}$, the linear incremental equations are simplified further.

Considering again Eq. (87) and recalling the assumption that incremental deformations are very small, the following approximation, as given by Novozhilov (1953), is incorporated.

$$dS_{ij} \approx d\tau_{ij} \tag{92}$$

Therefore, the new incremental equilibrium equation for small strain and large rotation is then,

$$\int_s (T_i + dT_i)\,\delta(du_i)\,ds + \int_v \rho_0 \{F_i(x) + dF_i\}\delta(du_i)\,dv$$

$$-\int_v \sigma_{ij}\delta(d\varepsilon_{ij})\,dv = \int_v \sigma_{ij}[-d\varepsilon_{ki}\delta(d\omega_{kj}) -$$

$$d\omega_{ki}\delta(d\varepsilon_{kj}) + d\omega_{ki}\delta(d\omega_{kj})]\,dv +$$

$$\int_v d\tau_{ij}\delta(d\varepsilon_{ij})\,dv \tag{93}$$

Thus, Eq. (93) leads to a set of symmetric equations under the assumption that incremental deformations are smaller than incremental rotations.

6.4 Incremental finite element equations

To present Eqs. (91) and (93) in a matrix form suitable for finite element analysis, the following column matrices are defined:

$$\{dU\},\ \{T\},\ \{F\},\ \{d\tau\},\ \{d\varepsilon\},\ \{d\tilde{\varepsilon}\}, \text{ and}$$

$$\{\sigma\} \tag{94}$$

where $\{dU\}$ is the matrix for incremental displacements of the body, $\{T\}$ is for surface traction forces, $\{F\}$ is for body forces, $\{d\tau\}$ is for incremental Cartesian stresses referred to the local coordinate frame in the subsequent configuration, $\{d\varepsilon\}$ is for incremental strains, $\{d\tilde{\varepsilon}\}$ is a combined column matrix in which the upper half is incremental strain components $d\varepsilon_{ij}$, and the lower half is incremental rotation components $d\omega_{ij}$, and $\{\sigma\}$ is for the Cartesian stresses referred to the global coordinates.

The incremental virtual work expression of Eq. (91) or (93) in matrix form is thus

$$\int_s \delta\{dU\}^t\{T+dT\}ds + \int_v \rho_0\delta\{dU\}^t\{F(x)+dF\}dv +$$

$$\int_v \delta\{d\varepsilon\}^t\{\sigma\}dv = \int_v \delta\{d\tilde{\varepsilon}\}^t[A]\{d\tilde{\varepsilon}\}dv -$$

$$\int_v \delta\{d\varepsilon\}^t(\{d\tau\}+[\psi]\{d\varepsilon\})\,dv \tag{95}$$

where the superscript "t" denotes matrix transpose and the $[\psi]$ matrix is neglected for the case of small strain and large rotation analysis (Eq. 93). The $[A]$ matrix includes the Cartesian stresses σ_{ij}. This matrix is symmetric under the assumption of small strain and large rotation. On the other hand, the $[A]$ matrix becomes asymmetric for large strain and large rotation analysis. Explicit expressions of $[A]$ for two-dimensional problems are given later.

Consider now a generic finite element. The relationship between the incremental displacement vector $\{dU\}$ and the vector of nodal displacement increments $\{dv\}$ referred to the global Cartesian frame is defined as

$$\{dU\} = [N]\{dv\} \tag{96}$$

where $[N]$ is a matrix of coordinate functions. In addition, the relation between vectors $\{d\varepsilon\}$ and $\{dv\}$, and between vectors $\{d\tilde{\varepsilon}\}$ and $\{dv\}$, can be defined using Eq. (96) and the kinematic equations. Thus,

$$\{d\varepsilon\} = [B]\{dv\} \tag{97a}$$

and

$$\{d\tilde{\varepsilon}\} = [\tilde{B}]\{dv\} \tag{97b}$$

where the $[B]$ and $[\tilde{B}]$ matrices consist of the quantities determined from the coordinates of an element.

Subsequently Eq. (95) can be rewritten for a single element. Although the equal sign will be used here, it should be noted that Eq. (95) is true only when the contribution of all elements is summed in the manner of the conventional direct stiffness method and that the element surface integrals are identically zero unless part of the element boundary is coincident with the body boundary. Thus, Eq. (95) is rearranged to give

$$\int_s \delta\{dv\}^t[N]^t\{T+dT\}ds + \int_v \rho_0\delta\{dv\}^t[N]^t\{F(x)+$$

$$dF)\,dv - \int_v \delta\{dv\}^t[B]^t\{\sigma\}dv$$

$$= \int_v \delta\{dv\}^t[\tilde{B}]^t[A][\tilde{B}]\{dv\}dv - \int_v \delta\{dv\}^t[B]^t$$

$$(\{d\tau\}+[\psi][B]\{dv\})\,dv \tag{98}$$

Since Eq. (98) must be satisfied for all kinematically admissible virtual nodal displacements, it follows that

$$\{dP\} = \int_v [\tilde{B}]^t[A][\tilde{B}]dv\{dv\} +$$

$$\int_v [B]^t(\{d\tau\}+[\psi][B]\{dv\})\,dv \tag{99}$$

where $\{dP\}$ is the column vector of the nodal forces given by

$$\{dP\} = \int_s [N]^t\{T+dT\}ds + \int_v \rho_0[N]^t\{F(x)+dF\}dv$$

$$-\int_v [B]^t\{\sigma\}dv \tag{100}$$

Introducing the symbolic elastic-plastic constitutive law

$$\{d\tau\} = [D]\{d\varepsilon\} = [D][B]\{dv\} \tag{101}$$

and substituting it into Eq. (99) gives

$$\{dP\} = \int_v [\tilde{B}]^t[A][\tilde{B}]dv\{dv\} +$$

$$\int_v [B]^t[\tilde{D}][B]dv\{dv\} \tag{102}$$

where $[\tilde{D}] = [D] + [\psi]$. Now the tangent stiffness matrix for an element can be written as

$$[K] = [K_m] + [K_g] \tag{103}$$

where $[K_m] = \int [B]^t [\tilde{D}] [B] dv$ and $[K_g] = \int [\tilde{B}]^t [A] [\tilde{B}] dv$. The $[K_m]$ matrix becomes symmetric for small strain and large rotation analysis and asymmetric for large strain and large rotation analysis. The $[K_g]$ matrix is the so-called geometric or initial stress stiffness matrix.

Introducing the constant strain triangle element, matrices $[A]$, $[B]$ and $[\tilde{B}]$ etc. are presented for two dimensional problems such as those under plane strain and plane stress conditions. It should be here noted that $\{dv\}$, $\{d\varepsilon\}$ and $\{d\tilde{\varepsilon}\}$ are $[du_x^1, du_y^1, du_x^2, du_y^2, du_x^3, du_y^3]^t$, $[d\varepsilon_{xx}, d\varepsilon_{yy}, d\gamma_{xy}]^t$ and $[d\varepsilon_{xx}, d\varepsilon_{yy}, d\gamma_{xy}, d\omega_{xy}]^t$ where du_x^i and du_y^i are the x and y displacement increment at node i of the triangle element respectively.

The $[A]$ matrix in the first term on the right hand side of Eq. (95) is determined from Eq. (91) for large strain and large rotation analysis.

$$[A] = \begin{pmatrix} \sigma_{xx} & 0 & \frac{1}{2}\sigma_{xy} & -\sigma_{xy} \\ 0 & \sigma_{yy} & \frac{1}{2}\sigma_{xy} & \sigma_{xy} \\ \frac{1}{2}\sigma_{xy} & \frac{1}{2}\sigma_{xy} & \frac{1}{4}(\sigma_{xx}+\sigma_{yy}) & \frac{1}{2}(\sigma_{xx}-\sigma_{yy}) \\ -\sigma_{xy} & \sigma_{xy} & \frac{1}{2}(\sigma_{xx}-\sigma_{yy}) & (\sigma_{xx}+\sigma_{yy}) \end{pmatrix} \tag{104}$$

Also, the matrix $[A]$ for small strain-large rotation analysis is derived from Eq. (93). In this case, $A_{11} = A_{12} = A_{13} = A_{21} = A_{22} = A_{23} = A_{31} = A_{32} = A_{33} = 0$, and the other part is the same as that in Eq. (104). Next, the $[B]$ and $[\tilde{B}]$ matrices are derived from Eq (96) and the kinematic equation between strains and displacements. Thus,

$$[B] = \frac{1}{2A} \begin{pmatrix} b_1 & 0 & b_2 & 0 & b_3 & 0 \\ 0 & c_1 & 0 & c_2 & 0 & c_3 \\ c_1 & b_1 & c_2 & b_2 & c_3 & b_3 \end{pmatrix} \tag{105a}$$

and

$$[\tilde{B}] = \frac{1}{2A} \begin{pmatrix} b_1 & 0 & b_2 & 0 & b_3 & 0 \\ 0 & c_1 & 0 & c_2 & 0 & c_3 \\ c_1 & b_1 & c_2 & b_2 & c_3 & b_3 \\ -c_1/2 & b_1/2 & -c_2/2 & b_2/2 & -c_3/2 & b_3/2 \end{pmatrix} \tag{105b}$$

where

$$a_1 = X_2 Y_3 - X_3 Y_2, \quad b_1 = Y_2 - Y_3, \quad c_1 = X_3 - X_2$$
$$a_2 = X_3 Y_1 - X_1 Y_3, \quad b_2 = Y_3 - Y_1, \quad c_2 = X_1 - X_3$$
$$a_3 = X_1 Y_2 - X_2 Y_1, \quad b_3 = Y_1 - Y_2, \quad c_3 = X_2 - X_1$$

and A, X_i and Y_i are the area of the triangle element, and X and Y coordinates of the i node of the triangle element respectively.

7 FINITE ELEMENT IMPLEMENTATION OF PLASTICITY MODELS

In this section, step-by-step computer implementation procedures for the Drucker-Prager perfectly plastic model and the advanced strain-hardening cap models are presented in a process where the incremental equilibrium equations can be numerically integrated. More specifically, these coding procedures have been implemented in a large deformation finite element analysis program "SOILSLP" at Purdue University for the analysis of earthquake-induced landslide problems (Mizuno and Chen, 1981).

7.1 Implementation of Drucker-Prager model

The mid-point integration method will be used here (Davidson and Chen, 1974). Consider an element whose equilibrium stresses and strains are respectively σ_{ij}^n and ε_{ij}^n at the beginning of the (n+1)-th increment. According to the stress state (elastic or plastic) in an element, either the elastic or the elastic-plastic constitutive matrix is used to calculate the initial tangent stiffness. Applying half of the incremental loads or displacements to the discretized body, incremental stresses $d\sigma_{ij}^{n+\frac{1}{2}}$ and strains $d\varepsilon_{ij}^{n+\frac{1}{2}}$ at the (n+½)-th increment are first estimated using the initial tangent stiffness.

At this stage, for an element which is previously plastic, the plastic unloading is checked because it may occur even if the incremental loads or displacements are monotonically increasing. Then, the positive scalar function $d\lambda$ defined in the flow rule equation is checked. If $d\lambda$ has a negative value, the element has undergone plastic unloading, and the elastic constitutive matrix is used to compute the subsequent tangent stiffness at mid-increment. On the other hand, if $d\lambda$ has a positive value, the element remains plastic, and the stress state $(\sigma_{ij}^n + d\sigma_{ij}^{n+\frac{1}{2}})$ is arbitrarily scaled back to the yield surface by assuming that the hydrostatic component I_1 and principal directions of the stress tensor σ_{ij} remain unchanged. This procedure is used because,

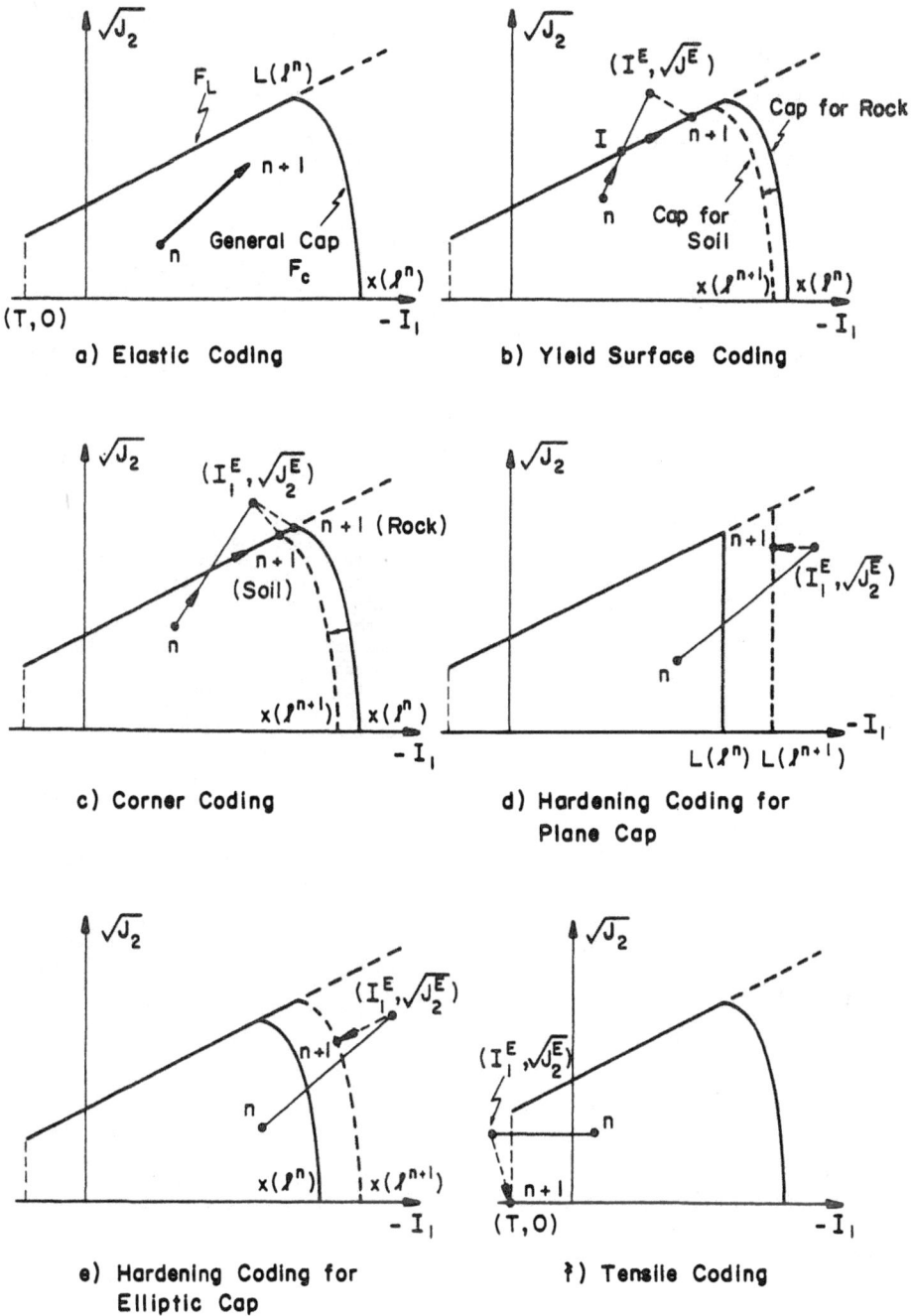

a) Elastic Coding

b) Yield Surface Coding

c) Corner Coding

d) Hardening Coding for Plane Cap

e) Hardening Coding for Elliptic Cap

f) Tensile Coding

Fig.10 Numerical implementation of cap models

after a number of such increments, the final stress would otherwise lie far away from the yield surface. The elastic-plastic constitutive matrix is used to compute the subsequent tangent stiffness at mid-increment.

For an element which was previously elastic, the value of k $(=\alpha I_1 + \sqrt{J_2})$ in the Drucker-Prager model at the end of the last n-th increment is evaluated. Thus,

$$k_n = \alpha I_1^n + \sqrt{J_2^n} \qquad (106)$$

where I_1^n and J_2^n are respectively the first and second invariants of the stress tensor and stress deviatoric tensor obtained at the n-th increment. The value of $k_{n+\frac{1}{2}}$ is then evaluated at mid-increment. To estimate k_{n+1} at the end of the increment, the following relationship is assumed:

$$k_{n+1} = k_n + 2(k_{n+\frac{1}{2}} - k_n) \qquad (107)$$

If k_{n+1} is less than the yield value k, it is assumed that the element remains in the elastic range during this increment. Otherwise, it is assumed that the element yields during this increment. In this latter case, a weighted average of the elastic and elastic-plastic constitutive matrix is used. To compute the averaged matrix, the stress state at mid-increment is scaled back to the yield surface. The averaged constitutive matrix is then computed as

$$[D_{av}] = m[D^e] + (1-m)[D^{ep}] \qquad (108)$$

where $m = (k - k_n)/2(k_{n+\frac{1}{2}} - k_n)$.

If the averaged matrix or elastic-plastic matrix is used to determine the mid-increment stiffness, the stress state in an element is assumed to be plastic at the end of the increment, regardless of whether the stress state lies inside or outside the yield surface. If an analysis is based on large deformation, the stress state at the end of the increment is first rotated from a local coordinate system to a global coordinate system. Then, the final stress state is scaled back to the yield surface. If the elastic constitutive matrix is used to compute the mid-increment stiffness, the stress state at the end of the increment is first rotated when a large deformation analysis is used. Then the yield condition is checked. If the stress state lies outside the yield surface, it is scaled back to the yield surface in preparation for the next increment. Otherwise, stress state remains unchanged and elastic constitutive matrix is used for the next increment.

Since the scaled stresses cannot in general be expected to satisfy equilibrium, the iteration procedure could be used in finite element coding to obtain a more accurate solution.

7.2 Implementation of cap model

Herein, the input quantities are the current stress components σ_{ij}^n and the hardening parameter ℓ^n obtained at the end of the n-th increment, and the components of the new strain increments $d\varepsilon_{ij}^{n+1}$ which are obtained from the current finite element calculation. Now our purpose is to determine the new quantities of the stress components σ_{ij}^{n+1} for the new strain components $(\varepsilon_{ij}^n + d\varepsilon_{ij}^{n+1})$. As a first step towards determining the corresponding stress increments $d\sigma_{ij}^{n+1}$, an elastic trial stress invariant I_1^E and elastic trial deviatoric stresses s_{ij} are computed as

$$I_1^E = I_1^n + 3K d\varepsilon_{kk}^{n+1} \qquad (109)$$

$$s_{ij}^E = s_{ij}^n + 2G de_{ij}^{n+1} \qquad (110)$$

where K, G, $d\varepsilon_{kk}^{n+1}$, and de_{ij}^{n+1} are the bulk modulus, the shear modulus, the volumetric strain increment and the deviatoric strain increments, respectively.

Since the cap models (elliptic cap model and plane cap model) have several loading surfaces, such as a Drucker-Prager type of yield surface, a hardening surface and a tension cut-off surface, the trial stresses are then checked with respect to each of these loading surfaces. In what follows, the finite element coding, including such items as "elastic coding", "yield surface coding", "hardening coding", "corner coding", and "tensile coding" is explained according to several cases of trial stress path (Baladi, 1979).

1. Elastic coding – If these trial stresses do not violate either of the loading functions, the behavior of an element is truly elastic, as shown in Fig. 10a. Thus, the hardening parameter ℓ^n remains unchanged, and the final stresses at the end of the (n+1)-th increment are written as

$$I_1^{n+1} = I_1^E \qquad (111)$$

$$s_{ij}^{n+1} = s_{ij}^E \qquad (112)$$

2. Yield surface coding – If the yield surface $f_L(I_1^E, \sqrt{J_2^E})$ is violated by the elastic trial stresses, as shown in Fig. 10b, i.e.,

$$I_1^E \geq L(\ell^n) \qquad (113)$$

and

$$f_L(I_1^E, \sqrt{J_2^E}) \geq 0 \qquad (114)$$

the trial stresses must be corrected such that the final stresses at the (n+1)-th increment satisfy the consistency condition, that is, $f_L(I_1^{n+1}, \sqrt{J_2^{n+1}}) = 0$. Since the elastic trial stresses generally lie outside the yield surface, the deviation from the yield surface ($f_L=0$) is thus, for a small strain increment,

$$df_L = f_L(I_1^E, \sqrt{J_2^E}) \qquad (115)$$

Substituting Eq. (115) into the flow rule equation, the plastic strain increment $d\varepsilon_{ij}^P$ is expressed as

$$d\varepsilon_{ij}^P = \frac{f_L(I_1^E, \sqrt{J_2^E})}{\dfrac{\partial f_L}{\partial \sigma_{kl}} D_{klmn}^e \dfrac{\partial g}{\partial \sigma_{mn}}} \frac{\partial g}{\partial \sigma_{ij}} \qquad (116)$$

where the function, g, is a plastic potential function. It should be noted that the calculation of Eq. (116) should be made with respect to the stresses which lie on the yield surface, as represented by point I in Fig. 10b. Consequently, the corresponding stress increment $d\sigma_{ij}^{n+1}$ is calculated by

$$d\sigma_{ij}^{n+1} = 3K(d\varepsilon_{kk}^{n+1} - d\varepsilon_{kk}^P)\delta_{ij} +$$
$$2G(d\varepsilon_{ij}^{n+1} - d\varepsilon_{ij}^P) \qquad (117)$$

where $d\varepsilon_{kk}^P$ and $d\varepsilon_{ij}^P$ are the increment of the plastic volumetric strain and the tensor of the plastic deviatoric strain increments respectively.

Furthermore, the hardening parameter ℓ^n, which is a function of plastic volumetric strain, must be updated. Thus,

$$\ell^{n+1} = \ell^n + \frac{\partial \ell}{\partial \varepsilon_{kk}^P}\bigg|_{\ell^n} d\varepsilon_{kk}^P \qquad (118)$$

This is the "yield surface coding" for a material such as soil in which the hardening cap is allowed to move back toward the origin. On the other hand, for a material like rock in which the hardening cap is not permitted to move back toward the origin, the hardening parameter ℓ^n is simply to set

$$\ell^{n+1} = \ell^n \qquad (119)$$

3. Corner coding - If the value of I_1^{n+1}, corrected from "yield surface coding", is found to be less than the present location of the cap (plane or elliptic cap), $L(\ell^n)$, the following "corner coding" must be utilized. For this case, the stress state must be on the corner at which the yield surface intersects the cap, as shown in Fig. 10c. It follows that

$$I_1^E = 3Kd\varepsilon_{kk}^P = I_1^{n+1}$$
$$= L(\ell^{n+1})$$
$$= \ell^{n+1}$$
$$= \ell^n + \frac{\partial \ell}{\partial \varepsilon_{kk}^P}\bigg|_{\ell^n} d\varepsilon_{kk}^P \qquad (120)$$

Eliminating $d\varepsilon_{kk}^P$ from Eq. (120) and solving for I_1^{n+1} leads to the final corner value

$$I_1^{n+1} = \ell^{n+1} = \frac{\dfrac{\partial \ell}{\partial \varepsilon_{kk}^P}\bigg|_{\ell^n} I_1^E + 3K\ell^n}{\dfrac{\partial \ell}{\partial \varepsilon_{kk}^P}\bigg|_{\ell^n} + 3K} \qquad (121)$$

Substituting I_1^{n+1} back into Eq. (120) and solving for plastic volumetric strain, $d\varepsilon_{kk}^P$, we obtain,

$$d\varepsilon_{kk}^P = \frac{I_1^E - \ell^n}{\dfrac{\partial \ell}{\partial \varepsilon_{kk}^P}\bigg|_{\ell^n} + 3K} \qquad (122)$$

Thus $d\lambda$ in a flow rule function is rewritten as

$$d\lambda = \frac{I_1^E - \ell^n}{3\dfrac{\partial \psi}{\partial I_1}(\dfrac{\partial \ell}{\partial \varepsilon_{kk}^P}\bigg|_{\ell^n} + 3K)} \qquad (123)$$

Next, $d\varepsilon_{ij}^P$ can be recalculated using $d\lambda$ in Eq. (123), and then $d\sigma_{ij}^{n+1}$ can be obtained from Eq. (117). On the other hand, for rocks the corner coding is simply to set

$$I_1^{n+1} = \ell^{n+1} = L(\ell^n) \qquad (124)$$

Although the cap does not move, the increment of plastic volumetric strain is obtained from Eq. (120), thus,

$$d\varepsilon_{kk}^P = \frac{I_1^E - \ell^n}{3K} \qquad (125)$$

Similarly, $d\lambda$ is obtained and then $d\sigma_{ij}^{n+1}$ is calculated from Eq. (117). Yield surface coding and corner coding complete the process for a stress state violating the yield surface. Now we are going to examine the case where the hardening cap, rather than the failure envelope, is violated.

4. hardening coding - The elastic trial stresses are checked against the hardening surface, f_c. If

$$f_c(I_1^E, \sqrt{J_2^E}, \varepsilon_{kk}^P) > 0 \text{ and } I_1^E < L(\ell^n) \qquad (126)$$

then the "hardening coding" is performed.

Now the procedure will be explained for a plane hardening cap, as shown in Fig. 10d. First, an increment of hardening parameter $d\ell^t$ is assumed, from which the corresponding increments, $d\varepsilon_{kk}^p$ and $d\varepsilon_{kk}^e$, are computed. Thus, the following relation should be satisfied by $d\ell^t$.

$$d\varepsilon_{kk}^{n+1} = \frac{\partial \varepsilon_{kk}^p}{\partial \ell}\bigg|_{\ell^n} d\ell^t + \frac{\ell^n + d\ell^t - I_1^{l_1}}{3K} \tag{127}$$

An iterative procedure for trial values of $d\ell^t$ is continued until Eq. (127) is satisfied. Once the value of $d\ell^t$ is determined, then $d\varepsilon_{kk}^p$ can be calculated. Similarly, $d\lambda$ is determined from the flow rule using $d\varepsilon_{kk}^p$, and $d\sigma_{ij}^{n+1}$ is evaluated from Eq. (117). Since the plane cap causes only plastic volumetric strain within the framework of the normality flow rule, the new s_{ij}^{n+1} can be taken directly as s_{ij}^E.

Implementation of the elliptic cap is performed in a different way from that of the plane cap coding (Fig. 10e). First, a trial value of $d\ell^t$ is assumed. Next, the corresponding trial value of $L(\ell^n + d\ell^t)$, $X(\ell^n + d\ell^t)$ and $d\varepsilon_{kk}^{pt}$ are computed. Further, a trial value of I_1^t is calculated. Thus,

$$I_1^t = I_1^E - 3K d\varepsilon_{kk}^{pt} \tag{128}$$

Now, the final stress state must satisfy the following conditions:

$$s_{ij}^t + 2Gd\varepsilon_{ij}^{pt} = s_{ij}^E \tag{129}$$

$$f_c(I_1^t, \sqrt{J_2^t}, \varepsilon_{kk}^{pt}) = 0 \tag{130}$$

where s_{ij}^t, $d\varepsilon_{ij}^{pt}$, and ε_{kk}^{pt} are respectively the trial tensor of deviatoric stresses, trial tensor of increments of plastic deviatoric strain and trial increments of plastic volumetric strain. Since $d\varepsilon_{kk}^{pt}$ is known, the trial value of the positive scalar function in the flow rule equation, $d\lambda$, is given by

$$d\lambda^t = \frac{d\varepsilon_{kk}^{pt}}{3 \dfrac{\partial f_c}{\partial I_1}\bigg|_{I_1^t, J_2^t}} \tag{131}$$

Therefore, $d\varepsilon_{ij}^{pt}$ is calculated as

$$d\varepsilon_{ij}^{pt} = \frac{s_{ij}^t \, d\varepsilon_{kk}^{pt} \dfrac{\partial f_c}{\partial \sqrt{J_2}}}{6\sqrt{J_2^t} \dfrac{\partial f_c}{\partial I_1}}\bigg|_{I_1^t, J_2^t} \tag{132}$$

Substituting Eq. (132) into Eq. (129), we finally have,

$$\sqrt{J_2^t} + \frac{Gd\varepsilon_{kk}^{pt} \dfrac{\partial f_c}{\partial \sqrt{J_2}}\bigg|}{3 \dfrac{\partial f_c}{\partial I_1}\bigg|_{I_1^t, J_2^t}} = \sqrt{J_2^E} \tag{133}$$

This iterative procedure (Baladi, 1979) is repeated until Eq. (133) is satisfied. After iteration, $d\varepsilon_{ij}^p$ is similarly obtained from the flow rule. Then, $d\sigma_{ij}^{n+1}$ is calculated from Eq. (117).

5. Tensile coding – Let us consider the case in which the elastic trial stresses violate the region of tension cut-off, as shown in Fig. 10f. Here, it is assumed that the new stress state is now shifted from the previous stress state $(I_1^n, \sqrt{J_2^n})$ to the point $(T,0)$ which is located on the I_1-axis. Therefore,

$$\sigma_{ij}^{n+1} = \frac{T}{3}\delta_{ij} \tag{134}$$

In addition, the dilatancy caused by tensile coding is considered as follows: If ℓ^n is less than zero,

$$d\varepsilon_{kk}^p = d\varepsilon_{kk}^{n+1} - \frac{T - I_1^n}{3K} \tag{135}$$

and if ℓ^n is greater than zero, $d\varepsilon_{kk}^p$ is simply set to zero. Thus the hardening parameter ℓ^{n+1} is determined as

$$\ell^{n+1} = \ell^n + \frac{\partial \ell}{\partial \varepsilon_{kk}^p}\bigg|_{\ell^n} d\varepsilon_{kk}^p \tag{136}$$

So far the implementation of several codings have been discussed. Here it should be noted that these codings are performed within the small strain increments. However, the new stress state σ_{ij}^{n+1} obtained from "yield surface coding," "corner coding" and "hardening coding" do not necessarily lie exactly on these loading surfaces. Therefore the procedure for scaling new stresses back onto these loading surfaces is now recommended. If a large deformation analysis is employed, then the stresses are rotated from the local Cartesian system to the global Cartesian system for the next calculation.

8 SEISMIC ANALYSES OF SLOPES

In this section, plasticity models such as Drucker-Prager model and cap model, which have been so far explained within the context of theoretical and numerical treatment, will be applied to seismic slope stability problems. Emphasis is placed here on the effect of a large deformation on the evaluation of overall slope stability problems. Further, of particular concern is

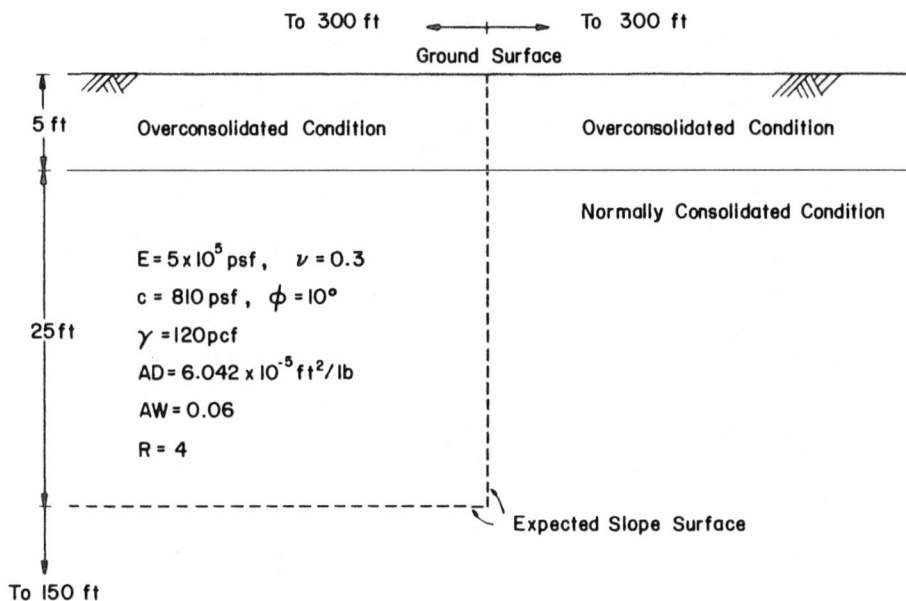

Fig.11 Initial condition of the ground

the comparison of failure modes and limit loads with those assumed in the limit analysis method (Mizuno and Chen, 1984).

8.1 Analyses of slopes prior to seismic loadings

In order to simulate the stress condition inside the slope after the completion of ground excavation, an elastic-plastic effective stress analysis of the slope is performed. Sequential loading to simulate a cut-down or built-up process is not considered here. Instead, the final configuration of the slope, and the spreading of yielded zones are investigated qualitatively by increasing the internal force due to the weight of soil from zero to the natural weight of $\gamma = 120$ pcf (18.85KN/m^3). A 30 ft (9.15 m) high vertical slope, as shown in Fig. 11, is considered. Here the vertical boundaries and bottom boundary are placed respectively at 300 and 150 ft away from the toe of the expected slope surface. Movement on the vertical boundaries is constrained horizontally only, and movement along the bottom boundary is constrained in both directions. The same mesh as used by Snitbhan and Chen (1978), consisting of 250 nodes and 216 rectangular elements, is utilized in the present finite element analy-

sis. The following material constants of soil are assumed: Young's modulus E = 5 x 10^5 psf (2.4×10^7 Pa), Poisson's ratio ν = 0.3. cohesion c = 810 psf (3.88×10^4 Pa), internal friction angle $\phi = 10^0$ and unit weight of soil γ = 120 pcf (18.85 KN/m^3). The material constants in the Drucker-Prager model are obtained from matching the model with the Coulomb criterion under the plane strain condition. Further, the material constants, W and D, in the hardening function of cap model, $\varepsilon_{kk}^p = W[\text{Exp}(Dx) - 1]$, are assumed to be 0.06 (the maximum compaction of plastic volumetric strain ε_{kk}^p and 6.042 x 10^{-5} psf^{-1} (1.26×10^{-6} Pa^{-1} respectively. The location of the elliptic cap surface, x, is determined according to the initial ground condition which is illustrated in Fig. 11. The value of x is assumed to be -6700 psf (compression) for soil elements above 5 ft (1.53 m) in depth, where the stress states in the ground are within the elliptic cap surface. On the other hand, for soil elements below 5 ft level, the locations of the cap surfaces are determined according to the initial stress states, so that the stress states lie on the corresponding cap surfaces. As for the yield surface in the elliptic cap model, the same surface as that of the Drucker-Prager model assumed above is utilized. The shape ratio of the elliptic cap surface, R, is assumed to be 4. Note that the initial stress state for all elements is set to zero at the beginning of both model analyses, while the elliptic cap surfaces for all elements remain fixed at the location determined above.

Here, large deformation analysis is employed to reflect the effect of slope configuration on the limit load due to earthquake loading. In addition, the initial stress method (Zienkiewicz et.al., 1969) is utilized in the numerical analysis.

In the case of the Drucker-Prager model, the ground surface behind the crest is found to settle approximately 3.2 ft (0.98 m) and the bulging extends 0.604 ft (0.184 m) from the original vertical slope line. For the elliptic cap model case, a larger amount of ground settlement (3.8 ft or 1.16 m) and bulging (0.804 ft or 0.245 m) in the horizontal direction are observed. However, deformed geometries of the two slopes are very similar at this load level, regardless of the relatively large difference in ground settlement.

8.2 Analyses of slopes during seismic loadings

Based on the analysis described previously, the seismic large deformation analyses of the vertical slopes are now performed here by employing the pseudo-static method. The 1934 El Centro (S-N direction) accelerogram is used as the input for the seismic loadings which are obtained as the product of soil mass, m, and acceleration, a. In the present analysis, the horizontal acceleration data between time t = 1.5 seconds and t = 4.5 seconds in the accelerogram are applied to the deformed shape of the slope after excavation. For simplicity, the distribution of acceleration is here assumed to be uniform throughout the slope. It acts to the right when its sign is positive. The response of the slope during the seismic loading and the velocity fields at the collapse stage are discussed as follows:

1. Responses of slopes - In Fig. 12, the acceleration-time-displacement curves are presented for both model cases. Note that the horizontal displacement at the nodal point above the toe is taken as the reference displacement in the figure, and that displacement is measured from the deformed shape of the slope after excavation, not from the original shape. The displacement response curves with solid and broken lines are results from the analyses with the Drucker-Prager and elliptic cap models respectively. Further, the zones of yielding during seismic loading are presented in Fig. 13 for both model cases. The yielded zones in the Drucker-Prager model and elliptic

cap model cases are illustrated in parallel in the upper and lower halves of the figure respectively. Note that the yield zones in the elliptic cap model case are taken to be the combination of the Drucker-Prager yielded zones and corner zones, but not the hardening zones.

The seismic loadings with a small magnitude act horizontally and uniformly to the right during the incipient period. This can be seen from the accelerogram in Fig. 12. Consequently the direction of horizontal displacement at the nodal point above the toe is to the right, and it prevents the slope from sliding. The displacements predicted by both models during the period between step 1 and step 3, as marked in Fig. 12, are almost the same. Further, the yielded zone predicted by the Drucker-Prager model contracts dramatically at step 1, as shown in Fig. 13b, compared with the yielded zone at the beginning of the seismic loading (Fig. 13a). In this case, the yielded zone is located just to the right of the toe and oriented diagonally downward. At step 2, it grows in size and then contracts again at step 3 (Fig. 13c and d). The yielded zones for the elliptic cap model case show a similar response during this period.

During the period between step 3 and step 4, the seismic load changes directions and acts to the left so that sliding of the slope is likely to occur. At step 4 with a = -0.065 g, the horizontal displacements are predicted to be 0.58 ft (0.177 m) and 0.74 ft (0.223 m) by the Drucker-Prager and elliptic cap models respectively. For the Drucker-Prager model case, the yielding spreads into the area below the toe and along the slip surface as would be predicted in the limit analysis method (Fig. 13e). For the elliptic cap model case, it spreads over this same general area but also at the ground surface behind the crest. In this case, the yield zones show a clearer picture of potential sliding of the slope. At step 5, however, the important spreading of the yielded zones, which could cause possible sliding of the slope, is not observed for both model cases because the load now acts to the right (Fig. 13f). As a result, the zones of yielding contract. At step 6, where the direction of loading is again reversed, the growth of the yielded zone is limited to a small area around the toe for the Drucker-Prager model case. However, for the elliptic cap model case, the yielded zone spreads along the entire slip surface of the limit analysis method (Fig. 13g).

During the period between step 7 and step 8, yielded zones spread over a small area near the vertical slope line for the Drucker-Prager model case and inside the slope for the elliptic cap model case (Fig.

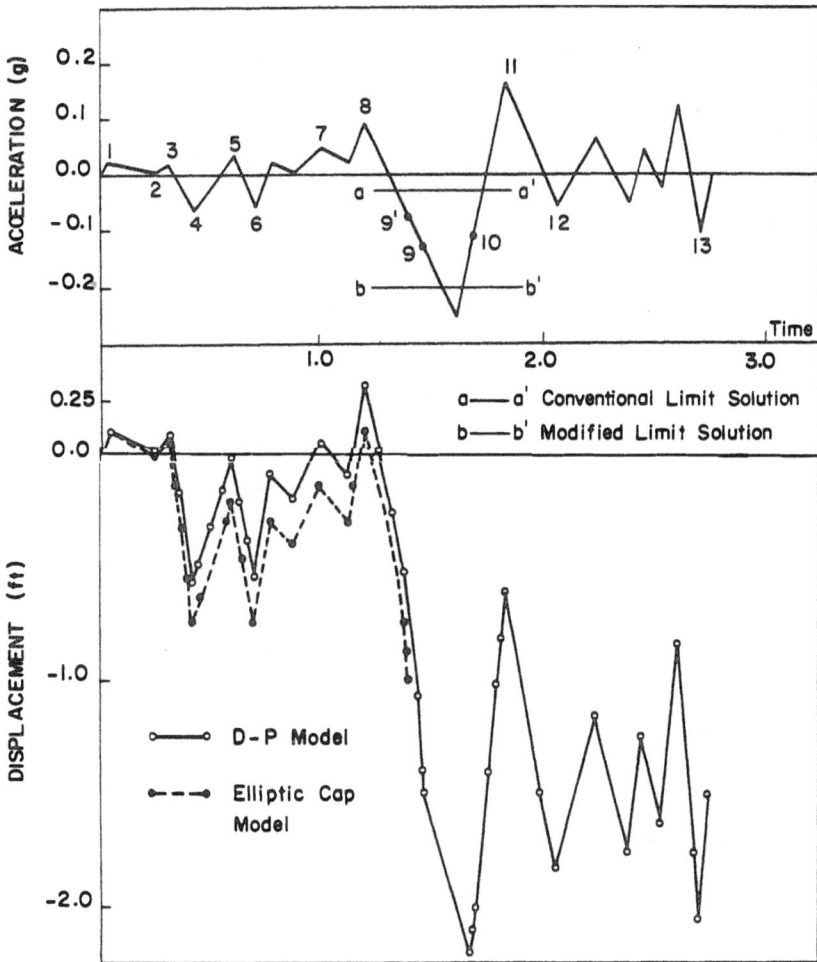

Fig.12 Acceleration-Time-Displacement curves by the Drucker-Prager and elliptic cap models

13h and i). These yielded zones do not appear to affect the critical behavior of the slope.

Beyond step 8, the seismic load acts to the left and directs toward a level corresponding to an acceleration of a = -0.225g. Since potential sliding of the slope is observed at step 4 (Fig. 13e), with a = -0.064 g, it is expected that the sliding mechanism would develop during this loading. Based on the finite element calculation (Fig. 12), it can be estimated that the sliding of the slope will occur at approximately a = -0.135 g and -0.075 g, for the Drucker-Prager and elliptic cap model cases respectively. On the other hand, the limit analysis method predicts the occurrence of sliding at a = -0.0287 g and -0.196 g by applying the full slope height of 30 ft (9.15 m) and the modified slope height of 25 ft (7.63 m), respectively. Thus, the finite element solutions lie between the two extreme solutions of the limit analysis method.

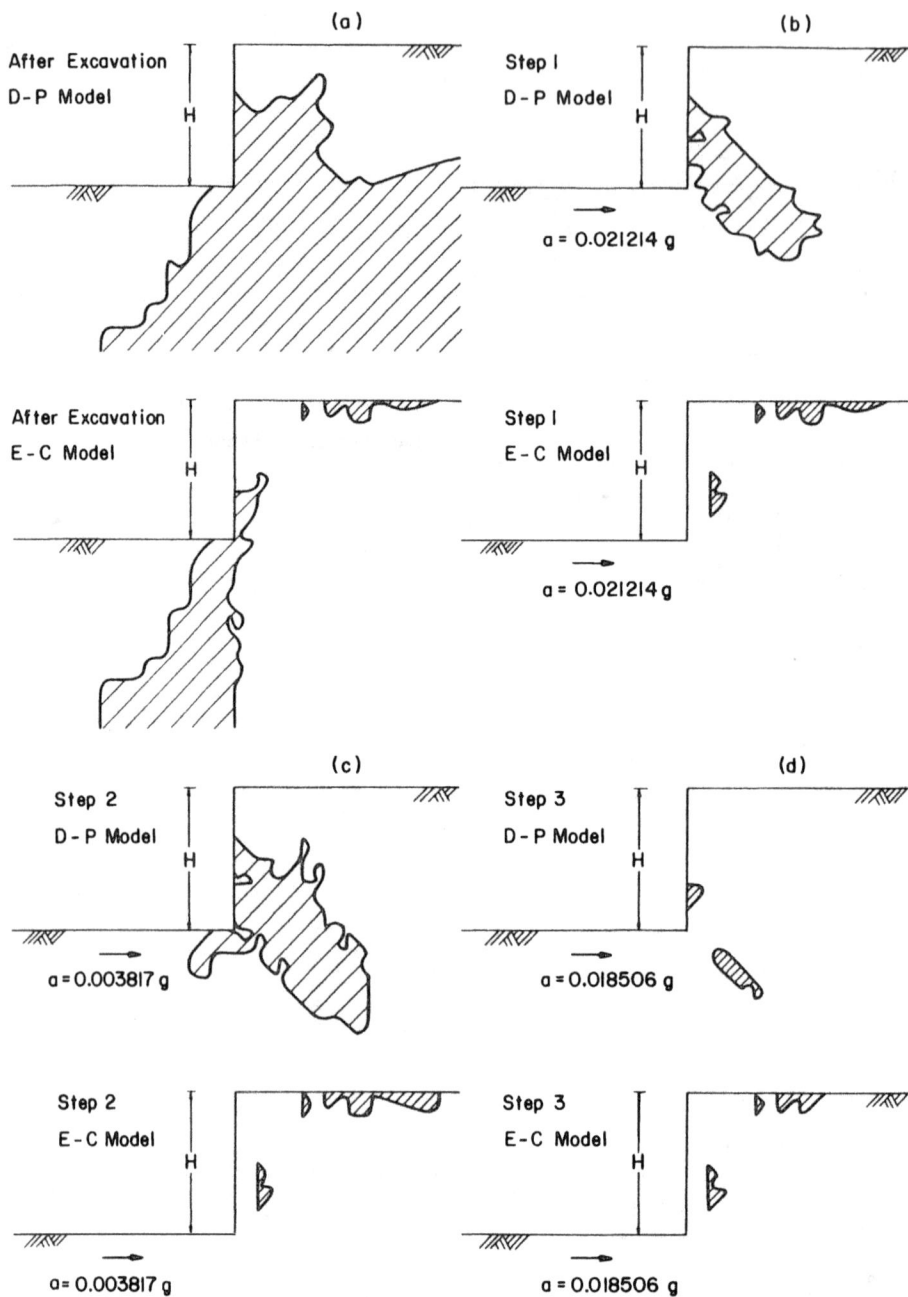

Fig.13   Yielded zones in the Drucker-Prager perfectly plastic material and elliptic cap hardening material slopes during seismic loading

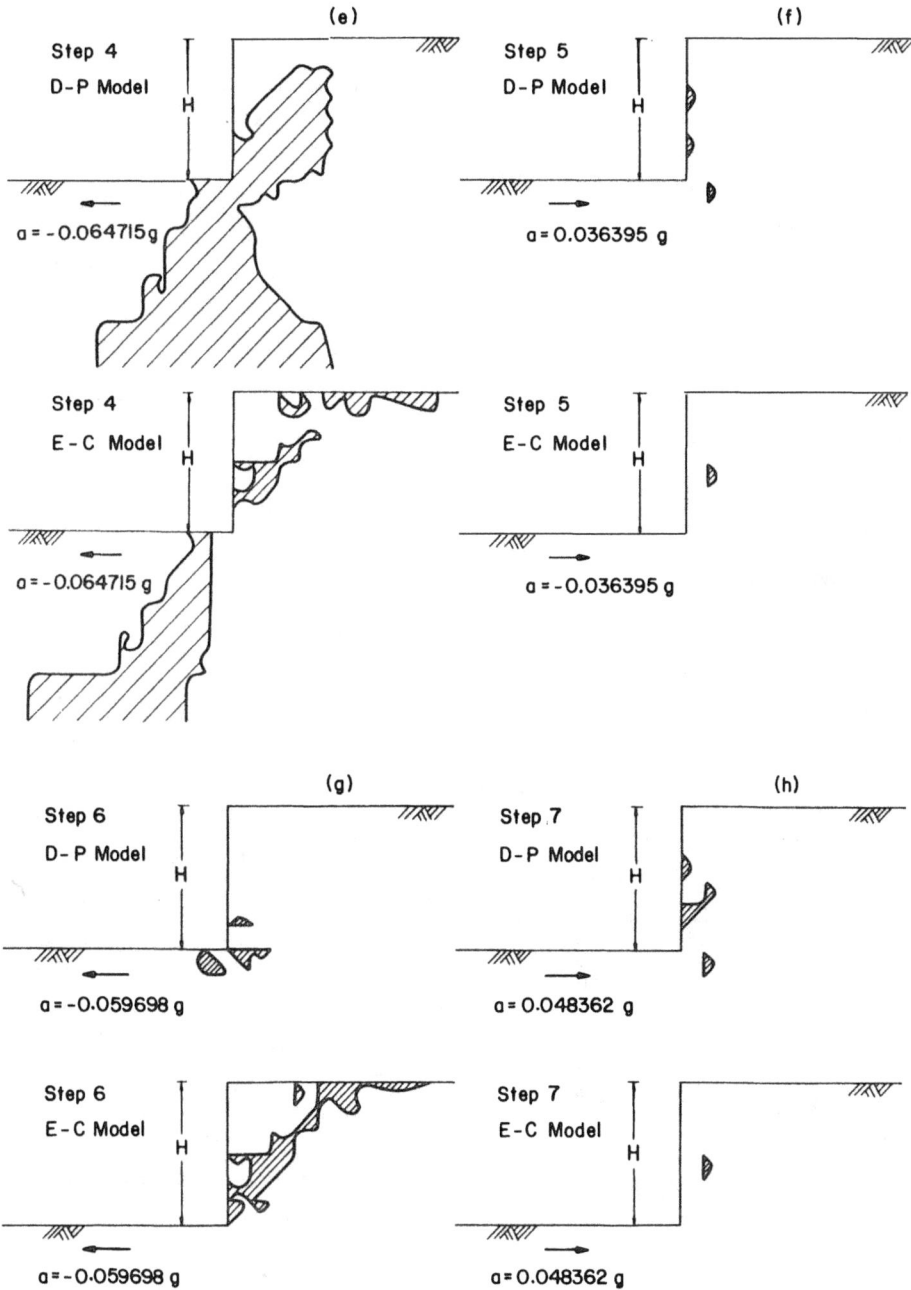

(e)

Step 4
D-P Model
H
a = -0.064715g

(f)

Step 5
D-P Model
H
a = 0.036395 g

Step 4
E-C Model
H
a = -0.064715 g

Step 5
E-C Model
H
a = -0.036395 g

(g)

Step 6
D-P Model
H
a = -0.059698 g

(h)

Step 7
D-P Model
H
a = 0.048362 g

Step 6
E-C Model
H
a = -0.059698 g

Step 7
E-C Model
H
a = 0.048362 g

Fig.13 Continued

420

(i)

Step 8

D-P Model

H

a = 0.094636 g

(j)

Step 9

D-P Model

H

a = - 0.135 g

Step 8

E-C Model

H

a = 0.094636 g

Step 9′

E-C Model

H

a = -0.075 g

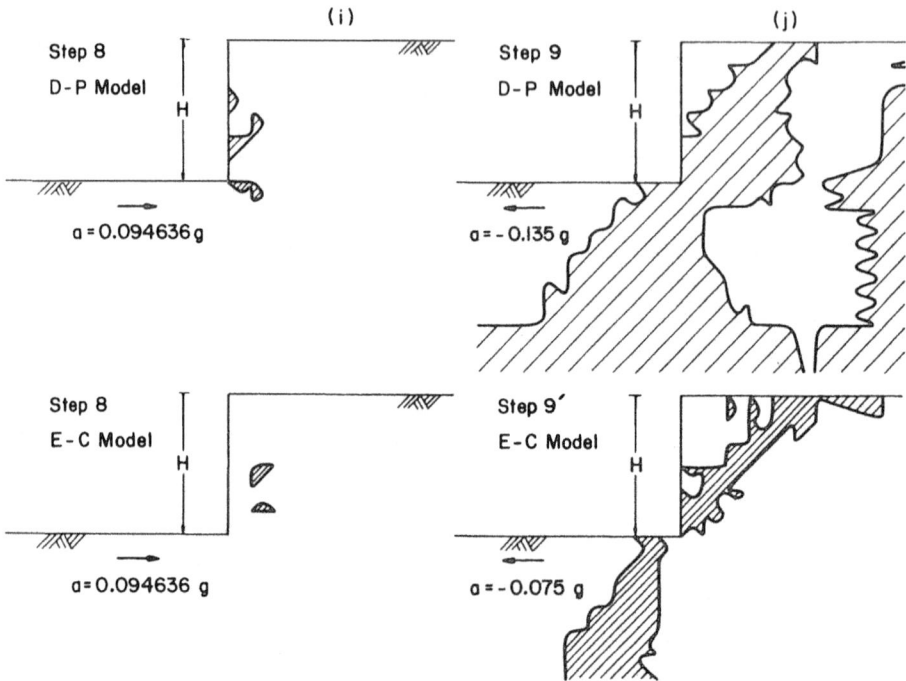

Fig.13 Continued

Further, at this sliding stage the horizontal displacement is predicted to be 1.49
ft (0.454 m) and 0.99 ft (0.302 m) for
the Drucker-Prager and elliptic cap model
cases respectively, as can be seen in Fig.
12. The yielded zone predicted by the Drucker-Prager model spreads downward into the
ground below the toe, diagonally upward
along portions of the slip surface of the
limit analysis method, and vertically in
the area approximately 40 ft behind the
crest (Fig. 13j). On the other hand, for
the elliptic cap model case, it spreads
into the ground below the front of the toe,
within a narrow area along the slip surface,
and at the ground surface behind the crest
(Fig. 13j).

2. Relative velocity fields - The relative velocity is defined here as the difference between the displacement increment
at a nodal point and that at the toe. The
relative velocity fields at the sliding
stage are shown in Fig. 14 for both model
cases. The magnitude and direction of each
relative velocity vector is shown by an
arrow, and the resultant of the relative-
displacement increment at the nodal point

above the toe is taken as a normalized unit
length. Further, the modified and conventional log-spiral slip surface of the limit
analysis method are illustrated with
solid and broken lines respectively (Chen
et al, 1971 and 1975). The former and latter
slip surfaces are obtained by using slope
heights of 25 ft (7.63 m) and 30 ft (9.15 m)
in the limit analysis solutions. Also, note
that the modified and conventional slip surfaces correspond to the failure mechanism at
a = -0.196 g and -0.0287 g respectively.
Since the loss of ground appears to occur
at a distance of approximately one-sixth
the total height above the toe, the relative velocity fields of both models agree
quite well with the modified slip surfaces.

The angle between the modified slip surface and the relative velocity vector at
the nodal point above the toe is measured
from the figures. For the Drucker-Prager
model case, it has an angle of 9°. It is
expected that if the seismic load is increased slightly, this angle would approach
10°, as would be predicted by the limit
analysis method. On the other hand, the angle in the elliptic cap model case is near-

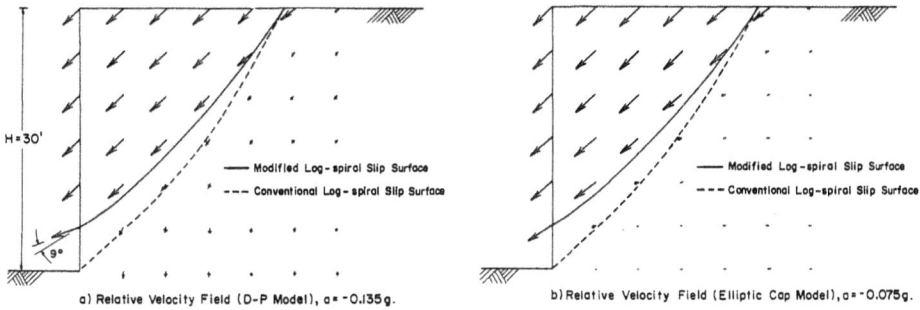

a) Relative Velocity Field (D-P Model), a = -0.135g.

b) Relative Velocity Field (Elliptic Cap Model), a = -0.075g.

Fig.14 Relative velocity field at sliding

ly zero. This means that the dilatancy has not been developed along the slip surface. Referring to Fig. 13j, most of the stress states along the slip surface are in the corner zone. Consequently, dilatancy on the slip surface is prevented to some extent.

8.3 Analyses of slopes after slidings

It is usually difficult to continue an elastic-plastic analysis with the finite element beyond the loading which causes instability of the slope, because the stiffness matrix itself becomes unstable. However, we propose here a method to calculate the post-sliding behavior of the slope under seismic loading conditions.

As can be observed from Fig. 14, the relative velocity vectors above the modified log-spiral slip surface are almost uniform in magnitude and have almost the same direction of approximately 45° downward to the left, except the vector at the point above the toe. Consequently, a rigid body type of sliding is assumed to continue until the magnitude of acceleration becomes again the same as that at the beginning of sliding. The pseudo-velocities at step 9 and step 9' in Fig. 12 are approximated as 6.3 ft/s (1.92 m/s) and 18.2 ft/s (5.5 m/s) for the Drucker-Prager and elliptic cap model cases, respectively. Thus, the corresponding pseudo-displacements are calculated as 1.05 ft (0.32 m) and 4.55 ft. (1.39 m).

This pseudo-displacement is now introduced into the configuration of the slope system at the beginning of sliding, in order to resume our finite element analysis. To perform this analysis, we note the basic theorem of the limit analysis, that is, "when the limit load is reached, the deformation proceeds under the constant

load, all stresses remain constant; only plastic (not elastic) increments of strain occur" (Drucker. et.al., 1952). Thus, the pseudo-displacement is added here as an plastic displacement into the configuration of the slope with the stress states held at the beginning of sliding. However, note that a direct addition of the full pseudo-displacement to the slope configuration causes numerical problems, because the sudden change of updated Lagrangian system generates large unequal nodal forces along the slip surface. Herein, we gradually add this displacement by dividing it into several increments. If numerical iteration causes problems during this process, then the acceleration at the beginning of sliding is slightly reduced. At each new stage, the remaining quantity of pseudo-displacement is again added into the latest configuration of the slope.

In this way, the addition of the pseudo-displacement of 1.05 ft (0.32 m) into the slope configuration corresponding to the Drucker-Prager model case, is completed at load step 10 with a = -0.11 g (Fig. 12).

For the elliptic cap model case, it has been estimated that "pseudo displacement" of 4.55 ft (1.39 m) causes a complete collapse of the slope.

Using this procedure, the post-sliding behavior has been analyzed for the Drucker-Prager model case. Figure 15 presents the zones of yielding during the seismic loading after load step 10. After the addition of the pseudo-displacement, the yielded zone spreads over a very narrow area along the slip surface, as shown in Fig. 15a. This spreading of the yielded zone is very interesting because it is consistent with the assumption of the limit analysis method. As the seismic load acts to the right during step 11, the yielded zone contracts, as shown in Fig. 15b. The yield zones at step 12 and 13 are illustrated in Figs. 15c and d.

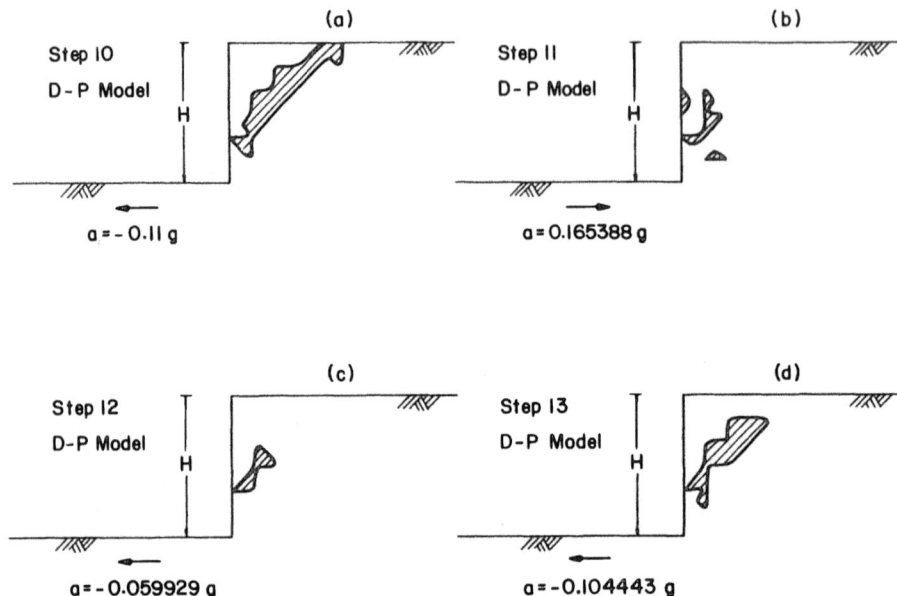

Fig.15 Yielded zone in the Drucker-Prager perfectly plastic material slope after sliding

As the loading acts to the left, the yield-ing begins to spread along the slip sur-face towards the ground surface behind the crest. However, the response of displace-ment is found to be almost elastic from step 10 to step 13, in spite of the exist-ence of the yielded zones.

9 SUMMARY AND CONCLUSION

This paper gives a summary of recent devel-opments and achievements in theoretical soil plasticity, with emphasis on their applica-tion to earthquake induced slope progressive failure and landslide analysis. The con-stitutive relations of geotechnical mater-ials play an important role in the analysis of soil-structure problems. The basic re-quirements for a soil model are:
1. The mathematical formulation should result in a unique and stable stress-strain relationship.
2. The constitutive equation should re-flect the key characteristics of experi-mental data.
3. The mathematical relationship should be defined by a very small number of param-eters which can be determined from standard test data.

4. The mathematical model should encom-pass the well known Coulomb criterion as a special case.
The plasticity models adopted for comput-er implementation in this paper are the Drucker-Prager model and generalized cap models which satisfy all these requirements. Furthermore, at the present time, it appears from the numerical application of these mod-els that there is little need for further re-finements of this type of model, since the present uncertainty in experimental meas-urements makes it desirable that constitu-tive models be kept relatively simple while their numerical implications on practical applications are examined more throughly. These analytical results should be compared with field data to assess the adequacy of these models in reproducing the important features of soil-structure interaction, and to identify areas in which improved modell-ing of the geotechnical material properties is necessary and essential.

ACKNOWLEDGEMENTS

This material is based upon work supported by the National Science Fundation under Grant No. PFR-7809326 to Purdue University.

REFERENCES

Atkinson, J.H. & Bransby, P.L. 1978. The Mechanics of Soils - An Introduction to Critical State Soil Mechanics. Maidenhead, England: McGraw-Hill Book Company, Inc.

Baladi, G.Y. 1979. Lecture Series on Constitutive Equations given at Purdue University.

Baladi, G.Y. & Rohani, B. 1979. Elastic-Plastic Model for Saturated Sand. J. Geotech. Engng. Div. ASCE, Vol.105 (GT4): 465-480.

Bishop, A.W. 1966. The Strength of Soils as Engineering Materials. Geotechnique, Vol. 16, No.2: 91-130.

Chen, W.F. 1975. Limit Analysis and Soil Plasticity. Amsterdam: Elsevier.

Chen, W.F. 1980. Plasticity in Soil Mechanics and Landslides. J. Engng. Mech. Div. ASCE, Vol.106 (EM3): 443-464.

Chen, W.F. 1982a. Plasticity in Reinforced Concrete. New York: McGraw-Hill Book Company, Inc.

Chen, W.F. 1982b. Soil Mechanics, Plasticity and Landslides. Purdue University Report, CE-STR-82-20.

Chen, W.F. & Giger, M.W. 1971. Limit Analysis of Stability of Embankments. J. Soil Mech. and Foun. Div. ASCE, Vol.97 (SM1): 19-26.

Chen, W.F. & Mizuno, E. 1979. On Material Constants for Soil and Concrete Models. Third ASCE/EMD Specialty Conference: 539-542.

Chen, W.F. & Saleeb, A.F. 1982. Constitutive Equations for Engineering Materials Vol.1 - Elasticity and Modeling. N.Y.: John Wiley Interscience.

Chen, W.F. & Saleeb, A.F. 1983. Constitutive Equations for Engineering Materials Vol.2 - Plasticity and Modeling. N.Y.: John Wiley Interscience.

Chen, W.F., Snitbhan, N. & Fang, H.Y. 1975. Stability of Slopes in Anisotropic Non-Homogenous Soils. Canadian Geotech. J., Vol.12, No.1: 146-152.

Dafalias, Y.F. & Herrmann, L.R. 1980. A Bounding Surface Soil Plasticity Model. International Symposium on Soils under Cyclic and Transient Loading, Vol.1: 335-345.

Dafalias, Y.F. & Herrmann, L.R. 1982. A Generalized Bounding Surface Constitutive Model for Clays. In R.N. Yong and E.T. Selig (eds.), Application of Plasticity and Generalized Stress-Strain in Geotechnical Engineering, pp.78-95. New York: ASCE.

Dafalias, Y.F. & Popov, E.P. 1975. A model of Nonlinearly Hardening Materials for Complex Loadings. Acta Mechanica, Vol.21: 173-192.

Davidson, H.L. & Chen, W.F. 1974. Elastic-Plastic Large Deformation Response of Clay to Footing Loads. Rep. No.355.18, Fritz Eng. Lab. Lehigh University, Pa.

Desai, C.S. & Christian, J.T. (eds.) 1979. Numerical Methods in Geotechnical Engineering. New York: McGraw-Hill Book Co., Inc.

Desai, C.S., Phan, H.V. & Perumpral, J.V. 1982. Mechanics of Three-Dimensional Soil-Structure Interaction, J. Engng. Mech. Div. ASCE, Vol.108 (EM5):731-747.

DiMaggio, F.L. & Sandler, I.S. 1971. Material Models for Granular Soils. J. Engng. Mech. Div. ASCE, Vol.122 (EM3): 936-950.

Drucker, D.C. 1966. Concepts of Path Independence and Material Stability for Soils. In. J. Kravtchenko and P.M. Sirieys (eds.), Rheol. Mecan. Soils. Proc. IUTAM Symp., pp.23-43. Berlin: Springer.

Drucker, D.C., Gibson, R.E. & Henkel, D.J. 1957. Soil Mechanics and Work-Hardening Theories of Plasticity. Transactions ASCE, Vol.122: 338-346.

Drucker, D.C. & Prager, W. 1952. Soil Mechanics and Plastic Analysis of Limit Design. Quarterly of Applied Mathematics, vol.10, No.2: 157-175.

Iwan, W.D. 1967. On a Class of Models for the Yielding Behavior of Continuous and Composite Systems. J. Applied Mechanics, Vol.34: 612-617.

Krieg, R.D. 1975. A Practical Two-Surface Plasticity Theory. J. Applied Mechanics, Vol.42: 641-646.

Lade, P.V. 1977. Elasto-Plastic Stress-Strain Theory for Cohesionless Soil with Curved Yield Surfaces. Int. J. Solids and Structures, Vol.13: 1014-1035.

Lade, P.V. 1979. Stress-Strain Theory for Normally Consolidated Clays. Proceedings of the 3rd International Conference on Numerical Methods in Geomechanics, pp. 1325-1337. Rotterdam: Balkema.

Lade, P.V. & Duncan, J.M. 1975. Elastoplastic Stress-Strain Theory for Cohesionless Soil. J. Geotech. Engng. Div. ASCE, Vol.101 (GT10): 1037-1053.

Masing, G. 1926. Eigenspannungen and Verfertigung heim Messing. Proceedings of the 2nd International Congress of Applied Mechanics. Zurich.

Mizuno, E. & Chen, W.F. 1981. Plasticity Models and Finite Element Implementation. In C.S. Desai and S.K. Saxena (eds.), Implementation of Computer Procedures and Stress-Strain Laws in Geotechnical Engineering, pp.519-534. Durham: Acorn Press.

Mizuno, E. & Chen, W.F. 1982a. Large-Deformation Finite Element Implementation of Soil Plasticity Models. Purdue University Report, CE-STR-82-4.

Mizuno, E. & Chen, W.F. 1982b. Plasticity

Modeling for Geological Media. Purdue University Report, CE-STR-82-15.

Mizuno, E. & Chen, W.F. 1983a. Plasticity Analysis of Slope with Different Flow Rules. Computers & Structures, Vol.17, No.3: 375-388.

Mizuno, E. & Chen, W.F. 1983b. Cap Models for Clay Strata to Footing Loads. Computers & Structures, Vol.17, No.4: 511-528.

Mizuno, E. & Chen, W.F. 1984. Plasticity Models for Seismic Analyses of Slopes. Soil Dynamics and Earthquake Engineering, Vol.3, No.1: 2-7.

Mroz, Z. 1967. On the Description of Anisotropic Hardening. J. Mech. Phys. Soils, Vol.15: 163-175.

Mroz, Z., Norris, V.A. & Zienkiewicz, O.C. 1978. An Anisotropic Hardening Model for Soils and Its Application to Cyclic Loading. Int. J. Num. Analy. Meth. Geomech., Vol.2: 203-221.

Mroz, Z., Norris, V.A. & Zienkiewicz, O.C. 1979. Application of an Anisotropic Hardening Model in the Analysis of Elastoplastic Deformation of Soils. Geotechnique, Vol.29, No.1: 1-34.

Mroz, Z., Norris, V.A. & Zienkiewicz, O.C. 1981. An Anisotropic, Critical State Model for Soils subjected to Cyclic loading. Geotechnique, Vol.31, No.4: 451-469.

Mroz, Z. & Pietruszczak, ST. 1983. A Constitutive Model for Sand with Anisotropic Hardening Rule. Int. J. Num. Analy. Meth. Geomech., Vol.7: 305-320.

Nelson, I. 1978. Consitutive Models for Use in Numerical Computations. Proc. of the International Symposium on Dynamic Method in Soil and Rock Mechanics, Vol.2, pp.45-97. Rotterdam: Balkema.

Nelson, I. & Baladi, G.Y. 1977. Outrunning Ground Shock Computed with Different Models. J. Engng. Mech. Div. ASCE, Vol.103 (EM3): 377-393.

Novozhilov, V.V. 1953. Foundations of the Nonlinear Theory of Elasticity. N.Y.: Graylock Press.

Palmer, A.C. (ed.) 1973. Proceedings of the Symposium on the Role of Plasticity in Soil Mechanics. Cambridge.

Parry, R.H.G. 1958. correspondence on "On the Yielding of Soils". Geotechnique, Vol.8: 185-186.

Parry, R.H.G. (ed.) 1972. Stress-Strain Behavior of Soils, Roscoe Memorial Symposium. Cambridge.

Pietruszczak, ST. & Mroz, Z. 1983. On Hardening Anisotropy of K_0-Consolidated Clays. Int. J. Num. Analy. Meth. Geomech., Vol.7: 19-38.

Prevost, J.H. 1977. Mathematical Modeling of Monotonic and Cyclic Undrained Clay Behavior. Int. J. Num. Analy. Meth. in Geomech., Vol.1: 195-216.

Prevost, J.H. 1978a. Anisotropic Undrained Stress-Strain Behavior of Clays. J. Geotech. Engng. Div. ASCE, Vol.104 (GT8): 1075-1090.

Prevost, J.H. 1978b. Plasticity Theory for Soil Stress Behavior. J. Engng. Mech. Div. ASCE, Vol.104 (EM5): 1177-1194.

Prevost, J.H., Cuny, B., Hughes, T.J.R. & Scott, R.F. 1981. Offshore Gravity Structures; Analysis. J. Geotech. Engng. Div. ASCE, Vol.107 (GT2): 143-165.

Prevost, J.H. & Hoeg, K. 1975. Effective Stress-Strain-Strength Model for Soils. J. Geotech. Engng. Div. ASCE, Vol.101 (GT3): 259-278.

Roscoe, K.H. & Burland, J.B. 1968. On the Generalized Stress-Strain Behavior of 'Wet' Clay. In J. Heyman and F.A. Leckie (eds.), Engineering Plasticity, pp.535-609. Cambridge: Cambridge University Press.

Roscoe, K.H., Schofield, A.N. & Thurairajah, A. 1963. Yielding of Clays in State Wetter than Critical. Geotechnique, Vol. 13, No.3: 211-240.

Roscoe, K.H., Schofield, A.N. & Wroth, C.P. 1958. On the Yielding of Soils. Geotechnique, Vol.8, No.1: 22-53.

Sandler, I.S. & Baron, M.L. 1979. Recent Development in the Constitutive Modeling of Geological Materials. Proceedings of the 3rd International Conference on Numerical Methods in Geomechanics, Vol.1, pp.363-376. Rotterdam: Balkema.

Sandler, I.S., DiMaggio, F.L. & Baladi, G. Y. 1976. Generalized Cap Model for Geological Materials. J. Geotech. Engng. Div. ASCE, Vol.102 (GT7): 683-699.

Schofield, A.N. & Wroth, C.P. 1968. Critical State Soil Mechanics. N.Y: McGraw-Hill Book Company, Inc.

Shield, R.T. 1955. On Coulomb's Law of Failure in Soils. J. Mech. Phys. Solids, 4(1): 10-16.

Snitbhan, N. & Chen, W.F. 1978. Elastic-Plastic Large Deformation Analysis of Soil Slopes. Computers & Structures, Vol.9: 567-577.

Van Eekelen, H.A.M. & Potts, D.M. 1978. The Behavior of Drammen Clay under Cyclic Loading. Geotechnique, Vol.28, No.2: 173-196.

Vermeer, P.A. 1978. A Double Hardening Model for Sand. Geotechnique, Vol.28, No.4: 413-433.

Zienkiewicz, O.C., Humpheson, C. & Lewis, R.W. 1975. Associated and Non-Associated Visco-Plasticity and Plasticity in Soil Mechanics. Geotechnique, Vol.25, No.4: 671-689.

Zienkiewicz, O.C., Norris, V.A. & Naylor, D.J. 1977. Plasticity and Viscoplasticity in Soil Mechanics with Special Reference to Cyclic Loading Problems. In P.G. Bergan,

et.al. (eds.), Finite Elements in Non-
Linear Mechanics, Vol.2, pp.455-485.
Tapir, Trondheim.
Zienkiewicz, O.C., Valliapan, S. & King, I.
P. 1969. Elastoplastic Solutions of Engi-
neering Problems: Initial Stress, Finite
Element Approach. Int. J. Num. Meth.
Engng., Vol.1: 75-100.

5.7.2

Mechanics of Material Behavior,
edited by G.J. Dvorak and R.T. Shield, 1984
Elsevier Science Publishers B.V., Amsterdam — Printed in the Netherlands

Soil Mechanics, Plasticity and Landslides

W.F. CHEN

School of Civil Engineering, Purdue University, West Lafayette, IN 47907 (U.S.A.)

Abstract

After a brief historical introduction, the impact of D.C. Drucker's ideas on the modern development of the theory of soil plasticity is discussed. Recent achievements in plasticity as applied to soil mechanics are summarized with emphasis on applications to earthquake-induced landslide problems.

To this end, a detailed description of the following three subjects is given: (1) stress—strain relations for soil; (2) limit analysis for seismic stability of slopes; and (3) finite element analysis for progressive failure behavior of slopes under earthquake loading. In this way, some of the interrelationships between the limit analysis of perfect plasticity and the finite element analysis of work-hardening plasticity are demonstrated, and their power and their relative merits and limitations for practical applications are evaluated within the context of their use in the seismic analyses of soil mass involving slope failures and landslides.

1. Introduction

In the 1950s major advances were made in the theory of metal plasticity by the development of (i) the fundamental theorems of limit analysis; (ii) Drucker's postulate or definition of stability of material; and (iii) the concept of normality condition or associated flow rule. The theory of limit analysis of perfect plasticity leads to practical methods that are needed to estimate the load-carrying capacity of structures in a more direct manner. The concept of a stable material provides a unified treatment and broad point of view of the stress—strain relations of plastic solids. The normality condition provides the necessary connection between the yield criterion or loading function and the plastic stress—strain relations. All these have led to a rigorous basis for the theory of classical plasticity, and laid down the foundations for subsequent notable developments.

The initial applications of the classical theory of plasticity were almost

32

exclusively concerned with perfectly plastic metallic solids such as mild steel which behaves approximately like a perfectly plastic material [62]. For these materials, the angle of internal friction ϕ is zero, no plastic volume change occurs and the only material property is the shear strength k or cohesion c in the terminology of soil mechanics. Numerical calculations were restricted to the method of characteristics based on the theory of the plane slip-line field analysis to derive the stress and velocity distribution in the plastic region [36]. Since the plane slip-line field analysis is rarely applicable to structures, exact and approximate calculations of the plastic collapse load were made exclusively by the methods of limit analysis [29].

The development of the modern theory of soil plasticity, as the new field was called, was strongly influenced by the well-established theory of metal plasticity. Soil mechanics specialists have been preoccupied with extending these concepts to answer the complex problems of soil behavior. Tresca's yield condition, used widely in metal plasticity, can be regarded as a special case of the condition of Coulomb on which the important concept of the limiting equilibrium of a soil medium had been firmly established in soil mechanics [80].

It is a relatively straightforward matter to extend the method of characteristics to cover Coulomb material where c and ϕ can either remain constant [79] or vary throughout the stress field in some specified manner [3]. In the theory of limit equilibrium, the introduction of stress—strain relations was obviated by the restriction to the consideration of equations of equilibrium and a yield condition. This produces what appears to be and sometimes is static determinacy for the solutions of slip line field equations. However, in many soil—structure interaction problems, the boundary conditions involve rates of displacement and the slip line equations are generally statically indeterminate. The key to obtaining a valid solution for such cases requires the basic knowledge of the stress—strain relations. Otherwise, a so-called solution is merely a guess.

The general theory of limit analysis, developed in the early 1950s, considers the stress—strain relation of a soil in an idealized manner. This idealization, termed normality or the associated flow rule, establishes the limit theorems on which limit analysis is based. Although the applications of limit analysis to soil mechanics are relatively recent, there have been an enormous number of practical solutions available [13]. Many of the solutions obtained by the method are remarkably good when comparing with the existing results for which satisfactory solutions already exist. As a result of this development, the meaning of the limit equilibrium solutions in the light of the upper- and lower-bound theorems of limit analysis becomes clear.

The first major advance in the extension of metal plasticity to soil plasticity was made in the paper "Soil Mechanics and Plastic Analysis or

Limit Design" by Drucker and Prager [32]. In this paper, the authors extended the Coulomb criterion to three-dimensional soil mechanics problems. The Coulomb criterion was interpreted by Drucker [28] as a modified Tresca as well as an extended von Mises yield criterion. The yield criterion obtained by Drucker and Prager for the later case is now known as the Drucker—Prager model or the extended von Mises model.

One of the main stumbling blocks in the further development of the stress—strain relations of soil based on the Drucker—Prager type or Coulomb type of yield surfaces to define the limit of elasticity and beginning of a continuing irreversible plastic deformation was the excessive prediction of dilation, which was the result of the use of the associated flow rule. It became necessary, therefore, to extend classical plasticity ideas to a "non-associated" form in which the plastic potential and yield surfaces are defined separately [24]. However, this modification eliminated the validity of the use of limit theorems for bounding collapse loads and created doubts about the uniqueness of solutions. Attempts have been made to revise the bounding theorems and to resolve the uniqueness problem, but to date not much success has been achieved through this route [58].

In 1957, an important advance was made in the paper "Soil Mechanics and Work-Hardening Theories of Plasticity" by Drucker, Gibson and Henkel [30]. In this paper the authors introduced the concept of work-hardening plasticity into soil mechanics. There are two important innovations in the paper. The first is the introduction of the idea of a work-hardening cap to the perfectly plastic yield surface such as the Coulomb type or Drucker—Prager type of yield criterion. The second innovation is the use of current soil density (or voids ratio, or plastic compaction) as the state variable or the strain-hardening parameter to determine the successive loading cap surfaces.

These ideas have led in turn to the generation of many soil models, most notably the development of the critical state soil mechanics at Cambridge University, U.K. These new soil models have grown increasingly complex as additional experimental data have been gathered, interpreted, and matched. This extension marks the beginning of the modern development of a consistent theory of soil plasticity [13].

Instead of tracing the historical development of the theory of soil plasticity in this paper, I shall attempt to survey its present status. The present survey deals with three equally important aspects of progress: stress—strain modeling of soils, the application of limit analysis to stability problems in soil mechanics, and new numerical solutions to specific boundary value problems. In stressing the new solutions in soil mechanics, three broad groups of typical problems in soil mechanics should be tackled: (1) earth retaining structures (active and passive pressures); (2) foundation (bearing capacity); and (3) soil slopes. However, the discussion of most of these problems would require the presentation of a

34

greater amount of details than is practical. Furthermore, almost all the interesting aspects of recent achievements can be brought out in the class of problems related to slopes. Accordingly, this paper is devoted primarily to a consideration of the concept of soil plasticity in relation to its applications to stability and progressive failures of slopes under static and earthquake loading conditions.

2. Stress—strain relations

Soil mechanics along with all other branches of mechanics of solids requires the consideration of geometry or compatibility and of equilibrium or dynamics. The essential set of equations that differentiate the soil from other solids is the relation between stress and strain. The behavior of soils is very complicated. The attempt to incorporate the various features of soil properties in a single mathematical model is not likely to be successful, but even if such a model could be constructed, it would be far too complex to serve as the basis for the solution of practical geotechnical engineering problems. Simplifications and idealizations are essential in order to produce simpler models that can represent those properties that are essential to the considered problem. Thus, any such simpler models should not be expected to be valid over a wide range of conditions [33, 35, 71].

The need for mathematical simplicity in the description of the mechanical properties of solids is understood quite well for metals where so much research effort has been expended by so many investigators [29]. Yet even for metals, the simple idealizations such as perfect plasticity, isotropic hardening, kinematic hardening, and mixed hardening are frequently used in solving practical problems [12, 21]. The same situation is to be expected for the stress—strain modeling of soil which is a far more complex material [34, 35, 70].

Drastic idealizations are valuable not only for the ease of treatment of practical engineering problems but also conceptually for a clear physical understanding of the essential features of the complex behavior of a material under certain conditions. Therefore, for soils, as for metals, perfect plasticity is still an excellent design assumption, while very complex stress—strain relations of soil which require an ever increasing elaboration in detail of a mathematical description may be approximated crudely by simple isotropic, kinematic, or mixed hardening models. Thus, the isotropic hardening cap models and Cambridge models, the kinematic hardening nested yield surfaces models, or the mixed hardening bounding surface models that have been proposed and developed in recent years are all within the realm of this simplification. In the sections that follow, these developments are described and, hopefully, unified within the same framework of physically and mathematically well-established theory of work-hardening plasticity.

It is important to note here that the path to reach the present state-of-the-art in soil plasticity is by no means easy and that much of the classical theory of plasticity for metals has been considerably modified in order to obtain reasonable results for soils. These achievements have been summarized in two earlier symposia [58, 61], as well as in two recent ASCE workshop and symposium proceedings [59, 81, 82]. An up to date summary will appear in the forthcoming book "Constitutive Equations for Engineering Materials — Vol. 2 Plasticity and Modeling" by Chen and Saleeb [21] and the Proceedings on "Constitutive Laws for Engineering Materials: Theory and Application" edited by Desai and Gallagher [25].

The use of strain- or work-hardening plasticity theories in soil mechanics has been developed for about twenty-five years, since publication of the classical paper by Drucker, Gibson and Henkel [30]. Most of the research has been conducted by engineers working in the area of soil statics. Recently, attention has been focused on the use of these models in soil dynamics [11]. The objective of this section is to set forth the state-of-the-art with respect to elastic-plastic stress–strain relations of soils. In doing so, it achieves not only the purpose of surveying the current research activity that has been going on very actively in this field in recent years, but also the survey gives the best indications of future problems that may result from the observations of the trend of recent developments.

One of the main problems in the theory of plasticity is to determine the nature of the subsequent yield surfaces. This post-yielding response is described by the hardening rule which specifies the rule for the evolution of the loading surfaces during the course of plastic deformations. Indeed, the assumption made concerning the hardening rule introduces a major distinction among various plasticity models developed for soils in recent years.

2.1 HARDENING (SOFTENING) RULES

There are several hardening rules that have been proposed to describe the growth of subsequent yield surfaces for strain-hardening (softening) materials. The choice of a specific rule depends primarily on the ease with which it can be applied and its ability to represent the hardening behavior of particular material. In general, three types of hardening rules have been commonly utilized [12]. These are: (1) Isotropic hardening; (2) Kinematic hardening; (3) Mixed hardening. In an isotropic hardening model, the initial yield surface is assumed to expand (or contract) uniformly without distortion as plastic flow continues. On the other hand, the kinematic hardening rule assumes that, during plastic deformations, the loading surface translates without rotation as a rigid body in the stress space, maintaining the size and shape of the initial yield surface. This rule provides a means of accounting for the Bauschinger effect, which refers to

36

one particular type of directional anisotropy induced by plastic deformations; namely that an initial plastic deformation of one sign reduces the resistance of the material with respect to a subsequent plastic deformation of the opposite sign. Therefore, kinematic hardening models are particularly suitable for materials with pronounced Bauschinger effect such as soils under cyclic and reversed types of loading.

A combination of isotropic and kinematic hardening models leads to a more general hardening rule, and therefore provides for more flexibility in describing the hardening behavior of the material. For a mixed (combined) hardening model, the loading surface experiences translation as well as expansion (contraction) in all directions, and different degrees of Bauschinger effect may be simulated. Kinematic and mixed types of hardening rules are generally known as anisotropic hardening models.

In the last few years, several plasticity models with more complex hardening rules combining the concepts of kinematic and isotropic hardening have been developed and applied to describe the behavior of soils under cyclic loading. Some of these models will be discussed in the sections that follow.

2.2 PERFECT PLASTICITY MODELS

Perfect plasticity is an appropriate idealization for a structural metal because it captures the essential features of its behavior. This includes small tangent modulus when compared with elastic modulus, when loading in the plastic range, and the unloading response is elastic. However, perfect plasticity is not nearly appropriate for soils. Some of the troubles and their justifications for adoption of this idealization for practical use were discussed in the paper "Concepts of Path Independence and Material Stability for Soils" by Drucker [27].

For the most part, the concept of perfect plasticity has been used extensively in the past in conventional soil mechanics in assessing the collapse load in stability problems. Different widely known techniques have been employed to obtain numerical solutions in these cases; such as the slip-line method [79], and the limit equilibrium method [80]. For the later case, the simple ideas of perfect plasticity have found direct application in many practical geotechnical engineering problems.

In addition to these classical methods, the more rigorous approach of modern limit analysis of perfect plasticity has been applied to a wide variety of practical stability problems. Using the well-known Coulomb yield criterion and its associated flow rule, many solutions have been obtained [13]. Recently, the stability analysis has been extended to include the earthquake loading, employing the pseudo-static force method [11, 14, 15, 19]. It should be emphasized here that the useful application of these techniques has not been exhausted. New and striking applications are not only possible but to be encouraged strongly, because of their

37

simplicity and power in helping us reach an understanding of, and feel for, a problem. Further, some predictions of this enormous idealization are very good. Much more of value will be uncovered as engineers who have need for particular results apply the methods of limit analysis and design to their own special problems. The power and simplicity of this method will be brought out in the subsequent discussions through the example of seismic stability analysis of slopes. This example will show not only the power and simplicity of the method, but also add insight to the practicing engineers' understanding of the complicated landslide problems [9, 42].

2.3 ISOTROPIC HARDENING MODELS (CONCEPT OF HARDENING CAP)

Drucker, Gibson and Henkel [30] were the first to suggest that soil might be modeled as an elastic-plastic isotropic hardening material. They proposed that successive yield surfaces might resemble extended von Mises (or Drucker–Prager) cones with convex end spherical caps (Fig. 1). As the soil strain-hardens, both the cone and the end cap expand. As mentioned previously, there are two important innovations in this work. The first is the introduction of the idea of a spherical cap fitted to the cone. The second is the use of current soil density (specific volume or void ratio) as the strain-hardening parameter to determine the successive loading surfaces, such as the surface marked I in Fig. 1 for a particular value of soil density. There will be a succession of such surfaces, all geometrically similar, but of different sizes, for different densities. This strain- or work-hardening model was a major step forward toward a more realistic representation of soil behavior. Two specific isotropic hardening models are discussed in the following.

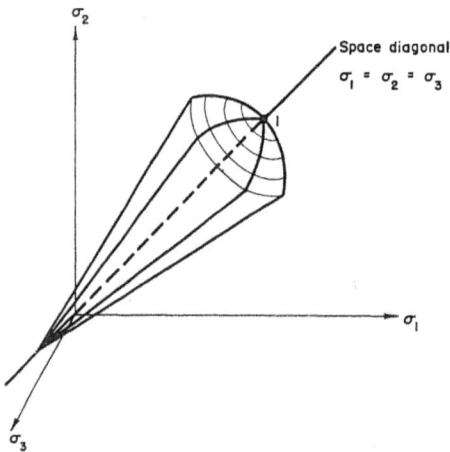

Fig. 1. A Drucker–Prager type of strain-hardening cap model.

38

(i) Cambridge models (concept of critical state)

The introduction of isotropic hardening plasticity into soil mechanics led in turn to the generation of the family of soil models of the strain-hardening cap type developed at Cambridge University by the late Professor Roscoe and his co-workers. Two of these models are now widely used. The first is known as the Cam-clay model [75] and was originally formulated by Roscoe, Schofield, and Thurairajah [69] for normally and lightly over-consolidated clays in the triaxial test. The second, known as the Modified Cam-clay model, was developed by Roscoe and Burland [67] as a modification and extension of the Cam-clay model to a general three-dimensional stress state. In both models, associated flow rule was used. The modified Cam-clay model is an isotropic, nonlinear elastic-plastic strain-hardening model. Only the volumetric strain is assumed to be partially recoverable; i.e. elastic shearing strain is assumed to be identically zero. The elastic volumetric strain is non-linearly dependent on the hydrostatic stress and is independent of the deviatoric (shear) stresses. For certain stress histories, strain-softening may occur. Extensive reviews of various types of Cambridge models have been given in two symposia [58, 61].

The important feature that has been the integral part of all Cambridge models is the concept of critical states proposed by Roscoe, Schofield, and Wroth [68], and independently conceived by Parry [60]. The critical state line is the locus of the failure points of all shear tests under both drained and undrained conditions. Its crucial property is that failure of initially isotropically compressed samples will occur once the states of stress in the samples reach the line, irrespective of the test path followed by the samples on their way to the critical state line [1]. At the critical state, large shear deformations occur with no change in stress or plastic volumetric strain.

Experimental results have indicated that, at failure, the behavior of soil is governed by the Coulomb failure criterion [2]. The failure (yield) surface corresponding to the Coulomb criterion in principal stress space is an irregular hexagon pyramid whose apex lies on the space diagonal as shown in Fig. 2. A strain-hardening cap intersects with the Coulomb surface in a line $ABCDEFA$. This line is called the critical state locus, and it is simply a line which separates the two surfaces corresponding to one fixed value of the specific volume (or void ratio). The complete failure surface is termed a state boundary surface [1, 67]. There will be similar surfaces of different sizes, but of the same shape, corresponding to different values of the specific volume. The geometric representation of the shape of the generalized state boundary surface requires four dimensions (i.e., three principal stress and one specific volume).

Since the development of Cambridge models, many attemps have been made to use various versions of these models in numerical solutions of

39

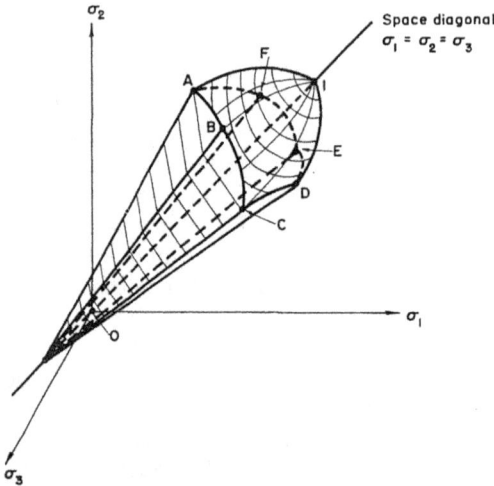

Fig. 2. The complete state boundary surface in principal stress space — Cambridge type
of strain-hardening cap model [1].

boundary value problems [4] and predictions of soil behavior in the field
(see for example, Palmer [58].)

(ii) Cap models

Dimaggio and Sandler [26] proposed a generalized cap model. The
yield function they used consisted of a perfectly-plastic (failure) portion
fitted to a strain-hardening elliptical cap. Associated flow rule was
employed for the failure and cap functions. In this model, the functional
forms for both the perfectly-plastic and the strain-hardening portions may
be quite general and would allow for the fitting of a wide range of
material properties. A simple version of the cap model will be used in the
later part of the paper for a complete progressive failure stress analysis of
earthquake-induced landslide problems.

The movement of the cap is controlled by the increase or decrease of
the plastic volumetric strain. Strain-hardening in this model can therefore
be reversed. It is this mechanism that leads to an effective control on
dilatancy, which can be kept quite small (effectively zero) as required for
many soils. Their model has also been adopted for rocks [73, 74] by allowing only expansion of the cap (i.e. hardening). In this variation of the
model, the cap movement is assumed to depend only on the maximum
previous value of the plastic volumetric strain, and consequently the cap is
not reversible. In this way, the model allows representation of a relatively
large amount of dilatancy, which is often observed during failure of rocks
at low pressures.

These generalized cap models have also been expanded to include rate

40

effects, and anisotropic behavior within the yield surface and viscoplastic behavior during yielding [54, 72]. Many of these variations of the generalized cap models are now widely used in ground shock computations [54—56, 72].

Using the idea of a cap fitted to the yield surface, Lade [44, 45] has developed a strain-hardening model to describe the behavior of different sands and normally consolidated clays. He used a spherical cap together with a modified version of the conical yield surface suggested earlier by Lade and Duncan [46]. In this model, both the conical yield surface and the cap are allowed to harden isotropically. However, for the cap, associated flow rule was employed, while non-associated rule was used in connection with the conical portion.

2.4 KINEMATIC HARDENING MODELS (CONCEPT OF NESTED YIELD
    SURFACE)

An alternative approach to the isotropic hardening type of models described above is provided by the kinematic type of strain-hardening rules. Recently, this approach has been employed by several researchers to provide for a more realistic representation of soil behavior under reversed, and particularly cyclic, loading conditions.

Iwan [38], following the related work of Masing [47], proposed one-dimensional plasticity models which consist of a collection of perfectly elastic and rigid-plastic or slip elements arranged in either a series—parallel or a parallel—series combination. The model can contain a very large number of elements, and the properties of these elements can be distributed such that they can match the particular form of hysteretic behavior of a certain type of soil. Such models are known as the overlay or mechanical sublayer models [83].

In order to extend the one-dimensional model to three-dimensional situations, an extended formulation of the classical incremental theory of plasticity has been proposed by Iwan [38]. Instead of using a single yield surface in stress space, he postulated a family (nest) of yield surface (Fig. 3) with each surface translating independently in a pure kinematic manner, or individually obeying a linear work-hardening model. Their combined action, in general, gives rise to a. nonlinear work-hardening behavior for the material as a whole. The approach leads to a realistic Bauschinger effect of a type that could not be obtained by using a single yield surface and a nonlinear work-hardening rule even with kinematic hardening. The same concept of using a field of nested yield surfaces was also proposed independently by Mroz [51]. These models have been used recently for soils and are usually known as multi-surface plasticity models. In all the proposed models of this type, associated flow rule has been utilized.

Figure 3 demonstrates the qualitative behavior of a multisurface model

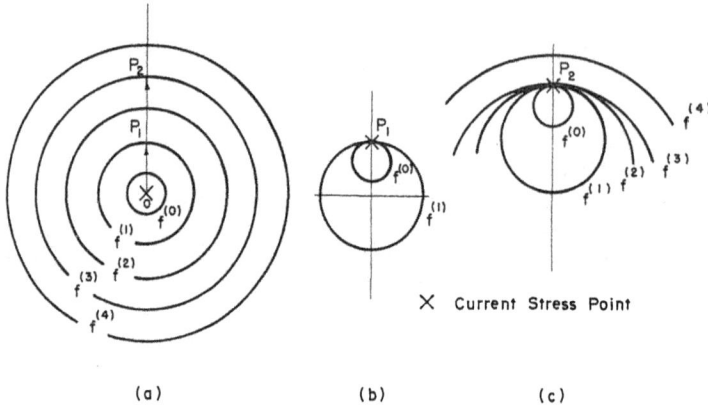

(a)   (b)   (c)

Fig. 3. Series of nested yield surfaces in Mroz/Iwan type of strain-hardening models.

with pure kinematic hardening. The initial positions of the yield surface $f^{(0)}$, $f^{(1)}$, $f^{(2)}$, and $f^{(3)}$ are shown in Fig. 3(a). When the stress point moves from O to P_1, elastic strains first occur until the surface $f^{(0)}$ is reached, where the plastic flow begins and the surface $f^{(0)}$ starts to move towards the surface $f^{(1)}$. Before their contact, the hardening modulus $H^{(0)}$ associated with $f^{(0)}$ governs the plastic flow according to the normality flow rule. However, when $f^{(0)}$ engages $f^{(1)}$ at P_1 (Fig. 3(b)), the first nesting surface $f^{(1)}$ becomes the active surface and, upon further loading, the hardening modulus $H^{(1)}$ applies in the flow rule. Both $f^{(0)}$ and $f^{(1)}$ are then translated by the stress point, and they remain tangent to each other on the stress path until they touch the yield surface $f^{(2)}$ which then becomes the active surface. For subsequent contacts of consecutive surfaces, the process is repeated with new corresponding values of hardening moduli applying. The situation when $f^{(3)}$ is reached is illustrated in Fig. 3(c).

Applications of the Iwan/Mroz model to study the seismic response of two-dimensional configurations of soils have been made by Joyner and Chen [40]. The model has been found to be particularly promising for use in calculating the response of earth dams subjected to earthquake ground shaking.

Recently, Prevost [63–65] has extended the Iwan/Mroz model for the undrained behavior of clays under monotonic and cyclic loading conditions. In this development, the basic strain-hardening rule is still of the kinematic type but a simultaneous isotropic hardening (or softening) is allowed. The rule used to govern the translation of the nested yield surfaces during plastic loading was that suggested by Mroz [51]. According to this rule, when the stress point P (Fig. 4) reaches the yield surface $f^{(m)}$, then, upon further loading (i.e. when the stress increment $\dot{\sigma}_{ij}$ is applied), the instantaneous translation of $f^{(m)}$ towards the next yield

42

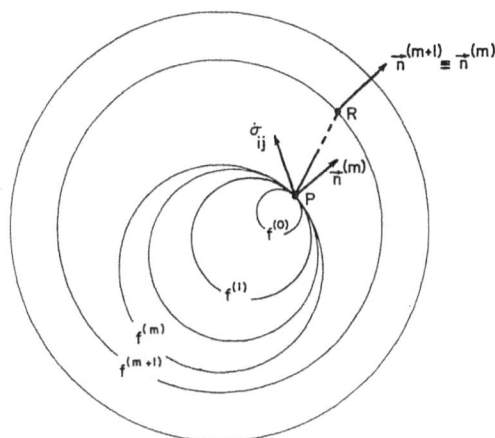

P ≡ Current Stress State

$\overrightarrow{PR}$ ≡ Direction of instantaneous
translation of yield surfaces

$f^{(0)}, \ldots, f^{(m)}$

Fig. 4. Translation rule of nested yield surfaces in stress space — Prevost type of strain-hardening models.

surface $f^{(m+1)}$ will be along $\overrightarrow{PR}$, where R (known as the conjugate point) is the point on $f^{(m+1)}$ with outward normal in the same direction as the normal to $f^{(m)}$ at P (i.e., $\bar{n}^{(m+1)} = \bar{n}^{(m)}$ in Fig. 4).

This model has been adopted very recently by Prevost et al. [66] in the finite element analyses of soil-structure interaction of centrifugal models under both monotonic and cyclic loadings simulating the situation encountered in the analysis of offshore gravity structure foundations under wave forces. The results obtained from the analysis agree quite well with those measured experimentally in the centrifuge. This study has demonstrated the ability of the multi-surface model to provide realistic representation of soil behavior under complex loadings.

2.5 MIXED HARDENING MODELS (CONCEPT OF BOUNDING SURFACE)

Various types of strain-hardening plasticity models have been recently employed for soils based on the bounding (consolidation or limiting) surface concept introduced earlier for metals [22, 43]. A two-surface model of this type was proposed by Mroz, Norris and Zienkiewicz [52, 53] for clays. A bounding surface, $F = 0$, representing the consolidation history of the soil, and a yield surface, $f = 0$, defining the elastic domain within the bounding surface (Fig. 5) were employed in the model.

The bounding surface was assumed to expand or contract isotropically, but the yield surface was allowed to translate, expand or contract within

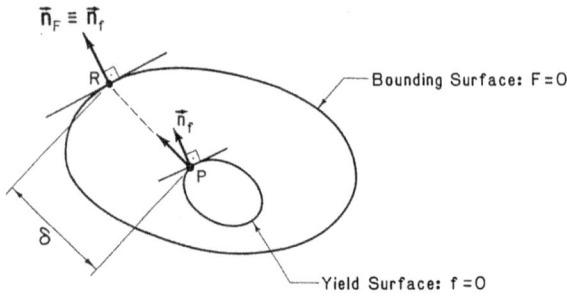

Fig. 5. Yield and bounding surfaces in stress space.

the domain enclosed by the bounding surface. The translation of the yield surface is governed by the same rule as the multi-surface models described earlier (i.e., f will translate towards the bounding surface along $\overline{PR}$ in Fig. 5).

The hardening modulus on the yield surface is assumed to vary depending on the relative configuration of the yield and bounding surfaces. It was this last assumption that distinguished the present formulation from the previous multi-surface models. Here, instead of using a field of nested yield surfaces with associated hardening moduli, an interpolation rule is utilized to define the variation of the hardening moduli between the yield and the bounding surfaces. In the interpolation rule used by Mroz et al. [52, 53], the hardening modulus, H, was taken as a function of the distance δ (Fig. 5) between the current stress point, P, on the yield surface and its conjugate point, R, on the bounding surface. Very detailed discussions of the present model and its application to represent the behavior of clays under monotonic and cyclic triaxial test conditions have been given by Mroz et al. [53].

Dafalias and Herrmann [23] have also used the bounding surface formulation to describe the behavior of clay under cyclic loadings. However, no explicit yield surface was postulated. Associated flow rule was utilized for the bounding surface. In this model, the variation of the plastic (hardening) modulus within the boundary surface is defined based on a very simple radial mapping rule. For each actual stress point within (or on) the bounding surface, a corresponding "image" point on the surface is specified as the intersection of the surface with the straight (radial) line connecting the origin with the current stress point (the origin was assumed to be always within the bounding surface). The actual hardening modulus is then assumed to be a function of the hardening modulus on the bounding surface, at the "image" point, and the distance between the actual stress point and its "image".

3. Limit analysis of perfect plasticity

Limit analysis is concerned with the development of efficient methods

44

for computing the collapse load in a direct manner. It is therefore of intense practical interest to practicing engineers. There have been an enormous number of applications in metal structures. Applications of limit analysis to reinforced concrete structures are more recent and are given in a recent book by Chen [12] as well as a colloquium proceedings [37]. Applications to typical stability problems in soil mechanics have been the most highly developed aspect of limit analysis so that the basic techniques and many numerical results have been summarized in the book by Chen [13]. Extensive references to the work before 1975 are also given in the book cited. An up-to-date reference to recent work on the applications of limit analysis to earth pressure, bearing capacity and slope stability problems can be found in the ASCE Proceedings [81, 82], theses [5—7, 41] and papers [8, 15], among others.

It is true, as in most fields of knowledge, that many of the basic ideas of perfect plasticity and limit analysis have been used extensively and fruitfully in the past in conventional soil mechanics through experimental studies and engineering intuition. Here, the standard and widely-known techniques of the slip-line method and the limit equilibrium method, among others, come to mind immediately, but these methods also have been mentioned previously.

The slip-line method uses the Coulomb criterion as the yield condition for soil. From the basic slip-line differential equations the slip-line network can be constructed and the collapse load determined. Examples of this approach are the solutions presented in the book by Sokolovskii [79].

The limit equilibrium method can be best described as an approximate approach to the construction of the slip-line field. It generally entails the assumption of the failure surface of various simple configurations from which it is possible to solve problems by simple statics. Terzaghi [80] cites some examples of this approach.

Although these methods are widely used in geotechnical practice, they neglect altogether the important fact that the stress—strain relations constitute an essential consideration in a complete theory of any branch of the continuum mechanics of deformable solids. Modern limit analysis methods take into consideration in an idealized manner the stress—strain relations of the continuum, the soil in the present case. This idealization, termed normality or flow rule, establishes the limit theorems on which limit analysis is based. Within the framework of perfect plasticity and the associated flow rule assumption, the approach is rigorous and the techniques are competitive with those of limit equilibrium approach. In several instances, especially in slope stability analysis, such a level of reliability and completeness has been achieved and firmly established in recent years that the limit analysis method can be used as a working tool for design engineers to solve everyday problems. Examples of this approach are the solutions presented in the book by Chen [13].

However, most of the applications of limit analysis of perfect plasticity to geotechnical problems have been limited to soil statics. The on-going work discussed here attempts to extend this method to the area of soil dynamics, in particular to earthquake-induced landslide problems. Further details of the method of attack are given in the sections that follow.

3.1 THE PROBLEM OF EARTHQUAKE-INDUCED LANDSLIDES

The conventional method for evaluating the effect of an earthquake load on the stability of a slope is the so-called "pseudo-static method of analysis". In this analysis, the inertia force is treated as an equivalent concentrated horizontal force (the "pseudo-static force") at some critical point (usually the center of gravity) of the critical sliding mass. The inadequacies of this method for slope stability analysis are discussed in various papers by many authors, most notably by H. Bolton Seed of the University of California [76].

(1) The pseudo-static force is applied as a permanent force whereas in reality the reduction in the stability of the slope exists only during the short period of time for which the unfavorable direction of the inertia force is induced. As a result, the factor of safety may drop below unity for a very short duration and some permanent displacements can occur in this duration. But this may not cause the collapse of a slope.

(2) Since the soils are not rigid, the acceleration is not uniform throughout the slope. The distribution of seismic coefficients is therefore a function of the height of a slope. This has been the subject of study of several investigators, in particular, Seed and his colleagues who used the viscoelastic model together with the finite element method [76]. Suggested values for the seismic coefficients for design purposes are available but have not been incorporated into the present pseudo-static method of analysis.

(3) There is no strong basis for the value of the seismic coefficient for the pseudo-static force as it is specified in the building codes. The empirical value ranges from 0.05 to 0.15 g for the design of earth dams in the United States. Somewhat higher values, 0.12–0.25 g, are used in Japan. For building design, the specified value in Los Angeles County is 0.15 g. In some parts of Japan, the value is as high as 0.30 g. California is currently considering a inertia force of 1.0 g.

Despite these limitations, the pseudo-static method continues to be used by consulting geotechnical engineers because it is required by the building codes, it is easier and less costly to apply, and satisfactory results have been obtained since 1933. This method will continue to be popular until an alternative method can be shown to be a more reasonable approach.

46

3.2 PROPOSED PROCEDURES FOR SEISMIC SAFETY ANALYSIS OF SLOPES

Limit analysis of perfect plasticity has been shown to be a powerful tool for geotechnical engineers in practice. Many solutions to static problems obtained by this method have been substantiated numerically by comparing these solutions with other existing results [13]. Herein, we suggest the following extension and modifications to the pseudo-static method of slope stability analysis, using the limit analysis techniques:

(1) Using the concept of superposition, the variation of the horizontal pseudo-static force throughout the depth of the slope can be incorporated. The factor of safety of a slope against collapse with nonuniform seismic coefficients can be determined. This work is currently in progress at Purdue University, where satisfactory results have been obtained [5–7]. Further extension to include such important items as pore water pressure, seepage forces, stratification, and three-dimensional effects will be made.

(2) In addition to the calculation of the factor of safety of a slope after a given earthquake shock, the effects of earthquakes on the displacements of a slope can be assessed. This is done following the concept of Newmark [57], again using the limit analysis techniques. This can be achieved in the following steps:

(a) Calculate the yield acceleration at which slippage will just begin to occur (or for which the factor of safety of the slope determined in (1) would reduce to 1.0).

(b) Instead of applying a permanent single, uniform value or several non-uniform values of the pseudo-static force to the slope, apply values from a discretized accelerogram of an actual or simulated earthquake.

(c) When the induced acceleration exceeds the yield acceleration, rigid body type of slope movement will occur. The magnitude of displacements can be evaluated by double integration of the part of the acceleration history above the yield acceleration. The acceleration and deceleration of the sliding mass are functions of the soil dynamics strength parameters.

(d) Determine the "stability" of the slope on the basis of this estimated total displacement by rigid body sliding.

(3) After several cycles of shaking, the potential failure surface may not be the one as originally assumed, because of the possible occurrence of liquefaction in the earth slopes. It is therefore necessary to find the most critical failure surface that reflects the change of shear strength of soil due to liquefaction. This can be achieved in the following steps.

(a) Determine the effective normal and shear stresses along the potential slip surfaces before and during a given earthquake shaking using the limit equilibrium method of slices.

(b) Evaluate the liquefaction potential for all points along the selected sliding surfaces by comparing the calculated shear stresses during earthquake shaking with the critical shear stresses required for liquefaction corresponding to a given initial stress condition and a given number of cycles.

(c) Calculate the total shear resistance against sliding for each of the selected potential failure surfaces.

(d) Determine the "reduced" yield acceleration corresponding to the critical slip surface that has been partially liquefied.

(e) Compute the total displacement by rigid body sliding of the critical mechanism.

Computer programs for the computation of safety factor, yield acceleration and displacements of slopes corresponding to a given earthquake loading have been developed at Purdue University [6]. A typical numerical example will be given in the following.

3.3 AN ILLUSTRATIVE EXAMPLE OF EARTHQUAKE-INDUCED LANDSLIDE PROBLEMS

A typical numerical example based on the present formulation is given below with the following input data (Fig. 6):

Design earthquake = El Centro, Dec. 30, 1934: c = cohesion = 200 psf, ϕ = friction angle = 24.5°, γ = unit weight of soil = 60 psf (submerged), β = slope angle = 44.5°, α = upper slope angle = 0, p = surcharge boundary load on AB = 120 psf, H = height of slope = 50 feet.

The yield acceleration of the slope at the beginning of the design earthquake is found to be $0.269\,g$. To obtain the updated or reduced yield acceleration at the end of each specified cycle of stress application, the empirical relationship proposed by Seed et al. [77] for liquefaction potential calculation is used. This results in the following information using a total of 20 slices in a typical limit equilibrium stress calculation.

(a) After 2 cycles of loading

Location of failure surface starting at $L = 4H$ on the top of the slope
Number of slices liquefied = 15
Yield acceleration after liquefaction = $0.192\,g$.

(b) After 4 cycles of loading

Location of failure surface starting at $L = 4H$ on the top of the slope
Number of slices liquefied = 17
Yield acceleration after liquefaction = $0.178\,g$

(c) After 7 cycles of loading

Location of failure surface starting at $L = 3H$ on the top of the slope
Number of slices liquefied = 18
Yield acceleration after liquefaction = $0.169\,g$

48

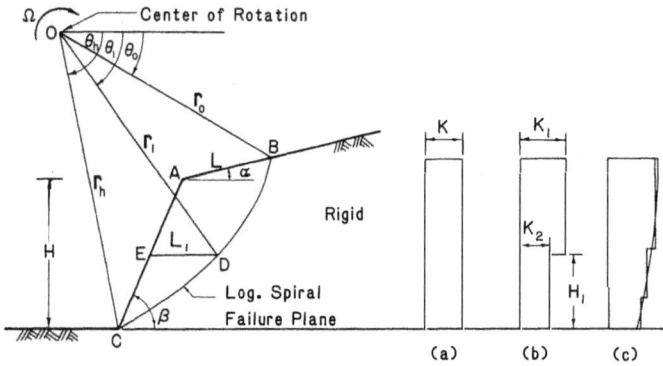

Fig. 6. Failure mechanism for the stability of an embankment and distributions of seismic coefficient.

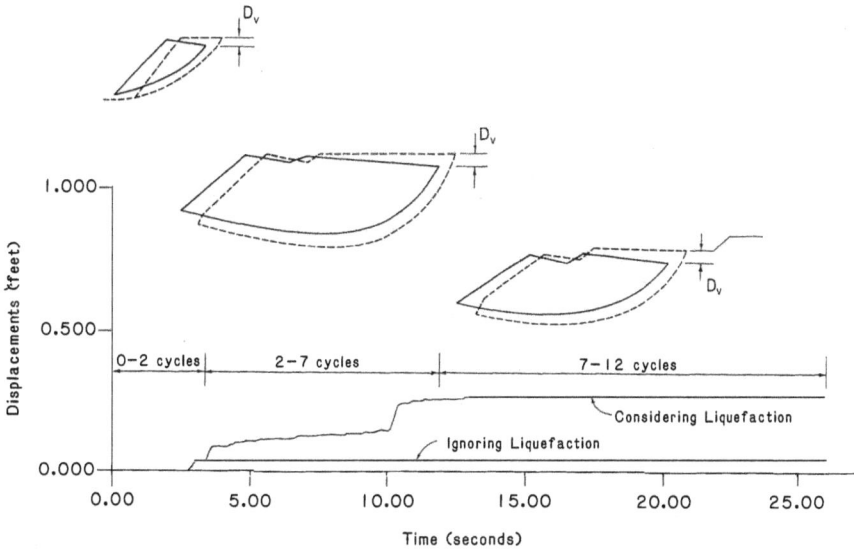

Fig. 7. Vertical displacement, D_v of point A of failure section (logspiral slip surface of local failure).

Once the critical sliding mechanism is identified and its reduced yield acceleration calculated, the rigid body type of slope movement can then be proceeded in a rather straightforward manner by double integration of that part of the acceleration history above the yield acceleration. A typical result is shown in Fig. 7 where the displacement—time relationships corresponding to the two cases: liquefaction considered and liquefaction ignored, are given. For the case of slope stability analysis ignoring the occurrence of liquefaction, the slope movement is generally small. However, including the effect of liquefaction, the slope movement can be significantly increased. The need for the determination of the "seismic stability" of the slope on the basis of this estimated total displacement is clearly demonstrated in this example calculation.

The influence of progressing liquefaction on the location of slip surfaces is shown in Fig. 8 where the displacements of sliding sections at the end of each cycle of loading are sketched. These calculated displacement patterns are seen to be similar to those observed in actual landslides due to earthquake shaking [57].

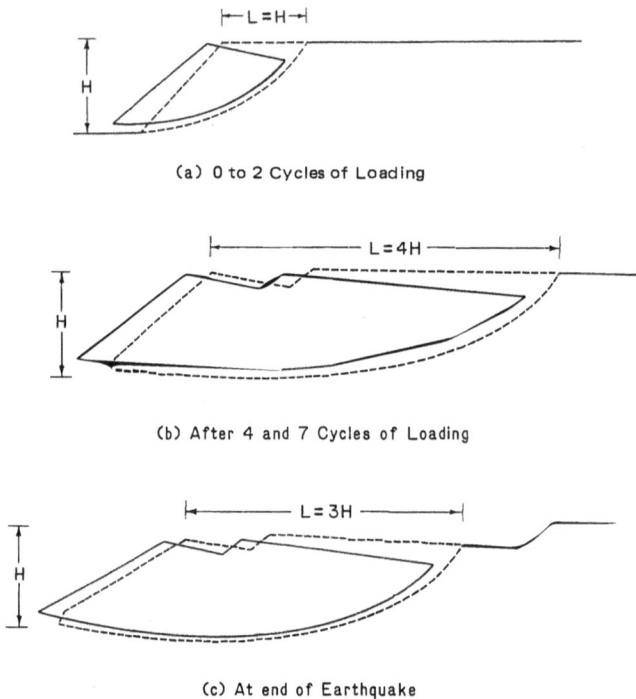

(a) 0 to 2 Cycles of Loading

(b) After 4 and 7 Cycles of Loading

(c) At end of Earthquake

Fig. 8. Displacement patterns of failure sections after specified cycles of loading.

4. Finite element analysis of work-hardening plasticity

It is well-known that a complete analysis of stress and strain in a structure, as the load is increased to failure, is generally very complicated. This is particularly true in soil mechanics and soil–structure interaction problems where, unlike traditional structural engineering, the analysis almost always involves two- or three-dimensional continua. With the

50

present state of development of finite element computer programs, we can confidently say that an almost unlimited range of solutions can now be obtained. These are not limited to linear elastic solid mechanics but can be extended to include problems of various kinds involving material and geometric non-linearities [10, 12, 16].

As has been indicated previously, a notable achievement in the stress strain modeling of soils based on work-hardening theories of plasticity has been made in recent years. What is really needed now in soil mechanics is to implement these theories in a finite element program for a proper study of the load--deformation behavior of soils at all stress levels leading up to failure in typical basic configurations of mixed boundary value problems under static and dynamic loading conditions. The objective of the work described in the forthcoming is therefore to show how the Drucker–Prager model with or without a hardening cap has been used for the solution of a complicated nonlinear boundary value problem. Attempts to obtain such computed solutions have shed much needed light on the suitability of such a drastic but extremely useful idealization of the soil as a perfectly plastic or isotropic hardening solid.

In the sections that follow, the development and verification of the "pseudo-static" method using finite element analysis with the elliptic cap model as well as the Drucker–Prager model are presented, and a comparative study is made with the limit analysis method. Details of the model formulations together with their computer implementations are given elsewhere [48–50]. Emphasis is placed here on the effect of large deformation on the evaluation of the overall slope stability problems. Further, of particular concern is the comparison of failure modes and limit loads with those assumed in the limit analysis method [18, 20].

4.1 ANALYSES OF SLOPES PRIOR TO SEISMIC LOADINGS

In order to simulate the stress condition inside the slope after the completion of the ground excavation, an elastic-plastic effective stress analysis of the slope is performed in this section. Sequential loading to simulate a cut-down or build-up process is not considered here. Instead, the final configuration of the slope, and the spreading of yielded zones are investigated qualitatively by increasing the internal force due to the weight of soil from zero to the natural weight of $\gamma = 120 \, \text{pcf}$ $(18.85 \, \text{kN/m}^3)$. A 30 ft (9.15 m) high vertical slope, as shown in Fig. 9 is considered. Here, the vertical boundaries and bottom boundary are placed respectively at 300 and 150 ft away from the toe of the expected slope surface. Movement on the vertical boundaries is constrained horizontally only and that along the bottom boundary is constrained in both directions. The same mesh as used by Snitbhan and Chen [78] consisting of 250 nodes and 216 rectangular elements is utilized in the present finite element analysis. Further, large deformation analysis is employed to

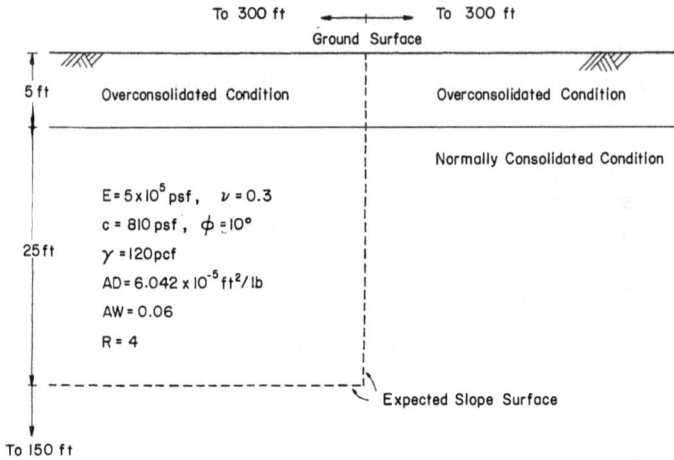

Fig. 9. Initial condition of ground.

reflect the effect of slope configuration on the limit load due to earth-quake loading. In addition, the initial stress method [84] is utilized in the numerical analysis. Details of the analysis are given elsewhere [48].

In the Drucker–Prager model case, the ground surface behind the crest is found to settle approximately 3.2 ft (0.98 m) and the bulging extends 0.604 ft (0.184 m) from the original vertical slope line. For the elliptic cap model case, larger amounts of ground settlement (3.8 ft or 1.16 m) and bulging (0.804 ft or 0.245 m) in the horizontal direction are observed. However, the deformed geometries of the two slopes are very similar at this load level, regardless of the relatively large difference in ground settlement.

4.2 ANALYSES OF SLOPES DURING SEISMIC LOADINGS

Based on the analysis described in the previous section, the seismic large deformation analyses of the vertical slopes are now performed in this section by employing the pseudo-static method. The 1934 El Centro (S–N direction) accelerogram is used as input for the seismic loadings which are obtained as the product of soil mass, m, and acceleration, a. In the present analysis, the horizontal acceleration data between time $t = 1.5$ s and $t = 4.5$ s in the accelerogram are applied to the deformed shape of the slope after excavation. For simplicity, the distribution of acceleration is here assumed to be uniform throughout the slope. It acts to the right when its sign is positive. The response of the slopes during seismic loading, and the velocity fields at the collapse stage are discussed briefly as follows:

52

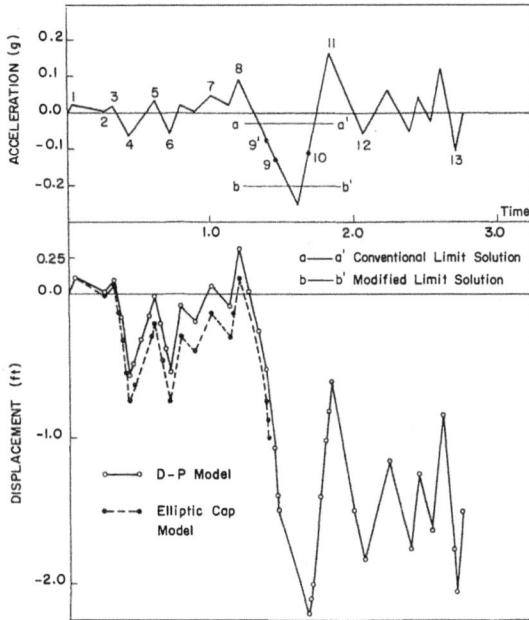

Fig. 10. Acceleration—time—displacement curves by the Drucker—Prager and elliptic cap models.

Response of Slopes

In Fig. 10, the acceleration—time displacement curves are presented for both model cases. Note that the horizontal displacement at the nodal point above the toe is taken as the reference displacement in the figure, and that displacement is measured from the deformed shape of the slope after excavation, not from the original shape. The displacement response curves with the solid and broken lines are results from the analyses with the Drucker—Prager and elliptic cap models, respectively.

Based on the finite element calculation (Fig. 10), the sliding of the slope is estimated to occur at approximately $a = -0.135\,g$ and $-0.075\,g$, for the Drucker—Prager and elliptic cap model cases, respectively. On the other hand, the limit analysis method predicts the occurrence of sliding at $a = -0.0287\,g$ and $-0.196\,g$ by applying the full slope height of 30 ft (9.15 m) and the modified slope height of 25 ft (7.63 m), respectively. Thus, the finite element solutions lie between the two extreme solutions of the limit analysis method.

Relative velocity fields

The relative velocity is defined here as the difference between the displacement increment at a nodal point and that at the toe. The relative

a) Relative Velocity Field (D-P Model), a = -0.135g.

b) Relative Velocity Field (Elliptic Cap Model), a = -0.075g.

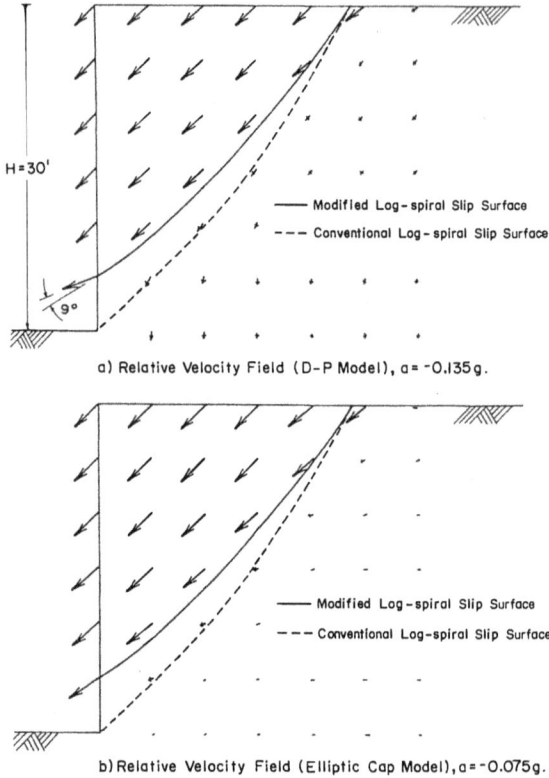

Fig. 11. Relative velocity fields at sliding.

velocity fields at the sliding stage are shown in Fig. 11, for both model cases. The magnitude and direction of each relative velocity vector is shown by an arrow, and the resultant of the relative-displacement increment at the nodal point above the toe is taken as a normalized unit length. Further, the modified and conventional log-spiral slip surface of the limit analysis method are illustrated with the solid and broken lines, respectively [17, 20]. The former and latter slip surfaces are obtained by using slope heights of 25 ft (7.63 m) and 30 ft (9.15 m) in the limit analysis solutions. Also, note that the modified and conventional slip surfaces correspond to failure mechanisms at $a = -0.196 g$ and $-0.0287 g$, respectively. Since the loss of ground appears to occur at a distance of approximately one-sixth of the total height above the toe as discussed previously, the relative velocity fields of both models agree quite well with the modified slip surfaces.

The angle between the modified slip surface and the relative velocity vector at the nodal point above the toe is measured from the figures. For the Drucker–Prager model case, it has an angle of 9°. It is expected that if

54

the seismic load is increased slightly, this angle would approach $10°$ as would be predicted by the limit analysis method. On the other hand, the angle in the elliptic cap model case is nearly zero. This means that the dilatancy has not been developed along the slip surface. Referring to Fig. 11, most of the stress states along the slip surface are in the corner zone. Consequently, the dilatancy on the slip surface is prevented to some extent.

5. Conclusion

This paper gives a summary of recent progress and achievement of theoretical soil plasticity with emphasis on applications to earthquake-induced slope failure and landslide analysis. It is also intended to give a brief summary of the origins and development of soil plasticity from about 1950 onwards. To this end, I should like to conclude by emphasizing again the great impact of D.C. Drucker's ideas on the evolution and modern development of the theory of soil plasticity.

The mechanical behavior of soils is much more complex than that of ductile metals. Accordingly, a more elaborate plasticity theory is required for soils than is required for metals. The extension of metal plasticity to problems involving plastic deformations in soils was first made by Drucker and Prager [32] based on the concept of perfect plasticity. This formed the central and most extensively developed part of the early theory of soil plasticity. The general theory of limit analysis, due to Drucker, Greenberg and Prager [31], had shed much needed light on the foundation of the much older theory of earth pressure. An important defect in the theory of earth pressure lay in its development without reference to stress—strain relations, the theory being based upon the concept of limiting equilibrium satisfying Coulomb's failure criterion.

The modern development of soil plasticity did not copy this unsatisfactory feature of the theory of earth pressure nor continued the simplicity of perfect plasticity, but instead, pursued the difficult course of work-hardening theory of plasticity. As a result of this, the theory of soil plasticity is now in a position to lead that of metal plasticity theory, especially in connection with cyclic loading applications. It is therefore appropriate to conclude here that the pioneer work of Drucker et al. [30] on work-hardening theories of soil plasticity and also that of the subsequent developments of Roscoe and his co-workers (1958–1963) mark the beginning of the modern developments of a consistent theory of soil plasticity.

Acknowledgment

This material is based in part upon work supported by the National Science Foundation under Grant No. PFR-7809326 to Purdue University.

55

References

1 J.H. Atkinson and P.L. Bransby, The Mechanics of Soils — An Introduction to Critical State Soil Mechanics. McGraw-Hill, Maidenhead, England, 1978.

2 A.W. Bishop, The strength of soils as engineering materials, Geotechnique, 22 (1972) 509—513.

3 J.R. Booker and E.H. Davis, A note on a plasticity solution to the stability of slopes in homogeneous clay. Geotechnique, 22 (1972) 509—513.

4 J.B. Burland, A method of estimating the pore pressures and displacements beneath embankments on soft, natural clay deposits. In: R.H.G. Parry (Ed.), Proc. Roscoe Memorial Symp., Cambridge Univ., March 29—31, 1971, Foulis, Oxfordshire, U.K., 1972, pp. 505—536.

5 S.W. Chan, Perfect Plasticity Upper Bound Limit Analysis of the Stability of a Seismic-Infirmed Earthslope. M.S. Thesis, Sch. of Mech. Eng., Purdue Univ., Aug. 1980.

6 C.J. Chang, Seismic Safety Analysis of Slopes. Ph.D. Thesis, Sch. of Civ. Eng., Purdue Univ., Aug. 1981.

7 M.F. Chang, Static and Seismic Lateral Earth Pressures on Rigid Retaining Structures, Ph.D. Thesis, Sch. of Civ. Eng., Purdue Univ., Aug. 1981.

8 M.F. Chang and W.F. Chen, Lateral Earth Pressures on Rigid Retaining Walls Subjected to Earthquake Forces. Solid Mech. Archives, 7, Sijthoff & Noordhoff, The Netherlands, Sept., 1982, pp. 315—362.

9 W.F. Chen, Mechanics of slope failures and landslides. In: J. Penzien, S.T. Mau and Z. Yang (Eds.), Proc. Advisory Meeting on Earthquake Eng. and Landslides, U.S.-R.O.C. Coop. Sci. Prog., Taipei, August 29—September 2, 1977, Nat. Sci. Found., Washington, D.C., 1977, pp. 219—232.

10 W.F. Chen, Constitutive equations for concrete. In: Special Publ. of the Assoc. of Bridge and Str. Eng. Colloq. on Plasticity in Reinforced Concrete, Copenhagen, 1979, Vol. 2, pp. 11—34.

11 W.F. Chen, Plasticity in soil mechanics and landslides. J. Eng. Mech. Div., ASCE, 106 (1980) 443—464.

12 W.F. Chen, Plasticity in Reinforced Concrete. McGraw-Hill, New York, 1982.

13 W.F. Chen, Limit Analysis and Soil Plasticity, Elsevier, Amsterdam, 1975.

14 W.F. Chen, C.J. Chang and J.T.P. Yao, Limit analysis of earthquake-induced slope failure. In: R.L. Sierakowski, (Ed.), Proc. 15th Annual Meeting Soc. Eng. Sci., Dec. 4—6, Gainesville, Univ. of Florida, 1978, pp. 533—538.

15 W.F. Chen and M.F. Chang, Limit analysis in soil mechanics and its applications to lateral earth pressure problems. Solid Mech. Archives, 6 (3) (1981) 331—399.

16 A.C.T. Chen and W.F. Chen, Constitutive relations for concrete, J. Eng. Mech. Div., ASCE, 101 (1975) 465—481.

17 W.F. Chen, and M.W. Giger, Limit analysis of stability of embankments. J. Soil Mech. and Found. Div., ASCE, 97 (1971) 19—26.

18 W.F. Chen, M.W. Giger and H.Y. Fang, On the limit analysis of stability of slopes, soils and foundations. Jap. Soc. Soil Mech. and Found. and Eng., 9 (1969) 23—32.

19 W.F. Chen and S.L. Koh, Earthquake-induced landslide problems. Proc. Central American Conf. on Earthquake Eng., 1, Universidad Centroamericana Jose Simeon Canas and the Ministry of Public Works of El Salvador and Lehigh Univ., San Salvador, Jan. 9—12, 1978, Envo, PA., pp. 665—685.

20 W.F. Chen, N. Snitbhan and H.Y. Fang, Stability of slopes in anisotropic non-homogenous soils. Canadian Geot. J., 12 (1975) 146—152.

21 W.F. Chen and A.F. Saleeb, Constitutive Equations for Engineering Materials, Vol. 1 — Elasticity and Modeling, Feb., 1982; Vol. 2 — Plasticity and Modeling, 1984. John Wiley Interscience, New York.

22 Y.F. Dafalias and E.P. Popov, A model of nonlinearly hardening materials for complex loadings. Acta Mech., 21 (1975) 173—192.

56

23 Y.F. Dafalias and R. Herrmann, A boundary surface soil plasticity model. Int. Symp. Soils Under Cyclic and Transient Loading, Swansea, U.K., 1980, Vol. 1, pp. 335—345.

24 E.H. Davis, Theories of plasticity and the failure of soil masses. In: I.K. Lee (Ed.), Soil Mechanics: Selected Topics, Butterworths, London, 1968, pp. 341—380.

25 C.S. Desai and R.H. Gallagher (Eds.), Constitutive Laws for Engineering Materials: Theory and Application. John Wiley, U.K., 1983.

26 F.L. Dimaggio and I.S. Sandler, Material model for granular soils. J. Eng. Mech. Div., ASCE, 97 June (1971) 935—950.

27 D.C. Drucker, Concepts of path independence and material stability for soils. In: J. Kravtchenko and P.M. Sirieys (Ed.), Rheol. Mecan. Soils Proc. IUTAM Symp. Grenoble, Springer, Berlin, 1966, pp. 23—43.

28 D.C. Drucker, Limit analysis of two- and three-dimensional soil mechanics problems. J. Mech. and Phys. Solids, 1 (1953) 217—226.

29 D.C. Drucker, Plasticity. In: J.N. Goodier and N.J. Hoff (Eds.), Structural Mechanics, Pergamon Press, London, 1960, pp. 407—455.

30 D.C. Drucker, R.E. Gibson and D.J. Henkel, Soil mechanics and work hardening theories of plasticity. Trans. ASCE, 122 (1957) 338—346.

31 D.C. Drucker, J.H. Greenberg and W. Prager, Extended limit design theorems for continuous media. Quart. Appl. Math., 9 (1952) 381—389.

32 D.C. Drucker and W. Prager, Soil mechanics and plastic analysis or limit design. Quart. Appl. Math. 10 (1952) 157—165.

33 J.M. Duncan and C.Y. Chang, Nonlinear analysis of stress and strain in soils. J. Soil Mech. Found. Div. ASCE, 96 (Sept. 1970) 1629—1653.

34 B.O. Hardin, The nature of stress—strain behavior of soils, Earthquake Eng. and Soil Dynamics, ASCE, 1 (1978) 3—90.

35 B.O. Hardin and V.P. Drnevich, Shear modulus and damping in soils: design equation and curves. J. Soil Mech. Found. Div., ASCE, 98 (1972) 667—692.

36 R. Hill, The Mathematical Theory of Plasticity, Oxford, Clarendon Press, 1950.

37 IABSE Proc. Colloq. on Plasticity in Reinforced Concrete, Copenhagen, May 21—23, IABSE Pub., Zurich, 1979.

38 W.D. Iwan, On a class of models for the yielding behavior of continuous and composite systems. J. Appl. Mech., 34 (1967) 612—617.

39 N. Janbu, Soil compressibility as determined by odometer and triaxial tests. Proc. Europ. Conf. on Soil Mech. and Found. Eng., Wiesbaden, Germany, 1963, Vol. 1, pp. 19—25.

40 W.B. Joyner and A.T.F. Chen, Calculation of nonlinear ground response in earthquakes. Bulletin Seismological Soc. of America, 65 (5) (1976).

41 K. Karal, Energy Method for Soil Stability Analyses. River and Harbour Lab., Norwegian Inst. Tech., Trondheim, Norway, 1979.

42 S.L. Koh and W.F. Chen, The Prevention and Control of Landslides. Presented at Sept. 1977, Joint U.S.—Southeast Asia Symp. on Eng. for Natural Hazards Protection, Manila.

43 R.D. Krieg, A practical two-surface plasticity theory. J. Appl. Mech., 42 (1975) 641—646.

44 P.V. Lade, Elasto-plastic stress—strain theory for cohesionless soil with curved yield surfaces. Int. J. Solids and Str., 13 (1977) 1014—1035.

45 P.V. Lade, Stress—strain theory for normally consolidated clay. Proc. 3rd Int. Conf. on Numer. Methods in Geomech., Aachen, Germany, 1979, Vol. 4, pp. 1325—1337.

46 P.V. Lade and J.M. Duncan, Elastoplastic stress—strain theory for cohesionless soil. J. Geot. Eng. Div., ASCE, 101 (1975) 1037—1053.

47 G. Masing, Eigenspannungen and Verfestigung beim Messing. Proc. 2nd Int. Cong. of Applied Mech., Zurich, 1926.

57

48 E. Mizuno, Plasticity Modeling of Soils and Finite Element Applications. Ph.D. Thesis, Sch. of Civ. Eng., Purdue Univ., Dec. 1981.

49 E. Mizuno and W.F. Chen, Analysis of soil response with different plasticity models. In: R.N. Yang and E.T. Selig (Eds.), Proc. of the Symp. Applications of Plasticity and Generalized Stress—Strain in Geotechnical Engineering, Hollywood, Florida, Oct. 27--31, 1980, ASCE, New York, 1982, pp. 115—138.

50 E. Mizuno and W.F. Chen, Plasticity models and finite element implementation. Proc. Symp. on Implementation of Computer Procedures and Stress—Strain Laws in Geot. Eng., Chicago, Aug. 3—6, 1981, Acorn Press, Durham, N.C., 1981, pp. 519—534.

51 Z. Mroz, On the description of anisotropic hardening. J. Mech. Phys. of Solids, 15 (1967) 163—175.

52 Z. Mroz, V.A. Norris and O.C. Zienkiewicz, An anisotropic hardening model for soils and its application to cyclic loading. Int. J. for Numer. and Anal. Meth. in Geomech., 2 (1978) 203—221.

53 Z. Mroz, V.A. Norris and O.C. Zienkiewicz, Application of an anisotropic hardening model in the analysis of elastoplastic deformation of soils. Geotech. 29 (1979) 1—34.

54 I. Nelson, Constitutive models for use in numerical computations. Proc. Int. Symp. on Dynamic Methods in Soil and Rock Mech., Balkema, Rotterdam, 1978, Vol. 2, pp. 45—97.

55 I. Nelson and G.Y. Baladi, Outrunning ground shock computed with different models, J. Eng. Mech. Div. ASCE, 103 (1977) 377—393.

56 I. Nelson, M.L. Baron and I. Sandler, Mathematical Models for Geological Materials for Wave Propagation Studies, Shock Waves and the Mechanical Properties of Solids. Syracuse Univ. Press, Syracuse, N.Y., 1971.

57 N.M. Newmark, Effects of earthquakes on dams and embankments. Geotech. 15 (1965) 139—160.

58 A.C. Palmer (Ed.), Proc. Symp. on the Role of Plasticity in Soil Mechanics, Cambridge Univ. Press, 1973.

59 G.N. Pand and O.C. Zienkiewicz (Eds.), Proc. Int. Symp. Soils Under Cyclic and Transient Loading, Balkema, Rotterdam, 1980.

60 R.H.G. Parry, correspondence on "On the Yielding of Soils". Geotech. 8 (1958) 185—186.

61 R.H.G. Parry (Ed.), Roscoe Memorial Symp.: Stress—Strain Behavior of Soils, Henly-on-Thames, Cambridge Univ., 1972.

62 W. Prager and P.G. Hodge, Theory of Perfectly Plastic Solids, Wiley, New York, 1950.

63 J. Prevost, Mathematical modelling of monotonic and cyclic undrained clay behavior. Int. J. for Numer. and Anal. Meth. in Geomech., 1 (1977) 195—216.

64 J.H. Prevost, Anisotropic undrained stress—strain behavior of clays. J. Geotech. Eng. Div. ASCE, 104 (1978) 1075—1090.

65 J.H. Prevost, Plasticity theory for soil stress behavior, J. Eng. Mech. Div. ASCE, 104 (1978) 1177—1194.

66 J.H. Prevost, B. Cuny, T.J.R. Hughes and R.F. Scott, Offshore gravity structures: analysis. J. Geotech. Div. ASCE, 107 (1981) 143—165.

67 K.H. Roscoe and J.B. Burland, On the generalized stress—strain behavior of "wet" clay. In: J. Heyman and F.A. Leckie (Eds.), Engineering Plasticity, Cambridge Univ. Press, 1968, pp. 535—609.

68 K.H. Roscoe, A.N. Schofield and C.P. Wroth, On the yielding of soils. Geotech. 8 (1958) 22—52.

69 K.H. Roscoe, A.N. Schofield and A. Thurairajah, An Evaluation of Test Data for Selecting a Yield Criterion for Soils. ASTM Spec. Tech. Pub. No. 361, ASTM, 1963, pp. 111—128.

58

70 A.F. Saleeb, Constitutive Models for Soils in Landslides, Ph.D. Thesis, Sch. Civ. Eng. Purdue Univ., May, 1981.

71 A.F. Saleeb and W.F. Chen, Nonlinear hyperelastic (Green) constitutive models for soils, Part I — Theory and Calibration; Part II — Predictions and Comparisons. In: R.K. Yong and H.Y. Ko (Eds.), Proc. North American Workshop on Limit Equilibrium, Plasticity, and Generalized Stress—Strain In Geotech. Eng., McGill Univ., Montreal, May 28—30, 1980; ASCE, New York, 1981.

72 I.S. Sandler and M.L. Baron, Recent development in the constitutive modeling of geological materials. In: Proc. 3rd Int. Conf. on Num. Methods in Geomechanics, Aachen, Germany, 1979, pp. 363—376.

73 I.S. Sandler and M.L. Baron, Material models of geological materials in ground shock. In: C.S. Desai (Ed.), Proc. of the Second Int. Conf. on Num. Methods in Geomech., Blacksburg, Virginia, June, 1976, ASCE, 1976, pp. 219—231.

74 I.S. Sandler, F.L. DiMaggio and G.Y. Baladi, Generalized cap model for geological materials. J. Geotech. Eng. Div. ASCE, 102 (1976) 683–-699.

75 A. Schofield and P. Wroth, Critical State Soil Mechanics, McGraw-Hill, New York, 1968.

76 H.B. Seed, A method for earthquake resistant design of earth dams. J. Soil Mech. and Found. Div. ASCE, 92 (1966) 13—41.

77 H.B. Seed, K.L. Lee and I.M. Idriss, Analysis of Sheffield dam failure. J. Soil Mech. Found. Div., ASCE, 95 (1969) 1453—1490.

78 N. Snitbhan and W.F. Chen, Elastic-plastic large deformation analysis of slopes. Computers and Structures, 8 (1978) 567—577.

79 V.V. Sokolovskii, Statics of Granular Media. Pergamon, New York, 1965.

80 K. Terzaghi, Theoretical Soil Mechanics. John Wiley, New York, 1943.

81 R.K. Yong and H.Y. Ko, (Eds.), Limit Equilibrium, Plasticity and Generalized Stress—Strain in Geotechnical Engineering. ASCE, New York, 1981.

82 R.N. Yong and E.T. Selig (Eds.), Application of Plasticity and Generalized Stress—Strain in Geotechnical Engineering. ASCE, New York, 1982.

83 O.C. Zienkiewicz, V. Norris and D.J. Naylor, Plasticity and viscoplasticity in soil mechanics with special reference to cyclic loading problems. In: P.G. Bergan et al. (Eds.), Finite Elements in Non-Linear Mechanics, Tapir, Trondheim, 1977, Vol. 2, pp. 455—485.

84 O.C. Zienkiewicz, S. Valliapan and I.P. King, Elasto-plastic solutions of engineering problems: initial stress, finite element approach. Int. J. Numer. Meth. Eng. 1 (1969) 75—100.

5.8 Structural Stability (結構穩定)

5.8.1

Columns with End Restraint and Bending in Load and Resistance Design Factor

W. F. CHEN and E. M. LUI

1. INTRODUCTION

Elastic Stability—Mathematical

The problem of structural stability has long been the subject of research for a number of researchers. Early in the 18th century, Euler[1] investigated the elastic stability of a centrally loaded isolated strut using the bifurcation approach. The bifurcation or eigenvalue approach is basically a mathematical approach. Under the assumptions that (1) the member is perfectly straight, (2) the material remains fully elastic and obeys Hooke's Law and (3) the deflection is small, a linear differential equation can be written based on a slightly deformed geometry of the member.

The eigenvalue solution to the characteristic equation of this differential equation will give the buckling load of the strut. This load corresponds to the state at which bifurcation of equilibrium takes place. At this load, the original straight position of the member ceases to be stable. Under this load, a small lateral disturbance will produce a large lateral displacement which will not disappear when the disturbance is removed. This buckling load is referred to as the critical load or Euler load given by

$$P_e = \frac{\pi^2 EI}{(KL)^2} \tag{1}$$

where

I = moment of inertia of the cross section
L = unbraced length of the column
K = effective length factor to account for the end conditions of the column

W. F. Chen is Professor and Head, Structural Engineering Department, School of Engineering, Purdue University, West Lafayette, Indiana.

E. M. Lui is a Graduate Assistant in that same department.

This paper, the T. R. Higgins Lectureship Award winner for 1985, was first presented at the Structural Stability Research Council Annual Technical Session and Meeting on April 16, 1985 in Cleveland, Ohio.

This formula gives a good prediction of the behavior of long columns so far as the axial stresses in the member remain below the proportional limit, i.e., if the member remains fully elastic. For short or intermediate columns, the assumption of fully elastic behavior will be questionable. Under the action of the applied force, some fibers of the cross section will yield. Consequently, only the elastic core of the cross section will be effective in resisting the additional applied force. Thus, the Euler load will overestimate the strength of the column.

Plastic Buckling—Physical

To account for the effect of inelasticity, two theories were proposed:[2,3] the double modulus theory and the tangent modulus theory. In the double modulus theory (also known as the reduced modulus theory), the axial load is assumed constant during buckling. Consequently, at buckling, the bending deformation of the column will produce strain reversal on the convex side of the member with the result that the elastic modulus E will govern the stress-strain behavior of the fibers. The concave side of the column, on the other hand, will continue to load and so the tangent modulus E_t will govern the stress-strain behavior of the fibers (Fig. 1). The critical load obtained based on this concept is referred to as the reduced modulus load given by

$$P_r = \frac{\pi^2 E_r I}{(KL)^2} = \frac{E_r}{E} P_e \tag{2}$$

where E_r is the reduced modulus.

The reduced modulus is a function of the tangent modulus and the geometry of the cross section. Hence the reduced modulus load depends on both the material property and the geometry of the cross section. The reduced modulus load is lower than the Euler load because the ratio E_r/E in Eq. 2 is always less than unity. It should be pointed out that the reduced modulus load can only be reached if the column is artificially held in a straight position when the tangent modulus load (to be discussed later) has been exceeded. The reduced modulus load can

P_r

SECTION A–A

A   A

$\Delta\epsilon_2$ — ϵ_r $\Delta\epsilon_1$

STRAIN DIAGRAM

N.A.

σ

E_t

σ_r $\Delta\sigma_1 = E_t \Delta\epsilon_1$

$\Delta\sigma_2 = E \,\Delta\epsilon_2$

E

$\Delta\epsilon_1$

$\Delta\epsilon_2$ ϵ_r

ϵ

— σ_r

$\Delta\sigma_2 = E \Delta\epsilon_2$
(unloading)

$\Delta\sigma_1 = E_t \Delta\epsilon_1$
(loading)

P_r

STRESS DIAGRAM

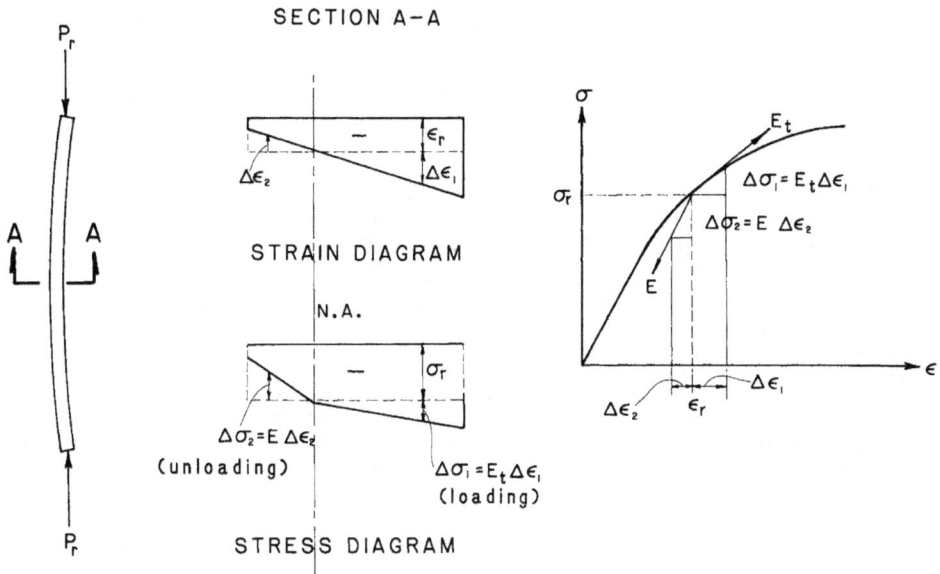

Fig. 1.   *Double (reduced) modulus theory*

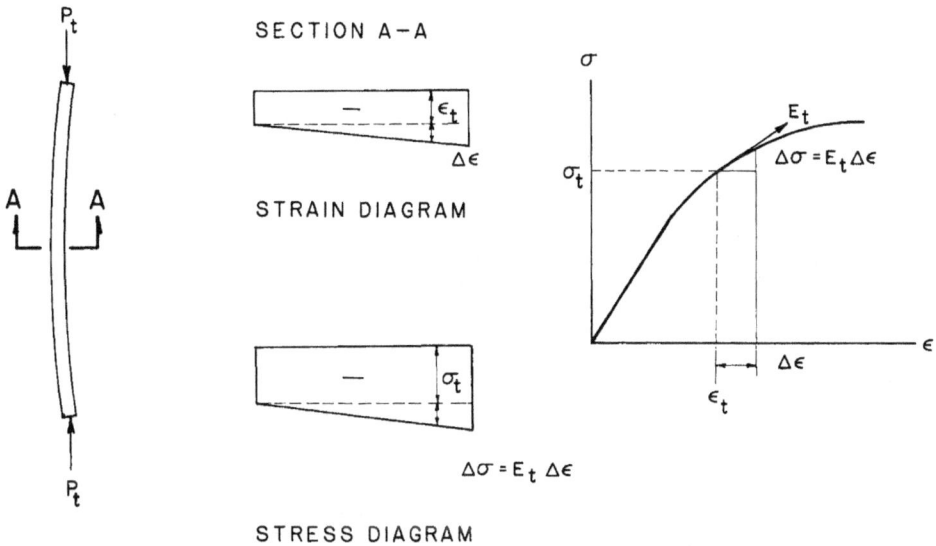

P_t

SECTION A–A

A   A

— ϵ_t $\Delta\epsilon$

STRAIN DIAGRAM

σ

E_t

σ_t $\Delta\sigma = E_t \Delta\epsilon$

$\Delta\epsilon$

ϵ_t

ϵ

— σ_t

$\Delta\sigma = E_t \,\Delta\epsilon$

P_t

STRESS DIAGRAM

Fig. 2.   *Tangent modulus theory*

never be reached even if the slightest geometrical imperfection is present in the column.

In the tangent modulus theory, the axial load is assumed to increase during buckling. The amount of increase is such that strain reversal will not take place and so the tangent modulus E_t will govern the stress-strain behavior of the entire cross section (Fig. 2). The critical load obtained is known as the tangent modulus load given by

$$P_t = \frac{\pi^2 E_t I}{(KL)^2} = \frac{E_t}{E} P_e \qquad (3)$$

The tangent modulus load, unlike the reduced modulus load, is independent of the geometry of the cross section. It depends only on the material property. For a steel column, the nonlinearity of the average stress-strain behavior of the cross-section is due to the presence of residual stress. Residual stresses arise as a result of the manufacturing process. When a compressive axial force is applied to a stub column (very short column), the fi-

bers that have compressive residual stresses will yield first. The fibers that have tensile residual stresses will yield later. As a result, yielding over the cross section of the column is a gradual process, as shown in Fig. 3.

The slope of the stub column stress-strain curve is the tangent modulus E_t of the member. Also shown in the figure is the stress-strain behavior of a coupon. A coupon, unlike a stub column, is free of residual stress. Therefore, its stress-strain relationship exhibits an elastic-perfectly plastic behavior.

The tangent modulus load marks the point of bifurcation of a perfectly straight inelastic column. The tangent modulus load is lower than the Euler and the reduced modulus loads and so it also represents the lowest load at which bifurcation of equilibrium can take place (Fig. 4).

Experiments on columns have demonstrated the failure loads of columns fall nearer to the tangent modulus loads than the reduced modulus loads. The theoretical justification for this observation was given by Shanley,[4] who, in 1947, investigated the buckling behavior of col-

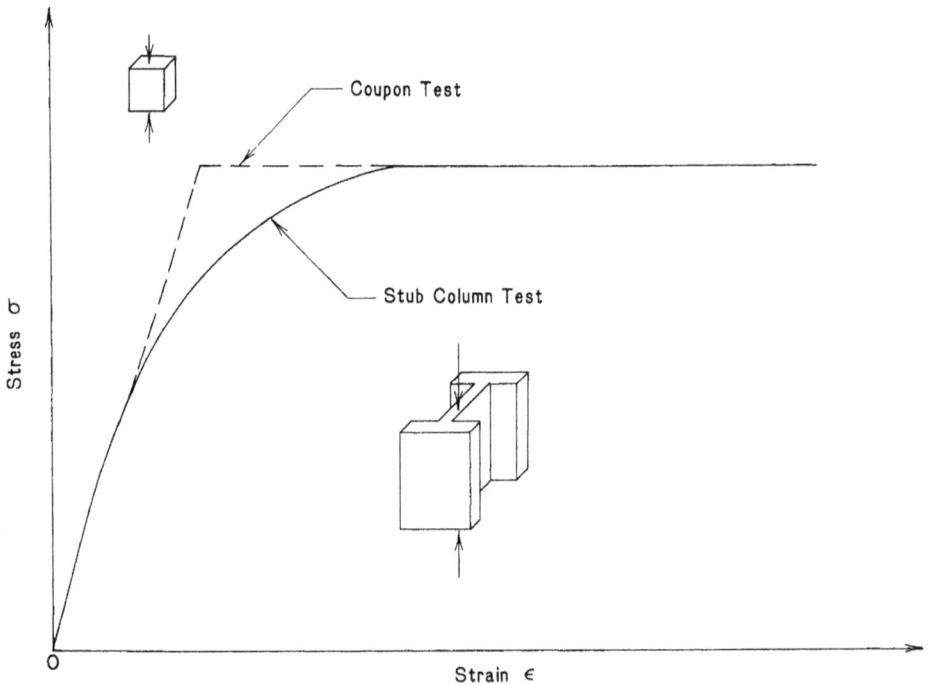

Fig. 3.   *Stress-strain relationship for steel*

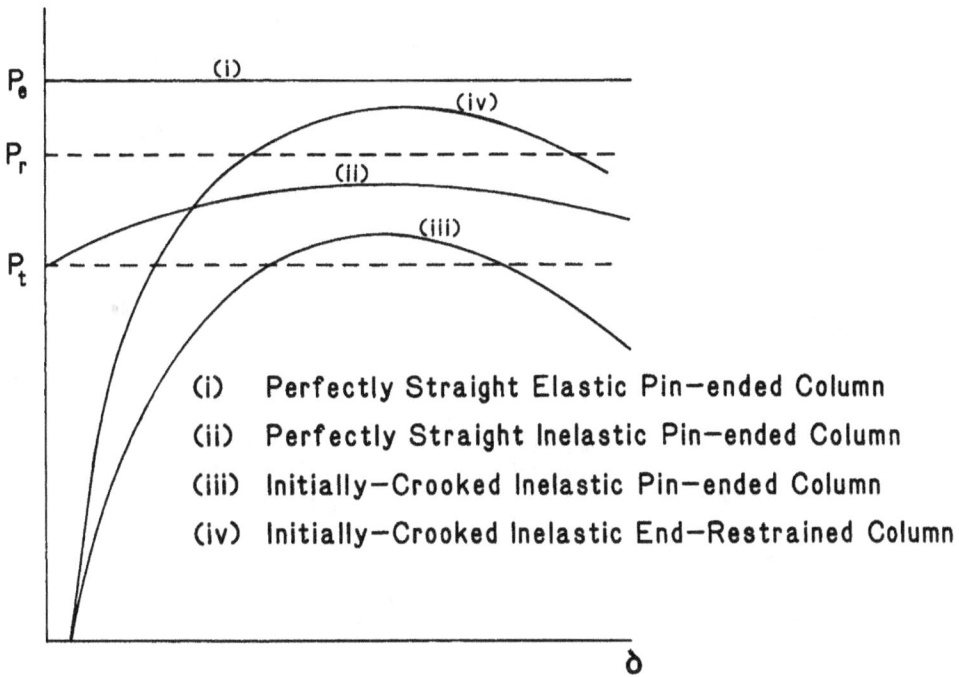

(i)   Perfectly Straight Elastic Pin—ended Column

(ii) Perfectly Straight Inelastic Pin—ended Column

(iii) Initially—Crooked Inelastic Pin—ended Column

(iv) Initially—Crooked Inelastic End—Restrained Column

Fig. 4.   Load-deflection behavior of columns

umns above the tangent modulus load. Using a simplified physical model, Shanley showed that bifurcation of equilibrium will take place when the applied load reaches the tangent modulus load. After bifurcation, increase in lateral deflection is accompanied by a slight increase in load above the tangent modulus load. Thus the maximum load is really slightly larger than the tangent modulus load, provided the column is perfectly straight. Extensions of Shanley's model to describe the buckling behavior of columns above the tangent modulus load were reported by Duberg and Wilder[5] and Johnston.[6]

In Ref. 5, it was shown that if a column were artificially held in a straight position up to a load somewhere in between the tangent modulus and reduced modulus loads, then released, it would start to bend with an increase in axial load. The magnitude of the increase, however, was less than that of the tangent modulus load. If the column was held in a straight configuration up to the reduced modulus load, then released, it would bend with no increase in axial load. Reference 6 demonstrates that when a column buckles at the tangent modulus load there is no strain reversal only for an infinitesimal increment of axial load.

For any finite increase of axial load above the tangent modulus load, the column assumes equilibrium positions with increasing deflection accompanied by a strain reversal on the convex side of the column. Nevertheless, the amount of strain reversal is less than that of the reduced modulus theory. The readers are referred to a paper by Johnston[7] for a more thorough discussion of the historic highlights of the column buckling theory.

The discussion so far pertains to columns which are perfectly straight. Columns in reality are rarely perfectly straight. Geometrical imperfection in a column tends to lower the maximum load of the member. As a result, the Structural Stability Research Council (formerly the Column Research Council) recommended the tangent modulus load be the representative failure load of a centrally loaded column.

The reduced modulus theory and the tangent modulus theory, as well as the Shanley's concept of inelastic column, are all based on physical reasoning. They provide solutions and explanations to the behavior of perfectly straight inelastic columns. The mathematical theory of elastic stability and the concepts of inelastic buckling are well explained in Refs. 8 and 9.

Plastic Stability—Numerical

As pointed out earlier, real steel columns not only exhibit inelasticity due to the presence of residual stresses, but also they possess initial crookedness. The analysis of columns with residual stresses and initial crookedness is rather complicated. The eigenvalue approach, which is valid only for perfectly straight columns, can not be used here. Instead, a different approach known as the stability approach must be utilized. In the stability approach, the load-deflection behavior of the column is traced from the start of loading to failure. The procedure is often carried out numerically using the computer because the differential equation governing the behavior of inelastic-crooked columns are often intractable, so closed form solutions are very difficult, if not impossible, to obtain. Various methods to obtain numerical solutions are presented in Refs. 10 and 11.

In addition to inelasticity and initial crookedness, the end conditions of a column also play an important role in affecting its behavior. The analyses of columns taking into consideration inelasticity, initial crookedness and end restraint were reported by a number of researchers in the past few years. The results are summarized in Ref. 12.

Structural Stability—Engineering

Columns in real structures seldom exist alone. The behavior of a column as an integral part of a structure is affected by the behavior of other structural members. In particular, in addition to carrying axial force, the column must be able to resist bending moments induced by the beam, so the column in reality behaves as a beam-column resisting both axial load and bending moments. The moment transfer mechanism between beams and columns is different depending on whether the connection is rigid or flexible. In other words, the behavior of the frame and its structural members is dependent on the rigidity of the connections. The stability analysis of frameworks with flexible connections has been a popular research topic in recent years. In particular, the recently published Load and Resistance Factor Design (LRFD) Specification[13] designates two types of construction in its provision: Type FR (fully restrained) and Type PR (partially restrained) constructions. Type PR construction requires explicit consideration of connection flexibility in proportioning structural members.

The stability analysis of flexibly connected frames requires connection modeling. Since connection moment-rotation behavior is usually nonlinear, the inclusion of a connection as a structural element in a limit state analysis requires the use of nonlinear structural theory. With the advent of computer technology, great advancement has been made in computer-aided analysis and design of structures. At the present time, first- and second-order elastic analyses of structures can conveniently be performed for nearly all types of structures. Analysis of

structures loaded into the inelastic range can also be performed for certain types of structures.

The continued development in computer hardware and software has made it possible for engineers and designers to predict structural behavior rather accurately. The advancement in structural analysis techniques coupled with the increased understanding of structural behavior has made it possible for engineers to adopt the limit state design philosophy. A limit state is defined as a condition at which a structural member or its component ceases to perform its intended function under normal condition (serviceability limit state) or failure under severe condition (ultimate limit state). Load and Resistance Factor Design is based on the limit state philosophy and thus it represents a more rational approach to the design of structures.

This paper attempts to summarize the state-of-the-art methods in the analysis and design of columns as individual members and as members of a structure. A second objective is to introduce to engineers the stability design criteria of members and frames in LRFD. Highlights of recent research as well as directions of further research will be discussed.

2. PIN-ENDED COLUMN

A pin-ended column is the most fundamental case of a column. The behavior of a pin-ended column represents an anchorpoint for the study of all other columns. For columns with long slenderness ratio, the Euler formula (Eq. 1) will provide a good estimate of their behavior. For intermediate or short columns, the Euler formula has to be modified according to the reduced modulus concept or the tangent modulus concept (Eqs. 2 and 3) to account for yielding (or plastification) over the cross section due to the presence of residual stresses. As mentioned earlier, the tangent modulus theory gives a better prediction of inelastic column behavior and hence it is adopted for design purposes.

CRC Curve

Based on the study of idealized columns with linear and parabolic residual stress distribution, as well as the test results of a number of small and medium-size, hot-rolled, wide-flange shapes of mild structural steel, the Column Research Council recommended in the first edition of the Guide[14] a parabola of the form

$$F_{cr} = F_y - B \left(\frac{KL}{r} \right)^2 \tag{4}$$

to represent column strength in the inelastic range. This parabola was chosen because it represented an approximate median between the tangent modulus strength of a W column buckled in the strong and weak directions. The column strength in the elastic range, however, is represented by the Euler formula. The point of demarcation between inelastic and elastic behavior was chosen to be $F_{cr} = 0.5\,F_y$. The number 0.5 was chosen as a conservative measure of the maximum value of compressive residual stress present in hot-rolled wide-flange shapes which is about $0.3\,F_y$. To obtain a smooth transition from the parabola to the Euler curve, the constant B in Eq. 4 was chosen to be $F_y^2/4\,\pi^2E$. The slenderness ratio that corresponds to $F_{cr} = 0.5\,F_y$ is designated as C_c in which

$$C_c = \sqrt{\frac{2\pi^2E}{F_y}} \qquad (5)$$

Thus, for columns with slenderness ratios less than or equal to C_c, the CRC curve assumes the shape of a parabola and for slenderness ratio exceeding C_c, the CRC curve takes the shape of a hyperbola, i.e.

$$F_{cr} = \begin{cases} F_y\left[1 - \dfrac{(KL/r)^2}{2C_c^2}\right] & \dfrac{KL}{r} \leq C_c \\[2ex] \dfrac{\pi^2E}{(KL/r)^2} & \dfrac{KL}{r} > C_c \end{cases} \qquad (6)$$

For comparison purposes, Eq. 6 is rewritten in its load form in terms of the nondimensional quantities P/P_y and λ_c in which P_y is the yield load given by $P_y = AF_y$ and λ_c is the slenderness parameter given by $\lambda_c = (KL/r)\sqrt{F_y/\pi^2E}$

$$\frac{P}{P_y} = \begin{cases} 1 - 0.25\lambda_c^2 & \lambda_c \leq \sqrt{2} \\[1ex] \lambda_c^{-2} & \lambda_c > \sqrt{2} \end{cases} \qquad (7)$$

The CRC curve is plotted in Fig. 5 in its nondimensional form (Eq. 7).

AISC/ASD Curve

The CRC curve divided by a variable factor of safety of

$$\frac{5}{3} + \frac{3}{8}\left(\frac{KL/r}{C_c}\right) - \frac{1}{8}\left(\frac{KL/r}{C_c}\right)^3$$

$$\left[= \frac{5}{3} + \frac{3}{8}\left(\frac{\lambda_c}{\sqrt{2}}\right) - \frac{1}{8}\left(\frac{\lambda_c}{\sqrt{2}}\right)^3\right]$$

in the inelastic range and a constant factor of safety of 23/12 in the elastic range gives the AISC Allowable Stress

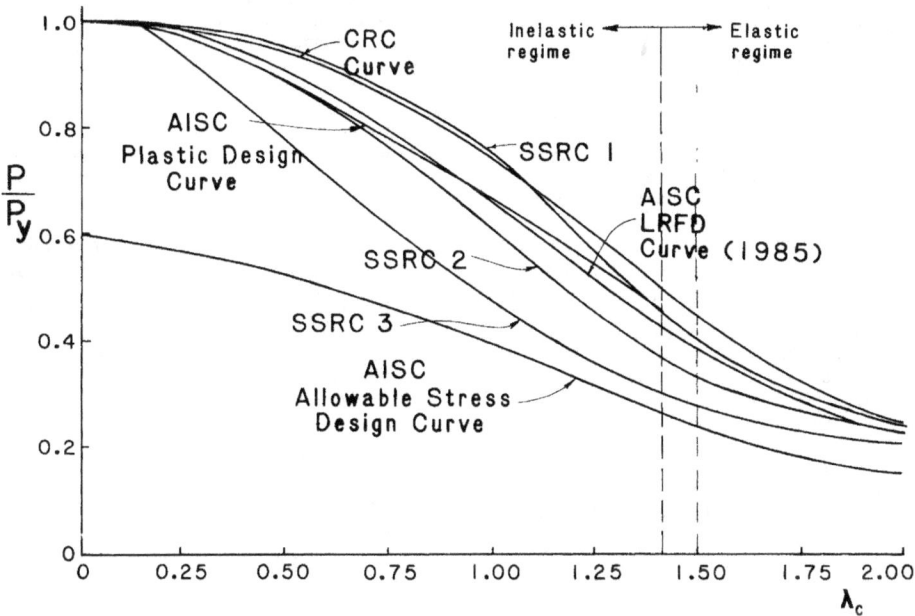

Fig. 5. Column design curves

Design (ASD) curve. The factors of safety are employed to account for geometrical imperfections and load eccentricities which are unavoidable in real columns. The AISC/ASD curve is also plotted in Fig. 5. The ASD column curve is used in conjunction with the ASD format given by

$$\frac{R_n}{F.S.} \geq \sum_{i=1}^{j} Q_{ni} \qquad (8)$$

where

R_n = nominal resistance
(for column design, $R_n/F.S.$ is represented by the ASD column curve)
Q_n = service loads

AISC/PD Curve

The ASD curve multiplied by a factor of 1.7 forms the AISC Plastic Design (PD) curve (Fig. 5). In plastic design only the inelastic regime of the curve is utilized because of the slenderness requirement. The design format for plastic design of columns is thus

$$\frac{1.7R_n}{F.S.} \geq \gamma \sum_{i=1}^{j} Q_{ni} \qquad (9)$$

where γ is the load factor used in the present AISC/PD Specification. The values for γ are: $\gamma = 1.7$ for live and dead loads only and $\gamma = 1.3$ for live and dead loads acting in conjunction with wind or earthquake loads.

SSRC Curves

Before proceeding any further, it should be stated that both the ASD curve and PD curve are originated from the CRC curve which was developed based on the bifurcation concept which postulates that the column is perfectly straight. Although residual stress is explicitly accounted for, the effect of geometrical imperfections is only accounted for implicitly by applying a variable factor of safety to the basic curve. Analysis of columns which explicitly take into consideration the effects of both residual stresses and initial crookedness was reported.[15] The stability approach was used in the analysis and a set of three curves referred to as the SSRC multiple column curves[16] was developed. Detailed expressions for these curves are given in Ref. 16. Approximate formulas for these curves based on physical reasoning which are useful for design are also reported.[17,18]

For comparison purposes, the three SSRC curves are plotted with the CRC, ASD and PD curves in Fig. 5. These curves belly down in the intermediate slenderness range ($0.75 < \lambda < 1.25$) due to the combined maximum detrimental effects of both residual stresses and initial crookedness on column strength in the numerical analysis. Tests of real columns have demonstrated the det-

rimental effects of residual stresses and initial crookedness are not always synergistic, so the SSRC curves which belly down in the intermediate slenderness range will be too conservative for most columns in building frames.

AISC/LRFD Curve

To provide a compromise between the CRC curve (developed based on the tangent modulus concept) and the SSRC curves (developed based on the stability concept), the 1985 AISC/LRFD Specification[13] adopted a curve of the form

$$\frac{P}{P_y} = \begin{cases} \exp(< -0.419\lambda_c^2) & \lambda_c \leq 1.5 \\ 0.877\lambda_c^{-2} & \lambda_c > 1.5 \end{cases} \qquad (10)$$

to represent basic column strength.

The LRFD curve is plotted on Fig. 5, together with the other curves described above. Note the LRFD curve lies between the CRC curve and the SSRC curve 2.

The LRFD format is

$$\phi R_n \geq \sum_{i=1}^{j} \gamma_i Q_{ni} \qquad (11)$$

where

R_n = nominal resistance
Q_n = nominal load effects
ϕ = resistance factor
γ = load factor

Note the LRFD format has the features of both the ASD and PD formats in that factors of safety are applied to both the load and resistance terms to account for the variabilities and uncertainties in predicting these values. Furthermore, these load and resistance factors (ϕ, γ) are evaluated based on first order probabilistic approach. Since different types of loads have different degrees of uncertainties, different load factors are used for different types of loads (e.g. 1.6 for live load, 1.2 for dead load, etc.). Therefore, the LRFD format represents a more rational design approach.

The expressions for various column curves described above and the three state-of-the-art design formats (ASD, PD, LRFD) are summarized in Tables 1 and 2.

3. COLUMNS WITH END RESTRAINT

Eigenvalue Analysis

In addition to residual stresses and initial crookedness, the end conditions of a column have a significant influence on column behavior. For perfectly straight elastic columns with idealized end conditions (ideally pinned or fully rigid), an eigenvalue analysis can be carried out to determine the critical load P_{cr}. The effective length fac-

Table 1. Summary of Column Curves

Column Curves	Column Equations	
CRC Curve	$\dfrac{P}{P_y} = 1 - \dfrac{\lambda_c^2}{4}$	$\lambda_c \leq \sqrt{2}$
	$\dfrac{P}{P_y} = \dfrac{1}{\lambda_c^2}$	$\lambda_c > \sqrt{2}$
AISC	$\dfrac{P}{P_y} = \dfrac{1 - \dfrac{\lambda_c^2}{4}}{\dfrac{5}{3} + \dfrac{3}{8}\left(\dfrac{\lambda_c}{\sqrt{2}}\right) - \dfrac{1}{8}\left(\dfrac{\lambda_c}{\sqrt{2}}\right)^3}$	$\lambda_c \leq \sqrt{2}$
Allowable Stress Design Curve	$\dfrac{P}{P_y} = \dfrac{12}{23}\dfrac{1}{\lambda_c^2}$	$\lambda_c > \sqrt{2}$
AISC Plastic Design Curve	$\dfrac{P}{P_y} = \dfrac{1.7\left(1 - \dfrac{\lambda_c^2}{4}\right)}{\dfrac{5}{3} + \dfrac{3}{8}\left(\dfrac{\lambda_c}{\sqrt{2}}\right) - \dfrac{1}{8}\left(\dfrac{\lambda_c}{\sqrt{2}}\right)^3} \leq 1.0$	$\lambda_c \leq \sqrt{2}$
AISC LRFD Curve (1985 Version)	$\dfrac{P}{P_y} = \exp(-0.419\lambda_c^2)$	$\lambda_c \leq 1.5$
	$\dfrac{P}{P_y} = \dfrac{0.877}{\lambda_c^2}$	$\lambda_c > 1.5$

Table 2. Summary of Design Formats

Allowable Stress Design (ASD)	$\dfrac{R_n}{F.S.} \geq \Sigma\, Q_{ni}$
Plastic Design (PD)	$R_n \geq \gamma \Sigma\, Q_{ni}$
Load and Resistance Factor Design (LRFD)	$\phi R_n \geq \Sigma\, \gamma_i\, Q_{ni}$

tor K for the column with the particular set of end conditions can be obtained by

$$K = \sqrt{\dfrac{P_e}{P_{cr}}} \qquad (12)$$

where P_e is the Euler load given by $P_e = \pi^2 EI/L^2$ in which L is the length of the column.

The effective length factor multiplied by the true length L of the column gives the effective length of the column which can be used for design. Table 3[19] gives the theoretical and recommended K values for columns with

Table 3. Theoretical and Recommended K Values for Idealized Columns

	(a)	(b)	(c)	(d)	(e)	(f)
Buckled shape of column is shown by dashed line						
Theoretical K value	0.5	0.7	1 0	1 0	2.0	2.0
Recommended design value when ideal conditions are approximated	0 65	0 80	1 2	1 0	2 10	2 0
End condition code	Rotation fixed and translation fixed / Rotation free and translation fixed / Rotation fixed and translation free / Rotation free and translation free					

various types of idealized end conditions. Since fully rigid supports are seldomly realized in real life, the recommended K values for cases with fixed support idealization are slightly higher than their theoretical values.

Numerical Analysis

It should be remembered that eigenvalue analysis can only be carried out for perfectly straight columns. For columns with initial crookedness, the stability or load-deflection approach must be used. In the load-deflection approach, the load-deflection behavior of the column is traced from the start of loading to collapse. The maximum load the column can carry is the peak point of the load-deflection curve. The analyses of non-sway columns with residual stresses, initial crookedness and small end restraint using the load-deflection approach have been reported by a number of researchers. The important results are summarized by the authors.[12] Some of the important findings are:

1. Comparing with pin-ended columns, the maximum load-carrying capacity of end-restrained columns increases as the degree of end restraint (as measured by the rotational stiffness of the connections connecting beams and columns) increases.
2. The increase in load-carrying capacity of end-restrained columns is more pronounced for slender columns when stability is the limit state than for short columns when yielding is the limit state.
3. The end-restraining effect on column strength is more noticeable for columns bent about their weak axes than for columns bent about their strong axes.
4. While residual stresses and initial crookedness have a destabilizing effect on columns strength, end restraint will provide a stabilizing effect which counteracts the detrimental effects of residual stresses and initial crookedness. However, the strengthening effect of end restraint is highly dependent on the slenderness of the column.

Practical Design of Initially Crooked Column with End Restraints

For design purposes, it is convenient to use the effective length factor approach in which the actual column with end restraints is converted to an equivalent pin-ended column by multiplying the actual unbraced length of the column by the effective length factor K, so the pin-ended column curves described in the preceding section can be utilized directly. The procedure to determine the effective length factor for initially crooked end-restrained columns with residual stresses is more involved than that of perfectly straight elastic columns with idealized end conditions. Equation 12 is not applicable anymore for the determination of the effective length factor K. Instead, a number of load-deflection curves, each corresponding to a specific slenderness ratio L/r (or slenderness parameter λ), are generated numerically. The peak points of these load-deflection curves are then plotted with the associated slenderness ratios (or slenderness parameters) to form a column curve (see Fig. 6). Each column curve is unique for a specific value of initial crookedness, a specific distribution of residual stress and a specific end restraint characteristic. To get the effective length factor, the end-restrained column curve (Fig. 6b) is compared with the corresponding pin-ended column curve and the K factor at any load level is given (Fig. 7),

$$K = \frac{\lambda_{ab}}{\lambda_{ac}} \tag{13}$$

where λ_{ab}, λ_{ac} are depicted in the figure.

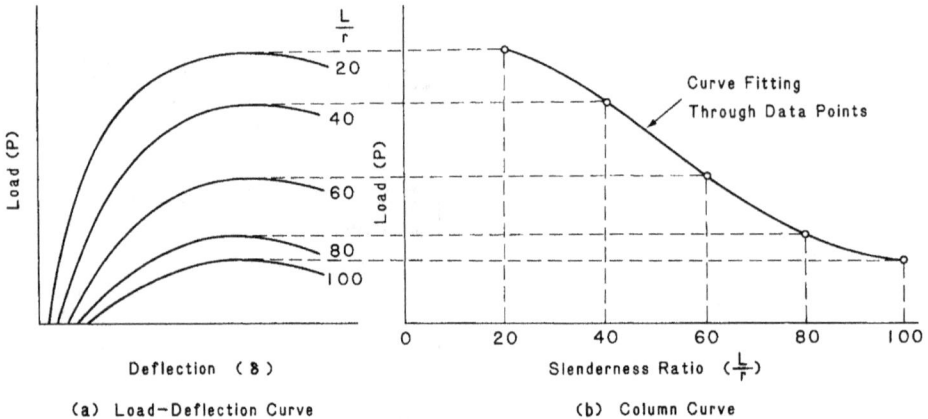

(a) Load-Deflection Curve   (b) Column Curve

Fig. 6. Determination of column-strength curve from load-deflection curves for an initially crooked end-restrained column

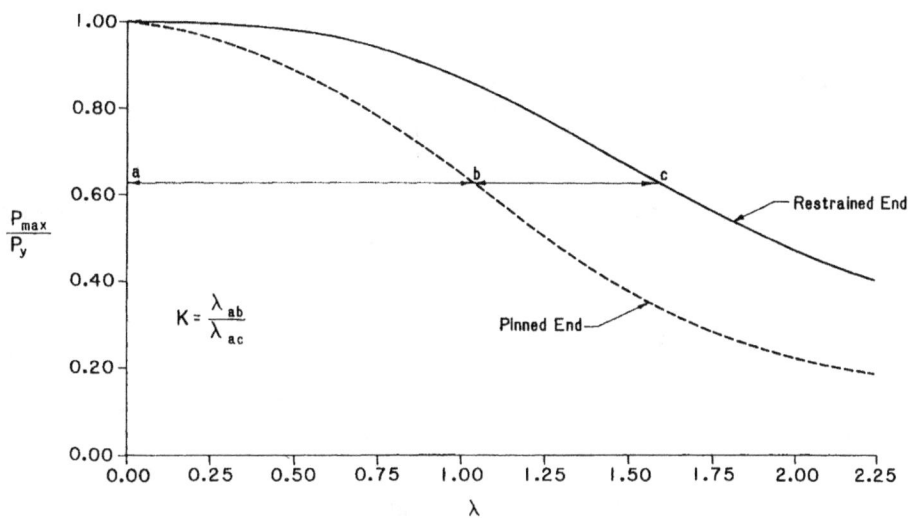

Fig. 7. Determination of effective length factor. K

Upon investigations of 83 end-restrained columns,[20] the values of K for each curve do not vary significantly over the load levels. Thus, a relationship between the K-factor and the magnitude of end restraint can be established. In particular, the expression

$$K = 1.0 - 0.017\bar{\alpha} \geq 0.6 \qquad (14)$$

where

$$\bar{\alpha} = \frac{2EI_g}{(M_p)_c L_g} \left[\frac{1}{1 + \dfrac{2EI_g}{L_g R_{ki}}} \right] \qquad (15)$$

in which
I_g = moment of inertia of the girder connected to the column
L_g = length of the girder
$(M_p)_c$ = plastic moment capacity of the column
R_{ki} = initial connection stiffness of the connection joining the beam to the column (Fig. 8)

was proposed[12] for non-sway columns with initial crookedness, residual stresses and small end restraints, taking into account the effect of beam flexibility. Procedures for the design of such columns have been reported in Refs. 12, 21 and discussed in Ref. 22.

At this point, it is interesting to compare the effective length factor K as described by Eq. 14 with the elastic effective length factor K_{el} determined by an eigenvalue

analysis assuming perfectly straight columns with end restraints provided by linear elastic rotational springs having spring stiffness R_{ki} at the ends. Such comparison is shown in Fig. 9. The dotted line is a plot of K versus $\bar{\alpha}$ whereas the solid lines are plots of K_{el} versus $\bar{\alpha}$. As can be seen, K_{el} gives a conservative estimate of column strength provided that λ is relatively low and $\bar{\alpha}$ is relatively high.

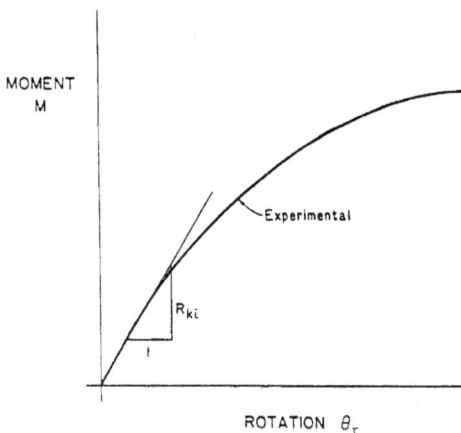

Fig. 8. Determination of R_{ki}

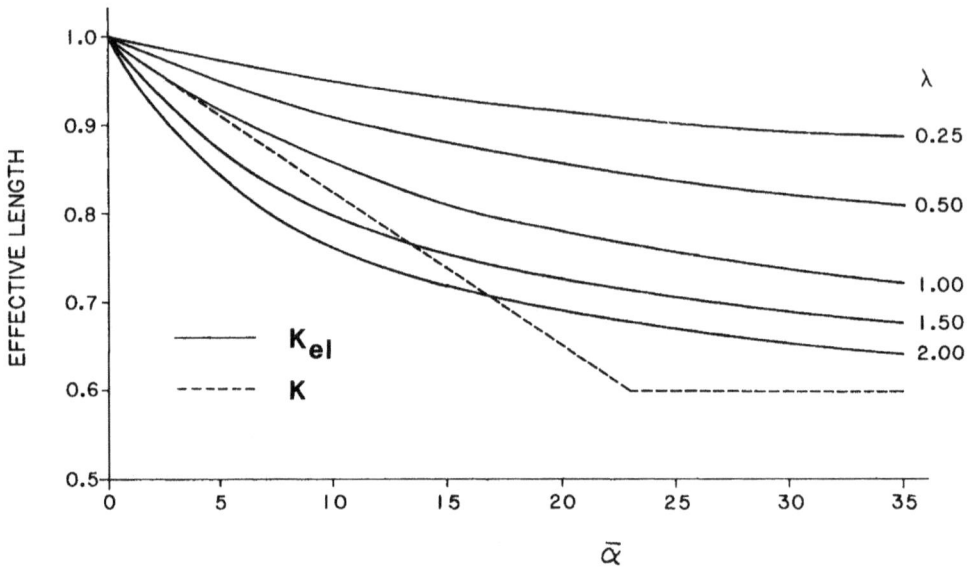

Fig. 9. *Comparison of K_{el} and K*

4. COLUMNS IN FRAMES

As mentioned earlier, columns in real structures usually exist as part of a frame. A column in a frame is usually subjected to the combined action of bending moments and axial thrust. As a result, part of the strength of the member is required to resist the bending moment and only the remaining part of the strength is available to resist the axial force. Thus, most columns in frames must be treated as beam-columns.

Columns in Braced Frames—B_1 Factor

A phenomenon associated with a beam-column is the secondary effect. When a braced member is subjected to both bending moments and axial force, the axial force acts through the deflection caused by the primary moments (moments arised from transverse loads and end moments acting on the member) to produce additional moment referred to as secondary or P-δ moment. Figure 10 shows schematically these two types of moments. The moment acting along the member is thus the algebraic sum of the primary and secondary moments. To obtain the exact value of this moment, a second-order analysis of the member is necessary. However, in lieu of such analysis, a simplified approach to obtain the total moment can be used.

Using the assumptions that

1. The deflection is small

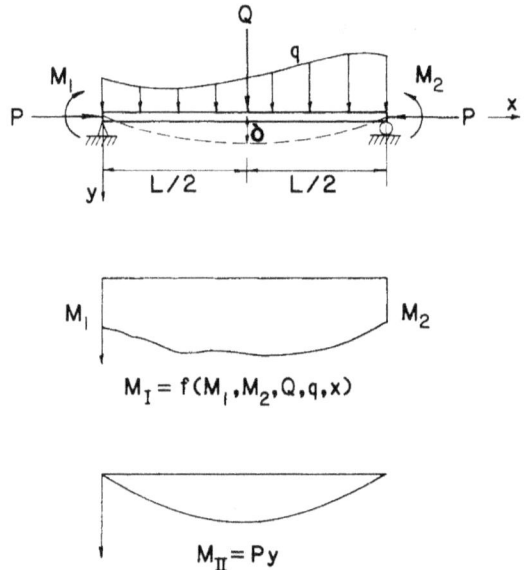

$$M_I = f(M_1, M_2, Q, q, x)$$

$$M_{II} = Py$$

Fig. 10. *P-δ effect*

2. The secondary moment M_{II} is in the form of a half sine wave
3. The maximum deflection δ occurs at midspan
4. The maximum primary moment M_{Imax} occurs at or near midspan

an approximate expression for the maximum moment can be derived.

Because of assumptions 1 to 3, we can relate the curvature caused by the secondary moment to the maximum deflection as

$$y''_{II} = -\frac{M_{II}}{EI} = -\frac{P\delta}{EI} \sin \frac{\pi x}{L} \qquad (16)$$

Integrating Eq. 16 twice and enforcing the boundary conditions $y(0) = 0$ and $y(L) = 0$ it can easily be shown that the secondary deflection (deflection caused by the P-δ effect) can be written as

$$y_{II} = \frac{P\delta}{EI} \left(\frac{L}{\pi}\right)^2 \sin \frac{\pi x}{L} \qquad (17)$$

from which the secondary deflection at midspan is

$$\delta_{II} = y_{II} \,|x = \frac{L}{2} = \delta \frac{P}{P_e} \qquad (18)$$

Since the total deflection at midspan is the sum of the primary and secondary deflections, i.e.

$$\delta = \delta_I + \delta_{II} \qquad (19)$$

we can eliminate δ_{II} by substituting Eq. 18 into Eq. 19. The result is

$$\delta = \frac{\delta_I}{\left(1 - \dfrac{P}{P_e}\right)} \qquad (20)$$

From assumption 4, we can write

$$M_{max} = M_{Imax} + P\delta \qquad (21)$$

If we substitute Eq. 20 into Eq. 21 and rearrange, we can write

$$M_{max} = \left[\frac{C_m}{1 - P/P_e}\right] M_{Imax} = B_1 M_{Imax} \qquad (22)$$

where

$$B_1 = \frac{C_m}{1 - P/P_e} \geq 1 \qquad (23)$$

$$C_m = 1 + \psi P/P_e$$

in which

$$\psi = \frac{\delta_I P_e}{M_{Imax}} - 1 \qquad (24)$$

Equation 22 shows the maximum moment in the member can be obtained by multiplying the maximum primary moment M_{Imax} by an amplification factor B_1 (the factor in parenthesis). Note this amplification factor must be greater than unity if it is of any importance. This is because if this factor is less than unity, then from Eq. 22 it is clear that $M_{Imax} > M_{max}$ and the designer will use M_{Imax} rather than M_{max} in proportioning the members. The condition that B_1 must be greater than unity is adopted in the present AISC/LRFD Specification which was not the case for the AISC/ASD Specification.

Figure 11 shows the value of ψ and C_m for several different load cases. It is important to point out that because of assumption 4, Eq. 24, which is derived from Eq. 21, is only applicable to the two simply supported cases (Cases 1 and 4). For the other cases in which the maximum primary moment M_{Imax} occurs at the end(s) (Cases 2, 3, 5) or occurs at midspan as well as at the ends (Case 6), the exact values of the maximum moments are first evaluated; the values for ψ are then obtained from calibration.[23]

Case	ψ	C_m
1 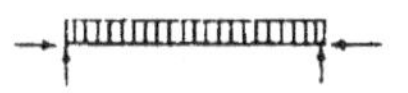	0	1.0
2 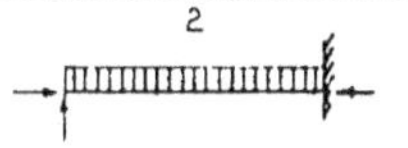	-0.4	1 -0.4 P/P$_e$
3 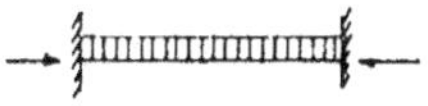	-0.4	1 -0.4 P/P$_e$
4 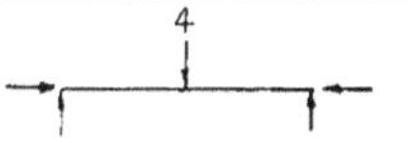	-0.2	1 -0.2 P/P$_e$
5 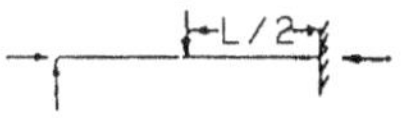	-0.3	1 -0.3 P/P$_e$
6 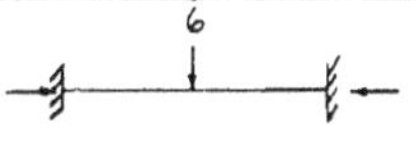	-0.2	1 -0.2 P/P$_e$

Fig. 11. Values of ψ for beam-columns under transverse loadings

116

For the cases in which end-restraint(s) is (are) present (Cases 2, 3, 5, 6), the value of C_m for usual P/P_e ratios is only slightly less than unity and a conservative value of 0.85 is thus suggested for C_m. For the two simply supported cases (Cases 1, 4), a value of 1 is suggested for C_m in the AISC/LRFD Specification.[13]

A special case arises when in-span transverse loads are absent in the member. For this case, the primary moment in the member is caused by end moments acting on the member ends. Since the maximum primary moment in members subjected to axial load and end moments seldom exists at midspan, C_m defined in Eq. 23 is invalid since it was developed based on assumption 4 postulating that M_{Imax} occurs at or near midspan. Instead, C_m is redefined as

$$C_m = 0.6 - 0.4\,(M_1/M_2) \geq 0.4 \qquad (25)$$

where M_1/M_2 is the ratio of the smaller to larger end moments of the unbraced length of the member and it is positive when the member is bent in reverse curvature and negative when the member is bent in single curvature.

Equation 25 was developed based on the equivalent moment concept. In the equivalent moment model, a pair of equal and opposite end moments are applied to the member which, when amplified by the amplification B_1, will give the same maximum moment as will the actual unequal end moments. It is obvious that the location of maximum moment will be distorted, but this is ignored for simplicity. C_m expressed in Eq. 25 was proposed by Austin[24] based on a more accurate expression derived by Massonnet.[25]

Columns in Unbraced Frames—B_2 Factor

The above discussion on moment amplification pertains to members in braced frames in which sidesway is prevented. For members in an unbraced frame, in addition to P-δ effect there is another effect known as the P-Δ effect. The P-Δ effect arises when the gravity loads of a frame act through the drift of the frame thus producing additional overturning moment and additional drift (Fig. 12). Since this is a destabilizing effect, it should be considered in design. Both the P-δ and P-Δ effects can be taken into account by using second-order analysis. The AISC/LRFD Specification[13] recommends the use of P-Δ moment amplification factor B_2 to account for the P-Δ effect in lieu of a second-order analysis.

Two expressions for B_2 are given in the specification

$$B_2 = \cfrac{1}{1 - \cfrac{\Sigma P_u \Delta_{oh}}{\Sigma HL}} \qquad (26)$$

and

$$B_2 = \cfrac{1}{1 - \cfrac{\Sigma P_u}{\Sigma P_e}} \qquad (27)$$

where

ΣP_u = design axial forces on all columns of a story, in kips

Δ_{oh} = translation deflection on the story under consideration based on a first-order analysis, in inches

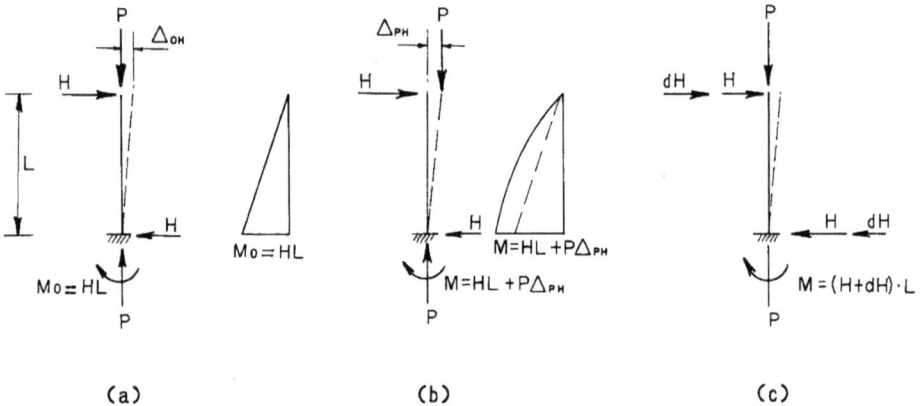

(a)   (b)   (c)

Fig. 12. Cantilever column

ΣH = sum of all story horizontal forces producing Δ_{oh}, in kips

L = story height, in inches

P_e = Euler load (Eq. 1)

Equation 26 was developed based on the story stiffness concept.[26,27,28] By assuming that

1. each story behaves independently of other stories, and
2. the additional moments in the columns caused by the P-Δ effect is equivalent to that caused by a lateral force of $\Sigma P_u \Delta/h$ where ΣP_u is the sum of all vertical forces on the story, Δ is the total frame drift including the P-Δ effect and h is the height of the story,

the sway stiffness of the story can be defined as:

$$S_F = \frac{\text{horizontal force}}{\text{lateral displacement}} \qquad (28)$$

$$= \frac{\Sigma H}{\Delta_{oh}} = \frac{\Sigma H + \Sigma P \Delta / h}{\Delta}$$

Solving the above equation for Δ gives

$$\Delta = \left(\frac{1}{1 - \dfrac{\Sigma P \Delta_{oh}}{\Sigma H h}} \right) \Delta_{oh} \qquad (29)$$

If rigid connections and elastic behavior are assumed, the magnified moment induced in the member as a result of sway M_{lt} will be proportional to the lateral deflections. Therefore, we can write

$$(M_{lt})_{max} = \left(\frac{1}{1 - \dfrac{\Sigma P \Delta_{oh}}{\Sigma H h}} \right) M_{lt} = B_2 M_{lt} \qquad (30)$$

where

M_{lt} = moment due to lateral translation determined from a first-order analysis

The alternative expression for the moment amplification factor B_2 is obtained as a direct extension of Eq. 20. Under the assumption that when sidesway instability is to occur in a story, all columns in that story will become unstable simultaneously, the P/P_e term in Eq. 20 is replaced by $\Sigma P_u/\Sigma P_e$ in which the summation is carried through all columns in a story.[29] Using the same argument as before that if elastic behavior and rigid connections are assumed, the story sway moment will be proportional to the lateral deflection. As a result, the maximum end moment accounting for the P-Δ effect can be written as

$$(M_{lt})_{max} = \left(\frac{1}{1 - \dfrac{\Sigma P_u}{\Sigma P_e}} \right) M_{lt} = B_2 M_{lt} \qquad (31)$$

The P-Δ moment amplification factor B_2 described above and recommended in the AISC/LRFD Specification represents an improvement over that recommended in the AISC/ASD Specification[19] in which the P-Δ moment amplification factor is expressed as $0.85/(1 - f_a/F'_e)$. The reason is that the B_2 factor in the LRFD Specification magnifies only the sway moment M_{lt}, whereas the moment amplification factor in the ASD Specification magnifies the total moment. If the bulk of the column moment does not produce sidesway, the approach recommended in the ASD Specification will be unduly conservative.

Column Design in LRFD for Type FR Construction

As mentioned earlier, for an unbraced frame, both the P-δ and P-Δ effects are important, so both effects have to be accounted for in design. The AISC/LRFD Specification recommends a superposition technique in which the P-δ (sometimes called the member instability) effect and the P-Δ (sometimes called the frame instability) effect are summed together algebraically to obtain the maximum design moment, i.e.

$$M_u = B_1 M_{nt} + B_2 M_{lt} \qquad (32)$$

where B_1 and B_2 are the P-δ and P-Δ moment amplification factors respectively and M_{nt} is the moment in the member assuming there is no lateral translation in the frame and M_{lt} is the moment in the member as a result of lateral translation of the frame.

In the actual design, M_{nt} is determined from a first-order analysis of the frame braced against lateral translation under the applied loads. M_{lt} is determined from a first-order analysis of the frame acted on by the reverse of the bracing forces (Fig. 13). It is important to note Eq. 32 is a conservative approach, since the maximum P-δ moment and the maximum P-Δ moment may not necessarily coincide at the same location. Furthermore, it should be remembered the expressions for B_1 and B_2 are only valid if the joints are rigid. In other words, Eq. 32 is only applicable to Type FR (fully restrained) construction in the LRFD Specification. Finally, it should also be mentioned that, depending on how the frame is braced against sway, the moments M_{nt} and M_{lt} calculated will be different for different arrangements of fictitious supports. However, for regular rectangular frames used in most building construction, the difference is insignificant for design purpose.

The validity of Eq. 32 has been checked by comparing the maximum moment calculated using Eq. 32 with the exact maximum moment calculated using a second-order elastic analysis.[30] It is concluded that for rectangular frames in which the P-Δ effect is not too significant, good correlation between the two calculated moments is observed.

Furthermore, if the P-Δ effect is not too significant, the two different expressions for B_2 (Eqs. 26 and 27) give comparable results provided that all the beams and columns in the story are rigidly connected.

If the P-Δ effect is significant, viz, if $B_2 > 1.5$, the approach suggested by LeMessurier[31] provides more accurate results. In his approach, the moment magnification factor is expressed as

$$B_2 = \cfrac{1}{1 - \cfrac{\Sigma P_u}{\Sigma P_L - \Sigma C_L P}} \qquad (33)$$

where

$$\Sigma P_L = \frac{\Sigma H h}{\Delta_{oh}} \qquad (34)$$

C_L is a factor accounting for the decrease in stiffness of the column due to the presence of the axial force.

Note that if C_L is insignificant, Eq. 33 reduces to Eq. 26.

5. CONNECTION RESTRAINT CHARACTERISTICS

The analyses and design of Type FR (fully restrained) and Type PR (partially restrained) frames differ in that for Type PR construction, the effect of connection flexibility must be taken into account. Since a connection is a highly statical indeterminate element, a rigorous analytical study of its behavior is quite a formidable task. In view of this, a special Task Group (TG25) of the Structural Stability Research Council was set up to investigate theoretically and experimentally connection behavior.

The behavior of a connection is best described by its moment-rotation relationship. Since most connection moment-rotational relationships are nonlinear almost from the start of loading, the analysis of structures including the effect of connection flexibility is an inherent nonlinear problem. To simplify the analysis technique, a number of simplified models have been proposed.

Connection Modeling

Figure 14 shows two simple linear models. The first model[32] uses the initial stiffness R_{ki} of the connection to represent the behavior of the connection for the entire range of loading. As can be seen, the validity of this

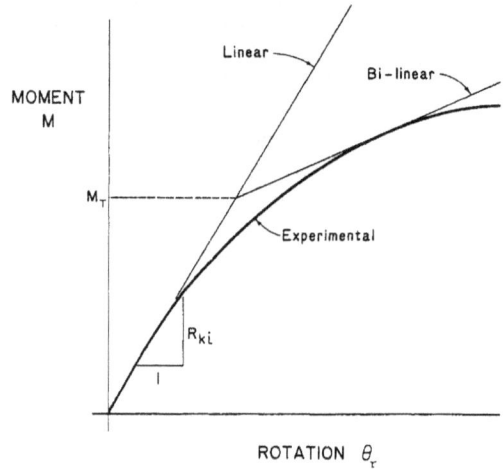

Fig. 14.   *Linear M-θ_r models*

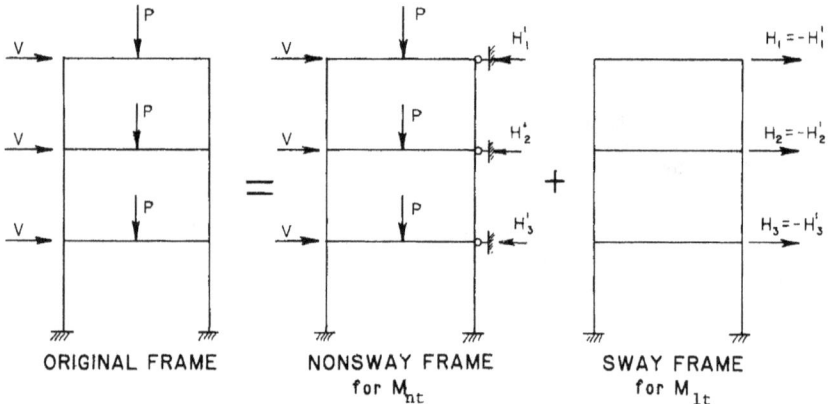

Fig. 13.   *Two fictional frames for M_{nt} and M_{lt}*

linear model deteriorates as the moment increases. To get a better representation of the connection stiffness, a bilinear model[33] was used. In the bilinear model, the initial slope of the moment-rotational line was replaced by a shallower line at a certain transition moment M_T. A direct extension of the bilinear model is the piecewise linear model[34] in which the nonlinear M-θ_r curve of the connection is represented by a series of straight line segments. Although the linear, bilinear or piecewise linear models are easy to implement, the inaccuracies and sudden jump in stiffness which are inherent in these models make them undesirable to be used in a limit state analysis routine.

To this end, Frye and Morris[35] proposed a polynomial model in which a polynomial is used to represent the connection M-θ_r behavior (Fig. 15). However, there is a major drawback in this model. Since the nature of a polynomial is to peak and trough within a certain range, the stiffness of the connection (as represented by the first derivative of the polynomial) may be negative, which is physically unjustifiable. To overcome this, Jones et al[36] uses a cubic B-spline curve fitting technique to improve the polynomial model (Fig. 15). In the cubic B-spline

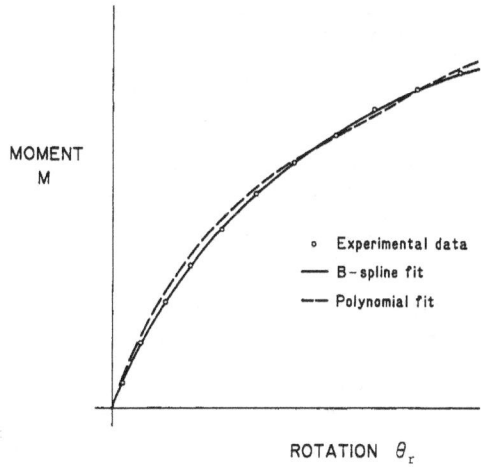

Fig. 15.   *B-spline and polynomial curve fit models*

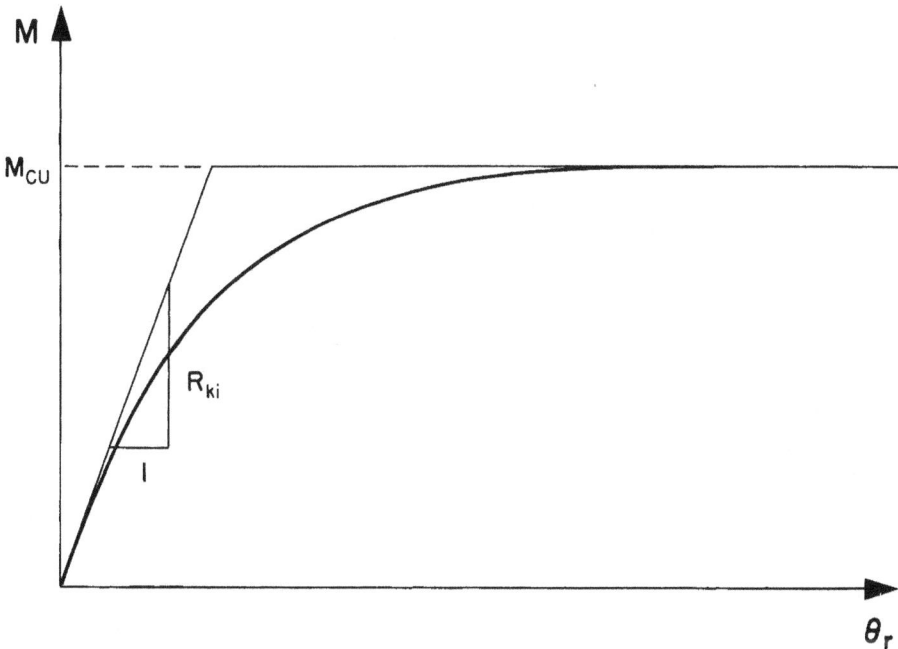

Fig. 16.   *Connection moment-rotation idealization used in the power model*

model, a cubic polynomial is used to curve-fit segments of a curve. Continuity between the first and second derivatives of each segment of curve are enforced. Although the cubic B-spline model gives a good representation of the connection behavior and circumvents the problem of negative stiffness, a large number of data are necessary for the curve-fitting process. To overcome this, the power model proposed by Colson[37] and the exponential model proposed by Lui[38] can be used.

In the power model,[37] a power function is used to represent the connection M-θ_r behavior. It has the form

$$\theta_r = \frac{|M|}{R_{ki}} \frac{1}{1 - \left|\dfrac{M}{M_{cu}}\right|^a} \qquad (35)$$

where (refer to Fig. 16)

R_{ki} = initial connection stiffness
M_{cu} = ultimate moment capacity of the connection
a = a parameter to account for the curvature of the M-θ relationships

In the exponential model,[38] the connection M-θ_r behavior is represented by an exponential function of the form

$$M = \sum_{j=1}^{n} C_j \left(1 - e^{-|\theta_r|/2j\,\alpha}\right) + M_o + R_{kf}|\theta_r| \qquad (36)$$

where

M_o = initial moment
R_{kf} = final or strain-hardening connection stiffness
α = scaling factor
C_j = connection model parameters

The connection model parameters are merely curve-fitting constants which can be obtained by using an optimization technique.

To demonstrate the validity of the exponential model, two experimentally obtained moment-rotation curves are curve-fitted with Eq. 36 using four curve-fitting constants and 10 sets of data from each curve. The results are shown in Figs. 17 and 18 respectively. The connection used in Fig. 17 was a double web angle connection tested by Lewitt, Chesson and Munse.[39] The connection used in Fig. 18 was a T-stub connection tested by Rathbun.[32] As can be seen, the exponential model gives an excellent representation of the test curves.

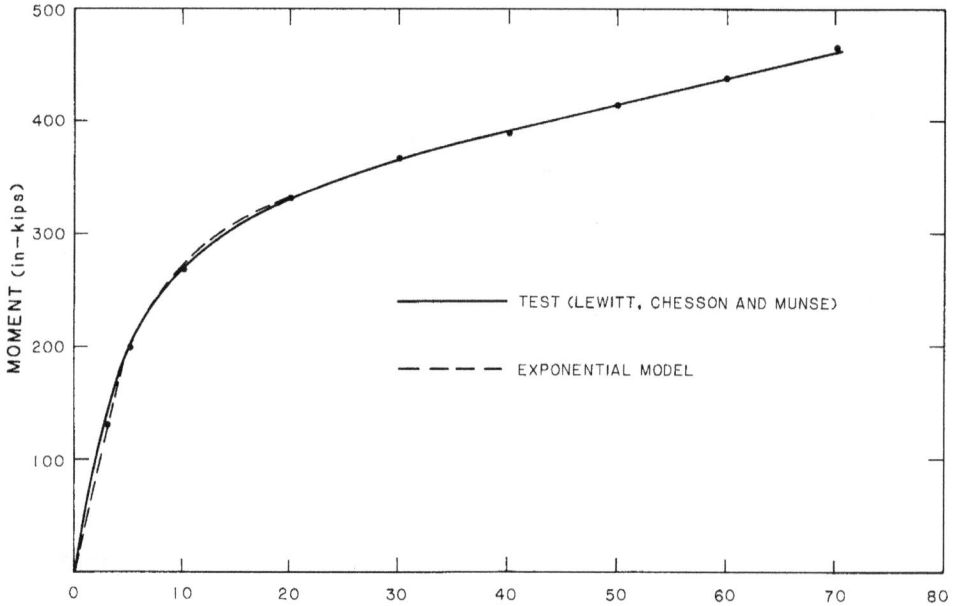

Fig. 17. *Comparison of exponential connection model with test by Lewitt, Chesson and Munse*

Research Need

As mentioned earlier, because of the complex geometries and stress distributions of most connections, most $M\text{-}\theta_r$ curves are nonlinear and thus almost all existing $M\text{-}\theta_r$ curves available today are obtained from experiments. Since most of these experiments were performed on connections which have become obsolete, it is essential that additional analytical and experimental investigations on commonly used connections be conducted in view of the advancement made on the limit state approach to analysis and design of steel structures.

6. BEHAVIOR OF COLUMNS WITH RIGID AND SEMI-RIGID CONNECTIONS

For columns in frames, another important phenomenon which the engineers should be aware of is the moment transfer mechanism between the beams and columns. One commonly posed question is: how can a beam restrain a column if at the same time it is inducing moment to the column? Whether a beam restrains or induces moment to the column depends on a number of factors. Some of the important ones are (1) the rigidity of the connections, (2) the relative stiffness of the beams and columns and (3) the load patterns and load sequences on the frame.

Moment Transfer—Rigid Connection

To study the moment transfer mechanism between the beam and the column, it is advantageous to look at the behavior of some simple subassemblages. Figure 19[40] shows a T-shaped subassemblage consisting of two beams and a column rigidly connected to one another. A concentrated load Q equal to half the yield load of the beam is applied to the midspan of each beam, an axial load P is then applied to an imperfect column with the influence of residual stress as well as with an initial out-of-straightness of $0.001L$. The moment distributions of the joint are plotted as P increases. By assuming that the beams behave elastically for the entire range of loading and the joint is rigidly connected, the moment at the joint is considered to consist of three parts:

1. Bending moment M_1 due to lateral load Q with joint fixed.
2. Bending moment M_2 due to joint translation as column buckles.
3. Bending moment M_3 due to joint rotation.

These three components of bending moments are shown schematically in Fig. 20.

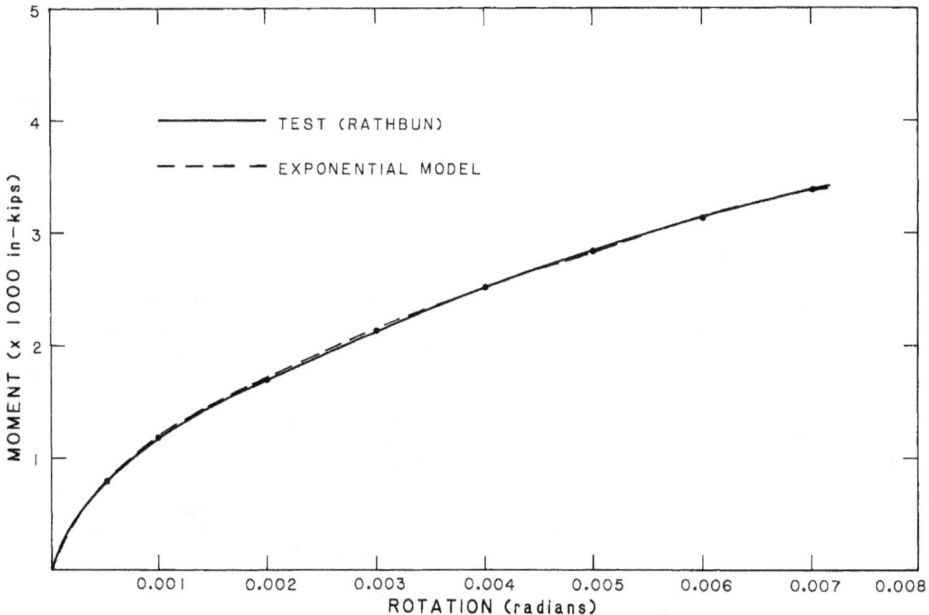

Fig. 18. Comparison of exponential connection model with test by Rathbun

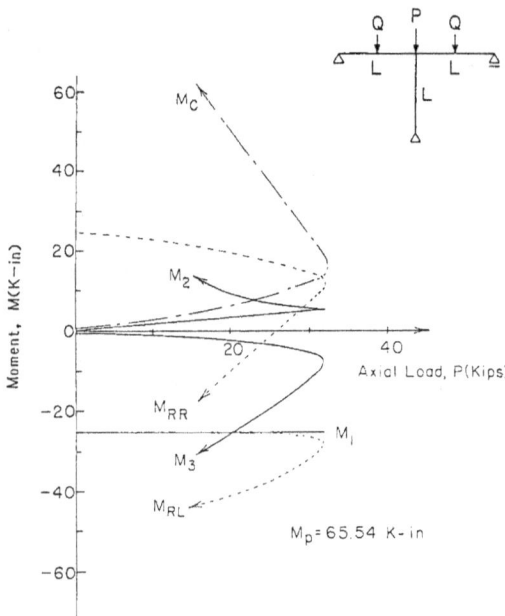

Fig. 19. Moment distributions at joint of subassemblage

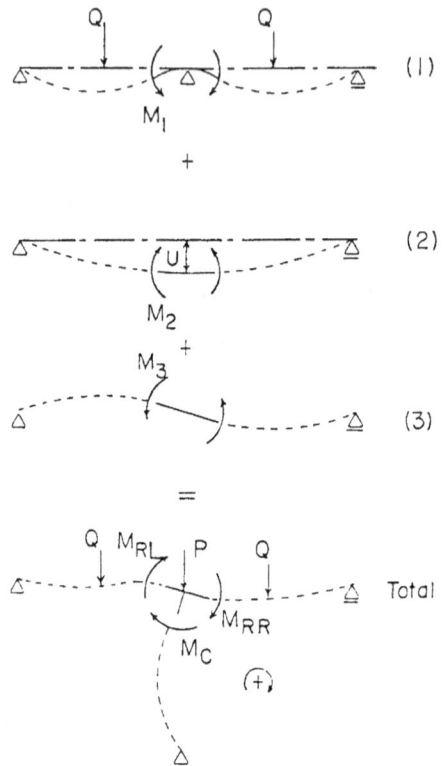

Fig. 20. Components of bending moments at joint of subassemblage

The bending moments in the left beam M_{RL} and in the right beam M_{RR} can be expressed as follows:

$$M_{RL} = -M_1 + M_2 - M_3 \qquad (37)$$

$$M_{RR} = M_1 - M_2 - M_3 \qquad (38)$$

The bending moment in the column M_C can be obtained by considering joint equilibrium.

$$M_C = -(M_{RL} + M_{RR}) = 2M_3 \qquad (39)$$

The variation of these bending moments with the axial load P is plotted in Fig. 19. It can be seen from the figure that the moment due to the buckling of the column is not negligible. Not only does it reduce the moment of the left beam, but, together with the moment arising from joint rotation, restrains the column during the final stage of loading. The moment of the right beam M_{RR}, at first inducing moment to the column, decreases gradually and at $P = 26$ kips reverses sign and becomes a restraining moment to the column. On the other hand, the moment of the left beam M_{RL} is always negative and thus always restrains the column.

Moment Transfer—Flexible Connection

If the connections are not rigid, the moment transfer mechanism between the beam and the column are more complicated because of the loading/unloading characteristic of the connections. To demonstrate this characteristic schematically, the readers are referred to Fig. 21. For this subassemblage (Fig. 21a), the beams are connected to the column by semi-rigid connections. Beam loads w_L, w_R are first applied to simulate the dead load of the structure. Figure 21b shows the directions of moments acting on the left- and right-hand side of the joint of the subassemblage. The corresponding M-θ_r curves for the left and right connections are also shown. The left connection will follow curve OA' and the right connections will follow curve OA''. The moment acting on the column will be M_{1L} on the left side of the joint and M_{1R} on the right side of the joint.

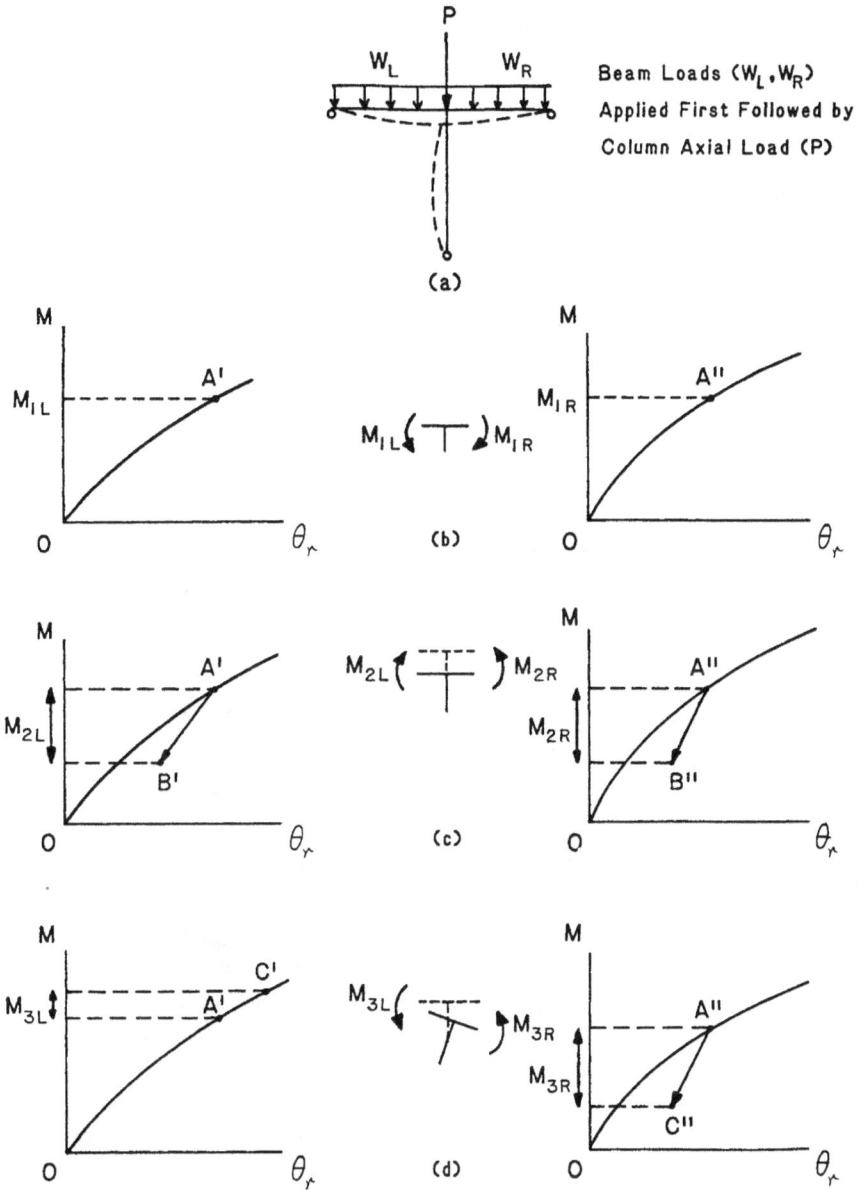

Fig. 21. Schematic representation of moment transfer mechanism of a flexibly connected subassemblage

Now, a column axial load P is applied to the subassemblage to simulate the live load. Under the action of P, the column will shorten and bend as shown by the dashed line in Fig. 21a. The moment induced due to shortening of the column is shown in Fig. 21c. Note that the directions of moments on both sides of the column are opposite to that of Fig. 21b. Therefore, unloading of the connections will result. As a result, the M-θ_r curve of the left connection will follow path $A'B'$ and that of the right connection will follow path $A''B''$. The slopes of $A'B'$ and $A''B''$ are parallel to the initial slopes to the corresponding M-θ_r curves.

In addition to column shortening, there is bending deformation in the column. As a result of bending, the joint will rotate. If rotation is in the direction as shown in Fig. 21a, the direction of moment induced will be that as shown in Fig. 21d. The induced moment to the left of the joint has the same direction as that of Fig. 21b but the direction of the induced moment to the right of the joint has opposite direction to that of Fig. 21b. In other words, the connection to the left of the columns will load while the connection to the right of the column will unload as a result of joint rotation.

Since the two column deformations, shortening and bending, occur simultaneously as P applies, the phenomenon depicted in Figs. 21c and 21d are concurrent events. Consequently, the connection on the left-hand side of the column may follow path $A'B'$ or $A'C'$ (i.e. unload or load) depending on whether M_{2L} is greater than or smaller than M_{3L}. On the other hand, the connection on the right-hand side of the column will always unload and so it will always exhibit a restraining effect to the column.

To study the behavior of flexibly connected frames, recourse to numerical methods is inevitable because of the inherent nonlinear nature of the problem. To give the reader an insight into the restraint characteristic between members of flexibly-connected frames, the behavior of the following subassemblage will be discussed.

To study the behavior of flexibly connected frames, the subassemblages shown in Figs. 22 and 23 are analyzed with the two load sequences applied as shown. The difference between the subassemblage of Fig. 22 and Fig. 23 is that rigid connections are assumed in the subassemblage of Fig. 22 and flexible connections with a moment-rotation behavior of Fig. 24 are used in the subassemblage of Fig. 23. The loadings for the two subassemblages are identical. The two beams are first loaded at midspan with a 5-kip concentrated load in load sequence 1. The column is also loaded with a 5-kip concentrated load in load sequence 1. The column loads (vertical at the joint and horizontal at quarter-points of the columns) are then applied monotonically in load sequence 2 until a plastic hinge formed in the column.

The distribution of joint moments for the rigidly connected and flexibly connected subassemblages are shown

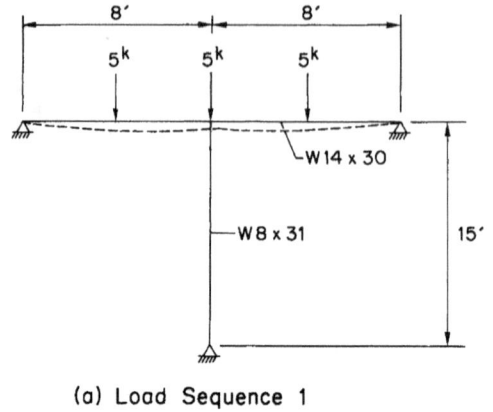

(a) Load Sequence 1

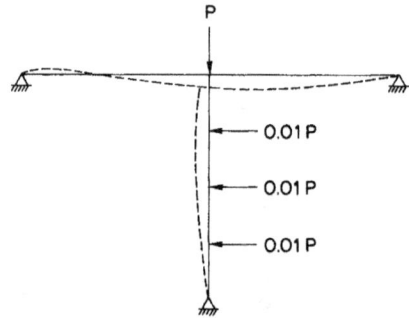

(b) Load Sequence 2

Fig. 22. Rigidly connected T-shaped subassemblage

in Figs. 25a and b respectively. The following observations can be made from the plots:

1. The column is restrained against buckling even though the beams are preloaded. For the rigidly connected subassemblage, restraint is offered by the left beam until at $P = 38$ kips the right beam starts to provide the restraint. At $P = 54$ kips, restraint is offered solely by the right beam. For the flexibly connected subassemblage both beams provide the restraint to the column.

2. The restraining effect is more pronounced for rigid connections than for flexible connections.

The difference in moment distribution around the joint

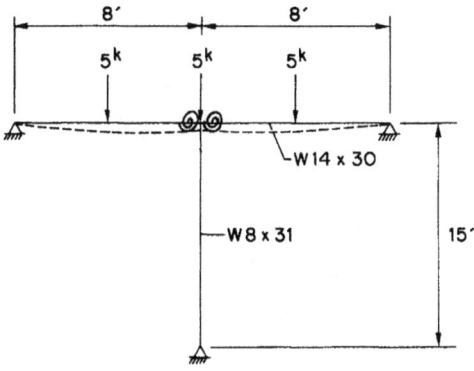

(a) Load Sequence 1

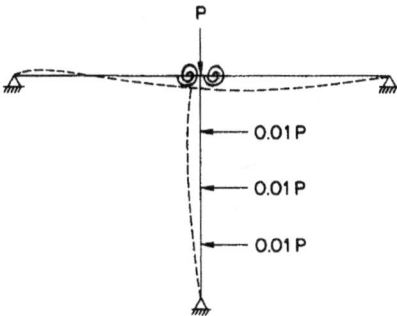

(b) Load Sequence 2

Fig. 23. Flexibly connected T-shaped subassemblage

Fig. 24. Connection moment-rotation behavior used for the T-shaped subassemblage

is apparent in Fig. 25. Of particular interest is the direction of M_{BL}. For the flexibly connected subassemblage, M_{BL} is always negative whereas for the rigidly connected subassemblage M_{BL} is only negative at low values of P but becomes positive at high values of P. The reason for this can be explained by reference to Fig. 26 in which the beam end moments at the joint are decomposed. At the end of load Sequence 1, M_{BL} is negative (i.e. counterclockwise, Fig. 26a). However, as load sequence 2 commences, the induced moment M_{BL} may be positive (i.e. clockwise, Fig. 26b) as a result of joint translation or negative (i.e. counterclockwise, Fig. 26c) as a result of joint rotation. Whether the final value of M_{BL} is positive or negative depends on whether joint translation or joint rotation dominates.

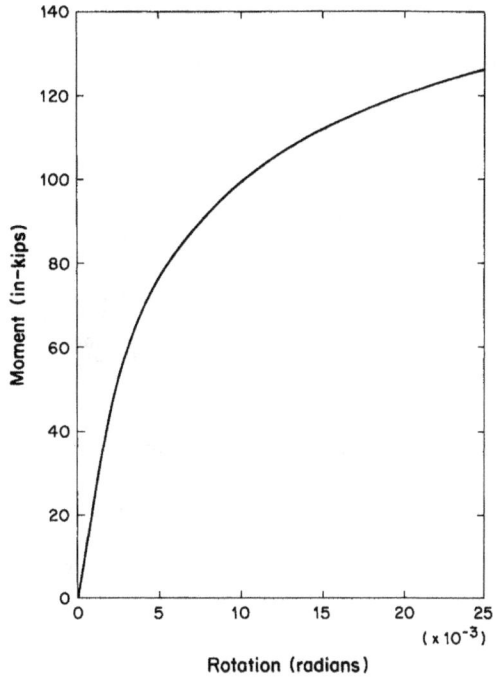

In Fig. 27, the magnitude of joint translation and joint rotation as a function of the applied force P are plotted. Although the magnitude of joint translation for both the rigidly connected and flexibly connected subassemblages are comparable, the joint rotation of the flexibly connected subassemblage is significantly larger than that of the rigidly connected subassemblage. As a result, the moment induced as a result of joint rotation will outweigh that of joint translation, hence the final value of M_{BL} for the flexibly connected frame is negative.

As for the right beam, regardless of whether joint translation or joint rotation dominates, the induced M_{BR} is almost always negative. As a result, this beam, except at the initial loading stage for the rigidly connected subassemblage, will always provide restraint to the column regardless of whether the connection is rigid or flexible. It should be mentioned that unloading occurs at the connection which connects the right beam to the column as load sequence 2 commences because the direction of moment at this location is opposite for load sequence 1

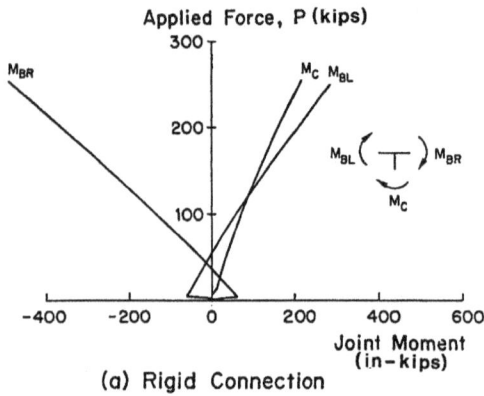

(a) Rigid Connection

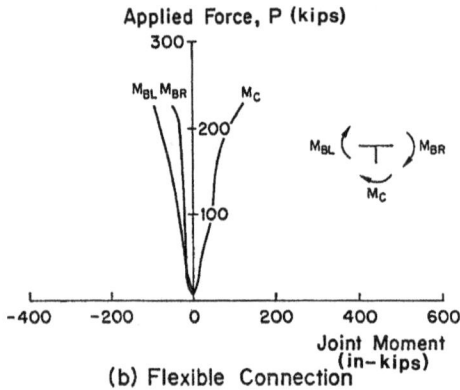

(b) Flexible Connection

Fig. 25. *Applied force vs. joint moment relationships*

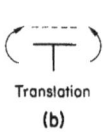

(a)

Joint moment at the end of load sequence 1

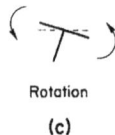

(b) Translation   (c) Rotation

Joint moment induced as a result of load sequence 2

Fig. 26. *Decomposition of joint moments*

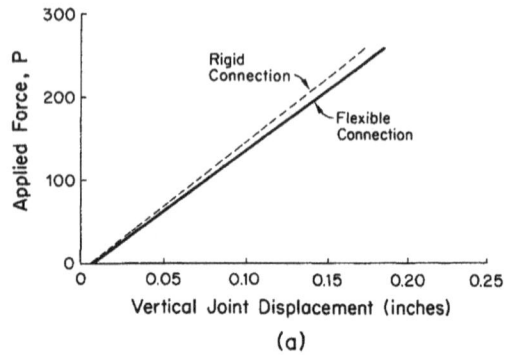

(a)

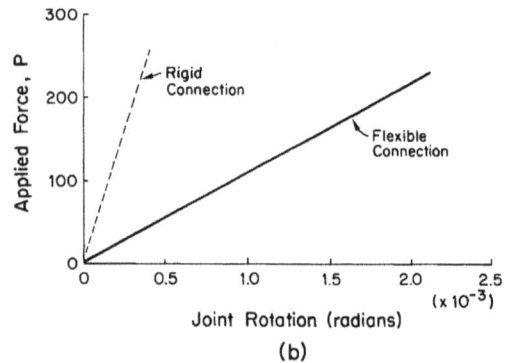

(b)

Fig. 27. *Joint displacement and rotation of the T-shaped subassemblage*

and load sequence 2 (see Fig. 26). Consequently, the left connection is offering tangent stiffness restraint to the column whereas the right connection is offering initial stiffness restraint.

A more detailed analysis and discussion of the behavior of subassemblages with flexible connections are given elsewhere.[38]

7. DESIGN OF COLUMNS WITH SEMI-RIGID CONNECTIONS

For design purposes, if the connections are rigid, one can just perform a first-order analysis on a trial frame. With the end moments and axial force known for each member, an interaction formula (to be discussed in the next section) can be used to check the trial sizes of the members. However, if the connections are not rigid, then care must be exercised in using the interaction equations in the proposed LRFD Specifications,[13] because the maximum moment in the member as determined by Eq. 32 will not be valid anymore, since the moment amplification factors B_1 and B_2 are only defined for rigidly connected frames.

One plausible solution is to use computer-aided analysis and design in which a second-order analysis is performed on the flexibly connected frame to determine the maximum moment including the P-δ and P-Δ effects directly. However, in lieu of such analysis, simplified design methods based on idealized connection behavior have been proposed.[21,38,41,42]

Additional Studies

Additional studies on the role of connections in effecting the strength and stiffness of frames have been reported by Ackroyd,[43] Moncarz and Gerstle[44] and Simitses et al.[45,46,47] In Ref. 43, it was shown an increase in connection stiffness does not always result in an increase in frame strength. For long-span frames only a few stories high where the lateral loads effects are small compared with gravity loads, an increase in connection stiffness may cause a decrease in frame strength. A parameter was defined which can be used as an index to determine whether providing overstiff connections will be beneficial or detrimental.

In addition to connection flexibility, another important factor that affects the limit state behavior of a frame is panel zone deformation. The study of panel zone deformation on frame behavior has been reported by Fielding et al,[48,49] Becker,[50] Kato[51] and Krawinkler.[52] In these studies, attention was given to the modeling of the shear deformation of the panel zone. The important results of these studies were summarized by the authors.[53] Recently, a finite element model of the panel zone which can represent all modes (extension, shear, bending) of deformation of the panel zone has been reported.[38] Generally speaking, the strength and stiffness of frames will be reduced if the effect of panel zone deformation is taken into account in the analysis procedures.

It should be mentioned that since both connection flexibility and panel zone deformation have detrimental effects on frame strength and stiffness, particular attention must be given by the designers to ensure that the strength of the frame is adequate and that the stability and drift will not be a problem for Type PR frames.

8. BEAM-COLUMN INTERACTION FORMULAS

With the end moments and axial force of a member known, the member can be proportioned so that it can resist these applied forces without premature failure. For design purposes, the proportion of the member is facilitated by the use of interaction formulas. The general form of an interaction formula is

$$f\left(\frac{P_u}{P_n}, \frac{M_{ux}}{M_{nx}}, \frac{M_{uy}}{M_{ny}}\right) \leq 1.0 \tag{40}$$

where P_u, M_{ux}, M_{uy} are the design axial load and bending moments (required strength) about the principal axes respectively and P_n, M_{nx} and M_{ny} are the corresponding ultimate axial force and moment capacities of the section. Interaction equations can be linear or nonlinear. A linear interaction equation is an equation in which the terms P_u/P_n, M_{ux}/M_{nx}, M_{uy}/M_{ny} are combined together linearly. A nonlinear interaction equation is one in which these terms are combined together nonlinearly.

Linear Interaction Equations

The AISC/LRFD Specification, based on the exact inelastic solution of 82 beam-columns,[54] proposed the following interaction equations for sway and nonsway beam-columns

$$\text{for } \frac{P_u}{\phi_c P_n} \geq 0.2$$

$$\frac{P_u}{\phi_c P_n} + \frac{8}{9}\left(\frac{M_{ux}}{\phi_b M_{nx}} + \frac{M_{uy}}{\phi_b M_{ny}}\right) \leq 1.0 \tag{41}$$

$$\text{for } \frac{P_u}{\phi_c P_n} < 0.2$$

$$\frac{P_u}{2\phi_c P_n} + \frac{M_{ux}}{\phi_b M_{nx}} + \frac{M_{uy}}{\phi_b M_{ny}} \leq 1.0 \tag{42}$$

The above interaction formulas are a simplification and clarification of interaction Formulas 1.6.1a and 1.6.1b used in the present AISC/ASD Specification.[19] Equation 1.6.1a is a check for stability and Eq. 1.6.1b is a check for strength. Since the amplification factor B_1 can be less than unity in the AISC/ASD Specification, both equations are required to be checked in the design. However, in the AISC/LRFD Specification, since the B_1 factor must be greater than unity, only one equation needs to be checked. The applicable equation is governed by the value $P_u/\phi_c P_n$, where P_u is the axial force in the member, P_n is the axial strength capacity of the member and ϕ_c is the the column resistance factor and has a value of 0.85.

M_{ux} and M_{uy} are the maximum moment (including the P-δ and P-Δ effects) in the member which may be determined from a second-order elastic analysis. However, in lieu of such analysis, their values may be determined from the simplified approach described in the section on Columns in Frames.

Nonlinear Interaction Equations

Equations 41 and 42 are applicable to members in both braced and unbraced frames. For members in braced frames, the new specification also recommended a nonlinear interaction equation based on the work in Ref. 55 for the design of I and wide-flange shapes. The equation has the form

$$\left(\frac{M_{ux}}{\phi_b M'_{px}}\right)^{\xi} + \left(\frac{M_{uy}}{\phi_b M'_{py}}\right)^{\xi} \leq 1.0 \qquad (43)$$

$$\left(\frac{C_{mx} M_{ux}}{\phi_b M'_{nx}}\right)^{\eta} + \left(\frac{C_{my} M_{uy}}{\phi_b M'_{ny}}\right) \leq 1.0 \qquad (44)$$

where

$$\xi = 1.6 - \frac{P_u P_y}{2[\ln(P_u P_y)]} \quad \text{for } 0.5 \leq b_f/d \leq 1.0 \qquad (45)$$

$$\eta = \begin{cases} 0.4 + \dfrac{P_u}{P_y} + \dfrac{b_f}{d} \geq 1.0 & \text{for } b_f/d \geq 0.3 \\[2ex] 1 & \text{for } b_f/d < 0.3 \end{cases} \qquad (46)$$

b_f = flange width, in.
d = member depth, in.

$$M'_{px} = 1.2 M_{px} [1 - (P_u/P_y)] \leq M_{px} \qquad (47)$$

$$M'_{py} = 1.2 M_{py} [1 - (P_u/P_y)] \leq M_{py} \qquad (48)$$

$$M'_{nx} = M_{nx} [1 - (P_u/\phi_c P_n)][1 - (P_u/P_{ex})] \qquad (49)$$

$$M'_{ny} = M_{ny} [1 - (P_u/\phi_c P_n)][1 - (P_u/P_{ey})] \qquad (50)$$

Design Evaluation

At this point, it is of interest to compare design carried out using the ASD format and the new LRFD format. A comprehensive comparative study was reported by Zhou and Chen.[56] In the study, a number of nonsway beam-columns under a specific set of force conditions were designed using both the ASD and LRFD formats. For the LRFD format, both the linear (LRFD-Linear) and nonlinear (LRFD-Non) interaction equations were used. The result of the comparison was expressed by the weight ratios of the sections chosen using the various design formats and interaction equations.

Figure 28 shows one such comparison. It can be seen that for single curvature bending ($C_{mx} = C_{my} = 1$), the LRFD approach is generally more liberal than the ASD approach. Figure 29 shows the distributions of the weight ratios for the beam-columns designed using live load to dead load (L_n/D_n) ratios of 1, 3 and 5 respectively. A constant roof live load to dead load (L_r/D_n) ratio of 0.2 was used for all three cases. From the figure, it can be seen that

1. The smaller the L_n/D_n ratio is, the more liberal is the LRFD approach as compared to the ASD approach.
2. The LRFD nonlinear equations will give the lightest section (except in cases when b_f/d ratio is small and the ratio of moment to axial force is large when the

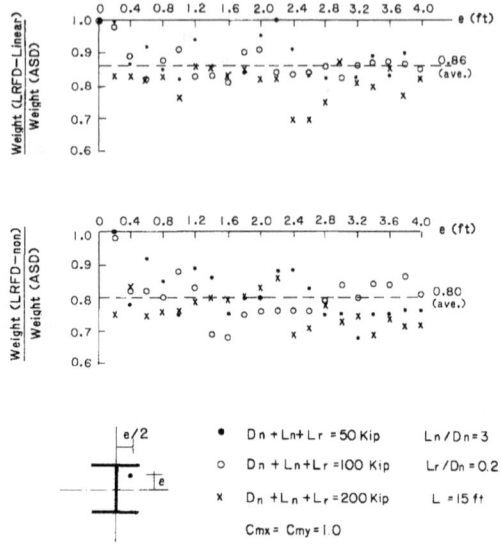

Fig. 28.   Weight ratio

LRFD linear equations give a more economical section).

The readers are referred to Ref. 56 for a more thorough discussion of the comparison.

Nonlinear Interaction Equations for Box Columns

Before leaving the subject, it is necessary to discuss recent development of the interaction equations for rectangular box beam-columns under biaxial loading. Based on a study by Zhou and Chen,[57] the following interactions are proposed as an extension of the expressions proposed previously by Chen and McGraw[58] for welded square box columns under biaxial loading to the present more general case of rectangular box cross section.

For short members

$$\left(\frac{M_{ux}}{M'_{px}}\right)^{\xi} + \left(\frac{M_{uy}}{M'_{py}}\right)^{\xi} \leq 1.0 \qquad (51)$$

where

$$\xi = 1.7 - \frac{P_u/P_y}{\ln (P_u/P_y)} \qquad (52)$$

$$M'_{px} = 1.2 (1 - P_u/P_y) M_{px} \leq M_{px} \qquad (53)$$

$$M'_{py} = 1.2 (1 - P_u/P_y) M_{py} \leq M_{py} \qquad (54)$$

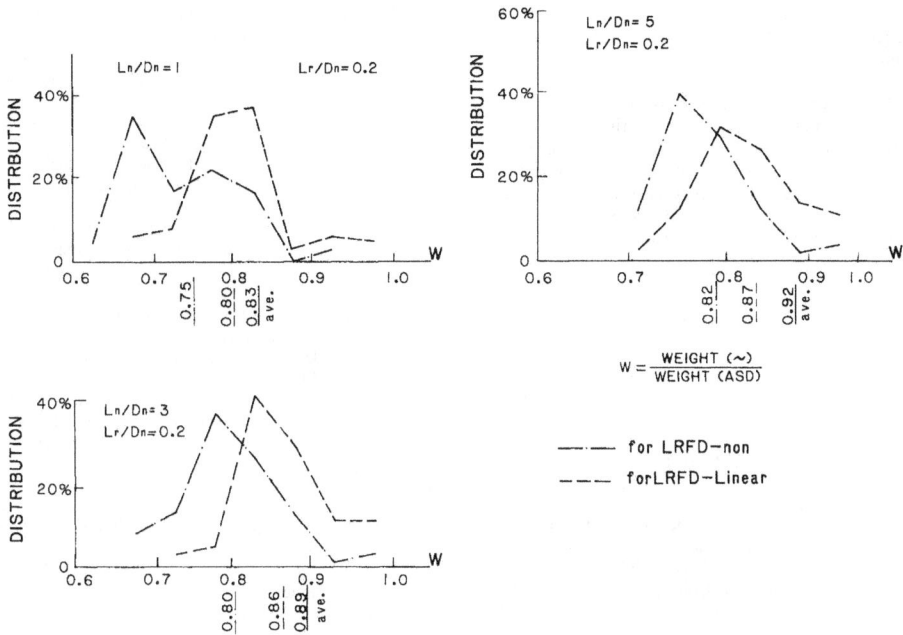

$$W = \frac{\text{WEIGHT } (\sim)}{\text{WEIGHT (ASD)}}$$

—··— for LRFD-non

— — — for LRFD-Linear

Fig. 29. Distribution of weight ratio

For long members

$$\left(\frac{M_{ux}}{M'_{nx}}\right)^{\alpha} + \left(\frac{M_{uy}}{M'_{ny}}\right)^{\alpha} \leq 1.0 \qquad (55)$$

where

$$\alpha = 1.7 - \frac{P_u/P_y}{\ln(P_u/P_y)} - a\frac{KL}{r}(P_u/P_y)^b \geq 1.1$$

$$M'_{nx} = M_{nx}(1 - P_u/P_n)\left[1 - \frac{P_u}{P_{ex}}\frac{1.25}{(B/H)^{1/3}}\right] \qquad (56)$$

$$M'_{ny} = M_{py}(1 - P_u/P_n)\left[1 - \frac{P_u}{P_{ex}}\frac{1.25}{(B/H)^{1/2}}\right]$$

in which

B = width and H = depth of the cross section

$$a = \begin{cases} 0.06 \text{ if } P_u/P_y \leq 0.4 \\ 0.15 \text{ if } P_u/P_y > 0.4 \end{cases} \qquad (57)$$

$$b = \begin{cases} 1.0 \text{ if } P_u/P_y \leq 0.4 \\ 2.0 \text{ if } P_u/P_y > 0.4 \end{cases} \qquad (58)$$

These nonlinear interaction formulas have been shown[57] to compare favorably with computer solutions and some experimental results.

9. SUMMARY AND CONCLUSIONS

In this paper, the state-of-the-art design philosophies with particular emphasis on the Load and Resistance Factor Design (LRFD) format are discussed. The background and relevant development of the design methods relating to columns and beam-columns are presented.

LRFD is a limit-state design method. A valid limit-state analysis and design of structures or structural members requires more understanding of structural behavior and more demand on structural analysis techniques. For example, in the AISC/LRFD Specification, a direct second-order analysis of the structure to determine the maximum design forces in the member is recommended. Furthermore, an extensive use of stability theory is employed in the development of the specification equations.

Although first-order theory is still extensively used by engineers and designers in proportioning members, the use of second-order analysis will become more and more popular in the near future. The use of more sophisticated analysis techniques is enhanced by the rapid development in computer hardware and software. In particular, the great advancement in microcomputers has enabled engineers and designers to perform fast and more effi-

130

cient analysis and design of most structural members. Consequently, computer-aided analysis and design of structure will continue to gain popularity as time proceeds.

In addition to theoretical investigations of structural behavior, the continued need for experimental investigations is inevitable, especially in the area of connection restraint characterization. Theoretical and experimental work must go in parallel paths to ensure the continued development of more economical and rational design procedures in view of recent great advancement in microcomputers.

REFERENCES

1. *Euler, L.* De Curvis Elasticis *Lausanne and Geneva 1744, pp. 267–268. The Euler formula was derived in a later paper, Sur le Forces des Colonnes Memoires de l'Academie Royale des Sciences et Belles Lettres, Vol. 13, Berlin, 1759.*

2. *Engesser, F.* Zeitschrift fur Architektur und Ingenieurwesen *35, 1889, p. 455. Schweizerische Bauzeitung Vol. 26, 1895, p. 24.*

3. *Considere, A.* Resistance des Pieces Comprimees *Congres International des Procedes de Construction Paris, Vol. 3, 1891, p. 371.*

4. *Shanley, F. R.* Inelastic Column Theory *Journal of the Aeronautical Sciences, Vol. 14, No. 5, May 1947, pp. 261–264.*

5. *Duberg, J. E. and T. W. Wilder* Column Behavior in the Plastic Stress Range *Journal of the Aeronautic Science, Vol. 17, No. 6, 1950.*

6. *Johnston, B. G.* Buckling Behavior Above the Tangent Modulus Load *Journal of the Engineering Mechanics Division, ASCE, Vol. 87, No. EM6, December 1961, pp. 79–99.*

7. *Johnston, B. G.* Column Buckling Theory: Historic Highlights *ASCE, Journal of Structural Engineering, Vol. 109, No. 9, September 1983, pp. 2086–2096.*

8. *Timoshenko, S. P. and J. M. Gere* Theory of Elastic Stability *2nd Ed., Engineering Societies Monographs, McGraw-Hill, New York, N.Y., 1961.*

9. *Bleich, F.* Buckling Strength of Metal Structures *Engineering Societies Monographs, McGraw-Hill, New York, N.Y., 1952.*

10. *Chen, W. F. and T. Atsuta* Theory of Beam-Columns: Vol. 1—In-plane Behavior and Design *McGraw-Hill, New York, N.Y., 1976.*

11. *Chen, W. F. and T. Atsuta* Theory of Beam-Columns: Vol. 2—Space Behavior and Design *McGraw-Hill, New York, N.Y., 1977.*

12. *Lui, E. M. and W. F. Chen* End Restraint and Column Design Using LRFD *AISC Engineering Journal, Vol. 20, No. 1, 1st Qtr., 1983, pp. 29–39.*

13. *American Institute of Steel Construction, Inc.* Proposed Load and Resistance Factor Design Specification for Structural Steel Buildings *Chicago, Ill., January 1985.*

14. *Johnston, B. G., Ed.* Guide to Design Criteria for Metal Compression Members *Column Research Council, 1960.*

15. *Bjorhovde, R.* Deterministic and Probabilistic Approaches to the Strength of Steel Columns *Ph.D. Dissertation, Department of Civil Engineering, Lehigh University, Bethlehem, Pa., 1972.*

16. *Johnston, B. G., Ed.* SSRC Guide to Stability Design Criteria for Metal Structures *3rd Ed., John Wiley, New York, N.Y., 1976.*

17. *Rondal, J. and R. Maquoi* Single Equation for SSRC Column Strength Curves Technical Notes *ASCE Journal of the Structural Division, Vol. 105, No. ST1, New York, N.Y., January 1979, pp. 247–250.*

18. *Lui, E. M. and W. F. Chen* Simplified Approach to the Analysis and Design of Columns with Imperfections *AISC Engineering Journal, Vol. 21, No. 2, 2nd Qtr., 1984.*

19. *American Institute of Steel Construction, Inc.* Specification for the Design, Fabrication and Erection of Structural Steel for Buildings *Chicago, Ill., November, 1978.*

20. *Lui, E. M. and W. F. Chen* Strength of H-Columns with Small End Restraints *The Journal of the Institute of Structural Engineers, Vol. 61B, No. 1, London, March 1983, pp. 17–26.*

21. *Bjorhovde, R.* Effect of End Restraint on Column Strength—Practical Applications *AISC Engineering Journal, Vol. 21, No. 1, 1st Qtr., 1984, pp. 1–13.*

22. *Lui, E. M. and W. F. Chen* Discussion on Effect of End Restraint on Column Strength—Practical Applications by R. Bjorhovde *to appear, 1985.*

23. *Iwankiw, N.* Note on Beam-Column Moment Amplification Factor *AISC Engineering Journal, Vol. 21, No. 1, 1st Qtr., 1984, pp. 21–23.*

24. *Austin, W. J.* Strength and Design of Metal Beam-Columns *ASCE Journal of Structural Division, Vol. 87, No. ST4, April 1961, pp. 1–32.*

25. *Massonnet, C.* Stability Considerations in the Design of Steel Columns *ASCE Journal of Structural Division, Vol. 85, No. ST7, September 1959, pp. 75–111.*

26. *Rosenblueth, E.* Slenderness Effects in Buildings *ASCE Journal of the Structural Division, Vol. 91, No. ST1, February 1965, pp. 229–252.*

27. *Stevens, L. K.* Elastic Stability of Practical Multistory Frames *Proceedings, Institute of Civil Engineers, Vol. 36, London, England, 1967.*

28. *Cheong-Siat-Moy, F.* Consideration of Secondary Effects in Frame Design *ASCE Journal of Structural Division, Vol. 103, No. ST10, October 1972, pp. 2,005–2,019.*

29. *Yura, J. A.* The Effective Length of Columns in Unbraced Frames *AISC Engineering Journal, Vol. 8, No. 2, April 1971, pp. 37–42.*

30. *McGuire, W.* Geometrical Nonlinear Analysis and Amplification Factors *Private Communication, August 1984.*

31. *LeMessurier, W. J.* A Practical Method of Second-Order Analysis, Part 2—Rigid Frames *AISC Engineering Journal, Vol. 14, No. 2, 2nd Qtr., 1972, pp. 49–67.*

32. *Rathbun, J. C.* Elastic Properties of Riveted Connections *Transactions of the American Society of Civil Engineers, Vol. 101, 1936, pp. 524–563.*

33. *Romstad, K. M. and C. V. Subramanian* Analysis of Frames with Partial Connection Rigidity *ASCE Journal of the Structural Division, Vol. 96, No. ST11, November, 1970, pp. 2,283–2,300.*

34. *Razzaq, Z.* End Restraint Effect on Steel Column Strength *ASCE Journal of the Structural Division, Vol. 109, No. ST2, February 1983, pp. 314–334.*

35. *Frye, M. J. and G. A. Morris* Analysis of Flexibly Connected Steel Frames *Canadian Journal of Civil Engineers, Vol. 2, No. 3, Canada, September 1975, pp. 280–291.*

36. *Jones, S. W., P. A. Kirby and D. A. Nethercot* Columns with Semirigid Joints *ASCE Journal of the Structural Division, Vol. 108, No. ST2, February 1982, pp. 361–372.*

37. *Colson, A. and J. M. Louveau* Connections Incidence on the Inelastic Behavior of Steel Structures *Euromech Colloquium 174, October 1983.*

38. *Lui, E. M.* Effects of Connection Flexibility and Panel Zone Deformation on the Behavior of Plane Steel Frames *Ph.D. Dissertation, Department of Structural Engineering, School of Civil Engineering, Purdue University, West Lafayette, Ind., 1985.*

39. *Lewitt, C. S., E. Chesson and W. H. Munse* Restraint Characteristics of Flexible Riveted and Bolted Beam-To-Column Connections *Engineering Experiment Station Bulletin No. 500, University of Illinois at Urbana-Champaign, Ill., January 1969.*

40. *Sugimoto, H.* Study of Offshore Structural Members and Frames *Ph.D. Dissertation, Department of Structural Engineering, School of Civil Engineering, Purdue University, West Lafayette, Ind., 1983.*

41. *Disque, R. O.* Directional Moment Connections—A Proposed Design Method for Unbraced Steel Frames *AISC Engineering Journal, 1st Qtr., 1975, pp. 14–18.*

42. *Driscoll, G. C.* Effective Length of Columns with Semi-Rigid Connections *AISC Engineering Journal, 4th Qtr., 1976, pp. 109–115.*

43. *Ackroyd, M. H.* Nonlinear Inelastic Stability of Flexibly-Connected Plane Steel Frames *Ph.D. Dissertation, Department of Civil, Environmental and Architectural Engineering, University of Colorado, Boulder, Colo., 1979.*

44. *Moncarz, P. D. and K. H. Gerstle* Steel Frames with Non-Linear Connections *ASCE Journal of the Structural Division, Vol. 107, No. ST8, August, 1981, pp. 1,427–1,441.*

45. *Simitses, G. J. and A. S. Vlahinos* Stability Analysis of a Semi-Rigidly Connected Simple Frame *Journal of Constructional Steel Research, Vol. 2, No. 3, September 1982, pp. 29–32.*

46. *Simitses, G. J. and J. Giri* Non-Linear Analysis of Unbraced Frames of Variable Geometry *International Journal of Non-Linear Mechanics, Vol. 17, No. 1, 1982, pp. 47–61.*

47. *Simitses, G. J., J. D. Swisshelm and A. S. Vlahinos* Flexibly-Jointed Unbraced Portal Frames *Journal of Constructional Steel Research, Vol. 4, 1984, pp. 27–44.*

48. *Fielding, D. J. and J. S. Huang* Shear in Beam-To-Column Connections *The Welding Journal, Vol. 50, July 1971.*

49. *Fielding, D. J. and W. F. Chen* Frame Analysis and Connection Shear Deformation *ASCE Journal of the Structural Division, Vol. 99, No. ST1, January 1973, pp. 1–18.*

50. *Becker, R.* Panel Zone Effect on the Strength and Stiffness of Steel Rigid Frames *AISC Engineering Journal, Vol. 12, No. 1, 1st Qtr., 1975, pp. 19–29.*

51. *Kato, B.* Beam-To-Column Connection Research in Japan *ASCE Journal of the Structural Division, Vol. 108, No. ST2, February 1982, pp. 343–360.*

52. *Krawinkler, H.* Shear in Beam-Column Joints in Seismic Design of Steel Frames *AISC Engineering Journal, 3rd Qtr., 1978, pp. 82–91.*

53. *Chen, W. F. and E. M. Lui* Effects of Connection Flexibility and Panel Zone Shear Deformation on the Behavior of Steel Frames Seminar on Tall Structures and Use of Prestressed Concrete in Hydraulic Structures *Indian National Group of the International Association for Bridge and Structural Engineering, Srinagar, May 24–26, 1984, pp. 155–176.*

54. *Kanchanalai, T.* The Design and Behavior of Beam-Columns in Unbraced Steel Frames *AISI Project No. 189, Report No. 2, Civil Engineering/Structures Research Lab, University of Texas at Austin, October 1977.*

55. *Tebedge, N. and W. F. Chen* Design Criteria for H-Columns Under Biaxial Bending *ASCE Journal of the Structural Division, Vol. 104, No. ST9, September 1978, pp. 1,355–1,370.*

56. *Zhou, S. P. and W. F. Chen* A Comparative Study of Beam-Columns in ASD and LRFD *Structural Engineering Report No. CE-STR-84-53, School of Civil Engineering, Purdue University, West Lafayette, Ind., 1984.*

57. *Zhou, S. P. and W. F. Chen* Design Criteria for Box Beam-Columns Under Biaxial Loading *Structural Engineering Report No. CE-STR-85-2, School of Civil Engineering, Purdue University, West Lafayette, Ind., 1985.*

58. *Chen, W. F. and J. McGraw* Behavior and Design of HSS-Columns Under Biaxial Bending *Proceedings of 2nd Engineering Mechanics Division Specialty Conference, Raleigh, N.C., May 23–25, 1977.*

5.8.2

ELSEVIER   Engineering Structures 22 (2000) 116–122

**ENGINEERING
STRUCTURES**

www.elsevier.com/locate/engstruct

Structural stability: from theory to practice

W.F. Chen [*]

School of Civil Engineering, Purdue University, West Lafayette, IN 47907, USA

Received 27 July 1997; accepted 9 January 1998

Abstract

Over the past 40 years drastic improvements in our knowledge regarding the behavior, strength and design of steel building frames has been achieved. This paper provides several specific examples in which new knowledge has been implemented and better design methods have been advanced in engineering practice. The directions of possible immediate implementation of some recent developments in advanced analysis for practical frame design are outlined. © 1999 Elsevier Science Ltd. All rights reserved.

Keywords: Advanced analysis; Buildings; Design; Effective length factor; Plasticity; Stability; Steel; Structural engineering

1. Introduction

What measuring stick should be used to assess the accomplishment of structural stability research for steel building frames in the past 40 years? Should it be the volume of papers presented, or the number of journal articles published, or the number of PhD theses produced or the number of new courses in the universities offered? I believe that the 'bottom line' for the structural engineering profession should be the amount of research which finds its way into practice. The profession has developed in the past decades in important ways as a direct result of these intensive research activities worldwide. The SSRC-related 'success stories' that can be attributed to these developments fall into a number of broad categories. For the Symposium honoring Professor T.V. Galambos, it is the most appropriate occasion to summarize these categories in this paper and provide several specific examples within each category where new knowledge has been implemented and, in some measure, a better understanding of the behavior of structural members and systems has been developed and better design methods have been advanced. Directions of possible immediate implementations of some recent developments for engineering practice are outlined here.

* Tel: + 1-765-494-2254; Fax: + 1-765-496-1105

0141-0296/00/$ - see front matter © 1999 Elsevier Science Ltd. All rights reserved.
PII: S0141-0296(98)00100-X

2. Behavior and design of structural members

Perhaps in no other area has there been such drastic improvement in our knowledge regarding the behavior, strength and design of structural members including columns, beams and beam-columns using the mainframe computing and finite element methods in the 1960s and 1970s. There has been a steady flow of results from SSRC research into the development of improved codes and standards governing the design of structural members in building codes. Major changes were made, for example, in the design of biaxially loaded columns. The studies ranged from full-scale tests to complex finite-element analysis of beam-columns under various load combinations in plane and in space. The information produced has been implemented in AISC building codes and Euro-codes and has become standard practice in the 1980's. These and other related developments were summarized in a two-volume book by Chen and Atsuta in 1976–77 [1] as well as the 1988 SSRC Guides edited by Galambos [2].

Another area where significant advances have been made is in the design of large fabricated cylindrical members as used in deep-water offshore structures. Several specific areas where the research results were instrumental in bringing about major changes in API codes and other practices for engineering in offshore structures include the effect of hydrostatic pressure on column strength, the beam-column strength and behavior considering dent damage effects, and the assessment of the

W.F. Chen/Engineering Structures 22 (2000) 116–122 117

strength of internally grout-repaired damaged members. This information and related developments were summarized in the books for example by Chen and Han [3] and Chen and Toma [4] among others.

3. Structural design with K factors

In current engineering practice, the interaction between the structural system and its members is represented by the effective length factor (Fig. 1). This classical approach to structural design is described clearly in the 1981 SSRC Technical Memorandum No. 5 [5], which provides the basis for the development of modern steel design methods including the popular load resistance factor design (LRFD) and allowable stress design (ASD) methods [6]. The effective length method generally provides a good method for the design of framed structures. However, despite its popular use in the past and present as a basis for design, the approach has major limitations.

The first of these is that it does not give an accurate indication of the factor against failure, because it does not consider the interaction of strength and stability between the member and structural system in a direct manner. It is a well recognized fact that the actual failure mode of the structural system often does not have any resemblance whatsoever to the elastic buckling mode of the structural system that is the basis for the determination of the effective length factor K.

The second and perhaps the most serious limitation is probably the rationale of the current two-stage process in design: elastic analysis is used to determine the forces acting on each member of a structural system, whereas inelastic analysis is used to determine the strength of each member treated as an isolated component. There is no verification of the compatibility between the isolated member and the member as part of a frame. The individ-

ual member strength equations as specified in specifications are not concerned with system compatibility. As a result, there is no explicit guarantee that all members will sustain their design loads under the geometric configuration imposed by the framework.

The other limitations of the effective length method include the difficulty of computing a K factor, which is not user-friendly for a computer-based design, and the inability of the method to predict the actual strength of a framed member, among many others. To this end, there is an increasing awareness of the need for practical analysis/design methods that can account for the compatibility between the member and system. With the rapid increase in the power of desktop computers and user-friendly software in recent years, the development of an alternative method to a direct design of structural system without the use of K factors becomes more attractive and realistic. The real challenge is making this type of new approach to design work and competitive in engineering practice. An extensive research on this topic, now known as advanced analysis to design, has been made at several universities around the world for many years, and significant advancements have been made, although much more remains to be done.

4. Advanced analysis to design

Extensive research has been devoted to the development and validation of several advanced analysis methods. A promising technique of developing high-order beam elements (making only one or two necessary to describe the behavior along a members' length) is in progress at Cornell University [7]. An intensely rigorous method using workstations and super-computers to solve thousands of degrees of freedom has been in development around the world for many years [8,9]. Simple calibration techniques and practical approaches have been researched here at Purdue [10]. Intermediate solutions include plastic-zone, quasi-plastic hinge, elastic–plastic hinge methods and various modifications thereof. All in some way account for residual stresses, geometric imperfections, non-linearities and moment redistribution throughout a structure. Briefly they are outlined below:

1. Plastic zone [8,9]:
   a. Discretized finite elements along the length and through the cross-section.
   b. Captures the incremental load-versus-deflection response considering the second-order geometric distortion.
   c. A constant residual stress pattern is assumed.
   d. The spread of plasticity is traced.

2. Quasi-plastic zone [7]:
   a. A compromise between plastic zone and elastic plastic hinge methods.

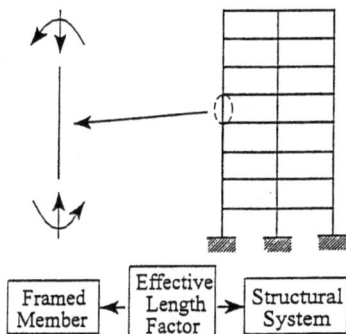

Fig. 1. Interaction between a structural system and its component members.

118   *W.F. Chen/Engineering Structures 22 (2000) 116–122*

b. The spread of plasticity is considered by flexibility coefficients.
c. A simplified residual stress pattern is used.
d. The fully plastic cross-section is calibrated to the plastic zone solution.
e. There is no potential to upgrade this from its current two-dimensional restriction.

3. Elastic–plastic hinge [11]
a. Zero length plastic hinges.
   b. No spread of yielding through the cross-section, or along the length.
   c. No consideration of residual stresses.
   d. Second-order geometric effects can be considered.

4. Refined plastic hinge method [12]:
   a. A step up from the elastic plastic model for two dimensions.
   b. Distributed plasticity-smooth stiffness degradation of a hinge.
   c. Inelasticity is considered indirectly by forces rather than strains. Tangent (E_t) modulus is used to describe the effect of residual stresses.
   d. Stiffness degradation function is used for gradual yielding.
   e. Connection flexibility can be modelled using rotational springs.

5. Practical refined plastic hinge method [13]:
   a. The refined model (4. above) is made practical by calibration to the LRFD empirical code equations.
   b. A separate modification of tangent modulus (E_t) is imposed to consider geometric imperfections.
   c. The CRC tangent modulus model is used allowing residual stresses to be considered separately.

For the last method to work effectively on popular commercial programs in use in design offices today further changes are needed. The Purdue method [14] was developed to perform designs, using simple modifications to elastic parameters familiar to LRFD users, comparable to those achieved by traditional code procedures—but a unique program was required. The advanced capabilities included two modifications for material non-linearity and one for geometric imperfections. The degradation of stiffness due to gradual yielding of a cross-section subject to flexural moments was defined as shown in Fig. 3. Residual stresses were accounted for by reducing the tangent modulus E_t as plotted in Fig. 2. The last non-linear effect, geometric imperfections, was considered by applying a further reduction to E_t. Details of this development will be summarized in the following section.

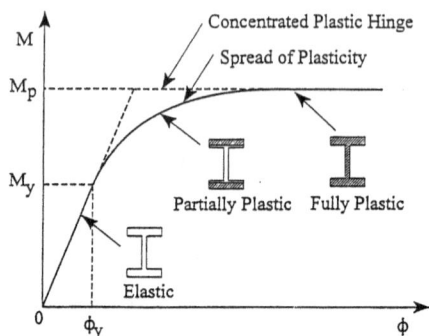

Fig. 2. Moment-curvature relationship for a perfect plastic and work hardening hinge.

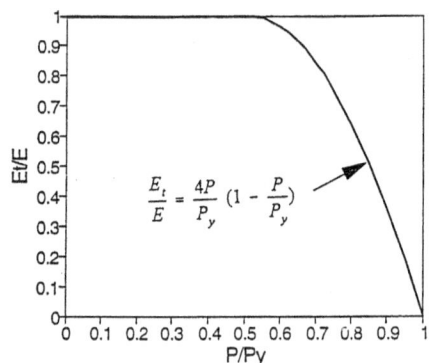

Fig. 3. Member tangent stiffness degradation derived from the CRC column curve.

Two difficulties were noted with regard this method, which, being changed, would result in a practical method of using current software to achieve the same types of analysis.

1. The choice of using stability functions to account for the P-Δ effect does not translate well into the finite element world. Stability functions use small deformations theory implicitly to capture an effect that many finite element analyses account for explicitly. The advantage of this technique is only one element is needed per member. This is becoming less and less of a driving issue with the speed of computer analysis today, and the price of having more elements along the length (while capturing behavior taking place out-of-alignment with the centre-line of the members) is small. This allows the finite element method (as it exists in most all structural analysis packages) to be used as they stand.

W.F. Chen/Engineering Structures 22 (2000) 116–122 119

2. The degradation in stiffness modification factors was applied by the Purdue team to coefficients in the stiffness matrix of each element. It was required that the fundamental solution algorithms be re-coded in order to consider this non-linear behavior 'simply'. This difficulty will be overcome with this method by taking advantage of the iterative capabilities inherent (or soon to be implemented) in common analysis packages [15]. Non-linearities due to material behavior fast becoming available in commercial codes, and using this to define a failure criterion would present a convenient and practicable 'advanced analysis' method.

5. Structural design without K factor

In the following, I shall briefly summarize a practical solution to the problem by simply modifying an elastic program with a modified tangent modulus based on the familiar CRC column strength equation together with a refined plastic hinge concept. These modifications consider the following key behavioral effects of a steel member: second-order, gradual yielding associated with residual stresses and flexure and geometric imperfections. To meet the current LRFD requirements, these modifications have been calibrated against the LRFD specification.

5.1. Second-order effects

To capture second-order effects, the simplified stability functions reported by Chen and Lui [16] is adopted. The incremental force-displacement relationship of a member may be written as, in the usual notations

$$\begin{bmatrix} \dot{M}_A \\ \dot{M}_B \\ \dot{P} \end{bmatrix} = \frac{EI}{I} \begin{bmatrix} S_1 & S_2 & 0 \\ S_2 & S_1 & 0 \\ 0 & 0 & A/I \end{bmatrix} \begin{bmatrix} \dot{\theta}_A \\ \dot{\theta}_B \\ \dot{e} \end{bmatrix} \tag{1}$$

where S_1 and S_2 are stability functions, for in-plane bending of a prismatic beam-column.

5.2. Cross-section plastic strength

The LRFD cross-section plastic strength curves are adopted for both strong and weak-axis bending

$$\frac{P}{\varphi_c P_y} + \frac{8}{9}\frac{M}{\varphi_b M_p} = 1.0 \quad \text{for } \frac{P}{\varphi_c P_y} \geq 0.2 \tag{2a}$$

$$\frac{P}{2\varphi_c P_y} + \frac{M}{\varphi_b M_p} = 1.0 \quad \text{for } \frac{P}{\varphi_c P_y} < 0.2 \tag{2b}$$

The reduction factors φ are selected as 0.85 for axial strength and 0.9 for flexural strength just as the LRFD specification does.

5.3. Residual stresses

The CRC tangent modulus is employed to account for the gradual yielding effect due to residual stresses along the length of members under axial loads. In this approach, the elastic modulus E instead of the moment of inertia I is reduced to account for the reduction of the elastic portion of the cross-section, because the reduction of elastic modulus is easier to implement than that of the moment of inertia for different sections. The reduction rate in stiffness for both strong and weak axis is taken to be the same and this reduction is reflected by the CRC E_t as (Fig. 2).

$$E_t = 1.0E \qquad \text{for } P \leq 0.5P_y \tag{3a}$$

$$E_t = 4\frac{P}{P_y}E\left(1 - \frac{P}{P_y}\right) \text{ for } P > 0.5P_y \tag{3b}$$

5.4. Distributed plasticity

When idealized plastic hinges are formed at the member ends, the elastic stiffness at the ends will be reduced abruptly to zero (Fig. 3). To represent a gradual transition from the elastic stiffness at the onset of yielding to the stiffness associated with a full plastic hinge at the ends, the parameter η representing a gradual stiffness reduction associated with flexure is introduced with $0 \leq \eta \leq 1.0$ according to the parabolic expression (Fig. 4)

$$\eta = 4\alpha(1 - \alpha) \text{ for } \alpha > 0.5 \tag{4}$$

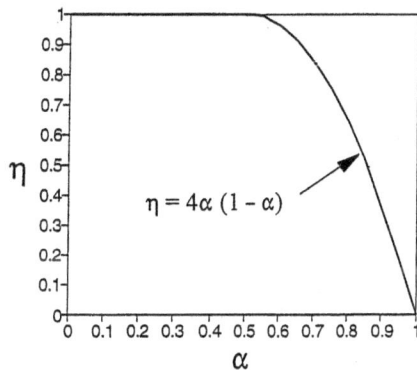

Fig. 4. Parabolic plastic hinge stiffness degradation function with $\alpha_o = 0.5$ based on LRFD sectional strength equation.

where α is the force-state parameter obtained from the limit state surface corresponding to the member ends (Fig. 5)

$$\alpha = \frac{P}{P_y} + \frac{8}{9}\frac{M}{M_p} \qquad \text{for } \frac{P}{P_y} \geq \frac{2}{9}\frac{M}{M_p} \tag{5a}$$

$$\alpha = \frac{P}{2P_y} + \frac{M}{M_p} \qquad \text{for } \frac{P}{P_y} \leq \frac{2}{9}\frac{M}{M_p} \tag{5b}$$

similar to the LRFD plastic sectional strength expressions. This is known as the refined plastic hinge concept. It reflects the distributed plasticity effects associated with bending actions at the member ends. When these refined plastic hinges are present at both ends of a member, the incremental elastic force–displacement relationship as given in Eq. (1) can now be modified to include both the inelasticity within the member by using E_t instead of E and the distributed plasticity at the ends by using the refined plastic hinge concept using the η parameter as

$$\begin{bmatrix} \dot{M}_A \\ \dot{M}_B \\ \dot{P} \end{bmatrix} = \tag{6}$$

$$\frac{E_t I}{L} \begin{bmatrix} \eta_A \left[S_1 - \frac{S_2^2}{S_1}(1-\eta_B) \right] & \eta_A \eta_B S_2 & 0 \\ \eta_A \eta_B S_2 & \eta_B \left[S_1 - \frac{S_2^2}{S_1}(1-\eta_A) \right] & 0 \\ 0 & 0 & \frac{A}{I} \end{bmatrix} \begin{bmatrix} \dot{\theta}_A \\ \dot{\theta}_B \\ \dot{e} \end{bmatrix}$$

where η_A and η_B are stiffness reduction factor as given in Eq. (4) at end A and end B, respectively. Details of this development are given elsewhere [14].

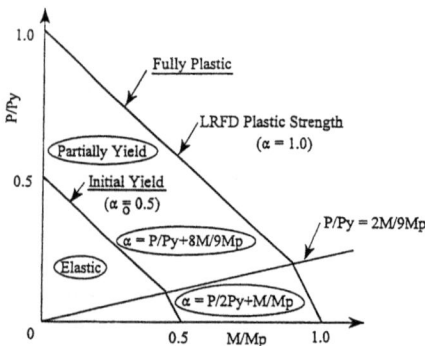

Fig. 6. CRC and reduced tangent modulus for members with geometrical imperfection.

5.5. Geometric imperfections

The degradation of member stiffness due to geometric imperfections may be simulated by a further reduction of member stiffness. This may be achieved simply by a further reduction of the tangent modulus E_t as (Fig. 6)

$$E_t' = 0.85 \, E_t \tag{7}$$

Herein, the reduction factor 0.85 is used to reduce further the CRC E_t as given in Eq. (3a) and (3b) to include the effect of geometric imperfections (Fig. 6). Thus, if the modulus E in the Euler buckling formula is replaced by E_t', the column strength curve as specified by the LRFD specification will be obtained within a maximum error of no more than 5% (Fig. 7). The further reduced modulus E_t' is applicable for both braced and unbraced members and frames.

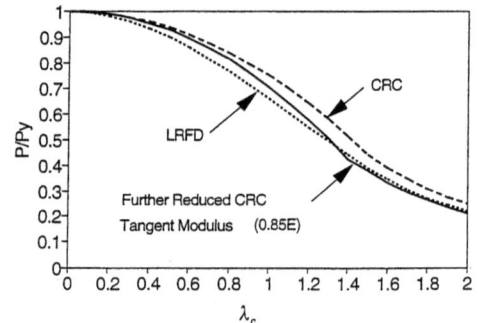

Fig. 5. Smooth stiffness degradation for a work-hardening plastic hinge based on LRFD sectional strength curve.

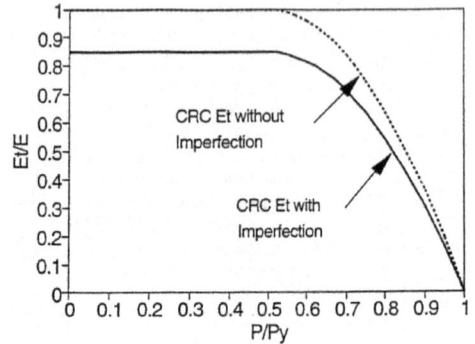

Fig. 7. Comparison of column strength curves with further reduced tangent modulus method.

W.F. Chen/Engineering Structures 22 (2000) 116–122 121

6. Behavior and design of structural system

When I first began to work in structural stability over 35 years ago, evaluation of the first-order response of a structural system was a significant problem. This includes linear elastic analysis and simple plastic analysis, and the progress made to the present state-of-the-art, which deals routinely with second-order inelastic analysis of complicated structural systems having hundreds of thousands of degrees of freedom, is miraculous. The modelling of all types of structural systems of high-rise buildings can now be handled quickly and efficiently on relatively inexpensive computers. The primary limitation is a sufficient understanding of the response of some secondary structural elements such as concrete floor slabs, composite joints and walls that make up that system to develop simple but realistic models that can be incorporated into the analysis programs.

The advent of personal computers, particularly in the computing and graphics performance of engineering workstations, has made more sophisticated methods of analysis feasible in design practice. While the use of first-order analysis for elastic or plastic design is still the norm of engineering practice, a new generation of codes has emerged that recommends the second-order theory as the preferred method of analysis. The basic theory for second-order inelastic analysis is well established and documented in open literature. The real challenge is making this type of analysis work in engineering practice. The advanced analysis approach to design as illustrated in Fig. 8 can predict more accurately the possible failure modes of a structure, exhibit a more uniform level of safety, and provide a better long-term serviceability and maintainability.

7. Seismic design with structural fuse

The analytical capability of tracing the performance of a structure into the non-linear range required by seismic loads is also available [17]. Advanced analysis combines the theory of stability with the theory of plasticity and traces the gradual plastification of members with

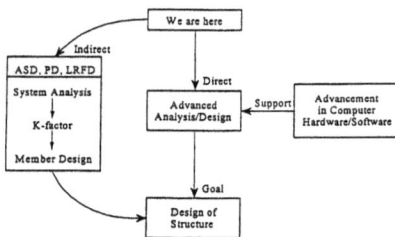

Fig. 8. Analysis and design methods.

rigid or flexible joints in a steel frame (Chen and Sohal, [18]). The power of this tool is as follows: with the ability to predict the actual moment distribution at load levels that require members to sustain their plastic moment capacity, 'breaks' can be strategically located throughout a structure. These structural 'fuses' can be designed to fail themselves without the risk of the building as a whole falling down, while leaving the majority of the connections in satisfactory condition (AISC, [19]). This would not only limit the amount of post-quake repair necessary, but would also indicate where the failed connections were and thus greatly reduce the expense of 'exploratory procedures'.

Some of the new aspects that can be further considered in design practice when performance based (via advanced analysis) becomes standard practice are:

Semi-rigid connections and thus partly restrained members have been researched extensively but this knowledge has not been able to be easily implemented into code-based design practice. Unless computer methods are adopted, small practical use can be garnered from the new knowledge of connection rotation characteristics and their effect on the global behavior of frames.

Furthermore, three-dimensional behavior is a natural extension. Even only partly considering out-of-plane behavior (e.g. lateral torsional buckling) becomes possible.

8. Summary

The practical design method presented here introduces the potential (and emphasizes the necessity) of advanced analysis procedures. Long accustomed to isolated member-by-member capacity checks, one analysis now considers all components and their interdependence. The global analysis provides information on the failure mode and thus allows an assessment of damage sustained at collapse loads. If the damage can be predicted, it can be controlled by design procedures calibrated to maintain adequate performance criteria (as opposed to the Codes' traditional safety levels). Only when design engineers are assured of the validity and convinced on the practicality of performance-based analysis and design will this and other advanced capabilities be implemented in day-to-day offices and the results of many years of research be granted their true fulfillment: a place in Practice.

References

[1] Chen WF, Atsuta T. Theory of beam-columns, Vol. 1, In-plane behavior and design, 1976, Vol. 2, Space behavior and design, 1977. New York: McGraw–Hill.

[2] Galambos TV, editor. Guide to stability design criteria for metal structures, 4th edn. Wiley–Interscience, 1988.

122   *W.F. Chen/Engineering Structures 22 (2000) 116–122*

[3] Chen WF, Han DJ. Tubular members in offshore structures. London: Pitman, 1985.

[4] Chen WF, Toma S. Analysis and software of cylindrical members. Boca Raton, FL: CRC Press, 1996.

[5] SSRC. General principles for the stability design of metal structures. Technical Memorandum No. 5, Civil Engineering, ASCE, February 1981, pp. 53–54.

[6] Chen WF, Lui EM. Structural stability: theory and implementation. New York: Elsevier, 1987.

[7] Deierlein GG. Steel-framed structures. Progress in structural engineering and materials, Vol. 1, No. 1. London: CRC Ltd., September 1997.

[8] Clarke MJ, Bridge RQ, Hancock GJ, Trahair NJ. Advanced analysis of steel building frames. Journal of Constructional Steel Research 1992;23(1–3):1–30.

[9] McGuire W. Computer-aided analysis. In: Dowling PJ, Harding JE, Bjorhovde R, editors. Constructional steel design—and international guide. Elsevier Applied Science, 1992:915–932.

[10] Chen WF, Toma S. Advanced analysis of steel frames. Boca Raton, FL: CRC Press, 1994.

[11] White DW, Chen WF. Plastic-hinge based methods for advanced analysis and design of steel frames. Bethlehem, PA: Structural Stability Research Council, Lehigh University, 1993.

[12] Liew JYR, White DW, Chen WF. Second-order refined plastic hinge analysis for frame design: Parts 1 and 2. Journal of Structural Engineering, ASCE 1993;119(11):3196–237.

[13] Kim SE, Chen WF. Practical advanced analysis for braced steel frame design and practical advanced analysis of unbraced steel frame design. Journal of Structural Engineering, ASCE 1996;122(11):1259–74.

[14] Chen WF, Kim SE. LRFD steel design using advanced analysis. Boca Raton, FL: CRC Press, 1997.

[15] Chen WF, Han DJ. Plasticity for structural engineers. New York: Springer–Verlag, 1988.

[16] Chen WF, Lui EM. Stability design of steel frames. Boca Raton, FL: CRC Press, 1992.

[17] White DW, Chen WF, editors. Proceedings of the US–Japan Seminar on Innovations in Stability Concepts and Methods for Seismic Design in Structural Steel [special issue]. Engineering Structures 1997;20(4–6).

[18] Chen WF, Sohal I. Plastic design and second-order analysis of steel frames. New York: Springer-Verlag, 1995.

[19] AISC. Load of resistance factor design specification for structural steel buildings. Chicago, IL: American Institute of Steel Construction, 1997.

5.9 Concrete Building Construction (混凝土建築施工)

5.9.1

Construction Loads on Supporting Floors

Recent analytical studies have reported that construction loads on supporting floors may appreciably exceed slab design loads. This article quantifies the errors incurred when using the simplified method to predict maximum shore and slab loads and recommends modifications to account for these errors. Modification coefficients are developed by comparing the results using the simplified method to results using a two-dimensional matrix structural analysis computer program and by comparison to field measurements.

by Xi-La Liu, Wai-Fah Chen, and Mark D. Bowman

Keywords: concrete construction; flat concrete slabs; floors; formwork (construction); **loads (forces)**; reinforced concrete; **shoring**; structural analysis; wood.

In this, the age of rapid construction of high-rise reinforced concrete buildings, the freshly placed concrete floor is supported temporarily by shores, which are in turn supported by previously cast floors. Recent analytical studies have reported that the construction loads on the supporting floors may appreciably exceed the slab design loads.[1] The problem is particularly troublesome in multistory reinforced concrete buildings in which the live load is small compared to the dead load.

Formwork should not be removed until the concrete attains adequate strength to carry both the dead load and the construction loads. From the contractor's viewpoint, formwork removal should be scheduled at the earliest possible time, since formwork represents a high proportion of the total construction cost of concrete structures. In practice, the schedule for formwork removal is often left to the discretion of the contractor. The recommendations suggested by ACI Committee 347[2] provide general guidelines regarding formwork removal. Nonetheless, the contractor has the tendency to minimize his costs by using less formwork and by removing that formwork as soon as he believes it is safe.

A number of concrete construction disasters have occurred as a result of early formwork removal.[3]

a comparison of the simplified method to field measurements and to two-dimensional refined analysis

On March 2, 1972, premature removal of shores supporting a 5-day-old immature slab led to the progressive collapse of an entire structure at Bailey Crossroads, near Alexandria, Virginia. Fourteen workers died.

Common causes of formwork disasters include excessive loads applied to the falsework, premature removal of shores and falsework, and inadequate lateral support for the shoring members. To assess structural safety during the construction phase, one must possess a thorough understanding of load distribution during the construction of concrete structures.

A few studies have been conducted on the development of analytical models that simulate the construction loading process. In 1952, Nielsen[4] presented a detailed analysis of the interaction during construction of formwork and floor slabs. Nielsen's procedures, however, are lengthy and inconvenient for practical use. In 1963, Grundy and Kabaila[5] developed a simplified method to determine the construction loads imposed on slabs and formwork. Neither of these methods have been

chosen by the present ACI specifications as requirements for scheduling formwork removal. Still, ACI Committee 347[2] does require that the minimum design load for shores be no less than one and one-half times the weight of a given floor of concrete plus formwork and construction loads. This coincides with the loads specified by Grundy and Kabaila.[5]

This article attempts to quantify the errors incurred when using the simplified method to predict the maximum shore and slab loads. These predicted loads are compared to measured loads reported in the literature and to loads predicted using a refined analysis method. Modifications of the loads predicted by the simplified method are recommended to conservatively account for modeling errors.

Simplified analysis

The analysis method developed by Grundy and Kabaila[5] to determine shore and slab loads is based on three assumptions:

• The axial stiffnesses of shores and reshores are assumed to be infinite. This assumption is based on the axial shore stiffness being much greater than the flexural stiffness of the slabs. Thus, all slabs interconnected by shores are assumed to deflect equally when a new load is added.

- All slabs are assumed to possess equal flexural stiffness, such that all loads are carried equally by the interconnected slabs. Grundy and Kabaila[5] found that the assumed concrete slab stiffness, from floor to floor, whether all equal or varying due to different ages, had little effect on the numerical results of the load distribution — 5 to 10 percent of the maximum slab load.
- The ground level floor or other base support is assumed to be rigid. This assumption is essentially true if the ground floor slab is in place to support the formwork. The foregoing assumption is not valid, however, if the ground level shores are supported on earth using mud sills. For that type of construction, the soil underneath the

mud sills will compress and cause the slabs to pick up additional load. The additional slab load can be predicted using empirical assumptions for compression of the soil or using the refined analysis technique.[6]

Two excellent examples illustrating the use of the simplified analysis on a step-by-step basis are given by Hurd.[7] The simplified method is used to determine loads carried by the shores and slabs of a multistory structure with two levels of shoring and one level of reshoring. In one example, only the weight of the slab *(D)* is taken into account; in the second example the weight of the shores $(0.10D)$, the reshores $(0.05D)$, and a construction live load $(0.5D)$ are also included. In both examples, the maximum shore and slab loads occur at the same location, although the values are different.

Field measurements

Agarwal and Gardner[8] used field measurements to check the accuracy of different analysis methods

for estimating the shore and slab load distribution. The objective of their investigation was to directly measure the load ratios experienced by shores and reshores during construction, and from that to determine the load carried by the floor slabs at different stages of construction.

The shore and reshore loads for two buildings were measured by Agarwal and Gardner: Alta Vista Towers in Ottawa, Ontario, and Place du Portage in Hull, Quebec. Steel shores were used for both structures. Construction of the 22-story Alta Vista Towers flat slab structure used three floors of shores and four floors of reshores. Measurement of the shore loadings, for the shore and reshore arrangements shown in Fig. 1, was limited to the 7th through the 13th story.

The Place du Portage structure is a 27-story flat slab office building. The construction used three levels of shores, with no reshores. The shoring arrangement is shown in Fig. 2. Shore measurements

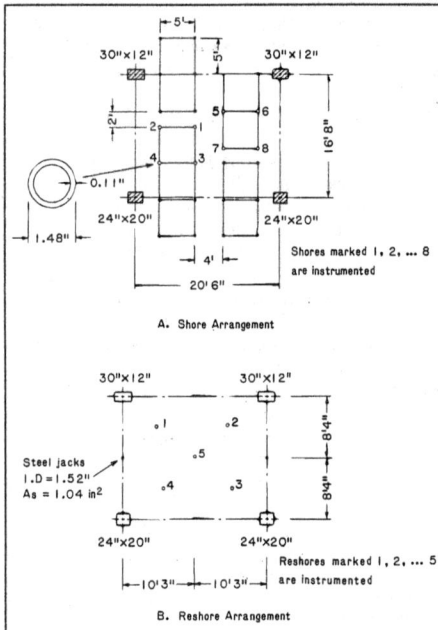

Fig. 1 (left) — Shore and reshore arrangements of Alta Vista Towers.[8]

Fig. 2 (above) — Shoring arrangements of Place du Portage.[8]

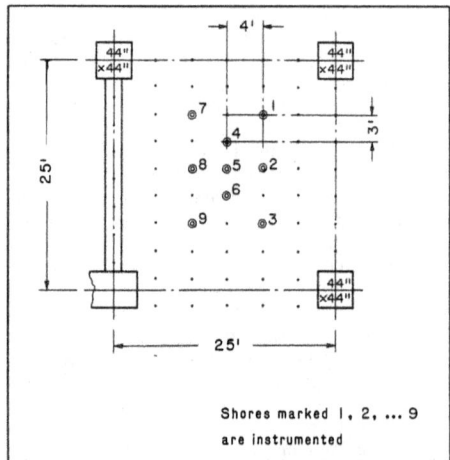

were taken from the 19th through the 22nd story.

Three factors concerning the field measurement data should be noted:

• Field measurements were not taken from ground level, so the actual slab and shore loads of the entire system during construction are not available.

• Typical shores and reshores chosen for instrumentation were located in the central areas of the slab (see Fig. 1 and 2); thus, the influence of the surrounding boundary beams and columns was less pronounced. From the recorded data, shores or reshores in the central region of the slab are different; the coefficient of variation of the shore axial force varies from 0.04 to 0.23, depending on the distance of the shore from the boundary line.[8]

• The values of E_r and the shore height are not reported in Reference 8. It is not possible, without knowing the values for E_r and the shore height, to use the measured shore load data to check other methods further.

Comparisons between the experimental results for both buildings, and the slab and shore loads predicted using the simplified analysis, are reported in Reference 8. In order to check the assumption of equal slab stiffness, the slab and shore loads were calculated both by using equal slab stiffnesses and by varying the slab stiffness for each floor to account for an increase in the modulus of elasticity of concrete, E_c, with age. Table 1 compares the field measurements for each story in both buildings to the calculated maximum slab and shore loads.

Generally, good agreement is observed between the predictions made by the simplified analysis and the experimental results. Use of variable slab stiffnesses in the simplified analysis produced an improved accuracy: the maximum error for variable slab stiffness is less than 10 percent, while for

constant slab stiffness it is as high as 16 percent. Furthermore, as shown in Reference 8, the calculated results based on either constant or variable slab stiffness consistently predict the correct construction step and the location of the maximum shore and slab loads.

The simplified analysis method is straightforward and easy to handle in practice. It is necessary, however, to examine a more refined analysis to check the applicability of the assumptions made in the simplified method, especially when flexible wooden shores are used. An improved analysis technique is also needed to examine the influence of slab boundary conditions on the distribution of shore and slab loads.

Refined analysis

The refined analysis technique uses elementary matrix computer modeling to deal with the shore-slab interaction in a more realistic fashion than does the simplified method. A structural analysis program (SAP V2) for static and dynamic response of linear systems was chosen for performing the refined analysis. Unlike the simplified method, which assumes infinitely rigid shores and equal slab stiffnesses, the refined method

considers the actual rigidity of shores and the time-dependent variation in slab stiffness due to concrete maturity.[9]

The refined analysis technique treats the shore-slab interaction as a two-dimensional problem; a comparison of the two-dimensional model with a three-dimensional model is discussed in Reference 6. The two-dimensional refined analysis model makes the following assumptions:

• the slabs behave elastically and their stiffnesses are time-dependent;

• the shores and reshores behave as continuous uniform elastic supports and their axial stiffnesses are finite and time-independent;

• the foundation is rigid;

• the joints between the shores and slabs are pinned connections; and

• the slab edges are either fixed or simply supported.

An example calculation was chosen similar to that of Nielsen's[i] example (Fig. 3), but which follows the construction procedure described in Reference 7. For this calculation a 7.1 in. x 35.4 in. (180 mm x 900 mm) rectangular slab-beam section is assumed to exist along each column line. Also, the following material properties are

TABLE 1—A comparison between field measurements and calculated maximum slab and shore loads for buildings tested by Agarwal and Gardener.[8]

Building	Level	Maximum slab load ÷ slab weight					Maximum shore load ÷ slab weight				
		Field measurement (1)	Calculated with E_c constant (2)	Calculated with E_c variable (3)	Comparisons (2) (1)	Comparisons (3) (1)	Field measurement (4)	Calculated with E_c constant (5)	Calculated with E_c variable (6)	Comparisons (4) (5)	Comparisons (4) (6)
Alta Vista Towers	7	1.88	1.72	1.83	0.91	0.97	1.65	1.40	1.58	0.85	0.96
	8	1.93	1.70	1.85	0.88	0.96	1.51	1.46	1.57	0.97	1.04
	9	1.91	1.72	1.74	0.90	0.91	1.70	1.42	1.58	0.84	0.93
	10	2.02	1.72	1.99	0.85	0.99	1.37	1.43	1.46	1.04	1.07
	11	1.07	1.13	0.99	1.06	0.93	1.68	1.44	1.62	0.89	1.00
	12						1.64	1.43	1.61	0.87	0.98
	Mean				0.92	0.95				0.90	1.00
	Std. dev.				0.08	0.03				0.09	0.05
Place du Portage	19	2.11		2.11		1.00	1.48		1.41		0.95
	20	1.34		1.30		0.97	1.44		1.44		1.00
	21						1.45		1.41		0.97
	Mean					0.99					0.97
	Std. dev.					0.02					0.03

TABLE 2—Fixed-end slab edges; a comparison of results using the simplified method to results using the refined method.

Construction step	Floor level	Slab load ÷ slab weight						Shore load ÷ slab weight					
		Simplified method (1)	Refined method max. (2)	min. (3)	avg. (4)	$\frac{(1)}{(2)}$	$\frac{(1)}{(4)}$	Simplified method (5)	Refined method max. (6)	min. (7)	avg. (8)	$\frac{(5)}{(6)}$	$\frac{(5)}{(8)}$
1	1							1.00	1.00	1.00	1.00	1.00	1.00
2	1*	1.00	0.37	0.28	0.34	2.70	2.94	2.00	1.83	1.49	1.60	1.09	1.25
	2							1.00	1.00	1.00	1.00	1.00	1.00
3,4	1	1.00						0.00	0.54	0.22	0.43	—	
	2	1.00	0.67	0.59	0.62	1.48	1.62						
5	1	1.00	1.60	1.54	1.58	0.62	0.63	1.00	0.44	0.24	0.31	2.27	3.23
	2	1.00	1.09	1.08	1.08	0.92	0.93	1.00	0.94	0.90	0.93	1.06	1.07
	3							1.00	1.00	1.00	1.00	1.00	1.00
6	1	1.33	1.78	1.75	1.77	0.75	0.75						
	2	1.34	1.19	1.16	1.17	1.12	1.14	0.33	0.84	0.70	0.79	0.39	0.42
	3	0.33	0.06	0.05	0.06	5.39	5.96	0.67	0.93	0.96	0.95	0.72	0.71
7,8	1	1.00	1.00	1.00	1.00	1.00	1.00						
	2	1.50	1.67	1.63	1.66	0.90	0.91						
	3	0.50	0.37	0.33	0.36	1.34	1.40	0.50	0.73	0.57	0.68	0.68	0.73
9	1	1.34	1.21	1.18	1.19	1.11	1.12						
	2	1.83	1.95	1.94	1.95	0.94	0.94	0.34	0.25	0.14	0.17	1.38	1.95
	3	0.83	0.87	0.85	0.86	0.96	0.96	1.17	1.18	1.10	1.13	0.99	1.04
	4							1.00	1.00	1.00	1.00	1.00	1.00
10	1	1.00	1.00	1.00	1.00	1.00	1.00						
	2'	1.95	2.06	2.05	2.06	0.95	0.95						
	3	0.94	0.92	0.91	0.91	1.02	1.03	0.95	1.07	1.04	1.05	0.89	0.90
	4	0.11	0.03	0.03	0.03	3.20	3.54	0.89	0.98	0.96	0.97	0.91	0.92

*Step and level of maximum shore loading.
'Step and level of maximum slab loading.

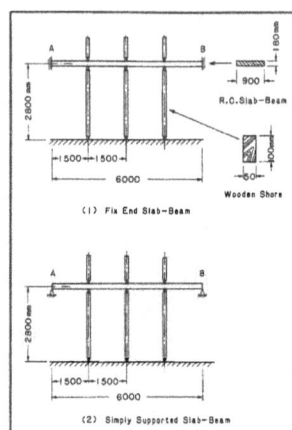

Fig. 3 — Example calculation model similar to Nielsen's example.[4]

assumed: the modulus of elasticity for concrete is $E_c = 5.1 \times 10^6$ psi $(3.5 \times 10^4$ MPa); the cylinder strength of concrete is $f'_c = 5900$ psi (41 MPa); Poisson's ratio for concrete is $\nu_c = 0.20$; the modulus of elasticity for wooden shores is $E_w = 1.1 \times 10^6$ psi $(7.75 \times 10^3$ MPa); and the compression strength f_w and Poisson's ratio ν_w of wood are 810 psi (5.6 MPa) and 0.3, respectively. The development of E_c and f'_c relative to their 28-day values are shown in Fig. 4.

A comparison between the shore and slab loads predicted using the refined analysis and the loads given in Table 5-2 of Reference 7, is presented in Table 2 for fixed-end slab edges and in Table 3 for simply supported slab edges. The maximum, minimum, and average shore load values predicted by the refined method for each slab level are reported in Tables 2 and 3. The axial shore loads predicted by the two methods often differ because in the simplified method all shores carry an equal load, whereas in the refined method the predicted load on each shore of a given slab depends on the relative

interstory deflection that occurs at a shore joint.

The results in Tables 2 and 3 also indicate that the simplified and refined methods both predict the identical location and construction step at which the maximum loads in the shores and slab occur (denoted respectively by * and ' in Tables 2 and 3), although the magnitudes of the loads are somewhat different. Two features concerning the effect of slab stiffness on the load distribution can be observed from the results reported in Tables 2 and 3. First, the influence of concrete age on the slab stiffness is not considered in the simplified method. Consequently, the simplified method underestimates the hardened slab load and overestimates the slab load from freshly placed concrete. Second, the difference between the results obtained using the refined and simplified methods is smaller in Table 3 than in Table 2. This is reasonable because the flexural stiffness of slabs with simply supported edges is less than that of slabs with fixed-end edges. Consequently, the assumption of infinite shore stiffness that

is used in the simplified method is more realistic for slabs with simply supported edges than for slabs with fixed-end edges.

The maximum load ratios for slabs and shores are summarized in Table 4. For the fixed-end slab case, analysis by the refined method predicted maximum shore loads less than those calculated by the simplified method, with a maximum difference of about 9 percent. However, for simply supported slab edges, the refined method predicted maximum shore loads 4 percent greater than those given by the simplified method.

The maximum load in the slabs predicted by the refined method is roughly 5 percent larger than the corresponding maximum load predicted by the simplified method for both end restraint conditions. The reason for differences in the results given by the two analysis techniques is that at step 2, while the level 2 concrete is being placed, the additional load does not all go through the shores to the ground level because the stiffness of the shores is not infinite, and the first floor slab picks up a portion of the additional load as it

TABLE 3—Simply supported slab edges; a comparison of results using the simplified method to results using the refined method.

Construction step	Floor level	Slab load ÷ slab weight			Shore load ÷ slab weight						
		Simplified method (1)	Refined method (2)	Comparison (1)/(2)	Simplified method (5)	Refined method max. (6)	min. (7)	avg. (8)		Comparison (5)/(6)	(5)/(8)
1	1				1.00	1.00	1.00	1.00		1.00	1.00
2	1*	0.00	0.07	—	2.00	2.07	1.79	1.88		0.97	1.06
	2				1.00	1.00	1.00	1.00		1.00	1.00
3,4	1	1.00	1.12	0.89							
	2	1.00	0.88	1.13	0.00	0.25	-0.01	0.16		—	—
5	1	1.00	1.20	0.83	1.00	0.90	0.64	0.73		1.11	1.38
	2	1.00	1.03	0.97	1.00	0.98	0.97	0.98		1.02	1.03
	3				1.00	1.00	1.00	1.00		1.00	1.00
6	1	1.33	1.52	0.87							
	2	1.34	1.27	1.05	0.33	0.60	1.44	0.55		0.55	0.60
	3	0.33	0.20	1.62	0.67	0.83	0.76	0.81		0.81	0.83
7,8	1	1.00	1.00	1.00							
	2	1.50	1.55	0.97							
	3	0.50	0.45	1.11	0.50	0.61	0.49	0.57		0.82	0.88
9	1	1.34	1.31	1.02							
	2	1.83	1.88	0.97	0.34	0.37	0.27	0.30		0.92	1.13
	3	0.83	0.80	1.04	1.17	1.23	1.17	1.19		0.95	0.98
	4				1.00	1.00	1.00	1.00		1.00	1.00
10	1	1.00	1.00	1.00							
	2†	1.95	2.02	1.03							
	3	0.94	0.90	1.04	0.95	1.02	1.01	1.02		0.93	0.94
	4	0.11	0.08	1.30	0.89	0.93	0.90	0.92		0.96	0.97

*Step and level of maximum shore load.
†Step and level of maximum slab load.

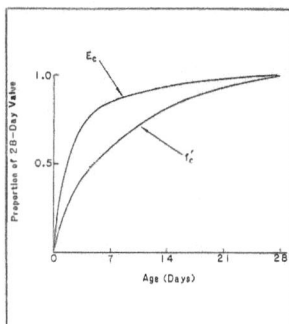

Fig. 4 — Development of E_c and f'_c with age.

deflects by an amount equal to the shortening of the shores. This also happens at step 5. Consequently, it can be noted that, when compared to the simplified method, the refined method predicts a higher maximum load for slabs; comparison of the maximum shore load depends upon the slab edge condition.

It is interesting to note that the foregoing comparison between the refined and simplified analyses is consistent with the comparison observed by Agarwal and Gardner[8] between field measurements and the simplified method. Increasing the slab and shore loads predicted by the simplified method by roughly 5 percent produces mean load values that are roughly consistent with the measured values. Additional field measurement studies are needed before any generalizations can be drawn. It appears, however, based on the results of the refined analysis and present field data, that increasing the shore and slab loads predicted by the simplified method by 5 to 10 percent, sufficiently adjusts the simplifying assumptions to provide more realistic values.

Conclusions

From comparisons between field measurements, the simplified analysis, and the refined analysis, it was found that:

• The simplified method developed by Grundy and Kabaila[5] is straightforward and easy to apply in practice. The simplicity of this method is very attractive for practical use.

• Comparison of maximum loads computed by the simplified method with those predicted by the refined method indicates that the maximum relative differences in the two methods varies between -5 and +9 percent. Because the effect of shore stiffness is not considered, the calculated results using the simplified method have larger errors than the results predicted using the refined method.

• The simplified method reliably predicts the construction step and location where the maximum slab and shore loads will occur but generally underestimates the actual load ratios. Consequently, the maximum slab and shore loads predicted by the simplified method can be corrected using a modification coefficient. The value of the modification coefficient for the simplified method should vary from 1.05 to 1.10 for design purposes.

TABLE 4—Maximum load ratios for slabs and shores.

	Slab			Shore		
	Simplified method (1)	Refined method (2)	(1) - (2) / (1) (percent)	Simplified method (3)	Refined method (4)	(3) - (4) / (3) (percent)
Fixed-end slab	1.95D	2.06D	-6	2.0D	1.83D	9
Simply supported slab	1.95D	2.02D	-4	2.0D	2.07D	-4

D = slab weight

References

1. Chen, W. F., and Liu, X. L., "Study of Concrete Frame Structures During Construction," *Structural Engineering Report* No. CE-STR-82-41, Purdue University, West Lafayette, Dec. 1982.

2. ACI Committee 347, "Recommended Practice for Concrete Formwork (ACI 347-78)," American Concrete Institute, Detroit, 1978, 37 pp.

3. Feld, Jacob, "Reshoring of Multistory Concrete Buildings," *Concrete Construction*, V. 19, No. 5, May 1974, pp. 243-248.

4. Neilsen, Knud E. C., "Loads on Reinforced Concrete Floor Slabs and Their Deformations During Construction," *Bulletin* No. 15, Final Report, Swedish Cement and Concrete Research Institute, Royal Institute of Technology, Stockholm, 1952.

5. Grundy, Paul, and Kabaila, A., "Construction Loads on Slabs with Shored Formwork in Multistory Buildings," ACI JOURNAL, *Proceedings* V. 60, No. 12, Dec. 1963, pp. 1729-1738.

6. Liu, Xila; Chen, Wai-Fah; and Bowman, Mark D., "Construction Load Analysis for Concrete Structures," *Journal of Structural Engineering*, ASCE, V. 111, No. 5, May 1985, pp. 1019-1036.

7. Hurd, M. K., *Formwork for Concrete*, 4th Edition, SP-4, American Concrete Institute, Detroit, 1981, 464 pp.

8. Agarwal, R. K., and Gardner, Noel J., "Form and Shore Requirements for Multistory Flat Slab Type Buildings," ACI JOURNAL, *Proceedings*, V. 71, No. 11, Nov. 1974, pp. 559-569.

9. Liu, X. L.; Chen, W. F.; and Bowman, M. D., "Load Distribution in Concrete Framed Structures During Construction," *Structural Engineering Report* No. CE-STR-83-38, Purdue University, West Lafayette, Jan. 1984.

Received and reviewed under Institute publication policies.

Acknowledgments

This paper is based on work supported by the National Science Foundation under a grant to Purdue University.

W. F. Chen is head of Structural Engineering in the school of Civil Engineering at Purdue University, West Lafayette, Indiana. He is a graduate of Cheng-Kung University in China, Lehigh University, and Brown University. Chen has taught at Lehigh and at Purdue since 1966. He has received numerous awards including the 1985 T. R. Higgins Lectureship Award from the American Institute of Steel Construction.

Xi-L. Liu is a graduate research assistant in civil engineering at Purdue University, West Lafayette, Indiana. He previously worked at the Institute of Building Research in Cheng Du, China, where he conducted research on the stability of concrete compression members and the development of reinforced concrete pipe columns.

*ACI member **Mark D. Bowman**, is an assistant professor of Structural Engineering in the School of Civil Engineering at Purdue University. He received degrees from Purdue University and from the University of Illinois, and has worked as a design engineer for a prestressed concrete manufacturer. He is a member of ACI Committee 215, Fatigue of Concrete Structures.*

5.9.2

Design considerations for formwork in multistorey concrete buildings

Khalid Mosallam and Wai-Fah Chen

School of Civil Engineering, Purdue University, West Lafayette, IN 47907, USA
(Received July 1989; revised October 1989)

In a multistorey reinforced concrete (RC) building construction, the formwork system and the partially completed structure support all construction loads occuring during the construction process. Field investigations and recent analytical studies have shown that construction loads on a supporting system including slabs, shores and reshores may appreciably exceed their load-carrying capacity and contribute to a significant portion of the disasters of RC buildings during construction.

This paper will examine many factors relating to the analysis of multistorey reinforced concrete buildings during construction. In particular, we shall examine causes of falsework failures, review existing codes and national standards as they affect safety in construction, discuss factors to be considered in the analysis, and summarize several construction procedures used in practice.

This paper also describes an analytical method using the current Load and Resistance Factor Design (LRFD) approach to check the slab adequacy during construction.

Keywords: multistorey buildings, reinforced concrete, formwork (construction)

In general, the primary causes of formwork disasters are (1) excessive loads, (2) premature removal of forms or shores, and (3) inadequate lateral support for the shoring members.

Hadipriono and Wang[1] examined the causes that resulted in 85 major falsework failures over the past 23 years. Three causes of failure were identified: triggering, enabling, and procedural causes (*Table 1*). Most failures occurred because of interaction of the triggering and enabling events that were, in many cases, produced by inadequacies in the procedural methods.

Inadequate falsework cross-bracing or lacing was the primary source of several falsework accidents (see *Table 1*, section b), such as the collapses of the Arroyo Seco Bridge in California, the Skyline Center Complex in Virginia, the highway ramp in East Chicago, and the Coliseum in New York.

National codes and standards

For concrete construction, many sets of requirements and standards exist ranging from those of federal standards to those of state and local building codes. The following four documents have established national applicability in the USA.

(1) The ACI Committee 347(1988), 'Guide to Formwork for Concrete'[2], which replaces a previous standard, ACI 347-78, 'Recommended Practice for Concrete Formwork'.
(2) ANSI A10.9-83, the American National Standards Institute 'American National Standard for Construction and Demolition Operations—Concrete and Masonry Work—Safety Requirements'[3].
(3) Chapter 6 of ACI 318-83 'Building Code Requirements for Reinforced Concrete', which deals with construction practices[4].
(4) OSHA Subpart Q (1988) of the federal Construction Safety and Health Regulations[5].

The new 'Guide to Formwork for Concrete', prepared by ACI Committee 347, and the ANSI A10.9 standard are more detailed voluntary consensus documents that establish much of the state-of-the-art technology which serves as a basis for safe construction practices. An ACI publication, SP-4 'Formwork for Concrete'[6] which serves as a commentary to ACI 347, has abundant detailed information relative to formwork practices including design aids, design examples and illustrations.

The revised OSHA Subpart Q called 'Concrete and Masonry Construction'[5] is simpler and more per-

0141-0296/90/030163-10/$03.00
© 1990 Butterworth–Heinemann Ltd

Formwork in multi-storey concrete buildings: K. Mosallam and W.-F. Chen

Table 1   Causes of falsework failure (Hadipriono and Wang[1])

Number of occurrence	Causes of failure
(a) Triggering cause of failure	
3	Heavy rain causing falsework foundation slippage
1	Strong river current causing falsework foundation slippage
1	Strong winds
4	Fire
5	Failure of equipment for moving formwork
4	Effects of formwork component failure
1	Concentrated load due to improper prestressing operation
2	Concentrated load due to construction material
2	Other imposed loads
27	Impact loads from concrete debris and other effects during concreting
3	Impact load from construction equipment/vehicles
5	Vibration from nearby equipment/vehicles or excavation work
6	Effect of improper/premature falsework or formwork removal
20	Other causes or not available
(b) Enabling causes of failure	
17	Inadequate falsework cross-bracing/lacing
14	Inadequate falsework component
9	Inadequate falsework connection
7	Inadequate falsework foundation
8	Inadequate falsework design
4	Insufficient number of shoring
1	Inadequate reshoring
4	Failure of movable falsework/formwork components
2	Improper installation/maintenance of construction equipment
1	Failure of permanent structure component
4	Inadequate soil foundation
2	Inadequate design/construction of permanent structure
30	Other causes or not available
(c) Procedural causes of failure	
23	Inadequate review of falsework design/construction
22	Lack of inspection of falsework/formwork during concreting
2	Improper concrete test prior to removing falsework/formwork
4	Employment of inexperienced/inadequately trained workmen
1	Inadequate communication between parties involved
5	Change of falsework design concept during construction
38	Other causes or not available

formance-oriented, leaving the contractor options in just how to meet the requirements.

While standards, numerous guidelines and manuals for the design of formwork exist, there is considerable variation in the requirements for design loads and in the safety factors required for construction. At the present time, no codes and standards provide a uniform approach to the design of the partially completed slab/shore system as a consistent system with the design standard and procedures used for permanent structures.

Theoretical review

Neilsen[7] was one of the first who studied the distribution of loads on shores and slabs in multistorey structures. He made an extensive theoretical analysis of the interaction between the formwork and the slabs supporting the falsework loads. In the analysis, he assumed that: (1) the slabs behave elastically; (2) shrinkage and creep of the concrete can be neglected; (3) shores are represented by uniform elastic support; and (4) the torsional moments and the shearing forces in the formwork are neglected. Neilsen's procedures seem lengthy and inconvenient for practical use.

Grundy and Kabaila[8] developed a simplified method for determining the construction loads imposed on the supporting slabs and formwork. They assumed that the rigidity of shores is infinite compared with that of the slabs. Blakey and Beresford[9] and Feld[10] extended Grundy and Kabaila's method to consider different slab shore systems and various construction rates.

Agarwal and Gardner[11] compared field measurements of the construction loads from two high-rise buildings with the loads predicted by the Grundy–Kabaila simplified method. The field measurements of construction loads agreed to an acceptable accuracy with those predicted by the simplified method of Grundy and Kabaila.

Liu, Chen and Bowman[12,13] developed refined two- and three-dimensional computer models to check the major assumptions used by the Grundy–Kabaila simplified method. They concluded that the simplified method is adequate for predicating the construction step and the location of the maximum slab moments and shore loads. However, the numerical values of slab moments and shore loads should be corrected by using a modification coefficient.

Liu, Lee and Chen[14] developed the mathematical formulation of the simplified method for all possible construction operations. This formulation was implemented in a computer program which can be run on a personal computer to analyse construction loads on shores and slabs during different construction steps. Also, slab safety at any construction stage can be checked by comparing the applied loads with the available slab strength.

El-Sheikh and Chen[15] studied the effect of fast construction rate on short- and long-term deflections of RC buildings. They concluded that a fast construction rate will not affect the short- or long-term deflections provided that high strength concrete and high early strength cement are used.

It should be noted that:

(1) up to now, none of the foregoing methods has been chosen by the present codes and standards for the analysis of RC buildings during construction, albeit they are referenced as acceptable analysis techniques by ACI Committee 347;
(2) all the aforementioned methods did not include the effect of horizontal (or lateral) loading in the distribution of loads on shores and slabs, or the effect of lateral bracing on the maximum slab or shore load distribution.

For safe and economical design, critical combinations of construction loads should be considered and the

Formwork in multi-storey concrete buildings: K. Mosallam and W.-F. Chen

transmission of construction loads to the interconnected slabs by shores and reshores must be analysed. Against the background of this information, the development of an analytical model simulating the concrete construction process is urgently needed. This model should examine the effect of horizontal loads on the distribution of loads on shores and slabs, and the effect of different bracing systems on the horizontal stiffness of falsework and on the maximum slab and shore load distribution.

As a first step, a thorough understanding of the different bracing systems used in practice is necessary. This is described below.

Bracing and lacing

The formwork system should be designed to transfer all horizontal loads to the ground or to the completed construction in such a manner as to ensure safety at all times.

Cross-bracing (diagonal bracing) functions primarily to resist lateral loads and to prevent instability of individual members. A major proportion of resistance against lateral movement can be provided by placing the columns at least a day ahead of the floor slab. The function of horizontal lacing is to prevent the shores and reshores from buckling by reducing the unsupported length. The brace system should be anchored in a manner to ensure stability of the total system.

Much of the bracing in use for formwork today is erected on the basis of experience and judgement of the superintendent or foreman on the job. Some good bracing practices (wooden and steel single-pole shores)[6,16] are listed here.

(1) The slenderness ratio is the relationship of unsupported length to the cross-sectional dimension in the face under consideration. If the shore is unbraced, use the dimension of the narrower face in determining the ratio. For wooden shores, this ratio must not exceed 50. For steel shores, the slenderness ratio is expressed as Kl/r, where l = unsupported length in inches, r = radius of gyration and K = effective length factor ($K = 1$ is used for steel shores, which are conservatively assumed to be pin ended). For steel shores, l/r should not exceed 200.

(2) The bracing system must be tied to solid ground or permanent construction, unless it is multidirectional with sufficient X-bracing to give it internal rigidity. For one-piece shores, horizontal strut bracing may be adequate. *Figures 1, 2* and *3* show three alternative schemes for bracing slab shoring using (1) braced bays, (2) braced lines, and (3) tying to the completed columns or walls. Minimum lateral loads used in designing components of this bracing are given by Hurd[6].

(3) For a specific job, as single-post shores are installed, the necessary stability lacing and bracing can be added to the shores. The lacing must be continuous horizontally on shores and in both directions in the horizontal plane. This lacing is to be located at approximately mid-height and is to be attached to the shore. Bracing or lacing connections should develop the strength of the bracing or lacing member, but at least 10d double-headed form nails should be used at each nailing plate driven from the

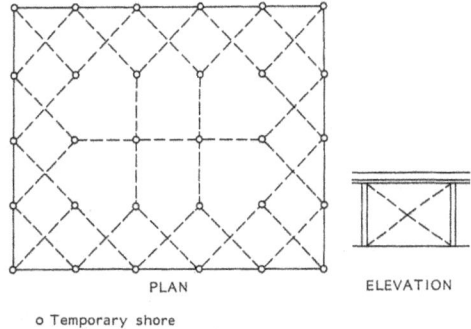

PLAN   ELEVATION

o Temporary shore

Figure 1 Braced bays; bays marked with dashed X-lines have complete X-bracing system on vertical lines in both directions as well as horizontal X-bracing. Centre shores tied in with strut bracing[6]

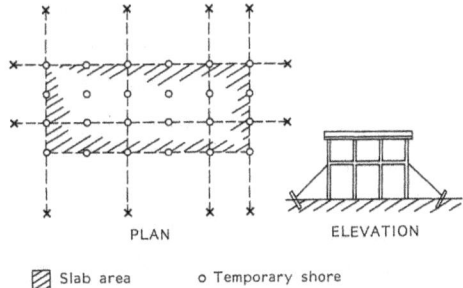

PLAN   ELEVATION

▨ Slab area   o Temporary shore

Figure 2 Braced lines; for short heights, strut bracing is used. X-bracing for spliced shores[6]

Form design

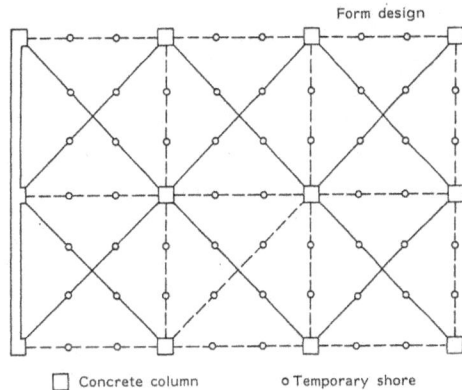

□ Concrete column   o Temporary shore

Figure 3 Use of completed columns or walls for bracing; solid lines represent X-bracing, dashed lines indicate one line of strut braces. Intermediate lines are needed if shores are spliced[6]

shore side through a two by six (preferred) or two by four. The nail should be driven all the way through and then clinched (see *Figure 4*).

(4) Diagonal bracing parallel to the stringer lines and perpendicular to stringer lines must be installed at

Formwork in multi-storey concrete buildings: K. Mosallam and W.-F. Chen

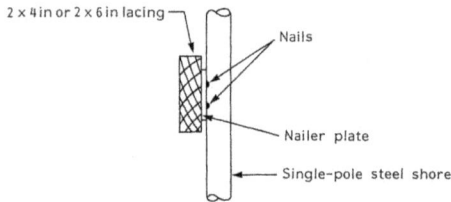

Figure 4   Position of nails and nailer plate when afixed to a steel shore[16]

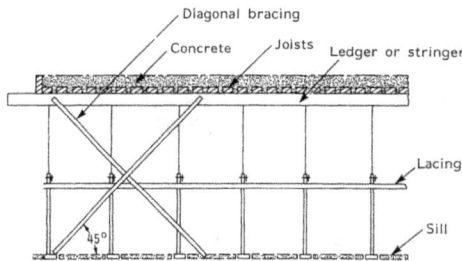

Figure 5   Proper configuration for vertical shoring[16]

both ends of the lines, from ledger to sill, to form a cross (X). If the stringer lines are long, this should be done at intermediate points (see *Figure 5*). Diagonal bracing should be installed at an angle not greater than 45° or not less than 30°. Diagonal bracing should be provided between lines of post shores to meet the expected lateral forces or at least every 10 to 15 feet (3 to 4.5 m) as minimum lateral stability requirements.

(5) Bracing between stringer lines should be done at the ends of the lines and at intermediate points. If the ratio of the height to the width of the stringer is two to one or less (four by eight or four by six), horizontal bracing is not needed for the stringer. If the height or width of the stringer is more than two to

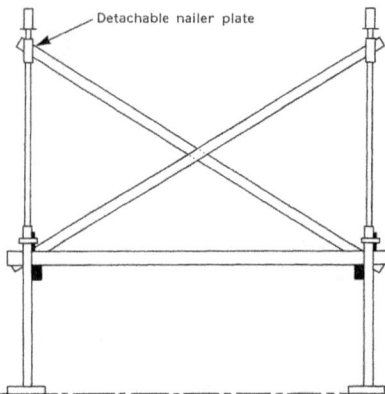

Figure 6   Proper configuration for lacing and diagonal bracing[16]

one (three by eight or four by ten), diagonal bracing or bridging must be provided between stringers for lateral stability. This bracing can take the form of either tying the lacing together and installing diagonal bracing between the top of the shore and the lacing ties (see *Figure 6*), or bracing diagonally from the top stringer line to the sills of another line.

Factors to be considered in the analysis

In determining the number of floors to be shored or reshored and to determine the loads transmitted to the floors, shores and reshores as a result of the construction sequence, ACI Committee 347 recommends the consideration of 10 factors.

(1) Structural design load of the slab or member including live load, partition loads, and other loads for which the engineer designed the slab. Where the engineer included a reduced live load for the design of certain members and allowances for construction loads, such values should be sown on the structural drawings and be taken into consideration when performing this analysis.
(2) Dead load weight of the concrete and formwork.
(3) Construction live loads, such as placing crews and equipment or stored materials.
(4) Design strength of concrete specified.
(5) Cycle time between placement of successive floors.
(6) Strength of concrete at time it is required to support shoring loads from above.
(7) The distribution of loads between floors, shores and reshores at the time of placing concrete, stripping formwork and removal of reshoring.
(8) Span of slab or structural member between permanent supports.
(9) Type of formwork systems, i.e., span of horizontal formwork components, individual shore loads, etc.
(10) Minimum age where appropriate.

Determination of shore and reshore adequacy

Adequacy of the shores and reshores can be determined simply by checking that the uniformly distributed loads calculated from the slab/shore analysis multiplied by an appropriate tributary area are less than the safe working loads of the shores.

Determination of slab adequacy

To ensure a safe construction scheme, the capacity of the slabs (taking into account the reduced strength of the immature slabs) must be compared with the slab loads determined by a slab-shore analysis at various stages of construction. If a slab load exceeds its capacity, then the proposed construction procedure cannot be used and one or more measures must be taken to either strengthen the slab or to reduce the load on the slab at the critical age. Several possibilities for doing this include increasing concrete strength at the time of loading by using heat, higher strength concrete or high early strength cement, changing the construction cycle, or modifying the number of shores or reshores storeys to produce a more favourable distribution of construction loads. The strength capability of a reinforced concrete slab is a function of the design ultimate load, age of the slab, construction temperature, type of cement and the critical modes of the slab behaviour. The method described herein can be used to check the slab adequacy. The method is similar to that of Reference 17 except that it uses the simplified method and it accounts for construction live loads. A Load and Resistance Factor Design (LRFD) approach is used whereby the nominal strength must be at least equal to the maximum effect of the design loads multiplied by appropriate load factors. The nominal strength available at the early ages at which the slab experiences construction loads, R_M, is less than the 28-day available strength, or

$$R_M = \beta R_n \tag{1}$$

in which β is a modification coefficient which depends on the concrete age, temperature, type of cement and the critical mode the slab behaviour. R_n is the nominal design strength.

If the design loads consist of dead and live loads, the LRFD strength requirement for service conditions may be written as

$$\gamma_D D + \gamma_L L \le \phi R_n \tag{2}$$

in which γ_D and γ_L are the dead and live load factors, respectively; ϕ is the strength reduction factor; and D and L are the dead and live loads, respectively.

Using the LRFD approach, one may write an analogous expression for construction rather than service load effects. This may writen as

$$\gamma_C C_t \le \phi R_M \tag{3}$$

where γ_C is the construction load factor; C_t is the construction slab load at age t; ϕ is the strength reduction factor which varies with concrete strength. Since there is not sufficient data to determine ϕ, it will be assumed constant and will be taken equal to ϕ given in equation (2).

If the inequality in equation (2) is replaced by an equality, then, solving for R_n

$$R_n = (\gamma_D D + \gamma_L L)/\phi \tag{4}$$

Substitution of this expression for R_n into equation (1) yields

$$R_M = \beta(\gamma_D D + \gamma_L L)/\phi \tag{5}$$

Substituting R_M from equation (5) into equation (3), and solving for C_t/D, one obtains, after some rearrangement

$$C_t/D \le \frac{\beta(\gamma_D + \gamma_L L/D)}{\gamma_C} \tag{6}$$

ACI 318-83[4] gives the design dead and live factors as $\gamma_D = 1.4$ and $\gamma_L = 1.7$. The American National Standard ANSI A10.9-1983[3] requires the use of a load factor of 1.3 for all construction loads. Using these values in the above equation, one obtains

$$C_t/D \le \frac{\beta(1.4 + 1.7L/D)}{1.3} \tag{7}$$

Using the simplified method and a modification coefficient of 1.1 (to allow for the error in the theory), the above equation can be rewritten as

$$C_t/D \le \frac{\beta(1.4 + 1.7L/D)}{1.1 \times 1.3} \tag{8}$$

which can be approximated as

$$C_t/D \le \beta(1.0 + 1.2L/D) \tag{9}$$

Table 2 Development of concrete strength (Liu et al.[14])

Age: (days)	Type I			Type II cement		
	73°F (22.8°C)	55°F (12.8°C)	40°F (4.4°C)	73°F (22.8°C)	55°F (12.8°C)	40°F (4.4°C)
1	0.31	0.15	0.03	0.54	0.33	0.11
2	0.47	0.28	0.11	0.65	0.50	0.30
3	0.59	0.40	0.18	0.74	0.62	0.43
4	0.66	0.49	0.24	0.78	0.66	0.54
5	0.72	0.57	0.32	0.81	0.70	0.63
6	0.76	0.63	0.39	0.83	0.73	0.70
7	0.79	0.68	0.44	0.85	0.75	0.77
8	0.81	0.72	0.48	0.86	0.77	0.80
9	0.83	0.75	0.52	0.88	0.79	0.82
10	0.85	0.77	0.56	0.89	0.81	0.84
11	0.86	0.80	0.59	0.90	0.82	0.86
12	0.88	0.82	0.62	0.91	0.84	0.88
13	0.89	0.84	0.64	0.92	0.85	0.89
14	0.90	0.86	0.67	0.92	0.86	0.90
21	0.96	0.94	0.80	0.97	0.93	0.99
28	1.00	1.02	0.88	1.00	0.96	1.07

Note: For a given treatment, β can be obtained by a linear interpolation

Formwork in multi-storey concrete buildings: K. Mosallam and W.-F. Chen

C_t/D is the maximum permissible slab load ratio determined from an analysis using the simplified method accounting for construction dead and live loads. According to the work reported by Gardner[18], β can be determined easily using *Table 2*. β can also be calculated from a maturity-based model proposed by Lew[17], using the following equation

$$\beta = [0.007M/(1 + 0.007M)] \times (4550/4000) \qquad (10)$$

where M is the maturity defined as the integral over time of the temperature of the concrete above a datum temperature, or

$$M = \int (T - T_0)dt \qquad (11)$$

in which T is the temperature of the concrete and T_0 is the datum temperature, i.e., that temperature below which there is no further strength gain with time. Equation (10) is based on results of compression tests conducted at 7 and 28 days under standard curing conditions ($T = 73°F$). The specified compressive strength for design was 4000 psi with maturity computed from equation (11) using $T_0 = 32°F$. Using equations (10) and (11), β is computed for different temperatures for Type I cement. This is shown in *Table 3* and can be compared with *Table 2*[14]. Both tables give close results for Type I cement when using a temperature of 73°F but different results when using a temperature of 40°F (this may be due to the assumption of $T_0 = 32°F$).

Examples

Several examples to check the adequacy of slabs during construction are described here. The examples illustrate the effects of varying the construction cycle (C), the influence of several curing temperature conditions (T), and the ability to affect the age at which a maximum slab load occurs by changing the construction scheme (M and N).

In all these examples, the simplified method was used to find the slab load ratio using the computer program developed by Liu et al.[14]. The adequacy of slabs was determined using equations (9), (10) and (11). The specified concrete strength is 4000 psi and the live-to-dead load ratio is taken to be 1.25 for the first two examples and 1.50 for the third example.

Example 1. *Figure 7* shows the graph resulting from load analysis for a construction scheme using two levels of shores and one level of reshores ($M = 2$ and $N = 1$). Two construction cycles, at 4 and 7 days per floor ($C = 4$ and $C = 7$), at a temperature of 73°F ($T = 73°F$) were used. The graph shows that the 4-day cycle is not acceptable and that a casting cycle of 7 days per floor is required.

Example 2. *Figure 8* shows the graph resulting from load analysis for a construction scheme using two levels of shores and one level of reshores ($M = 2$ and $N = 1$) and a casting cycle of 7 days per floor ($C = 7$). Two curves are shown, one for 73°F ($T = 73°F$) and one for 55°F ($T = 55°F$) curing temperature. This graph indicates that, if the concrete can be maintained at only 55°F, the slab load will exceed the capacity at the age of 14 days.

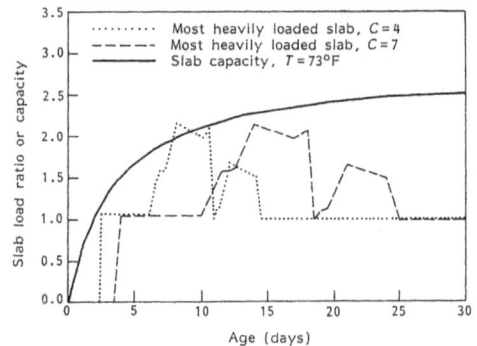

Figure 7 Comparison of construction cycles (C); $M = 2$, $N = 1$, $T = 73°F$

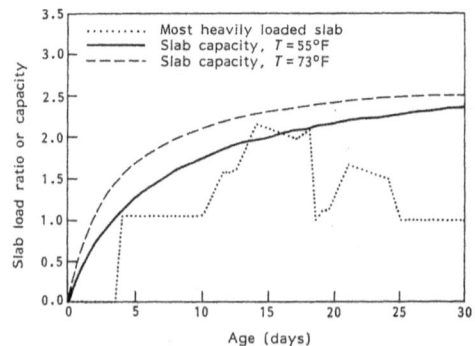

Figure 8 Comparison of curing temperatures (T); $M = 2$, $N = 1$, $C = 7$

Table 3 Development of concrete strength*

Age (days)	Type I cement		
	73°F (22.8°C)	55°F (12.8°C)	40°F (4.4°C)
1	0.25	0.16	0.06
2	0.41	0.28	0.11
3	0.53	0.37	0.16
4	0.61	0.44	0.21
5	0.67	0.51	0.25
6	0.72	0.56	0.29
7	0.76	0.60	0.32
8	0.79	0.64	0.35
9	0.82	0.67	0.38
10	0.84	0.70	0.41
11	0.86	0.73	0.43
12	0.88	0.75	0.46
13	0.90	0.77	0.48
14	0.91	0.79	0.50
21	0.98	0.88	0.61
28	1.00	0.93	0.69

*Using equations (10) and (11) (Lew[17])

Example 3. *Figure 9* shows the graph resulting from load analysis for a different construction scheme using one, two or three levels of shores and one level of reshores ($M = 1$, 2 or 3 and $N = 1$). A casting cycle of 5 days per floor ($C = 5$) at a temperature of 60°F ($T = 60°F$) was used. As can be seen, if three levels of shores are used, the maximum slab load, while larger than that for the other two schemes, is less than the predicted slab capacity.

The assumption that there is a constant ratio between concrete strengths at different ages is not always correct. For example, 7-day strength has often been estimated to be about 76% of the 28-day strength (see *Table 3*). In reality, for a given mix design, the ratio varies depending on the early strength. *Figure 10*[19], which was generated from actual test data, shows that as the 7 day strength increases, the ratio of 7-day strength to 28-day strength increases. Also, because concrete is a variable material, its strength will change from day to day. Changing climatic conditions can induce upward or downward trends as shown in *Figure 11*[19]. For these reasons, in a field situation, the strength of concrete should be verified by test as construction proceeds, not be based on typical curves or tables, as shown in this paper.

Construction procedures

Some of the difficulties associated with making a precise analysis have to do with varying practices and definitions of shoring and reshoring. For purposes of this discussion, the following definitions apply[2].

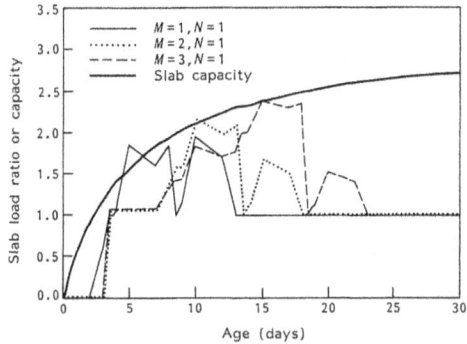

Figure 9   Comparison of construction schemes (M and N); $C = 5$, $T = 60°F$

- *Shores*—vertical or inclined support members designed to carry the weight of formwork, concrete and construction loads above.
- *Reshores*—shores placed snugly under a stripped concrete slab or structural member after the original forms and shores have been removed from a large area, thus requiring the new slab or structural member to deflect and support its own weight and existing construction loads applied prior to the installation of the reshores.
- *Backshores*—shores placed snugly under a stripped concrete slab or structural member after the original

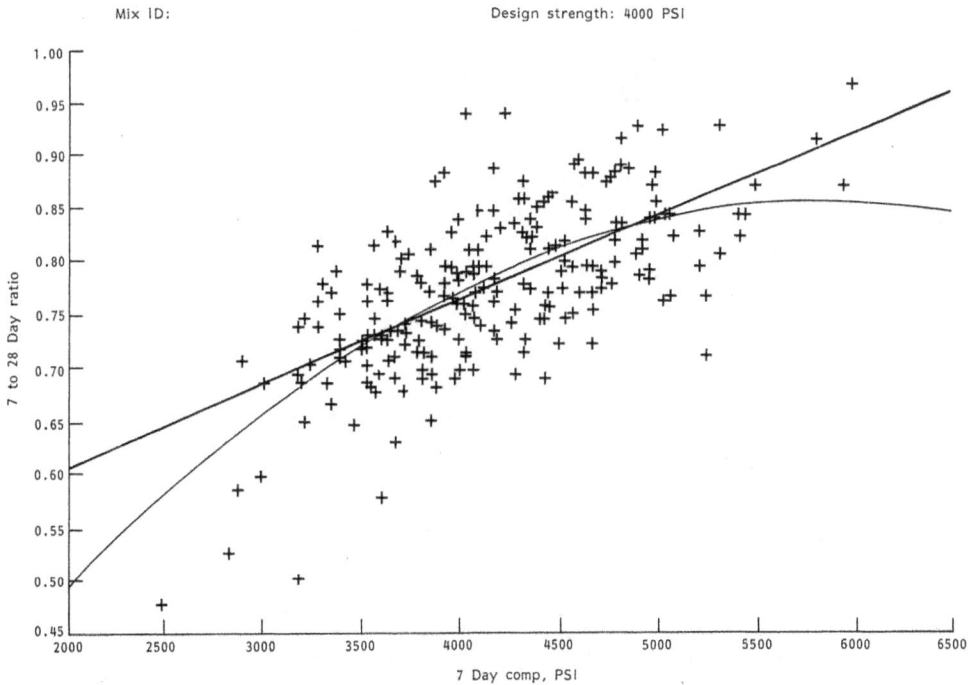

Figure 10   For a given mix, ratio of 7-day strength to 28-day strength varies depending on early strength[19]

Formwork in multi-storey concrete buildings: K. Mosallam and W.-F. Chen

Mix ID: Design strength: 4000 PSI

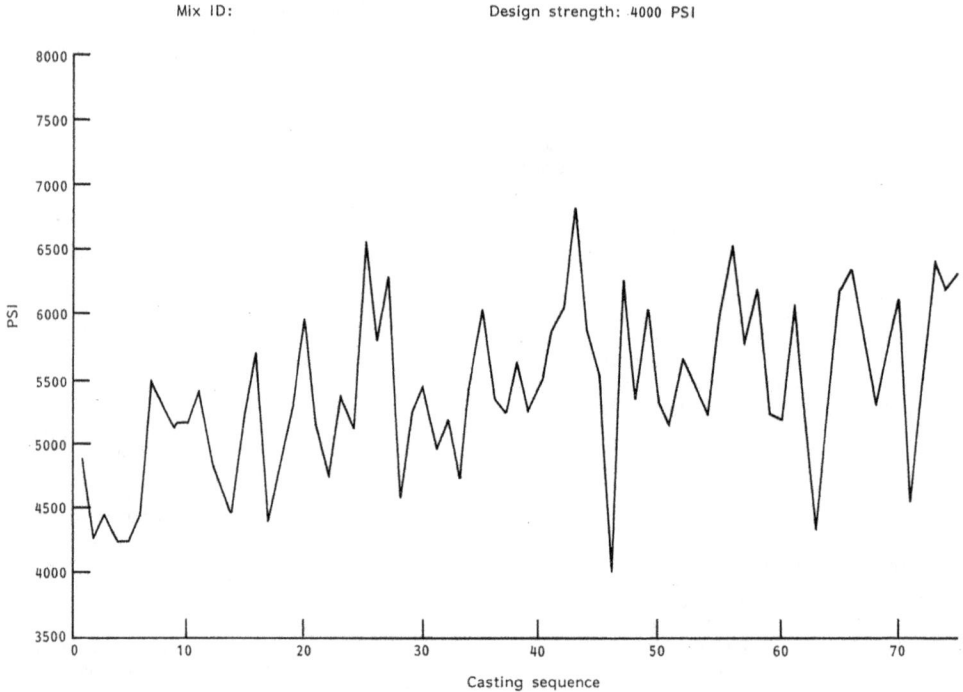

Figure 11   Changes in concrete strength over time[19]

forms and shores have been removed from a small area without allowing the slab to deflect or support its own weight or existing construction loads from above.

● *Preshores*—a refined version of the reshoring method. It is the technique of scheduled reshoring whereby the unsupported slab is reduced and controlled[20].

The loads occuring on the supporting slabs can be large compared with the slabs' designed capacity (as shown in the previous section) and uniquely determined by the construction procedure, namely *shores only, shores and reshores, shores and backshores*, and *shores, preshores and reshores*. These methods are described briefly below.

Shoring

The typical construction procedure for slabs involves casting the new slab on shores (formwork) and removal of supports from the lowest level at prehaps four to five days after casting the slab. Shoring is carried to several floors so the load of the fresh concrete will be shared among several supporting slabs. Increasing the number of shores will delay the occurrence of the peak load, thus benefiting from the strength growth with time; however, it will increase the maximum slab load and the maximum shore load[21].

Shoring, reshoring

The shore/reshore technique involves using typically only one level of shores and several levels of reshores. Basically, the forms/shores are removed from beneath a slab allowing it to deflect and carry its own weight; reshores are then installed allowing the load during concrete placement to be shared between the various slabs in the systems. The slab loads imposed using reshoring construction are much less than those using shores but the loads are imposed at earlier ages. Reshoring usually requires fewer levels of interconnected slabs, thus freeing more areas for other trades. Near-capacity loads in slabs usually occur for shorter periods. The disadvantage of the reshoring technique is that, while allowing the slab to deflect reduces the slab maximum loads applied to the slab, the slab deflection can be excessive as the concrete is too young to have developed much stiffness.

Shoring, backshoring

With backshoring, so long as the first level shores remain in place in contact with grade, each tier of shores must carry the weight of all concrete and construction loads above it. Once the tier of backshores in contact with grade has been removed, the assumption is made that the system of slabs behaves elastically. The slabs interconnected by backsores will deflect equally during addition or removal of loads. Loads will be distributed among the slabs in proportion to their developed stiffness. Using backshores,

Table 4 Comparison of reshoring and backshoring (ACI 347)

Backshoring	Reshoring
Strip small areas	Strip several entire bays
Do not let slab deflect	Allow slab to deflect
Install backshores before any further stripping occurs	Install reshores without removing deflection
Slab does not carry its own weight	Slabs carry their weight
Backshores have an initial load	Reshores have no initial load

stripping of forms may be accomplished at an early age because large areas of concrete are not required to carry their own weight. New slabs carry less load, thus reducing the effects of early creep. Backshoring requires knowledgeable supervision and extreme caution. Care must be exercised to ensure that individual shores are not overloaded during stripping. *Table 4* compares key features of reshoring and backshoring[2].

Shoring, preshoring and reshoring

In preshoring construction, specified shores in the shore removal process are removed leaving other shores in place; reshores are then installed at the location(s) of the removed shores; the remaining shores are then removed and replaced by reshores. The method is illustrated in *Figure 12*[20]. *Figure 12a* represents the casting of the new floor onto formwork supported by shores a, b and c. *Figure 12b* shows shores a and c removed, allowing the slab to deflect over a span of half the bay. *Figure 12c* represents the reshores placed at a and c beneath the deflected slab

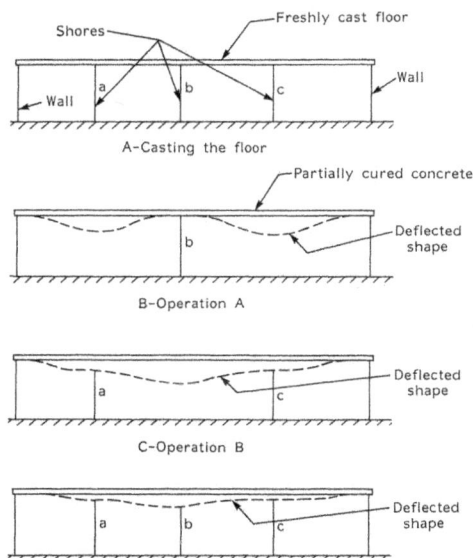

Figure 12 Schematic representation of the preshoring technique[20]

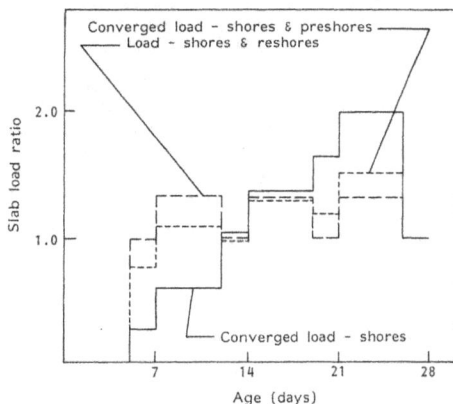

Figure 13 Comparison of load histories for shored, preshored and reshored construction[20]

and shore b removed. Finally in *Figure 12d* reshores are placed at location b. Generally operations A and B would be carried out during the same day.

The advantage of preshoring is that the unsupported span lengths are reduced, thus reducing slab deflections. One disadvantage of preshoring is that to realize the benefits close control of the construction process is needed. Also, preshoring is more complicated to model than reshoring in that the analysis has to consider slab and shore deformations in addition to static equilibrium of the loads[20]. *Figure 13*[20] shows ratios calculated for shoring, reshoring and preshoring schedules for the converged solution in each case. Shoring gives the highest load ratio, preshoring the middle load ratio and reshoring the smallest load ratio.

Conclusions

An attempt has been made to examine many factors relating to the analysis of multistorey reinforced concrete buildings during construction, including causes of falsework failures, a review of existing codes and national standards as they affect safety in construction, and a review of the state-of-the-art knowledge. Based on this study, the following conclusions can be drawn.

(1) A survey of reported failures in concrete buildings under construction indicates that almost half of the building falsework failures were due to deficiences in regular vertical shores, and one out of two falsework collapses was due to inadequate lateral bracing.

(2) To date, only a few studies have been conducted on the development of analytical models to simulate the construction process; moreover, none of these models has been chosen or confirmed by code or standard writing bodies.

(3) All the previous analytical models *have not* included the effect of lateral loads in the distribution of loads on shores and slabs or the effect of lateral bracing on the maximum slab or shore load distribution. For safe and economical design, critical combinations of construction loads should be considered and the

Formwork in multi-storey concrete buildings: K. Mosallam and W.-F. Chen

transmission of construction loads to the interconnected slabs by shores and reshores must be analysed.

(4) It is impossible to recommend a single general
procedure for shoring and reshoring multistorey structures, since the interrelationships and variables differ
widely from one job to another. Shoring, reshoring,
backshoring and preshoring are different construction
procedures used to distribute construction loads
through the lower floors. Though load distribution
analysis is similar, there are significant differences
in magnitude, duration, and timing of floor and shore
loads for each procedure.

References

1 Hadipriono, F.C. and Wang, H.K. 'Analysis of causes of formwork
   failures in concrete structures,' *J. Construction Engng and Management, ASCE* 1986, *112*, No. 1, 112–121
2 ACI Committee 347, 'Guide to Formwork for Concrete', *ACI Structural Journal* September–October 1988, 22643–22646
3 American National Standards Institute 'American National Standard
   for Construction and Demolition Operations—Concrete and Masonry
   Work—Safety Requirements (ANSI A10.9-1983)', New York, 1983,
   22 pp
4 ACI Committee 318, 'Building Code Requirements for Reinforced Concrete (ACI 318–83)', American Concrete Institute, Detroit, MI, 1983,
   111 pp
5 Hurd, M.K. 'OSHA revises concrete construction safety rules', *Concrete Construction* December 1988, 1079–1090
6 Hurd, M.K. 'Formwork for concrete', American Concrete Institute,
   Detroit, MI, 1979, 464 pp
7 Neilsen, K.E.C. 'Load on reinforced concrete slabs and their deformation during construction', *Bulletin No. 15*, Final Report, Swedish
   Cement and Concrete Research Institute of Technology, Stockholm,
   1952, 112 pp
8 Grundy, P. and Kabaila, A. 'Construction loads on slabs with shored
   formwork in multistory buildings', *ACI Journal* 1963, *60*, No. 12,
   1729–1738
9 Blakey, F.A. and Beresford, F.D. 'Stripping of formwork for concrete buildings in relation to structural design', *Civil Engineering
   (Trans. of Instn. Engrs, Australia)* 1965, CE7, No. 2, 92–96
10 Feld, J. 'Reshoring of multistory concrete buildings', *Concrete Construction* May 1974, 243–248
11 Agarwal, R.K. and Gardner, N.J. 'Form and shore requirements for
   multistory flat slab type buildings', *ACI Journal* 1974, *71*, No. 11,
   559–569
12 Liu, X.L., Chen, W.F. and Bowman, M.D. 'Construction load analysis
   for concrete structures', *J. Struct. Engng* 1985, *111*, No. 5, 1019–1036
13 Liu, X.L., Chen, W.F. and Bowman, M.D. 'Shore slab interaction
   in concrete buildings', *J. Construction Engng and Management, ASCE*
   1986, *112*, No. 2, 227–244
14 Liu, X.L., Lee, H.M. and Chen, W.F. 'Analysis of construction loads
   on slabs and shores by personal computer', *Concrete International:
   Design & Construction* June 1988, 21–30
15 El-Sheikh, M. and Chen, W.F. 'Effects of fast construction rate on
   deflections of R.C. buildings', *J. Struct. Engng* 1988, *114*, No. 10,
   2225–2238
16 National Safety Council 'Vertical shoring of concrete formwork', *National Safety News* 1980, *122*, No. 5, 51–59
17 Lew, H.S. and Gross, J.L. 'Analysis of shoring loads and capacity
   for multistory concrete construction', *ACI Second International Conference on Forming Economical Concrete Buildings*, Chicago, IL,
   November 1984, 109–130
18 Gardner, N.J. 'Shoring, reshoring, and safety', *Concrete International:
   Design & Construction* 1985, *7*, No. 4, 28–34
19 Shilstone, J.M. 'Evaluating concrete potential and performance', *ACI
   Third International Conference on Forming Economical Concrete
   Buildings*, Dallas, TX, November 1986, 147–168
20 Gardner, N.J. and Chan, C.S. 'Comparison of preshore and reshore
   procedures for flat slabs', *ACI Second International Conference on Forming Economical Concrete Buildings*, Chicago, IL, November 1984,
   157–174
21 Liu, X.L., Lee, H.M. and Chen, W.F. 'Shoring and reshoring of high-
   rise buildings', *Concrete International: Design & Construction* January
   1989, 64–68
22 Chen, W.F. and Liu, X.L. 'Study of concrete framed structures during construction', special session on Safety Considerations during
   Construction, *ASCE Annual Convention*, New Orleans, 25–29 Oct
   1982, 10 pp
23 Feld, J. 'Concrete formwork failures', *ACI Journal*, 1975, *72*, No.
   7, 351–355
24 Lew, H.S. 'Safety during construction of concrete buildings—a status
   report', *NBS Building Science Series No. 80*, National Bureau of Standards, Washington, D.C., 1976, 48 pp
25 Liu, X.L. and Chen, W.F. 'Analysis of reinforced concrete structures
   during construction', *Struc. Engng Rep.*, CE-STR-33, Purdue Univ.,
   West Lafayette, IN, 1985, 218 pp

5.10 Advanced Analysis (高等分析)

5.10.1

ELSEVIER

Journal of Constructional Steel Research 55 (2000) 245–265

JOURNAL OF
CONSTRUCTIONAL
STEEL RESEARCH

www.elsevier.com/locate/jcsr

Advanced inelastic analysis of frame structures

J.Y. Richard Liew [a,*], W.F. Chen [b], H. Chen [a]

[a] *Department of Civil Engineering, National University of Singapore, 10 Kent Ridge Crescent, Singapore 119260, Singapore*
[b] *School of Civil Engineering, Purdue University, West Lafayette IN 47907, USA*

Abstract

This paper provides a state-of-the-art summary of recent advances in inelastic analysis of space frame structures. Particular attentions are devoted to inelastic modelling of framework components for accurate representation of frame behaviour and the applications of plastic hinge analysis for large-scale framework. Issues related to inelastic buckling and post-buckling unloading of struts, modelling of gradual yielding in steel beam-columns, inelastic modelling of composite floor beams subject to sagging and hogging moments, modelling of building core walls and semi-rigid beam-to-column connections in three-dimensional frameworks are discussed. Numerical examples are provided to illustrate the acceptability of the use of the inelastic models in predicting the ultimate strength and inelastic behaviour of spatial frameworks. © 2000 Elsevier Science Ltd. All rights reserved.

Keywords: Advanced analysis; Beam-column; Connection model; Inelastic analysis; Limit-state design; Nonlinear analysis; Plastic hinge; Semi-rigid connection; Thin-walled element

1. Introduction

The conventional approach for teaching advanced structural analysis courses is to treat theory of structural stability and plastic analysis/design as two separate topics. Little time has been spent to formulate solution methods and solve problems combining the theories of plasticity and stability. This philosophy of teaching combined with a need to simplify practical problems has led to an analysis/design approach in which forces and deformation demands are estimated from an elastic analysis and acceptability of a structure is assessed by comparing the demands with the component capacities defined in traditional limit-states checks. In other words, the design

* Corresponding author.

0143-974X/00/$ - see front matter © 2000 Elsevier Science Ltd. All rights reserved.
PII: S 0 1 4 3 - 9 7 4 X (9 9) 0 0 0 8 8 - 7

246   *J.Y.R. Liew et al. / Journal of Constructional Steel Research 55 (2000) 245–265*

specifications provide guidance to perform ultimate strength checks of structural components based on elastic forces obtained from global analysis. They are not adequate in addressing behavioural issues related to system limit-states. With the rapid advancement of computer technology (computer speed is doubled almost every one and a half years), research works are currently in full swing to develop the advanced inelastic analysis methods which can sufficiently represent the behavioural effects associated with member primary limit states such that the separated specification member capacity checks are not required. In other words, there is a potential for direct handling of limit states within analysis models.

This paper summarizes the recent work on second-order plastic hinge analysis of space frames with an ultimate aim of implementing advanced analysis for frame design. In the present work, space trusses are modelled using the strut-and-tie model, and rigid space frames are modelled using the beam-column model. Cross section plastification is modelled using the plastic hinge theory. To capture inelastic buckling of a member, plastic hinges are allowed to form at the ends and/or along the length of the member. To allow for gradual yielding effect, a two-surface plastic hinge model is proposed. The initial yield surface bounds the region of elastic sectional behaviour, while the bounding surface defines the state of full plastification of the section. Smooth degradation of stiffness is assumed as the force state moves from the initial yield surface towards the bounding surface. Composite floor beams are modelled by spread-of-plasticity approach considering full composite action between floor slabs and steel beams. Structural core-walls are modelled using "simplified" thin-walled line elements. Significant twisting action in the core walls is analysed to include both warping and torsional effects. Connections are modelled as a rotational spring having the moment–rotation relationship described by the four-parameter power model. The advanced inelastic analysis program is used to investigate the limit-states behaviour of space truss and tubular braced frame. Lastly, nonlinear inelastic analyses are carried on a core-braced framework considering the effects of composite beams and semi-rigid connections.

2. Inelastic analysis of space trusses

A space truss is a special kind of frame structure in which members are pin-connected at their ends. The inelastic large-displacement behaviour of space trusses can be modelled using the strut and tie model developed by Liew et al. [1]. The strut model includes three distinct axial load-shortening curves as shown in Fig. 1. The elastic loading stage is associated with the compression of an initially imperfect column until its maximum strength, P_{max}, is reached. The post-buckling stage is developed by assuming that a plastic hinge forms at the mid-span of the member. The post-buckling unloading stage is characterised by the fact that the strut member is stretched due to the change in sign of the incremental force. It is assumed that the post-buckling unloading curve follows the elastic loading curve with the member's imperfection magnitude computed based on the initial shortening value Δ_f as shown in Fig. 1. The maximum strength of an axially loaded strut, P_{max}, can be made to

J.Y.R. Liew et al. / Journal of Constructional Steel Research 55 (2000) 245–265   247

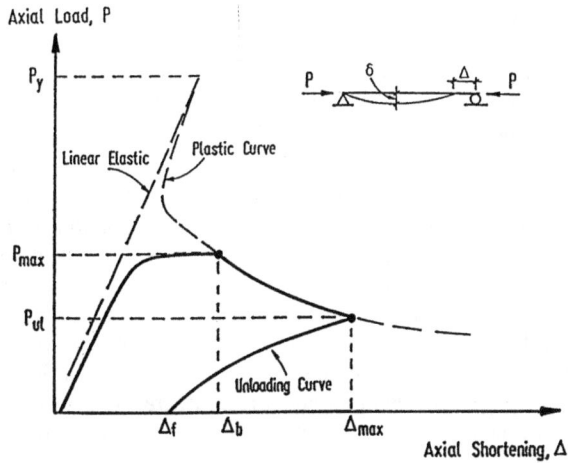

Fig. 1.   Axial load-shortening curve for a strut member.

conform to the code formulae for the design of columns by assigning an equivalent initial imperfection in the member. For tie members, the axial load–elongation relationship as shown in Fig. 2 is obtained based on the bilinear elastic–plastic stress–strain curve of the material. The main advantage of the proposed model is that member axial load–deformation relationships are developed based on the strut equations from the design specification. A separate check for member stability is therefore not required.

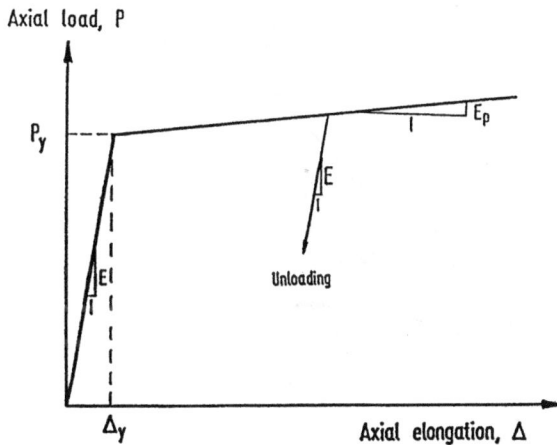

Fig. 2.   Axial load-elongation curve for a tie member.

248 *J.Y.R. Liew et al. / Journal of Constructional Steel Research 55 (2000) 245–265*

3. Modelling of inelasticity in beam-columns

A three-dimensional (3-D) beam-column formulation proposed by Liew and Tang [2] provides a realistic representation of geometric and material nonlinear effects in the member. The analysis operates on element stress resultants, i.e., forces and moments. The motion of material is described by an updated Lagrangian formulation and referred to a Cartesian co-ordinate system. The basic element formulation is based on exact solution of the fourth-order differential equation of a beam-column subjected to end forces [3]. It consists of three transitional and three rotational degrees of freedom at each node as shown in Fig. 3. The effects of large displacements and coupling between lateral deflection and axial strain are included by using nonlinear strain relations. The tangent stiffness matrices are derived in a consistent manner from energy principles. The elastic tangent stiffness matrices are calculated from closed form expressions, with no numerical integration over the element cross-section or over the element length. They contain the influence of axial on lateral deformations in the form of stability functions.

The proposed analysis can capture the formation of a plastic hinge between the element ends with minimum computational effort. Based on the member initial out-of-straightness, the deformed element shape and the forces at element ends and the force state within the element length can be established by taking the equilibrium of axial force and moment at the internal cross-section. The element length is divided

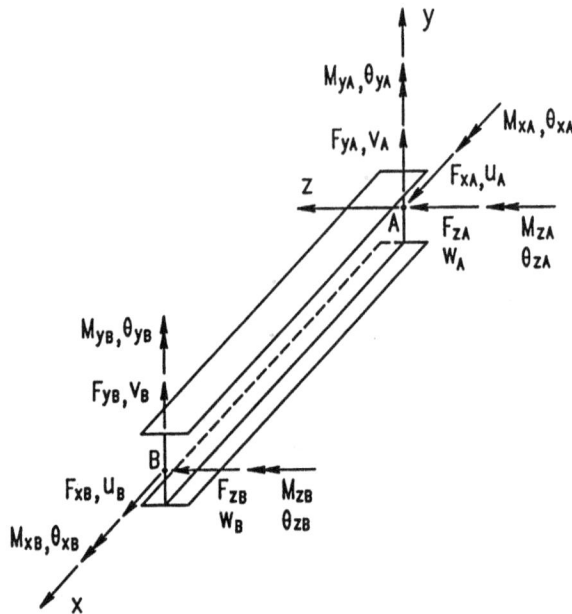

Fig. 3.   Beam-column element.

J.Y.R. Liew et al. / Journal of Constructional Steel Research 55 (2000) 245–265 249

into six segments with equal length. The cross-sectional forces are then checked at five points between the element ends. A plastic hinge is said to have formed when the plastic strength is reached at any of these points. The analysis will automatically subdivide the original element into two sub-elements at the plastic hinge location. The internal hinge is then modelled by an end hinge at one of the sub-elements. The stiffness matrices for the two sub-elements are determined. The inelastic stiffness properties for the original element are obtained by static condensation of the "extra" node at the location of the internal plastic hinge.

A plasticity model that accounts for partial yielding and hardening is formulated according to the boundary surface concept proposed by Hilmy et al. [4]. This model employs two interaction surfaces—one yield surface and one bounding surface—as shown in Fig. 4. The yield surface bounds the region of elastic cross-sectional behaviour, while the bounding surface defines the state of full plastification of the cross-section. The bounding surface encloses the cross-sectional force state and the yield surface at any stage during the yielding process. The size of the initial yield surface is described by Z_y which is the surface extension parameter. A suitable Z_y value can be selected to model the effect due to initial residual stress in the cross-section. Once yielding is initiated, the yield surface will translate so that the cross-section force state remains on the yield surface during subsequent loading. This type of approach is termed "refined" plastic hinge analysis in [5].

4. Modelling of steel–concrete composite beams

When a building frame is subject to both gravity and lateral loads, the distribution of bending moment in the composite floor beam varies along the length. In the

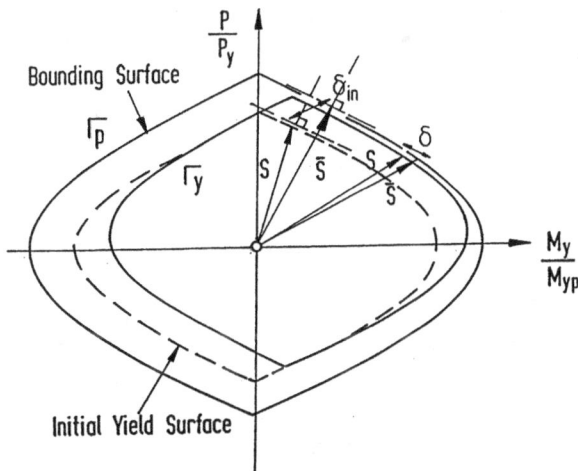

Fig. 4. Two-surface plasticity model

250   J.Y.R. Liew et al. / Journal of Constructional Steel Research 55 (2000) 245–265

hogging moment region, concrete in tension may crack and the steel reinforcements in the slab may yield. In the sagging moment region, a large bending moment may cause the yielding of the steel section and crushing of the concrete. Consequently, the cross-section properties and the flexural stiffness of a composite beam will vary along the length of each span. The nonlinear inelastic behaviour of a composite beam should be considered in the limit state analysis of composite structures.

In the proposed approach, a composite beam is subdivided into a finite number of segments to capture the varying flexural stiffnesses along the member length as shown in Fig. 5. The instantaneous flexural stiffness can be derived using the simpli-fied expressions for the moment–curvature relationship, and is used in the inelastic stiffness formulation to consider the yielding of the cross-section. Hence, section discretization is not required. The instantaneous flexural stiffness of a segment depends on the force state at its midpoint. The composite beam is a non-prismatic member because of the variation of instantaneous flexural stiffness along the length. In order to reduce the computational efforts, length-wise discretization can be minim-ized by reducing the number of internal nodes. This can be achieved by performing a yield check at all internal nodes along the member length. If several continuous segments are found to be in elastic state as shown in Fig. 5, they are combined to form one internal element. Convergence studies indicate that not more than 16 seg-ments per member are required to accurately capture the spread-of-plasticity effect of a composite beam member.

Li et al. [6] investigated a large number of variables required to fully describe the moment–curvature relationship of a composite beam section as shown in Fig. 6 in both sagging and hogging moment regions. If detailed information on rebars is available, Li's moment–curvature equations can be used to calculate the flexural stiffness of a composite beam under hogging moment. If major reinforcement bars are not present in the slab, the composite beam under hogging moment behaves like a steel section. Attalla's moment–curvature–thrust equation [7] may be used to determine the flexural stiffness of a steel section because it can predict the cross-sectional behaviour more accurately than other forms of moment–curvature–thrust equations.

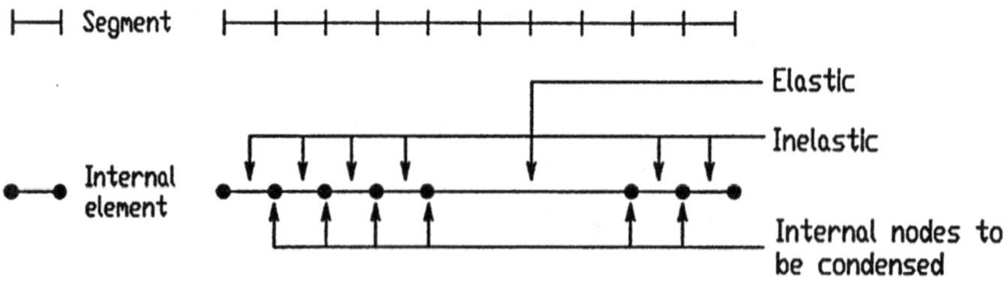

Fig. 5. Spread-of-plasticity analysis for a composite beam.

J.Y.R. Liew et al. / Journal of Constructional Steel Research 55 (2000) 245–265 251

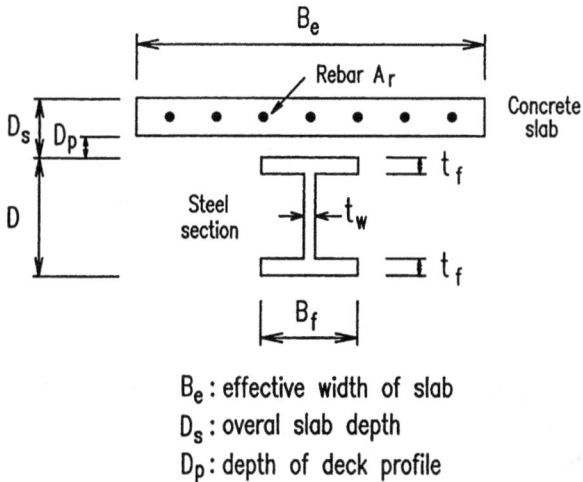

B_e : effective width of slab
D_s : overal slab depth
D_p : depth of deck profile

Fig. 6. Composite beam section.

5. Modelling of core walls

Core-walls are modelled by the thin-walled beam-column element for their pro-portional similarity to Vlasov's thin-walled beams and for their computational efficiency in the inelastic analysis [8,9]. As shown in Fig. 7, the thin-walled beam-column element has an additional warping degree-of-freedom over the beam-column element at each end. The local coordinate is chosen: axis x lies on the shear centre axis, and the y and z axes are parallel to the principal $\bar{y}$ and $\bar{z}$ axes. Some force and displacement components are referred to the shear centre, whereas the remaining ones are referred to the centroid of the section. However, before the element stiffness matrices are transformed into the global coordinate, it is necessary that all the forces and displacements are referred to a single point. The shear centre can be selected as the reference point. The detailed derivation for the elastic and geometric matrices of the thin-walled beam-column element is given in [8,10]. Because the height-to-width ratio of core-walls is large and the axial force with respect to the sectional area is small in practical buildings, material nonlinearity of core-walls maybe considered approximately by assuming that the plastic resistance is controlled by bending action only. The locations of the shear centre and the centroid of the cross-section are assumed not to change due to the inelastic effects.

6. Diaphragm action

For normal building frameworks, a floor slab may be modelled as a rigid dia-phragm, which is assumed to provide infinite in-plane stiffness and without any out-of-plane stiffness. The lateral response of a floor slab is characterised by two trans-

252   *J.Y.R. Liew et al. / Journal of Constructional Steel Research 55 (2000) 245–265*

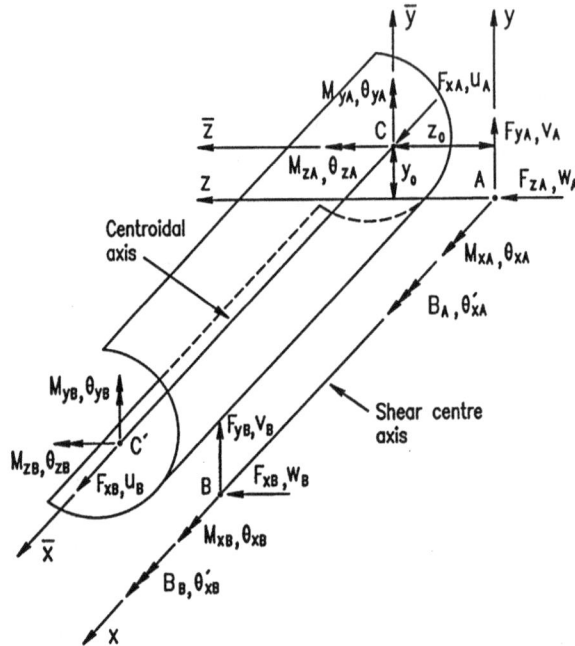

Fig. 7.   Thin-walled beam-column element.

lational and one rotational degrees of freedom located at the floor master node. The detailed formulation that incorporates the diaphragm action is given in [8]. The analytical methods described in this paper can be used to analyse 3-D frameworks with or without a rigid floor diaphragm. However, it is important to model the flexibility of the floor slab for building frames with complex floor geometry or with lateral braces. Currently only a limited number of frame modelling packages allow for a flexible floor slab.

7. Modelling of semi-rigid connections

Beam-to-column connections may be modelled as zero-length rotational spring elements in the nonlinear analysis of semi-rigid frames [11]. The proposed formulation allows the relative torsional and flexural rotations between the member end and the connection. When a semi-rigid connection is specified at one end of a member, the corresponding local rotational degrees-of-freedom between the member end and the connection with respect to the local member coordinate are treated as the additional global unknowns in the structure stiffness equation. Thus there is no need to modify the tangent stiffness matrix of framing members.

A four-parameter power model proposed by Hsieh [12] is adopted to represent

J.Y.R. Liew et al. / Journal of Constructional Steel Research 55 (2000) 245–265   253

the moment–rotation relationship of typical connections due to its simplicity and robustness for representing the basic behaviour of connections, and ease of implementation in the nonlinear inelastic analysis program. This connection model has the following form:

$$M = \frac{(K_e - K_p)\theta}{[1 + |(K_e - K_p)\theta / M_0|^n]^{1/n}} + K_p\theta \tag{1}$$

in which K_e is the initial stiffness of the connection, K_p is the strain-hardening stiffness of the connection, M_0 is a reference moment, and n is a shape parameter (see Fig. 8). By differentiating Eq. (1) with respect to θ, the tangent connection stiffness, K_t, can be obtained. To allow for the unloading of the connection associated with non-proportional loading and inelastic force redistribution, the tangent connection stiffness is equal to the initial stiffness K_e.

The four parameters may be determined through curve-fitting if the experimental data are available, or by using analytic formulations if the connection details are known. However, in the structural design practice, it is unlikely that specific connection details will be known during the preliminary design until the structural members have been sized in the final design. Since connection flexibility will affect the structural response and therefore the required member sizes, there is a need to develop some means to account for connection behaviour during the analysis and design process before the final member sizes are selected. One solution is to use the standard connection reference curves which are based on the connection test database. An optimisation approach utilising the conjugate-gradient method is first used to find a set of parameters (M_0, K_e, K_p, and n) which gives the best curve-fit to the experi-

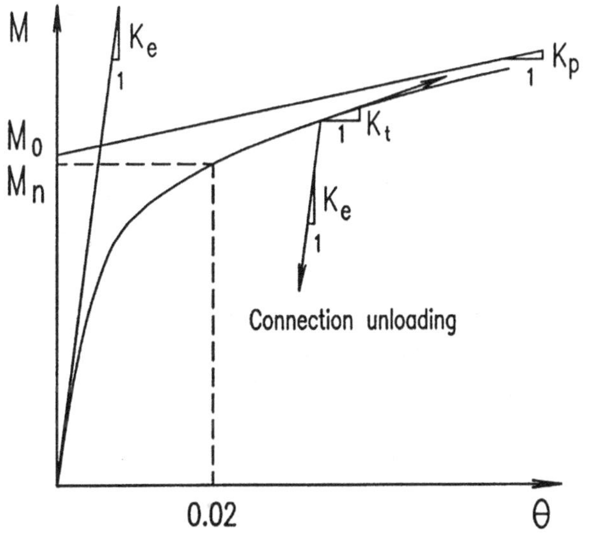

Fig. 8. Four-parameter power model.

Table 1
Parameters and M_n/M_{pb} values for connections under in-plane bending[a]

Connection type	$M_0'=M_0/M_n$	$K_e'=K_e/M_n$	$K_p'=K_p/M_n$	n	M_n/M_{pb}	
					At the beam framing about the major-axis of column (see Fig. 9)	At the beam framing about the minor-axis of column (see Fig. 9)
DWA	1.03	301	5.0	1.06	0.05	0.025
TSAW	0.94	363	6.9	1.11	0.4	0.2
EEP	0.97	309	5.5	1.20	1.0	0.5

[a] DWA: Double web-angle connection; TSAW: Top- and seat-angle connections with double web angles; EEP: Extended end-plate connection without column stiffeners.

mental connection response data. The moment–rotation curves are then normalised with respect to the nominal connection capacity M_n, which equals to the moment at a rotation of 0.02 radian as shown in Fig. 8. The standard reference curve is calibrated by fitting a curve through the average of the normalised curves. The average values of $M'=M/M_n$, $K'_e=K_e/M_n$, $K'_p=K_p/M_n$ and n in the standard reference curves for nine types of commonly used connections under in-plane bending have been established [12]. Then, for the analysis of the overall structure, only the connection type and nominal connection capacity would need to be defined without unnecessary concern over the final connection details. Based on the connection test database, a survey of the ratio of M_n/M_{pb} for different types of connections have been carried out, in which M_{pb} is the plastic bending capacity of a beam where the semi-rigid connection is located. The standard reference curve parameters and values of M_n/M_{pb} for three typical connections are listed in Table 1 and Fig. 9.

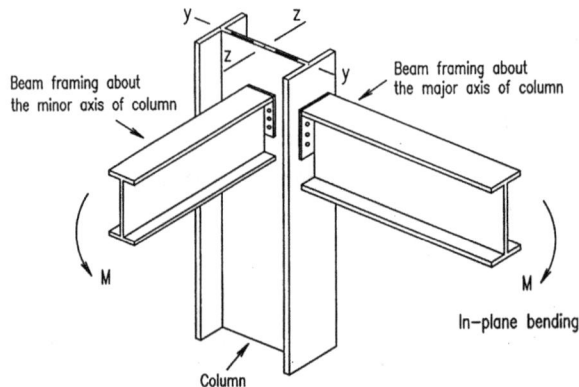

Fig. 9.   Beam-to-column connections.

J.Y.R. Liew et al. / Journal of Constructional Steel Research 55 (2000) 245–265   255

8. Analysis of a space truss system

Fig. 10 shows a star-shaped space truss system subjected to a single downward concentrated load applied at the crown joint [1]. Member cross-sectional area is $A=10$ mm^2, and moment of inertia is $I=41.7$ mm^4. Elastic modulus is $E=2.034\times10^5$ N/mm^2, plastic modulus is $E_p=E/1000$, and yield strength is $f_y=400$ N/mm^2. All supports are restrained against translations and the remaining nodes are free to translate in space. For the advanced analysis, the axial load-shortening curves for the strut members are obtained based on the column curve "b" of BS 5950 [13] and the tension members are modelled as linearly elastic with strain hardening defined by the plastic modulus E_p. The axial load–deformation curves are shown in Fig. 12.

The structure is analysed by elastic large displacement analysis and by the proposed advanced analysis. The load–displacement curves for the crown joint are shown in Fig. 11. The one-to-one correspondence between the behaviours of structure and members are labelled as A, B, C and D in Figs. 11 and 12. The limit load predicted by the elastic large displacement analysis and advanced analysis are 633 N and 372 N, respectively. The elastic large displacement analysis overestimates the system strength and stability by 70%. By observing point "B" in Figs. 11 and 12, it can be seen that the limit load of the truss corresponds to the geometric instability of the system and not due to any member buckling. Based on member capacity check using design specification BS 5950, the vertical components of the axial capacity of six top members 1 to 6 (1132 N) yields the limit load of 543 N as against 372 N obtained by advanced analysis making an overestimation of about 46%. Additional

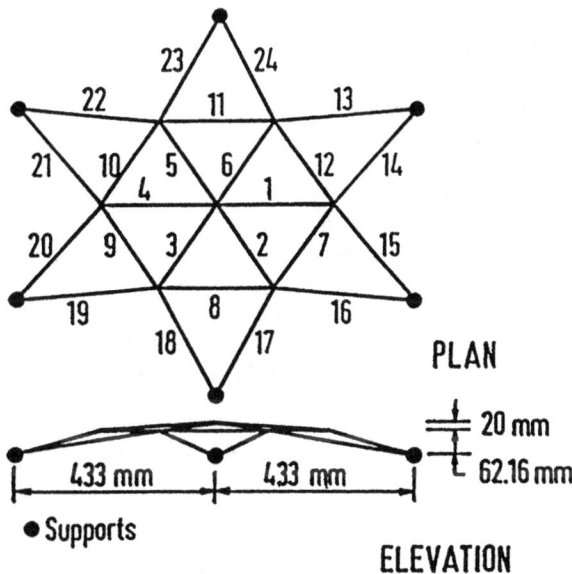

Fig. 10.   Space truss system.

256 *J.Y.R. Liew et al. / Journal of Constructional Steel Research 55 (2000) 245–265*

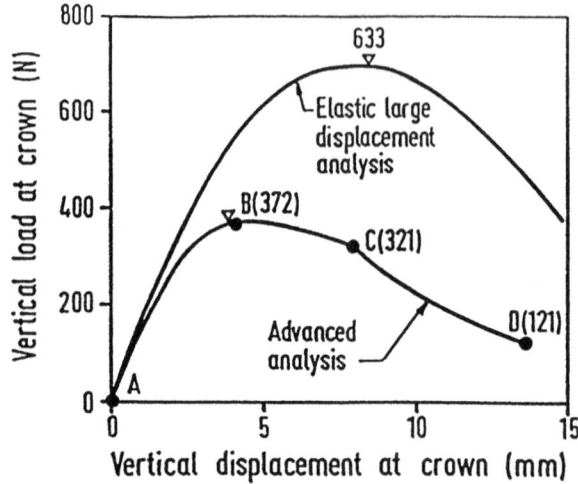

Fig. 11. Load–displacement curves of space truss.

numerical studies carried out by Liew et al. [1] show that if the limit load of the structure is governed by the buckling of individual members and not because of the accelerated geometric instability of the system, then the overestimation of limit load by conventional design is at an acceptable level.

9. Tubular braced frame

Fig. 13 shows a space frame consisting of X- and inverted V-braces in the x- and y-directions [2]. Member sizes and yield strength are given in Table 2. The frame is fixed at the column bases, and it is subjected to gravity load and lateral load in the negative y-direction acting at the top of the frame. The lateral load of 1000 kN is applied in equal proportions to the column tops above the X-braced frame "ABCD". Each tubular member is modelled as one inelastic beam-column element. Non-proportional loading is assumed, i.e., the gravity load (factored self-weight) is applied first and then followed by the lateral load. Analysis is also carried on the 2-D X-braced frame, denoted as frame "ABCD" in Fig. 13, and the results are compared with those obtained from a full 3-D analysis so that the strength enhancement due to the 3-D effect can be evaluated.

The lateral load–displacement curves obtained from 2-D and 3-D analyses are shown in Fig. 14. For the 3-D frame with perfectly straight members, first yield takes place at the lateral load ratio of 1.415. Member buckling first occurs at the diagonal member of the X-brace at a load ratio of 2.070. The frame reaches its maximum resistance at a load ratio of 2.445 when an additional compression member in the X-braces buckle as shown in the inserted diagram in Fig. 14. In the post-collapse range, more compression members including those in the V-braces buckle.

J.Y.R. Liew et al. / Journal of Constructional Steel Research 55 (2000) 245–265   257

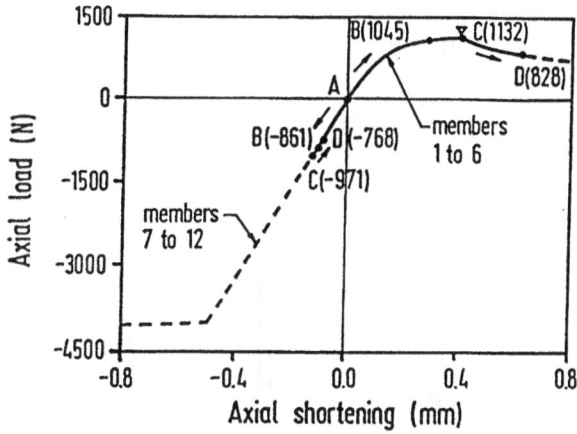

(a)

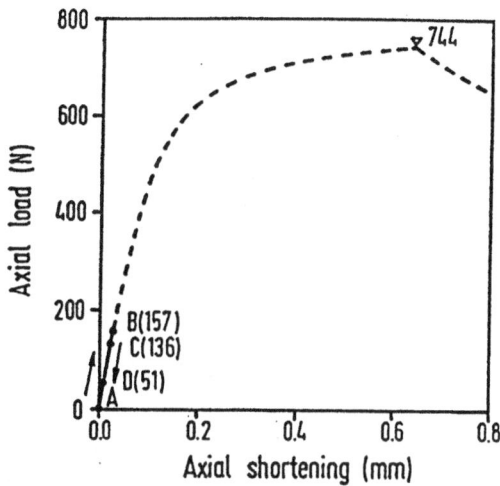

(b)

Fig. 12.   Axial load–deformation curves for truss members: (a) 1 to 12; (b) 13 to 24.

This is because the horizontal force acting in the y–z plane produces torsion forces and induces lateral force in the V-braced frame in the x-z plane. The limit load factor obtained from the analysis of the 2-D X-braced frame "ABCD" is 1.484 and the corresponding load-displacement curve is shown in Fig. 14. The strength enhance-

258   *J.Y.R. Liew et al. / Journal of Constructional Steel Research 55 (2000) 245–265*

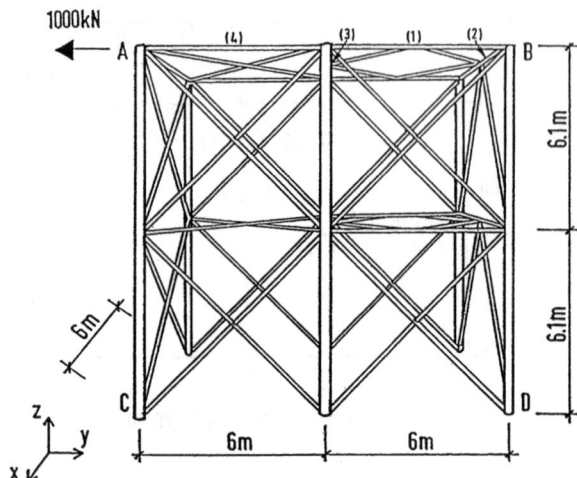

Fig. 13. Tubular braced frame.

Table 2
Member sizes and yield strength of a tubular braced frame

Member attribute	Tubular section D(mm)×t(mm)	Yield strength f_y (N/mm²)	Member attribute	Tubular section D(mm)×t(mm)	Yield strength f_y (N/mm²)
Horizontal member (1)	168.3×4.5	250	External vertical V-braces	168.3×5.6	250
Horizontal member (2)	273×5.6	240	Internal vertical V-braces	168.3×4.5	250
Horizontal member (3)	273×9.3	358	Vertical X-braces	168.3×4.5	250
Horizontal member (4)	168.3×18.3	358	Horizontal X-braces	168.3×4.5	250
All Columns	355.6×25.4	358			

ment due to the 3-D behaviour over the 2-D capacity is about 65%. The 3-D frame action is achieved by load distribution through the plan bracing on the decks.

10. Composite core-braced frame

Figs. 15 and 16 show a 24-storey core-braced frame with total height of H=87.792 m [9]. Beam sections are given in Table 3. Thickness of the concrete core-walls is 0.254 m. Concrete lintel beams with depth of 1.219 m are rigidly connected to core-walls to resist the lateral and torsional loads efficiently. A36 steel is used for all sections. Thickness of the concrete slab is 127 mm. All floors are assumed to be rigid

J.Y.R. Liew et al. / Journal of Constructional Steel Research 55 (2000) 245–265   259

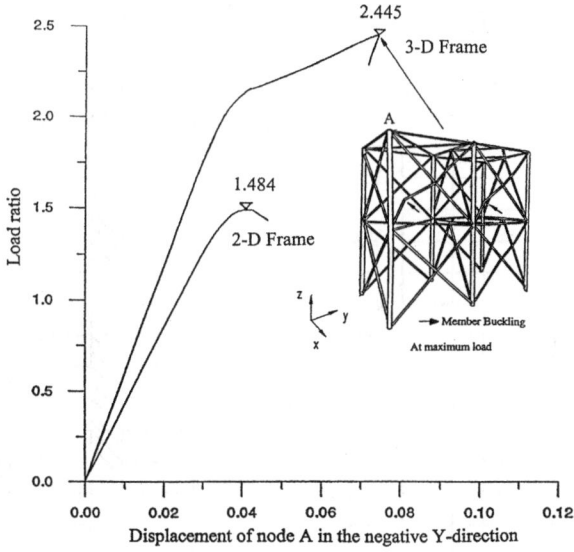

Fig. 14.   Load–displacement curves of a tubular frame.

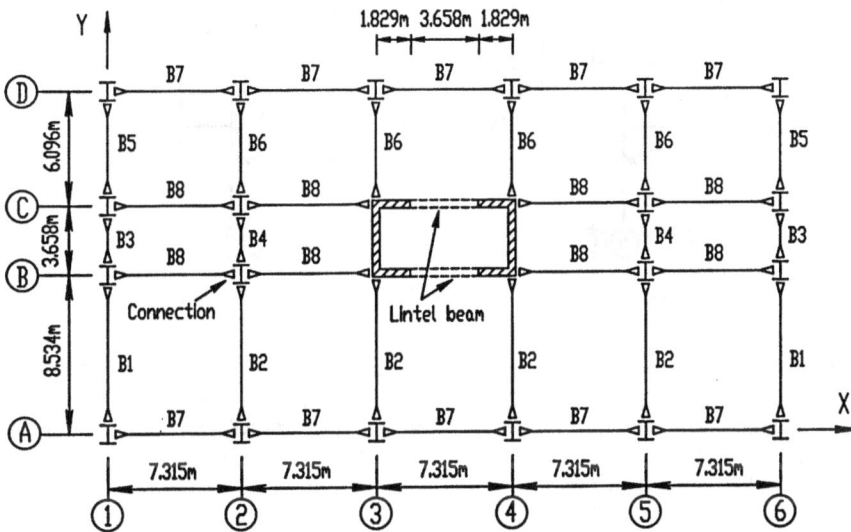

Fig. 15.   Plan view of a core-braced frame.

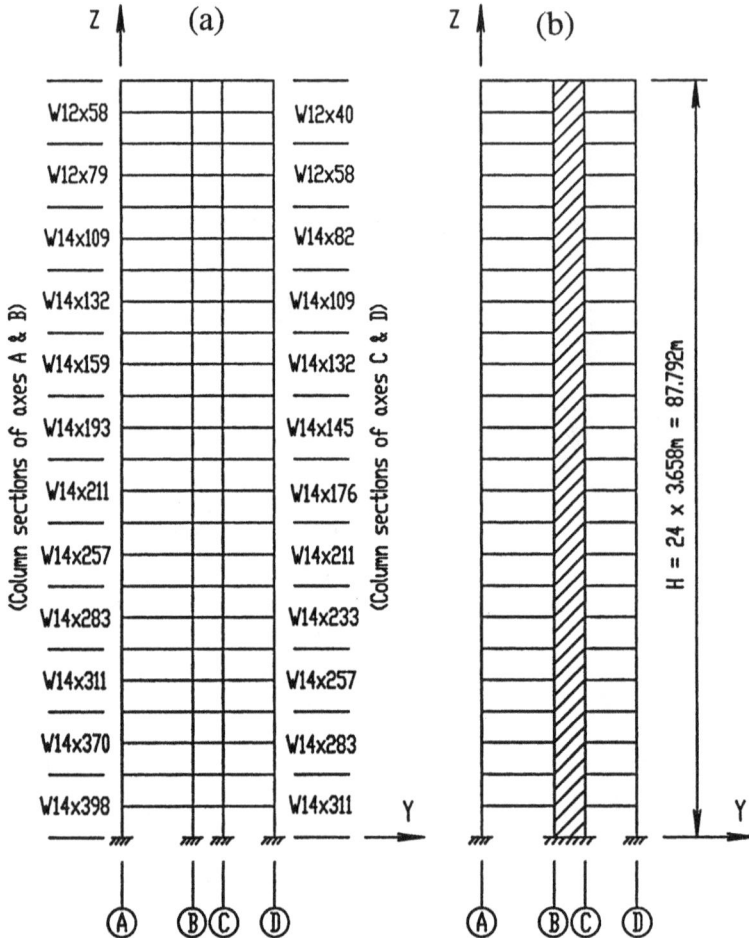

Fig. 16.   Elevation view of a core-braced frame: (a) at axes 1, 2, 5, 6; (b) at axes 3, 4.

in plane to account for the diaphragm action of concrete slabs. Material properties of concrete are: modulus of elasticity E_c=23,400 N/mm^2, and compressive strength f''_c =23.4 N/mm^2. The structure is analysed for the most critical load combination of gravity loads=4.8 kN/m^2 and wind loads=0.96 kN/m^2 acting in the y-direction.

Core-walls are modelled as thin-walled beam-column elements. In this example, core-walls are mainly subjected to the bending moment about the principle $\bar{z}$-axis, which is parallel to the global x-axis. The bending moment about the principle $\bar{y}$-axis is small. The plastic section modulus about the principle $\bar{z}$ axis of channel-shaped core-wall section is Z=2.549 m^3. The height-to-width ratio of core-walls is 24:1. It is assumed that the plastic resistance of core-walls is dominated by the plastic

J.Y.R. Liew et al. / Journal of Constructional Steel Research 55 (2000) 245–265 261

Table 3
Section properties of composite beam sections[a]

Beam No.	B_e (mm)	D_s (mm)	D_p (mm)	Steel section	Under sagging moment			Under hogging moment		
					M_u (kNm)	EI (kNm²)	α_y	M_p (kNm)	EI (kNm²)	α'_y
B1	1067	127	0	W18×60	852	1.72×10^5	0.76	500	8.19×10^4	0.61
B2	2134	127	0	W18×60	931	2.08×10^5	0.73	500	8.19×10^4	0.61
B3	457	127	0	W12×16	184	2.55×10^4	0.68	82	8.57×10^3	0.60
B4	915	127	0	W12×16	197	3.10×10^4	0.68	82	8.57×10^3	0.60
B5	762	127	0	W14×34	416	6.75×10^4	0.73	222	2.83×10^4	0.62
B6	1524	127	0	W14×34	452	8.17×10^4	0.71	222	2.83×10^4	0.62
B7	915	127	0	W14×30	382	6.39×10^4	0.72	192	2.42×10^4	0.62
B8	1830	127	0	W16×45	654	1.32×10^5	0.72	335	4.88×10^4	0.62

[a] M_u: ultimate moment of composite beam section under sagging moment; M_p: plastic moment of steel section; EI: elastic flexural stiffness; α_y and α'_y: ratios of yield moment.

262   *J.Y.R. Liew et al. / Journal of Constructional Steel Research 55 (2000) 245–265*

bending resistance about the principle $\bar{z}$-axis, $M_z=0.8Zf''_c=4.8\times10^4$ kNm, only. The plastic resistance of core-walls has been reduced to approximately account for the tensile cracking and some axial force interaction effect. Each steel column is modelled as one inelastic beam-column element.

Firstly, inelastic analyses are carried out on a rigid core-braced frame by using two beam models:

1. Pure steel beam model. Composite action between steel beams and concrete slab is ignored, and steel beams are modelled by the inelastic beam-column elements.
2. Composite beam model. Steel beam and concrete slab with effective width B_e are fully composite. Composite beams are analysed by the spread-of-plasticity approach. The properties of composite beam sections are given in Table 3. The flexural stiffness–moment relationship of a composite beam section under sagging moment is based on the moment–curvature relationship proposed by Li et al. [6]. By ignoring the contribution of rebars, a composite beam under hogging moment behaves like a steel section. The flexural stiffness–moment relationship of a steel section is obtained from the M–Φ–P relationship proposed by Attalla et al. [7]. As shown in Fig. 17, when the bending moment is less than the yield moment, the cross-section is elastic. When the moment increases from the yield moment to the moment capacity, the instantaneous flexural stiffness decreases rapidly.

When the loads are proportionally increased, many beams and columns enter the plastic state. Hence the lateral stiffness of the framework decreases and more of the lateral load is resisted by the core walls. When plastic hinges form at the bottom and top of the core-walls in the first storey, the whole structure collapses. The load–lateral deflection curves obtained from both analyses are shown in Fig. 18. It can be seen from Fig. 18 that if considering the composite action of a slab, the limit

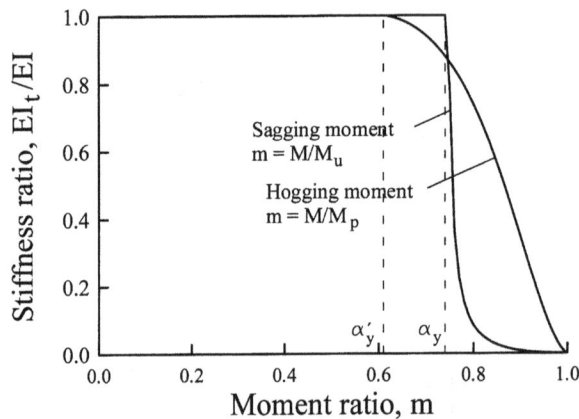

Fig. 17.   Flexural stiffness–moment relationship of a composite beam section.

J.Y.R. Liew et al. / Journal of Constructional Steel Research 55 (2000) 245–265 263

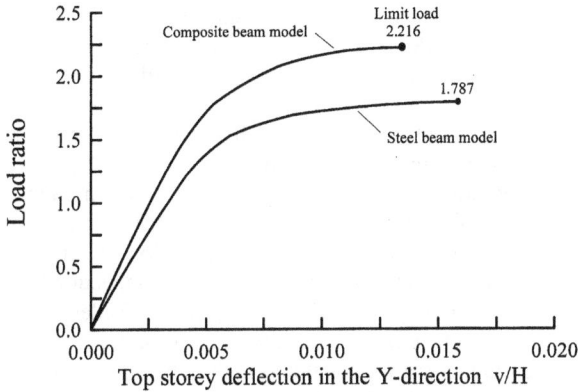

Fig. 18. Load–deflection curves of a rigid core-braced frame with different beam models.

load and initial lateral stiffness of the core-braced frame are 24% and 36% higher than those of the core-braced frame with a pure steel beam model.

To study the lateral resistance of the core-wall only, inelastic analysis is performed on the building assuming the surrounding framework is pin-connected to the RC core. In this case, the whole building relies on the core-walls to provide the lateral resistance. The limit load and initial lateral stiffness of the frame with pin-connections are 36% and 21% of those of the rigid frame (see Fig. 19). Similarly, to study the lateral resistance capacity of the pure steel framework, the elastic modulus and the compressive strength of concrete are assigned to be very small values. The frame collapses at a load ratio of 0.654, which is similar to that of the frame with pin-connections as shown in Fig. 19. The building cannot only rely on core-walls or

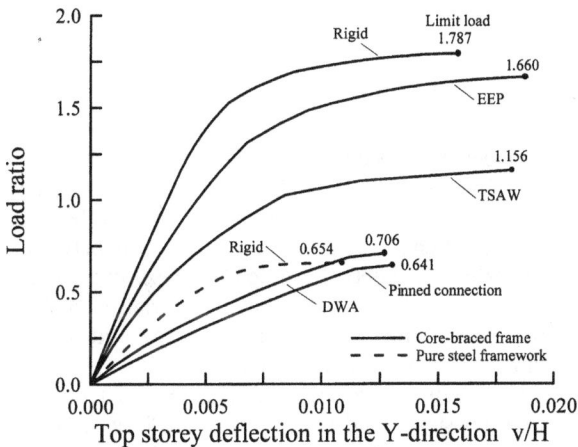

Fig. 19. Load–deflection curves of a core-braced frame with different connection types.

steel frameworks to provide the lateral load resistance. Core-walls and steel frameworks must interact to provide an efficient lateral load resistance system.

Semi-rigid construction is faster and cheaper than rigid construction. For high-rise building design, inter-storey deflection and overall frame drift are always the main concern. From an economical point of view, it is necessary to reduce the number of moment connections in high-rise building construction. The use of core-braced frames with semi-rigid connections may provide the optimum balance between the dual objectives of buildability and functionality [11]. Different types of connections as given in Table 1 are assumed to study the connection effect on the limit loads and lateral deflections of the frame. The arrangement of connection in the steel framework is shown in Fig. 15. Composite joint effect is not considered because the four-parameter power model is developed based on steel connections. The proposed semi-rigid formulation can model the torsional and both major- and minor-axis flexibility. However, in this analysis, only the relative rotations about the major-axis of the beam section are allowed at the semi-rigid connections. This is due to two reasons: (1) at present there is little experimental information on the torsional and out-of-plane behaviours of a semi-rigid connection, and (2) for typical framed structures with rigid floors, the torsional and out-of-plane effects of semi-rigid connections are not significant. Inelastic analyses are performed on core-braced frames with "DWA", "TSAW" and "EEP" connections with limit loads and load–deflection curves shown in Fig. 19. If "EEP" connections are adopted, the load and lateral stiffness can reach 93% and 81% of those of the rigid frame. The limit load and inelastic stiffness of the frame with "DWA" connections are only a little higher than those of the frame with pin-connections. The limit load and inelastic behaviour of the frame with "TSAW" connections are between those of the frame with "EEP" connections and the frame with "DWA" connections. It can be concluded that if proper semi-rigid connections are used, the frame can be constructed much faster and cheaper than the rigid frame, at the same time satisfying the system's strength and serviceability limit states.

11. Conclusions

Several inelastic models have been proposed for modelling the inelastic behaviour of framing components within the context of advanced inelastic analysis. The basic feature of these models is to capture the nonlinear inelastic behaviour of framing components so that the strength and stability interaction between members and framework can be assessed accurately. Limit load analyses carried on the star-shaped space truss and 3-D tubular frame show that the advanced inelastic analysis, in addition to predicting the limit load of the structure, helps to identify the load sharing and force distribution mechanism of individual members. The identification of the critical members in the structure enables the engineer to redesign the system to enhance the structural resistance and performance. Analyses carried out on a core-braced composite frame indicate that the limit load and lateral stiffness of the framework can be increased quite significantly if composite action between beam beams

J.Y.R. Liew et al. / Journal of Constructional Steel Research 55 (2000) 245–265 265

and floor slab is considered in the analysis. In steelworks construction, site assembly is easier with semi-rigid bolted connections than with rigidly welded connections. Analyses on a core-braced frame with semi-rigid connections show that proper bolted connections can be used to achieve the optimum balance between buildability and functionality. Thus, ease of erection and speed of construction will lead to early completion of a project. Labour cost saving and early return of investment can be realised eventually. The use of advanced analysis tools and system limit-state design philosophy would eventually enhance the understanding of 3-D frame behaviour and achieve a more rational and cost effective design.

References

[1] Liew JYR, Punniyakotty NM, Shanmugam NE. Advanced analysis and design of spatial structures. J Construct Steel Res 1997;42(1):21–48.

[2] Liew JYR, Tang LK. Nonlinear refined plastic hinge analysis of space frame structures. Research Report No. CE029/99. Department of Civil Engineering, National University of Singapore, 1998.

[3] Liew JYR, Chen H, Shanmugam NE. Stability functions for second-order inelastic analysis of space frames. In: Proceedings of the 4th International Conference on Steel and Aluminium Structures, Espoo, Finland, June 20–23, 1999:19–26.

[4] Hilmy AI, Abel JF. Material and geometric nonlinear dynamic analysis of steel frames using computer graphics. Comp Struct 1985;21(4):825–40.

[5] Chen WF, Toma S, editors. Advanced analysis of steel frames: theory, software, and applications. Boca Raton, FL: CRC Press, 1994.

[6] Li TQ, Choo BS, Nethercot DA. Moment curvature relations for steel and composite beams. J Singapore Struct Steel Soc, Steel Struct 1993;4(1):35–51.

[7] Attalla MR, Deierlein GG, McGuire W. Spread of plasticity: quasi-plastic hinge approach. ASCE J Struct Eng 1994;120(8):2451–73.

[8] Liew JYR, Chen H, Yu CH, Shanmugam NE, Tang LK. Second-order inelastic analysis of three-dimensional core-braced frames. Research Report No. CE024/97. Department of Civil Engineering, National University of Singapore, 1997.

[9] Liew JYR, Chen H, Yu CH, Shanmugam NE. Advanced inelastic analysis of thin-walled core-braced frames. In: Proceedings of the Second International Conference on Thin-Walled Structures, Singapore, December 2–4, 1998:485–92.

[10] Conci A. Large displacement analysis of thin-walled beams with generic open section. Int J Numer Meth Eng 1992;33:2109–27.

[11] Chen WF, Goto Y, Liew JYR. Stability design of semi-rigid frames. New York: John Wiley and Sons Inc, 1996.

[12] Hsieh SH. Analysis of three-dimensional steel frames with semi-rigid connections. Structural Engineering Report 90-1. School of Civil and Environmental Engineering, Cornell University, Ithaca, NY, 1990.

[13] BS 5950. Structural use of steelwork in building, Part 1: Code of practice for design in simple and continuous construction: hot rolled sections. UK, 1990.

5.10.2

Toward Practical Advanced Analysis for Steel Frame Design

Wai-Fah Chen
Dr, Prof. of Civil Eng.,
Dept. of Civil Engineering,
University of Hawaii,
Honolulu, USA
Contact: chenwf@eng.hawaii.edu

Summary

Extensive research on the topic of steel frame design with advanced analysis methods has been conducted over the past 10 to 15 years in recognition of the fact of rapid advancement in computing power in recent years. Despite significant research and development efforts, usage of advanced analysis for direct frame design has, nevertheless, been slow in coming to the profession. This is clearly and primarily caused by a lack of understanding and appreciation of the applications of second-order inelastic analysis methods on the part of the designers. Practical implementation of the advanced analysis methods for structural steel design is now a real possibility. This paper describes one of the practical advanced analysis methods that have been calibrated against the current American Institute of Steel Construction (AISC) load and resistance factor design specification, which requires only simple modifications of an existing in-house elastic program that is widely available in current engineering practice.

Keywords: design methods and aids; steel structures; stress analysis; structural frameworks.

Introduction

Before steel frame structures reach their limit states of strength or stability, they exhibit some forms of geometric and material nonlinearities. For decades, structural engineers and researchers have been exploring various approaches for assessing this structural behavior for the analysis and design of steel structures. The assessment methods, accordingly, have evolved over time, from hand calculation approach based on member capacity checks to computer-based approach based on advanced analysis to consider the interdependent effects between member and frame stability.

The strength and stability of a structural system and its members are related, but the interaction is treated separately in the current (AISC) load and resistance factor design specification,[1] among others.

In current engineering practice, the interaction between the structural system and its members is represented by the effective length factor. These indirect analysis and design methods include historically the allowable stress design (ASD), plastic design (PD), and load resistance factor design (LRFD). All these design approaches require an elastic analysis and separate specification member capacity checks including the calculation of the effective length factor, commonly known as the *K-factor*.

Advanced analysis can be defined as a method of second-order inelastic analysis that sufficiently represents the behavioral effects associated with member primary limit states such that separated specification member capacity checks are not required. The analysis must, therefore, take into account aspects that influence the behavior of the frame, which may include: material nonlinearity, second-order effects, residual stresses, geometrical imperfections, local buckling, connection behavior, end restraint, and interaction with the foundations.

With the present development of computer technology, two aspects, the stability of separate individual members, and the stability of their structural frameworks as a whole, can be treated rigorously for the determination of the maximum strength of the structures. The development of this direct approach to design is called *advanced analysis* or more specifically, *second-order inelastic analysis for frame design*. In this direct approach, there is no need to compute the effective length factor, because separate member capacity checks encompassed by the specification equations are not required. This design approach is illustrated in *Fig. 1* and marked as the direct analysis and design method.

In the United States, the term *advanced analysis* strictly means *second-order inelastic analysis* for frame design without the use of the effective length factor (*K*-factor). There are three stages of

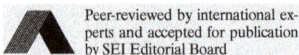

Paper received: January 7, 2009
Paper accepted: March 9, 2009

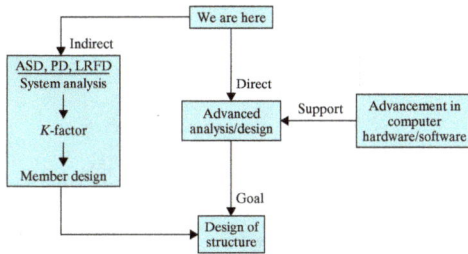

Fig. 1: Indirect and direct analysis and design methods

Fig. 2: Interaction between a structural system and its component members using the K-factor concept

progress to achieve the advanced analysis for frame design at the present time:

Stage 1: Direct second-order elastic analysis eliminating the use of amplification factors. This direct computation of second-order forces was encouraged in the 1986 AISC/LRFD specifications. This is LRFD design but not advanced analysis for design.

Stage 2: Direct second-order plastic analysis with the use of K-factor to do member-by-member capacity checks with code requirements. This is a transition to advanced analysis, but it is still not advanced analysis because the K-factor is still needed in the design process.

Stage 3: Direct second-order inelastic analysis for frame design without the use of K-factor to do member-by-member capacity checks with code requirements. The code requirements are met automatically in the advanced analysis for the structural system. This is called *advanced analysis*.

The purpose of this paper is to present a simple, concise, and reasonably comprehensive introduction to a practical, direct method of steel frame design, using advanced analysis that will produce almost identical member sizes as those of the traditional LRFD method. The direct method described in the following is limited to two-dimensional steel frames, so that the spatial behavior is not considered. Lateral torsional buckling is assumed to be prevented by adequate lateral braces. Compact plastic W sections are assumed so that sections can develop their full plastic moment capacity without buckling locally. All loads are statically applied.

Current Engineering Practice

In current engineering practice, there is a fundamental two stage process in the design operation:

– The forces acting on the structural members are determined by conducting an elastic structural system analysis.
– The sizes of various structural members are selected by checking against the ultimate strength equations specified in design codes.

As mentioned previously, the interaction behavior between individual members and their structural system is accounted for approximately by the use of the effective length factor K-concept as illustrated schematically in *Fig. 2*. However, despite its popular use in current practice as a basis for design, the effective length approach has the following major limitations:

– It cannot reflect the inelastic distributions of internal forces in a structural system.
– It cannot provide information on the failure mechanisms of a structural system.
– It is not easy to implement in an integrated computer design application.
– It is a time-consuming process by calculating every K-factor for each separate member capacity check.

Furthermore, some of these difficulties are more so on seismic designs because additional questions are frequently asked:

– How is the structure going to behave during an earthquake?
– Which part of the structure is the most critical area?
– What will happen if part of the structure yields or fails?
– What might happen if forces greater than the code has specified occur?

Considering these limitations and drawbacks and the rapid advancement of computing power, the second-order inelastic analysis approach or so-called *advanced analysis* approach provides an alternative approach to structural analysis and design. The practical

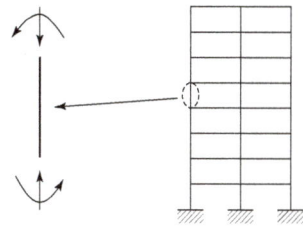

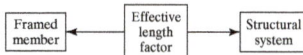

advanced analysis as presented in this paper is an elastic-plastic hinge-based analysis, modified to include the geometry imperfections, gradually yielding and residual stress effects, and semi-rigid connections. In this approach, all these drawbacks mentioned above were overcome. There is no need to compute the effective length factor, yet it will produce almost identical member sizes as those of the LRFD method.[1]

The Practical Theory

In the following text, an encapsulation of the practical theory will be presented that will enable incorporate advanced analysis of existing elastic analysis packages. The aim is to provide a smooth transition from the current indirect LRFD analysis and design practice to the direct advanced analysis method suitable for adoption in engineering practice. Details of the theory and step-by-step procedures are given in the book by Chen and Kim.[2]

To use an existing in-house elastic program, the practical theory makes simple modifications of the elastic modulus of material and plastic strength of cross section to account for key behavioral effects of a steel member: second-order, gradual yielding associated with residual stresses and flexure, and geometric imperfections. To meet the current LRFD requirements, these modifications have been calibrated against the LRFD specifications. This is described briefly in the following text.

Second-order Effects

To capture second-order effects, stability functions are recommended because they lead to large savings in modeling and solution efforts by using one or two elements per member. The simplified stability functions reported by Chen and Lui[3] may be used.

The incremental moment–rotation relationship of member ends may be written as, in the usual notations:

$$
\begin{bmatrix} \dot{M}_A \\ \dot{M}_B \\ \dot{P} \end{bmatrix} = \frac{EI}{L} \begin{bmatrix} S_1 & S_2 & 0 \\ S_2 & S_1 & 0 \\ 0 & 0 & A/I \end{bmatrix} \begin{bmatrix} \dot{\theta}_A \\ \dot{\theta}_B \\ \dot{e} \end{bmatrix}, \quad (1)
$$

where S_1, S_2 = stability functions; $\dot{M}_A$, $\dot{M}_B$ = incremental end moment; $\dot{P}$ = incremental axial force; $\dot{\theta}_A$, $\dot{\theta}_B$ = incremental joint rotation; $\dot{e}$ = incremental axial displacement; A, I, L = area, moment of inertia, and length of beam-column element; E = modulus of elasticity.

The stability functions for in-plane bending of a prismatic beam-column are defined as:

$$
S_1 = \frac{kL \sin kL - (kL)^2 \cos kL}{2 - 2\cos kL - kL \sin kL}, \quad (2a)
$$

$$
S_2 = \frac{(kL)^2 - kL \sin kL}{2 - 2\cos kL - kL \sin kL}. \quad (2b)
$$

For simplicity in computing, Lui and Chen[4] have proposed approximate expressions for the stability functions S_1 and S_2 as ($k^2 = P/EI$) if $-2,0 \le kL \le 2,0$.

$$
S_1 = 4 + \frac{2\pi^2 \rho}{15} - \frac{(0,01\rho + 0,543)\rho^2}{4 + \rho} \\ - \frac{(0,004\rho + 0,285)\rho^2}{8,183 + \rho}, \quad (3a)
$$

$$
S_2 = 2 - \frac{\pi^2 \rho}{30} + \frac{(0,01\rho + 0,543)\rho^2}{4 + \rho} \\ - \frac{(0,004\rho + 0,285)\rho^2}{0,183 + \rho}, \quad (3b)
$$

where the non-dimensional axial force P is defined as:

$$
\rho = \frac{P}{P_e} = \frac{P}{(\pi^2 EI)/L^2}. \quad (3c)
$$

Tangent Modulus Model Associated with Residual Stresses

The Column Research Council (CRC) tangent modulus concept is used to account for the gradual yielding effect due to residual stresses along the length of members under axial loads between

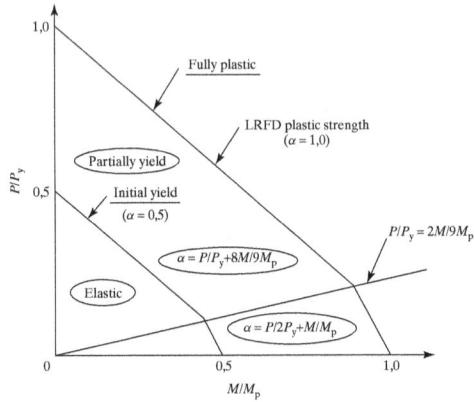

Fig. 3: Smooth stiffness degradation for a work hardening plastic hinge based on LRFD sectional strength curve as the limit or fully yielded surface

two plastic hinges. In this concept, the elastic modulus E instead of moment of inertia I is reduced to account for the reduction of the elastic portion of the cross section because the reduction of elastic modulus is easier to implement than that of moment of inertia for different sections. The reduction rate in stiffness between weak-axis and strong-axis is different, but this is not considered here. This is because rapid degradation in stiffness in the weak-axis strength is compensated well by the stronger weak-axis plastic strength. As a result, this simplicity will make the present methods practical.

The tangent modulus based on the CRC column strength formulas[5] can be presented as:

$$
\frac{E_t}{E} = 1,0 \text{ for } P \le 0,5 P_y, \quad (4a)
$$

$$
\frac{E_t}{E} = 4\frac{P}{P_y}\left(1 - \frac{P}{P_y}\right) \text{ for } P > 0,5 P_y, \quad (4b)
$$

where P_y = squash load.

Equation (4) is shown as dotted curve in *Fig. 4* in Section *Initial Geometric Imperfections* in which the modeling of geometrical imperfections is discussed.

Two-surface Stiffness Degradation Model Associated with Flexure

The tangent modulus model in Eq. (4) is suitable for $P/P_y > 0,5$, but it is not sufficient to represent the stiffness degradation for cases with small axial forces and large bending moments. A gradual stiffness degradation of

plastic hinge is required to represent the distributed plasticity effects associated with bending actions. Therefore, the hardening plastic hinge model is introduced to represent the gradual transition from elastic stiffness to zero stiffness associated with a fully developed plastic hinge.

To represent a gradual transition from the elastic stiffness at the onset of yielding to the stiffness associated with a full plastic hinge at the end, a parabolic model simulating the gradual degradation of the element stiffness due to the plastification of the steel section is used as shown in *Fig. 3*. The factor η, representing a gradual stiffness reduction associated with flexure, is proposed by Liew et al.[6]:

$$
\eta = 1 \text{ for } \alpha \le 0,5, \quad (5a)
$$

$$
\eta = 4\alpha(1 - \alpha) \text{ for } \alpha > 0,5. \quad (5b)
$$

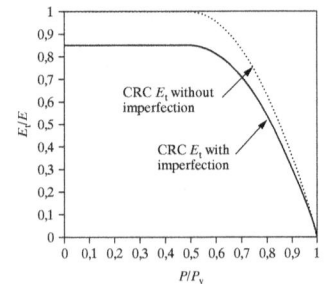

Fig. 4: Further reduced CRC tangent modulus $E'_t = 0,85 E_t$ for members with geometrical imperfections.

In this model, α is the force-state parameter obtained from the limit state surface corresponding to the member ends. The term α is based on the AISC-LRFD sectional strength curve and is expressed as:

$$\alpha = \frac{P}{P_y} + \frac{8}{9}\frac{M}{M_p} \text{ for } \frac{P}{P_y} \geq \frac{2}{9}\frac{M}{M_p}, \quad (6a)$$

$$\alpha = \frac{1}{2}\frac{P}{P_y} + \frac{M}{M_p} \text{ for } \frac{P}{P_y} < \frac{2}{9}\frac{M}{M_p}, \quad (6b)$$

where P, M = second-order axial force and bending moment at the cross section; M_p = plastic moment capacity.

Initial yielding is assumed to occur at $\alpha = 0,5$, and the yield surface function corresponding to $\alpha = 1,0$ represents the state of forces where the cross section has fully yielded and their corresponding state of forces (P, M) can only move along the yield surface under continuous loading condition.

Incremental Force–Displacement Relationship

In advanced analysis, the incremental load concept is applied to trace the force–displacement relationship of the structure and its components. When the work hardening plastic hinges are presented at both ends of a beam-column element, the incremental force–displacement relationship can be expressed as, in the usual notations:

$$\begin{Bmatrix} \dot{M}_A \\ \dot{M}_B \\ \dot{P} \end{Bmatrix} = \frac{E_t I}{L} \begin{bmatrix} \eta_A \left[S_1 - \frac{S_2^2}{S_1}(1-\eta_B) \right] \\ \eta_A \eta_B S_2 \\ 0 \end{bmatrix}$$

$$\times \begin{bmatrix} \eta_A \eta_B S_2 & 0 \\ \eta_B \left[S_1 - \frac{S_2^2}{S_1}(1-\eta_A) \right] & 0 \\ 0 & A/I \end{bmatrix} \begin{Bmatrix} \dot{\theta}_A \\ \dot{\theta}_B \\ \dot{e} \end{Bmatrix}$$

(7)

where η_A and η_B are the stiffness reduction factors at ends A and B, respectively. Note that the modulus of elasticity E in Eq. (7) is replaced by the tangent modulus E_t to account for the effects of residual stresses.

The parameter η represents a gradual stiffness reduction associated with flexure at sections. The partial plastification at cross sections in the end of elements is denoted by $0 < \eta < 1$. The

η may be assumed to very according to the parabolic expression:

$$\eta = 4\alpha(1-\alpha) \text{ for } \alpha > 0,5. \quad (8)$$

The refined plastic-hinge analysis implicitly accounts for the effects of both residual stresses and spread of yielded zones. To this end, refined plastic-hinge analysis may be regarded as equivalent to the plastic-zone analysis.

Initial Geometric Imperfections

Geometric imperfection modeling combined with the CRC tangent modulus model is discussed in the following text. There are three approaches: the explicit imperfection modeling method, the equivalent notional load method, and the further reduced tangent modulus method. There are two types of initial geometric imperfections for steel members: out-of-straightness and out-of-plumbness. These imperfections create additional moments in the column members, causing further reduction of the member bending stiffness.

The ECCS,[7,8] AS,[9] and CSA[10,11] specifications recommend the out-of-straightness imperfection varying in parabolic shape with a maximum in-plane deflection at the mid-height. It is noted that this explicit modeling method in braced frames requires the inconvenient imperfection modeling at the center of columns, although the inconvenience is much lighter than that of the conventional LRFD method for frame design.

The ECCS[7,8] and the CSA[10,11] also introduced the equivalent load concept, which accounted for the geometric imperfections in unbraced frames, but not in braced frames. In Eurocode 3, the equivalent load concept may also be used for braced frames.[12] For braced frames, an equivalent notional load may be applied at mid-height of a column because the ends of the column are braced. For unbraced frames, the geometric imperfections of a frame may be replaced by the equivalent notional lateral loads expressed as a fraction of the gravity loads acting on the story. The drawback of this method for braced frames is that it requires tedious input of notional loads at the center of each column. Another drawback for unbraced frame is that the axial force in the columns must be known in advance to determine the notional loads before analysis, but these are often difficult to calculate for large structures that are subject to lateral wind loads. To avoid this difficulty, it is recommended that

the further reduced tangent modulus method be used. This is described in the following text.

The idea of using the reduced tangent modulus concept is to further reduce the tangent modulus, E_t, to account for further stiffness degradation due to geometrical imperfections. The degradation of member stiffness due to geometric imperfections may be simulated by an equivalent reduction of member stiffness. The further reduced tangent modulus approach was proposed by Chen and Kim[2] to avoid the drawbacks of the first two methods. By including the effects of stiffness degradation due to geometric imperfection, the reduction factor of 0,85, which is determined by calibrating with the almost exact plastic-zone solutions, is used to further reduce the CRC-E_t value as given previously in Eq. (4). The same reduction factor of 0,85 is used for both braced and unbraced structures including both out-of-straightness and out-of-plumbness imperfections as shown in *Fig. 4* by the solid curve marked with imperfection, or simply $E_t' = 0,85 E_t$. This further reduced tangent modulus 0,85 is equivalent to the geometric imperfection of $L_c/500$ of column length or story height.

The key considerations of the conventional LRFD method and the practical advanced analysis method are compared in *Table 1*. Although the LRFD method does account for key behavioral effects implicitly in its column strength and beam-column interaction equations, the advanced analysis method as described previously accounts for these effects explicitly in the form of stability function, stiffness degradation function, and stiffness reduction factor for geometric imperfections.

Frames with Partially Restrained Connections

Connections in real structures do not possess the idealized characteristics, typically used in design codes around the world, of being fully restrained (rigid) or simple (pin-ended). Most commonly used connections in steel buildings are semi-rigid partial strength connections (partially restrained). Several studies have demonstrated the potential for improving stability, strength, and serviceability for structures with semi-rigid partial strength connections.

If the reader is interested in the semi-rigid frame design with LRFD using the advanced analysis method, the

Key considerations	LRFD	Proposed advanced method
Second-order effects	Column curve, B_1, B_2 amplification factors	Stability function Necessary number of elements: – beam with uniform load: 2 – column in braced frame: 2 – column in unbraced frame: 1
Geometric imperfections	Column curve	Further reduced tangent modulus method – reduction factor– $E_t' = 0,85\ E_t$ – see *Fig. 4*
Stiffness degradation associated with residual stresses	Column curve	CRC tangent modulus curve – see Eq. 4
Stiffness degradation associated with flexure	Column curve, beam-column interaction equations	Parabolic degradation function – refined plastic hinge concept (work hardening plastic hinge) – see Eqs. 5 and 6 and *Fig. 3*
Connection nonlinearity	No procedure	Power model/rotational spring

Table 1: Comparison of key considerations of load and resistance factor design (LRFD) method with the proposed practical advanced analysis method

book by Chen *et al.*[13] entitled *Stability Design of Semi-Rigid Frame Design* provides an in-depth coverage of recent developments in the design of these frames. It also provides computer software that will enable the reader to perform planar frame analysis in a direct manner for a better understanding of behavior and a more realistic prediction of a system's strength and stability. Practical, analytical methods for evaluating connection flexibility and its influence on the stability of the entire framework are also described.

These methods range from simplified member-by-member technique to more sophisticated computer-based advanced analysis and design approaches including many sample problems and detailed design procedures for each type of method.

Six-story Frame

Vogel[14] presented the load–displacement relationships of a six-story frame using second-order plastic-zone analysis. The frame is shown in *Fig. 5*. The

stress–strain relationship is elastic-plastic with linear strain hardening. The geometric imperfections are $L_c/450$ of column length or story height.

For comparison, the out-of-plumbness of $L_c/450$ is used in the explicit modeling method. The notional load factor of 1/450 expressed as a fraction of the gravity loads acting on the story and the reduced tangent modulus factor of 0,85 are used. Because the further reduced tangent modulus is equivalent to the geometric imperfection of $L_c/500$, the geometric imperfection of $L_c/4500$ is additionally modeled in the further reduced tangent modulus method, where $L_c/4500$ is the difference between the Vogel's geometric imperfection of $L_c/450$ and the proposed geometric imperfection of $L_c/500$.

The load–displacement curves of the three proposed second-order elastic analysis methods together with Vogel's second-order plastic-zone analysis are compared in *Fig. 6*. The errors in strength prediction by the proposed methods are less than 1%. Explicit imperfection modeling and the equivalent notional load method under-predict lateral displacements by 3%, and the further reduced tangent modulus method shows a good agreement in displacement with the Vogel's exact solution. Vogel's frame is a good example of how the reduced tangent modulus method predicts lateral displacement well under reasonable load combinations.

Conclusion

Over 40 years ago, evaluation of the first-order elastic response of a

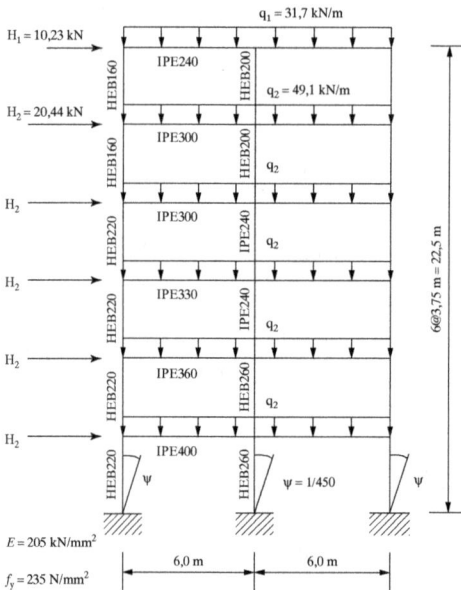

Fig. 5: Configuration and load condition of Vogel's six-story frame for verification study

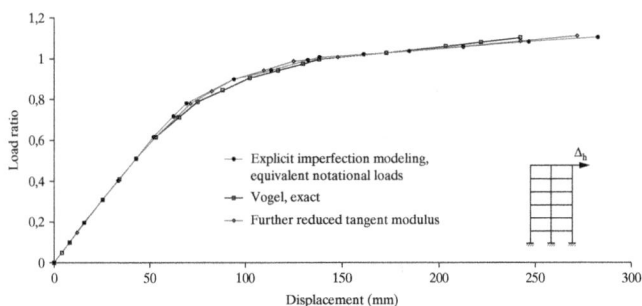

Fig. 6: Comparison of displacements for Vogel's six-story frame

structural system was a significant problem. The progress to the present state-of-the-art that deals routinely with second-order nonlinear behavior of complicated systems having hundreds of thousands of degrees of freedom is miraculous.

The basic theory for second-order inelastic analysis is well established and documented in the published literature.[15] The real challenge is making this type of analysis work in engineering practice. The advent of personal computers has made more sophisticated methods of analysis feasible in design practice. The proposed practical theory and analysis procedures using an existing in-house elastic program are aimed at providing a smooth transition from the current indirect LRFD analysis and design practice to the direct advanced analysis method suitable for use in the current engineering practice.

The practical advanced analysis method will enable the reader to perform planar frame analysis in a direct manner for a better understanding of the behavior and more realistic prediction of the strength and stability of the frames. This will help better understanding and appreciation of the applications of advanced analysis methods on the part of the designers. The potential applications of advanced analysis for steel frame design and performance are far reaching. It is a state-of-the-art methodology for the structural engineering profession for the 21st century.

References

[1] AISC. *Load and Resistance Factor Design, Manual of Steel Construction.* American Institute of Steel Construction: Chicago, 2005.

[2] Chen WF, Kim SE. *LRFD Steel Design Using Advanced Analysis.* CRC Press: Boca Raton, FL, 1997.

[3] Chen WF, Lui EM. *Stability Design of Steel Frames.* CRC Press: Boca Raton, FL, 1992.

[4] Lui EM, Chen WF. Steel frame analysis with flexible joints. *J. Constr. Steel Res. (Special Issue on Joint Flexibility in Steel Frames)* 1987; Vol. 8, 161–202.

[5] Galambos TV. *Guide to Stability Design Criteria for Metal Structures,* 4th edn. Wiley: New York, 1988.

[6] Liew JYR, White DW, Chen WF. Second-order refined plastic hinge analysis for frame design: part 1. *J. Struct. Eng., ASCE* 1993; **119**: 3196–3216.

[7] ECCS. *Essentials of Eurocode 3 Design Manual for Steel Structures in Building,* ECCS-Advisory Committee 5, No. 65, 1991.

[8] ECCS. *Ultimate Limit State Calculation of Sway Frames with Rigid Joints,* Technical Committee 8 – Structural Stability Technical Working Group 8.2 – System, Publication No. 33, 1984.

[9] Standards Australia. *AS4100-1990, Steel Structures,* Sydney, Australia, 1990.

[10] Canadian Standard Association. *Limit States Design of Steel Structures,* CAN/CSA-S16.1-M94, 1994.

[11] Canadian Standard Association. *Limit States Design of Steel Structures,* CAN/CSA-S16.1-M89, 1989.

[12] European Committee for Standardization. *Eurocode-3, 2005. EN 1993-1-1: Design of Steel Structures—General Rules and Rules for Buildings,* 2005.

[13] Chen WF, Goto Y, Liew JYR. *Stability Design of Semi-Rigid Frame Design.* Wiley: New York, 1996.

[14] Vogel U. Calibrating frames. *Stahlbau* 1985; **10**: 1–7.

[15] Chen WF, Toma S. *Advanced Analysis of Steel Frames.* CRC Press: Boca Raton, FL, 1994.

5.11 Structural Engineering (結構工程)

5.11.1

ENGINEERING MECHANICS IN STRUCTURAL ENGINEERING RESEARCH AND EDUCATION

W.F. Chen
School of Civil Engineering,
Purdue University, USA.

This paper was presented as the keynote lecture at the International Symposium on the Role of Engineering Mechanics in the Training of Engineers, held on 23-26 June, 1994 at Tsinghua University, Beijing. It has been published in the Proceedings.

1.   ENGINEERING MECHANICS

The words "engineering mechanics" imply a mathematical formulation of the engineering problem and of the basic equations to be used in its solution. Structural engineers earn their living by practising the art and science of applied mechanics to the design, analysis, and evaluation of all types of engineering structures. The purpose of an ultimate design is the production of a physical structure capable of withstanding the environmental conditions to which it may be subjected. Thus, structural engineers have to assess the loading the structure will be called on to resist during its life, and they must also know the properties of the materials used in the structure to resist this loading for maximum economy and safety as well as for aesthetic appeal. There is, therefore, a fundamental two-stage process in the engineering design operation: firstly, the forces acting on the structural members of elements must be determined, and secondly, the reactions of the materials of the members or elements to the forces must be defined. The first stage involves an analysis of the structural system; while the second involves a constitutive modelling of structural materials. The more comprehensive background a structural engineer has in engineering mechanics training, the better the structural engineer will be able to design his structure. This is illustrated in the present paper.

2.   CONSTITUTIVE MODELLING OF ENGINEERING MATERIALS

Structural mechanics, as the term implies is a branch of mechanics of solids. A valid solution to a solid mechanics problem must satisfy the equations of equilibrium or of motion, the equations of compatibility or geometry, and the relation between stress and strain or constitutive equations of materials. A rigorous mathematical analysis of these equations is extremely difficult and simplifi-

cations and idealizations must be made for further progress to engineering practice. The key to obtaining valid and practical engineering solutions is the establishment of the proper relation between forces or stress quantities (generalized stresses) and compatible or geometric quantities (generalized strains). Without such basic generalization of the stress-strain relations of materials to structural elements or their equivalent, a rigorous solution to an engineering problem is almost impossible. In the following sections, we will adopt a unified point of view by basing the development of structural engineering on the basic and narrower concept of stress-strain relations of materials and of their extension and generalization. In this way, stress-strain relations, strength of materials simplifications, finite element generalizations and high performance computing all are made more natural and understandable by this fundamental approach of engineering mechanics. To obtain a readily understandable and soluble mathematical model for an engineering material, drastic idealizations concerning its deformation characteristics are necessary. Thus, for example, for a simple mathematical model utilizing the theory of elasticity, the following assumptions are necessary: (Chen and Saleeb, 1982)

1.   The stress and strain relationship of the material is *linear*.

2.   The material is *homogeneous* and no discontinuities exist.

3.   The material is *isotropic*.

4.   The material is perfectly *elastic*.

These assumptions are, of course, not true for a real material, but they provide a close approximation in many actual applications. For a rigorous correct solution of an engineering problem, a three-dimensional stress analysis must be carried out to determine: six stress components (σ_x, σ_y, σ_z, τ_{xy}, τ_{yz}, τ_{zx}), six strain components (ε_x, ε_y, ε_z, γ_{xy}, γ_{yz}, γ_{zx}), and three displacement components (u, v, w). In the case of plane strain/stress problem, the more simple application involves only three components (σ_x, σ_y, τ_{xy}), three strain components (ε_x, ε_y, γ_{xy}), and two displacements components (u, v). They may be resolved by considering the various boundary conditions which must be fulfilled for solution of a boundary value problem. A simple two-dimensional elasticity formulation, so familiar in engineering mechanics, as represented and illustrated by the following mathematical analysis:

(i)   stress-strain relationship *(material)*:

$$\varepsilon_x = \frac{1}{E}\left(\sigma_x - v\sigma_y\right)$$
$$\varepsilon_y = \frac{1}{E}\left(\sigma_y - v\sigma_x\right) \tag{1}$$
$$\gamma_{xy} = \frac{\tau_{xy}}{G}$$

(ii)   strain-displacement relationship *(logic)*:

$$\varepsilon_x = \frac{\partial u}{\partial x}, \ \varepsilon_y = \frac{\partial v}{\partial y}, \ \gamma_{xy} = \frac{\partial u}{\partial y} + \frac{\partial v}{\partial x} \tag{2}$$

(iii) equations of equilibrium *(physics)*:

$$\frac{\partial \sigma_x}{\partial x} + \frac{\partial \tau_{xy}}{\partial y} + X = 0$$
$$\frac{\partial \sigma_y}{\partial y} + \frac{\partial \tau_{xy}}{\partial x} + Y = 0 \tag{3}$$

(iv)   equation of compatibility *(geometry)*:

$$\frac{\partial^2 \varepsilon_x}{\partial y^2} + \frac{\partial^2 \varepsilon_y}{\partial x^2} = \frac{\partial^2 \gamma_{xy}}{\partial x \partial y} \tag{4}$$

The extreme difficulties involved in the use of formal mathematics in the solution of the above simultaneous partial differential equations for any engineering structures lend particular attraction to the use of larger elements as a means of solving problems associated with engineering design in structures. This is achieved by devising the structure into large elements, so that the forces and resultant displacements at the boundary of each element may be obtained and assembled into a particular directional form. This is known as the *strength of materials* approach to structural elements and members, leading to the modern development of finite element approach to structural system, a branch of computer-based structural engineering, that has been greatly enhanced by the recent rapid development of *high performance computing* in software engineering. [8]

3.   STRENGTH OF MATERIALS APPROACH TO STRUCTURAL ELEMENTS

For structural members and frames, an element or segment with full cross section A can be taken as the basic building block for mathematical analysis instead of a material point as illustrated previously for an elasticity problem. The element may be subject to axial force, twist moment, or bending moment separately or in combination. These loads are called the *generalized stresses* of the problem. The corresponding *generalized strains* are the elongation per unit length, the angle of twist per unit length, or the relative rotation of the end cross sections per unit length about the axis of bending (or curvature). The very simple but powerful kinematic assumption must be made to relate the strain to generalized strain from which the *generalized stress-strain relationship* for the structural element can be properly derived. This approach, which underlies the well-established branch of engineering mechanics of solids known as *strength of materials* is extremely effective in providing close approximate solutions to many structural engineering problems involving such structures as frames, plates and shells. A simple beam bending problem, so familiar in practice, is illustrated in the forthcoming by the derivation of its generalized stress-strain relation or the moment-curvature (M-ϕ) relation.

(i)   The strain ε to the generalized strain ϕ is established by the linear kinematic assumption *(kinematic)*

$$\varepsilon = y\phi \tag{5}$$

(ii)   The stress σ to the generalized stress M is obtained by the equilibrium consideration *(equilibrium)*

$$M = \int_A \sigma y \, dA \tag{6}$$

(iii) The use of the stress-strain relation *(material)*

$$\sigma = E\varepsilon \tag{7}$$

leads to be desired relationship between the generalized stress M to the generalized strain ϕ in the simple form *(element)*

$$M = EI \, \phi \tag{8}$$

where the EI value is the beam bending stiffness.

This approach reduces many structural engineering problems, which are intractable in the exact sense, to elementary problems with the strength of materials assumption. This approach has lead naturally to the modern development of the very powerful computer-based method of structural analysis known as *finite element analysis*. Indeed, the finite element analysis has provided structural engineers with a tool of very wide applicability. In principle, at least, numerical solution can be achieved for virtually any engineering structures. The development of finite element analysis process is described in what follows.

4.   FINITE ELEMENT APPROACH TO STRUCTURAL SYSTEMS

The finite-element method can be thought of as a

generalization of the strength of materials approach to structural system by selecting finite elements as the basic building blocks for the system or the continuum. The basic steps in the finite element process follows closely the process of strength of materials approach described in the preceding section. Here, as in the strength of materials case, the basic step in the simplification is the derivation of the generalized stress and generalized strain relationship that relates the nodal force $\{F\}$ to the nodal displacement $\{\delta\}$ for an individual element. To derive this relation, we follow the identical steps as in the beam bending case given previously.

(i) The strain $\{\varepsilon\}$ to the generalized strain $\{\delta\}$ is established by the kinematic assumption or the *shape-function* $\{N\}$ *(kinematic)*

$$\{\varepsilon\} = [B]\{\delta\} \tag{9}$$

the matrix $[B]$ generally being composed of derivatives of the shape functions.

(ii) The stress $\{\sigma\}$ to the generalized stress $\{F\}$ is obtained by applying the virtual work equation *(equilibrium)*

$$\{F\} = \int_V [B]^T \{\sigma\}\, dV \tag{10}$$

(iii) The use of the stress-strain relation *(material)*

$$\{\sigma\} = [D]\{\varepsilon\} \tag{11}$$

leads to the desired relationship between the generalized stress $\{F\}$ to the generalized strain $\{\delta\}$ in the simple form (element)

$$\{F\} = [k]\{\delta\} \tag{12}$$

where the element stiffness matrix has the value

$$[k] = \int_V [B]^T [D][B]\, dV \tag{13}$$

generally, except for special cases, the integration indicated in Eq. (13) must be carried out numerically in the computer.

Once the element stiffness matrices have been calculated and transformed from local to global coordinates, the structure stiffness matrix $[K]$, which relates the load $\{R\}$ to the nodal-displacement $\{\delta\}$ of the complete structure, can now be formed by the systematic addition of element stiffness

$$\{R\} = [K]\{\delta\} \tag{14}$$

This load-displacement relation can also be viewed as the generalized stress $\{R\}$ and generalized strain $\{\delta\}$ relationship of the structural system. Once the nodal-point displacement vector $\{\delta\}$ known, the element stresses within each element are found

from Eq. (11), using Eq. (9)

$$\{\sigma\} = [D][B]\{\delta\} \tag{15}$$

Generally, the solution of the simultaneous equations (14) must be carried out numerically and incrementally by an iteration process in the computer. Thanks to the rapid development of the high speed computers and modern numerical techniques, incremental inelastic analysis of virtually any structural problems can now be carried out by this powerful finite-element method. With the recent development of high performance computing in software engineering, it has greatly benefitted the field of structural engineering, in particular, its finite element applications. The high performance computing has been the subject of intense research in recent years. A brief description of this development currently underway at Purdue University is given in the forthcoming.

5. HIGH PERFORMANCE COMPUTING IN STRUCTURAL ENGINEERING

High performance computing, i.e. the integration of *parallel processing, object-oriented design and analysis, and computer graphics*, provides the possibility to speed up the implementation process of *engineering mechanics theory* and widen the use of engineering research in practice. Advances in structural engineering have created a demand for high-quality software systems which utilize new hardware capabilities effectively and efficiently. However, due to the difficulties frequently encountered in maintaining existing software and in developing new software, the tremendous increases in computing power offered by modern computers cannot always be fully utilized to meet the demand. At the present time, software development and maintenance are the main barriers to the infusion of advanced computer technology in structural engineering research and education. In computationally intensive applications, structural engineering researchers have been opting for the use of supercomputers, especially parallel computers. More specifically, they have been focusing in the area of development of concurrent algorithms for specific applications. A major problem encountered by those researchers is not only due to the various types of parallel architectures existent in the market, but also because there is no unified way in which parallel computer's compilers are written. As a consequence, the implementation of concurrent algorithms in actual parallel computers can be very cumbersome. Therefore, many of the advances achieved in the area of concurrent algorithms are not being taken advantage of by the structural engineering community at large. The research in progress at Purdue University is an attempt to address the challenges mentioned above. [13] The authors believe that parallel processing technology is a major breakthrough in engineering computing. They also believe that the application of the object-oriented paradigm to domain-specific design of software provides the greatest promise for facilitating reuse and rapid prototyping of con-current software applications. The research being described here is that of a prototype object-oriented concurrent software development environment. The final goal of this research is to demonstrate its potential for improvement of quality and productivity in research and education development for large-scale computing in structures domains. It has been found that structural engineers with a strong engineering mechanics background are the most desirable persons to carry out such tasks and to achieve the main goal of this research: an object-oriented concurrent software development environment. This is described in the forthcoming.

The two critical issues addressed by this research work are:

• The under-utilization of the parallel computer technology by the structural engineering community;

• The efforts made on the development and extension of software for structural engineering computing often cannot be accumulated, and thus become wasted.

The main two tasks in this work are, therefore, the creation and testing of a parallel programming environment and of a development environment that promotes software reuse in the specific structures domain. This is to be achieved by applying the object-oriented paradigm to the design of concurrent software applications. The environment should provide *systematic* support to the development of application software, as well as serve as a crucial layer between structural engineering applications and evolving computer technology. The improvement of software quality and productivity for research and education activity will be investigated through the use of the environment in the development of a number of structural engineering applications in research and education. This domain-specific environment is referred to in the discussion below as **Structural Engineering concurrent Software Development Environment (SECSDE).** This research attempts to build the basic framework of the SECSDE, and to establish a model of such an environment for other civil engineering area [7].

The final version of the SECSDE will consist of a concurrent software development tool reusable software components (including existing parallel algorithms) and CASE (Computer Aided Software Engineering) tools which support software reuse. Reusable software components will include both structural engineering specific components, and general-purpose components whose use is not limited to structural engineering. The CASE tools of the SECSDE will be utilized for managing the software components and helping programmers to find and integrate components into applications. The concurrent software portability tool will permit the development of portable parallel codes for the several multiprocessing machines available in the system. Its extension to future parallel architectures should be straightforward, since it will include the necessary abstractions to deal with the parallel programming idiosyncrasies.

Object-Oriented Programming (OOP) is the key methodology in the development of the reusable components of the SECSDE. This is because software *reusability* is the central objective of OOP. Object-oriented programming appears to be the most promising technique at the present time for attaining the goal of software reusability. This programming methodology facilitates a new style of software development: software development based on a large number of prefabricated reusable software components. This new style of software development should be more productive than previous styles. Software developed in this way should be less error-prone, more readily modified, and more extendible.

As mentioned previously, object-oriented and parallel programming are the major methodologies for the software components development in the SECSDE. The C++ language is the major implementation language. However, the environment will not force applications to follow an object-oriented methodology or to use the C++ language. Where possible, interfaces for conventional languages such as C and FORTRAN will be provided for major software components of the environment. In Figure 1, the class libraries of the SECSDE are shown. In Figure a depiction of an example finite element subsystem is given. Also, as an example, in Figure 2, one of the objects *(element)* and its corresponding subsystem are shown in some detail.

6. STRUCTURAL ENGINEERING RESEARCH

It would be impossible to investigate the usefulness of this software development environment unless a number of application test cases were studied. The first application to be tested will be a Finite Element platform, whose initial thrust will focus on *building structural performance assessment*. This will be a two fold departure from conventional approaches. Firstly, structural analysis and design are currently largely based on the study of members and connections, or of simple structural systems. In the envisioned environment, by utilizing high performance computing, the analysis of whole structural systems will be made possible. Secondly,

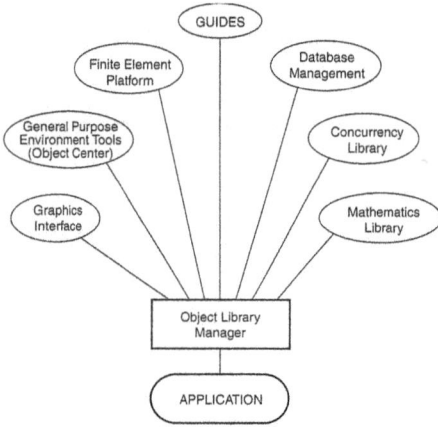

Figure 1
The class libraries for the envisioned environment

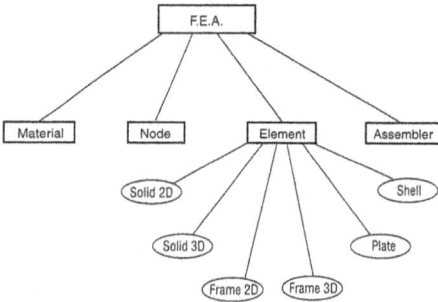

Figure 2 *A Finite Element Subsystem in the environment*

presently finite element analysis software is either application specific or so general that it is too time consuming for the implementation of real applications. In this work, because of the use of an object-oriented design, only portions of the software needed for the analysis of a particular problem will be linked together to form an executable program. This will result in improvements in usage of memory and in run-time efficiency. Furthermore, when this platform is combined with the CASE tools for the envisioned environment, the ability for collaborative work among finite element researchers will be greatly enhanced. Many research topics fall in the category of building structural performance assessment. Research on several of these topics is currently underway at Purdue University. They are described briefly in the following paragraphs [9].

6.1 Life-Cycle Analysis of Structural Systems

The first application of interest is that of life-cycle analysis of structural systems. The goal of this

analysis is to stimulate the behavior of a structural system at the various stages of its useful life. The life of a structural system can be subdivided into three major phases: construction, service and maintenance. At present, structural analysis and design are based on the service phase of the life-cycle of a structural system.. However, it is often during construction when structures are subjected to most problems. Figure 3 illustrates the risk of failure of a system in the various phases of its life. Life-cycle analysis has a high demand for com-putational power. Thus, high quality software systems are essential to the progress of structural engineering in this research area. Research on the construction sequencing of concrete buildings is currently underway as a test case of the SECSDE. This ties directly with research works that have been done at Purdue University since 1982 [12]. These studies have helped provide information to techniques that can be used to simulate shoring and reshoring loads and the load distribution during the construction phase. However, much research remains to be done. Topics such as: construction variability, lateral loading resisting systems, and classification of construction loads have still not been studied.

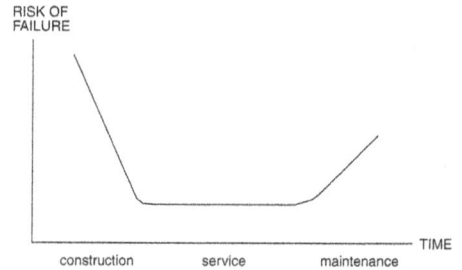

Figure 3
Phases in the Life-cycle of a Structural System

6.2 Constitutive Behavior of Concrete Materials

A second application focuses on the modelling of the inelastic constitutive behavior of concrete materials. Constitutive equations are of central importance to concrete mechanics and the engineering design of reinforced concrete structures. A better understanding of the behavior of concrete materials is necessary to precisely analyze concrete structures by numerical procedures, and it may lead to a more efficient use of the material. However, despite extensive research in this area, the behavior of concrete materials has still not been fully understood or properly modelled, especially during the post-failure regime.

Research is currently underway on an unified elastic-plastic-damage theory, which is based on recent fundamental understanding of concrete micro-mechanics at Purdue University [1]. The environment described here provides the best setting for the development of this constitutive model of concrete materials. It will eliminate the cumbersome task of implementing such a model in an existing general purpose finite element analysis. Also, because of its parallel processing tools, SECSDE will allow for the analysis of mechanisms of deformation obtained from micro-mechanics studies, Figure 4, which is a computationally intensive task.

6.3 Advanced Analysis for Steel Frame Design

Finally, an application on the design of steel frame structures considering both geometric and material nonlinearities in a direct manner, [14] Figure 5, is being researched. Much work has been done,

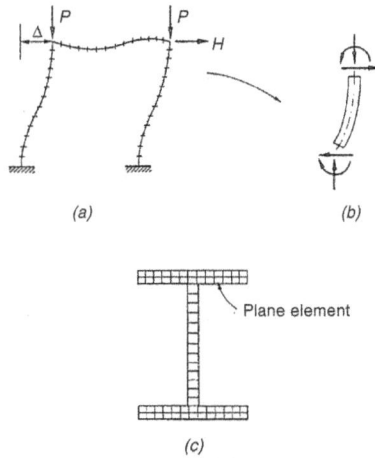

(A) – A concrete mass subjected to compressive loading. Concrete is composed of three phases:
1. mortar
2. coarse aggregate, and 3. mortar-course aggregate interface. Also shown are some mechanical loading-induced microcracks in the interface

(B) – Magnification of mortar portion of concrete in (A). Microcracking can be intense in paste-sand particle interface because of high surface area of sand particles

(C) – Further magnification of mortar in (B) showing details of the interface which includes stacks of well-developed calcium hydroxide crytals, some entrapped-air which is remnance of bleeding, and a thin layer of amoorphous calcium hydroxide uniformly surrounding the sand particle

Figure 4 *Details of concrete mass and magnification of its mortar portion for (A) concrete; (B) mortar portion of concrete; (C) further magnification of mortar. (After Chen and Cohen, 1992)*

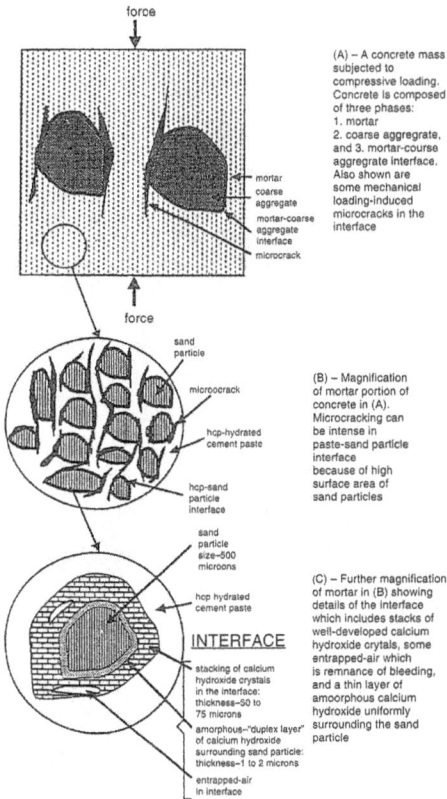

Figure 5 *Plastic zone theory for frame analysis.*
(a) frame; (b) beam-column element;
(c) section. (After Chen and Toma, 1994).

which have validated the utilization of second-order inelastic analysis procedures [4]. This type of research suites very nicely into the SECSDE, whose final architecture will allow for the testing of various sophisticated second-order inelastic element models. It will also, due to its high performance computing capabilities permit the analysis of the structural system as a whole. These features will permit the research of several possible failure mechanisms both at the local and global levels.

7. COMPUTER-AIDED INSTRUCTION

Computer technologies have made a great impact on almost every aspect of our life. To engineering education, computers offer the possibility of individualized instruction and simulation of the real world. [9]. In doing so, the learning process can be made more interesting, challenging and effective. Infusing computer technology into educational curricula will prepare students for life in an information society. We are facing the challenge now and for the 21st century to infuse the rapidly advancing computer technologies into the curricula of higher engineering education. Currently, computer hardware and software potential has given rise to an unprecedented opportunity to infuse computer technology into engineering education and to given momentum to the advancement of education technology.

7.1 CAI for Structural Engineering

Computer-Aided Instruction (CAI) software or courseware for structural engineering falls basically into three categories: (1) courseware for courses in basic theory such as *Statics and Dynamics, Elasticity, and Plasticity*; (See for example, Chen and Zhang, 1990) (2) course courseware for structural analysis; [3] and (3) courseware for structural design. [10]. Courseware for basic theory attempts to help students understand the principles and fundamental methods. [11]. Structural analysis courseware is aimed at helping students to grasp concepts, to appreciate the structural behavior of elements and systems, to realize the significance of assumptions in basic theorems, and to develop the student's "structural sense" [4]. Structural design courseware is tied to the solution of realistic problems in the real world. It enables students to learn more about design philosophy and design considerations as well as to become familiar, with recommendations of codes of practice [2]. CAI software can also be categorized as either *problem-solving courseware systems or intelligent computer tutoring systems* (which are also referred to as *Intelligent Computer-Aided Instruction (ICAI)* courseware systems).

7.2   Problem-Solving CAI Courseware

Problem-solving courseware encompasses tools aimed at helping students to master the concepts and methods taught in a course by solving problems. The executive of problem-solving courseware is usually a student-initiated learning process: *problem-definite, problem-solving,* and *result-interpretation.* Taking full advantage of the high speed, vast storage, and high-resolution graphics of workstations, the problem-solving CAI courseware helps student gain technical competence through repeated hands-on analysis. Students learn the concepts, and understand the process of structural modelling better. An important benefit of CAI courseware is that it helps students to gain a sense of the scale up from basic engineering courses to the realistic problems in engineering practice. In traditional structural engineering education, the students usually do not have the opportunity to participate in engineering practice until they hold their first position in business or industry. They are often surprised to find their knowledge is *incomplete* and rather limited for solving practical problems. With the help of the CAI courseware, the students will be able to understand and solve more realistic problems in the educational environment.

A notable advance in this development is the SOCRATES project, *the Study of Complementary Research and Teaching in Engineering Science* [11]. It was created by a grant from the U.S. Department of Education to Cornell University in 1986. Since then, a large number of problem-solving courseware programs have been developed and distributed to several engineering schools in the United States. This is pioneering work in this area. Similar work is currently underway at Purdue University, using SESDE.

7.3   ICAI Courseware

Problem-solving courseware, which provides a foundation for the development of ICAI courseware, can be incorporated with artificial intelligence to develop more powerful and more attractive ICAI courseware. The purpose of ICAI courseware is to combine the experience of problem-solving and the motivation of discovery learning with the effective guidance of tutorial interactions. With an intelligent tutoring system, students have more control over their own learning through the selection of work time, location, and pace of learning among other things. Students have more flexibility to choose instructional materials to meet individual needs. The various teaching strategies embodied in the system can better suit individual learning styles. With the tremendous growth in the development of expert systems and expert system tools, it is of great interest to introduce knowledge-based tutoring in engineering education. This is a wide open area with huge potential.

ICAI courseware can be formed by adding certain ICAI-specific or course-specific knowledge bases which contain problem-solving expertise and teaching strategies, and facilities such as problem-generating systems and diagnostic systems for evaluating students. Experienced faculty should be involved with formulating student diagnostic systems and teaching strategies. These ICAI courseware systems hold significant promise as a crucial mechanism for responding to the learning, education and training needs of highly industrialized nations. Although this learning technology is presently still in its infancy, the integration and development of the rich and diverse tools of the SECSDE will make the development of ICAI courseware practical to accomplish.

8.   CONCLUSIONS

Structural engineers apply the art and science of engineering mechanics to engineering practice by developing the mathematical tools to be used in its solution. The basic tool to obtaining valid solutions is the establishment of the relation between force or stress quantities, called *generalized stress*, and compatible or geometric quantities, called *generalized strain*, for various structural elements. Without such basic knowledge of the stress-strain relations or their equivalent generalized stress-strain relations, a so-called solution is merely a guess. The development of these basic relationships is in the realm of engineering mechanics. Once this basic relationship is established for a structural engineering problem, the solution can always be achieved with the aid of computers.

The first part of this paper therefore discusses this unified point of view by basing the development of structural engineering on the basic concept of stress-strain relations of materials and their extension and generalization leading to the modern development of the very powerful computer-based method of structural analysis known as *finite element analysis*.

Thanks to the rapid development of the high speed computers, any structural engineering problems can now be carried out rigorously and solved by this powerful computer-based method. With the recent development of *high performance computing* in software engineering, an evolution of the structural engineering profession is currently underway. This evolution is driven by the rapid advances in computer technology. Research and instruction in structural engineering are becoming more software dependent and more software intensive. The current trend is expected to continue into the 21st century. The success and pace of this evolution depends on the rapid and economic development of structural engineering software. The need for the development of a domain-specific software development environment for structural engineering computing is therefore described here. Such an environment is expected to lead to a higher software productivity, and great improvements in software quality, reliability, maintainability, flexibility for modification and extendibility. The domain-specific environment currently under development in the School of Civil Engineering at Purdue University, referred to as Structural Engineering Concurrent Software Development Environment (SECSDE), is presented.

High Performance Computing, i.e. the integration of parallel processing, object design and analysis, and computer graphics, as a part of SECSDE development, will make hitherto unrealizable approaches feasible, including such challenging subjects as life-cycle analysis of structural systems; and micromechanical study of concrete materials connecting material science to structural engineering. The increasing utilization of computer power through SECSDE will also demand further progress in many other areas of structural engineering. Structural engineers with strong backgrounds in computer science and engineering mechanics are in high demand and are the only few who are able to contribute to this development and to meet such a challenge.

Computer-Aided Instruction will also be in demand in the 21st century. The current development of computing environment such as SECSDE will give rise to an unprecedented opportunity for educators to develop CAI for engineering education. Although some progress has been made in this area at Purdue and elsewhere, much more remains to be done. Some of this development relating to structural engineering is outlined near the end of this paper.

9. REFERENCES

[1] Chen, W.F., Concrete Plasticity: Macro- and Micro Approaches, International Journal of Mechanical Science, Vol. 35, No. 12, 1993, pp. 1097-1109.

[2] Chen, W.F. and Lui, E.M., Stability Design of Steel Frames, CRC Press, Boca Raton, Florida, 1991, 380 pp.

[3] Chen, W.F. and Mosallam, K.H., Concrete Buildings: Analysis for Safe Construction, CRC Press, Boca Raton, Florida, 1991, 260 pp.

[4] Chen, W.F. and Toma, S. Editors, Advanced Analysis of Steel Frames: Theory, Software and Applications, CRC Press, Boca Raton, Florida, 1994, 384 pp.

[5] Chen, W.F. and Saleeb, A.F., "Constitutive Equations for Engineering Materials," Vol. 1 – Elasticity and Modelling, John Wiley Interscience, New York, 1982, 580 pp.

[6] Chen, W.F. and Zhang, H., Structural Plasticity: Theory, Problems and CAE Software, Springer-Verlag, New York, 1990, 250 pp.

[7] Chen, W.F. and Dunsmore, H.E., White, D.W. and Zhang, H., "SESDE – An Envisioned Software Development Environment for Structural Engineering," Structural Engineering Technical Report, CE-STR-90-6, Purdue University, West Lafayette, IN, 1990, 20 pp.

[8] Chen, W.F., Sotelino, E.D. and White, D.W., High Performance Computing in Civil Engineering Applications, Conference on Computer Applications in Civil and Hydraulic Engineering, October 15-16, Tainan, Taiwan, 1993, pp. 1-18.

[9] Chen, W.F., White, D.W. and Zhang, H., "Preparing Structural Engineering Research and Education for the 21st Century," Journal of the Chinese Institute of Civil Engineering and Hydraulic Engineering, Vol. 2, No. 2, 1990, pp. 95-106.

[10] ELRED, EDRFD Reference Manual and User's Guide Visual Edge Software Ltd., Release 10, St. Laurent, Quebec, 1990.

[11]   Ingraffea, T. and Mink, KK., "Project SOCRA-
        TES: Fostering a New Collegiality," Academic
        Computer, pp. 20-21, 60-62 (1988).

[12]   Liu, X.L., Lee, H.M. and Chen, W.F., "Analysis
        of Construction Loads on Slabs and Shores
        by Personal Computer," Concrete Inter-
        national: Design and Construction, 1988,
        pp. 20-21.

[13]   Sotelino, E.D., White, D.W. and Chen, W.F.,
        Domain-Specific Object-Oriented Environment
        for parallel Computing, Steel Structures,
        Journal of Singapore Structural Steel Society,
        Vol. 3, No. 1, December, 1992, pp. 47-60.

[14]   White, D.W., Liew, J.Y.R., and Chen, W.F.,
        "Application of Second-Order Inelastic
        Analysis for Frame Design: A Report to SSRC
        Task Group 29 on Recent Researcha and
        the Perceived State-of-Art," Structural
        Engineering Technical Report, CE-STR-91-12,
        Purdue University, West Lafayette, IN, 1991,
        116 pp.

5.11.2

Proceedings of ICE
Civil Engineering 162 May 2009
Pages xxxxx Paper 08-00022

doi: 10.1680/cien.2009.162.2.0

Keywords
buildings, structures & design; design
methods & aids; mathematical modelling

Wai-Fah Chen
PhD

is professor of civil engineering at
the University of Hawaii, USA

Seeing the big picture in structural engineering

This paper sets out to provide a 'big-picture' guide to the major advances in structural engineering design that have taken place over the last five decades. Rapid advances in computer technology during this period have spurred the development of structural calculations, ranging from the simple strength-of-materials approach in the 1950s, to the finite-element type of structural analysis for design in more recent years, and to the modern development of scientific simulation and visualisation for structural problems in the years to come. The paper concludes that the continually emerging nature of structural engineering and its associated codes of practice offers an exciting career, balancing idealisations of scientific theory with engineering reality and employing the latest computing technology.

Structural engineering is one of the many divisions of applied mechanics. In structural engineering, elasticity and plasticity in particular, and mechanics of materials and continuum mechanics in general, are studied and employed because they are powerful mathematical tools and provide rich physical content.

Here, as in every division of applied mechanics, the engineer operates with ideal material models and ideal structural systems. The theories of reinforced-concrete design, for example, do not deal with real reinforced concrete. They operate with an ideal composite material consisting of concrete and steel, the design properties of which have been approximated from those of real reinforced concrete by a process of drastic idealisation and simplification.

The same statement also applies to the formulation of the basic equations on equilibrium and compatibility of a real structural system. The magnitude of difference between the actual performance of a real structure in the real world and the performance predicted on the basis of drastically simplified theory can only be ascertained by long-term experience and observation, as realistically reflected in building codes supplemented with a variety of safety factors to account for the difference.

This paper focuses on the theories which have stood the test of experience and have been used widely in the actual design of practical structural engineering solutions.

Mechanics and engineering design

The mechanics analysis of a given structural problem or a proposed structural design involves mathematical formulation with three sets of basic equations and solutions

- equilibrium equations or motion reflecting law of physics (Newton's law or physics)

ice
Institution of Civil Engineers

CIVIL **ENGINEERING**

AUTHORS •

- constitutive equations or stress–strain relations reflecting material behaviour (materials or experiments)
- compatibility equations or kinematical relations reflecting the geometry (continuity or logic).

The interrelationship of these three sets of basic equations is shown in Figure 1 for the case of static analysis.

For an elastic material, the constitutive equations are embodied in Hooke's law and the mechanics of elastic structures are well understood. The construction of the analytical tools for calculating stresses and deformations in a strained elastic structure is the dominant concern with the theory of elasticity, which provides familiar methods for solutions of the basic field equations.[1]

Of greater practical importance is the determination of the load-carrying capacity of structures composed of elements which yield on reaching a certain stress state. The construction of the analytical tools for calculating stress and deformations in a structure deformed beyond the elastic range is the dominant concern with the theory of plasticity, which predicts the conditions under which the structure will fail and what type of failure mode will occur. In general, as a matter of practicality in engineering practice, it is the irreversible, inelastic range of deformation that is of major concern in structural engineering.

To achieve a realistic and practical engineering-design solution, the following must be performed with the most advanced computers: drastic idealisations and simplifications for the field equations of equilibrium, of material behaviour and

of kinematics. This paper highlights several key breakthroughs in concepts to reach the present state of analysis and design that is familiar to structural engineers. These breakthroughs include, most notably, the following concepts and theorems.

- The generalised-stresses and generalised-strains concept connects the conventional strength-of-materials approach to a continuum-mechanics-based theory of plasticity leading to the modern development of finite-element solutions in structural engineering.
- The simple plastic-hinge concept enables the direct application of simple plastic theory to steel-frame design, in particular, leading to the modern development of advanced analysis for design of steel building structures.
- The proof of the limit theorems of perfect plasticity provides rational principles for preliminary structural design via simple equilibrium or kinematical processes that are consistent with the engineer's intuitive approach to design, leading to modern development of strut–tie models for design of reinforced-concrete structures in particular.

Generalised stresses and strains

To simplify the field equations for a realistic engineering solution, it is more convenient to formulate the elastic or plastic relations in terms of elements from which the parts of the structure are composed rather than for the material point as defined elegantly in the concept of continuum mechanics.

For example, for a reinforced-concrete beam in a building framework, the basic element or segment can be obtained by cutting through the entire thickness of length Δs as shown in Figure 2. Thanks to this approach, it is then possible to replace the six stress components acting on the element by one dominant normal stress resultant – the bending moment, M (generalised stress). Similarly, the corresponding six deformational strain components can be reduced to one dominant strain resultant – the angle of relative rotation, Δφ (generalised strain).

The relationship between the value of bending moment M and the angle of relative rotation Δφ for the ends of the section represents the material behaviour of the segment (generalised stress–strain relation). The relationship is linear and reversible before the formation of a first crack, becomes linear again but somewhat less stiff, and then irreversible when it reaches a certain value, at which the magnitude of the rotational deformation increases almost without change in the loading as shown by the horizontal line a–b in Figure 2 (nearly elastic / perfectly plastic relation).

In reality, of course, the stresses and strains in the segment are much more complicated. As the loading increases, the concrete in the tension side of the segment starts to crack, progressively open up and move upward, diminishing the compression zone. When the loading increases further, more cracks develop, while the bonding between the reinforcing steel and its surrounding concrete starts to deteriorate and to crack. Finally, the reinforcement starts to yield, resulting in an almost constant moment capacity with nearly an

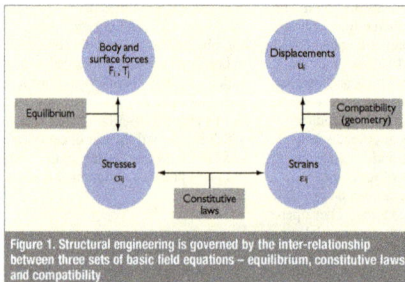

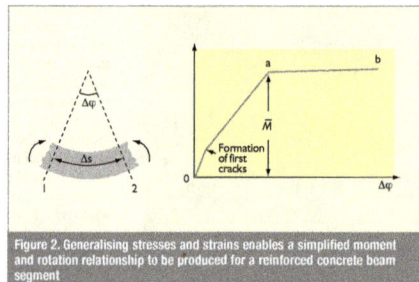

Figure 1. Structural engineering is governed by the inter-relationship between three sets of basic field equations – equilibrium, constitutive laws and compatibility

Figure 2. Generalising stresses and strains enables a simplified moment and rotation relationship to be produced for a reinforced concrete beam segment

unlimited rotational deformation.

The concept of using the generalised stresses and generalised strains for inelastic structural analysis and design was employed for the first time in 1952 by Prager[2] in establishing his general theory of limit design and, later in 1959, utilised prominently by Hodge[3] in his popular text on the plastic analysis of structures. It took great insight fully to understand the impact of unifying the conventional strength-of-materials approach with the modern theory of plasticity and limit design in a consistent manner.

Thanks to the approach, the complex local stress and strain states in an element are avoided and the field of application of the theory of plasticity to engineering structures can be broadened significantly. This expansion and generalisation resulted in the development of modern structural theories, among them several structural elements including beam elements, plate elements, shell elements and finite elements.

The study and mathematical formulation of engineering structures have led to a fundamental three-stage process in mechanics operation.

■ First, the stresses in a structural element to the generalised stresses acting on the element are determined by the use of equilibrium equations.
■ Second, the deformations of the material in the element to that stresses to the generalised strains of the element are established through a kinematical assumption such as 'plane section remains plane after bending'.
■ Finally, the generalised stresses

and generalised strains are derived through the use of stress–strain relations on the material.

This mechanics operation becomes particularly clear when this process is applied to the segment of a reinforced concrete beam in bending shown in Figure 2 (see Appendix). Since the solution for the simultaneous equation (see Equation 10 in the Appendix) must be carried out numerically in incremental form by computer, many numerical procedures were developed in the period from 1970 to the 1980s, including the necessity of developing an efficient iterative process to deal with load–path dependency of the inelastic material behaviour.

As a result of the progress, together with the rapid advancement in computing power, large amounts of numerical data were generated in a variety of structural engineering applications during this period. Table 1 briefly summarises some structural engineering impacts resulting from applications of finite-element methods with plasticity theory.

Plastic-hinge concept

The generalised stress and generalised strain relation of a structural element must be further idealised in order to develop simple plastic theory for engineering practice. This leads, for example, to ignoring strain-hardening and also to eliminating entirely the effect of time from the calculations.

For materials such as hot-rolled wide flange sections, the residual stress effects on its initial yielding of the cross-section are ignored. Also ignored are the possible local buckling effects on its maximum

plastic moment capacity. For materials such as reinforced concrete, the actual relation between the deformations and internal stresses is exceedingly complex, as explained previously.

To operate effectively in the real world, structural engineers must deal with idealisations of idealisations. In a structural concrete element, a single time-independent stress–strain curve in simple compression has already been designated as an idealisation of reality (dashed curve in Figure 3) that already ignores the variability of concrete material along with time effects.

To derive the moment–curvature relation of the type shown in Figure 2, additional idealisation of the material properties is required so that calculations are not excessively complex. To this end, the real stress–strain curves of steel and concrete are further simplified to be elastic / perfectly plastic, as represented by the solid lines in Figure 3.

With the idealisations of idealisations, it is possible to develop a simple plastic theory for frame design by introducing the powerful concept of a plastic hinge to deal with the complex plastic behaviour. The following characteristics are attributed to the moment–rotation relation of the beam element

■ in the elastic state before yield, either in concrete or steel, the generalised stress–strain behaviour of the element is linearly elastic with a constant elastic stiffness
■ in the state of yield, the generalised strain (deformation) can increase indefinitely for a constant value of the corresponding generalised stress with a full plastic moment capacity
■ the influences of axial force, residual stresses, local stress concentration

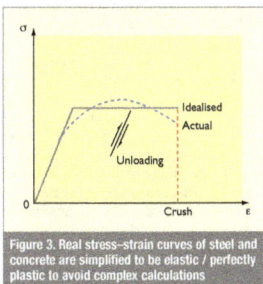

Figure 3. Real stress–strain curves of steel and concrete are simplified to be elastic / perfectly plastic to avoid complex calculations

Table 1. Structural engineering developments resulting from applications of finite-element methods with plasticity theory in the 1970s and 1980s			
Decade	Key theme	Developments	Further information
1970s	Numerical studies of member-strength equations	■ Beam-strength equation leading to beam-design curve. ■ Column-strength equation leading to column-design curve. ■ Beam–column-strength equation leading to beam–column-interaction design curve. ■ Bi-axially-loaded-column strength equation for plastic design in steel building frames.	Two-volume beam–columns treatise by Chen and Atsuta[4,5] and in the SSRC guide edited by Galambos[6]
1980s	Limit states to design	■ Development of reliability-based codes. ■ Publication of 1986 AISC/LRFD specification in USA[7] and Europe.[8] ■ Introduction of second-order elastic analysis to design codes. ■ Explicit consideration of semi-rigid connections in frame design (now known as 'partially restrained construction') in USA[9] and in Europe.[10]	Structural stability books by Chen and Lui[11] and Chen[12]

AUTHORS •

and partial yielding during the plastification process on the beam element's stiffness and rotation capacity are all ignored.

These characteristics of a beam element can be considered as defining an ideally plastic body or element. In the previous example of the reinforced-concrete beam bent in one plane (Figure 2), the moment–rotation curve for the ideal element is similar to the idealised stress–strain curve shown in Figure 3.

Despite the complexity of the real response of the beam element in details, all the local complexities exhibit relatively low influence and resistance to increasing load close to the maximum strength because of ductile or cracking behaviour of the materials. In fact, for a wide flange steel section, the dominant key feature of the element is captured nicely by this further idealisation of perfect plasticity concept.

Simple plastic theory

With the introduction of the plastic-hinge concept, simple plastic methods of structural analysis were rapidly developed in the 1960s for the purpose of calculating the plastic-collapse loads of frame structures. It was a remarkable achievement that it was possible to calculate the collapse load directly without considering the intervening elastic–plastic range.

Thanks to the simple plastic-hinge concept, the plastic methods of analysis may now be said to be fully developed as a sequence of elastic analysis. The plastic methods of analysis have since become

an important part of the modern theory of structures for practising structural engineers and an integral part of the undergraduate teaching curriculum.

Refined plastic hinge

The simple plastic hinge used in the preceding section may be called a concentrated plastic hinge where the zero-length plastic hinge can form suddenly from the limit of elasticity of a member. This concentrated plastic-hinge concept accounts for inelasticity but not the spread of yielding or plasticity at sections, or the influence of residual stress.

Depending on the geometry used to form the equilibrium equations, the elastic-plastic-hinge method may be divided into first-order and second-order plastic analyses. For the first-order elastic-plastic analysis, the un-deformed geometry is used and non-linear geometry effects are neglected. As a result, the predicted collapse is the same as the simple plastic analysis or rigid-plastic analysis. In second-order elastic-plastic-hinge analysis, the deformed shape is considered and geometry non-linearities can be included with the use of stability function for a beam–column element to capture the second-order effect. Load–displacement curves for various analysis methods are shown in Figure 4.

Desktop computing has now made it possible to combine the theory of stability and the theory of plasticity and apply them directly for frame design. This is known as the second-order inelastic analysis for frame design or, simply put, advanced analysis for frame design. The 2005 American

Institute of Steel Construction (AISC) load and resistance factor design (LRFD) specifications were issued to permit the use of this new analysis procedure for structural design.[7]

Various methods have been proposed in literature in order to make the advanced analysis applicable to engineering practice. Among them, the notional-load plastic hinge is one of the most widely known. The notional-load plastic-hinge method is achieved by applying additional fictitious equivalent lateral loads to account for the influences of residual stresses, member imperfections and distributed plasticity that are not included in the elastic-plastic hinge method.[13] With certain modifications, this method is accepted in the European Convention for Constructional Steelwork.[14]

To account for the spread of plasticity, a refined plastic-hinge method has been proposed to improve the accuracy of the notional-load plastic-hinge method. To meet the current specification requirements in the USA, further modifications and calibration of the refined plastic hinge are made against the current AISC LRFD codes. Details of this development are given in the book by Chen and Kim.[15]

Toward advanced analysis

In the USA, the term 'advanced analysis' strictly means 'second-order inelastic analysis' for frame design without the use of effective length factor (K-factor). There are three stages of progress to achieve the advanced analysis for frame design at the present time.

■ *Stage 1.* Direct second-order elastic analysis eliminating the use of amplification factors. This direct computation of second-order forces was encouraged in the 1986 AISC/LRFD specifications.[7] This is load-and-resistance-factor design but not advanced analysis for design.
■ *Stage 2.* Direct second-order plastic analysis with the use of K-factor to do member-by-member capacity checks with code requirements. This is a transition to advanced analysis, but it is still not advanced analysis because the K-factor is still needed in the design process.
■ *Stage 3.* Direct second-order inelastic analysis for frame design without the

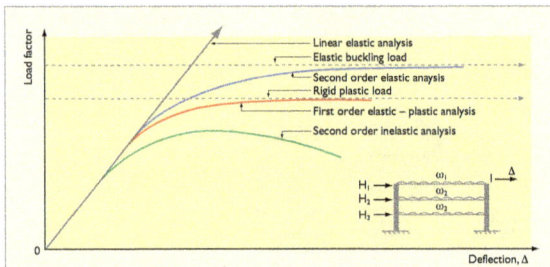

Figure 4. Load-deflection results of various analysis methods for structural frames – second-order inelastic analysis has been made possible through advances in desktop computing and is now accepted in codes

use of K-factor to do member-by-member capacity checks with code requirements. The code requirements are met automatically in the advanced analysis for the structural system. This is called advanced analysis.

As described previously, the introduction of the notional-load, refined-plastic-hinge method has reduced the complex second-order inelastic analysis to a second-order elastic analysis problem. Thus, the existing computer program for a second-order elastic analysis can be used directly for structural system design without the use of the effective length factor K. This notional-load, refined-plastic-hinge concept has made the advanced analysis work in engineering practice, since it only requires the use of existing second-order elastic analysis software.[15]

The advanced analysis can be further extended to include semi-rigid connections (partially restrained construction or simply 'PR construction' in the terminology of AISC specifications) as a separate structural element in a steel frame design. This is now under active development.

Limit theorems of perfect plasticity

The further idealisation of perfect plasticity gives powerful limit theorems of plasticity,[16] which provide an excellent guide for preliminary design as well as for analysis of structure.

- *Lower-bound theorem*. If an equilibrium distribution of stress can be found which balances the applied load, and is everywhere below yield or at yield, the structure will not collapse or will just be at the point of collapse.
- *Upper-bound theorem*. The structure will collapse if there is a compatible pattern of plastic-failure mechanism for which the rate at which the external forces equals or exceeds the rate of internal dissipation.

The lower-bound theorem states that the structure will adjust itself to carry the applied load if at all possible. It gives lower bound or safe values of the collapse loading. The maximum lower bound is the collapse load itself. The upper bound states that if a mechanism of plastic failure exists,

the structure will not stand up. It gives an upper bound or unsafe values of the collapse loading. The minimum upper bound is the collapse load itself.

Since the limit theorems were proved by writing the equations of equilibrium in the original geometry of a structure, they are not valid for important applications such as buckling or stability and crack propagation. However, they do provide an excellent and quick preliminary estimate of a load-carrying capacity of a structure. The quick estimate of the collapse load of a structure is of great value, not only as a simple check for a more refined computer analysis, but also as a basis for a preliminary engineering design.

The structural applications of the limit theorems started with the development of the simple plastic theory for steel building design[17] and extended to the development of yield-line theory for reinforced-concrete slab design;[18] limit theorems have been explored carefully for applications to stability problems in soil mechanics,[19] complemented by applications to the metal forming process[20] and studied thoroughly in metal matrix composites applications,[21] among others.

Yield-line theory for slab design

For reinforced-concrete slabs with a load acting perpendicular to their plane, the classical yield-line theory (or rupture-line theory) has been successfully used in the analysis and design. Since the yield-line theory for slabs uses a mechanism-solution technique, it gives upper bounds on the collapse load according to the upper-bound theorem of limit analysis.

As in the simple beam theory, moment–curvature relation in two dimensions instead of one is used and the actual moment–curvature curve is idealised as two straight lines similar to the elastic perfectly plastic stress–strain curve for uniaxial tests shown in Figure 3. The plastic moment capacity in a slab is exactly the same as for the beam element with the plastic moment capacity now expressed in terms of strength per unit length.

Consider as an example an annular concrete slab supported along the inside and outside edges, with axial symmetric reinforcement above and below and loaded with axial symmetric vertical load as shown in Figure 5(a). Owing to the axial symmetry of the slab and loading, the possible

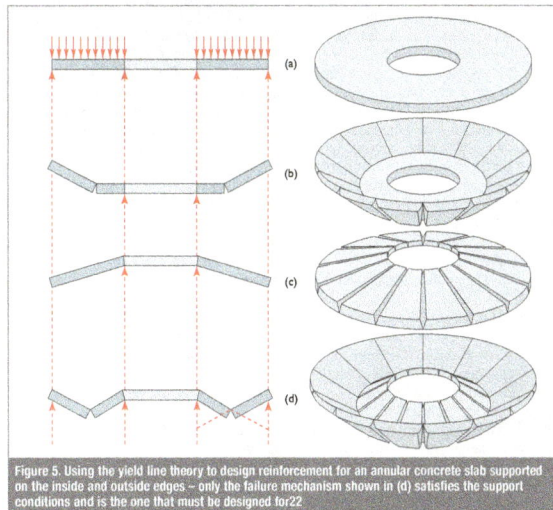

Figure 5. Using the yield line theory to design reinforcement for an annular concrete slab supported on the inside and outside edges – only the failure mechanism shown in (d) satisfies the support conditions and is the one that must be designed for22

AUTHORS •

plastic deformations of the slab element associated with the opening of cracks can be either in a circumferential direction or a radial direction as sketched in Figure 5(b) and (c). Only the combination of the two deformations satisfies the support conditions – shown in Figure 5(d).

Once the failure mechanism is sketched, it is a simple matter to write its work equation expressed in terms of the unknown position of the circumferential crack. The usual minimum process can then be applied to determine the critical position of the circumferential crack resulting in a minimum upper-bound solution or collapse load. The failure mechanism shown in Figure 5(d) may be somewhat unexpected but it does provide a clear physical picture of the failure process and helps structural engineers develop suitable designs for its reinforcement.

It is worth repeating here that in the application of limit analysis to structural engineering, it is presupposed that, up to the instant of collapse, the deformation and the displacements remain sufficiently small that one can ignore the change in the geometry of the deformed structure with the establishment and calculation of the element's equations of equilibrium as well as its kinematical relationships. This restriction is not a problem for reinforced-concrete structures since they are sufficiently rigid, but this may be an important limitation in the case of some thin-walled metal structures. However, for reinforced-concrete material, it is necessary to consider the possibility of brittle fractures prior to collapse due to excessive yielding of steel. Strengthening the elements and avoiding brittle fractures require individual testing and verification before design.

Drucker's simple beam model
Drucker's simple beam model[23] illustrates the basic concept of lower- and upper-bound techniques of limit analysis using a simple reinforced-concrete beam (Figure 6). For simplicity, it is assumed the concrete beam with negligible weight cannot carry tension, so it must act as a very flat arch. The outward thrust of the arch is shown in Figure 6(a) as being taken by a steel tension tie between two end-plates bearing on the concrete. The steel is unbonded.

Efficient use of material would seem

to dictate at first that, at the ultimate or collapse load, both steel and concrete should be at their yield stresses. Since the equilibrium distribution of stress in the concrete and the steel as shown in Figure 6(a) is nowhere tensile in the concrete and is everywhere at or below yield, the beam will not collapse at this load or will be just at the point of collapse, according to the lower-bound theorem. This approach so far focuses on the lower-bound equilibrium technique and thus the strength of the beam might be under-estimated.

Figure 6(b) is a kinematical picture associated with an assumed plastic failure mechanism, which gives an upper bound on the collapse load. The assumed failure mechanism as drawn shows the stretching or yielding of the steel tie and the crushing plastically of the shaded areas of concrete at the ends as well as in the centre. This failure mechanism results in an upper-bound solution which turns out to be equal to the lower-bound solution of Figure 6(a).

Thus, we obtain the correct answer for the idealised beam according to the limit theorem despite the fact that either the stress field as constructed in Figure 6(a) or the plastic collapse mechanism as assumed in Figure 6(b) is the real stress distribution or the real failure mechanism respectively. This simple example clearly illustrates the basic concept and power of limit analysis as applied to reinforced-concrete structures. It also physically shows how the load is carried in a composite structure through arching for tension-weak concrete and stretching for tension-strong steel to its supports or foundations.

Strut–tie model
The upper-bound technique of limit analysis, the yield-line theory for slab design, has been used in engineering practice but the application of stress fields to reinforced-concrete design based on the concept of lower-bound theorem of limit analysis is a more recent development.

One of the most important advances in reinforced concrete in recent years is the extension of lower-bound-limit-theorem-based design procedures to shear, torsion, bearing stresses and the design of structural discontinuities such as joints and corners. As illustrated in the preceding example, these techniques will have the advantage of allowing a designer to follow the forces

through a structure.[24, 25]

The strut–tie model is based on the lower-bound theorem of limit analysis. In this model, the complex stress distribution in the structure is idealised as a truss carrying the imposed loading through the structure to its supports, as illustrated in Drucker's simple beam model. Like a real truss, a strut-and-tie model consists of compression struts and tension ties interconnected at nodes.

Using the stress legs sketched in Figure 7, a lower-bound stress field that satisfies equilibrium and does not violate yield criteria at any point can be constructed to provide a safe estimate of capacity on the reinforced-concrete structures.

The strut–tie model has been well developed in the USA over the last decade and the subject was presented in several texts as a standard method for shear, joints and support bearing design.[27, 28] The strut–tie method was also introduced in the American Association of State Highway and Transportation Officials' LRFD specifications[29] as well as in the American Concrete Institute 318 building codes.[30] A typical example of a strut-and-tie model for a common structural joint design is shown in Figure 8.

Summary

Over the last 50 years, remarkable developments have occurred in computer hardware and software. Advancement in computer technology has spurred the development of structural calculations, ranging from the simple strength-of-

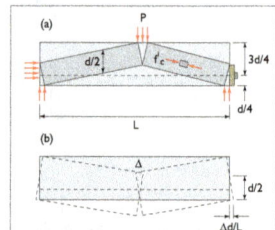

Figure 6. Drucker's simple model for a concrete beam reinforced with an unbonded tendon: (a) a lower-bound or equilibrium representation of arch action and (b) a kinematical view of the upper-bound collapse mechanism[23]

materials approach in early years, to the finite-element type of structural analysis for design in recent years and to the modern development of scientific simulation and visualisation for structural problems in the years to come.

Table 2 summarises briefly the 'major advances' of structural engineering that can be attributed to the 'breakthrough' of mechanics formulation, material modelling or computing power; where new knowledge has been implemented in structural engineering; and, in some measure, structural engineering practice has been fundamentally changed.

Concluding remarks

The determination of the load-carrying capacity of structures through the application of the theory of plasticity is topic on which significant progress has been made in recent years. This is in contrast to the earlier era of design, with undue emphasis on linear elastic analysis.

Engineering specifications contained rules that helped engineers avoid most of the errors of over-design or under-design with guidelines derived from experience and tests. However, rules based on past experience work well only for designs lying within the scope of that range. They cannot be depended upon outside that range. Ideally, design guidelines and rules should be derived from sound physical and mathematical principles.

Like the theory of elasticity in earlier eras, the theory of plasticity in recent years provides one of the success stories of applied mechanics that led to the development of modern design guidelines and rules. The mathematical theory of plasticity enables structural engineers to go beyond the elastic range in a time-independent but theoretically consistent way for inelastic structural analysis and design.[31]

The introduction of the concept on generalised force (or stress) and generalised displacement (or strain) and the establishment of the general theory of limit analysis and design in the 1950s laid the foundation for the revolution in structural engineering in subsequent years. The adoption of plas-

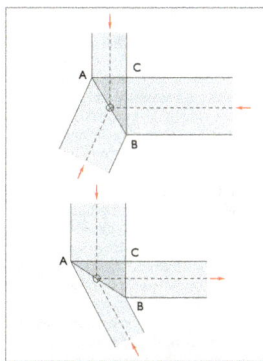

Figure 7. Using stress-legs to produce a stress field at a stress joint[26]

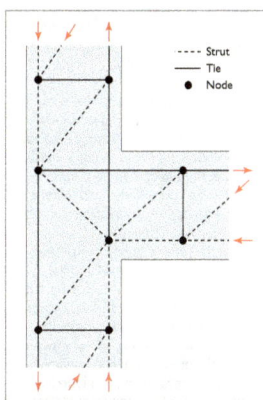

Figure 8. Application of the strut-and-tie model to reinforced concrete joint design

Mechanics	Materials	Computing	Structural analysis and design
Strength of materials formulation: closed-form solutions by series expansion, numerical solutions by finite difference	Linear elasticity	Slide rule and calculator environment	Strength of materials approach to structural engineering in the early years ▪ allowable stress design with K-factor ▪ amplification factor for second-order effect ▪ moment-distribution or slope-deflection methods for load distribution in framed structures ▪ member-by-member design process ▪ design rules based on allowable strength of members by tests.
Limit-analysis methods: mechanism method and equilibrium method, plastic-hinge concept	Perfect plasticity	Slide rule and calculator environment	Simple plastic analysis method for steel frame design in the early years ▪ plastic analysis and design with K-factor ▪ amplification factor for second-order effects ▪ upper- and lower-bound methods for frame design ▪ member-by-member design process ▪ design rules based on ultimate strength of members by tests.
Finite-element formulation using shape function and virtual work equation: generalised stresses and generalised strains concept	General plasticity	Mainframe computing environment	Finite-element approach to structural engineering in recent years ▪ development of member strength equations with probability and reliability theory ▪ development of reliability-based codes ▪ limit states to design with K-factor ▪ direct calculation of second-order effect ▪ member-by-member design approach ▪ design rules based on load factor and resistance factor concept by mathematical theory.
Advanced analysis: combining theory of stability with theory of plasticity	General plasticity	Desktop computing with object-oriented programming	Second-order inelastic analysis for direct frame design as current progress ▪ structural system approach to design without K-factor and amplification factor ▪ explicit consideration of the influence of structural joints in the analysis/design process. ▪ development of performance-based codes ▪ consideration of 'structural fuse' concept in design. ▪ design based on maximum strength of the structural system without member-by-member strength check.
Model-based simulation based on the integration of mechanics, computing, physics and materials science	Deterioration science or aging	High-performance computing	Large-scale simulation of structural system over its life-cycle performance analysis ▪ numerical challenges: proper modelling of discontinuity and fracture or crack for tension-weak materials ▪ software challenges: radically different scales in time and/or space ▪ material challenges: from time-independent elastic and inelastic material model to time-dependent modelling reflecting material degradation and deterioration science ▪ design process includes modelling (physics), simulation (computing), virtualisation (software) and verification (experiment).

Table 2. Interaction of mechanics, materials and computing on the impact and advancement of structural engineering practice

AUTHORS •

tic-analysis methods in steel specifications started the revolution in the 1960s.

Thanks to the rapid advancement of computing power beginning in the 1970s, the study of mechanics and mathematical formulation subsequently focused on the study of structural elements, of which the parts of the structure are composed, rather than the material itself. Thanks to this approach, the field of application of the theory of plasticity to structural engineering has broadened appreciably.

In more recent years, various analysis approaches to the estimation of stress, strain and displacement, including analytical, numerical, physical and analogue techniques, have advanced and are readily available to structural engineers. In particular, the finite-element technique is the most versatile and popular. As a result of this success, design specifications around the world have been undergoing several stages of revolutionary change, from the allowable stress design, to plastic design, to load-resistance factor design and to the more recent performance-based design – as exemplified by the new Eurocodes.

We are now in a desktop environment for unlimited computing. Computer simulation has now joined the theory and experimentation as a third path for engineering design and performance evaluation. Simulation is computing, theory is modelling, and experimentation is validation of the result. The emerging areas of model-based simulation in structural engineering will include, most notably, the following topics.

- From the present structural-system approach to the life-cycle structural analysis and design, covering: construction sequence analysis during construction; performance analysis during service; and degradation and deterioration analysis during maintenance, rehabilitation and demolition.
- From the present finite-element modelling for continuous media to the finite-block types of modelling for tension-weak materials, which will develop cracks and subsequently change the geometry and topology of the structure.
- From the present time-independent elastic and inelastic material modelling to the time-dependent modelling reflecting material degradation and deterioration science.

A busy, productive, and exciting future lies ahead for all who wish to participate in these emerging areas of research and application. These areas are inherently interdisciplinary in science and engineering, where computation plays the key role. Scientists provide a consistent theory for application, but structural engineers must continue to face the reality of dealing with the idealisations of idealisations of these theories in order to make them work and applicable to the real world of engineering.

Seeing the big picture will enable all structural engineers to make a difference in the further advancement of their art in the years to come.

Appendix

This appendix shows how the fundamental mechanics operation is applied to the segment of a reinforced-concrete beam in bending, shown in Figure 2.

Step 1. The stress σ in the beam element is related to the generalised stress M acting on the beam element through the application of equilibrium equation

$$1. \qquad M = \int \sigma y \, dA$$

Step 2. The strain ε in the beam element is related to the generalised strain or curvature φ of the element through the application of the kinematical assumption that the plane section remains plane after bending

$$2. \qquad \varepsilon = y\phi$$

Step 3. The generalised stress M can now be related to the generalised strain φ through the application of the simple linear elastic stress–strain relation σ = Eε

$$3. \qquad M = \int \sigma y \, dA = E\phi \int y^2 \, dA = EI\phi$$

where E is Young's modulus and I is the moment of inertia of the section.

Similar processes can be applied to other structural elements in an elastic as well as an inelastic range with a general state of stresses and strains. This process will be

illustrated again in the following section, which deals with the modern development of finite-element application.

Finite-element formulation
The general governing equation of the modern finite-element method for a general static inelastic analysis can be derived from the basic mechanics operation as outlined above in a straightforward manner.

The basic step in any finite-element analysis is the derivation of the element-stiffness matrix $[k]$, which relates the increment-of-nodal-displacement vector $\{d\delta\}$ (generalised strain) to the increment-of-nodal-force vector $\{dF\}$ (generalised stress). This is known as the incremental generalised-stress–generalised-strain relation of an individual element. To derive this relation, the three basic sets of field equations described previously must be satisfied.

In the following, the basic steps outlined in the preceding section are used to derive the basic stiffness relation valid for any type or shape of finite element selected. Details of this derivation can be found in the book by Chen,[24] among others.

Kinematical condition
Step 1. Express the increment of internal displacement vector $\{du\}$ at any point within the element in terms of the increment-of-nodal-displacement vector $\{d\delta\}$ by means of assumed shape functions $[N]$, which replace the kinematical assumption that the plane section remains plane after bending in the case of simple beam theory

$$4. \qquad \{du\} = [N]\{d\delta\}$$

Step 2. With displacement known at all points within the element, the increment-of-strain vector $\{d\varepsilon\}$ at any point can be obtained by taking suitable derivatives of Equation 4. This establishes relationship of the incremental strain $\{d\varepsilon\}$ to the incremental generalised strain $\{d\delta\}$ as the counterpart of Equation 2 in the case of simple beam theory

$$5. \qquad \{d\varepsilon\} = [B]\{d\delta\}$$

The matrix $[B]$ generally is composed of derivatives of the shape functions.

Equilibrium condition

Step 3. The equilibrium relation between the increment-of-nodal-force vector $\{dF\}$ and the internal stress vector $\{d\sigma\}$ at any point can be established directly by applying the principle of virtual work with an arbitrarily imposed virtual compatible set of the type given in Equation 5. This establishes the relationship of the incremental stress $\{d\sigma\}$ to the incremental generalised stress $\{dF\}$ as the counterpart of Equation 1 in the case of simple beam theory

6.
$$\{dF\} = \int [B]^* \{d\sigma\} \, dv$$

Constitutive relations

Using the incremental theory of plasticity, the incremental stress–strain relation can be expressed in the general form

7.
$$\{d\sigma\} = [D] \{d\varepsilon\}$$

where $[D]$ is the elastoplastic tangent stiffness matrix for the current stress state. Substituting Equation 7 in Equation 6 and using Equation 5 leads to the desired element stiffness matrix $[k]$

8.
$$\{dF\} = [k] \{d\delta\}$$

where

9.
$$[k] = \int [B]^* [D] [B] \, dv$$

Modern finite-element solutions

Once the element stiffness matrices have been calculated and transformed from local to global coordinates, the structure stiffness matrix $[K]$, which relates the load increment $\{dR\}$ to the generalised strain increment or nodal-displacement increment $\{d\delta\}$ of the complete structure, can be formed by the systematic addition of element stiffness

10.
$$\{dR\} = [K] \{d\delta\}$$

Since the external load vector increment $\{dR\}$ is known, the unknown nodal-displacement vector increment $\{d\delta\}$ can be found by solving Equation 10. With the

known $\{d\delta\}$, the element stress within each element can be found from Equation 7 using Equation 5

11.
$$\{d\sigma\} = [D] [B] \{d\delta\}$$

References

1. SOKOLNIKOFF I. S. *Mathematical Theory of Elasticity*, 2nd ed. McGraw-Hill, New York, NY, USA, 1956.
2. PRAGER W. Sectional address. *Eighth International Congress of Theoretical and Applied Mechanics*, Istanbul, 1952.
3. HODGE P. G. *The Plastic Analysis of Structures.* McGraw-Hill, New York, NY, USA, 1959.
4. CHEN W. F. and ATSUTA T. *Theory of Beam–Columns, Vol. 1 – In-Plane Behavior and Design.* McGraw-Hill, New York, NY, USA, 1976, reprinted by J. Ross, Orlando, FL, USA, 2007.
5. CHEN W. F. and ATSUTA T. *Theory of Beam–Columns, Vol. 2 – Space Behavior and Design.* McGraw-Hill, New York, NY, USA, 1977, reprinted by J. Ross, Orlando, FL, USA, 2007.
6. GALAMBOS T. V. (ed.). *Guide to Stability Design Criteria for Metal Structures*, 4th ed. Wiley, New York, NY, USA, 1988.
7. AMERICAN INSTITUTE OF STEEL CONSTRUCTION. *Load and Resistance Factor Design Specification for Structural Steel Buildings.* AISC, Chicago, IL, USA, 1986, 2005.
8. EUROPEAN CONVENTION FOR CONSTRUCTIONAL STEELWORK. *Ultimate Limit State Calculation of Sway Frames with Rigid Joints.* ECCS, Brussels, Belgium, 1984, Publication No.33.
9. CHEN W. F. and KIM Y. S. *Practical Analysis for Partially Restrained Frame Design.* Structural Stability Research Council, Lehigh University, Bethlehem, PA, USA, 1998.
10. EUROPEAN CONVENTION FOR CONSTRUCTIONAL STEELWORK. *Analysis and Design of Steel Frames with Semi-Rigid Joints.* ECCS, Brussels, Belgium, 1992, Publication No. 61.
11. CHEN W. F. and LUI E. M. *Stability Design of Steel Frames.* CRC Press, Boca Raton, FL, USA, 1991.
12. CHEN W. F. (ed.). *Semi-Rigid Connections in Steel Frames.* Council on Tall Buildings and Urban Habitat, McGraw-Hill, New York, NY, USA, 1993.
13. LIEW J. Y. R., WHITE D. W. and CHEN W. F. Notional-load plastic-hinge method for frame design. *Journal of Structural Engineering, ASCE*, 1994, **120**, No. 5, 1434–1454.
14. EUROPEAN CONVENTION FOR CONSTRUCTIONAL STEELWORK. *Essential of Eurocode 3 Design Manual for Steel Structures in Buildings.* ECCS, Brussels, Belgium, 1991, Advisory Committee 5, No. 65.
15. CHEN W. F. and KIM S. E. *LRFD Design Using Advanced Analysis.* CRC Press, Boca Raton, FL, USA, 1997.
16. DRUCKER D. C., PRAGER W. and GREENBERG H. Extended limit design theorems for continuous media. *Quarterly of Applied Mathematics*, 1952, **9**, No. 4, 381–389.
17. NEAL B. G. *The Plastic Methods of Structural Analysis.* Chapman & Hall, London, 1957.
18. NIELSEN M. P. Limit analysis of reinforced concrete slabs. *Acta Polytechnica Scandinavica, Civil Engineering Building Construction Series*, 1964, No. 26.
19. CHEN W. F. *Limit Analysis and Soil Plasticity.* Elsevier, Amsterdam, The Netherlands, 1975, reprinted by J. Ross, Orlando, FL, USA, 2007.
20. JOHNSON W. The mechanics of metal working plasticity. In *Applied Mechanics Update* (STEELE C. R. and SPRINGER G. S. (eds)). American Society of Mechanical Engineers, New York, NY, USA, 1986.
21. DVORAK G. J. and BAHEI-EL-DIN Y. A. Plasticity analysis of fibrous composites. *Journal of Applied Mechanics*, 1982, **49**, No. 2, 327–335.
22. GVOZDEV A. A. The determination of the value of the collapse load for statically indeterminate systems undergoing plastic deformation. *International Journal of Mechanical Science*, 1960, **1**, No. 4, 322–335. (Translation of paper in *Proceedings of Conference on Plastic Deformations*, December, 1936, Akademia Nauk SSSR, Moscow-Leningrad, 1938).
23. DRUCKER D. C. On Structural Concrete and the Theorems of Limit Analysis. International Association for Bridge and Structural Engineering, Zurich, Switzerland, 1961, Publication No. 21.
24. CHEN W. F. *Plasticity in Reinforced Concrete.* McGraw-Hill, New York, NY, USA, 1982, reprinted by J. Ross, Orlando, FL, USA, 2007.
25. NIELSEN M. P. *Limit Analysis and Concrete Plasticity*, 2nd edition. CRC Press, Boca Raton, FL, USA, 1999.
26. CHEN W. F. and HAN D. J. *Plasticity for Structural Engineers.* Springer-Verlag, New York, NY, USA, 1988, reprinted by J. Ross, Orlando, FL, USA, 2007.
27. SCHLAICH J. and SCHAFER K. Design and detailing of structural concrete using strut-and-tie models. *The Structural Engineer*, 1991, **69**, No. 6, 113–125.
28. ASCE–ACI COMMITTEE 445 ON SHEAR AND TORSION. Recent approaches to shear design of structural concrete. *Journal of Structural Engineering, ASCE*, 1998, **124**, No. 12, 1375–1417.
29. AMERICAN ASSOCIATION OF STATE HIGHWAY AND TRANSPORTATION OFFICIALS. *AASHTO LRFD Bridge Specifications*, 2nd ed. AASHTO, Washington, D. C., USA, 1998.
30. ACI COMMITTEE 318. *Building Code Requirements for Structural Concrete (ACI 318-02) and Commentary (ACI 318R-02).* American Concrete Institute, Farmington Hills, MI, USA, 2002.
31. HILL R. *The Mathematical Theory of Plasticity.* Oxford University Press, New York, NY, USA, 1950.

Acknowledgement

This paper is based on the author's keynote lecture presented at the eleventh east Asia-Pacific conference on structural engineering and construction (EASEC-11) entitled *Building a sustainable environment* November 19-21, 2008, Taipei, Taiwan.

What do you think?

If you would like to comment on this paper, please email up to 200 words to the editor at journals@ice.org.uk.

If you would like to write a paper of 2000 to 3500 words about your own experience in this or any related area of civil engineering, the editor will be happy to provide any help or advice you need.

CHAPTER 6
ACADEMIC MASTERPIECE BOOKS
學術專著

Dr. Chen is the author or co-author of 23 books, 33 edited books and 57 contributing chapters.

6.1 Books and Reviews

1. (With Viest IM, Rentschler GP.) (1972) *Current Research on Tall Buildings*. ASCE-IABSE Joint Committee on Planning and Design of Tall Buildings, ASCE Publication, August, 126 pp.
2. (1975) *Limit Analysis and Soil Plasticity*. Amsterdam, Elsevier, 638 pp. (Translated into Chinese in 1995 by People's Communication Publishing House in Beijing, China.)
3. (With Atsuta T.) (1976) *Theory of Beam-Columns, Volume 1 In-Plane Behavior and Design*. New York, McGraw-Hill, 513 pp.
4. (With Atsuta T.) (1977) *Theory of Beam-Columns, Volume 2 Space Behavior and Design*. New York, McGraw-Hill, 732 pp. (Both volumes were translated into Chinese in 1994–1995 by Science/Technology Book Co. in Taipei, Taiwan, and in 1996 by People's Communication House in Beijing, China.)
5. (1982) *Plasticity in Reinforced Concrete*. New York, McGraw-Hill, 474 pp. (Translated into Japanese in 1985 by Maruzen Co., Tokyo, Japan.)
6. (With Saleeb AF.) (1982) *Constitutive Equations for Engineering Materials, Volume 1 — Elasticity and Modeling*. New York, Wiley Interscience, 580 pp. (Translated by Yu TQ *et al.* into Chinese in 2001 and published by Huazhong University of Science and Technology Press in China, 450 pp.)
7. (1994) *Constitutive Equations for Engineering Materials, Volume 2 — Plasticity and Modeling*. Amsterdam, Elsevier, 1096 pp. (Translated by Yu TQ *et al.* into Chinese in 2001 and published by Huazhong University of Science and Technology Press in China, 451 pp.) (Both volumes were translated and edited into two graduate texts by Yu TQ *et al.* — *Elasticity and Plasticity*, 335 pp.

and *Constitutive Equations for Materials of Concrete and Soil*, 427 pp. China Architecture & Building Press, 2004.)

8. (With Han DJ.) (1985) *Tubular Members in Offshore Structures*. London, Pitman, 271 pp.

9. (With Baladi GY.) (1985) *Soil Plasticity: Theory and Implementation*. Amsterdam, Elsevier, 231 pp.

10. (With Lui EM.) (1987) *Structural Stability: Theory and Implementation*. New York, Elsevier, 486 pp.

11. (With Han DJ.) (1988) *Plasticity for Structural Engineers*. New York, Springer-Verlag, 600 pp.

12. (With Zhang H.) (1990) *Structural Plasticity: Theory, Problems and CAE Software*. New York, Springer-Verlag, 250 pp.

13. (With Liu XL.) (1990) *Limit Analysis in Soil Mechanics*. Amsterdam, Elsevier, 477 pp.

14. (With Mizuno E.) (1990) *Nonlinear Analysis in Soil Mechanics*. Amsterdam, Elsevier, 661 pp.

15. (With Lui EM.) (1999) *Stability Design of Steel Frames*. Boca Raton, FL, CRC Press, 1991, 380 pp. (Translated by Zhou SP into Chinese and published by Shanghai-World Book Company in Shanghai, China, 420 pp.)

16. (With Mosallam KH.) (1991) *Concrete Buildings: Analysis for Safe Construction*. Boca Raton, FL, CRC Press, 260 pp.

17. (With Sohal IS.) (1995) *Plastic Design and Second-Order Analysis of Steel Frames*. New York, Springer-Verlag, 509 pp.

18. (With Goto Y, Liew JYR.) (1996) *Stability Design of Semi-Rigid Frames*. New York, John Wiley and Sons, 468 pp.

19. (With Kim SE.) (1997) *LRFD Steel Design Using Advanced Analysis*. Boca Raton, FL, CRC Press, 448 pp.

20. (With Kim YS.) (1998) *Practical Analysis for Partially Restrained Frame Design*. Structural Stability Research Council, Lehigh University, Bethlehem, PA, 82 pp.

21. (2007) *My Life's Journey: Reflections of an Academic*. Singapore, World Scientific Publishing Company, 449 pp. (Chinese translation by Yu TQ, 2009, 450 pp.)

22. (With El-Metwally S.) (2011) *Understanding Structural Engineering: From Theory to Practice*. Boca Raton, FL, CRC Press, 255 pp.

23. (With El-Metwally S.) (2017) *Structural Concrete: Strut-and-Tie Models for Unified Design*. Boca Raton, FL, CRC Press, 230 pp.

J Ross Publishing Classics
Ross Publishing Classics are world-renowned texts and monographs written by preeminent scholars.

These books have been widely read, discussed and cited in the literature since their inaugural publications but unavailable for some time to students, researchers, professionals and libraries. J Ross Publishing is proudly making these valuable references and texts available once again.

Theory of Beam-Columns, Volume 1
In-Plane Behavior and Design
Wai-Fah Chen and Toshio Atsuta
Special Direct Price: $54.95

This first volume of a two-volume work presents the basic theoretical principles, methods of analysis in obtaining the solutions of beam-columns, and developments of theories of biaxially loaded beam-columns, and shows how these theories can be used in the solution of practical design problems.

    After presenting the basic theory the authors proceed to solutions of particular problems. The book discusses both refined and simplified design procedures, along with their limitations. It is left to the engineer to choose among them as he sees fit.

Theory of Beam-Columns, Volume 2
Space Behavior and Design
Wai-Fah Chen and Toshio Atsuta
Special Direct Price: $54.95

This second volume of a two-volume work discusses systematically the complete theory of space beam-columns. The book presents principles and methods of analysis for beam columns in space, which should be the basis for structural design, and shows how these theories are applied for the solution of practical design problems. With the importance of the role of beam-columns in modem structures, this set of books will be invaluable to structural engineers, designers, specification writing bodies, and researchers.

Plasticity for Structural Engineers
Wai-Fah Chen and Da-Jian Han
Special Direct Price: $54.95

This comprehensive text addresses the elastic and plastic behavior of general structural elements under combined stress. It sets out to examine the stress strain behaviors of materials under simple test conditions and proceeds to show how these behaviors can be generalized under combined stress. The topic of structural plasticity is presented in a manner that is simple, and concise, encompassing the classical theory of metal plasticity as well as concrete plasticity.

Plasticity in Reinforced Concrete
Wai-Fah Chen
Special Direct Price: $49.95

This indispensable reference presents a unified treatment of mathematical models of concrete structural analysis. In Part I, the author considers the experimental data regarding stress and strain characteristics of concrete under biaxial and multiaxial stress states and presents empirical equations for modulus and fracture strength. Part II discusses concrete elasticity, generalized failure, and fracture criteria, while the final part addresses concrete plasticity with applications of limit analysis and finite element analysis to concrete and reinforced structures.

Limit Analysis and Soil Plasticity
Wai-Fah Chen
Special Direct Price: $54.95

Devoted to the theory and applications of limit analysis as applied to soil mechanics, this text also contains information on soil plasticity and rock-like material such as concrete. The first part of the book describes the techniques of limit analysis in detail and are illustrated by many examples. The second part deals with the applications of limit analysis to "classical soil mechanics problems" and the third part presents advances on bearing capacity problems of concrete blocks or rock. The final part discusses the modern development of the theory of soil plasticity.

Book Reviews

Limit Analysis and Soil Plasticity, Amsterdam, Elsevier, 1975

The Civil Engineer in South Africa, **February 1976**

In spite of attending and lecturing at three post-graduate soil mechanics courses, I think it is fair to say that one reading this new volume I perceived for the first time what may well be termed an underlying unity of approach or a basic philosophy of understanding for analysis applicable to the whole sphere of the mechanics of soil "structure".

    ... This volume is likely to become a classic on the subject and standard required reading for all who take soil mechanics and engineering seriously enough to desire to keep abreast of current developments ...

Indeed, this volume is definitely of interest to all engineers, not only geotechnical, who are seeking a simple and practical analytical method for solutions in problems in the mechanics of solids, but also to those who try to understand and perceive the philosophy of engineering and the essential unity of all physical process.

Applied Mechanics Reviews, Vol. 29, May 1976, W Prager, Switzerland

The exceedingly well written treatise can serve not only as a valuable reference work on the use of limit analysis in soil mechanics, but also as an excellent text on limit analysis in general, even though the vast majority of examples is taken from soil mechanics.

Geotechnical Engineering, December 1976, EW Brand

Dr Chen has been active for some years in the field of plasticity as applied to geotechnical engineering, and by writing this treatise *Limit Analysis and Soil Plasticity* he has shown himself to be one of the foremost exponents of the powerful mathematical tool.

... This book is a work of scholarship, which is more concerned with providing explanation for the fundamental stress-strain behavior of materials than it is with providing design methods. In this book Dr Chen has succeeded admirably; his treatise will stimulate researchers for many years to come.

Canadian Civil Engineering, February 1977, M Cleary

Soil Mechanics theory has been brought up to date with this fine reference book and it is highly recommended ...

The more familiar the engineer gets with this reference book, the greater the value it is worth ordering.

Geoderma, Vol. 15, No. 5, June 1976, AJ Koolen

From reading this book it appears that the author has a near encyclopedic knowledge of the huge amount of scientific results on soil mechanics problems that have been published during the last two centuries.

The book may be thought of as being a "milestone" in the growing tendency by soil mechanics practitioners to adopt limit analysis, because it is the first single reference book on the subject. It can be recommended to anybody who is interested in the calculation of collapse loads .... The book will surely find its way into various collections of soil mechanics books.

Applied Mechanics Reviews, **Vol. 32, No. 1, January 1979, A Kezdi, Hungary**

It is gratifying to see how the author commands even the most difficult problems; he produced a clear and didactic treatment.

It can be recommended to every civil engineer who is more or less engaged with soil mechanics.

Geotechnique, **March 1976, RGJ**

This book is a most welcome addition to the soil mechanics literature. It contains a wealth of material, which is of great interest to all engineers concerned with the calculation of limit loads in civil engineering.

This book, although somewhat expensive for 630 pages, should nevertheless be enthusiastically received by the university teacher, the researcher worker, and the civil engineer who practice limit analysis.

Book Reviews

Theory of Beam-Columns — Volume 1 In-Plane Behavior and Design
McGraw-Hill, 1976

Concrete International, **April 1979**

... and the latest developments of theories of biaxially loaded beam-columns. The books shows how these theories can be used in the solutions of practical design problems. The student, researcher, specification writer, or practitioner can all benefit from the use of this volume.

Book Reviews

Theory of Beam-Columns — Volume 2 Space Behavior and Design
McGraw-Hill, 1977

Concrete International, **April 1979**

... Structural engineers, researchers, specification writers, and designers will find this book of interest.

New Civil Engineer, **Canada, 25 May 1978, Jonathan Wood**

The book will be of particular value to the research worker and specialist analyst and programmer. However, clear style and presentation make it suitable for the working engineer or student who wishes to get to the root of the theory that underlies codes of practice and computer programs.

Book Reviews

Soil Plasticity: Theory and Implementation
Amsterdam, Elsevier, 1985

***Applied Mechanics Review*, Vol. 39, No. 9, September 1986, Mihail Popescu**

Computers are becoming more accessible daily. The need for assistance from computer analysis techniques in designing geotechnical problems is a fact of life. The publication of this volume will greatly facilitate the appropriate use of these techniques.

    The volume is a quality publication, with numerous illustrations to compliment the text. As both an educator and practicing engineer, ... strongly recommend it to anyone interested in learning the basics of computer methods in soil plasticity.

Book Reviews

Constitutive Equations for Engineering Materials, Volume 1 — Elasticity and Modeling
Wiley Interscience, 1982

Constitutive Equations for Engineering Materials, Volume 2 — Plasticity and Modeling
Amsterdam, Elsevier, 1994

***Studies in Applied Mechanics*, V.37A–37B**

This two-volume reference-text both provides the necessary foundations of the theory of elasticity and plasticity in civil engineers in general, and structural, materials, and geotechnical engineers in particular; and present recent results in the development of constitutive models for metals, concrete, and soil, their numerical implementation to computer program, and some finite element solutions for typical problems in structural and geotechnical engineering applications.

Volume 1 deals with the development of stress-strain model for metals, concrete, and soils based on the principles of elasticity.

Volume 2 extends the elasticity-based stress-strain models to the plastic range and develops plasticity-based models for engineering application.

Book Reviews

Structural Stability: Theory and Implementation
New York, Elsevier, 1987

Applied Mechanics Review, Vol. 41, No. 3, March 1988, P Spanos

This is an excellent book which is written by two experts in the field of structural stability .... The treatment of the various subjects is clear and thorough, and is suitable for a one semester first course in structural stability.

JF Camey iii, Vanderbilt University, January 1988

... The treatment compares favorably with the works of SP Timoshenko and JM Gere, *Theory of Elastic Stability* (1961), and F Bleich, *Buckling Strength of Metal Structures* (1952). Engineering libraries should include this important work in their civil engineering collections.

TV Galambos, University of Minnesota, 8 May 1987

... Your book *"Structural Stability"*, Beautiful! Just right for an introductory course of graduate students.

Kurt H Gerstle, University of Colorado, 16 December 1987

Your section "General Principles" is deceptively simple, but seems to contain all the essential thinking.

Book Reviews

Plasticity for Structural Engineers
New York, Springer-Verlag, 1988

Applied Mechanics Reviews, Vol. 42, 1989

… This book must be a treasure in the bookshelf of those who seriously commit themselves in the study of structural plasticity. The book is worth the price right down to the last penny.

Bautechnik, BRD, 67.1990.1

… Excellent treatment of the plasticity theory for structural engineers.

Technische Mechanik, DDR, 10.1989.3

… This book will contribute to reducing the level of "chaos" in treatment of this subject typical of other books.

Beton und Stahlbetonbau, BRD, 1989.9

… For English speaking students, the book will become a standard for those interested in plasticity theory …. The book will find great interest in Germany in research institutions and schools teaching steel and concrete design as well as soil engineering.

Book Reviews

Structural Plasticity: Theory, Problems and CAE Software
New York, Springer-Verlag, 1990

Applied Mechanics Review, Vol. 44, No. 8, August 1991, Alexandra M Vinogradov

This book is to help students and structural engineers learn and practice how to solve typical engineering plasticity problems in general and, more importantly, how to use computers to solve plasticity problems in structural engineering in particular. Clearly, the authors have been successful in attaining this goal.

Structural Plasticity: Theory, Problems and CAE Software is written in a concise and logical manner. It can be recommended as an excellent teaching aid in various structural engineering courses, as well, and it will be of immediate practical assistance to engineers and researchers dealing with problems of structural analysis and design.

Zentralblatt Fur Mathematik, und ihre Grenzgebiete, BRD, 711.1991, A Bertram

This book is meant as a supplement to the textbook on *Plasticity for Structural Engineers* by WF Chen and DJ Han. The book may serve well to educational purpose with students in engineering …

Book Reviews

Nonlinear Analysis in Soil Mechanics, Amsterdam, Elsevier, 1990

Bulletin of the Association of Engineering Geologists, Vol. XXIX, No. 2, June 1992

This is an excellent reference book for anyone who is engaged in the numerical analysis of geotechnical problems and the characterization of geotechnical materials from laboratory tests. Overall, the book is very clearly written and provides the kind of details, examples and illustrations that will make the book a very useful reference.

Book Reviews

Stability Design of Semi-Rigid Frames
John Wiley and Sons, New York, 1996

Steel News, Engineering Journal, AISC, Robert F Lorenz

For those of you who have enjoyed, during the past decade or so, nibbling on the technical morsels offered by WF Chen and his associates at Purdue University, you can now enjoy sitting down to the whole meal, Chen has just published a new book, *Stability Design of Semi-Rigid Frames* (January 1996), and let me assure you that it is an enjoyable feast.

In many ways, the book provides a receipt for discerning between several advanced analysis techniques for designing the steel frames of the future .... These new methods eventually could eliminate the need for specification checks, K-factor iterations, inelastic stiffness reduction and other error-plagued miseries.

This book is recommended to those who truly want a taste of the impact of the electronic revolution on steel frame analysis during the upcoming decade. I think it provides both map and compass ... to help you survive whatever the technological future holds.

Book Review

My Life's Journey: Reflections of an Academic
Singapore, World Scientific Publishing Company, 2007
Reviewed by SL Chan

In Chinese, "seeing a successful scholar stimulates us to up-grade" is a popular statement reminding us to struggle. Reading this book enables us to appreciate the route of a successful academic. This 449-page book is about the life of a successful engineering professor, which should be of great interest to younger and experienced academics about the career of a renounced and senior academic, whose achievements can be seen on p. 291 of the book. The author, Professor WF Chen, has successfully established himself as a giant of research in Engineering and contributed much to the fields of steel and concrete structures, structural and soil mechanics and others and has published very extensively in the fields. This book is about the personal and non-technical side of the great academic.

The book describes the very beginning of his life in Nanjing, which was the capital of several dynasties in China. The book continues to describe his study in Taiwan after the Chinese civil war and later in the USA. It contains seventeen chapters with the first one about his childhood, the second one on his middle school, next on his voyage overseas, then on family and love story, academic career at Lehigh and at Purdue, his life story, his academic contributions, his view on this changing time, his life in the rising China, his students, his teacher, the transformation of college, reform in higher education in China, his advisors and link with National Taiwan University and finally with the last chapter was written by one of Professor Chen's successful student, Professor Toshio Atsuta, about his career and the relationship with Professor WF Chen. We should remember the 2-volume state-of-the-art book on beam-columns written jointly by Chen and Atsuta and this chapter is about the background of these two authors.

Professor Chen has supervised students from many countries including the USA, China, Japan, Korea and others. His research interest spans across numerous topics like beam-columns, steel structures, connections, soil plasticity, concrete plasticity, polymer concrete and computing application in engineering design. One of his talents is on the vision and projection. One can see much of his work is not only pioneering, but also paving the path for research and development in the field. For example, in pace with the development of high-speed computer technology and development of low-cost and high speed personal computers, he, independently and in parallel with some other professors in Australia, China, USA and UK, led the development of engineering design from member-based to system-based design, which was finally adopted in various design codes and named as "Advanced Analysis".

This book was written in an easy-to-read manner with explanation of some terms in Chinese. It contains many photographs of great historical value. The biography is strongly recommended for academics and researchers at different stages.

Book review

"My Life's Journey: Reflections of an Academic" by Wai-Fah Chen

World Scientific, Singapore; 2007; Price $54, 449 pages

Most books written by engineering professors deal with theories, designs, or technologies. To meet the rapidly changing, diversifying world, engineers are required to possess competencies not limited to the technical capabilities. The problem is that few books written by engineering professors have gone beyond the domain of science and engineering. It is in this sense that the book by Prof. Wai-Fah Chen, a respected professor in engineering and education, on his life journey is timely. This book contains not only a narrative of the life story of the author himself, but also a reflection of the evolvement of structural engineering and engineering education in the past four decades, in which the author was highly involved.

This book contains 17 chapters. The first seven chapters, arranged in a chronological order, were devoted to the academic growth of the author. The next seven chapters were given to specific topics, such as the evolvement of structural engineering, life in the rapidly changing world, eyewitness of recent rise of China, previous students at a glance, the teacher D. C. Drucker, engineering college in transformation, and higher education reform in China. The remaining three chapters were given by his three students, describing their early connections with the author and personal achievements each in China, Taiwan and Japan.

This book can find readers in three major groups. For those who grew up around the time of the author, this book gave a vivid description of the stories of the author's life struggle through the Sino-Japanese War (1937-1945) and the Chinese Civil War (1945-1949). Two things appear rather unimaginable to young people living in the present days. One is the author's story on how to escape from China to Taiwan via the strait following the retreat of the government in 1949, and the other is how to overcome all the obstacles to apply for a visa for advanced study and to take a voyage via the Pacific Ocean to the America in the 1960s, to pursue the golden dream as an immigrant.

For those working in structural engineering, this book contains a rich history of evolution of structural

mechanics, structural engineering and steel structural design, in particular with relation to the AISC specifications, via his careers at Lehigh and Purdue Universities. A sufficient coverage is given of the development of plastic design for steel structures in the 1960s, based primarily on the research at the Fritz Engineering Laboratory of Lehigh University under the leadership of Lynn S. Beedle. The author was the key person in extending the concept of limit analysis to soil mechanics and concrete materials.

And for those young people growing in the rise of China, it provides a perspective of the history of the inevitable changes of China in the past half a century through his life journey and eye witness, especially for the Tiananmen Event in China on June 4, 1989. Though China's higher education has expanded rapidly following the economic booming in the past two decades, the author pointed out that there will be a long way to go for any university in China to be considered of world-class quality and reputation, if they do not change the government-controlled system to a market-driven system.

As I am a professor educated in the traditional Chinese environment, obtaining an advanced degree from the U.S.A. in 1984, and teaching in structural engineering in a university in Taiwan, I belong mainly to the second group of readership and partially to the third group due to the cultural connection. The following are the points of view I would like to elaborate on this book from my own perspective:

First, the author was highly productive in technical publications. For example, he has supervised a total of 56 Ph.D. students, and co-authored with his students on most of their publications, which include 19 books and some 350 refereed journal articles. Thus, the stories given in this book offer a good reference for graduate students to interact with the supervisor, and for young faculty members to select a teaching job, a research topic, and even to supervise the graduate students. All these issues are instructive to young faculty members who take a teaching job shortly after finishing their

Ph.D. degrees. It is generally understood that most young faculty members can only learn by doing, with basically little or limited assistance offered to them by their institutions for career-building.

Second, the author was renowned for his achievements in three seemingly distinct areas, i.e., steel structures, soil mechanics, and concrete materials. From this book, one can have some ideas about how he links the three areas together through the same concept of limit analysis. The ambition underlying such an effort, a good example of integration, is really amazing to most of us centering on a narrow part of a single area nowadays. Even so, we should not forget that the basic principle adopted by the author in all his research is "nothing can be more practical than a simple theory," or "just make things simple."

Third, the author has successfully merged his job of teaching and research with his family life and education of his children. When reading this memoir, I got a strong feeling of family value in the author's mind, especially through the poems and various articles he wrote for the children, classmates, and junior relatives in different occasions. We all know that under the present stringent environment, the family value has receded in competition with other requirements such as publications and promotion for faculty members. The author offers a good example for combining both.

Fourth, the book offers many valuable ideas for enhancement of engineering education and for transformation of engineering colleges to a top-tier institution, based on his experience as Dean of College of Engineering of University of Hawaii at Manoa. It is amazing to see that by the time he stepped down as the dean, the College's performance has been greatly improved, in terms of fund-raising, faculty retention and recruiting, and student recruiting, partly via adoption of the merit-raised salary adjustment system for the faculty. Besides, the author pays lots of attention to higher education in China through his personal involvement as early as in 1981 right after the announcement of the open-door policy by Deng Xiaoping. His suggestions for the higher education reform in China may be critical, but are just to the point.

I would like to take Prof. Ted Galambos' comments on this book, as given in his e-mail to the author, as part of the concluding remarks of this review: "....The book is not only enjoyable because it describes your personal and professional history, but also because it contains the history of structural mechanics and structural engineering in the past half Century.....". It is my belief that young faculty members and graduate students can benefit a great lot from reading this book, and a valuable contribution for engineering professors at their stage of retirement to the profession is to write books beyond their technological domains.

Y. B. Yang
Department of Civil Engineering,
National Taiwan University, Taipei, Taiwan 10617

Modern Steel Construction, April 2008
Ted Galambos, University of Minnesota

The Life and Times of an Engineering Giant

Professor Wai-Fah Chen is one of the major figures in the history of modern civil engineering. He is known to the steel construction community for his many contributions to structural analysis, stability theory and its applications, and behavior of connections. Many of the results of his research have found their way into several specifications of the AISC. He was a member of the AISC Committee of Specifications for many years, and he is presently an emeritus member of this group. After his recent retirement as the Dean of Engineering at the University of Hawaii, he sat down and wrote a 450-page book, *My Life's Journey: Reflections of an Academic*, that delightfully combines his personal life experiences and family history; a history of the intellectual and practical developments in structural engineering, engineering mechanics, plasticity theory, and geo-engineering in the past half-century; experiences in engineering education; reminiscences about teachers, colleagues, and friends; reviews of his research contributions; and general philosophic comments on the past, present, and future of the civil engineering profession in general and the academic teaching/research community in particular.

Wai-Fah Chen had a very adventurous youth during World War II and the following Chinese Civil War, eventually ending up in Taiwan where he obtained his BCCE degree. He describes these early years through a lively narrative, setting the tone for the rest of the book. Many of the readers of MSC will find their name, or the names of their teachers and colleagues, in the book. They will also be able to trace the origin of many of the criteria that are now a part of their daily design life. Engineering teachers will appreciate learning about the origins of steel design, concrete design, and soil mechanics concepts from the many books authored by Chen and his students.

Professor Chen is truly a giant in our professional field. Readers will not only be delighted and educated by this book, but will also feel proud to be part of a dynamic and creative profession.

6.2 Edited Books

1. (1980) Beam-to-column building connections: State-of-the-art. ASCE Portland Convention, 14–18 April, Preprint Volume No. 80–179, ASCE Publication, New York, 260 pp.

2. (With Ting EC.) (1980) Fracture in concrete. Proceedings of a Session at the ASCE Hollywood Convention, 27–31 October, FL, ASCE, New York, 110 pp.

3. (With Lewis ADM.) (1983) *Recent Advances in Engineering Mechanics and Their Impact on Civil Engineering Practice*. Two Volumes, ASCE, New York, 1326 pp.

4. (1985) Connection flexibility and steel frames. Proceedings of a Session at the ASCE Detroit Convention, 21–25 October, ASCE Publication, No. 482, New York, 122 pp.

5. (1987) Joint flexibility in steel frames. *Journal of Constructional Steel Research*, Special Issue, Vol. 8, Elsevier Applied Science, London, 290 pp.

6. (1988) Steel beam-to-column building connections. *Journal of Constructional Steel Research*, Special Issue, Vol. 10, Elsevier Applied Science, London, 482 pp.

7. (With Sohal IS.) (1988) Cylindrical members in offshore structures. *Thin-Walled Structures*, Special Issue, Vol. 7, Elsevier Applied Science, London, pp. 153–285.

8. (1993) *Semi-Rigid Connections in Steel Frames. Council on Tall Buildings and Urban Habitat*. New York, McGraw-Hill, 318 pp.

9. (With White DW.) (1993) Plastic hinge based methods for advanced analysis and design of steel frames: An assessment of the state-of-the-art. Structural Stability Research Council, Bethlehem, PA, 299 pp.

10. (With Toma S.) (1994) *Advanced Analysis of Steel Frames*. Boca Raton, FL, CRC Press, 384 pp.

11. (1995) *The Civil Engineering Handbook*. Boca Raton, FL, CRC Press, 2609 pp.

12. (With Toma S.) (1996) *Analysis and Software of Cylindrical Members*. Boca Raton, FL, CRC Press, 309 pp.

13. (1997) *Handbook of Structural Engineering*. Boca Raton, FL, CRC Press, 1600 pp.

14. (With White DW.) (1998) Innovations in stability concepts and methods for seismic design in structural steel. Proceedings, US-Japan Seminar, Special Issues of Engineering Structures, Vol. 20, Nos. 4–6, April–June, 569 pp.

15. (With King WS, Zhou SP, Duan L.) (1999) *Structural Steel Design: LRFD Method* (in Chinese). Taiwan, Science/Technology Book Company, 268 pp.

16. (With Duan L.) (2000) *Bridge Engineering Handbook*. Boca Raton, FL, CRC Press, 2020 pp. (Winner of *Choice* [January 2001] magazine's Outstanding Academic Title award for 2000.)

17. (2000) *Practical Analysis for Semi-Rigid Frame Design*. Singapore, World Scientific Publishing Company, 465 pp.

18. (With Liew JYR.) (2001) High performance computing. *Computer-Aided Civil and Infrastructure Engineering*, Special Issue, Vol. 16, No. 5, 78 pp.
19. (With Scawthorn C.) (2002) *Earthquake Engineering Handbook*. Boca Raton, FL, CRC Press, 1376 pp.
20. (With Liew JYR.) (2002) *The Civil Engineering Handbook*. 2nd Edition. Boca Raton, FL, CRC Press, 2904 pp.
21. (With Duan L.) (2003) *Bridge Engineering: Substructure Design*. Boca Raton, FL, CRC Press, 272 pp. (Translated into Chinese by China Machine Press, 2008, 260 pp.)
22. (With Duan L.) (2003) *Bridge Engineering: Seismic Design*. Boca Raton, FL, CRC Press, 480 pp. (Translated into Chinese by China Machine Press, 2008, 508 pp.)
23. (With Duan L.) (2003) *Bridge Engineering: Construction and Maintenance*. Boca Raton, FL, CRC Press, 256 pp.
24. (With Lui EM.) (2005) *Handbook of Structural Engineering*. 2nd Edition. Boca Raton, FL, CRC Press, 1768 pp.
25. (With Lui EM.) (2006) *Earthquake Engineering far Structural Design*. Boca Raton, FL, CRC Press, 264 pp. (Translated into Chinese by Chongqing University Press, Chongqing, China, 2009.)
26. (With Lui EM.) (2006) *Principles of Structural Design*. Boca Raton, FL, CRC Press, 528 pp.
27. (With Kishi N, Komuro M.) (2011) *Handbook of Semi-rigid Connections*. Fort Lauderdale, FL, 1248 pp.
28. (With Duan L.) (2014) *Handbook of International Bridge Engineering*. Boca Raton, FL, CRC Press, 1394 pp.
29. (With Duan L.) (2014) *Bridge Engineering Handbook, Second Edition: Fundamentals*. Boca Raton, FL, CRC Press, 591 pp.
30. (With Duan L.) (2014) *Bridge Engineering Handbook, Second Edition: Superstructure Design*. Boca Raton, FL, CRC Press, 736 pp.
31. (With Duan L.) (2014) *Bridge Engineering Handbook, Second Edition: Substructure Design*. Boca Raton, FL, CRC Press, 370 pp.
32. (With Duan L.) (2014) *Bridge Engineering Handbook, Second Edition: Seismic Design*. Boca Raton, FL, CRC Press, 740 pp.
33. (With Duan L.) (2014) *Bridge Engineering Handbook, Second Edition: Construction and Maintenance*. Boca Raton, FL, CRC Press, 661 pp.

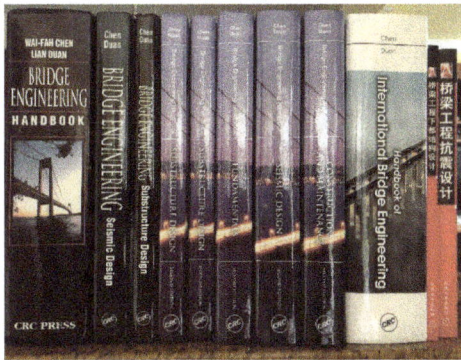

Bridge engineering handbooks.

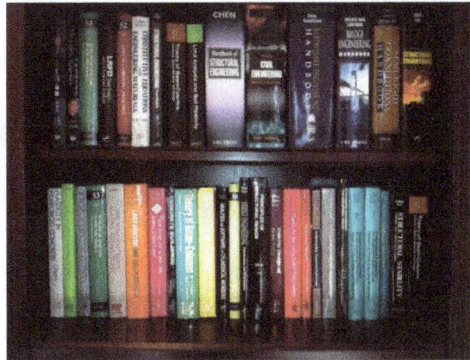

Technical books at a glance.

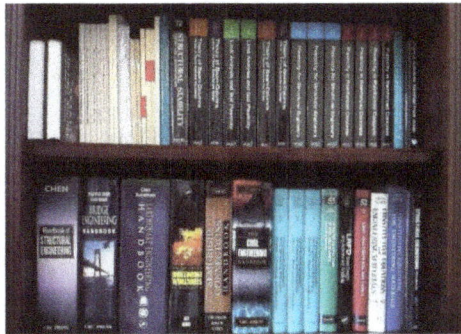

Handbooks at a glance.

Translated technical books at a glance.

6.3 Contributing Chapters

1. (With Drucker DC.) (1968) On the use of simple discontinuous fields to bound limit loads. In *Engineering Plasticity*, Editors, Heyman J, Leckie FA. London, Cambridge University Press, pp. 129–145.

2. (1971) Additional design considerations, Chapter 6. In *Plastic Design in Steel — A Guide and Commentary*, ASCE – Manual and Report on Engineering Practice — No. 41, 2nd Edition, New York, ASCE, pp. 61–84.

3. (1976) Foundation stability-theory and applications. In *Analysis and Design of Building Foundations*, Editor, Fang HY. Bethlehem, PA, Envo Publishing Company, pp. 37–102.

4. (1976) Beam-columns, Chapter 8. In *SSRC Guide to Stability Design Criteria for Metal Structures*, 3rd Edition, Editor, Johnston BG, New York, Wiley-Interscience, pp. 189–226.

5. (1978) Connections, Chapter 7. In *ASCE Monograph on Structural Design of Tall Steel Buildings*, Editor-In-Chief, Beedle LS, New York, ASCE, pp. 485–575.

6. (With Bazant ZP, Darwin D, Buyukozturk O, Willam KJ *et al.*) (1981) Constitutive relations and failure theories, Chapter 2. In *Finite Element Analysis of Reinforced Concrete Structures*, ASCE Special Publication, New York, pp. 34–148.

7. (1982) Box and cylindrical columns under biaxial bending, Chapter 2. In *Vol. 1 — Axially Compressed Structures, in Developments in the Stability and Strength of Structures*, Editor, Narayanan, R. London, Applied Science Publishers, pp. 83–127.

8. (1983) Effective length of columns with simple connections. In *Developments in Tall Buildings*, Editor-in-Chief, Beedle LS. Council on Tall Buildings and Urban Habitat. Stroudsburg, PA, Hutchinson Ross Publishing Company, pp. 513–520.

9. (1984) The continuum theory of rock mechanics, Chapter 3. In *Mechanics of Oil Shale*, Editors, Chong KP, Smith JW. London, Applied Science Publishers, pp. 71–126.

10. (1984) Constitutive modeling in soil mechanics, Chapter 5. In *Mechanics of Engineering Materials*, Editors, Desai CS, Gallagher RH. London, John Wiley and Sons, pp. 91–120.

11. (1984) Soil mechanics, plasticity and landslides. In *Special Anniversary Volume on Mechanics of Material Behavior to Honor Dean Daniel C. Drucker*, Editors, Dvorak GJ, Shield RT. Amsterdam, Elsevier, pp. 31–58.

12. (With Saleeb AF.) (1984) Plasticity modeling for engineering materials. In *Special Anniversary Volume (Verba Volant, Scripta Manent) for Professor Massonnet*, Belgium, Liege, pp. 117–132.

13. (With Lui EM.) (1985) Beam-to-column moment-resisting connections, Chapter 6. In *Vol. 4 — Steel Framed Structures (Stability and Strength), in Developments in the Stability and Strength of Structures*, Editor, Narayanan R. London, Applied Science Publishers, pp. 153–203.

14. (With Sugimoto H.) (1985) Fabricated tubular columns used in offshore structures, Chapter 5. In *Vol. 5 — Shell Structures (Stability and Strength), in Developments in the Stability and Strength of Structures*, Editor, Narayanan R. London, Applied Science Publishers, pp. 137–184.

15. (1985) Constitutive relations for concrete, rock and soils: Discusser's report. In *Mechanics of Geomaterials*, Editor, Bazant Z. New York, John Wiley and Sons, pp. 65–86.

16. (With Mizuno E.) (1986) Plasticity modeling and its applications to geomechanics. In *Recent Developments in Laboratory and Field Tests and Analysis of Geotechnical Problems*, Editors, Balasubramaniam AS, Chandra S, Bergado DT. Rotterdam, AA Balkema, pp. 391–426.

17. (With Lui EM.) (1986) Generalized column equation — A physical approach. In *Advances in Tall Buildings*, Editor-In-Chief, Beedle LS. New York, Van Nostrand Reinhold, pp. 323–352.

18. (With Lui EM.) (1986) Recent developments in structural connections. In *Advances in Tall Buildings*, Editor-In-Chief, Beedle LS. New York, Van Nostrand Reinhold, pp. 353–365.

19. (With McCarron WO.) (1986) Modeling of soils and rocks based on concepts of plasticity. In *Recent Developments in Laboratory and Field Tests and Analysis of Geotechnical Problems*, Editors, Balasubramaniam AS, Chandra S, Bergado DT. Rotterdam, AA Balkema, pp. 467–510.

20. (1986) Semi-rigid connections in steel frames. In *High-Rise Buildings: Recent Progress*, Editor, Beedle LS. Council on Tall Buildings and Urban Habitat, Lehigh University, Bethlehem, PA, pp. 171–189.

21. (1988) Connection flexibility in steel structures. In *Second Century of the Skyscraper*, Editor-In-Chief, Beedle LS. Council on Tall Buildings and Urban Habitat, Lehigh University, Bethlehem, PA, pp. 857–884.

22. (1989) Analysis of steel frames with flexible joints, Chapter 10. In *Structural Connections: Stability and Strength*, Editor, Narayanan R. London, Applied Science Publishers, pp. 335–444.

23. (With McCarron WO.) (1991) Bearing capacity of shallow foundations, Chapter 4. In *Foundation Engineering Handbook*, 2nd Edition, Editor, Fang HY. New York, Van Nostrand Reinhold, pp. 144–165.

24. (1991) Beam-columns. North America Chapter Editor, Chapter 5. In *Stability of Metal Structures — A World View*, Editor-in-Chief, Beedle LS. Structural Stability Research Council, Lehigh University, Bethlehem, PA, pp. 318–330, 358, and 370–376.

25. (With White DW, Liew JYR.) (1992) Beam-columns, Chapter 2.5. In *Constructional Steel Design: An International Guide*, Editors, Dowling PJ, Harding JG, Bjorhovde R. London, Elsevier Science Publishers, pp. 105–132.

26. (With White DW.) (1993) Introduction. Plastic hinge based methods for advanced analysis and design of steel frames: An assessment of the state-of-the-art, Structural Stability Research Council, Lehigh University, Bethlehem, PA, pp. 1–22.

27. (With White DW, Liew JYR.) (1993) Toward advanced analysis in LRFD. In *Plastic Hinge Based Methods for Advanced Analysis and Design of Steel Frames: An Assessment of the State-of-the-Art*, Structural Stability Research Council, Lehigh University, pp. 95–173.

28. (With Lui EM.) (1993) Analysis of members in semi-rigid unbraced (sway) frames, Chapter 4. In *Semi-Rigid Connections in Steel Frames*, Editor-in-Chief, Beedle LS. New York, McGraw-Hill, pp. 73–100.

29. (With Kishi N.) (1993) Creating design application models from historical experimental database, Appendix A. In *Semi-Rigid Connections in Steel Frames*, Editor-in-Chief, Beedle LS. New York, McGraw-Hill, pp. 211–232.

30. (With Wu FH.) (1993) Moment-rotation relationship of semi-rigid steel beam-to-column connections, Appendix B. In *Semi-Rigid Connections in Steel Frames*, Editor-in-Chief, Beedle LS. New York, McGraw-Hill, pp. 233–268.

31. (With Yamaguchi E, Kotsovos MD, Pan AD.) (1993) Constitutive models, Chapter 2. In *Finite Element Analysis of Reinforced Concrete Structures II*, Editor, Isenberg J. ASCE Proceeding Publication, New York, pp. 36–117.

32. (With Liew JYR, White DW.) (1993) Beam-columns. In *Constructional Steel Design: An International Guide*, Editors, Dowling PJ, Harding JE, Bjorhovde R. London, Elsevier Applied Science, pp. 105–132.

33. (With Liew JYR.) (1994) Trends toward advanced analysis, Chapter 1. In *Advanced Analysis of Steel Frames*, Editors, Chen WF, Toma S. Boca Raton, FL, CRC Press, pp. 1–45.

34. (With Liew JYR.) (1994) Second-order plastic hinge analysis of frames, Chapter 4. In *Advanced Analysis of Steel Frames*, Editors, Chen WF, Toma S. Boca Raton, FL, CRC Press, pp. 139–194.

35. (With Zhou SP.) (1994) Plastic-zone analysis of beam-columns and portal frames, Chapter 5. In *Advanced Analysis of Steel Frames*, Editors, Chen WF, Toma S. Boca Raton, FL, CRC Press, pp. 195–258.

36. (With Toma S.) (1994) Benchmark problems and solutions, Chapter 7. In *Advanced Analysis of Steel Frames*, Editors, Chen WF, Toma S. Boca Raton, FL, CRC Press, pp. 321–371.

37. (With Luan M.) (1995) Stability analysis in soil mechanics. In *Geomechanics and Geotechnical Engineering*, Editors, Li JJ, Luan M, Song YF. Dalian University of Technology, Dalian, China, pp. 113–123.

38. (With Toma S.) (1996) Introduction, Chapter 1. In *Analysis and Software of Cylindrical Members*, Editors, Chen WF, Toma S. Boca Raton, FL, CRC Press, pp. 1–4.

39. (With Toma S.) (1996) Centrally and eccentrically compressed columns, Chapter 2. In *Analysis and Software of Cylindrical Members*, Editors, Chen WF, Toma S. Boca Raton, FL, CRC Press, pp. 5–53.

40. (With Toma S.) (1996) Approximate analysis of beam-columns, Chapter 3. In *Analysis and Software of Cylindrical Members*, Editors, Chen WF, Toma S. Boca Raton, FL, CRC Press, pp. 55–104.

41. (With Toma S.) (1996) Cyclic behavior and modeling, Chapter 4. In *Analysis and Software of Cylindrical Members*, Editors, Chen WF, Toma S. Boca Raton, FL, CRC Press, pp. 105–165.

42. (With Sohal IS.) (1996) Analysis considering local buckling effects, Chapter 5. In *Analysis and Software of Cylindrical Members*, Editors, Chen WF, Toma S. Boca Raton, FL, CRC Press, pp. 167–207.

43. (With Duan L.) (1996) Analysis considering dent damage effects, Chapter 6. In *Analysis and Software of Cylindrical Members*, Editors, Chen WF, Toma S. Boca Raton, FL, CRC Press, pp. 105–165.

44. (With Liew JYR.) (1997) LRFD-limit design of frames, Chapter 6. In *Steel Design Handbook — LRFD Method*, Editor, Tamboli AR. New York, McGraw-Hill, pp. 6-1 to 6-83.

45. (With Toma S, Duan L.) (1997) Bridge structures, Chapter 10. In *Handbook of Structural Engineering*, Editor-in-Chief, Chen WF. Boca Raton, FL, CRC Press, 10-1 to 10-103.

46. (With Liew JYR, Balendra T.) (1997) Multistory frame structures, Chapter 12. In *Handbook of Structural Engineering*, Editor-in-Chief, Chen WF. Boca Raton, FL, CRC Press, pp. 12-1 to 12-73.

47. (With Duan L.) (1997) Effective length factors of compression members, Chapter 17. In *Handbook of Structural Engineering*, Editor-in-Chief, Chen WF. Boca Raton, FL, CRC Press, pp. 17-1 to 17-52.

48. (With Kim SE.) (1997) An innovative design for steel frame using advanced analysis, Chapter 28. In *Handbook on Structural Engineering*, Editor-in-Chief, Chen WF. Boca Raton, FL, CRC Press, pp. 28-1 to 28-56.

49. (With Kim SE.) (1998) Design of steel structures with LRFD using advanced analysis. In *Stability and Ductility of Steel Structures*, Editors, Usami T, Itoh Y. Pergamon/Elsevier, pp. 153–166.

50. (With Duan L.) (2000) Effective length of compression members, Chapter 52. In *Bridge Engineering Handbook*, Editors, Chen WF, Duan L. Boca Raton, FL, CRC Press, pp. 52-1 to 52-22.

51. (With Duan M, Perdrikaris PC.) (2000) Impact effect of moving vehicles, Chapter 56. In *Bridge Engineering Handbook*, Editors, Chen WF, Duan L. Boca Raton, FL, CRC Press, pp. 56-1 to 56-19.

52. (With Duan L.) (2002) Bridges, Chapter 18. In *Earthquake Engineering Handbook*, Editors, Chen WF, Scawthorn C. Boca Raton, FL, CRC Press, pp. 18-1 to 18-56.

53. (With Kim SE.) (2005) Steel frame design using advanced analysis, Chapter 5. In *Handbook of Structural Engineering*, 2nd Edition, Editors, Chen WF, Lui EM. Boca Raton, FL, CRC Press, pp. 5-1 to 5-49.

54. (With Duan L, Reno M, Unjoh S.) (2005) Seismic design of bridges, Chapter 20. In *Handbook of Structural Engineering*, 2nd Edition, Editors, Chen WF, Lui EM. Boca Raton, FL, CRC Press, pp. 20-1 to 20-55.

55. (With Duan L.) (2005) Effective length factors of compression members, Chapter 31. In *Handbook of Structural Engineering*, 2nd Edition, Editors, Chen WF, Lui EM. Boca Raton, FL, CRC Press, pp. 31-1 to 31-57.

56. (With Toma S, Duan L.) (2005) Bridge structures, Chapter 25. In *Handbook of Structural Engineering*, 2nd Edition, Editors, Chen WF, Lui EM. Boca Raton, FL, CRC Press, 25-1 to 25-49.

57. (With Duan L, Lei HG.) (2014) Effective length of compression members, Chapter 18. In *Bridge Engineering Handbook, 2nd Edition: Fundamentals*, Editors, Chen WF, Duan L. Boca Raton, FL, CRC Press, 427–449.

CHAPTER 7
ACADEMIC PUBLICATIONS
期刊論文目錄

Dr. Chen is the author or co-author of more than 622 articles in various refereed Technical Journals (359), Conference Proceedings and Symposium Volumes (263). He has also been speaker for invited lectures, seminars and conferences (110).

7.1 Journal Articles (359)

1968
1. Chen WF. (1968) On the rate of dissipation of energy in soils. *Soils and Foundations*, **8** (4): 48–51. https://doi.org/10.3208/sandf1960.8.4_48
2. Chen WF. (1968) Discussion on application of limit plasticity in soil mechanics by Liam Finn. *Journal of the Soil Mechanics and Foundations Division*, **94** (SM2): 608–613.
3. Chen WF, Santathadaporn S. (1968) Review of column behavior under biaxial loading. *Journal of the Structural Division*, **94** (ST12): 2999–3021.

1969
1. Chen WF, Santathadapom S. (1969) Curvature and the solutions of eccentrically loaded columns. *Journal of the Engineering Mechanics Division*, **95** (EM1): 21–40.
2. Chen WF. (1969) Methods of computing geometric relations in frames. *Journal of the Structural Division*, **95** (ST8): 1789–1794.
3. Chen WF, Giger MW, Fang HY. (1969) On the limit analysis of stability of slopes. *Soils and Foundations*, **9** (4): 23–32. https://doi.org/10.3208/sandf1960.9.4_23
4. Chen WF. (1969) Soil mechanics and the theorems of limit analysis. *Journal of the Soil Mechanics and Foundations Division*, **95** (SM2): 493–518.
5. Chen WF. (1969) Thoughts of the present and future role of plasticity in concrete or rock mechanics. *Journal of Civil Engineering*, **10**: 5–11.
6. Chen WF, Drucker DC. (1969) Bearing capacity of concrete blocks or rock. *Journal of the Engineering Mechanics Division*, **95** (EM4): 955–978.

1970

1. Chen WF. (1970) Plastic indentation of metal blocks by a flat punch. *Journal of the Engineering Mechanics Division*, **96** (EM3): 353–363.
2. Chen WF. (1970) Effects of initial curvature on column strength. *Journal of the Structural Division*, **96** (ST12): 2685–2691.
3. Chen WF. (1970) Extensibility of concrete and theorems of limit analysis. *Journal of the Engineering Mechanics Division*, **96** (EM3): 341–352.
4. Chen WF, Scawthorn C. (1970) Limit analysis and limit equilibrium solutions in soil mechanics. *Soils and Foundations*, **10** (3): 13–49. https://doi.org/10.3208/sandf1960.10.3_1
5. Hyland MW, Chen WF. (1970) Bearing capacity of concrete blocks. *ACI Journal*, **67** (3): 228–236. DOI: 10.14359/7264 https://www.concrete.org/publications/internationalconcreteabstractsportal. aspx?m=details&id=7264
6. Chen WF. (1970) Discussion on circular and logarithmic spiral slip surfaces, by Eric Spencer. *Journal of the Soil Mechanics and Foundations Division*, **96** (SM1): 324–326.
7. Chen WF. (1970) General solution of inelastic beam-column problem. *Journal of the Engineering Mechanics Division*, **96** (EM4): 421–441.
8. Chen WF. (1970) Double-punch test for tensile strength of concrete. *ACI Journal*, **67** (12): 993–995 (Closure, Vol. 68, No. 6, 1971, 480–481).
9. Chen WF, Santathadaporn S. (1970) Discussion on approximate yield surface equations, by GA Morris and SJ Fenves. *Journal of the Engineering Mechanics Division*, **96** (EM2): 188–190.
10. Santathadaporn S, Chen WF. (1970) Interaction curves for sections under biaxial bending and axial force. *WRC Bulletin*, 148: 1–11.

1971

1. Chen WF, Carson JL. (1971) Stress-strain properties of random wire reinforced concrete. *ACI Journal*, **68** (12): 933–936. DOI: 10.14359/7249 https://www.concrete.org/publications/internationalconcreteabstractsportal. aspx?m=details&id=7249
2. Chen WF, Covarrubias S. (1971) Bearing capacity of concrete blocks. *Journal of the Engineering Mechanics Division*, **97** (EM5): 1413–1431.
3. Chen WF, Giger MW. (1971) Limit analysis of stability of slopes. *Journal of the Soil Mechanics and Foundations Division*, **97** (SM1): 19–26.
4. Chen ACT, Huang T, Chen WF. (1971) Discussion on flexural rigidity of concrete column sections, by IC Medland and DA Taylor. *Journal of the Structural Division*, **97** (ST11): 2763–2766.

5. Chen WF. (1971) Further studies of inelastic beam-column problem. *Journal of the Structural Divisions*, **97** (ST2): 529–544.
6. Chen WF. (1971) Approximate solution of beam-columns. *Journal of the Structural Divisions*, **97** (ST2): 741–751.

1972

1. Dismuke TD, Chen WF, Fang HY. (1972) Tensile strength of rock by the double-punch method. *Rock Mechanics*, **4** (2): 79–87.
   https://doi.org/10.1007/BF01239138
2. Chen WF, Trumbauer BE. (1972) Double-punch test and tensile strength of concrete. *Journal of Materials*, ASTM, **7** (2): 148–154.
3. Chen WF, Atsuta T. (1972) Interaction equations for bi-axially loaded sections. *Journal of the Structural Division*, **98** (ST5): 1035–1052 (Closure, Vol. 99, No. ST12, 1973, 2488–2493).
4. Chen WF. (1972) Discussion on stability and load capacity of members with no tensile strength, by Felix Y Yokel. *Journal of the Structural Division*, **98** (ST5): 1193–1197.
5. Chen WF, Atsuta T. (1972) Simple interaction equations for beam-columns. *Journal of the Structural Division*, **98** (ST7): 1413–1426 (Closure, Vol. 99, No. ST10, 1973, 2210–2211).
6. Santathadaporn S, Chen WF. (1972) Tangent stiffness method for biaxial bending. *Journal of the Structural Division*, **98** (ST1): 153–163 (Closure, Vol. 99, No. ST3, 1973, 578–579).
7. Chen WF, Atsuta T. (1972) Column curvature curve method for analysis of beam-columns. *The Structural Engineer*, ISE, **50** (6): 233–240.

1973

1. Huang JS, Chen WF, Beedle LS. (1973) Behavior and design of steel beam-to-column moment connections. *WRC Bulletin*, **188**: 1–23.
2. Regec JE, Huang JS, Chen WF. (1973) Test of a fully-welded beam-to-column connection. *WRC Bulletin*, **188**: 24–35.
3. Santathadaporn S, Chen WF. (1973) Analysis of bi-axially loaded steel h-columns. *Journal of the Structural Division*, **99** (ST3): 491–509.
4. Chen WF, Newlin DE. (1973) Column web strength in beam-to-column connection. *Journal of the Structural Division*, **99** (ST9): 1978–1984.
5. Fielding DJ, Chen WF. (1973) Steel frame analysis and connection shear deformation. *Journal of the Structural Division*, **99** (ST1): 1–18.

6. Chen WF, Atsuta T. (1973) Strength of eccentrically loaded walls. *International Journal of Solids and Structures*, **9** (10): 1283–1300. https://doi.org/10.1016/0020-7683(73)90115-7

7. Chen WF. (1973) Bearing strength of concrete blocks. *Journal of the Engineering Mechanics Division*, **99** (EM6): 1309–1321.

8. Chen WF, Chen CH. (1973) Analysis of concrete filled steel tubular beam-columns. *IABSE Publication* 33-11, 37–52.

9. Chen WF, Davidson HL. (1973) Bearing capacity determination by limit analysis. *Journal of the Soil Mechanics and Foundations Division*, **99** (SM6): 433–449.

10. Chen WF, Rosenfarb J. (1973) Limit analysis solutions of earth pressure problems. *Soils and Foundations*, **13** (4): 45–60. https://doi.org/10.3208/sandf1972.13.4_45

11. Chen WF, Atsuta T. (1973) Ultimate strength of bi-axially loaded steel H-columns. *Journal of the Structural Division*, **99** (ST3): 469–489 (Closure, Vol. 100, No. ST10, 1974, 2149).

12. Chen WF, Atsuta T. (1973) Inelastic response of column segments under biaxial loads. *Journal of the Engineering Mechanics Division*, **99** (EM4): 685–701.

13. Manson JA, Chen WF, Vanderhoff JW, Liu YN. (1973) Stress-strain behavior of polymer impregnated concrete. *Polymer Preprints*, ACS Publication, **14** (2): 1203–1208.

1974

1. Chen WF, Dahl-Jorgensen E. (1974) Polymer-impregnated concrete as a structural material. *Magazine of Concrete Research*, **26** (96): 16–20. https://doi.org/10.1680/macr.1974.26.86.16

2. Chen WF, Oppenheim IJ. (1974) Web buckling strength of beam-to-column connections. *Journal of the Structural Division*, **100** (ST1): 279–285.

3. Chen WF, Colgrove TA. (1974) Double-punch test for tensile strength of concrete. *Transportation Research Record* 504, National Research Council, pp. 43–50.

4. Chen WF, Atsuta T. (1974) Interaction curves for steel sections under axial load and biaxial bending. *The Engineering Journal*, **17** (A-3): I–VIII.

5. Tebedge N, Chen WF. (1974) Design criteria for H-Columns under biaxial loading. *Journal of the Structural Division*, **100** (ST3): 579–598 (Closure, Vol. 101, No. ST8, 1975, 1705–1708).

1975

1. Dahl-Jorgensen E, Chen WF, Manson JA, Liu YN, Vanderhoff JW. (1975) Polymer-impregnated concrete: Laboratory studies. *Journal of Transportation Engineering*, **101** (TE1): 29–45.

2. Chen WF, Manson JA, Mehta HC, Vanderhoff JW. (1975) Innovations in impregnation techniques for highway concrete. *Polymer Concrete*, Transportation Research Record 542, TRB, Washington, DC, pp. 29–40.

3. Mehta HC, Vanderhoff JW, Chen WF, Manson JA. (1975) Polymer-impregnated concrete: Field studies. *Journal of Transportation Engineering*, **101** (TE1): 1–27.

4. Chen WF, Snitbhan N. (1975) On slip surface and slope stability analysis. *Soils and Foundations*, **15** (3): 41–49. https://doi.org/10.3208/sandf1972.15.3_41

5. Chen ACT, Chen WF. (1975) Constitutive relations for concrete. *Journal of the Engineering Mechanics Division*, **101** (EM4): 465–481.

6. Chen ACT, Chen WF. (1975) Constitutive equations and punch-indentation of concrete. *Journal of the Engineering Mechanics Division*, **101** (EM6): 889–906.

7. Chen WF, Snitbhan N, Fang HY. (1975) Stability of slopes in anisotropic, non-homogeneous soil. *Canadian Geotechnical Journal*, **12** (1): 146–152. https://doi.org/10.1139/t75-014

8. Chen WF, Shoraka MT. (1975) Tangent stiffness method for biaxial bending of reinforced concrete columns. *IABSE Publications*, **35-1**: 23–44.

1976

1. Mehta HC, Chen WF, Manson JA, Vanderhoff JW. (1976) High-temperature drying of thick concrete slabs. *Journal of Transportation Engineering*, **102** (TE2): 185–200.

2. Chen WF, Mehta HC, Slutter RG. (1976) Sulfur- and polymer-impregnated brick and block prism. *Journal of Testing and Evaluation*, ASTM, **4** (4): 283–292.

3. Parfitt J, Chen WF. (1976) Tests of welded steel beam-to-column moment connections. *Journal of the Structural Division*, **102** (STl): 189–202.

4. Davidson HL, Chen WF. (1976) Two elastic-plastic soil models for numerical analysis. *Soils and Foundations*, **16** (2): 43–50. https://doi.org/10.3208/sandf1972.16.2_43

5. Chen ACT, Chen WF. (1976) Nonlinear analysis of concrete splitting tests. *Computers and Structures*, **6** (6): 451–457. https://doi.org/10.1016/0045-7949(76)90039-0

6. Ross DA, Chen WF. (1976) Design criteria for steel I columns under axial load and biaxial bending. *Canadian Journal of Civil Engineering*, **3** (3): 466–473. https://doi.org/10.1139/l76-047

7. Standig KF, Rentschler GP, Chen WF. (1976) Tests of bolted beam-to-column flange moment connections. *WRC Bulletin* 218.

1977

1. Chen WF, Vanderhoff JW, Manson JA, Mehta HC. (1977) Field impregnation techniques for highway concrete. *Journal of Transportation Engineering*, **103** (TE3): 355–368.

2. Davidson HL, Chen WF. (1977) Nonlinear response of undrained clay to footings. *Computers and Structures*, **7** (4): 539–546. https://doi.org/10.1016/0045-7949(77)90017-7

3. Chen WF, Ross DA. (1977) The axial strength and behavior of cylindrical columns. *Journal of Petroleum Technology*, AIME, pp. 239–241.

4. Chen WF, Ross DA. (1977) Tests of fabricated tubular columns. *Journal of the Structural Division*, **103** (ST3): 619–634 (Closure, Vol. 104, No. ST9, 1978, pp. 1536–1538).

5. Liu YN, Manson JA, Chen WF, Vanderhoff JW. (1977) Polymer-impregnated mortars I. Effect of polymer state on mechanical behavior. *Polymer Engineering and Science*, **17** (5): 325–334. https://doi.org/10.1002/pen.760170511

1978

1. Davidson HL, Chen WF. (1978) Nonlinear response of drained clay to footings. *Computers and Structures*, **8** (2): 281–290. https://doi.org/10.1016/0045-7949(78)90035-4

2. Snitbhan N, Chen WF. (1978) Elastic-plastic large deformation analysis of soil slopes. *Computer and Structures*, **9** (6): 567–577. https://doi.org/10.1016/0045-7949

3. Chen WF, Chang TYP. (1978) Plasticity solutions for concrete splitting tests. *Journal of the Engineering Mechanics Division*, **104** (EM3): 691–704 (Closure, Vol. 105, No. EM6, 1979, 1064–1067).

1979

1. Suzuki H, Chen WF, Chang TYP. (1979) Implosion analysis of concrete cylindrical vessels. *Journal of Pressure Vessel Technology*, ASME, **101** (1): 98–102. https://doi.org/10.1115/1.3454605

2. Chen WF, Chang TYP, Mehta HC. (1979) Experiments on axially loaded concrete shells. *Journal of the Structural Division*, **105** (ST8): 1673– 1688.

3. Toma S, Chen WF. (1979) Analysis of fabricated tubular columns. *Journal of the Structural Division*, **105** (ST11): 2343–2366 (Errata, Vol. 106, No. ST1, 1980, 2357).

1980

1. Chen WF, Ting EC. (1980) Constitutive models for concrete structures. *Journal of the Engineering Mechanics Division*, **106** (EMl): 1–19.
2. Rentschler GP, Chen WF, Driscoll GC. (1980) Tests of beam-to-column web moment connections. *Journal of the Structural Division*, **106** (ST5): 1005–1022.
3. Chen WF, Yuan RL. (1980) Tensile strength of concrete: Double-punch test. *Journal of the Structural Division*, **106** (ST8): 1673–1693.
4. Chen WF. (1980) End restraint and column stability. *Journal of the Structural Division*, **106** (ST11): 2279–2295 (Closure, Vol. 108, No. ST8, 1982, 1929–1933).
5. Chen WF, Suzuki H. (1980) Constitutive models for concrete. *Computers and Structures*, **12** (1): 23–32. https://doi.org/10.1016/0045-7949(80)90091-7
6. Ross DA, Tall L, Chen WF. (1980) Fabricated tubular steel columns. *Journal of the Structural Division*, **106** (ST1): 265–282.
7. Cheong-Siat-Moy F, Chen WF. (1980) Limit states design of steel beam-columns, a state-of-the-art review. *Solid Mechanics Archives*, Noordhoff, The Netherlands, **5** (1): 29–73.
8. Yuan RL, Chen WF. (1980) Behavior of sulfur-infiltrated concrete in sodium chloride solution. *Special Publication*, SP-65-17, pp. 291–308.
   DOI: 10.14359/6359
   https://www.concrete.org/publications/internationalconcreteabstractsportal.
   aspx?m=details&id=6359
9. Chen WF. (1980) Plasticity in soil mechanics and landslides. *Journal of the Engineering Mechanics Division*, **106** (EM3): 443–464.
10. Chen WF, Suzuki H, Chang TYP. (1980) Nonlinear analysis of concrete cylinder structures under hydrostatic loading. *Computers and Structures*, **12** (4): 559–570. https://doi.org/10.1016/0045-7949(80)90131-5
11. Chen WF, Suzuki H, Chang TYP. (1980) End effects of pressure-resistant concrete shells. *Journal of the Structural Division*, **106** (ST4): 751–771.
12. Chen WF, Koh SL. (1980) Plasticity approach to landslide problems. *Engineering Geology*, **16** (1–2): 125–133. https://doi.org/10.1016/0013-7952(80)90012-5

1981

1. Graff WJ, Chen WF. (1981) Bottom-supported concrete platforms: Overview. *Journal of the Structural Division*, **107** (ST6): 1059–1081.

2. Chen WF, Patel KV. (1981) Static behavior of beam-to-column moment connections. *Journal of the Structural Division*, **107** (ST9): 1815–1838.
3. Chen WF, Chang MF. (1981) Limit analysis in soil mechanics and its applications to lateral earth pressure problems. *Solid Mechanics Archives*, Noordhoff, The Netherlands, **6** (3): 331–399.
4. Saleeb AF, Chen WF. (1981) Elastic-plastic large displacement analysis of pipe. *Journal of the Structural Division*, **107** (ST4): 605–626.

1982

1. Yuan RL, McLelland GR, Chen WF. (1982) Experiments on closing reinforced concrete corners. *Journal of the Structural Division*, **108** (ST4): 771–779.
2. Hsieh SS, Ting EC, Chen WF. (1982) A plastic-fracture model for concrete. *International Journal of Solids and Structures*, **18** (3): 181–197. https://doi.org/10.1016/0020-7683(82)90001-4
3. Al-Noury SI, Chen WF. (1982) Behavior and design of reinforced and composite concrete sections. *Journal of the Structural Division*, **108** (ST6): 1266–1284.
4. Toma S, Chen WF. (1982) Cyclic analysis of fixed-ended steel beam-columns. *Journal of the Structural Division*, **108** (ST6): 1385–1399.
5. Toma S, Chen WF, Finn LD. (1982) External pressure and sectional behavior of fabricated tubes. *Journal of the Structural Division*, **108** (ST1): 177–194.
6. Chang MF, Chen WF. (1982) Lateral earth pressures on rigid retaining walls subjected to earthquake forces. *Solid Mechanics Archives*, **7** (4): 315–362.
7. Rentschler GP, Chen WF, Driscoll GC. (1982) Beam-to-column web connection details. *Journal of the Structural Division*, **108** (ST2): 393–409.
8. Sugimoto H, Chen WF. (1982) Small end restraint effects on strength of H-columns. *Journal of the Structural Division*, **108** (ST3): 661–681 (Closure, Vol. 109, No. 4, 1983, 1073–1077).
9. Al-Noury SI, Chen WF. (1982) Finite segment method for biaxially loaded RC columns. *Journal of the Structural Division*, **108** (ST4): 780–799.
10. Toma S, Chen WF. (1982) Inelastic cyclic analysis of pinned-ended tubes. *Journal of the Structural Division*, **108** (ST10): 2279–2294.

1983

1. Toma S, Chen WF. (1983) Design of vertical chords in deepwater platform. *Journal of Structural Engineering*, **109** (11): 2733–2746. https://doi.org/10.1061/(ASCE)0733-9445(1983)109:11(2733)
2. Han DJ, Chen WF. (1983) Buckling and cyclic inelastic analysis of steel tubular beam-columns. *Engineering Structures*, **5** (2): 119–132. https://doi.org/10.1016/0141-0296(83)90025-1

3. Sawada T, Chen WF. (1983) Earthquake-induced slope failure in non-homogeneous, anisotropic soils. *Soils and Foundations*, **23** (2): 125–139. https://doi.org/10.3208/sandf1972.23.2_125

4. Han DJ, Chen WF. (1983) Behavior of portal and strut types of beam-columns. *Engineering Structures*, **5** (1): 15–25. https://doi.org/10.1016/0141-0296(83)90036-6

5. Lui EM, Chen WF. (1983) Strength of H-columns with small end restraints. *Journal of the Institution of Structural Engineers*, **61B** (1), Part B Quarterly, London, pp. 17–26.

6. Lui EM, Chen WF. (1983) End restraint and column design using LRFD. *Engineering Journal*, **20** (1): 29–39. https://www.aisc.org/End-Restraint-and-Column-Design-Using-LRFD

7. Mizuno E, Chen WF. (1983) Plasticity analysis of slope with different flow rules. *Computers and Structures*, **17** (3): 375–388. https://doi.org/10.1016/0045-7949(83)90130-X

8. Mizuno E, Chen WF. (1983) Cap models for clay strata to footing loads. *Computers and Structures*, **17** (4): 511–528. https://doi.org/10.1016/0045-7949(83)90046-9

9. Toma S, Chen WF. (1983) Post-buckling behavior of tubular beam-columns. *Journal of Structural Engineering*, **109** (8): 918–1932. https://doi.org/10.1061/(ASCE)0733-9445(1983)109:8(1918)

10. Chen WF. (1983) A rapid method of computing geometric relations in structural analysis. *Civil Engineering*, **53** (5): 51–53.

11. Suzuki H, Chen WF. (1983) Elastic-plastic-fracture analysis of concrete structures. *Computers and Structures*, **16** (6): 697–705. https://doi.org/10.1016/0045-7949(83)90061-5

12. Toma S, Chen WF. (1983) Cyclic inelastic analysis of tubular column sections. *Computers and Structures*, **16** (6): 707–716. https://doi.org/10.1016/0045-7949(83)90062-7

1984

1. Mizuno E, Chen WF. (1984) Plasticity models for seismic analyses of slopes. *Journal of Soil Dynamics and Earthquake Engineering*, **3** (1): 2–7. https://doi.org/10.1016/0261-7277(84)90021-4

2. Lui EM, Chen WF. (1984) Simplified approach to the analysis and design of columns with imperfections. *Engineering Journal*, **21** (2): 99–117. https://www.aisc.org/Simplified-Approach-to-the-Analysis-and-Design-of-Columns-with-Imperfections

3. Toussi S, Yao JTP, Chen WF. (1984) A damage indicator for reinforced concrete frames. *ACI Journal*, **81** (3): 260–267. DOI: 10.14359/10682 https://www.concrete.org/publications/internationalconcreteabstractsportal. aspx?m=details&id=10682

4. Liu XL, Chen WF. (1984) Reinforced concrete pipe columns: Behavior and design. *Journal of Structural Engineering*, **110** (6): 1356–1373. https://doi.org/10.1061/(ASCE)0733-9445(1984)110:6(1356)

5. Liu XL, Chen WF. (1984) Reinforced concrete centrifugal pipe columns. *Journal of Structural Engineering*, **110** (7): 1665–1678. https://doi.org/10.1061/(ASCE)0733-9445(1984)110:7(1665)

6. Chang CJ, Chen WF, Yao JTP. (1984) Seismic displacements in slopes by limit analysis. *Journal of Geotechnical Engineering*, **110** (7): 860–874. https://doi.org/10.1061/(ASCE)0733-9410(1984)110:7(860)

7. Sohal IS, Chen WF. (1984) Moment-curvature expressions for fabricated tubes. *Journal of Structural Engineering*, **110** (11): 2738–2757. https://doi.org/10.1061/(ASCE)0733-9445(1984)110:11(2738)

8. Yener M, Chen WF. (1984) On in-place strength of concrete and pullout tests. *Journal of Cement, Concrete and Aggregates*, **6** (2): 90–99. https://doi.org/10.1520/CCA10361J

9. Patel KV, Chen WF. (1984) Nonlinear analysis of steel moment connections. *Journal of Structural Engineering*, **110** (8): 1861–1874. https://doi.org/10.1061/(ASCE)0733-9445(1984)110:8(1861)

1985

1. Chen WF, Lui EM. (1985) Stability design criteria for steel members and frames in the United States. *Journal of Constructional Steel Research*, **5** (1): 31–74. https://doi.org/10.1016/0143-974X(85)90019-7

2. Liu XL, Chen WF, Bowman MD. (1985) Construction load analysis for concrete structures. *Journal of Structural Engineering*, **111** (5): 1019–1036. https://doi.org/10.1061/(ASCE)0733-9445(1985)111:5(1019)

3. Chen WF, Sugimoto H. (1985) Moment-curvature-axial-compression-pressure relationship of structural tubes. *Journal of Constructional Steel Research*, **5** (4): 247–264. https://doi.org/10.1016/0143-974X(85)90023-9

4. Sohal IS, Chen WF. (1985) Large bending of pipes. *Engineering Structures*, **7** (2): 121–130. https://doi.org/10.1016/0141-0296(85)90022-7

5. Patel KV, Chen WF. (1985) Analysis of a fully bolted moment connection using NONSAP. *Computers and Structures*, **21** (3): 505–511. https://doi.org/10.1016/0045-7949(85)90129-4

6. Sugimoto H, Chen WF. (1985) Inelastic post-buckling behavior of tubular members. *Journal of Structural Engineering*, **111** (9): 1965–1978. https://doi.org/10.1061/(ASCE)0733-9445(1985)111:9(1965)

7. Zhou SP, Chen WF. (1985) Design criteria for box-columns under biaxial loading. *Journal of Structural Engineering*, **111** (12): 2643–2658. https://doi.org/10.1061/(ASCE)0733-9445(1985)111:12(2643)

8. Chen WF, Lui EM. (1985) Columns with end restraint and bending in load and resistance factor design. *Engineering Journal*, **22** (3): 105–132. https://www.aisc.org/Columns-with-End-Restraint-and-Bending-in-Load-and-Resistance-Factor-Design

9. Yener M, Chen WF. (1985) Evaluation of in-place flexural strength of concrete. *ACI Journal*, **82** (6): 788-796. https://www.concrete.org/publications/internationalconcreteabstractsportal.aspx?m=details&id=10389

10. Han DJ, Chen WF. (1985) A nonuniform hardening plasticity model for concrete materials. *Mechanics of Materials*, **4** (4): 1–20. https://doi.org/10.1016/0167-6636(85)90025-0

11. Liu XL, Chen WF, Bowman MD. (1985) Construction loads on supporting floors. *Concrete International*, **7** (12): 21–26. https://www.concrete.org/publications/internationalconcreteabstractsportal.aspx?m=details&id=9369

1986

1. Liu XL, Chen WF, Bowman MD. (1986) Shore-slab interaction in concrete buildings. *Journal of Construction Engineering and Management*, **112** (2): 227–244. https://doi.org/10.1061/(ASCE)0733-9364(1986)112:2(227)

2. Han DJ, Chen WF. (1986) Strain-space plasticity formulation for hardening-softening materials with elastoplastic coupling. *International Journal of Solids and Structures*, **22** (8): 935–950. https://doi.org/10.1016/0020-7683(86)90072-7

3. Lui EM, Chen WF. (1986) Analysis and behavior of flexibly-jointed frames. *Engineering Structures*, **8** (2): 107–118. https://doi.org/10.1016/0141-0296(86)90026-X

4. Chen WF, Lui EM. (1986) Steel beam-to-column moment connections — Part I: Flange moment connections. *Solid Mechanics Archives*, **11** (4): 257–316.

5. Lui EM, Chen WF. (1986) Frame analysis with panel zone deformation. *International Journal of Solids and Structures*, **22** (12): 1599–1627. https://doi.org/10.1016/0020-7683(86)90065-X

1987

1. Zhang XJ, Chen WF. (1987) Stability analysis of slopes with general nonlinear failure criterion, *International Journal for Numerical and Analytical Methods in Geomechanics*, **11** (1): 33–50. https://doi.org/10.1002/nag.1610110104

2. Han DJ, Chen WF. (1987) Constitutive modeling in analysis of concrete structures. *Journal of Engineering Mechanics*, **113** (4): 577–593. https://doi.org/10.1061/(ASCE)0733-9399(1987)113:4(577)

3. McCarron WO, Chen WF. (1987) Application of a bounding surface model to Boston blue clay. *Computers and Structures*, **26** (6): 887–897. https://doi.org/10.1016/0045-7949(87)90105-2

4. Larralde J, Chen WF. (1987) Estimation of mechanical deterioration of highway rigid pavements. *Journal of Transportation Engineering*, **113** (2): 193–208. https://doi.org/10.1061/(ASCE)0733-947X(1987)113:2(193)

5. Liu XL, Chen WF. (1987) Probability distribution of maximum wooden shore loads in multistory R.C. buildings. *Journal of Structural Safety*, **4** (3): 197–215. https://doi.org/10.1016/0167-4730(87)90013-0

6. Liu XL, Chen WF. (1987) Effect of creep on load distribution in multistory reinforced concrete buildings during construction. *ACI Structural Journal*, **84** (3): 192–200. DOI: 10.14359/2629
https://www.concrete.org/publications/internationalconcreteabstractsportal.aspx?m=details&id=2629

7. Chen WF, Zhou SP. (1987) Inelastic analysis of steel braced frames with flexible joints. *International Journal of Solids and Structures*, **23** (5): 631–649. https://doi.org/10.1016/0020-7683(87)90023-0

8. Chen WF, Zhou SP. (1987) Design of beam-columns using allowable stress design and load and resistance factor design. *Engineering Structures*, **9** (3): 201–209. https://doi.org/10.1016/0141-0296(87)90016-2

9. Sohal IS, Chen WF. (1987) Local buckling and sectional behavior of fabricated tubes. *Journal of Structural Engineering*, **113** (3): 519–533. https://doi.org/10.1061/(ASCE)0733-9445(1987)113:3(519)

10. Lui EM, Chen WF. (1987) Steel frame analysis with flexible joints. *Journal of Constructional Steel Research*, **8**: 161–202. https://doi.org/10.1016/0143-974X(87)90058-7

11. Goto Y, Chen WF. (1987) On the computer-based design analysis for flexibly jointed frames. *Journal of Constructional Steel Research*, **8**: 203–231. https://doi.org/10.1016/0143-974X(87)90059-9

12. Chen WF, Lui EM. (1987) Effects of joint flexibility on the behavior of steel frames, *Computers and Structures*, **26** (5): 719-732. https://doi.org/10.1016/0045-7949(87)90021-6

13. Chen WF, Lui EM. (1987) Steel beam-to-column connections part II — Web moment connections, *Solid Mechanics Archives*, **12** (1): 327–378.

14. Chang TY, Taniguchi H, Chen WF. (1987) Nonlinear finite element analysis of reinforced concrete panels. *Journal of Structural Engineering*, **113** (1): 122–140. https://doi.org/10.1061/(ASCE)0733-9445(1987)113:1(122)

15. Chen WF, Zhou SP. (1987) C_m factor in load and resistance factor design. *Journal of Structural Engineering*, **113** (8): 1738–1754. https://doi.org/10.1061/(ASCE)0733-9445(1987)113:8(1738)

16. Goto Y, Chen WF. (1987) Second-order elastic analysis for frame design. *Journal of Structural Engineering*, **113** (7): 1501–1519. https://doi.org/10.1061/(ASCE)0733-9445(1987)113:7(1501) (Closure, Vol. 115, No. 2, 1989, 503–506).

17. McCarron WO, Chen WF. (1987) A capped plasticity model applied to Boston blue clay. *Canadian Geotechnical Journal*, **24** (4): 630–644. https://doi.org/10.1139/t87-077

18. Ohtani YC, Chen WF. (1987) Hypoelastic-perfectly plastic model for concrete materials. *Journal of Engineering Mechanics*, **113** (12): 1840–1860. https://doi.org/10.1061/(ASCE)0733-9399(1987)113:12(1840)

19. Sohal IS, Chen WF. (1987) Local buckling and inelastic cyclic behavior of tubular members. *Thin-Walled Structures*, **5** (6): 455–475. https://doi.org/10.1016/0263-8231(87)90033-4

20. Chen WF, Sugimoto H. (1987) Analysis of tubular beam-columns and frames under reversed loading. *Engineering Structures*, **9** (4): 233–242. https://doi.org/10.1016/0141-0296(87)90022-8

1988

1. Chen WF, Lui EM. (1988) Static flange moment connections. *Journal of Constructional Steel Research*, **10**: 39–88. https://doi.org/10.1016/0143-974X(88)90027-2

2. Chen WF, Lui EM. (1988) Static web moment connections. *Journal of Constructional Steel Research*, **10**: 89–131. https://doi.org/10.1016/0143-974X(88)90028-4

3. Nethercot DA, Chen WF. (1988) Effects of connections on columns. *Journal of Constructional Steel Research*, **10**: 201–239. https://doi.org/10.1016/0143-974X(88)90031-4

4. Sohal IS, Chen WF. (1988) Local and post-buckling behavior of tubular beam-columns. *Journal of Structural Engineering*, **114** (5): 1073–1090. https://doi.org/10.1061/(ASCE)0733-9445(1988)114:5(1073)

5. Kato B, Chen WF, Nakao M. (1988) Effects of joint-panel shear deformation on frames. *Journal of Constructional Steel Research*, **10**: 269–320. https://doi.org/10.1016/0143-974X(88)90033-8

6. Liu XL, Lee HM, Chen WF. (1988) Analysis of construction loads on slabs and shores by personal computer. *Concrete International*, **10** (6): 21–30. https://www.concrete.org/publications/internationalconcreteabstractsportal. aspx?m=details&id=3553

7. Chen WF. (1988) Evaluation of plasticity-based constitutive models for concrete materials. *Solid Mechanics Archives*, **13** (1): 1–63.

8. Hsieh SS, Ting EC, Chen WF. (1988) Applications of a plastic-fracture model to concrete structures. *Computers and Structures*, **28** (3): 373–393. https://doi.org/10.1016/0045-7949(88)90077-6

9. El-Metwally SE, Chen WF. (1988) Moment-rotation modeling of reinforced concrete beam-column connections. *ACI Structural Journal*, **85** (4): 384–394. DOI:10.14359/2656
    https://www.concrete.org/publications/internationalconcreteabstractsportal. aspx?m=details&id=2656

10. Sohal IS, Chen WF. (1988) Local buckling and inelastic cyclic behavior of tubular sections. *Thin-Walled Structures*, **6** (1): 63–80. https://doi.org/10.1016/0263-8231(88)90026-2

11. Chen WF, Sohal IS. (1988) Cylindrical members in offshore structures. *Thin-Walled Structures*, **6** (3): 153–285. https://doi.org/10.1016/0263-8231(88)90010-9

12. Duan L, Chen WF. (1988) Effective length factor for columns in braced frames. *Journal of Structural Engineering*, **114** (10): 2357–2370. https://doi.org/10.1061/(ASCE)0733-9445(1988)114:10(2357)

13. El-Sheikh M, Chen WF. (1988) Effects of fast construction rate on deflections of R.C. buildings. *Journal of Structural Engineering*, **114** (10): 2225–2238. https://doi.org/10.1061/(ASCE)0733-9445(1988)114:10(2225)

14. Lui EM, Chen WF. (1988) Behavior of braced and unbraced semi-rigid frames. *International Journal of Solids and Structures*, **24** (9): 893–913. https://doi.org/10.1016/0020-7683(88)90040-6

15. Duan L, Chen WF. (1988) Design rules of built-up members in load and resistance factor design. *Journal of Structural Engineering*, **114** (11): 2544–2554. https://doi.org/10.1061/(ASCE)0733-9445(1988)114:11(2544)
    (Closure, Vol. 117, No. 1, 1991, 303–304) https://doi.org/10.1061/(ASCE)0733-9445(1991)117:1(303.2)

16. Ohtani Y, Chen WF. (1988) Multiple hardening plasticity for concrete materials. *Journal of Engineering Mechanics*, **114** (11): 1890–1910.
https://doi.org/10.1061/(ASCE)0733-9399(1988)114:11(1890)

1989

1. Sohal IS, Duan L, Chen WF. (1989) Design interaction equations for steel members. *Journal of Structural Engineering*, **115** (7): 1650–1665.
https://doi.org/10.1061/(ASCE)0733-9445(1989)115:7(1650)
(Closure, Vol. 117, No. 7, 1991, 2193–2196)
https://doi.org/10.1061/(ASCE)0733-9445(1991)117:7(2193.2)
2. Duan L, Sohal IS, Chen WF. (1980) On beam-column moment amplification factor. *Engineering Journal*, **26** (4): 130–135.
https://www.aisc.org/On-Beam-Column-Moment-Amplification-Factor
3. El-Sheikh M, Chen WF. (1989) Maximum probabilistic shore load in multistory R/C buildings. *Computers and Structures*, **32** (6): 1347–1357.
https://doi.org/10.1016/0045-7949(89)90311-8
4. Duan L, Chen WF. (1989) Effective length factor for columns in unbraced frames. *Journal of Structural Engineering*, **115** (1): 149–165.
https://doi.org/10.1061/(ASCE)0733-9445(1989)115:1(149)
Errata, Vol. 122, No. 2, 1996, 224–225.
https://doi.org/10.1061/(ASCE)0733-9445(1996)122:2(224)
5. Chen WF, Kishi N. (1989) Semirigid steel beam-to-column connections: Data base and modeling. *Journal of Structural Engineering*, **115** (1): 105–119.
https://doi.org/10.1061/(ASCE)0733-9445(1989)115:1(105)
6. Liu XL, Lee HM, Chen WF. (1989) Shoring and reshoring of high-rise buildings. *Concrete International*, **11** (1): 64–68.
https://www.concrete.org/publications/internationalconcreteabstractsportal.aspx?m=details&id=3371
7. Goto Y, Chen WF. (1989) On the validity of Wagner hypothesis. *International Journal of Solids and Structures*, **25** (6): 621–634.
https://doi.org/10.1016/0020-7683(89)90029-2
8. Duan L, Chen WF. (1989) Design interaction equation for steel beam-columns. *Journal of Structural Engineering*, **115** (5): 1225–1243.
https://doi.org/10.1061/(ASCE)0733-9445(1989)115:5(1225)
(Closure, Vol. 117, No. 8, 1991, 2554–2561)
https://doi.org/10.1061/(ASCE)0733-9445(1991)117:8(2554.2)

9. Van Wijk AJ, Larralde J, Lovell CW, Chen WF. (1989) Pumping predicting model for highway pavements. *Journal of Transportation Engineering*, **115** (2): 161–175.
    https://doi.org/10.1061/(ASCE)0733-947X(1989)115:2(161)

10. Duan L, Wang FM, Chen WF. (1989) Flexural rigidity of reinforced concrete members. *ACI Structural Journal*, **86** (4): 419–427 (Closure, Vol. 87, No. 3, 1990, 364–365). DOI: 10.14359/9228
    https://www.concrete.org/publications/internationalconcreteabstractsportal.
    aspx?m=details&id=9228

11. Chen WF, Bubenik TA, Sohal IS. (1989) Effect of pressure on tubular beam-column capacity. *Journal of Constructional Steel Research*, **13** (1): 23–42.
    https://doi.org/10.1016/0143-974X(89)90003-5

12. El-Metwally SE, Chen WF. (1989) Nonlinear behavior of R/C frames. *Computers and Structures*, **32** (6): 1203–1209.
    https://doi.org/10.1016/0045-7949(89)90297-6

13. Ohtani Y, Chen WF. (1989) A plastic-softening model for concrete materials. *Computers and Structures*, **33** (4): 1047–1055.
    https://doi.org/10.1016/0045-7949(89)90440-9

14. Kishi N, Matsuoka K, Nomachi S, Chen WF. (1989) Formulation of initial connection stiffness and ultimate moment capacity of steel beam-to-column angle connections. *Journal of Structural Engineering*, **35A**: 97–105.

15. Kishi N, Matsuoka K, Nomachi S, Chen WF. (1989) Data base of steel beam-to-column connection tests and its applications. *Journal of Structural Engineering*, **35A**: 75–82.

16. Liu XL, Wang QY, Chen WF. (1989) Layered analysis with generalized failure criterion. *Computers and Structures*, **33** (5): 1117–1124.
    https://doi.org/10.1016/0045-7949(89)90448-3

17. El-Metwally SE, Chen WF. (1989) Load-deformation relations for reinforced concrete sections. *ACI Structural Journal*, **86** (2): 163–167 (Closure, Vol. 87, No. 1, 1990, 117–118). DOI: 14359/2662
    https://www.concrete.org/publications/internationalconcreteabstractsportal.
    aspx?m=details&id=2662

1990

1. Huang TK, Chen WF. (1990) Simple procedure for determining cap-plasticity-model parameters. *Journal of Geotechnical Engineering*, **116** (3): 492–513.
    https://doi.org/10.1061/(ASCE)0733-9410(1990)116:3(492)

2. Kishi N, Chen WF. (1990) Moment-rotation relations of semirigid connections with angles. *Journal of Structural Engineering*, **116** (7): 1813–1834. https://doi.org/10.1061/(ASCE)0733-9445(1990)116:7(1813)

3. Duan L, Chen WF. (1990) Design interaction equations for cylindrical tubular beam-columns. *Journal of Structural Engineering*, **116** (7): 1794–1812. https://doi.org/10.1061/(ASCE)0733-9445(1990)116:7(1794)

4. Barakat M, Chen WF. (1990) Practical analysis of semi-rigid frames. *Engineering Journal*, **27** (2): 54–68.
   https://www.aisc.org/Practical-Analysis-of-Semi-Rigid-Frames

5. Zhou SP, Duan L, Chen WF. (1990) Comparison of design equations for steel beam-columns. *Structural Engineering Review*, **2** (1): 45–53.

6. Al-Mashary F, Chen WF. (1990) Elastic second-order analysis for frame design. *Journal of Constructional Steel Research*, **15** (4): 303–322. https://doi.org/10.1016/0143-974X(90)90052-I

7. Mosallam K, Chen WF. (1990) Design considerations for formwork in multistory concrete buildings. *Engineering Structures*, **12** (3): 163–172. https://doi.org/10.1016/0141-0296(90)90003-B
   See also *Construction and Building Materials*, Vol. 6, No. 1, 1992, pp. 23–30, reprinted from *Engineering Structures*.

8. Yamaguchi E, Chen WF. (1990) Cracking model for finite element analysis of concrete materials. *Journal of Engineering Mechanics*, **116** (6): 1242–1260. https://doi.org/10.1061/(ASCE)0733-9399(1990)116:6(1242)

9. Wu FH, Chen WF. (1990) A design model for semi-rigid connections. *Engineering Structures*, **12** (2): 88–97. https://doi.org/10.1016/0141-0296(90)90013-I

10. El-Metwally SE, Ashour AF, Chen WF. (1990) Instability analysis of eccentrically loaded concrete walls. *Journal of Structural Engineering*, **116** (10): 2853–2872. https://doi.org/10.1061/(ASCE)0733-9445(1990)116:10(2862)

11. Duan L, Chen WF. (1990) A yield surface equation for doubly symmetrical sections. *Engineering Structures*, **12** (2): 114–119. https://doi.org/10.1016/0141-0296(90)90016-L

12. Yamaguchi E, Chen WF. (1990) Post-failure behavior of concrete in compression. *Engineering Fracture Mechanics*, **37** (5): 1011–1023. https://doi.org/10.1016/0013-7944(90)90024-B

13. Chen WF, Duan L, Zhou SP. (1990) Second-order inelastic analysis of braced portal frames: Evaluation of design formulae in LRFD and GBJ specifications. *Steel Structures, Journal of Singapore Structural Steel Society*, **1** (1): 5–15.

14. El-Metwally SE, El-Shahhat AM, Chen WF. (1990) 3-D nonlinear analysis of R/C slender columns. *Computers and Structures*, **37** (5): 863–872. https://doi.org/10.1016/0045-7949(90)90114-H

15. Chen WF, White DW, Zhang H. (1990) Preparing structural engineering research and education for the 21st century. *Journal of the Chinese Institute of Civil and Hydraulic Engineering*, **2** (2): 95–106.

1991

1. Mosallam KH, Chen WF. (1991) Determining shoring loads for reinforced concrete construction. *ACI Structural Journal*, **88** (3): 340–350.
DOI: 10.14359/2733
https://www.concrete.org/publications/internationalconcreteabstractsportal.aspx?m=details&id=2733

2. Goto Y, Suzuki S, Chen WF. (1991) Bowing effect on elastic stability of frames under primary bending moments. *Journal of Structural Engineering*, **117** (1): 111–127. https://doi.org/10.1061/(ASCE)0733-9445(1991)117:1(111)

3. Yamaguchi E, Chen WF. (1991) Microcrack propagation study of concrete under compression. *Journal of Engineering Mechanics*, **117** (3): 653–673.
https://doi.org/10.1061/(ASCE)0733-9399(1991)117:3(653)

4. Cai CS, Liu XL, Chen WF. (1991) Further verifications of beam-column strength equations. *Journal of Structural Engineering*, **117** (2): 501–513.
https://doi.org/10.1061/(ASCE)0733-9445(1991)117:2(501)

5. Chen WF, Yamaguchi E, Zhang H. (1991) On the loading criteria in the theory of plasticity. *Computers and Structures*, **39** (6): 679–683.
https://doi.org/10.1016/0045-7949(91)90210-D

6. Barakat M, Chen WF. (1991) Design analysis of semi-rigid frames: Evaluation and implementation. *Engineering Journal*, **28** (2): 55–64.
https://www.aisc.org/Design-Analysis-of-Semi-Rigid-Frames-Evaluation-and-Implementation

7. El-Metwally SE, Ashour AF, Chen WF. (1991) Behavior and strength of concrete masonry walls. *ACI Structural Journal*, **88** (1): 42–48. DOI:10.14359/3095
https://www.concrete.org/publications/internationalconcreteabstractsportal.aspx?m=details&id=3095

8. Liew JYR, White DW, Chen WF. (1991) Beam-column design in steel framework — Insights and current methods and trends. *Journal of Constructional Steel Research*, **18** (4): 269–308.
https://doi.org/10.1016/0143-974X(91)90009-P

9. Goto Y, Suzuki S, Chen WF. (1991) Analysis of critical behavior of semi-rigid frames with or without load history in connections. *Solids and Structures*, **27** (4): 467–483. https://doi.org/10.1016/0020-7683(91)90135-3

10. Lee HM, Liu XL, Chen WF. (1991) Creep analysis of concrete buildings during construction. *Journal of Structural Engineering*, **117** (10): 3135–3148. https://doi.org/10.1061/(ASCE)0733-9445(1991)117:10(3135)

11. Al-Mashary F, Chen WF. (1991) Simplified second-order inelastic analysis for steel frames. *The Structural Engineers*, **69** (23): 395–399.

12. Liew JYR, Chen WF. (1991) Refining the plastic hinge concept for advanced analysis/design of steel frames. *Steel Structures, Journal of Singapore Structural Steel Society*, **2** (1): 13–28.

1992

1. Zeng, JM, Duan L, Wang FM, Chen WF. (1992) Flexural rigidity of reinforced concrete columns. *ACI Structural Journal*, **89** (2): 150–158. DOI: 10.14359/2926 https://www.concrete.org/publications/internationalconcreteabstractsportal. aspx?m=details&id=2926

2. King WS, White DW, Chen WF. (1992) On second-order inelastic analysis methods for steel frame design. *Journal of Structural Engineering*, **118** (2): 408–428. https://doi.org/10.1061/(ASCE)0733-9445(1992)118:2(408)

3. Lee HM, Chen WF, Liu XL. (1992) Construction live load caused by powered buggies. *Concrete International*, **14** (1): 47–51. https://www.concrete.org/publications/internationalconcreteabstractsportal. aspx?m=details&id=1316

4. Chen WF. (1992) Design of beam-columns in steel frames in the United States. *Thin-Walled Structures*, **13** (1–2): 1–83. https://doi.org/10.1016/0263-8231(92)90003-F

5. King WS, White DW, Chen WF. (1992) A modified plastic hinge method for second-order inelastic analysis of rigid frames. *Structural Engineering Review*, **4** (1): 31–41.

6. Toma S, Chen WF. (1992) European calibration frames for second-order inelastic analysis. *Engineering Structures*, **14** (1): 7–14. https://doi.org/10.1016/0141-0296(92)90003-9

7. El-Shahhat AM, Chen WF. (1992) Improved analysis of shore-slab interaction. *ACI Structural Journal*, **89** (5): 528–537. DOI: 10.14359/2969 https://www.concrete.org/publications/internationalconcreteabstractsportal. aspx?m=details&id=2969

8. Huang TK, Chen WF, Chameau JL (1992) Application of cap-plasticity model to embankment problems. *Computers and Structures*, **44** (6): 1349–1370. https://doi.org/10.1016/0045-7949(92)90377-C

9. Mosallam K, Chen WF. (1992) Construction load distributions for laterally braced formwork. *ACI Structural Journal*, **89** (4): 415–424. DOI: 10.14359/3025 https://www.concrete.org/publications/internationalconcreteabstractsportal. aspx?m=details&id=3025

10. Sotelino ED, White DW, Chen WF. (1992) Domain-specific object-oriented environment for parallel computing. *Steel Structures, Journal of Singapore Structural Steel Society*, **3** (1): 47–60.

1993

1. Duan L, Chen WF, Loh JT. (1993) Analysis of dented tubular members using moment curvature approach. *Thin-Walled Structures*, **15** (1): 15–41. https://doi.org/10.1016/0263-8231(93)90011-X

2. Abdel-Ghaffar M, White DW, Chen WF. (1993) An error estimate and step size control method for nonlinear solution techniques. *Finite Elements in Analysis and Design*, **13** (2–3): 137–148. https://doi.org/10.1016/0168-874X(93)90053-S

3. King WS, Duan L, Zhou RG, Hu YX, Chen WF. (1993) K-factors of framed columns restrained by tapered girders in US codes. *Engineering Structures*, **15** (5): 369–378. https://doi.org/10.1016/0141-0296(93)90040-B

4. Zhou RG, Hu YX, Duan L, Chen WF. (1993) An approximate solution of creep of orthogonal anisotropic concrete thin plates. *Engineering Structures*, **15** (1): 61–66. https://doi.org/10.1016/0141-0296(93)90018-Y

5. Duan L, Loh JT, Chen WF. (1993) Moment-curvature relationships for dented tubular sections. *Journal of Structural Engineering*, **119** (3): 809–830. https://doi.org/10.1061/(ASCE)0733-9445(1993)119:3(809) (Closure, Vol. 120, No. 7, 1994, 2255–2257) https://doi.org/10.1061/(ASCE)0733-9445(1994)120:7(2256)

6. Goto Y, Suzuki S, Chen WF. (1993) Stability behavior of semi-rigid sway frames. *Engineering Structures*, **15** (3): 209–219. https://doi.org/10.1016/0141-0296(93)90055-9

7. Sawada T, Chen WF, Nomachi SG. (1993) Assessment of seismic displacements of slopes. *Soil Dynamics and Earthquake Engineering*, **12** (6): 357–362. https://doi.org/10.1016/0267-7261(93)90038-S

8. Hu YX, Zhou RG, King WS, Duan L, Chen WF. (1993) On effective length factor of frame columns in ACI code. *ACI Structural Journal*, **90** (2): 135–143.

DOI: 10.14359/4203

https://www.concrete.org/publications/internationalconcreteabstractsportal.
aspx?m=details&id=4203

9. Kishi N, Chen WF, Goto Y, Matsuoka KG. (1993) Design aid of semi-rigid connections for frame analysis. *Engineering Journal*, **30** (3): 90–107.
    https://www.aisc.org/Design-Aid-of-Semi-rigid-Connections-for-Frame-Analysis

10. Chen WF. (1993) Concrete plasticity: Macro and micro approaches. *International Journal of Mechanical Sciences*, **35** (12): 1097–1109.
    https://doi.org/10.1016/0020-7403(93)90058-3

11. Duan L, King WS, Chen WF. (1993) K-factor equation to alignment charts for column design. *ACI Structural Journal*, **90** (3): 242–248. DOI: 10.14359/4232
    https://www.concrete.org/publications/internationalconcreteabstractsportal.
    aspx?m=details&id=4232

12. El-Shahhat A, Chen WF. (1993) Deflection-based analysis for concrete buildings during construction. *Structural Engineering Review*, **5** (4): 285–299.

13. El-Shahhat A, Rosowsky DV, Chen WF. (1993) Construction safety of multistory concrete buildings. *ACI Structural Journal*, **90** (4): 335–341.
    DOI: 10.14359/3974
    https://www.concrete.org/publications/internationalconcreteabstractsportal.
    aspx?m=details&id=3974

14. Liew JYR, White DW, Chen WF. (1993) Second-order refined plastic-hinge analysis for frame design part I. *Journal of Structural Engineering*, **119** (11): 3196–3216. https://doi.org/10.1061/(ASCE)0733-9445(1993)119:11(3196)

15. Liew JYR, White DW, Chen WF. (1993) Second-order refined plastic-hinge analysis for frame design part II. *Journal of Structural Engineering*, **119** (11): 3217–3237. https://doi.org/10.1061/(ASCE)0733-9445(1993)119:11(3217)

16. King WS, Chen WF. (1993) LRFD analysis for semi-rigid frame design. *Engineering Journal*, **30** (4): 130–140.
    https://www.aisc.org/LRFD-Analysis-for-Semi-Rigid-Frame-Design

17. Kishi N, Chen WF, Goto Y, Matsuoka K. (1993) Analysis program for the design of flexibly jointed frames. *Computers and Structures*, **49** (4): 705–713.
    https://doi.org/10.1016/0045-7949(93)90073-M

18. Liew JYR, White DW, Chen WF. (1993) Limit states design of semi-rigid frames using advanced analysis part 1: Connection modeling and classification. *Journal of Constructional Steel Research*, **26** (1): 1–27.
    https://doi.org/10.1016/0143-974X(93)90065-Z

19. Liew JYR, White DW, Chen WF. (1993) Limit states design of semi-rigid frames using advanced analysis part 2: Analysis and design. *Journal of Constructional Steel Research*, **26** (1): 29–57. https://doi.org/10.1016/0143-974X(93)90066-2

1994

1. Resheidat M, Ghanma M, Numayr K, Sutton C, Chen WF. (1994) Improved EI estimate for reinforced concrete circular columns. *Materials and Structures*, **27** (9): 515–526. https://doi.org/10.1007/BF02473212

2. King WS, Chen WF. (1994) Practical second-order inelastic analysis of semirigid frames. *Journal of Structural Engineering*, **120** (7): 2156–2175. https://doi.org/10.1061/(ASCE)0733-9445(1994)120:7(2156)

3. Toma S, Chen WF. (1994) Calibration frames for second-order inelastic analysis in Japan. *Journal of Constructional Steel Research*, **28** (1): 51–77. https://doi.org/10.1016/0143-974X(94)90033-7

4. Chan SL, Chen WF. (1994) Second-order inelastic analysis of steel frames by personal computers. *Journal of Structural Engineering*, **21** (2): 99–106.

5. Kishi N, Goto Y, Komuro M, Chen WF. (1994) Sway analysis of tall building frames with mixed use of rigid and semi-rigid connections. *Journal of Constructional Steel*, **2**: 1–8.

6. Liew JYR, White DW, Chen WF. (1994) Notational-load plastic-hinge method for frame design. *Journal of Structural Engineering*, **120** (5): 1434–1454. https://doi.org/10.1061/(ASCE)0733-9445(1994)120:5(1434)

7. Liew JYR, Chen WF. (1994) Implications of using refined plastic hinge analysis for load and resistance factor design. *Thin-Walled Structures*, **20** (4): 17–47. https://doi.org/10.1016/0263-8231(94)90054-X

8. Duan L, Chen WF, Loh J. (1994) Ultimate strength of damaged members. *International Journal of Offshore and Polar Engineering*, **4** (2): 127–133.

9. Rosowsky DV, Huang YL, Chen WF, Yen T. (1994) Modeling concrete placement loads during construction. *Structural Engineering Review*, **6** (2): 71–84.

10. Rosowsky DV, Huston D, Fuhr P, Chen WF. (1994) Measuring formwork loads during construction. *Concrete International*, **16** (11): 21–25. https://www.concrete.org/publications/internationalconcreteabstractsportal.aspx?m=details&id=1574

11. Cohen MD, Goldman A, Chen WF. (1994) The role of silica fume in mortar: Transition zone versus bulk paste modification. *Cement and Concrete Research*, **24** (1): 95–98. https://doi.org/10.1016/0008-8846(94)90089-2

12. Sawada T, Nomachi SG, Chen WF. (1994) Seismic bearing capacity of a mounded foundation near a down-hill slope by pseudo-static analysis. *Soils and Foundations*, **34** (1): 11–17. https://doi.org/10.3208/sandf1972.34.11

13. Chen WF. (1994) Engineering mechanics in structural engineering research and education. *Steel Structures, Journal of Singapore Structural Steel Society*, **5** (1): 85–93.

14. Kishi N, Hasan R, Chen WF, Goto Y. (1994) Power model for semi-rigid connections, steel structures. *Steel Structures, Journal of Singapore Structural Steel Society*, **5** (1): 37–48

15. El-Shahhat AM, Rosowsky DV, Chen WF. (1994) Partial factor design for reinforced concrete buildings during construction. *ACI Structural Journal*, **91** (4): 475–485. DOI: 10.14359/4207
    https://www.concrete.org/publications/internationalconcreteabstractsportal. aspx?m=details&id=4207

1995

1. Toma S, Chen WF, White DW. (1995) A selection of calibration frames in North America for second-order inelastic analysis. *Engineering Structures*, **17** (2): 104–112. https://doi.org/10.1016/0141-0296(95)92641-K

2. Abdalla KM, Chen WF. (1995) Expanded database of semi-rigid steel connections. *Computers and Structures*, **56** (4): 553–564.
   https://doi.org/10.1016/0045-7949(94)00558-K

3. Chen WF, Chan SL. (1995) Second-order inelastic analysis of steel frame using element with midspan and end springs. *Journal of Structural Engineering*, **121** (3): 530–541. https://doi.org/10.1061/(ASCE)0733-9445(1995)121:3(530)

4. Shanmugam NE, Chen WF. (1995) An assessment of K factor formulas. *Engineering Journal*, **32** (1): 3–11.
   https://www.aisc.org/An-Assessment-of-K-Factor-Formulas

5. Resheidat M, Ghanma M, Sutton C, Chen WF. (1995) Flexural rigidity of biaxially loaded reinforced concrete rectangular column sections. *Computers and Structures*, **55** (4): 601–614. https://doi.org/10.1016/0045-7949(94)00493-M

6. Lu J, White DW, Chen WF, Dunsmore HE. (1995) A matrix class library in C++ for structural engineering computing. *Computers and Structures*, **55** (1): 95–112. https://doi.org/10.1016/0045-7949(94)00421-X

7. Liew JYR, Chen WF. (1995) Analysis and design of steel frames considering panel joint deformation. *Journal of Structural Engineering*, **121** (10): 1531–1540. https://doi.org/10.1061/(ASCE)0733-9445(1995)121:10(1531)
   (Closure, Vol. 123, No. 3, 1997, 381–383.)
   https://doi.org/10.1061/(ASCE)0733-9445(1997)123:3(382)

8. Chan SL, Zhou ZH, Chen WF, Peng JL, Pan AD. (1995) Stability analysis of semirigid steel scaffolding. *Engineering Structures*, **17** (8): 568–574.
   https://doi.org/10.1016/0141-0296(95)00011-U

9. Kim SE, Chen WF. (1995) Practical advanced analysis for frame design-case study. *Steel Structures, Journal of Singapore Structural Steel Society*, **6** (1): 61–73.

10. Luan M, Lin G, Guo Y, Chen WF, Pan AD, Wang YK. (1995) Generalized sliding wedge method and its application to stability analysis in soil mechanics. *Chinese Journal of Geotechnical Engineering*, **17** (4): 1–9.

11. El-Shahhat AM, Chen WF. (1995) Toward a life cycle analysis of concrete structures. *International Journal for Engineering Analysis and Design*, **2** (3): 35–54.

12. Yen T, Peng JL, Lin IC, Chen WF, Pan AD. (1995) Why frequent scaffold failures during construction? *Construction News Record*, No. 155, Taiwan, pp. 32–43. (In Chinese).

1996

1. Peng JL, Rosowsky DV, Pan AD, Chen WF, Chan SL, Yen T. (1996) Analysis of concrete placement load effects using influence surfaces. *ACI Structural Journal*, **93** (2): 180–186. DOI: 10.14359/1485
   https://www.concrete.org/publications/internationalconcreteabstractsportal.
   aspx?m=details&id=1485

2. Zhao XH, Chen WF. (1996) Stress analysis of a S and particle with interface in cement paste under uniaxial loading. *International Journal for Numerical and Analytical Methods in Geomechanics*, **20** (4): 275–285.
   https://doi.org/10.1002/(SICI)1096-9853(199604)20:4<275::AID-
   NAG819>3.0.CO;2-Q

3. Huang YL, Yen T, Lin YC, Chen WF. (1996) Loading behavior of form supports during concrete placing. *Journal of the Chinese Institute of Civil and Hydraulic Engineering*, **8** (2): 281–285. (In Chinese).

4. Chen WF, Yamaguchi E. (1996) Spotlight on steel moment frames. *Civil Engineering*, **66** (3): 44–46.

5. Zhao XH, Chen WF. (1996) The influence of interface layer on microstructural stresses in mortar. *International Journal for Numerical and Analytical Methods in Geomechanics*, **20** (3): 215–222.
   https://doi.org/10.1002/(SICI)1096-9853(199603)20:3<215::AID-
   NAG818>3.0.CO;2-2

6. Ramesh G, Sotelino ED, Chen WF. (1996) Effect of transition zone on elastic moduli of concrete materials. *Cement and Concrete Research*, **26** (4): 611– 622.
   https://doi.org/10.1016/0008-8846(96)00016-6

7. Resheidat M, Numayr K, Sutton C, Chen WF. (1996) Improved EI estimation for RC rectangular columns. *Journal of Structural Engineering*, **23** (3): 151–157.

8. Abdalla KA, Alshegeir A, Chen WF. (1996) Analysis and design of mushroom slabs with a strut-tie model. *Computer and Structures*, **58** (2): 429–434. https://doi.org/10.1016/0045-7949(95)00135-4

9. Kim SE, Chen WF. (1996) Practical advanced analysis for semi-rigid frame design. *Engineering Journal*, **33** (4): 129–141. https://www.aisc.org/Practical-Advanced-Analysis-for-Semi-rigid-Frame-Design

10. Kishi N, Chen WF, Goto Y, Hasan R. (1996) Behavior of tall buildings with mixed use of rigid and semi-rigid connections. *Computers and Structures*, **61** (6): 1193–1206. https://doi.org/10.1016/0045-7949(96)00052-1

11. Resheidat M, Ghanma M, Sutton C, Chen WF. (1996) Flexural rigidity of biaxially loaded RC rectangular column sections. *Journal of Structural Engineering*, **22** (4): 201–210.

12. Yen T, Huang YL, Lin YC, Chi RC, Chen WF. (1996) On the resistance capacity of steel scaffolds. *Journal of the Chinese Institute of Civil and Hydraulic Engineering*, **8** (1): 33–43. (In Chinese).

13. Yen T, Huang YL, Chen WF, Pan AD, Chen SL. (1996) Application of steel scaffolds and their failure modes. *Construction New Record*, Taiwan, **158**: 13–22. (In Chinese).

14. Peng JL, Pan AD, Rosowsky DV, Chen WF, Yen T, Chan SL. (1996) High clearance scaffold systems during construction — I. Structural modeling and modes of failure. *Engineering Structures*, **18** (3): 247–257. https://doi.org/10.1016/0141-0296(95)00144-1

15. Peng JL, Pan AD, Rosowsky DV, Chen WF, Yen T, Chan SL. (1996) High clearance scaffold systems during construction — II. Structural analysis and development of design guidelines. *Engineering Structures*, **18** (3): 258–267. https://doi.org/10.1016/0141-0296(95)00145-X

16. Duan MZ, Chen WF. (1996) Design guidelines for safe concrete construction. *Concrete International*, **18** (10): 44–49. https://www.concrete.org/publications/internationalconcreteabstractsportal. aspx?m=details&id=9426

17. Kim SE, Chen WF. (1996) Practical advanced analysis for braced steel frame design. *Journal of Structural Engineering*, **122** (11): 1266–1274. https://doi.org/10.1061/(ASCE)0733-9445(1996)122:11(1266)

18. Kim SE, Chen WF. (1996) Practical advanced analysis for unbraced steel frame design. *Journal of Structural Engineering*, **122** (11): 1259–1265. https://doi.org/10.1061/(ASCE)0733-9445(1996)122:11(1259)

19. Lan YM, Chen WF. (1996) Applications of advanced composite materials to civil substructures. *Journal of Civil Engineering Technology*, pp. 107–118. (In Chinese).

20. Kim SE, Chen WF. (1996) Reduced tangent modulus plastic-hinge method for steel structure design. *International Journal of Steel Structures*, **8** (1): 145–154.

21. Yen T, Peng JL, Chen WF, Pan AD, Chan SL. (1996) Applications of steel scaffolds and investigation of failure modes of formwork supports. *Construction News Record*, No. 158, Taiwan, pp. 13–22. (In Chinese).

22. Yen T, Peng JL, Chen WF, Pan AD, Rosowsky DV. (1996) Effects on safety of formwork supports due to concrete placement and load path. *Construction News Record*, No. 161, Taiwan, pp. 18–32. (In Chinese).

1997

1. Kishi N, Hasan R, Chen WF, Goto Y. (1997) Study of Eurocode 3 steel connection classification. *Engineering Structures*, **19** (9): 772–779. https://doi.org/10.1016/S0141-0296(96)00151-4

2. Kishi N, Chen WF, Goto Y. (1997) Effective length factor of columns in semi-rigid and unbraced frames. *Journal of Structural Engineering*, **123** (3): 313–320. https://doi.org/10.1061/(ASCE)0733-9445(1997)123:3(313) (Closure, Vol. 124, No. 10, 1998, pp. 1230–1231) https://doi.org/10.1061/(ASCE)0733-9445(1998)124:10(1230.2)

3. Peng JL, Yen T, Lin Y, Wu KL, Chen WF. (1997) Performance of scaffold frame shoring under pattern loads and load paths. *Journal of Construction Engineering and Management*, **123** (2): 138–145. https://doi.org/10.1061/(ASCE)0733-9364(1997)123:2(138)

4. Joh C, Chen WF. (1997) Application of fracture mechanics to steel connections in moment frames under seismic loading. *Advances in Structural Engineering*, **1** (1): 23–37. https://doi.org/10.1177/136943329700100104

5. Kim SE, Chen WF. (1997) Further studies of practical advanced analysis for weak-axis bending. *Engineering Structures*, **19** (6): 407–416. https://doi.org/10.1016/S0141-0296(96)00070-3

6. Peng JL, Pan ADE, Chen WF, Yen Y, Chan SL. (1997) Structural modeling and analysis of modular falsework systems. *Journal of Structural Engineering*, **123** (9): 1245–1251. https://doi.org/10.1061/(ASCE)0733-9445(1997)123:9(1245)

7. Hasan R, Kishi N, Chen WF, Komuro M. (1997) Evaluation of rigidity extended end-plate connections. *Journal of Structural Engineering*, **123** (12): 1595–1602. https://doi.org/10.1061/(ASCE)0733-9445(1997)123:12(1595)

8. Chen WF, Bowman MD, Yang WH. (1997) The behavior and load-carrying capacity of seated-beam connections. *Engineering Journal*, **34** (31): 89–103. https://www.aisc.org/The-Behavior-and-Load-Carrying-Capacity-of-Unstiffened-Seated-Beam-Connections

1998

1. Kim SE, Chen WF. (1998) A sensitivity study on number of elements in plastic hinge refined analysis. *Computers and Structures*, **66** (5): 665–673. https://doi.org/10.1016/S0045-7949(97)00081-3

2. Zhao XH, Chen WF. (1998) Solutions of multilayer inclusion problems under uniform field. *Journal of Engineering Mechanics*, **124** (2): 209–216. https://doi.org/10.1061/(ASCE)0733-9399(1998)124:2(209)

3. Zhao XH, Chen WF. (1998) Effective elastic moduli of concrete with interface layer. *Computers and Structures*, **66** (2–3): 275–288. https://doi.org/10.1016/S0045-7949(97)00056-4

4. Zhao XH, Chen WF. (1998) The effective elastic moduli of concrete and composite materials. *Composites Part B: Engineering*, **29** (1): 31–40. https://doi.org/10.1016/S1359-8368(97)80861-X

5. Chen WF. (1998) Implementing advanced analysis for steel frame design. *Progress in Structural Engineering and Materials*, **1** (3): 323–328. https://doi.org/10.1002/pse.2260010315

6. Kim YS, Chen WF. (1998) Practical analysis for partially restrained frame design. *Journal of Structural Engineering*, **124** (7): 736–749. https://doi.org/10.1061/(ASCE)0733-9445(1998)124:7(736)

7. Kim YS, Chen WF. (1998) Design tables for top- and seat-angle with double web-angle connections. *Engineering Journal*, **35** (2): 50–75. https://www.aisc.org/Design-Tables-for-Top-and-Seat-Angle-with-Double-Web-Angle-Connections

8. Peng JL, Rosowsky DV, Pan AD, Chen WF, Chan SL. (1998) Simplified modeling and analysis of pattern loading effects on shoring systems during construction. *Advances in Structural Engineering*, **1** (3): 203–218. https://doi.org/10.1177/136943329800100305

9. Kishi N, Chen WF, Goto Y, Komuro M. (1998) Effective length factor of columns in flexibly jointed and braced frames. *Journal of Constructional Steel Research*, **47** (1–2): 93–118. https://doi.org/10.1016/S0143-974X(98)80104-1

10. Hasan R, Kishi N, Chen WF. (1998) A new nonlinear connection classification system. *Journal of Constructional Steel Research*, **47** (1–2): 119–140. https://doi.org/10.1016/S0143-974X(98)80105-3

11. Ramesh G, Sotelino ED, Chen WF. (1998) Effect of transition zone on elastic stresses in concrete materials. *Journal of Materials in Civil Engineering*, **10** (4): 275–282. https://doi.org/10.1061/(ASCE)0899-1561(1998)10:4(275)

1999

1. Li G, Zhou RG, Duan L, Chen WF. (1999) Multiobjective and multilevel optimization for steel frames. *Engineering Structures,* **21** (6): 519–529. https://doi.org/10.1016/S0141-0296(97)00226-5
2. Kim SE, Chen WF. (1999) Design guide for steel frames using advanced analysis program. *Engineering Structures,* **21** (4): 352–364. https://doi.org/10.1016/S0141-0296(97)00209-5
3. Joh CB, Chen WF. (1999) Fracture strength of welded flange-bolted web connections. *Journal of Structural Engineering,* **125** (5): 565–571. https://doi.org/10.1061/(ASCE)0733-9445(1999)125:5(565)
4. Zeng DR, Zhou SP, Chen WF. (1999) The strength and stiffness analysis of the bird-shape plates. *Industrial Construction,* **29** (7): 4–6. (In Chinese).
5. Zhou SP, Chen WF. (1999) Local stability analysis of bird-shape plates. *Industrial Construction,* **29** (7): 7–11. (In Chinese).
6. Zhou SP, Cheng YC, Chen WF. (1999) Torsional buckling of bird-shape plates. *Industrial Construction,* **29** (7): 12–16. (In Chinese).
7. Yang WH, Bowman MD, Chen WF. (1999) Experimental study on bolted unstiffened seat angle connections. *Journal of Structural Engineering,* **125** (11): 1224–1231. https://doi.org/10.1061/(ASCE)0733-9445(1999)125:11(1224)
8. Kim SE, Chen WF. (1999) Guidelines to unbraced frame design with LRFD. *The Structural Design of Tall Buildings,* **8** (4): 273–288. https://doi.org/10.1002/(SICI)1099-1794(199912)8:4<273::AID-TAL135>3.0.CO;2-B

2000

1. Kim SE, Kim MK, Chen WF. (2000) Improved refined plastic hinge analysis accounting for strain reversal. *Engineering Structures,* **22** (1): 15–25. https://doi.org/10.1016/S0141-0296(98)00079-0
2. Chen WF. (2000) Structural Stability: From theory to practice. *Engineering Structures,* **22** (2): 116–122. https://doi.org/10.1016/S0141-0296(98)00100-X
3. Chen WF. (2000) Plasticity, limit analysis and structural design. *International Journal of Solids and Structures,* **37** (1–2): 81–92. https://doi.org/10.1016/S0020-7683(99)00079-7
4. Youakim SAS, El-Metewally SEE, Chen WF. (2000) Nonlinear analysis of tunnels in clayey/sandy soil with a concrete lining. *Engineering Structures,* **22** (6): 707–722. https://doi.org/10.1016/S0141-0296(99)00008-5
5. Liew JYR, Chen H, Chen WF. (2000) Advanced inelastic analysis of frame structures. *Journal of Constructional Steel Research,* **55** (1–3): 245–265. https://doi.org/10.1016/S0143-974X(99)00088-7

6. Chen IH, Chen WF. (2000) Major design impact of 1997 LRFD steel seismic code revision in USA. *Journal of Structural Engineering*, **27** (1): 1–16.

7. Liew JYR, Chen H, Shanmugam NE, Chen WF. (2000) Improved nonlinear plastic hinge analysis of space frame structures. *Engineering Structures*, **22** (10): 1324–1338. https://doi.org/10.1016/S0141-0296(99)00085-1

8. Huang YL, Chen WF, Chen HJ, Yen T, Kao YG, Lin CQ. (2000) A monitoring method for scaffold-frame shoring system for elevated concrete formwork. *Computers and Structures*, **78** (5): 681–690.
https://doi.org/10.1016/S0045-7949(00)00051-1

2001

1. Zhou SP, Duan L, Cheng YC, Chen WF. (2001) Global stability analysis for bird-shaped girders. *Journal of Structural Engineering*, **127** (3): 306–313.
https://doi.org/10.1061/(ASCE)0733-9445(2001)127:3(306)

2. Peng JL, Pan ADE, Chen WF. (2001) Approximate analysis method for modular tubular falsework. *Journal of Structural Engineering*, **127** (3): 256–263.
https://doi.org/10.1061/(ASCE)0733-9445(2001)127:3(256)

3. Kishi N, Ahmed A, Yabuki N, Chen WF. (2001) Nonlinear finite element analysis of top- and seat-angle with double wed-angle connections. *Structural Engineering and Mechanics*, **12** (2): 201–214.
https://doi.org/10.12989/sem.2001.12.2.201

4. Chen WF, Kim SE, Choi SH. (2001) Practical second-order inelastic analysis for three-dimensional steel frames. *Steel Structures*, **1**: 213–223.

5. Joh C, Chen WF. (2000) Seismic behavior of steel moment connections with composite slab. *Steel Structures*, **1**: 175–183.

2002

1. Cheng HL, Sotelino ED, Chen WF. (2001) Strength estimation for FRP wrapped reinforced concrete columns. *Steel and Composite Structures*, **2** (1): 1–20.
https://doi.org/10.12989/scs.2002.2.1.001

2. Wongkaew K, Chen WF. (2002) Consideration of out-of-plane buckling in advanced analysis for planar steel frame design. *Journal of Constructional Steel Research*, **58** (5–8): 943–965. https://doi.org/10.1016/S0143-974X(01)00091-8

2003

1. Lan YM, Sotelino ED, Chen WF. (2003) The strain-space consistent tangent operator and return mapping algorithm for constitutive modeling of confined concrete, *International Journal of Applied Science and Engineering*, **1** (1): 17–29.

2. Teng MH, Sotelino ED, Chen WF. (2003) Performance evaluation of RC bridge columns wrapped with fiber reinforced polymers. *Journal of Composite for Construction*, **7** (2): 83-92.
   https://doi.org/10.1061/(ASCE)1090-0268(2003)7:2(83)
3. Duan L, Chen WF. (2003) Basic design criteria for bridge crossing, open sea and bay area. *Marine Georesources and Geotechnology*, **21** (3–4): 289–305.
   https://doi.org/10.1080/713773403

2004

1. Abdalla KM, Chen WF. (2004) Effect of control fluid and surface treatments of high strength bolts. *International Journal of Applied Science and Engineering*, **2** (1): 1–15.
2. Cheng HL, Sotelino ED, Chen WF. (2004) Sensitivity study and design procedure for frp wrapped reinforced concrete circular columns. *International Journal of Applied Science and Engineering*, **2** (2): 148–162.
   DOI:10.6703/IJASE.2004.2(2).148
3. Hwa K, Chen WF. (2004) Survival time prediction of steel frames under elevated temperature using advanced analysis. *International Journal of Steel Structures*, **4**: 187–196.

2005

1. Joh CB, Chen WF. (2005) Effects of concrete slab on ductility of steel moment connections. *International Journal on Advanced Steel Construction*, **1** (1): 3–22.
   DOI:10.18057/IJASC.2005.1.1.1

2006

1. King WS, Chen CJ, Duan L, Chen WF. (2006) Plastic analysis of frames with tapered member. *Journal of Architecture and Civil Engineering*, **23** (2): 9–19.
2. Abdallah KM, Chen WF. (2006) Base plate design in steel structures — A new approach. *Journal of Structural Engineering*, Chennai, India.

2007

1. Yang CY, Xu MX, Chen WF. (2007) Reliability analysis of shotcrete lining during tunnel construction. *Journal of Construction Engineering and Management*, **133** (12): 975–981. https://doi.org/10.1061/(ASCE)0733-9364(2007)133:12(975)

2008

1. Chen WF. (2008) Advanced analysis for structural steel design. *Journal of Frontiers of Architecture and Civil Engineering in China*, **2** (3): 189–196.

2. Chen WF. (2008) Structural engineering — Seeing big picture. *KSCE Journal of Civil Engineering*, **12** (1): 25–29.
https://doi.org/10.1007/s12205-008-8025-7

3. King WS, Duan L, Chen WF, Pan CL. (2008) Education improvement in construction ethics. *Journal of Professional Issues in Engineering Education and Practice*, **134** (1): 12–19.
https://doi.org/10.1061/(ASCE)1052-3928(2008)134:1(12)

4. Iu CK, Chen WF, Chan SL, Ma TW. (2008) Direct second-order elastic analysis for steel frame design. *KSCE Journal of Civil Engineering*, **12** (6): 379–389.
https://doi.org/10.1007/s12205-008-0379-3

5. Chen WF. (2008) My life journey — Connecting the DOTS. *Magazine of the Chinese Institute of Civil and Hydraulic Engineering*, **35** (3): 1–6.

2009

1. Iu CK, Bradford MA, Chen WF. (2009) Second-order inelastic analysis of composite frames structures based on the refined plastic hinge method. *Engineering Structures*, **31** (3): 799–813.
https://doi.org/10.1016/j.engstruct.2008.12.007

2. Chen WF. (2009) Seeing the big picture in structural engineering. *Proceedings of the Institute of Civil Engineers — Civil Engineering*, **162** (2): 87–95.
https://doi.org/10.1680/cien.2009.162.2.87

3. Chen WF. (2009) Toward practical advanced analysis for frame design. *Structural Engineering International*, **19** (3): 234–239.
https://doi.org/10.2749/101686609788957847

2013

1. Peng JL, Yen T, Lu LC, Chou T, Chen WF. (2013) Mechanism investigation of falsework collapse for North Mountain Interchange Project on National Highway 5 — Part 1. *Structural Engineering* **28** (3): 61–93. (In Chinese).

2. Peng JL, Yen T, Lu LC, Chou T, Chen WF. (2013) Mechanism investigation of falsework collapse for North Mountain Interchange Project on National Highway 5 — Part 2. *Structural Engineering*, **28** (4): 15–36. (In Chinese).

2015

1. Peng JL, Ho CM, Lin CC, Chen WF. (2015) Load-carrying capacity of single-row steel scaffolds with various setups. *Advanced Steel Construction*, **11** (2): 185–210. DOI:10.18057/IJASC.2015.11.2.4
https://www.researchgate.net/publication/277399084

2017

1. Peng JL, Wang CS, Wu CW, Chen WF. (2017) Experiment and stability analysis on heavy-duty scaffold systems with top shores. *Advanced Steel Construction*, **13** (3): 293–317. DOI: 10.18057/IJASC.2017.13.3.6
   https://www.researchgate.net/publication/320233326
2. Peng JL, Ho CM, Chan SL, Chen WF. (2017) Stability study on structural systems assembled by system scaffolds. *Journal of Constructional Steel Research*, **137** (October): 137–151. https://doi.org/10.1016/j.jcsr.2017.06.004

7.2 Conference Proceedings Articles (263)

1969

1. (1969) The plastic identification of metal blocks by a flat punch. *ASCE Annual Meeting and National Meeting on Structural Engineering*, Louisville, KY, 14–18 April. Meeting Reprint 890.

1971

1. (1971) The application of limit analysis to a three-dimensional problem in concrete. *Proceedings of the Third Canadian Congress of Applied Mechanics*, The University of Calgary, Canada, 12–21 May, pp. 419–420.
2. (With Fang HY.) (1971) New method for determination of tensile strength of soils. *Highway Research Record*, No. 345, 50th Annual Meeting, Highway Research Board, 18–22 January, pp. 62–68.

1972

1. (With Fang HY.) (1972) Further study of double-punch test for tensile strength of soils. *Proceedings of the Third Southeast Asian Conference on Soil Engineering*, Hong Kong, pp. 211–215.
2. (With Tebedge N, Tall L.) (1972) Experimental studies on European heavy shapes. *Proceedings of the International Colloquium on Column Strength*, IABSE, Paris, France, 23–24 November, pp. 301–320.
3. (With Gilligan JA.) (1972) Connections, State-of-the-Art No. 5, Technical Committee 15. *Proceedings, International Conference on Planning and Design of Tall Buildings*, Vol. II, 21–26 August, pp. 327–342.
4. (With Fielding DJ.) (1972) Frame analysis considering connection shear deformation, Discussion No. 1, Technical Committee 15. *Proceedings, International Conference on Planning and Design of Tall Buildings*, Vol. II, 21–26 August, pp. 365–370.

5. (1972) Behavior of biaxially loaded columns, Discussion No. 6, Technical Committee 16. *Proceedings, International Conference on Planning and Design of Tall Buildings*, Vol. II, 21–26 August, pp. 575–578.
6. (1972) Behavior of concrete-filled steel tubular columns, Discussion No. 3, Technical Committee No. 23. *Proceedings, International Conference on Planning and Design of Tall Buildings*, Vol. III, 21–26 August, pp. 608–612.
7. (With Tebedge N, Tall L.) (1972) On the behavior of a heavy welded steel column. *Proceedings of the International Colloquium on Column Strength*, IABSE, Paris, France, 23–24 November, pp. 9–24.
8. (1972) Behavior of biaxially loaded columns, Discussion No. 6, Planning and Design of Tall Buildings. *Proceedings of ASCE-IABSE International Conference*, Vol. II-TC 16, New York, pp. 575–580.

1973

1. (1973) Plastic response for arbitrary histories of biaxial loads. *Proceedings, Fourth Canadian Congress of Applied Mechanics*, Montreal, 28 May to 1 June, pp. 289–290.
2. (1973) Plastic H-column under repeated biaxial loading. *Symposium on Resistance and Ultimate Deformability of Structures Acted on by Well-Defined Repeated Loads*, IABSE publication, Lisbon, pp. 1–7.
3. (With Dahl-Jorgensen E.) (1973) Stress-strain properties of polymer modified concrete. *Symposium on Polymers in Concrete*, Publication SP-40, 1 January, pp. 347–358. DOI: 10.14359/17398
   https://www.concrete.org/publications/internationalconcreteabstractsportal.aspx?m=details&id=17398
4. (With Rentschler GP.) (1973) Ultimate strength of concrete-filled steel tubular beam-columns. *Regional Conference on Tall Buildings*, Madrid, Spain, 17–19 September, pp. 79–98.

1974

1. (With Shoraka MT.) (1974) Analysis and design of reinforced columns under biaxial loading. *IABSE Symposium on Design and Safety of Reinforced Concrete Compression Members*, Quebec, August, pp. 187–195.
   http://doi.org/10.5169/seals-15733
2. (With Carson JL). (1974) Bearing capacity of fiber reinforced concrete. *International Symposium on Fiber Reinforced Concrete*, Publication SP-44, 1 January, pp. 209–220. DOI: 10.14359/17896
   https://www.concrete.org/publications/internationalconcreteabstractsportal.aspx?m=details&id=17896

3. (With Chen ACT.) (1974) On structural concrete and work-hardening theories of plasticity. *Proceedings of the Society of Engineering Science*, Duke University, Durham, North Carolina, 11–13 November, pp. 304–305.
4. (With Chen ACT.) (1974) Strength of laterally loaded reinforced concrete columns. *IABSE Symposium on Design and Safety of Reinforced Concrete Compression Members*, Quebec, August, pp. 315–322.
5. (With Huang JS, Beedle LS.) (1974) Recent results on connection research at Lehigh. *Regional Conference on Tall Buildings*, Bangkok, Thailand, January, pp. 799–813.

1975

1. (1975) Reinforced concrete constitutive relations for application to OTEC plant structures. *Proceedings, Third Workshop on Ocean Thermal Energy Conversion*, Houston, TX, 8–10 May, pp. 99–102.
2. (With Manson JA *et al.*) (1975) Polymer-impregnated concrete for highway and structural applications. *Proceedings of the First International Congress on Polymer Concretes*, The Concrete Society, London, 5–7 May, pp. 403–408.
3. (With Snitbhan N, Fang HY.) (1975) Slope stability analysis of layered soils. *Proceedings of the Fourth Southeast Asian Conference on Soil Engineering*, Kuala Lampur, Malaysia, 7–10 April, pp. 5-26 to 5-29.
4. (With Mehta HC.) (1975) Structural use of sulfur for impregnation of building materials. *Proceedings, Pan-Pacific Tall Building Conference*, Hawaii, 26–29 January, pp. 123–135.
5. (With Mehta HC, Manson JA, Vanderhoff JW.) (1975) Use of polymers in highway bridge slabs. *Proceedings, Inter-Associations Colloquium on Behavior in Service of Concrete Structures*, Liege, 4–6 June.

1976

1. (With Snitbhan N.) (1976) Plasticity solutions for slopes. *Numerical Methods in Geomechanics, Proceedings of the Second International Conference on Numerical Methods in Geomechanics*, Blacksburg, Virginia, June, pp. 731–743.
2. (With Snitbhan N.) (1976) Finite element analysis of large deformation in slopes. *Numerical Methods in Geomechanics, Proceedings of the Second International Conference on Numerical Methods in Geomechanics*, Blacksburg, Virginia, June, pp. 744–756.
3. (With Davidson HL.) (1976) Nonlinear analyses in soil and solid mechanics. *Numerical Methods in Geomechanics, Proceedings of the Second International Conference on Numerical Methods in Geomechanics*, Blacksburg, Virginia, June, pp. 205–216.

4. (1976) Behavior of bolted moment connections in steel buildings. *Proceedings of the Regional Conference on Tall Buildings*, Hong Kong, 20–22 September, pp. 26–47.

5. (With Dahl-Jorgensen E.) (1976) Stress-strain behavior of polymer-impregnated concrete. *Proceedings on the International Symposium on New Horizons in Construction Materials*, Bethlehem, PA, Envo Publishing Co., pp. 303–326.

6. (With Mehta HC, Manson JA, Vanderhoff JW.) (1976) Use of polymers in highway bridge slabs. *Proceedings on the International Symposium on New Horizons in Construction Materials*, Bethlehem, PA, Envo Publishing Co., pp. 327–343.

7. (With Mehta HC, Pepe AJ.) (1976) Split-cylinder test and double-punch test for tensile strength of concrete. *Proceedings on the International Symposium on New Horizons in Construction Materials*, Bethlehem, PA, Envo Publishing Co., pp. 625–642.

8. (With Rentschler GP.) (1976) Test and analysis of beam-to-column web connections. *Proceedings of the National Structural Engineering Conference on Methods of Structural Analysis*, Madison, Wisconsin, 22–25 August, pp. 957–976.

9. (With Liu YN, Manson JA, Vanderhoff JW.) (1976) Polymer-impregnated mortar: Effects of polymer composition, *Symposium on Polymer Blends and Composites for New Materials*. Material Engineering, and Science Division, American Institute of Chemical Engineers, Atlantic City, AIChe Meeting, 29 August to 1 September.

10. (With Ross DA.) (1976) The axial strength and behavior of cylindrical columns. *Proceedings of Eighth Annual Offshore Technology Conference*, Paper No. OTC 2683. Dallas, TX, 3–6 May, pp. 741–754.

11. (With Ross DA.) (1976) Tests of fabricated tubular columns. *ASCE National Water Resources and Ocean Engineering Convention*, San Diego, CA, 5–8 April. Reprint 2735.

12. (With Ross DA.) (1976) Behavior of fabricated tubular columns under biaxial bending. *ASCE National Water Resources and Ocean Engineering Convention*, San Diego, CA, 5–8 April. Reprint 2659.

1977

1. (With Chang TY, Suzuki H.) (1977) Analysis of concrete cylindrical hulls under hydrostatic pressure. *ASME Energy Technology Conference and Exhibit of the Pressure Vessels and Piping Division*, Houston, TX, 18–27 September, Preprint No. 77-PVP-42.

2. (1977) Studies of axially loaded fabricated tubular columns. *Preliminary Report, International Colloquium on Stability of Structures*, Liege, 13–15 April, pp. 61–70.
3. (1977) Design of box columns under biaxial bending. *Preliminary Report, International Colloquium on Stability of Structures*, Liege, 13–15 April, pp. 355–361.
4. (1977) Manufacturing process of fabricated tubular columns. *Final Report, Second International Colloquium on Stability of Steel Structures*, Liege, 13–15 April, pp. 57–60.
5. (1977) Theory of beam-columns — The state-of-the-art review. *Proceedings of the International Colloquium on Stability of Structures under Static and Dynamic Loads*, SSRC/ASCE, Washington, DC, 17–19 March, pp. 631–641.
6. (1977) Analytical studies for solution of soil structure interaction problems. *International Symposium on Soil Structure Interaction*, Roorkee, India, 3–7 January, pp. 556–575.
7. (With Davidson HL.) (1977) Large deformation response of clay to loads. *Proceedings of the International Conference on Finite Elements in Nonlinear Solid and Structural Mechanics*, Geilo, Norway, 29 August to 1 September.
8. (1977) Mechanics of slope failures and landslides. *Proceedings of the Advisory Meeting on Earthquake Engineering and Landslides*, US-Republic of China Cooperative Science Program, Taipei, Taiwan, 29 August to 2 September, pp. 219–232.
9. (1977) Analysis of soil structure interaction problems. *Proceedings of the 14th Annual Meeting*, Society of Engineering Science, Lehigh University, Bethlehem, PA, 14–16 November, pp. 1149–1164.
10. (With Koh SL.) (1977) Prevention and control of landslides. *Joint United States-Southeast Asia Symposium on Engineering for Natural Hazards Protection*, Manila, Philippines, 28 September.
11. (With McGraw JE.) (1977) Analysis and design of HSS-columns under biaxial bending. *Proceedings of the Second Annual Engineering Mechanics Division Specialty Conference*, Raleigh, NC, 23–25 May, pp. 568–571.
12. (With Davidson HL.) (1977) Elastic-plastic large deformation response of soft clay to footing load. *Proceedings of the International Symposium on Soft Clay*, Bangkok, Thailand, 5–6 July, pp. 629–646.

1978

1. (With Hsu DS, Yao JTP.) (1978) Risk analysis of seismic structures. *Proceedings of Central American Conference on Earthquake Engineering*, San Salvador, El Salvador, 9–14 January, pp. 37–51.

2. (With Dahl-Jorgensen, E.) (1978) Monomer impregnated through case-in perforated pipes. *Polymers in Concrete: International Symposium*, SP58, American Concrete Institute Detroit, 1 January, pp. 299–312.
   DOI: 10.14359/17798
   https://www.concrete.org/publications/internationalconcreteabstractsportal.
   aspx?m=details&id=17798

3. (With Yuan RL.) (1978) Study of sulphur-impregnated brick, block prisms and concrete pipes. *Proceedings of the International Conference on Sulphur in Construction*, Canada Centre for Mineral and Energy Technology, Calgary, Alberta, 12–15 September, pp. 91–106.

4. (With Patel KV.) (1978) Radius expansion of pressure-resistant vessels. *ASME/CSME Pressure Vessels and Piping Conference*, Montreal, Canada, 25–30 June. In Special Volume PVP-PB-028, *Inelastic Behavior of Pressure Vessel and Piping Components*, editors, Chang TY and Krempl E, pp. 143–163.

5. (With Mehta HC, Manson JA, Vanderhoff JW.) (1978) Stress-strain behavior of polymer-impregnated concrete beams, columns, and shells. *Polymers in Concrete: International Symposium*, SP58, American Concrete Institute, 1 January, pp. 161–186. DOI: 10.14359/17790
   https://www.concrete.org/publications/internationalconcreteabstractsportal.
   aspx?m=details&id=17790

6. (With Suzuki H.) (1978) Constitutive models for concrete. *ASCE National Convention*, Chicago, 16–20 October, pp. 51–79.

7. (With Chang CJ, Yao JTP.) (1978) Limit analysis of earthquake-induced slope failure. *Proceedings of the 15th Annual Meeting of the Society of Engineering Science*, Gainesville, FL, 4–6 December, pp. 533–538.

8. (With Yuan RL.) (1978) Abrasion tests for fiber reinforced concrete. *Proceedings of the International Symposium on Testing and Test Methods of Fiber Cement Composites*, University of Sheffield, England, 5–7 April, pp. 531–536.

9. (With Yuan RL.) (1978) Double punch test for tensile strength of fiber and polymer impregnated concretes. *Proceedings of the International Symposium on Testing and Test Methods of Fiber Cement Composites*, University of Sheffield, England, 4–7 April, pp. 511–523.

10. (With Koh SL.) (1978) Earthquake-induced landslide problems. *Proceedings of Central American Conference on Earthquake Engineering*, San Salvador, El Salvador, 9–14 January, pp. 665–685.

11. (With Rentschler GP.) (1978) Tests of beam-to-column web moment connections. *ASCE Spring Convention and Exhibit*, Pittsburgh, PA, 24–28 April. Reprint 3202.

1979 — Conference Proceedings

1. (1979) Nonlinear analysis of concrete structures. *Proceedings of the International Conference on Computer Applications in Civil Engineering*, Roorkee, India, pp. 11-17 to 11-26.

2. (With Hsieh SS, Ting EC.) (1979) An elastic-fracture model for concrete. *Proceedings of the Third ASCE/EMD Specialty Conference*, Austin, TX, 17–19 September, pp. 437–440.

3. (With Ting EC.) (1979) On constitutive models for analysis of concrete structures. *ASCE Convention and Exposition*, Boston, MA, 2–6 April. Preprint 3475.

4. (1979) Constitutive equations for concrete. *Introductory Report, Colloquium on Plasticity in Reinforced Concrete*, IABSE Publication, Copenhagen, Denmark, 21–23 May, pp. 11–34.

5. (1979) Nonlinear analysis of pressure-resistant concrete cylinder structures. *Proceedings of the International Conference on Engineering Applications of the Finite Element Method*, Hovik, Norway, 9–11 May, pp. 19-1 to 19-31.

6. (With Toma S.) (1979) Stability of tubular members in sea-based structures. *Proceedings of CANCAN 79*, Sherbrooke, 27 May to 1 June, pp. 235–236.

7. (With Yuan RL.) (1979) Sulphur-treated concrete slabs. *Proceedings of the International Conference on Advances in Concrete Slab Technology*, University of Dundee, Scotland, 3–6 April, pp. 75–83.

8. (With Haynes HH, Chang TY, Suzuki H.) (1979) External hydrostatic pressure loading of concrete cylinder shells. *ASME Pressure Vessels and Piping Congress*, San Francisco, 25–29 June. Preprint No. 79-PVP-125.

9. (With Mizuno E.) (1979) On material constants for soil and concrete models. *Proceedings of the Third ASCE/EMD Specialty Conference*, Austin, TX, 17–19 September, pp. 539–542.

10. (With Toma S, Yuan RL.) (1979) Strength of fabricated tubular columns in offshore structures. *Proceedings of the International Conference on Thin-Walled Structures*, University of Strathclyde, Glasgow, Scotland, 3–6 April, pp. 271–286.

11. (1979) Moment-curvature relation of fabricated tubular columns. *Proceedings of the 16th Midwestern Mechanics Conference*, Kansas State University, 19–21 September.

12. (1979) Influence of end restraint on column stability. *ASCE Convention and Exposition*, Atlanta, GA, 23–25 October. Preprint 3608.

13. (1979) Material behavior under various types of loading. *Proceedings of a Workshop on High Strength Concrete*, University of Illinois at Chicago Circle, 2–4 December, pp. 93–95.

1980

1. (With Fang HY, Atsuta T.) (1980) Optimum design of retaining walls on sloping hillside. *Proceedings of the Sixth Southeast Asian Conference on Soil Engineering*, Taipei, Taiwan, 19–23 May, pp. 391–406.
2. (With Hsieh SS, Ting EC.) (1980) A plastic-fracture model for concrete. *Fracture in Concrete, Proceedings of a Session at the ASCE National Convention*, Hollywood, FL, 27–31 October, editors, Chen WF and Ting EC, pp. 50–64.
3. (With Mizuno E.) (1980) Analysis of soil response with different plasticity models. *ASCE Convention and Exposition*, Florida, 27–31 October. Preprint 80–637.

1981

1. (With Mizuno E.) (1981) Plasticity models and finite element implementation. *Proceedings of the Symposium on Implementation of Computer Procedures and Stress-Strain Laws in Geotechnical Engineering*, Chicago, Illinois, 3–6 August, pp. 519–553.
2. (1981) Plasticity in reinforced concrete. *Proceedings of the Workshop on Constitutive Relations for Concrete*, Air Force Weapons Laboratory and New Mexico Engineering Research Institute, Albuquerque, NM, 28–29 April.
3. (With Yuan RL.) (1981) Plasticized sulphur impregnated masonry block, clay brick and concrete members. *European Conference on Building Materials*, Copenhagen, Denmark, 7–8 February, pp. 139–151.
4. (With Yuan RL.) (1981) Corrosion study of sulphur impregnated concrete. *Third International Congress on Polymers in Concrete*, College of Engineering, Nihon University, Koriyama, Japan, 13–15 May, pp. 1170–1184.
5. (1981) Recent advances on analysis and design of steel beam-columns in USA. *Proceedings of the US-Japan Seminar on Inelastic Instability of Steel Structures and Structural Elements*, Tokyo, Japan, pp. 224–293.
6. (With Mizuno E.) (1980) Plasticity models for soils-comparison and discussion. *Proceedings of the Workshop on Limit Equilibrium, Plasticity and Generalized Stress-Strain in Geotechnical Engineering*, Montreal, Quebec, Canada, 28–30 May, pp. 328–351.
7. (With Mizuno E.) (1980) Plasticity models for soils. *Proceedings of the Workshop on Limit Equilibrium, Plasticity and Generalized Stress-Strain in Geotechnical Engineering*, Montreal, Quebec, Canada, 28–30 May, pp. 553–591.
8. (With Saleeb AF.) (1980) Nonlinear hyperelastic (green) constitutive models for soils: Predictions and comparisons. *Proceedings of the Workshop on Limit Equilibrium, Plasticity and Generalized Stress-Strain in Geotechnical Engineering*, Montreal, Quebec, Canada, 28–30 May, pp. 265–285.

9. (With Saleeb AF.) (1980) Nonlinear hyperelastic (green) constitutive models for soils: Theory and calibration. *Proceedings of the Workshop on Limit Equilibrium, Plasticity and Generalized Stress-Strain in Geotechnical Engineering*, Montreal, Quebec, Canada, 28–30 May, pp. 492–538.

1982

1. (With Liu XL.) (1982) Study of concrete framed structures during construction. *Special Session on Safety Considerations During Construction*, Annual Convention, New Orleans, LA, 25–29 October.
2. (With Mizuno E.) (1980) Analysis of soil response with different plasticity models. *Application of Plasticity and Generalized Stress-Strain in Geotechnical Engineering, Proceedings of the Symposium on Limit Equilibrium, Plasticity and Generalized Stress-Station Applications in Geotechnical Engineering*, in conjunction with the 1980 ASCE Annual Convention and Exposition, Hollywood, Florida, 27–31 October, pp. 115–138.
3. (With Mizuno E.) (1980) Plasticity models for soils: Comparison and discussion. *Application of Plasticity and Generalized Stress-Strain in Geotechnical Engineering, Proceedings of the Symposium on Limit Equilibrium, Plasticity and Generalized Stress-Station Applications in Geotechnical Engineering*, in conjunction with the 1980 ASCE Annual Convention and Exposition, Hollywood, Florida, 27–31 October.
4. (With Saleeb AF.) (1980) Nonlinear hyperelastic (green) constitutive models for soils: Predictions and comparisons. *Application of Plasticity and Generalized Stress-Strain in Geotechnical Engineering, Proceedings of the Symposium on Limit Equilibrium, Plasticity and Generalized Stress-Station Applications in Geotechnical Engineering*, in conjunction with the 1980 ASCE Annual Convention and Exposition, Hollywood, Florida, 27–31 October.
5. (With Saleeb AF.) (1980) Nonlinear hyperelastic (green) constitutive models for soils: Theory and calibration, *Application of Plasticity and Generalized Stress-Strain in Geotechnical Engineering, Proceedings of the Symposium on Limit Equilibrium, Plasticity and Generalized Stress-Station Applications in Geotechnical Engineering*, in conjunction with the 1980 ASCE Annual Convention and Exposition, Hollywood, Florida, 27–31 October.
6. (With Saleeb AF.) (1982) Constitutive modeling of soils — An overview. *ASCE Convention and Exposition*, Las Vegas, NV, 26–30 April. Reprint 82-006.
7. (With Mizuno E.) (1982) Cap models in soil mechanics. *ASCE Convention and Exposition*, Las Vegas, NV, 26–30 April. Reprint 82-006.

8. (1982) Three-dimensional elastic-plastic fracture analysis for concrete structures. *ASCE Convention and Exposition*, New Orleans, LA, 25–29 October. Reprint 82-501.

9. (With Liu XL, Chang TY.) (1982) Simplified inelastic analysis of reinforced concrete panels. *ASCE Convention and Exposition*, New Orleans, LA, 25–29 October. Reprint 82-526.

1983

1. (1983) Constitutive modeling in soil mechanics. *Proceedings of the International Conference on Constitutive Laws for Engineering Materials-Theory and Application*, Tucson, AZ, 10–14 January, pp. 183–184.

2. (1983) Stability design of columns in North America. *Design Limit States of Steel Structures, Proceedings of the First International Correspondence Conference*, Technical University of BRNO, Czechoslovakia, Brno, pp. 301–319.

3. (With Bjorhovde R.) (1983) Design criteria for end restrained columns. *Proceedings of the Third International Colloquium on Stability of Metal Structures*, Paris, France, 16–17 November.

4. (With Bjorhovde R.) (1983) Behavior of columns — A comprehensive treatment. *WH Munse Symposium Behavior of Metal Structures*, Philadelphia, PA, 17 May, pp. 85–102.

5. (With Han DJ.) (1983) Failure criteria of concrete materials. *The CNRS International Colloquium on Failure Criteria of Structured Media*, Villard-de-Lans, Grenoble, France, 21–24 June, pp. 375–384.

6. (With Lui EM.) (1983) Design of beam-columns in North America. *Proceedings of the Third International Colloquium on Stability of Metal Structures*, Structural Stability Research Council, Bethlehem, PA, May, pp. 253–291.

7. (With Yener M, Nishikawa A.) (1983) In-place flexural strength evaluation of concrete. *Recent Advances in Engineering Mechanics and Their Impact on Civil Engineering Practice, Proceedings of Fourth Engineering Mechanics Division Specialty Conference*, editors, Chen WF and Lewis ADM. West Lafayette, IN, 23–25 May, pp. 1032–1035.

8. (With Chang TYP.) (1983) A plastic-fracture model for concrete materials. *Proceedings of the Symposium on the Interaction of Non-Nuclear Munitions with Structures*, Colorado Springs, Colorado, 10–13 May.

9. (With Wang SH, Yao JTP.) (1983) Reliability of antenna structures. *Recent Advances in Engineering Mechanics and Their Impact on Civil Engineering Practice, Proceedings of Fourth Engineering Mechanics Division Specialty Conference*, Vol. 2, editors, Chen WF and Lewis ADM. West Lafayette, IN, 23–25 May, pp. 855–858.

10. (With Sawada T, Nomachi SG.) (1983) Seismic stability of non-homogeneous, anisotropic slopes. *Recent Advances in Engineering Mechanics and Their Impact on Civil Engineering Practice, Proceedings of the Fourth Engineering Mechanics Division Specialty Conference*, Vol. 2, editors, Chen WF and Lewis ADM. West Lafayette, IN, 23–25 May, pp. 1009–1012.

11. (1983) Soil mechanics, plasticity and landslides. *Daniel C. Drucker Symposium on the Mechanics of Material Behavior*, University of Illinois at Urbana-Champaign. Also, In *Mechanics of Material Behavior* (1983), pp. 31–58.

12. (With Mizuno E.) (1983) Plasticity modeling and its application to geomechanics. *Proceedings of the International Symposium on Recent Development in Laboratory and Field Tests and Analysis of Geotechnical Problems*, editors, Balasubramaniam AS, Chandra S and Bergado DT. Bangkok, Thailand, 6–9 December, pp. 391–421.

13. (With McCarron WO.) (1983) Modeling of soils and rocks based on concepts of plasticity. *Proceedings of the International Symposium on Recent Development in Laboratory and Field Tests and Analysis of Geotechnical Problems*, Bangkok, Thailand, 6–9 December, pp. 467–509.

14. (With Toma S.) (1983) Analytical models of tubular beam-columns. *IABSE Proceedings P67/83*, Zurich, Switzerland, pp. 193–212.

15. (With Sugimoto H.) (1983) Inelastic post buckling behavior of tubular members. *Structure Congress*, Houston, TX, 17–19 October. Reprint SC-5.

16. Chen WF, Sugimoto H. (1983) Moment-curvature-axial compression-pressure relationship of structural tubes. *Structure Congress*, Houston, TX, 17–19 October. Reprint SC-6.

1984

1. (With Lui EM.) (1984) Steel column design using LRFD. *Proceedings of the International Conference on Steel Structures*, Singapore, 7–9 March, pp. 244–260.

2. (1984) Inelastic cyclic behavior of tubular members in offshore structures. *Proceedings of the Eighth World Conference on Earthquake Engineering*, San Francisco, CA, 21–28 July.

3. (With Sohal IS.) (1984) Design of tubular members in offshore structures. *Proceedings of the SSRC Technical Sessions*, San Francisco, 10–11 April, pp. 251–268.

4. (With Han DJ.) (1984) On constitutive modeling of concrete materials. *Proceedings of the RILEM-CEB-CNRS International Conference on Concrete Under Multiaxial Conditions*, Toulouse, France, 22–24 May, pp. 210–219.

5. (With Lui EM.) (1984) Effects of connection flexibility and panel zone shear deformation on the behavior of steel frames. *ING/IABSE Seminar on Tall Structures and Use of Prestressed Concrete in Hydraulic Structures*, Srinagar, India, 24–26 May, pp. I-55 to I-76.

6. (With Yuan RL.) (1984) Temperature stress in a large concrete foundation. *Proceedings of the Fifth ASCE-EMD Specialty Conference*, Laramie, Wyoming, 1–3 August, pp. 615–618.

7. (With Yuan RL.) (1984) Corrosion of reinforcing steel in fly-ash concrete. *Proceedings of the Fifth ASCE-EMD Specialty Conference*, Laramie, Wyoming, 1–3 August, pp. 1284–1287.

8. (With Chang TY.) (1984) Plasticity modeling of geotechnical materials. *Proceedings of the Fifth ASCE-EMD Specialty Conference*, Laramie, Wyoming, 1–3 August.

9. (With Sawada T, Nomachi SG, Takahashi Y.) (1984) Evaluation of limit seismic factor in anisotropic cohesionless slopes. *Proceedings of the Fifth ASCE-EMD Specialty Conference*, Laramie, Wyoming, 1–3 August, pp. 997–1000.

1985

1. (With Liu XL.) (1985) Study of concrete framed structures during construction. *Proceedings of the Conference on Use of Concrete-Steel in Structures*, Madras, India, 18–19 March, pp. 6-1 to 6-12.

2. (With Liu XL, Chang TY.) (1985) Inelastic analysis of reinforced concrete panels. *The Second Symposium on the Interaction of Non-Nuclear Munitions with Structures*, Panama City, FL, 15–19 April.

3. (With Chang CJ, Yao JTP.) (1985) Evaluation of seismic factor of safety of a submarine slope by limit analysis. *Proceedings of the 1983 Shanghai Symposium on Marine Geotechnology and Near Shore/Offshore Structures*, pp. 262–295.

4. (With Sawada T, Nomachi SG.) (1983) Stability of slopes with anisotropic cohesion strength against earthquakes. *Proceedings of the 33rd Japan National Congress for Applied Mechanics*. Also in *Theoretical and Applied Mechanics* (1985), pp. 417–432.

5. (With Han DJ.) (1983) A five-parameter mixed-hardening model for concrete materials. *International Centre for Mechanical Sciences, Symposium on Plasticity Today — Modeling, Methods and Applications*, Udine, Italy, pp. 587–602.

6. (1985) Constitutive relations for concrete rock and soils — Discusser's report, Chapter 5. *IUTAM Prager Symposium Book on Mechanics of Geomaterials: Rocks, Concrete and Soils*, London, pp. 65–86.

7. (With Larralde J.) (1985) Computer model for analysis of rigid pavements with fatigue. *Proceedings of the Third International Conference on Concrete Pavement Design and Rehabilitation*, Purdue University, West Lafayette, IN, 23–25 April, pp. 537–547.

8. (With Yener M.) (1985) Breakoff tests to measure flexural strength of concrete. *Proceedings of the 22nd Annual Technical Meeting, Society of Engineering Science*, University Park, PA, 7–9 October. Preprint No. 22.

1986

1. (With Yamaguchi E.) (1986) On constitutive modeling of concrete materials. *Proceedings of the US/Japan Joint Seminar on Finite Element Analysis of Reinforced Concrete Structures*, Tokyo, Japan, 21–24 May, pp. 48–71.

2. (With Toma S.) (1986) Analysis and design of fabricated tubular beam-columns. *Proceedings of the First Asian Conference on Structural Engineering and Construction*, 15–17 January, Bangkok, Thailand.

3. (With McCarron WO.) (1986) Modeling of soils and rocks based on concepts of plasticity. *AIT Symposium and Course on Laboratory and Field Tests and Analysis of Geotechnical Problems*, Rotterdam, The Netherlands, pp. 467–510.

4. (With Mizuno E.) (1986) Plasticity modeling and its application to geomechanics. *AIT Symposium and Course on Laboratory and Field Tests and Analysis of Geotechnical Problems*, Rotterdam, The Netherlands, pp. 391–426.

5. (With Sohal IS.) (1986) Effect of local buckling on the behavior of tubular braces. *Proceedings of Annual Technical Session, Structural Stability Research Council*, Washington, DC, 15–16 April, pp. 201–210.

6. (With Liu XL.) (1986) Safety analysis of high-rise R.C. buildings during construction. *Proceedings of the International Symposium on Fundamental Theory of Reinforced Concrete and Prestressed Concrete*, Nanjing Institute of Technology, Nanjing, China, September.

7. (With Lui EM.) (1986) Connection panel zone deformations in steel frames. *Proceedings of a Session at the ASCE New Orleans Convention*, 15–18 September, pp. 1–15.

1987

1. (With Yamaguchi E.) (1987) On micro-mechanics of fracture and constitutive modeling of concrete materials. *Proceedings of the Second International Conference on Constitutive Laws for Engineering Materials: Theory and Application*, Tucson, Arizona, 5–10 January, pp. 939–947.

2. (With Sohal IS.) (1987) Local buckling in tubular braces. *Proceedings of the Sixth International Offshore Mechanics and Arctic Engineering Symposium*, Houston, TX, 1–5 March, pp. 467–474.

3. (With Kishi N.) (1987) On steel connection data bank at Purdue University. *Proceedings of the Sessions at Structures Congress Related to Materials and Member Behavior*, Orlando, FL, 17–20 August, pp. 89–106.

4. (With McCarron WO.) (1987) Application of a bounding surface model. *Proceedings of the Second International Conference and Short Course on Constitutive Laws for Engineering Materials: Theory and Applications*, Tucson, Arizona, 5–10 January, pp. 1257–1264.

5. (With Yamaguchi E.) (1987) On softening behavior of concrete materials. *Proceedings of the 20th Midwestern Mechanics Conference*, Purdue University, West Lafayette, IN, 31 August to 2 September, pp. 272–1277.

6. (With Chan SW, Koh SL.) (1984) Upper bound limit analysis of stability of a seismic-infirmed earthslopes. *Proceedings of the Symposium and Specialty Sessions on Geotechnical Aspects of Mass and Material Transportation*, AIT, Bangkok, Thailand, 3–14 December, pp. 373–428.

7. (With Sohal IS.) (1987) Local buckling and cyclic behavior of tubular members. *Proceedings of the Sessions at Structures Congress '87 Related to Dynamics of Structures*, Orlando, FL, 17–20 August, pp. 411–426.

8. (With Lee HM, Liu XL.) (1987) A computational method for construction live load of multistory highrise buildings. *Proceedings of the First International Conference on Computer Application and Engineering*, Guangzhou, China, January.

1988

1. (With Kishi N, Matsuoka KG, Nomachi SG.) (1987) Moment-rotation relation of top- and seat-angle with double web-angle connections. *Proceedings of the Workshop on Connections and the Behavior, Strength and Design of Steel Structures*, Cachan, France, 25–27 May, pp. 121–134.

2. (With Kishi N, Matsuoka KW, Nomachi SG.) (1987) Moment-rotation relation of single/double web-angle connections. *Proceedings of the Workshop on Connections and the Behavior, Strength and Design of Steel Structures*, Cachan, France, 25–27 May, pp. 135–149.

3. (1988) Computers in stability research. *Proceedings of 1988 Structural Stability Research Council Annual Technical Session and Meeting*, Minneapolis, MN, 26–27 April.

4. (With McCarron WO.) (1987) An elastic-plastic two-surface model for non-cohesive soils. *Proceedings of the International Workshop on Constitutive Equations for Granular Non-Cohesive Soils*, Cleveland, OH, 22–24 July, pp. 427–446.
5. (With Mizuno E.) (1987) A multi-surface model for non-cohesive soils. *Proceedings of the International Workshop on Constitutive Equations for Granular Non-Cohesive Soils*, Cleveland, OH, 22–24 July, pp. 481–500.
6. (1987) Evaluation of constitutive models in soil mechanics. *Proceedings of the International Workshop on Constitutive Equations for Granular Non-Cohesive Soils*, Cleveland, OH, 22–24 July, pp. 687–693.

1989

1. (With Sohal IS.) (1989) Design interaction equations for offshore tubular members in deep water. *Proceedings of the Energy-Sources Technology Conference*, Houston, TX, January 22–25, pp. 1–4.
2. (With Kishi N.) (1989) Moment-rotation relation of top- and seat-angle connections. *Proceedings of the International Colloquium on Bolted and Special Connections*, Moscow, May 15–20.
3. (With Zhou SP, Duan L.) (1989) Second-order inelastic analysis of portal frames. *Proceedings of the Fourth International Colloquium on Structural Stability Asian Session*, Beijing, China, October, pp. 681–688.
4. (With Toma S, Honda Y.) (1989) Effect of initial deflection and residual stress on tubular column strength. *Proceedings of the Fourth International Colloquium on Structural Stability Asian Session*, Beijing, China, October, pp. 144–153.
5. (With McCarron WO.) (1989) Formulation and implementation of bounding surface model. *Proceedings of Symposium on Computer and Physical Modeling in Geotechnical Engineering*, Balkema, Rotterdam, The Netherlands, pp. 439–474.
6. Design of beam-columns in frames. (1989) *Proceedings of SSRC Fourth International Colloquium on Stability of Metal Structures: Code Difference Around the World*, Lehigh University, Bethlehem, PA, pp. 135–147.
7. (With Mizuno E.) (1989) A cyclic plasticity model for soils. *Proceedings of the Second International Symposium on Plasticity and Its Applications*, TSU, MIE, Japan, 31 July to 4 August.

1990

1. (With White DW.) (1990) Second-order inelastic analysis for frame design. *Proceedings of National Symposium on Advances in Steel Structures*, India Institute of Technology, Madras, India, 7–9 February, pp. 461–474.

2. (With Yamaguchi E.) (1990) Microcrack propagation and softening behavior of concrete materials. *International Conference on Micromechanics of Failure of Quasi-Brittle Materials*, Society for Experimental Mechanics, Albuquerque, New Mexico, 6–8 June.
3. (With Mizuno E.) (1989) A cyclic plasticity model for sands. *Proceedings of the Second International Symposium on Plasticity and Its Current Applications*, Mie University, Japan, 31 July to 4 August.
4. (1990) Practical methods of second-order inelastic analysis for frame design. Keynote Lecture. *Proceedings of the Fourth Rail Bridge Centenary Conference on Developments in Structural Engineering*, Edinburgh, Scotland, 21–23 August, pp. 681–695.
5. (With El-Sheikh M.) (1990) Behavior and strength of telescopic steel shores. *Proceedings of the ASCE Materials Engineering Congress*, Denver, CO, 13–15 August.

1991

1. (With Huang TK.) (1989) Cap plasticity model for embankment: From theory to practice. *Proceedings of the Second International Symposium on Environmental Geotechnology*, Tongji University, Shanghai, China, 15–17 May, pp. 1–40.
2. (With Aboussalah M.) (1991) Nonlinear finite element analysis of concrete cylindrical shells subjected to hydrostatic pressure. *Proceedings of the Second International Conference on Computer Methods and Water Resources*, Marrakesh, Morocco, 18–22 February.
3. (With Zhou SP, Duan L.) (1991) The P-DELTA effect in portal frames. *Proceedings of the International Conference on Steel and Aluminum Structures*, Singapore, 22–24 May, pp. 713–722.
4. (With Lui EM.) (1991) Analysis and design of steel frames in the USA — 90's and beyond. *Proceedings of the International Conference on Steel and Aluminum Structures*, Singapore, 22–24 May, pp. 1–19.
5. (With Pan AD.) (1991) Finite element and finite block methods in geomechanics. *Proceedings of the Third International Conference on Constitutive Laws for Engineering Materials: Theory and Applications and Workshop on Innovative Use of Materials in Industrial and Infrastructure Design and Manufacturing*, Tucson, AZ, 7–11 January, pp. 669–675.
6. (1991) Design analysis of semi-rigid frames with LRFD. *Proceedings of the Second International Workshop on Connections in Steel Structures*, Pittsburgh, 10–12 April, pp. 370–379.

7. (With Lee HM, Pan AD.) (1991) Analysis of post-cracking behavior of steel-concrete structures by the finite block method. *Proceedings of the Third International Conference on Steel-Concrete Composite Structures*, Fukuoka, Japan, 26–29 September.
8. (With Pan AD.) (1991) State-of-the-art future developments of the finite element method. *Proceedings of the International Symposium on Concrete Engineering: New Technologies and Developments*, Nanjing, China, 18–20 September, pp. 1–12.
9. (With Al-Mashary F.) (1991) K-Factor paradox resolution. *Proceedings of Structures Congress*, Indianapolis, IN, 29 April to 1 May, pp. 191–194.
10. (With Liew JYR, White DW.) (1991) On K-factor for beam-column design in steel frameworks. *Proceedings of Structures Congress*, Indianapolis, IN, 29 April to 1 May, pp. 494–497.
11. (With Pan AD.) (1991) The finite block method. *Proceedings of Structures Congress*, Indianapolis, IN, 29 April to 1 May, pp. 693–696.
12. (With Abdel-Ghaffar M, White DW.) (1991) Simplified second-order inelastic analysis for steel frame design. *Proceedings of the Sessions on Approximate Methods of Structural Analysis and Design, and Verification Procedures*, Structures Congress, Indianapolis, IN, pp. 47–62.
13. (With White DW, Liew JYR.) (1991) Consideration of inelastic stability in frame design. *Proceedings of Annual Technical Session*, Structural Stability Research Council, Chicago, 15–17 April, pp. 65–76.
14. (With White DW, Liew JYR.) (1991) Considerations of inelastic stability in frame design. *Proceedings of the William McGuire Symposium*, Cornell University, Ithaca, NY, pp. 27–41.

1992

1. (With Liew JYR.) (1991) Seismic resistant design of steel moment resisting frames considering panel-zone deformation. *Proceedings of the US-Japan Seminar on Stability and Ductility of Steel Structures under Cyclic Loading*, Osaka, Japan, 1–8 July, pp. 323–334.
2. (With Kishi N, Matsuoka KW.) (1992) A determination of moment-rotation curve of connection with angles. *Proceedings of Structures Congress '92*, San Antonio, TX, 13–15 April, pp. 85–88.
3. (With Abdel-Ghaffar M.) (1992) Plastic analysis/design with LRFD. *Proceedings of Structures Congress '92*, San Antonio, TX, 13–15 April, pp. 222–225.
4. (With Cohen MD.) (1992) Micro mechanical considerations in concrete constitutive modeling. *Proceedings of Structures Congress '92*, San Antonio, TX, 13–15 April, pp. 270–273.

5. (With Pan AD *et al.*) (1992) Constitutive models. *Proceedings of Structures Congress '92*, San Antonio, TX, 13–15 April, pp. 860–863.

6. (With Abdel-Ghaffar M, White DW.) (1992) An error estimate and step size control method for nonlinear solution techniques. *Proceedings of Structural Stability Research Council Annual Technical Session and Meeting*, Pittsburgh, 6–8 April, pp. 291–302.

7. (With Duan L, Loh JT.) (1992) Behavior and strength of damaged tubular members in offshore structures. *Proceedings of the Second International Offshore and Polar Engineering Conference*, San Francisco, CA, 14–19 June, pp. 400–407.

8. (With Nomachi SG, Kida T, Sawada T.) (1992) On slope displacement by a logarithmic spiral failure slide during earthquake. *Proceedings of the Sixth International Symposium on Landslides*, Christchurch, New Zealand, 10–14 February, pp. 1193–1198.

9. (With Sawada T, Nomachi SG, Kida T.) (1992) Model test and analysis for seismic displacement of mounded foundation near a down-hill slope. *Proceedings of the International Symposium on Soil Improvement and Pile Foundation*, ISSIPF, Nanjing, China, 25–27 March, pp. 316–321.

10. (With White DW.) (1992) Advanced analysis for frame design. *Proceedings of the 1992 NSF Structures, Geomechanics and Building Systems Grantees Conference*, San Juan, Puerto Rico, 10–12 June, pp. 100–103.

1993

1. (With Pan AD.) (1992) Construction failure analysis of multistory concrete buildings. *Structural Failure, Product Liability and Technical Insurance 4: Proceedings of the Fourth International Conference on Structural Failure, Product Liability and Technical Insurance*, Vienna, Austria, 6–9 July, pp. 381–387.

2. (With Lu J, White DW.) (1993) Applying object-oriented design to finite element programming. *Proceedings of the 1993 Symposium on Applied Computing*, Association for Computing Machinery, Indianapolis, IN, February, pp. 424–429.

3. Concrete Plasticity: Macro and micro approaches. (1993) *Advances in Engineering Plasticity and Its Applications: Proceedings of the Asia-Pacific Symposium on Advances in Engineering Plasticity and Its Applications*, Elsevier Amsterdam, 1 May.

4. (With Sotelino ED, White DW.) (1993) High performance computing in civil engineering applications. *Proceedings of the Conference on Computer Applications in Civil and Hydraulic Engineering*, Tainan, Taiwan, October 15–16, pp. 1–18.

5. Preparing civil engineering research and education for the 21st century. (1993) *Proceedings of the Workshop on Transportation Development and Construction Research*, National Cheng-Kung University, Tainan, Taiwan, October 12–14, pp. 163–184.

6. (With Pan AD, Yen T.) (1993) Construction safety of temporary structures. *Proceedings of the Workshop on Transportation Development and Construction Research*, National Cheng-Kung University, Tainan, Taiwan, October 12–14, pp. 207–239.

1994

1. (With Lu J, White DW, Dunsmore HE, Sotelino E.) (1994) F++: An object-oriented application framework for finite element programming. *The Second Annual Object-Oriented Numeric Conference*, Sunriver, OR, 24–27 April, pp. 438–447.

2. (With Liew JYR.) (1994) Advanced analysis for semi-rigid frame design. *Proceedings of the International Workshop and Seminar on Behavior of Steel Structures in Seismic Area*, Timisoara, Romania, 26 June to 1 July.

3. (With Sotelino E, White DW.) (1994) An automated environment for parallel computing. *Proceedings of the 11th Analysis and Computation Conference*, Atlanta, GA, 24–28 April, pp. 193–202.

4. (With Shanmugam NE.) (1994) Column and beam-column design — A world view. *ASCE Structures Congress XII*, Atlanta, GA, 24–28 April, pp. 1581–1586.

5. (With Liew JYR.) (1994) Design implications of advanced analysis in LRFD. *ASCE Structures Congress XII*, Atlanta, GA, 24–28 April, pp. 1167–1172.

6. (With Chan SL, Zhou ZW.) (1994) Towards robust, effective and efficient second-order analysis of practical steel frames. *ASCE Structures Congress XII*, Atlanta, GA, 24–28 April, pp. 1173–1178.

7. (With Huang TK.) (1994) Plasticity analysis in geotechnical engineering: From theory to practice. *Developments in Geotechnical Engineering*, Balkema, Rotterdam, pp. 49–79.

8. (With Kishi N, Goto Y, Matsuoka KI.) (1994) Behavior of tall frame combining rigid and semi-rigid connections. *ASCE Structures Congress XII*, Atlanta, GA, 24–28 April, pp. 1179–1184.

9. Second-order inelastic analysis for frame design. (1994) *Proceedings of the 50th Anniversary Conference*, Structural Stability Research Council, Bethlehem, PA, pp. 197–209.

10. (With Goto Y, Miyashita S.) (1994) Classification of semi-rigid connections based on the elastic-plastic mechanical behavior of frames. *ASCE Structures Congress XII*, 24–28 April, pp. 1191–1196.

11. Concrete plasticity: Recent developments. (1994) *Proceedings of the 12th US National Congress of Applied Mechanics*, Seattle, Washington, June, pp. S86–S90.

12. (With Sawada T, Nomachi SG.) (1994) Stability analysis and model test for seismic displacement of a slope. *Proceedings of the International Symposium on Pre-Failure Deformation Characteristics of Geomaterials*, Sapporo, Japan, 12–14 September, pp. 665–671.

13. (With Chang TY.) (1994) Engineering mechanics in structural engineering research and education. *Proceedings of the International Symposium on the Role of Engineering Mechanics in the Training of Engineers*, pp. 7–22.

14. (With Luan M, Lin G, Pan AD, Wang YK.) (1994) FEM-based formulation and interpretation of discontinuous deformation analysis model in geomechanics. *Proceedings of the Workshop for Young Chinese Scholars on Geotechnical Engineering and Mechanics and Their Applications*, Wuhan, China, December, pp. 57–65. (In Chinese).

15. (With Hsieh SH, Sotelino ED and White DW.) (1994) An object-oriented environment for domain partitioning in parallel finite element computations. *The Second International Conference on Computational Structures Technology*, Athens, Greece.

16. (With Kishi N, Goto Y and Komoro M.) (1994) Sway analysis of tall building frames with mixed use of rigid and semi-rigid connections. *Proceedings of Constructional Steel*, Japan, pp. 53–60.

1995

1. Steel Connection Failures During the Northridge Earthquake. (1995) *Proceedings of the Workshop on Urban Disaster Mitigation and the Role of Engineering and Technology*, Chicago, IL, 15–16 July, pp. 167–176.

2. Strength of damaged cylindrical members under combined axial load and bending. (1995) *Proceedings of the 1995 Annual Technical Session and Theme Conference on Stability Problems Related to Aging Damaged and Deteriorated Structures*, Kansas City, 28–29 March.

3. (With Duan L.) (1995) Design equations for cylindrical steel members in offshore structures. *Proceedings of ASCE Structures Congress XIII*, Boston, MA, 2–5 April, pp. 1383–1398.

4. (With Luan M, Lin G.) (1995) Lumped-parameter model and nonlinear DSSI analysis. *Proceedings of the Third International Conference on Recent Advances in Geotechnical Earthquake Engineering and Soil Dynamics*, St. Louis, MO, 2–7 April, pp. 355–360.

5. (With Liew JYR.) (1995) Design of semi-rigid sway frames. *Proceedings of the Third International Conference on Steel and Aluminum Structures*, Bogazici University, Istanbul, Turkey, 24–26 May, pp. 293–300.

6. (With Huston DR, Rosowsky DV, Fuhr PL.) (1995) Construction shoring load measurements. *Proceedings of ASCE Structures Congress XIII*, Boston, MA, 3–5 April, pp. 1373–1377.

7. (1995) Second-order plastic analysis for frame design. *Proceedings of Plasticity '95*, 17–21 July, pp. 447–450.

8. (With Hsieh SH, Sotelino ED, White DW.) (1995) Toward object-oriented parallel finite element computations. *High Performance Computing*, Taiwan, 16–18 September.

9. (With Maleck AE, White DW.) (1995) Practical applications of advanced analysis in steel design. *Proceedings of the Fourth Pacific Structural Steel Conference*, Singapore, pp. 119–126.

10. (With White DW.) (1995) Characteristic semi-rigid connection relationships for frame analysis and design. *Proceedings of the Third International Workshop: Connections in Steel Structures III: Behavior, Strength and Design*, Italy, Pergamon, 29–31 May, pp. 299–308.

1996

1. (With Huston DR, Fuhr PL, Rosowsky DV.) (1996) Load monitoring and hazard warning systems for buildings under construction. *Proceedings of the Symposium on Smart Structures and Materials*, SPIE 2719-10, San Diego, CA.

2. (With Kim SE.) (1996) Practical advanced analysis for steel frame design. *Proceedings of the 12th Conference on Analysis and Computation*, Chicago, IL, 15–18 April, pp. 19–30.

3. (With Huang YL, Yen T.) (1996) A monitoring system for high-clearance scaffold systems during construction. *Proceedings of Structures Congress XIV*, Chicago, IL, 15–18 April, pp. 719–726.

4. (With Kim SE.) (1996) An innovative design of steel frame using advanced analysis. *Proceedings of the Third Asian Pacific Conference on Computational Mechanics*, Seoul, Korea, 16–18 September.

5. (1996) Advanced plastic analysis for steel frame design. *Workshop on Elasticity and Plasticity in Design, 15th Congress*, International Association for Bridge and Structural Engineering, Copenhagen, 16–20 June, pp. 863–868.

6. (With Kishi N, Goto Y, Komuro M.) (1996) Column moment of semi-rigid frames by means of B1/B2 method. *Proceedings of the International Conference on Advances in Steel Structures*, Hong Kong, 11–14 December, pp. 239–244.

7. (With Peng JL, Yen T, Pan AD, Chan SL.) (1996) Structural modeling and analysis of scaffold system. *Proceedings of the International Conference on Advances in Steel Structures*, Hong Kong, 11–14 December.

8. (With Peng JL, Pan AD, Yen T, Chan SL.) (1996) Simplified modeling of scaffold support systems, *The Third Structural Engineering Conference*, Taiwan, 1–3 September.

9. (With Ramesh G, Sotelino ED.) (1996) Effect of transition zone on the pre-peak mechanical behavior of mortar. *Proceedings of the Materials Engineering Conference* Annual Convention and Exposition, Washington, DC, 11–13 November, pp. 1238–1245.

1997

1. (With Zeng DR, Zhou SP.) (1997) Analysis of bird-shape plate with 45 m span. *Proceedings of the Fourth International Kerensky Conference*, Hong Kong, 3–5 September.

2. (1997) Structural Stability: From theory to practice. *Proceedings of Innovations in Structural Design: Strength, Stability, Reliability, A Symposium Honoring TV Galambos*, Minneapolis, Minnesota, 6–7 June, pp. 29–37.

3. (1997) Moment Frame Connections revisited: A fundamental fracture mechanics approach. *Proceedings of ANS Symposium on Trends in Structural Mechanics*, Waterloo, Canada, 10–12 July, pp. 255–272.

4. (1997) Design of steel structures with LRFD using advanced analysis. *Proceedings of the Fifth International Colloquium on Stability and Ductility of Steel Structures*, Nagoya, Japan, 29–31 July, pp. 21–32.

5. (With Sotelino SE.) (1997) The development of object-oriented environment for parallel computing in structural engineering. *Proceedings of the Sixth International Conference on Education and Practice of Computational Methods in Engineering and Science*, Guangzhou, China, 4–7 August, pp. 3–25.

6. (With Zeng DR, Zhou SP.) (1997) Analysis of bird-shape plate with long span. *Proceedings of the Sixth International Conference on Education and Practice of Computational Methods in Engineering and Science*, Guangzhou, China, 4–7 August, pp. 122–127.

7. (With Duan MZ.) (1997) Improved simplified method for slab and shore load analysis during construction. *Material Modelle und Methoden zur wirklichkeitsnahen Berechnung von Beton-, Stahlbeton-und Spannbetonbauteilen*, A Special Volume Honoring G Mehlhorn, Kassel, University of Kassel, Germany, pp. 219–226.

8. (With Lan M, Sotelino ED.) (1997) 3D modeling of RC columns retrofit with advanced composite jackets. *Proceedings of the ASME Applied Mechanics Division*, Northwestern University, Evanston, IL, 29 June to 2 July.

1998
1. (1998) Concrete plasticity: Past, present and future. *Proceedings of the International Symposium on Strength Theory: Application, Development and Prospects for 21st Century*, Xian, China, pp. 7–49.
2. (With Chan SL.) (1998) Design of unbraced semi-rigid steel frames by advanced analysis. *Proceedings of the International Symposium on Strength Theory: Application, Development and Prospects for 21st Century*, Xian, China, 9–11 September, pp. 1001–1008.
3. (With Huang TK.) (1998) Strong motion analysis of geotechnical engineering. *Proceedings of Symposium on Taiwan Strong Motion Instrumentation Program*, Alishan, Taiwan, 10–11 June, pp. 1–10.
4. (1998) Structural engineering: Past, present and future. *Proceedings of the Fourth National Conference on Structural Engineering*, Taipei, Taiwan, 9–11 September, pp. 1–16.
5. (1998) Frames with partially restrained connections, *Proceedings of the Theme Conference Workshop on Frames with Partially Restrained Connections*, Annual Technical Session and Meeting, Structural Stability Research Council, Atlanta, GA, 21–23 September.
6. (With Kim SE.) (1998) Development of refined plastic hinge analysis accounting for strain reversal. *Proceedings of the Fifth Pacific Structural Steel Conference*, Seoul, Korea, 13–16 October, pp. 259–264.
7. (With Kishi N, Goto Y, Komuro M.) (1998) Applicability of *B2* method on sway analysis of flexibly joint frames. *Proceedings of the Fifth Pacific Structural Steel Conference*, Seoul, Korea, 13–16 October, pp. 271–276.
8. (With Joh CB.) (1998) Fracture strength of welded flange-bolted web connections with backing bars. *Proceedings of the Fifth Pacific Structural Steel Conference*, Seoul, Korea, 13–16 October, pp. 759–764.
9. (With Kim Y.) (1998) Practical analysis for partially restrained frame design. *Proceedings of the Fifth Pacific Structural Steel Conference*, Seoul, Korea, 13–16 October, pp. 807–812.
10. (With Sotelino ED.) (1998) Future challenges for simulation in structural engineering. *Proceedings of the Fourth World Congress on Computational Mechanics*, Buenos Aires, Argentina, 30 June to 3 July. (In CD-ROM.)

1999

1. (With Duan L.) (1999) Seismic design methodologies and performance-based criteria. *Proceedings of the International Workshop on Mitigation of Seismic Effects on Transportation Structures*, Taipei, Taiwan, 12–14 July, pp. 130–141.
2. (With Peng JL and Pan AD.) (1999) An approximate method of second-order inelastic analysis for modular tubular falsework. *Proceedings of the First International Conference on Structural Engineering*, Kunming, Yunnan, China, 18–20 October.
3. (With Sotelino ED.) (1999) Modeling and simulation of concrete materials. *Proceedings of the Seminar on Post Peak Behavior of RC Structures Subjected to Seismic Loads*, Lake Yamanaka, Japan, 25–29 October, pp. 1–21.

2000

1. (With Doo CS, Sotelino ED.) (2000) Practical approach to cracking of welded moment-resisting frames. *Proceedings of the Symposium Honoring Professor Alfredo H-S Ang's Career*, University of Illinois at Urbana-Champaign, 19–20 May. (In CD-ROM.)

2001

1. (With Oh SH, Joh CB, Chang IH.) (2001) Slab effects on the seismic behavior of steel moment connections. *Proceedings of the Sixth Pacific Structural Steel Conference*, Beijing, China, pp. 539–544.
2. (2001) Foreword. *Bulletin*, No. 2, Hong Kong Institute of Steel Construction, Hong Kong.
3. (With Komuro M, Kishi N, Yabuki N.) (2001) Dynamic response analysis of semi-rigid portal frames under harmonic wave excitation. *The First International Conference on Steel and Composite Structures*, Busan, Korea, 14–16 June, pp. 335–342.

2002

1. (2001) High performance computing in civil engineering: Challenges and opportunities. *Proceeding of the International Conference on Innovation and Sustainable Development of Civil Engineering in the 21st Century*, Beijing, China, 1–3 August, pp. K17–K21.

2003

1. (With Kishi N, Komuro M.) (2003) Seismic response analysis of steel frame with mixed use of rigid and semi-rigid connections. *Behavior of Steel Structures in Seismic Areas*, Naples, Italy, 9–12 June, pp. 527–532.

2004

1. (With Komuro M, Kishi N.) (2004) Elasto-plastic FE analysis on moment-rotation relations of top- and seat-angle connections. *Connections in Steel Structures*, Amsterdam, 3–4 June, pp. 111–120.
2. (With Kishi N, Komuro M.) (2004) Four-parameter power model for M-theta curves of end-plate connections. *Connections in Steel Structures*, Amsterdam, 3–4 June, pp. 99–110.

2005

1. (With Liu XL.) (2005) Developments of structural calculations. *Proceedings of the International Symposium on Innovation and Sustainability of Structures in Civil Engineering — Including Seismic Loading*, Nanjing, China, 20–22 November.

2006

1. (With Chan SL.) (2006) Advanced analysis of steel frames — From theory to practice. *Proceeding of the XI International Conference on Metal Structures: Progress in Steel, Composite and Aluminum Structures*, Rzeszow, Poland, 21–23 June, pp. 20–29.

2007

1. (With Iu CK, Ma TW, Chan SL.) (2007) Integrating design and analysis: A system approach. *Proceedings of the International Conference on Recent Developments in Structural Engineering*, Department of Civil Engineering, Manipal Institute of Technology, Manipal, India, 30–31 August, pp. 31–41.
2. (With Cho SH, Chan SL, Ma D.) (2007) Modern simulation-based structural design. *Proceedings of the International Conference on Recent Developments in Structural Engineering*, Department of Civil Engineering, Manipal Institute of Technology, Manipal, India, 30–31 August, pp. 54–62.
3. (With Xu MX, Yang CY.) (2007) Statistical uncertainties in estimation of distribution parameters. *Proceedings of the Second International Conference on Structural Condition Assessment, Monitoring and Improvement*, China, 19–21 November, pp. 108–115.

2008

1. (With Komuro M, Kishi N.) (2008) Experimental study on aseismic strengthening method of steel-pipe piers using aramid fiber sheet. *Proceedings of the Annual Stability Conference*, Structural Stability Research Council, Nashville, Tennessee, 2–5 April.

2. (2008) Seeing the big picture and making a difference in structural engineering. *Proceedings of the 11th East Asia Pacific Conference on Structural Engineering and Construction, Building a Sustainable Environment*, Taipei, Taiwan, 19–21 November.

3. (With Xu MX, Yang CY.) (2008) Adaptability of interpolated curves to the displacement-based method for reliability analysis. *Proceedings of the First International Symposium on Innovation and Sustainability of Modern Railway*, Nanchang, China, 16–17 October, pp. 278–286.

2014

1. (With Peng JL, Chan SL.) (2014) Leaning column effect on load capacity of building falsework used in Taiwan. *Proceedings of the Workshop of Practice of Building Demolition and Temporary Works*, The Hong Kong Polytechnic University, the Hong Kong Institute of Steel Construction and the Hong Kong Institution of Engineers, Hong Kong, 27 June, pp. 41–48.

7.3 Invited Lectures, Seminars and Conference Organizer (110)

1972

1. **Topical Lecture:** (1972) Connections, state-of-the-art. *International Conference on Planning and Design of Tall Buildings*, Bethlehem, PA, 22 August.

1977

1. **US Delegate:** (1977) Mechanics of slope failures and landslides. *US-ROC Joint Meeting on Earthquake Engineering and Landslides*, Taipei, Taiwan, 29 August to 2 September.

2. **Keynote Lecture:** (1977) Theory of beam-columns — State-of-the-art review. *International Colloquium on Stability of Structures under Static and Dynamic Loads*, Washington, DC, 17 March.

1979

1. **Theme Lecture:** (1979) Strength of fabricated tubular columns in offshore structures. *International Conference on Thin-Walled Structures*, Glasgow, Scotland, 3–6 April.

2. **Theme Reporter:** (1979) Material behavior under various types of loading. *Workshop on High Strength Concrete*, Chicago, 2–4 December.

3. **Keynote Lecture:** (1979) Constitutive equations for concrete. *IABSE Colloquium on Plasticity in Reinforced Concrete, International Association for Bridge and Structural Engineering*, Copenhagen, Denmark, 21 May.

4. **Topical Lecture:** (1979) Nonlinear analysis of pressure-resistant concrete cylinder structures. *International Conference on Engineering Applications of the Finite Element Method*, Hovik, Norway, 9–11 May.

1980

1. **US Delegate:** (1980) Constitutive modeling of concrete materials. *US-Japan Joint Seminar on Finite Element Analysis of Reinforced Concrete Structures*, Tokyo, May.

1981

1. **Model Predictor:** (1981) Plasticity models for soils and nonlinear hyperelastic models for soils. *Workshop on Limit Equilibrium, Plasticity and Generalized Stress-Strain in Geotechnical Engineering*, Montreal, Canada.

2. **US Delegate:** (1981) Recent advances on analysis and design of steel beam-columns in USA. *US-Japan Seminar on Inelastic Instability of Steel Structures and Structural Elements*, Tokyo.

3. **Keynote Lecture:** (1981) Plasticity in reinforced concrete. *Workshop on Constitutive Relations for Concrete*, Albuquerque, NM, April.

1983

1. **Topical Lecture:** (1983) A five-parameter mixed-hardening model for concrete materials. *Symposium on Plasticity Today, International Centre for Mechanical Sciences*, Udine, Italy, 28 June.

2. **Discusser's Report:** (1983) Constitutive relations for concrete, rock and soils. *IUTAM Prager Symposium on Mechanics of Geomaterials: Rocks, Concrete and Soils*, Chicago, 12 September.

3. **Topical Lecture:** (1983) Soil mechanics, plasticity and landslides. *Daniel C. Drucker Symposium on the Mechanics of Material Behavior*, Urbana, IL, 6 June.

4. **Chairman:** (1983) *ASCE/EMD Specialty Conference*, Purdue University, 23–25 May. (JTP Yao, Co-Chairman.)

5. **Topical Lecture:** (1983) Constitutive modeling in soil mechanics. *International Conference on Constitutive Laws for Engineering Materials*, Tucson, AR, 10–12 January.

6. **Topical Lecture:** (1983) Failure criteria of concrete materials. *CNRS International Colloquium on Failure Criteria of Structured Media*, Grenoble, France.

7. **Theme Reporter:** (1983) Design of beam-column in North America. *The Third International Colloquium on Stability of Metal Structures*, Toronto, Canada, 9 May.

8. **Seminar:** (1983) Design of columns in LRFD. *Department of Civil Engineering, Stanford University*, Stanford, CA, October.

9. **Topical Reporter:** (1983) On concrete constitutive modeling. *Symposium on Non-Nuclear Structures*, Colorado Springs, Colorado, 10–13 May.

1984

1. **Seminar:** (1984) Structural steel design in North America. Tam-Kang University, Taipei, Taiwan, 2 March.

2. **Keynote Lecture:** (1984) Steel column design using LRFD. *International Conference on Steel Structures*, Singapore, 7–9 March.

3. **Topical Lecture:** (1984) A nonuniform hardening plasticity model for concrete materials. *DOD/ONR Workshop*, Northwestern University, Chicago, 19–21 November.

1985

1. **TR Higgins Lecture:** (1985) Columns with end restraint and bending in load and resistance factor design. Structural Stability Research Council Annual Meeting, Cleveland, OH, 16 April.

2. **Seminar:** (1985) Columns with end restraint and bending in LRFD. Georgia Institute of Technology, Atlanta, 24 November.

3. **Seminar:** (1985) Design of columns under biaxial loading. National Bureau of Standards, Washington, DC, 6 March.

4. **Special Lecture:** (1985) Column design in LRFD. American Institute of Steel Construction and Association of Structural Engineers in Greater Chicago Area, May.

5. **Seminar:** (1985) On constitutive equations of concrete materials. University of Kassel, Kassel, West Germany, 19 June.

6. **Seminars:** (1985) Introduction to material modeling. Concrete Plasticity and Limit Analysis, and Soil Plasticity and Numerical Implementation, ETH, Zurich, 16–17 June.

7. **Seminar:** (1985) End restraint and column design using LRFD. University of Kassel, Kassel, West Germany, 26 June.

8. **Seminar:** (1985) Constitutive modeling of reinforced concrete. Japan Concrete Institute, Tokyo, 20 May.
9. **US Delegate:** (1985) On constitutive modeling of concrete materials. *US-Japan Joint Seminar on Finite Element Analysis of Reinforced Concrete Structures*, Tokyo, 21 May.

1987

1. **Topical Lectures:** (1987) Moment-rotation relation of single/double web-angle connections, and Moment-rotation relation of top- and seat-angle with double web-angle connections. *Workshop on Connections and the Behavior, Strength and Design of Steel Structures*, Cachan, France, 26 May.
2. **Theme Lectures:** (1987) Application of a bounding surface model, and On micro-mechanics of fracture and constitutive modeling of concrete materials. *The Second International Conference on Constitutive Laws for Engineering Materials*, Tucson, Arizona, 5–10 January.

1988

1. **Model Predictor:** (1988) Evaluation of constitutive models in soil mechanics, and An elastic-plastic two-surface model for non-cohesive soils, *Workshop on Constitutive Equations for Granular Non-Cohesive Soils*, Cleveland, OH.
2. **Theme Lecture:** (1988) Connection flexibility in steel structures. *International Conference on Second Century of the Skyscraper*, Chicago.
3. **Theme Reporter:** (1988) Computers in stability research. *Structural Stability Research Council Annual Technical Sessions and Meeting*, Minneapolis, MN, 26 April.

1989

1. **Seminar:** (1989) Finite element versus finite block theories in geomechanics. Department of Earth and Atmospheric Sciences, Purdue University, West Lafayette, IN, 4 October.
2. **Theme Reporter:** (1989) Design of beam-columns in frames. *The Fourth International Colloquium on Stability of Metal Structures: Code Difference around the World*, New York, 17 April.
3. **Keynote Lecture:** (1989) Cap plasticity model for embankment: From theory to practice. *The Second International Symposium on Environmental Geotechnology*, Shanghai, PRC, 15 May.

1990

1. **Seminar Tour:** (1990) New issues facing research and education computing in structural engineering, second-order inelastic analysis of steel buildings, finite element analysis, constitutive modeling of materials, and structural engineering. Some New Topics in Structural Engineering, and Preparing Structural Engineering Research and Education for 21st Century, at Kyushu University, Kumamoto University, Nagoya University, Nagoya Institute of Technology, Tokyo University, Hokkai-Gakuen University, and Muroran Institute of Technology. The Japan Society for Promotion of Science, Japan, 22 May to 22 June.

2. **Topical Lecture:** (1990) Microcrack propagation and softening behavior of concrete materials. *International Conference on Micromechanics of Failure of Quasi-Brittle Materials, Society of Experimental Mechanics,* Albuquerque, NM, 7 June.

3. **Keynote Lecture:** (1990) Second-order inelastic analysis for frame design. *National Symposium on Advances in Steel Structures,* Madras, India, 7 February.

4. **Keynote Lecture:** (1990) Practical methods of second-order inelastic analysis for frame design. *The Fourth Rail Bridge Centenary Conference on Developments in Structural Engineering,* Edinburgh, Scotland, 22 August.

1991

1. **Seminars:** (1991) Second-order inelastic analysis for frame design, and Finite block methods in structural engineering. National University of Singapore and Nanyang Technological Institute, Singapore, 22–24 May.

2. **Theme Lecture:** (1991) Constitutive models. *International Workshop on Finite Element Analysis of Reinforced Concrete,* Columbia University, New York, 4 June.

3. **Seminar:** (1991) Finite element and finite block analysis in structural engineering. Center for Advanced Technology for Large Structural Systems, Lehigh University, Bethlehem, PA, 24 October.

4. **Keynote Lecture:** (1991) Analysis and design of steel frames in the USA — 90's and beyond. *International Conference on Steel and Aluminum Structures,* Singapore, 22 May.

5. **Topical Lecture:** (1991) Finite element and finite block methods in geomechanics. *The Third International Conference on Constitutive Laws for Engineering Materials,* Tucson, 8 January.

6. **Topical Lecture:** (1991) Design analysis of semi-rigid frames with LRFD. *The Second International Workshop on Connections in Steel Structures,* Pittsburgh, 11 April.

7. **Keynote Lecture:** (1991) State-of-the-art future developments of the finite element method. *International Symposium on Concrete Engineering*, Nanjing, PRC, 18 September.
8. **US Delegate:** (1991) Seismic resistant design of steel moment resisting frames considering panel-zone deformation. *US-Japan Seminar on Stability and Ductility of Steel Structures under Cyclic Loading*, Osaka, Japan, 2 July.
9. **SSSS Annual Lecture:** (1991) New developments in stability design of steel structures. Singapore Structural Steel Society, Singapore, 20 May.
10. **Special Lectures:** (1991) Workshop on plasticity theory and structural stability. National Chung-Hsing University, Taichung, Taiwan, 9 Lectures, 11–15 May.
11. **Seminars:** (1991) Finite block method in civil engineering. National Cheng-Kung University, Tainan, and National Central University, Taipei, Taiwan, May.

1992

1. **Seminar:** (1992) *Finite Block Analysis in Structural Engineering*, Hong Kong Polytechnic, Hong Kong, 17 December.
2. **Keynote Lecture:** (1992) Concrete plasticity: Macro and micro approaches. *Asia-Pacific Symposium on Advances in Engineering Plasticity and Its Applications*, Hong Kong Polytechnic, Hong Kong, 15 December.
3. **Theme Lecture:** (1992) Beam-columns. *First World Conference on Constructional Steel Design*, Acapulco, Mexico, 7 December.
4. **Invited Lecture:** (1992) Construction failure analysis of multistory concrete buildings. *International Conference on Structural Failure, Product Liability and Technical Insurance*, Vienna, Austria, 8 July.
5. **Seminar:** (1992) Finite element and finite block analysis in civil engineering. Syracuse University, Syracuse, NY, 3 April.
6. **Theme Lecture:** (1992) Introduction and scope. *Workshop on Plastic Hinge Based Methods for Advanced Inelastic Analysis and Design of Steel Frames*, Pittsburgh, PA, 5 April.

1993

1. **Keynote Lecture:** (1993) High performance computing in civil engineering. *Workshop on Applications of Computer to Civil and Hydraulic Engineering*, National Cheng-Kung University, Tainan, Taiwan, 15 October.
2. **Theme Lecture:** (1993) Preparing civil engineering research and education for the 21st century. *Workshop on Transportation Development and Construction Research*, National Cheng-Kung University, Tainan, Taiwan, 14 October.

3. **Theme Lecture:** (1993) *Workshop on Transportation Development and Construction Research*, National Cheng-Kung University, Tainan, Taiwan, 14 October.

4. **Seminar:** (1993) Finite element and finite block methods in civil engineering. Northwestern University, Chicago, 2 April.

1994

1. **Keynote Lecture:** (1994) Engineering mechanics in structural engineering research and education. *International Symposium on the Role of Engineering Mechanics in the Training of Engineers*, Tsinghua University, Beijing, China, 23 June.

2. **Seminar:** (1994) Steel moment connection failures during Northridge Earthquake. Ministry of Transportation and Communications, Taipei, Taiwan, 6 July.

3. **Theme Lecture:** (1994) Concrete plasticity: Recent developments. *Topic Symposium on Geomechanics, 12th US National Congress of Applied Mechanics*, University of Washington, Seattle, Washington, 26 June to 1 July.

4. **Theme Lecture:** (1994) Toward life-cycle analysis of concrete structures. Council of Labor Affairs of Taiwan, Taipei, Taiwan, 4–5 July.

5. **Topical Lecture:** (1994) Second-order inelastic analysis. 50th Anniversary Meeting, Structural Stability Research Council, Lehigh University, Bethlehem, PA, 20–22 June.

6. **Seminars:** (1994) The role of engineering mechanics in structural engineering and education. Chinese Society of Structural Engineers, Taipei, Taiwan, 6 July. Also, NSF Center for Advanced Technology for Large Structural System, Lehigh University, Bethlehem, PA, 28 October.

1995

1. **Theme Lecture:** (1995) Second-order plastic analysis for frame design. *The Fifth International Symposium on Plasticity and Its Current Applications*, Osaka, Japan, 17–21 July.

1996

1. **Keynote Lecture:** (1996) From research to practice. *The Fourth International Symposium on Structural Engineering for Young Experts*, Beijing, China, 26 August.

2. **Organizer:** (With White DW.). (1996) US-Japan seminar on advanced stability concepts and methods for seismic design in structural steel. National Science Foundation and Japan Society for Promotion of Science, Honolulu, 15–18 July.

3. **Public Lecture:** (1996) From research to practice: A structural engineer's own reflection. Hong Kong Institution of Engineers, Hong Kong, 13 June.
4. **Seminars:** (1996) High performance computing in structural engineering, and Steel connection failures during the Northridge Earthquake. The University of Hong Kong, 10 and 14 June.
5. **Theme Speaker:** (1996) Advanced plastic analysis for steel frame design. *Workshop on Elasticity and Plasticity in Design, 15th Congress of the International Association for Bridge and Structural Engineering*, Copenhagen, Denmark, 17 June.
6. **Seminar:** (1996) Steel connection failure during Northridge Earthquake. Tsinghua University, Beijing, China, 27 August.

1997

1. **Keynote Lecture:** (1997) From research to practice: A structural engineer's own reflection. Sigma Xi Research Society Annual Banquet, Purdue University, 25 September.
2. **Seminar:** (1997) Advanced analysis made LRFD design easy. Washington University, St. Louis, 24 March.
3. **Theme Lecture:** (1997) Structural stability: From theory to practice. *A Symposium Honoring TV Galambos*, Minneapolis, Minnesota, 6 June.
4. **Invitation Lecture:** (1997) Design of steel structures with LRFD using advanced analysis. *The Fifth International Colloquium on Stability and Ductility of Steel Structures*, Nagoya, Japan, 29 July.
5. **Keynote Lecture:** (1997) The development of object-oriented environment for parallel computing in structural engineering. *The Sixth International Conference on Education and Practice of Computational Methods in Engineering and Science*, Guangzhou, China, 4 August.
6. **Invited Speaker:** (1997) Moment frame connections revisited: A fundamental fracture mechanics approach. *ANS Symposium on Trends in Structural Mechanics*, Waterloo, Canada, 10 July.

1998

1. **Keynote Lecture:** (1998) Concrete plasticity: Past, present and future. *International Symposium on Strength Theories, Applications and Developments*, Xian, China, 9–11 September.
2. **Keynote Lecture:** (1998) Structural engineering, past, present and future. *The Fourth National Conference on Structural Engineering*, Taipei, Taiwan, 9–11 September.

3. **Keynote Lecture:** (1998) Frames with partially restrained connections. *Theme Conference Workshop on Frames with Partially Restrained Connections*, Structural Stability Research Council, Atlanta, GA, 23 September.

4. **Keynote Lecture:** (With Sotelino E and White DW.) (1998) Future challenges for structural engineering simulation. *The Fourth World Congress on Computational Mechanics*, Buenos Aires, Argentina, 29 June to 2 July.

1999

1. **Keynote Lecture:** (With Liew R.) (1999) Advanced inelastic analysis of frame structures. *The Sixth International Colloquium on Stability and Ductility of Steel Structures*, Timisoara, Romania, 9–11 September.

2000

1. **Keynote Speech:** (2000) Telecommunication of the future. *The Public Safety Communications Conference 2000*, Honolulu Police Department, McCoy Pavilion, Honolulu, Hawaii, 27 July.

2. **Keynote Speech:** (2000) Vision of the new millennium: Career flexibility in rapidly changing times. *Association of Chinese Scholars in Hawaii's Annual Meeting and Symposium on the Beginning of a New Millennium: A Critical Path of Political, Social and Technological Transformation*, East-West Center, University of Hawaii at Manoa, Honolulu, Hawaii, 25 June.

3. **Banquet Speaker:** (2000) How to build a first-rate engineering College at UH. ASCE/SEAOH Banquet, Honolulu, Hawaii, 18 May.

4. **Colloquium Speaker:** (2000) Modeling and simulation: Linking HPC to engineering. Physics Colloquium, University of Hawaii at Manoa, Honolulu, Hawaii, 23 March.

5. **Luncheon Speaker:** (2000) The future direction for the University of Hawaii's College of Engineering. Engineers and Architects of Hawaii, Honolulu, Hawaii, 31 March.

6. **Keynote Lecture:** (2000) Model-based simulation in civil engineering: Challenges and opportunities. *Chinese Institute of Civil and Hydraulic Engineering Conference on Computer Applications to Civil and Hydraulic Engineering*, Taichung, Taiwan, 17 February.

7. **Special Seminar:** (2000) Modeling and simulation — Linking HPC to engineering. National Center for High-Performance Computing, Taipei, Taiwan, 16 February.

8. **Seminar:** (2000) Plasticity, limit analysis, and structural design. National Taiwan University, Taipei, Taiwan, 16 February.

9. **Special Lecture:** (2000) How to build a first-rate university. Chao-Yang University of Technology, Taichung, Taiwan, 17 February.

2001

1. **Theme Lecture:** (2001) High tech and new economy in Hawaii: Tomorrow land, today. *Association of Chinese Scholars in Hawaii's Annual Meeting and Symposium on The Vision of Chinese-Based Economy in the New Millennium*, East-West Center, University of Hawaii at Manoa, Honolulu, Hawaii, 23 June.

2002

1. **Keynote Speech:** (2002) High performance computing in civil engineering: Challenges and opportunities. *International Conference on Innovation and Sustainable Development of Civil Engineering in the 21st Century*, Beijing, China, 2 August.
2. **Luncheon Speech:** (2002) Engineering and education. *Society of American Military Engineers*, Pearl Harbor, Hawaii, 14 May.

2005

1. **Keynote Speech:** (With Kim SE.) (2005) Advanced analysis for steel design: From theory to practice. *The Fourth International Conference on Advances in Steel Structures*, Shanghai, China, 13–15 June.
2. **Keynote Speech:** (With Liu XL.) (2005) Development of structural calculations. *International Symposium on Innovation and Sustainability of Structures in 'Civil Engineering — Including Seismic Engineering*, Nanjing, China, 20–22 November.

2006

1. **Keynote Speech:** (With Chan SL.) (2006) Advanced analysis for steel design. *The XI International Conference on Metal Structures*, Rzeszow, Poland, 21–23 June.
2. **Keynote Speech:** (2006) Academic achievements and contributions of academician Chen. Bureau of High Speed Rails, Taiwan, 28 June. In *The Magazine of the Chinese Institute of Civil and Hydraulic Engineering*, Vol. 33, No. 4, August 2006, pp. 1–7.

2008

1. **Keynote Speech:** (2008) My life journey: Connecting the dots. Asian/Pacific American Heritage Month, the National Science Foundation, Washington DC, 14 May.

2. **Luncheon Lecture:** (2008) My life journey: Reflections of an academic. Center of Biographical Research, University of Hawaii, Honolulu, Hawaii, 31 January.
3. **Luncheon Lecture:** (2008) My life journey: Connecting the dots. Engineers and Architects of Hawaii, Honolulu, Hawaii, 18 April.
4. **Keynote Speech:** (2008) Seeing the big picture and making a difference in structural engineering. The 11th East Asia Pacific Conference on Structural Engineering and Construction, Taipei, Taiwan, 19 November.

2010
1. **Keynote Speech:** (2010) The art of engineering — Simplicity. *The 14th Construction Engineering and Management Conference*, Taipei, Taiwan, 2 July.

CHAPTER 8
PHD DISSERTATIONS
博士論文目錄

Dr Chen has been the advisor of 55 PhD students from Lehigh University, Purdue University and the University of Hawaii.

8.1 Lehigh University, 1966–1976

1. Santathadaporn, Sakda (1970) *Analysis of Biaxially Loaded Columns.*
2. Fielding, David J. (1971) *Structural Behavior of Welded Beam to-Column Connections.*
3. Atsuta, Toshio (1972) *Analyses of Inelastic Beam-Columns.*
4. Huang, Joseph S. (1973) *Behavior and Design of Steel Beam-to-Column Moment Connections.*
5. Chen, A.C.T. (1973) *Constitutive Relations of Concrete and Punch Indentation Problems.*
6. Davidson, Hugh L. (1974) *Elastic-Plastic Large Deformation Response of Clay to Footing Loads.*
7. Snitbhan, Nimitchai (1975) *Plasticity Solutions for Slopes in Anisotropic Inhomogeneous Soil.*
8. Ross, David Alexander (1978) *The Strength and Behavior of Fabricated Tubular Steel Columns.*
9. Rentschler, Glenn P. (1979) *Analysis and Design of Steel Beam-to-Column Web Connections.*

8.2 Purdue University, 1976–1999

1. Toma, Shouji (1980) *Analysis and Design of Fabricated Tubular Beam-Columns.*
2. Al-Noury, Soliman Ibrahom I. (1980) *A Study of Reinforced Concrete Compression Members under Biaxial Bending.*
3. Patel, Kirit Vallabhbhai (1981) *Analysis of Steel Beam-to-Column Moment Connections.*

4. Saleeb, AF. (1981) *Constitutive Models for Soils in Landslides.*
5. Mizuno, E. (1981) *Plasticity Modeling of Soils and Finite Element Applications.*
6. Chang, Chemg-Jung (1981) *Seismic Safety Analysis of Slopes.*
7. Chang, MF. (1981) *Static and Seismic Lateral Earth Pressure on Rigid Retaining Structures.*
8. Hsieh, SS. (1981) *Elastic-Plastic-Fracture Analysis of Concrete Structures.*
9. Sugimoto, Hironori (1983) *Study of Offshore Structural Members and Frames.*
10. Han, Dajian (1984) *Constitutive Model in Analysis of Concrete Structures.*
11. Larralde, Jesus Silverio (1984) *Structural Analysis of Rigid Pavement with Pumping.*
12. Lui, Eric Mun (1985) *Effects of Connection Flexibility and Panel Zone Deformation on the Behavior Plane Steel Frames.*
13. McCarron, Willian Oris (1985) *Soil Plasticity and Finite Element Applications (Cap Model, Bounding Surface, Boston Blue Clay, Embankment Reinforcement).*
14. Liu, Xila (1985) *Analysis of Reinforced Concrete Building During Construction (Computer Model, Creep, Random, Nonlinear).*
15. Sohal, Iqbal Singh (1986) *Local Buckling in the Analysis of Cylindrical Members.*
16. El-Metwally, Salah El-Din El-Said (1986) *Nonlinear Analysis of Reinforced Concrete Frames (Joint Flexibility).*
17. Ohtani, Yasuhiro (1987) *Constitutive Modeling of Concrete Materials for Engineering Applications.*
18. Yamaguchi, Eiki (1987) *Microcrack Propagation and Softening Behavior of Concrete Materials.*
19. Barakat, Munzer Abdel-Karim (1988) *Simplified Design Analysis of Frames with Semi-rigid Connections.*
20. Wu, Fu-Hsiang (1988) *Semi-rigid Connections in Steel Frames.*
21. El-Shiekh, Magdy (1989) *Effects of Construction Process on R/C Buildings.*
22. Al-Mashary, Faisal Abdullah (1989) *Simplified Nonlinear Analysis for Steel Frames.*
23. Aboussalah, Mohamed (1989) *Application of Constitutive Models to Concrete Structures.*
24. King, Wonsun (1990) *Simplified Second-order Inelastic Analysis for Frame Design.*
25. Huang, Tien-Kuen (1990) *The Application of Capped Plasticity Model to Embankment Problems.*
26. Duan, Lian (1990) *Stability Analysis and Design of Steel Structures.*
27. Zhang, Hong (1991) *A Structural Engineering Software Development Environment.*

28. Mossalam, KH. (1991) *Construction Loads and Load Effects in Concrete Building Construction.*

29. Liew, Jat-Yuen Richard (1992) *Advanced Analysis for Frame Design.*

30. Abdel-Ghaffar, Maheeb ME. (1992) *Post Failure Analysis for Steel Structures.*

31. El-Shahhat, Ashraf Mohamed (1993) *Safety of Concrete High-rise Buildings during Construction.*

32. Peng, Jui-Lin (1994) *Analysis Models and Design Guidelines for High-Clearance Scaffold Systems.*

33. Lu, Jun (1994) *An Object-oriented Application Framework for Finite Element Analysis in Structural Engineering.*

34. Huang, Yue-Lin (1995) *Gravity Load and Resistance Factor Design Guidelines for High Clearance Scaffold Systems.*

35. Kim, Seung-Eock (1996) *Practical Advanced Analysis for Steel Frame Design.*

36. Duan, Ming-Zhu (1996) *Design Guidelines for Safe Construction of Concrete Buildings.*

37. Gambheera, Rsmesh (1997) *Micromechanical Study of Concrete Materials with Interfacial Transition Zone.*

38. Yang, Wei-Hong (1997) *The Behavior and Design of Unstiffened Seated-Beam Connections.*

39. Joh, Changbin (1998) *Application of Fracture Mechanics to Steel Moment Connections.*

40. Kim, Yosuk (1998) *Practical Analysis for PR Frame Design.*

41. Lai, Chih-Hung (1998) *Analysis and Design of Jack-Up Chords.*

42. Chen, I-Hong (1999) *Practical Advanced Analysis in Seismic Design of Steel Building Frames.*

43. Doo, Chung-Soo (1999) *Fracture Design Criteria for Special Moment-Resisting Frames.*

44. Wongkaew, Kamaiton (2000) *Practical Advanced Analysis for Design of Laterally Unrestrained Steel Planar Frames Under In-plane Loads.*

45. Cheng, Hsiao-Lin (2000) *A Study of FRP Wrapped Reinforced Concrete Columns.*

8.3 University of Hawaii, 1999–2006

1. Hwa, Ken (2003) *Toward Advanced Analysis in Steel Frame Design.*

POSTSCRIPT
後記 - 師生情誼

It is my great honor, as a former student of Prof. Wai-Fah Chen, to edit Plasticity, Limit Analysis, Stability and Structural Design, an esteemed collection (陳惠發院士文集) of his academic works and accomplishments during his 40-year academic life journey.

Time flies! My friendship with Prof. Chen started in 1985. Professor Chen, a world-renowned authority on solid mechanics, structural engineering and geotechnical engineering, had been at the forefront of scientific research and education and made a profound contribution from theory to practice. His rigorous academics and teaching by example made me understand the soul of innovative and realistic research and led me through the research door from theory to practice. His optimism, calmness and gentleness made me fully understand the true meaning of loving life, creating the future, and flexibly adjusting my goals. He is humble, cautious and persevering, and is a role model of life in my heart.

I was born in Shanxi, China, in 1954. I earned my diploma in civil engineering and MS in structural engineering in 1975 and 1981, respectively, from the Taiyuan University of Technology (TYUT) (太原理工大學), China. In 1990, I obtained my PhD in structural engineering from Purdue University, West Lafayette, Indiana. I worked at the Northeastern China Power Design Institute from 1975 to 1978 and taught at TYUT from 1981 to 1985. I have been working with the California Department of Transportation (Caltrans) since 1991.

In May 1978, I took the first national graduate entrance examination after the Culture Revolution in China. In October 1978, I was admitted to the Taiyuan Institute of Technology (太原工學院) (later renamed TYUT) as a graduate student in structural engineering. During those passionate years, we studied very hard, diligently and actively, to learn the latest development of science and technology in order to realize the four modernizations: industry, agriculture, science and technology, and national defense. I often went to the Taiyuan Foreign Bookstore to look for the latest scientific research information. On 1 April 1981, I luckily purchased a two-volume book set entitled *Theory of Beam-Columns*, which was co-authored by Prof. Chen and T Atsuta. Theories of the elastoplastic analysis in this beam-column masterpiece greatly helped and guided me in my Master thesis "Investigation of elastoplastic behavior of reinforced concrete continuous beams". At that time, I heard that Prof. Chen was an internationally famous scholar and a professor at Purdue University. Though I was absorbing his book for nutrition and knowledge, I never expected to end up studying under his supervision at Purdue one day in the future.

In May 1985, I was selected as the first group of students in the Shanxi Province to study abroad. The thought of the possibility of studying at Purdue University came to mind immediately. I found out Prof. Chen's mailing address from *Theory of Beam-Columns*, and I took the liberty to write a letter to him on 15 June, expressing my heart's desire to study at Purdue. Prof. Chen was visiting Germany at the time, and my letter was transferred from the United States (US) to Germany. After receiving my letter, he wrote back on 28 June from Germany to encourage me to apply for the PhD graduate program. I was so excited and appreciative when I received Prof. Chen's letter on 6 July. Never had I imagined that the world-famous Prof. Chen would reply to an unknown foreign student so quickly! I excitedly prepared my transcript, asked for recommendation letters from my professors, and submitted my application to the Purdue University Graduate School. In December 1985, I received an admission letter from Purdue. On 17 January 1986, in my first time flying in an airplane, I flew across the Pacific Ocean from Beijing to San Francisco, then Chicago, and finally, West Lafayette, Indiana, Purdue University, where I became one of Prof. Chen's PhD students. To this day, *Theory of Beam-Columns* follows me wherever I go and is still on my bookshelf.

My student life at Purdue from January 1986 to December 1990 was intense, yet fulfilling. Uniform red-brick campus buildings, quiet and beautiful West Lafayette, autumn red leaves over hills and fields, the intoxicating Wabash River, active classroom teaching, various term papers, unfettered teacher-student exchanges, academic lectures and seminars, overnight research, lifelong beneficiaries from professors… all these experiences are vivid and fresh like scenes from a movie.

One of Prof. Chen's methods to advise students was to train them to read extensive literature, understand the state-of-art and the state-of-practice, find problems, ask questions, explore ideas, conduct research, find solutions, and implement to practice. I remember that Prof. Chen used to send me documents and materials almost every day!

In 1987, based on a test report on steel built-up members provided by Prof. Chen, I quickly developed a set of design formulas and drafted my first English paper "Design rules of built-up members in load and resistance factor design", which I then sent to him for review. During a discussion, he asked me two questions: "What is new?" and "What is its practical value?" I was unprepared and hurriedly answered.

Obviously, he was not satisfied with my answers because he advised that theoretical background and derivations should be explained first, the proposed design formulas must be validated by experimental data available, and then further simplification should be made for practical implementation. In my draft paper, Prof. Chen marked his technical and grammar comments in red throughout! He not only pointed out the directions for further improvement, but also revised my poor English writing, and even checked the references one by one. Professor Chen always returned my drafts with thoughtful comments, questions and recommendations in one or two days. His careful, critical and innovative thinking motivated me to work hard. Professor Chen would usually review a paper four to five times before marking it as "Final". This paper was published in the American Society of Civil Engineers (ASCE) *Journal of Structural Engineering* in November 1988. He repeatedly taught his students that in any technical paper, there must be something new, no matter how big or small. Also, the method proposed should be concise and practical to use, and the conclusions are clear. His pursuit of innovation truth-seeking, exploration of theoretical practice, rigorous scholarly attitude, resolute style, concise and clear philosophy, have deeply educated and influenced me greatly.

Oftentimes, Professor Chen arranged for us to attend various academic conferences and encouraged us to directly communicate with frontline experts and scholars to establish academic relationships. In May 1988, his PhD students consisting of Fu-Hsiang Wu, Hong Zhang, Tien-Kuen Huang and myself drove to Blacksburg, West Virginia, to participate in the ASCE Engineering Mechanics Specialty Conference. At this meeting, in addition to participating in various academic sessions, we met Dr Genhua Shi, a pioneer in discontinuous medium mechanics and Prof. Xila Liu, who graduated from Purdue in 1985 and returned to China to teach at the Tsinghua and Shanghai Jiaotong Universities. Their passion and determined career goals deeply influenced me. Professor Liu and I have had a wonderful friendship and learned from each other for the past thirty years.

Professor Chen unselfishly supported his students, allowing them to stand on his giant shoulders to grow. I co-authored with Prof. Chen and published 15 papers during my PhD studies at Purdue. He always listed me as the first author. Under his careful and kind supervision, I completed my doctoral dissertation "Stability analysis and design of steel structures" in three parts — "Effective length factor of framed columns", "Design interaction equations for beam-columns", and "Behavior and strength of dented tubular members", consisting of 10 published papers.

It was my dream to become a university faculty after I obtained my PhD degree in December 1990. However, I joined the Division of Structures, California Department of Transportation (Caltrans), the most prestigious and influential bridge design unit in all the State Departments of Transportation in the US, as a transportation engineer

(Civil). When I first came to Caltrans, I felt like I had returned to the Northeast China Power Design Institute, where I had worked from 1975 to 1978. I quickly found my balance position. I have completed bridge design projects successfully, explored my knowledge on inelastic analysis of reinforced concrete structures and the stability of steel structures to solve engineering problems and thus contribute to the engineering profession.

For nearly three decades after graduating from Purdue, Prof. Chen and I have maintained close personal contact, academic cooperation and exchanges. He has been providing memorable guidance and constructive recommendations. We have had many happy reunions in Indiana, Hawaii and California.

The San Francisco Oakland Bay Bridge (SFOBB) is one of the most important bridges in the US. During the 1989 Loma Prieta earthquake, one of the east spans collapsed, causing the bridge to close for more than a month, which caused great economic losses. The seismic strengthening of SFOBB was Caltrans' most important project in the 1990s. In late November 1996, I undertook the task of writing the "San Francisco-Oakland Bay Bridge west spans seismic retrofit design criteria". There was only one and a half months from the drafting of the design criteria to the Peer Review Panel meeting. I sent the first draft to Prof. Chen for his review in mid-December and hoped to have his comments before the New Year in 1997. What made it unforgettable for me was that Prof. Chen expedited his critical review and sent his valuable comments by express mail before Christmas in 1996. The 120-page design criteria were highly evaluated by the Peer Review Panel. Two new methods proposed in the design criteria: "Section properties for latticed members of San Francisco Oakland Bay Bridge" and "Effect of compound buckling on compression strength of built-up members" were later published in the ASCE *Journal of Bridge Engineering* in 2000 and the American Society of Steel Structures (AISC) *Engineering Journal* in 2002, respectively. I was extremely surprised that the former article was awarded the prestigious Arthur Mellen Wellington Prize from ASCE in 2001. The proposed methods of these two articles were adopted by the AISC Specification for Structural Steel Buildings and the American Association of State Highway Transportation Officials LRFD Bridge Design Specifications.

In 1998, I proposed to collaborate with Prof. Chen to edit the first-ever *Bridge Engineering Handbook*. After more than two years of hard work from more than 140 world-class bridge engineering scholars and experts, *Bridge Engineering Handbook*, a comprehensive reference and resource book, was published in 2000 by CRC Press. The 2000-page Handbook represents a milestone and masterpiece that presents bridge engineering from planning, design and construction to maintenance. It integrates ancient and modern bridge engineering practices worldwide, explains the engineering examples and calculation steps in detail, and emphasizes the theory application and practice. The handbook won the "Outstanding Academic Title Award" in 2000 from CHOICE magazine. In 2014, we worked together again to publish a five-volume set of the second edition of this handbook and another new handbook titled the *Handbook of International Bridge Engineering*.

During my 30-year Caltrans career, I have chaired the Structural Steel Committee since 2000. I was the lead engineer for the development of Caltrans Seismic Design Specifications for Steel Bridges and editor for Caltrans' Bridge Design Practice Manual. I also served as a member for several National Highway Cooperative Research Program Panels and was a Transportation Research Board Steel Committee member from 2000 to 2006.

I received the Professional Achievement Award from the Professional Engineers in California Government in 2007, the Distinguished Engineering Achievement Award from the Engineers' Council in 2010, and the Outstanding Management and Engineering in Transportation — James Robert Award — from California Department of Transportation in 2017.

Lian Duan (段煉)
Sacramento, California (加利福尼亞沙加緬度)
May 2020

INDEX

Part I: Technical Terms

Part II: Organizations and Institutions

Part III: Individual Names Cited in the Book